X-RAY LASERS 2002

Related Titles from AIP Conference Proceedings

To learn more about these titles, or the AIP Conference Proceedings Series, please visit the webpage **http://proceedings.aip.org/proceedings**

X-RAY LASERS 2002

8th International Conference on
X-Ray Lasers

Aspen, Colorado   27–31 May 2002

SPONSORING ORGANIZATIONS
National Science Foundation
U.S. Department of Energy
Lawrence Livermore National Laboratory
Colorado State University
Princeton University

EDITORS

Jorge J. Rocca
Colorado State University

James Dunn
*Lawrence Livermore
National Laboratory*

Szymon Suckewer
Princeton University

AMERICAN
INSTITUTE
OFPHYSICS

Melville, New York, 2002
AIP CONFERENCE PROCEEDINGS ■ VOLUME 641

The article on pp. 461-468 was authored by U.S. Government employees and is not covered by the below mentioned copyright.

L.C. Catalog Card No. 2002113721
ISBN 0-7354-0096-2
ISSN 0094-243X
Printed in the United States of America

CONTENTS

TRANSIENT COLLISIONAL X-RAY LASERS: EXPERIMENTS SIMULATIONS

THEORY OF X-RAY LASERS AND AMPLIFIED BEAMS

INVESTIGATION OF X-RAY LASER MEDIA

FREE ELECTRON LASERS AND OTHER ACCELERATION-DRIVEN X-RAY SOURCES

HIGH-ORDER HARMONICS AND OTHER HIGH-BRIGHTNESS X-RAY SOURCES

APPLICATIONS OF X-RAY LASERS AND OTHER BRIGHT X-RAY SOURCES

XUV OPTICS, INSTRUMENTATION, AND CHARACTERIZATION
OF X-RAY BEAMS

Preface

This volume contains the papers presented at the 8th International Conference on X-Ray Lasers held in the town of Aspen, Colorado, from May 26th to May 31, 2002, attended by participants from 15 countries. This meeting was part of a series that reviews the most recent advances in the development and application of X-Ray Lasers. The first meeting was held in Aussois in 1986 and was followed by conferences at York (1990), Schliersee (1992), Williamsburg (1994), Lund (1996), Kyoto (1998) and Saint Malo (2000). The next two conferences are scheduled to be held in Beijing in 2004 and in Berlin in 2006. The published proceedings of these conferences provide the reader with a good overview of the significant progress that the field of X-Ray Lasers has achieved over the years. This book will give some further insight on many of the scientific and technical issues related to the development and application of these short wavelength coherent sources.

The papers compiled in this volume give an updated view of the rapid progress in the development of X-Ray Lasers, and review several applications. A clear emphasis of the work summarized in these papers is the development of compact and practical X-Ray Lasers, driven by the success in pumping with table-top short-pulse optical lasers and with fast capillary discharges. This has produced table-top laser sources that have a peak spectral brightness several orders of magnitude larger than synchrotrons, that now allow the successful realization of unique table-top experiments such as the interferometry of very high density plasmas. The contents of these proceedings also show that the progress in source development is accompanied by the demonstration of new optics and x-ray instrumentation that are vital to realize the full potential of these new sources in novel scientific and technological applications.

The 8th International Conference on X-Ray lasers was sponsored by the U.S. National Science Foundation, the U.S. Department of Energy Division of Chemical Sciences, Geosciences and Biosciences Division of the Office of Basic Energy Sciences, and by Lawrence Livermore National Laboratory. Partial support was also generously provided by, Colorado State University, Princeton University, and Positive Light.

The organizers also would like to acknowledge the invaluable assistance of Cathy L.Z. Smith, Maria Julia Marconi, Carmen S. Menoni, Joy Dunn and Ana I. Rocca in the successful organization and running of the conference, and thank Dinesh Patel for the beautiful photograph of the Maroon Bells mountains near Aspen that was used in the conference poster.

Finally, the editors want to express their sincere appreciation to Jean-Luc Beaumont for the expert work done in the compilation of these proceedings.

Jorge J. Rocca
James Dunn
Szymon Suckewer

Conference Organization

Conference Chairman: Jorge J. Rocca, *Colorado State University*
Co-Chair: James Dunn, *Lawrence Livermore National Laboratory*
Co-Chair: Szymon Suckewer, *Princeton University*

Conference Secretary: Cathy L. Z. Smith, *Colorado State University*

International Advisory Board

H. Daido (Institute for Laser Engineering, Osaka University, Japan)
D. C. Eder (Lawrence Livermore National Laboratory, USA)
R. C. Elton (University of Maryland, USA)
E. Fill (Max-Planck Institut für Quantenoptik, Garching,, Germany)
P. Hagelstein (Massachusetts Institute of Technology, Cambridge, USA)
P. Jaeglé (Laboratoire de Spectroscopie Atomique et Ionique, Orsay, France)
G. Jamelot (Laboratoire de Spectroscopie Atomique et Ionique, Orsay, France)
Y. Kato (Japan Atomic Energy Research Institute, Kyoto, Japan)
M. H. Key (Lawrence Livermore National Laboratory, USA)
V. V. Korobkin (General Physics Institute, Moscow, Russia)
D. L. Mathews (Lawrence Livermore National Laboratory, USA)
P. Nickles (Max-Born Institut, Berlin, Germany)
G. J. Pert (University of York, UK)
J. J. Rocca (Colorado State University, Fort Collins, USA)
S. Suckewer (Princeton University, USA)
S. Svanberg (Lund Institute of Technology, Sweden)
G. J. Tallents (University of York, UK)
A. V. Vinogradov (Lebedev Physical Institute, Moscow, Russia)
Z. Xu (Shanghai Institute of Optics and Fine Mechanics, China)
J. Zhang (Institute of Physics, Beijing, China)

Local Organizing Committee

Cathy L. Z. Smith, *Colorado State University*
Carmen S. Menoni, *Colorado State University*
Siu Au Lee, *Colorado State University*
Margaret M. Murnane, *University of Colorado*
Neil C. Gallagher, *Colorado State University*
Henry C. Kapteyn, *University of Colorado*
Bradley M. Luther, *Colorado State University*
Maria Julia Marconi, *Colorado State University*

Sponsors

The conference was scientifically sponsored by

National Science Foundation (US)

US Department of Energy, Division of Chemical Sciences, Geosciences, and Biosciences, Office of Basic Energy Sciences

Lawrence Livermore National Laboratory

Partial support was generously provided by

Colorado State University

Princeton University

Positive Light, USA

8th International Conference on X-Ray Lasers Award to Professor Pierre Jaeglé

"For life-long achievements in the X-Ray laser Research and for leadership in the creation of strong European X-Ray Laser community"

Aspen, Colorado. May 30, 2002

The accompanying photograph shows Pierre Jaeglé in the beautiful Maroon Bells mountain area near the town of Aspen, where this conference was held. In 1986, Pierre organised in Aussois, France, the first meeting of this series. The mountains are a very appropriate scenario for a photograph of Pierre, who spent his childhood in Embrun, in the French Alps, where he acquired a taste for skiing and climbing. The Maroon Bells was the site of a mid-conference excursion, that provided both an appropriate break from the busy schedule of oral talks and poster sessions and a relaxed atmosphere for continued scientific discussions. Pierre, who is an expert mountaineer, was the leader of this excursion. He guided the group of the participants in a hike, in which many of us had to struggle to keep up with his pace. However, it is not for this reason that he was the recipient of the 8th International Conference on X-Ray Lasers Award. The award was conferred to him by the community in recognition for life-long contributions to the field of X-Ray Lasers (XRL). Pierre has completed through the years a series of very significant experiments in XUV spectroscopy, X-Ray Laser development and applications. He is also the inspiring mentor to an entire new generation of European XRL scientists. Besides his experimental activity on XRL, Pierre Jaeglé is curious in theoretical developments and he collaborated with a number of theoreticians, in particular with Alain Sureau. More generally, epistemology and philosophy also interest him, and he has written several books alone or in collaboration with philosophers and scientists. Finally it must be mentioned that besides a very successful scientific career, Pierre Jaeglé is also a citizen who has actively participated and continues to participate in the struggles for peace, democracy and freedom.

Pierre in the Colorado mountains near Aspen

TRANSIENT COLLISIONAL X-RAY LASERS:
EXPERIMENTS SIMULATIONS

Measurement of Gain Duration for Ne-like Ni and Ni-like Ag

Y. Abou-Ali[*], G J. Tallents[*], M H. Edwards[*], R. Keenan[†], S J. Topping[†], C L S. Lewis[†], O. Guilbaud[¶], A. Klisnick[¶] and D. Ros[¶]

[*]Department of Physics, University of York, Heslington, York, YO10 5DD
[†]School of Mathematics and Physics, Queens University of Belfast, Belfast, BT7 1NN, UK
[¶]LIXAM, Université Paris-Sud, Bât, 350, 91405 Orsay Cedex, France

Abstract. A simple streaked slit diagnostic was developed to record resonance line emission from the plasmas used to produce Ne-like nickel and Ni-like silver X-ray lasing. A Kentech streak camera with a 500 μm wide horizontal timing slit viewing the plasma transversely with a space resolving slit of width 270 μm positioned normal to the timing slit to give spatial resolution along the line focus length. An estimate of the duration of X-ray laser gain is obtained by temporally resolving resonance line emission from states near in energy to the upper lasing level. The emission duration was measured to be 33 - 40 ps for Ni and 19 - 22 ps for Ag. We may expect the duration of this emission to be an upper bound on the gain duration. The travelling wave velocity at the leading edge of the main pulse was measured to be close to the speed of light.

INTRODUCTION

The first demonstration of a soft X-ray laser with a high gain-length product was in 1984 at a wavelength of 21 nm. In this experiment, the NOVA laser was used as the pump laser, delivering about a kilojoule of energy [1]. Since then much effort has been made to reduce the pump energy requirement. Using pre-pulses with ~ 100 ps pulses has reduced the driver energy for a saturated soft X-ray laser at 7 nm to less than 100 J [2]. Transient collisional excitation using ~ ps pumping pulses superimposed on a long duration pre-pulse has since further reduced the pump laser energy needed to achieve saturated X-ray lasers to less than 10 J [3]. The initial demonstration of transient collisional excitation was obtained with the Ne-like Ti 3p→3s transition at 32.6nm [4], but was soon extended to the Ni-like Pd 4d→ 4p line at 14.6 nm [5], Ne-like Ge at 19.6 nm [6] and Ni-like Sm at 7.3 nm [7]. In transient collisional excitation the long pulse (~ 300 – 600 ps) is used to produce and ionize plasma with a large population of the desired ionization stage. The pre-plasma is allowed to expand and is than strongly heated by the short (~ ps) pulse. A large population inversion is obtained via collisional excitation before further ionization can occur. This leads to large gain coefficients. The consequent short gain duration t_G, requires that amplifier lengths cannot be longer than about c t_G, unless a traveling wave pump is used to keep the excitation in phase with the amplifying group of photons [7].

CP641, X-Ray Lasers 2002: 8th International Conference on X-Ray Lasers, edited by J. J. Rocca et al.
© 2002 American Institute of Physics 0-7354-0096-2/02/$19.00

Measurement of the duration of gain in a plasma medium is difficult as the upper quantum states directly involved in the laser transition are metastable to decay to the ground state. If they were not metastable, the creation of a population inversion would not occur. Consequently, it is not possible to determine the laser upper quantum state population by directly measuring the resonance line emission from the upper state. In this paper, an estimate of the duration of the X-ray laser gain is obtained by temporally resolving the spectrally integrated resonance line emission from states near in energy to the upper lasing level. We may expect the duration of this emission to be an upper bound on the gain duration.

EXPERIMENTAL ARRANGEMENT

The experiment was performed at the Vulcan Nd:glass laser at the Rutherford Appleton Laboratory, UK. Two beams from a 1.053 μm Ti-Sapphire-Nd:glass chirped pulse amplification laser system irradiate the solid silver or nickel slab target. A pre-plasma was formed with a background pulse of duration 280 ps and peak irradiance $\sim 10^{13}$ Wcm^{-2} in a line focus of length 16 mm and width 100 μm. The main pulse irradiance is enhanced by chirp pulse amplification (CPA) to give a duration of 1.2 ps and peak irradiance $\sim 10^{15}$ Wcm^{-2} in a line focus of length 12 mm and width 100 μm. The background pulse was produced from a component of the uncompressed main pulse and hence arrived on target without random jitter relative to the main pulse. The background pulse and main pulse were focused onto the targets with separate focusing systems comprising a refracting lens (the background pulse) or a parabolic shaped mirror (the main pulse) to produce a spot focus which was imaged in each case by an off-axis spherical mirror to produce the line focus on the target [8]. The experimental set-up is similar to that detailed earlier by Klisnick et al [9]. The peak-to-peak separation of the two pulses was controlled by varying the optical path length of the long pulse via a timing slide within the target area. For results reported in this paper, we used a delay between pulses of 200 ps (for Ni-like silver) or 300 ps (for Ne-like nickel).

As the X-ray laser pulse travels at a maximum of the vacuum speed of light c in the plasma (corresponding to a time of 33 ps/cm), a mis-match between the propagating X-ray laser pulse and the onset of gain can result. Travelling wave pumping is employed to ensure that the X-ray pulse and the gain onset coincide. There is an intrinsic travelling wave velocity for the energy pulsefront along the line focus in our experiment of 2.5c due to the geometry for production of the line focus. A grating is employed before the compression gratings to reduce the laser pulsefront velocity to c along the line focus. The travelling wave pulsefront velocity was fine tuned by adjustment of the second pulse compressor grating [10].

The X-ray laser output was recorded using a flat field grating spectrometer, coupled either with an X-ray CCD or with an Axis Photonique streak camera. A more detailed description of this diagnostic and measurement of the X-ray laser output are given in accompanying papers [11, 12].

A streaked slit diagnostic was developed to record resonance line emission from states near in energy to the upper laser levels. A Kentech streak camera with a thin 0.3

μm gold photocathode coated onto 1 μm thick aluminum foil viewed the plasma transversely from the front side at an angle of 82.5° to the focal line. The estimated temporal resolution of ≈ 6 ps with this streak camera is dominated by the 500 μm wide entrance slit. A space resolving slit of width 270 μm positioned normal to the timing slit gave spatial resolution ≈ 0.7 mm along the line focus length. The output signal from the streak camera was amplified by a standard 50/40 Kentech intensifier and recorded by a CCD camera with 27 μm pixels. A filter of 2.4 μm thick aluminum was employed so that emission in the 0.8 –2.1 nm wavelength range was recorded with the diagnostic (figure 1). The emission is dominated by resonance line emission for both Ni-like silver (resonance emission in the range 1.7 to 2.2 nm from states with electron configurations $3p^6 3d^9 4p$, $3p^6 3d^9 4f$, $3p^5 3d^{10} 4s$ and $3p^5 3d^{10} 4d$) and Ne-like nickel (resonance emission in the range 1 to 1.4 nm from states with electron configurations $2s^2 2p^5 3s$, $2s^2 2p^5 3d$ and $2s2p^6 3p$). Continuum emission at the electron temperatures present in the plasma does not constitute a major component of the recorded emission. Time and spectrally-resolved nickel spectra recorded over the 0.95 – 1.45 nm spectral range show that Ne-like emission is more intense than the F-like and continuum emission.

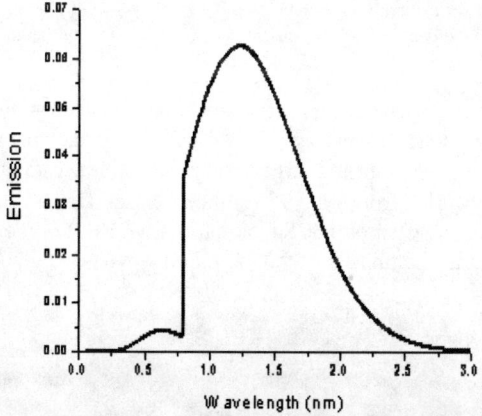

FIGURE 1. The bremsstrahlung emission of temperature 500 eV calculated to transmit through a 2.4 μm aluminum filter as used in the streaked slit diagnostic. The transmission is dominated by emission close in wavelength to the resonance lines of Ni-like silver and Ne-like nickel.

A crossed-slit camera with two slits monitored the line focus plasma uniformity. Slits of width 100 μm and 8 μm produced respective magnifications of 1.65 and 3.5 with spatial resolution of 160 μm and 10 μm in the directions along and across the line focus. These spatial resolutions were limited respectively by the slit width and CCD pixel size. The crossed-slit camera was filtered with 25 μm beryllium and 50 μm plastic (CH) so that emission at wavelengths below 0.6 nm was recorded. A KAP crystal spectrometer with space resolving slit measured the time integrated resonance line emission of Ne-like nickel with a spatial resolution along the plasma line of 300 μm.

RESULTS

A sample streaked slit output showing spatial resolution of 0.7 mm along the target length and temporal resolution of 6 ps is given in figure 2. Emission from the pre-plasma followed by the emission associated with the 1.2 ps main pulse is clearly apparent. This known time delay between the pulses has been used to calibrate the streaked slit camera streak speed.

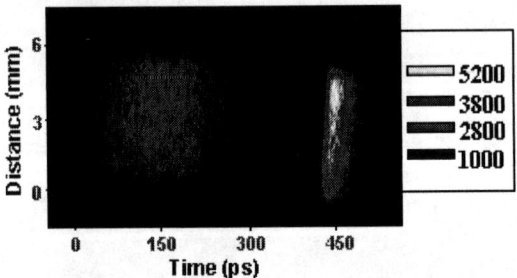

FIGURE 2. Streaked slit camera image showing the plasma emission filtered by a 2.4 μm aluminum filter (see figure 1) from the background and main laser pulses incident onto a nickel target. The relative intensity scale is also shown.

The main pulse emission occurs at times determined by the travelling wave irradiation. Assuming that the leading edge of the main pulse emission corresponds to the time of laser light reaching the target, figure 2 implies a travelling wave velocity for the laser energy pulse front of 0.95 (± 0.05) c (see figure 3) in agreement with measurements made using an optical streak camera viewing scattered laser light from the line focus during the setting-up of the travelling wave grating [10].

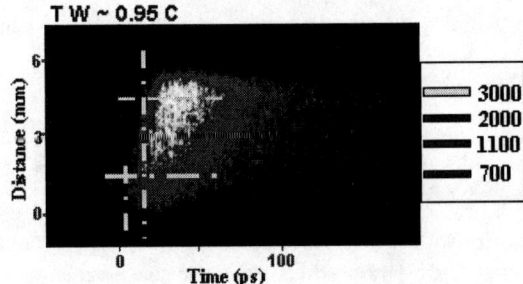

FIGURE 3. Emission at the leading edge of the main pulse emission recorded by the streaked slit camera can be used to measure the travelling wave excitation velocity (≈ 0.95 c). The relative intensity scale is also shown.

The variation along the target length as measured by the crossed-slit camera, crystal spectrometer and the emission integrated in time from the streaked-slit camera are shown for the same shot in figure 4. Similar emission variations for the three diagnostics are recorded.

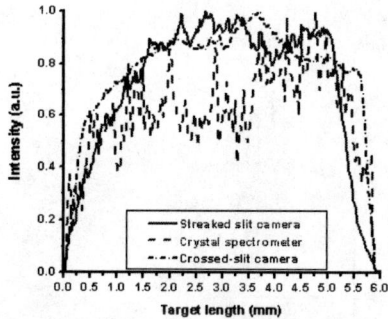

FIGURE 4. The variation along the target length as measured by the crossed-slit camera, crystal spectrometer for the so-called 3C or $2s^22p^53d\ ^1P_1 \rightarrow 2s^22p^6\ ^1S_0$ line at 1.243 nm and the emission integrated in time from the streaked slit camera. The nickel target was measured before the shot to have a length of 5.9 mm.

The spectrum recorded with the crystal spectrometer for the nickel target is shown in figure 5. The spectral lines have been identified following Boiko et al [13].

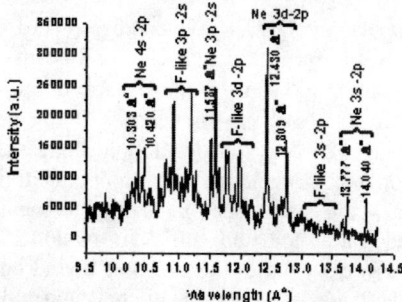

FIGURE 5. Time integrated Ne- and F-like nickel spectrum recorded with the crystal spectrometer.

Line-outs of emission recorded by the streaked slit camera for the silver and nickel targets are shown in figure 6. There is a sharp rise in emission when the laser pulse is incident followed by an approximately exponential decrease with time.

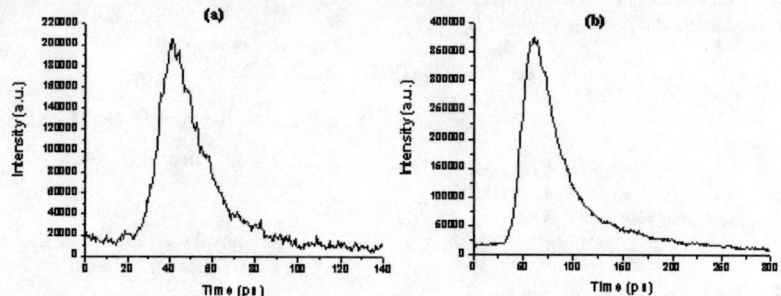

FIGURE 6. Line-out of main pulse emission recorded by the streaked slit camera for (a) a silver target (b) a nickel target.

The emission associated with the short pulse laser has measured fwhm duration of 19 - 23 ps for silver and 33 - 40 ps for nickel targets (figure 7).

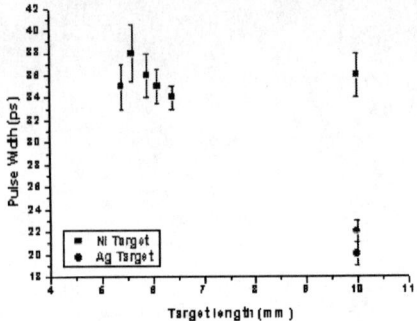

FIGURE 7. Measurements of plasma emission duration (fwhm) associated with the main pulse found using the streaked slit camera for different target lengths.

CONCLUSION

A simple streaked slit diagnostic was developed to record resonance line emission from the plasmas used to produce Ne-like nickel and Ni-like silver X-ray lasing. A Kentech streak camera with a 500 μm wide horizontal timing slit viewing the plasma transversely with a space resolving slit of width 270 μm positioned normal to the timing slit was employed to give spatial resolution along the line focus length. An estimate of the duration of X-ray laser gain is obtained by temporally resolving resonance line emission from states near in energy to the upper lasing level. The emission duration was measured to be 33 - 40 ps for Ni and 19- 23 ps for Ag. We may expect the duration of this emission to be an upper bound on the gain duration. The travelling wave velocity at the leading edge of the main pulse was measured to be close to the speed of light.

REFERENCES

[1] Matthews, D. L et al., *Phys. Rev. Lett* **54**, 110 (1985).
[2] Nilsen, J et al., *Phys. Rev A* **48**, 4682 (1993).
[3] Dunn, J et al., *PRL* **84**, 4834 (2000).
[4] Nickles, P. V et al., *Phys. Rev. Lett* **78**, 2748 (1997).
[5] Dunn, J et al., *Phys. Rev. Lett.* **80**, 2825 (1998).
[6] Kalachnikov, M. P et al., *Phys. Rev.A* **57**, 4778 (1998); Warwick, P. J et al., *J. Opt. Soc. Am. B* **15**, 1808 (1998).
[7] King, R. E et al., *Phys. Rev, A* **64**, 053810 (2001).
[8] Ross, I. N., *Appl. Optics* **36**, 9348 (1987).
[9] Klisnick, A et al., *Phys Rev. A* **65**, 033810 (2002).
[10] Collier, J.,Pepler, D., Danson, C., Warwick, J., Lewis, C., and Neely, D., *Central Laser Facility Annual Report* p209-10 (ISBN 0902376705) 1996/97; Klisnick, A., *J. Opt. Soc. Am. B*17, 1093 (2000).
[11] Tallents, G J et al., this proceeding.
[12] Guilbaud, O et al., this proceeding.
[13] Boiko, V. A., Faenov, A. Y., and Pikuz, S. A., *J. Quant. Spect. Radiat. Trans.* **19**, 11 (1978).

Injector-amplifier design for tabletop Ne-like x-ray lasers

J. Dunn [a], A.L. Osterheld [a], A. Ya. Faenov [b], T.A. Pikuz [b], and V.N. Shlyaptsev [c]

[a] Lawrence Livermore National Laboratory, Livermore, CA 94551-9900
[b] Multicharged Ions Spectra Data Center of VNIIFTRI, Mendeleevo, Moscow Region 141570 Russia
[c] University of California Davis-Livermore, Livermore, California 94551

Abstract. We report new results using the LLNL COMET laser to evaluate the effectiveness of different target architectures to improve the output and characteristics of the transient x-ray laser scheme. Surprising observations were found when the laser line focus irradiating a single slab Cr or Fe target was divided into two or three distinct plasma column sections with millimeter scale gaps between each plasma. The Ne-like 3p $^1S_0 \rightarrow$ 3s $^1P_1$ 28.5 nm and 25.5 nm x-ray laser lines, for Cr and Fe, respectively, were improved in beam divergence, by 2 – 3 times, and peak intensity, by up to one order of magnitude, when compared with a single plasma column of the same length or longer. This was contrary to expectations since these large-scale inhomogeneities introduced along the plasma, as well as attenuation from the cold plasma at the end of each section, would be detrimental to the x-ray propagation and amplification. Instead an injector-amplifier (IA) type process appears to be at work where the plasma gaps may be beneficially modifying the ray propagation and coupling through the high Ne-like ion gain regions. We present results showing the output of the amplifier stage with increasing length for the IA targets together with beam deflection and divergence measurements.

INTRODUCTION

The use of oscillator-amplifier or injector-amplifier designs, i.e. multiple- instead of single-stage designs, have been suggested as one way to improve the output characteristics for large, high power, laser-generated x-ray lasers. An injector-amplifier (IA) architecture was proposed in order to improve the output coherence by seeding an amplifier with a single-mode from an oscillator or injector [1]. Preliminary experiments investigating this effect on the NOVA laser were reported approximately 10 years ago [2]. Another important reason for using IA design was to mitigate against refraction effects bending the x-ray laser beam out of the gain region when target lengths exceeded 2 cm [3, 4]. This work demonstrated that with optimization of the target offset, the coupling efficiency between the two stages was maximized and the effect of refraction was reduced. More recently hybrid oscillator-amplifier schemes have been proposed by using higher order harmonics as a seed to be amplified in a fast capillary discharge gain medium [5] or a picosecond-heated solid target [6].

CP641, X-Ray Lasers 2002: 8th International Conference on X-Ray Lasers, edited by J. J. Rocca et al.
© 2002 American Institute of Physics 0-7354-0096-2/02/$19.00

In many respects, the high efficiency, transient gain, tabletop x-ray lasers can also benefit from injector-amplifier designs. While the required target lengths for saturated output are considerably shorter, typically less than 1 cm, refraction can still be a limiting effect because of the extremely high gains, g > 100 cm^{-1}, available in higher density regions [7]. Small deflection angles of the x-ray laser beam out of these regions may reduce the gain by 10 – 20 % or more which is sufficient to substantially decrease the x-ray laser output over a few millimeters of plasma length.

We report experimental results from a very simple approach where a single, flat slab target has been irradiated with two or three shorter line foci produced by placing a mask in the focusing beam. The millimeter-scale gaps in the line focus form multiple stages in the plasma column and are observed to substantially improve the beam divergence and intensity of several Ne-like ion x-ray lasers.

EXPERIMENTAL DESCRIPTION

The experiment was carried out on the Compact Multipulse Terawatt (COMET) laser system at LLNL [8]. This laser, operating at 1054 nm wavelength, utilizes chirped pulse amplification to produce the two laser beams to generate the x-ray laser. For this study on Ne-like Cr and Fe x-ray lasers, energy of 0.6 – 4.8 J in a 600 ps pulse was followed by a delay of 1.4 ns before utilizing a 1.2 ps excitation pulse with 5 J. The laser repetition rate was 1 shot every 4 minutes. The line focus length of 1.1 cm was achieved with a cylindrical lens and an on-axis paraboloid. The 600 ps beam was defocused to a width of ~150 μm (FWHM) while the 1.2 ps beam was focused to 80 μm [9]. A simple reflection echelon technique was adopted to produce a traveling wave line focus as described in previously related x-ray laser work [10]. The traveling wave optic consisted of five flat mirror segments placed before the focusing optics where each mirror segment was offset by 0.12 cm to introduce a traveling wave towards the spectrometer with a delay of 7.7 ps per step. This corresponded to a phase velocity of c along the line focus length with five steps.

The on axis x-ray laser output was observed with a 1200 line mm^{-1} variable-spaced flat-field grating spectrometer with a back-thinned 1024 × 1024 charge-coupled device (CCD). A 1 μm thick Al filter, determined to have a filter transmission of 0.032 at 28.54 nm, was used at highest x-ray laser intensities to prevent the CCD from saturating. Fiducial wires, placed in front of the spectrometer, were aligned relative to the target surface with a telescope in order to calibrate the angular deflection and beam divergence of the x-ray laser in the horizontal direction. Flat polished Cr and Fe slab targets were used in the experiment and tilted back by ~5 mrad in the horizontal direction to compensate for refraction of the x-ray laser in the plasma column. A CCD x-ray double-slit camera with 25 μm spatial resolution monitored the line focus plasma uniformity and overlap of the laser pulses. In addition to these diagnostics, several Focusing Spectrometers with Spatial Resolution in 1-Dimension (FSSR-1D) instruments using Kodak DEF film and CCD detector arrays were employed [11]. These gave $n = 3 \rightarrow 2, 4 \rightarrow 2$ Ne-like Fe and Cr resonance line emission measurements with spatial information along the length of the laser line focus.

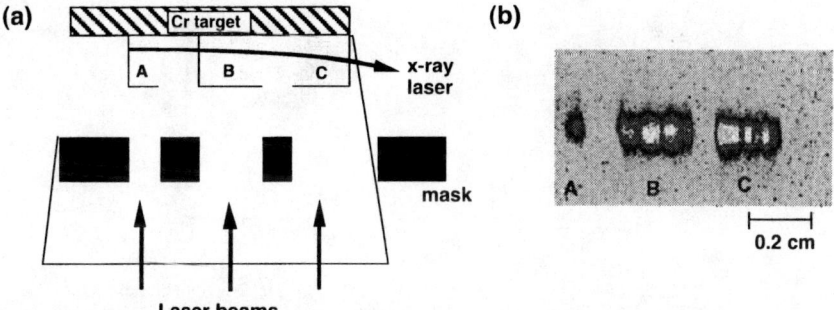

Fig. 1 (a) A mask placed in focusing beam generates three-stages A, B, and C along the Cr plasma. (b) X-ray double-slit camera image of x-ray emission showing three Cr stages clearly visible in the line focus. Magnification is 1× along the focus length and 14× across the focus width.

To compare the effect of the multi-stage target to the single stage target, it was important to maintain the same line intensity along the line focus but with the gaps. Although it would have been possible to create gaps in the line focus by simply separating the segments using the traveling wave stepped mirror, this may have changed the longitudinal intensity profile. Instead an alternative method was tried where a mask was placed in the focusing beam about 12 cm from the target. This created small gaps of approximately 1 mm length with minimal diffraction effects on the line focus. Figure 1 shows this experimental layout. The x-ray laser output could be studied with and without the gaps. A two-stage device was tried for Fe and a three-stage for the Cr x-ray laser with similar results. We report on a three-stage Cr x-ray laser using a 1 cm target irradiated with ~3 J, 600 ps and 4.8 J, 1.2 ps.

EXPERIMENTAL RESULTS

The conventional single-stage Cr targets have been reported recently [9]. Strong lasing was observed on the 28.54 nm $3p \rightarrow 3s$ line, with weaker lasing on both the 25.91 nm $3d \rightarrow 3p$ line and the short wavelength 24.03 nm $3p \rightarrow 3s$ line. An intensity versus length study was performed for single-stage target lengths of 0.2 cm to 1 cm, and small-signal gain of 31 cm^{-1} was determined for the 28.54 nm $3p \rightarrow 3s$ line for targets up to 0.4 cm. The output was found to increase at a lower exponential rate up to 1 cm. The overall gain length product of 16.7 ± 0.6 was above the predicted saturation intensity. Using this as a reference for single-stage Cr target output, the mask was placed in the beam to generate the three-stage target, as shown in Fig. 1. Similar laser pumping conditions were repeated. The x-ray laser intensity was found to be substantially higher for the combined lengths of the three stages when compared

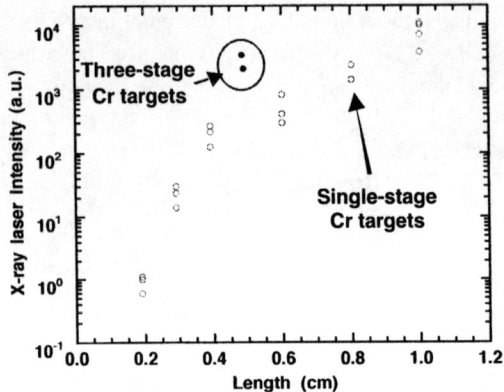

Fig. 2 X-ray laser output of the Cr 38.5 nm $3p \rightarrow 3s$ line as a function of target length for Cr single-stage (open circles) and three-stage targets (closed circles). Note that the three-stage target length is the combined length of all three stages as determined by the double-slit x-ray imaging camera.

with a single stage length. This is shown in Fig. 2 where the three-stage Cr target for 0.5 cm is higher than a 0.8 cm single-stage. For this situation, a close proximity injector-amplifier action is proposed as the explanation for the enhanced output of the multi-stage target. For nomenclature purposes, the first two stages, A and B in Fig. 1, are considered as the injector and stage C as the amplifier.

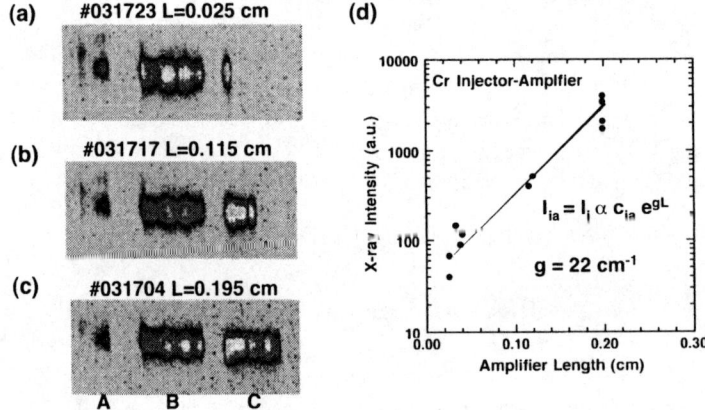

Fig. 3 (a), (b), (c) are x-ray images of the line focus where the amplifier section length, L, is varied from 0.025, 0.115 and 0.195 cm, respectively. (d) The x-ray laser output intensity of the Cr 28.5 nm line plotted as a function of amplifier length, L.

The length of the amplifier section, up to a maximum of 0.2 cm, could be varied by moving the target longitudinally in the line focus. The injector stage length (A and B) remained constant and it was possible to measure the total output from stage B, and therefore the injector intensity I_i coupled into the amplifier stage. Stage A length, a single-stage, was less than 0.1 cm and the output intensity was too low to measure. Figures 3(a) – (c) show x-ray double-slit images of the line focus plasma where the amplifier stage is being varied in length. The corresponding x-ray laser intensity of the Cr 28.5 nm line is plotted as a function of the amplifier length, Fig. 3(d). It would be expected that the injector-amplifier output, I_{ia}, can be written as $I_{ia} = I_i\ \alpha\ c_{ia}\ e^{gL} + I_a$ where I_i is the injector output, α the attenuation from cold plasma at the end of the injector and amplifier stages, c_{ia} coupling efficiency from injector to amplifier stages, g the small signal gain for amplifier length L, and I_a is the unseeded amplified spontaneous emission (ASE) from the amplifier stage. From Fig. 2 the ASE from a 0.2 cm target is ~1 count and so all of the amplifier output is due to seeding. The average small signal gain is estimated to be ~22 cm^{-1}.

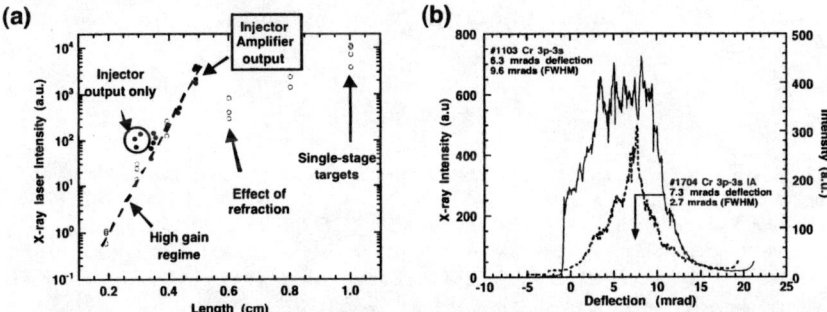

Fig. 4 (a) Cr x-ray laser intensity plotted against length for single-stage and multi-stage targets. (b) Angular pointing of a 1 cm single-stage plotted with a 0.5 cm multi-stage target.

Figure 4(a) shows the multi-stage output plotted together with the single-stage data. The overall output of the 0.5 cm multi-stage while not exceeding the 1 cm single stage output has an estimated gain length product of 15 and is close to saturation. Figure 4(b) shows the pointing angle and beam divergence of a 1 cm single-stage target and multi-stage 0.5 cm target. The single-stage deflection angle and beam divergence are similar to previous results for Ti [12, 13]. The multi-stage target although shorter has a substantially narrower beam divergence angle. Typically 2 – 3 times narrower beam divergence for the multi-stage targets was observed.

DISCUSSION

This simple injector-amplifier target architecture consisting of multiple stages demonstrates improved x-ray characteristics in beam divergence and output mainly by mitigating the effects of refraction, to be modeled in simulations. It was expected that the large scale non-uniformities in the line focus would be substantially detrimental to the x-ray laser propagation along the plasma column. This has not been the case and raises the question of beam uniformity in transient gain x-ray lasers. Secondly, at the ends of each stage regions of cold plasma will be created resulting in absorption losses by free-free inverse bremsstrahlung or photo-ionization of low Pd charge states. However, even though this setup is not ideal, the effect of creating a multiple stage target appears to maintain the x-ray laser propagating through the high gain region, Fig. 4(a), without the roll-off that affects single targets of length greater than 0.4 cm. The large spatial extent of the Ne-like gain region away from the target is expected to be important for this scheme.

ACKNOWLEDGMENTS

The authors would like to thank Jim Hunter for technical support in this research. This work was performed under the auspices of the US Department of Energy by the University of California Lawrence Livermore National Laboratory under Contract No. W-7405-Eng-48.

REFERENCES

1. M.D. Rosen, J.E. Trebes, and D.L. Matthews, *Comments Plasma Phys. Controlled Fusion* **10**, pp. 245 - 252 (1987).
2. G.M. Shimkaveg *et al.*, in *X-ray Lasers 1992*, edited by E.E. Fill, *IOP Conference Series* 125, Bristol and Philadelphia, 1992, pp.61 – 66.
3. C.L.S. Lewis, D. Neely, D.M. O'Neill, J.O. Uhomoibhi, M.H. Key, Y. Al Hadithi, G.J. Tallents, and S.A. Ramsden, *Optic. Comm.* **91**, pp. 71-76 (1992).
4. S. Wang *et al.*, *J. Opt. Soc. Am. B.* **9**, pp. 360-368 (1992).
5. J.J. Rocca, J.L.A. Chilla, S. Sakadzic, A. Rahman, J. Filevich, E. Jankowska, E.C. Hammarsten, and B. Luther, in *"Soft x-ray Lasers and Applications IV"*, ed. E.E. Fill and J.J. Rocca, *SPIE Proc.* **4505**, pp. 1-6 (2001).
6. N. Hasegawa, A.V. Kilpio, K. Nagashima, T. Kawachi, M. Kado, M. Tanaka, S. Namba, K. Takahashi, K. Sukegawa, P. Lu, H. Tang, M. Kishimoto, R. Tai, H. Daido, Y. Kato, *ibid* **4505**, pp. 204-210 (2001).
7. V.N. Shlyaptsev P.V. Nickles, T. Schlegel, M.P. Kalahnikov, and A.L. Osterheld, in *"Ultrashort Wavelength Lasers"*, ed. S. Suckewer, *SPIE Proc.* **2012**, pp. 111-118 (1993).
8. J. Dunn, J. Nilsen, A.L. Osterheld, Y. Li, and V.N. Shlyaptsev, *Opt. Lett.* **24**, pp. 101-3 (1999).
9. J. Dunn *et al.*, in "X-ray Lasers 2000", 7th International conference on X-ray Lasers, St. Malo, France June 19 – 23, ed. G. Jamelot, C. Möller, A. Klisnick, *J.Phys. IV* **11**, Pr2-19 (2001).
10. J.R. Crespo *et al*, *Proc. SPIE Int. Soc. Opt. Eng.* **2012**, pp.258-264 (1993).
11. A.Ya. Faenov, S.A. Pikuz, A.I. Erko, B.A. Bryunetkin, V.M. Dyakin, G.V. Ivanenkov, A.R. Mingaleev, T.A. Pikuz, V.M. Romanova, T.A. Shelkovenko. *Phys. Scr.* **50**, pp. 333 (1994).
12. M.P. Kalachnikov, P.V. Nickles, M. Schnürer, W. Sandner, V.N. Shlyaptsev, C. Danson, D. Neely, E. Wolfrum, J. Zhang, A. Behjat, A. Demir, G.J. Tallents, P.J. Warwick, and C.L.S. Lewis, *Phys. Rev. A* **57**, pp. 4778-4783 (1998).
13. J. Dunn, A.L. Osterheld, Y. Li, J. Nilsen, and V.N. Shlyaptsev, in "Short Wavelength Lasers and Applications", ed. J.G. Eden and J.J. Rocca, *IEEE Journal of Selected Topics in Quantum Electronics* **5**(6), 1441 – 1446 (1999).

Time-Resolved Measurements of the Transient X-ray Laser Emission

M. Edwards[1], Y. Abou-Ali[1], S. J. Pestehe[1], F. Strati[1], G. Tallents[1], S. Hubert[2], R. Keenan[2], S. Topping[2], C.Lewis[2], O. Guilbaud[3], A. Klisnick[3],D. Ros[3], R. Clarke[4], D. Neely[4], M. Notley[4]

[1]*Department of physics, York University, Heslington, YO10 5D, United Kingdom*
[2]*School of mathematics and physics, Queen's University of Belfast, Belfast BT7 1NN,United Kingdom*
[3]*LIXAM, Universite Paris-Sud, Bâtiment 350, 91405 Orsay Cedex, France*
[4]*RAL, Chilton Didcot, Oxfordshire OX11 0QX, United Kingdom*

Abstract. A measurement of the time-resolved emission of transient X-ray laser pulses is described. An ultra-fast X-UV streak camera set at the focal plane of a flat field Spectrometer was used to obtain the temporal evolution of X-UV spectra of Ni-like Ag and Ne-like Ni plasmas in a small wavelength range covering the lasing lines. The total time resolution of the device was of 1,1 ps. The FWHM duration of the X-ray laser pulse was measured to be 3.5 ps for Ni-like Ag $(3d^9 4d(3/2,3/2)_{J=0} \rightarrow 3d^9 4p(5/2,3/2)_{J=1}$, $\lambda=13.9$ nm) and 13 ± 2 ps for Ne-like Ni $(2p^5 3p(1/2,1/2)_{J=0} \rightarrow 2p^5 3s(1/2,1/2)_{J=1}$, $\lambda=23.1$ nm). Lasing signal was also observed on the 4f-4d laser line $(\lambda=16$nm) in Ni-like Ag,, in both time-integrated and time-resolved spectra.

INTRODUCTION

Since the first demonstration [1] of laser action in neon-like ions in the X-UV range there has been determined effort to improve the characteristics of such devices to make them suitable for various applications. In particular, effort has been invested to produce very intense and saturated soft x-ray lasers operating at shorter wavelength and with shorter pulse durations. The early X-ray lasers, based on collisional pumping of Ne-like ions, pumped with nanosecond laser pulses, showed relatively short-lived lasing ($\approx$100ps) on J=0-1 3p-3s line[2]. Shorter X-ray laser pulse durations were obtained employing multipulse and prepulse techniques with a main pulse of $\approx$100ps on Ne-like [3] and Ni-like ions[4] and pulse duration of 35 ps were routinely demonstrated [5]. With the advent of X-ray laser pumping using the transient excitation scheme and CPA pump lasers ($\approx$1 ps), shorter X-UV pulse durations were expected [6-9]. Direct observation with streak cameras of limited temporal resolution gave the first estimates of this duration [9-12].

Measurements of the X-ray laser pulse duration is important for a better understanding of amplification and propagation conditions in the plasma gain region of the X-ray laser. Moreover the X-UV pulse duration defines the temporal resolution in potential experiments using X-ray lasers.

CP641, *X-Ray Lasers 2002: 8ᵗʰ International Conference on X-Ray Lasers*, edited by J. J. Rocca et al.
© 2002 American Institute of Physics 0-7354-0096-2/02/$19.00

The duration of transient X-ray lasing has been recently measured by Klisnick et al [13] to be as short as 2 ps with picosecond pumping. However, the instrument temporal resolution for this measurement was of comparable magnitude. In this contribution, the duration of X-ray lasing is measured with an X-ray streak camera with 700 fs temporal resolution [14, 15]. Combined with a temporal smearing due to the spectrometer employed, we have measured X-ray laser pulse durations for Ni-like silver and Ne-like nickel with a total time resolution estimated at 1.1 ps.

The measurement of transient X-ray laser pulses of picosecond duration is difficult as most X-ray streak cameras have temporal resolution > 2 ps [14]. It is desirable to spectrally resolve the X-ray laser line from possible long-lived, broad band background plasma emission using a spectrometer. Spectrometer designs produce optical path length differences and hence temporal smearing. It is important to carefully minimise this temporal smearing in the operation of the spectrometer.

EXPERIMENT

X-ray lasing was produced in Ni-like silver at 13.9 nm and Ne-like nickel at 23.1 nm by irradiating solid silver or nickel slabs with two laser beams of wavelength 1.06 μm from the VULCAN glass laser at the Rutherford Appleton Laboratory. A plasma was formed with a background pulse of duration 280 ps and peak irradiance 2×10^{13} W.cm^{-2} in a line focus of length 16 mm and width 100 μm. The main pulse irradiance is enhanced by chirp pulse amplification (CPA) to give a duration of 1.2 ps and peak irradiance 7×10^{15} Wcm^{-2} in a line focus of length 12 mm and width 100 μm. The background pulse was produced from a component of the uncompressed main pulse and hence arrived on target without random jitter relative to the main pulse. Delay between the two pulses after optimising the output energy was 200 ps for Ni-like silver and 300 ps for Ne-like nickel. The background pulse and main pulse were focussed onto the targets with separate focussing systems comprising a refracting lens (the background pulse) or a parabolic shaped mirror (the main pulse) to produce a spot focus which was imaged in each case by an off-axis spherical mirror to produce the line focus on the target [16].

With ~ ps pumping, simulations and our measurements show that the gain duration is ~ 10 – 20 ps for Ni-like ions and 30 – 40 ps for Ne-like ions. As the X-ray laser pulse travels at a maximum of the vacuum speed of light c in the plasma (corresponding to a time of 33 ps/cm), this can result in a mis-match between the propagating X-ray laser pulse and the onset of gain. Travelling wave pumping is employed [17] to ensure that the X-ray pulse and the gain onset coincide. There is an intrinsic travelling wave velocity for the energy pulsefront along the line focus in our experiment of 2.5c due to the geometry for production of the line focus. A grating is employed before the compression gratings to reduce the laser pulsefront velocity to c along the line focus. Intrinsic and nominal travelling wave velocities were measured with a Hammamatsu IR streak camera with a time resolution of 1 ps.

The X-ray laser output was recorded using a flat field grating spectrometer [18] with an Axis Photonique streak camera [19] positioned at the spectrometer detection plane. The flat field spectrometer was positioned so that the X-ray laser output was incident at a grazing angle of 3.55° onto a $n_g \approx 1200$ line/mm grating of radius of curvature 5.649 m. An aperture near the spectrometer grating defined the vertical angular acceptance and ensured that only a length $d \approx 16$ mm of the grating in the direction along the grating surface normal to the grating rulings was illuminated. The distance d determines the temporal smearing Δt_g on the X-ray laser pulse produced by the grating.

We have

$$\Delta t_g = \frac{n_g d\lambda}{c} \qquad (1)$$

where λ is the X-ray laser wavelength. For Ni-like silver ($\lambda = 13,9$ nm) $\Delta t_g = 0,9$ ps. Ray tracing [20] shows that the aperiodic rulings on the grating [18] produce an approximate planar focus at distance 237 mm from the grating. The grating was positioned horizontally, so that the variation of the X-ray laser beam intensity with horizontal angle from the vertical target surface could be observed at the spectrometer detection plane. The spectrometer was initially operated with a CCD detection system for each target material to check the divergence and the deflection of the X-ray laser beam. The streak camera was subsequently positioned with a vertical entrance aperture at the position of peak X-ray laser output recorded by the CCD camera. The peak-to-peak delay between the pre-pulse and main pulse was varied to optimise the X-ray laser output using the flat field spectrometer with CCD detection (e.g. figure 1).

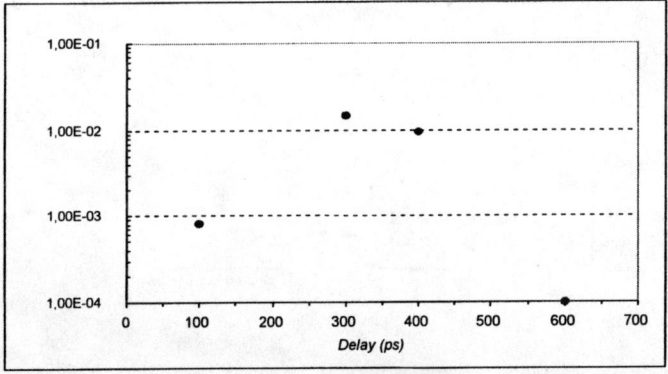

FIGURE 1. Time integrated energy output of the Ne-like Ni X-ray laser line at 23.1 nm from a target of length 4.3 mm as a function of the peak-to-peak separation between the CPA main 1.2 ps pulse and 280 ps background pulse. The background pulse is incident first.

The streak camera was triggered using an Auston switch irradiated by a component of the chirped laser pulse separately compressed using a grating pair. This minimised jitter between the main laser pulse on target and the streak trigger. The streak camera was operated below saturation so that space charge effects in the streak tube due to the production of too many photo-electrons did not degrade the temporal resolution. The streak camera temporal resolution and sweep speed were calibrated using frequency tripled pulses (wavelength 270 nm) from a 60 fs Ti:sapphiire laser operating at INRS, University of Quebec in Montreal [20]. With operation below saturation, the temporal resolution with a KI photocathode is estimated at 700 fs. Combining this resolution quadratically with Δt_g due to the spectrometer gives a total instrument temporal resolution of 1.1 ps. The streak camera sweep rate is 5 ($\pm$ 0.5) ps/mm. For some shots a Kentech 50/40 image intensifier was employed at the streak camera output phosphor screen and coupled to a Photometrics CH350 CCD detector. For other shots, we recorded the streak output directly onto the CCD detector.

RESULTS

The temporal variation of the Ni-like silver output at 13.9 nm and Ne-like nickel output at 23.1 nm as recorded with the streaked spectrometer using an image intensifier are shown in figure 3. The Ni-like silver X-ray laser output shows a sharp rise with a slow fall in output as is expected from modelling studies [21]. The X-ray laser pulse duration (full-width at half-maximum FWHM) for silver is 3 ps. Not employing an image intensifier gave a silver output with an approximately Gaussian shape, but again with a 3 ps pulse duration. Without the image intensifier, we believe that the streak camera may have been operating just into saturation (with a slight excess of photo-electrons) which degraded the temporal resolution so that the structure of the X-ray laser output could no longer be seen. When the estimated temporal resolution was comparable to the X-ray laser pulse duration, a Gaussian shape for the nickel X-ray laser pulse was also observed by Klisnick et al [13].

(a)

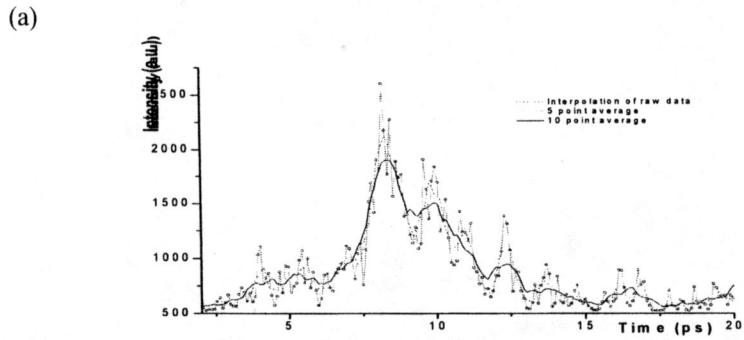

FIGURE 2a.   The temporal variation of X-ray lasing at 13.9nm for Ni-like silver. An image intensifier was used with the streak camera. A 10mm long silver target was irradiated with peak irradiance of 3.3×10^{12} Wcm^{-2} in the 280 ps background pulse and 1.3×10^{15} Wcm^{-2} in the 1.2 ps main pulse with peak-to-peak separation of 200 ps. The output was filtered with 1.2 μm plastic (CH). The FWHM X-ray laser duration was measured to be 3 ps.

(b)

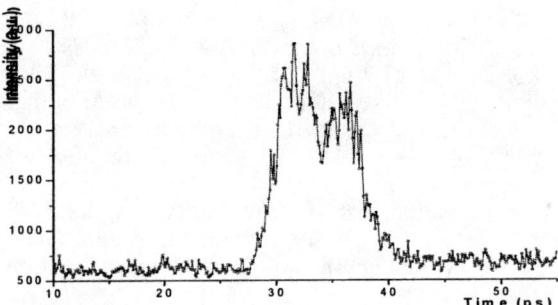

FIGURE 2b. The temporal variation of X-ray lasing at 23.1nm for Ne-like nickel. An image intensifier was used with the streak camera. A 6 mm long nickel target was irradiated with peak irradiance 6×10^{13} Wcm^{-2} in the 280 ps background pulse and 5×10^{15} Wcm^{-2} in the 1.2 ps main pulse with peak-to-peak separation of 300 ps. The output was filtered by 0.2 μm plastic (CH). The FWHM x-ray laser duration was measured to be 13 ps.

A pulse duration of 3.5± ps (full width at half maximum) was observed for the Ni-like silver line (4f $^1P_1$ $\rightarrow$ 4d $^1P_1$) at 16.1 nm, first observed by Kuba et al [12] (see figure 3). The Ne-like nickel X-ray laser output at 23.1 nm shows a sharp rise with a slow decrease with a pulse duration of 13±2 ps (full width at half maximum) . A double humped temporal modulation is also observed, probably caused by saturation of the streak camera.

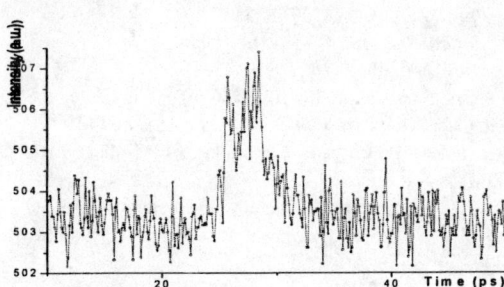

FIGURE 3. The temporal variation of the X-ray lasing line at 16.1nm for Ni-like silver. A 10 mm long silver target was irradiated with peak irradiance of 5.5×10^{12} Wcm^{-2} in the 280 ps background pulse and 1.7×10^{15} Wcm^{-2} in the 1.2 ps main pulse with peak-to-peak separation of 200 ps. The output was filtered by 0.6 μm of plastic (CH). The FWHM x-ray laser duration was measured to be 3.5 ps.

19

CONCLUSIONS

Measurements of the duration of X-ray lasing pumped with 1.2 ps pulses from the VULCAN optical laser have been obtained. We have measured X-ray laser pulse durations for Ni-like silver at 13.9 nm and Ne-like nickel at 23.1 m with a total time resolution of 1.1 ps. For Ni-like silver, the X-ray laser output has a steep rise following by an approximately exponential temporal decay with full width at half maximum (fwhm) of 3 ps. For Ne-like nickel, the duration of lasing is ≈ 13 ps (fwhm).

We plan further measurements of laser and gain duration with comparable temporal resolution to search for gain narrowing and saturation re-broadening of the X-ray laser pulse durations and the effect of the main pulse duration.

ACKNOWLEDGMENTS

This work is financially supported by the U.K. Engineering and Physics Science Research Council, by the E.U. Training and Mobility of Researchers programme and by the Laboratoire Europeen Associe. The assistance of the laser, target manufacture, engineering and other staff at the Central Laser Facility is gratefully acknowledged.

REFERENCES

[1] D.L. Matthews et al., *Phys. Rev. Lett.* **54**, 110 (1985)

[2] B.Rus et al., *J. Opt. Soc. Am. B* **11**, 564 (1994)

[3] R Kodama et al., *Optics Commun.* **90**, 95 (1992)

[4] H. Daido et al., *Phys. Rev. Lett.* **75**, 1074 (1995)

[5] J. Zhang et al., *Phys. Rev. Lett.*, **78**, 3856 (1997)

[6] P.V. Nickles et al., *Phys Rev. Lett.* **78**, 2748 (1997)

[7] P.J. Warwick et al., *J. Opt. Soc. Am. B* **15**, 1808 (1998)

[8] J. Dunn et al , *Phys. Rev. Lett.* **80**, 2825 (1998)

[9] A. Klisnick et al., *J. Opt. Soc. Am. B.* **17**, 1093 (2000)

[10] P.A. Simms, S. Mc Cabe, G.J. Pert, *Optics commun.* **153**, 164 (1998)

[12] J Kuba, A Klisnick, D Ros, P Fourcade, G Jamelot, J L Miquel, N Blanchot and J F Wyart 2000 Phys. Rev. A **62**, 043808

[13] A Klisnick, J Kuba, D Ros, R Smith, G Jamelot, C Chenais-Popovics, R Keenan, S J Topping, C L S Lewis, F Strati, G J Tallents, D Neely, R Clarke, J Collier, A G MacPhee, F Bortolotto, PV Nickles and K A Janulewicz 2002 Phys Rev A **65**, 033810.

[14] P Gallant et al, *Rev. Sci. Inst.*, **71** 10 3627-33 (2000)

[15] F Strati, D.Phil. thesis, University of York (unpublished) (2002)

[16] I N Ross, *Appl. Optics* **36**, 9348 (1987)

[17] J Collier, D Pepler, C Danson, J Warwick, C Lewis and D Neely 1997 Central Laser Facility Annual Report 1996/97 p209-10 (ISBN 0902376705)

[18] T Kita, T Harada, N Nakano, H Kuroda, *Appl. Optics*, **22**, 512 (1983)

[19] Axis-Photonique Inc., Varennes, Canada

[20] E Constant, E Mevel, A Zair, V Bagnoud and F Salin, *X-ray Lasers 2000* J Physique IV, **11**, Pr 2 p537-40 (2000)

[21] L Casperson and A Yariv, *Phys. Rev. Lett.* **26**, 6, 293-5 (1971)

Investigations on soft x-ray lasers with a picosecond-laser-irradiated gas puff target

H. Fiedorowicz[a], A. Bartnik[a], R. Jarocki[a], R. Rakowski[a], M. Szczurek[a],

J. Dunn[b], R.F. Smith[b], J. Hunter[b], J. Nilsen[b], and V.N. Shlyaptsev[c]

[a] *Institute of Optoelectronics, Military University of Technology, 00-908 Warsaw, Poland*
[b] *Lawrence Livermore National Laboratory, Livermore, CA 94551-9900*
[c] *University of California Davis-Livermore, Livermore, CA 94551*

Abstract. We present results of experimental studies on transient gain soft x-ray lasers with a picosecond-laser-irradiated gas puff target. The target in a form of an elongated gas sheet is formed by pulsed injection of gas through a slit nozzle using a high-pressure electromagnetic valve developed and characterized at the Institute of Optoelectronics. The x-ray laser experiments were performed at the Lawrence Livermore National Laboratory using the tabletop Compact Multipulse Terawatt (COMET) laser to irradiate argon, krypton or xenon gas puff targets. Soft x-ray lasing in neon-like argon on the $3p$-$3s$ transition at 46.9 nm and the $3d$-$3p$ transition at 45.1 nm have been demonstrated, however, no amplification for nickel-like krypton or xenon was observed. Results of the experiments are presented and discussed.

INTRODUCTION

Efficient laser-driven soft x-ray lasers are strongly required for many applications in science and technology. Substantial advances in the field have been achieved during the last few years using different variants of the prepulse technique and the travelling wave excitation [1-5]. In this technique a solid target is irradiated with two laser pulses. The first pulse creates a large-scale length preformed plasma, and the subsequent pulse heats the plasma to lasing conditions. Strong soft x-ray lasing with neon-like and nickel-like ions has been demonstrated with less than 10 J of laser pumping energy when the second pulse is a picosecond laser pulse from the CPA (chirped-pulse amplification) laser [6-9]. This opens the way for practical table-top soft x-ray lasers.

In this work we present investigations on soft x-ray lasers using a picosecond-laser-irradiated gas puff target instead of a solid. The gas puff target in a form of an elongated gas sheet was formed by pulsed injection of gas from a high-pressure solenoid valve through a slit nozzle. The gas puff valve was developed and characterized at the Institute of Optoelectronics. Gas puff targets have been used to obtain soft x-ray lasers pumped with sub-nanosecond high-energy lasers [10] and pose a number of substantial differences and advantages when compared to solid targets.

CP641, *X-Ray Lasers 2002: 8th International Conference on X-Ray Lasers*, edited by J. J. Rocca et al.
© 2002 American Institute of Physics 0-7354-0096-2/02/$19.00

A fundamental difference between gas puff and solid targets is that the gas puff provides the gain medium at a density closer to the lasing conditions. The laser-plasma coupling is also much different than with a solid, since the plasma starts below critical density. Moreover, use of a gas puff target instead of a solid target offers the advantage of developing a high-repetition rate x-ray laser with no target debris production. The gas puff target provides better control over the density and minimizes gradients in the plasma.

SOFT X-RAY LASER EXPERIMENTS

The x-ray laser experiments were performed at the Lawrence Livermore National Laboratory using the Compact Multipulse Terawatt (COMET) laser system. Two laser pulses of time duration 0.6 ns and 6 ps with a total energy of 10 J, were focused onto an elongated, sheet-like gas puff target of the maximum length of 0.9 cm. The line focus with length of 1.6 cm was achieved by using a cylindrical lens in combination with an on-axis paraboloid. The target was irradiated using the travelling wave geometry. The soft x-ray spectrometer equipped with a variable-spaced flat-field grating and a CCD camera was used to measure x-ray laser spectra along the axis of the line focus. Additionally, the plasma column was imaged using a x-ray camera with two crossed slits. The details of the experimental setup are given elsewhere [11-12].

Two series of experiments have been carried out. In the first series, performed in January 2000, the collisionally excited transient gain soft x-ray lasers with nickel-like krypton and nickel-like xenon were studied. Although the computer simulations showed high gains for both lasers, no lasing was observed in those experiments. The results of the studies were presented at the 7th X-ray Laser Conference at Saint Malo [13]. The reason for the lack of lasing was not clear, however, one reason may be related to optimizing the correct Kr gas density in the interaction region. In the case of the xenon laser the energy of the pumping laser pulse may not be sufficient for lasing. The x-ray laser with nickel-like xenon ions has been demonstrated recently at the Advanced Photon Research (APR) Center in Japan using the gas puff target created with the same valve but irradiated with higher laser energies [14].

In the second series of experiments performed at Livermore in March 2001 we studied soft x-ray lasing with neon-like argon ions, and strong amplification both on the $3p$-$3s$ transition at 46.9 nm and the $3d$-$3p$ transition at 45.1 nm was demonstrated with gain coefficients of 10.6 cm^{-1} for both lines on a maximum target length of 0.9 cm [15]. The output of both lines was stable and reproducible as a result of optimizing the experimental conditions. Spectral and spatial characteristics of the x-ray laser beams have been measured for various parameters of the gas puff targets and the pumping laser pulses [11, 12]. The beam divergence was measured to be 9 - 12 mrad (FWHM), containing narrow 1.2 – 3.0 mrad (FWHM) features, for the x-ray laser beams [12]. A typical on-axis spectrum from an irradiated gas-puff is shown in Fig. 1(a). Both x-ray

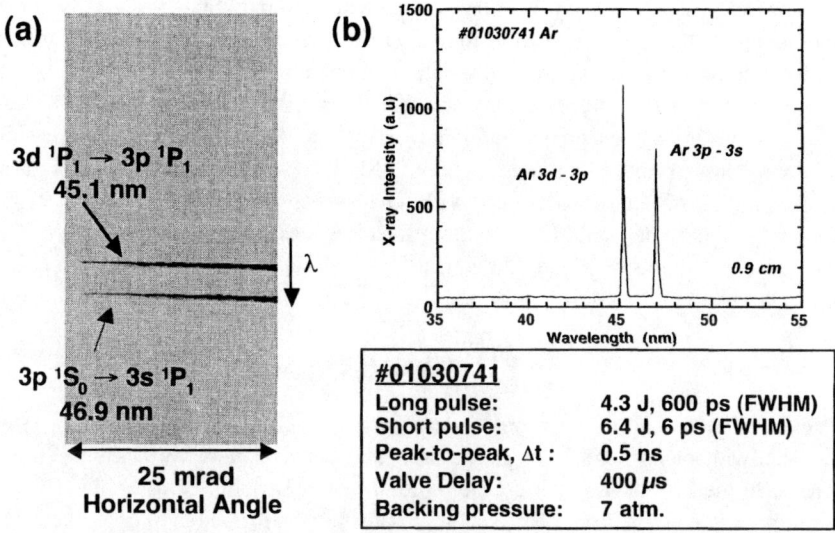

Fig. 1 (a) On-axis spectrum from an Ar gas puff target showing intense x-ray laser lines at 45.1 nm and 46.9 nm. The orthogonal direction to the dispersed axis is the horizontal deflection angle of the x-ray laser lines in the plasma blow-off direction. (b) An intensity lineout through the spectrum (a). Laser and experimental conditions are tabulated.

lines are labeled and the intensity of two x-ray laser lines are many orders of magnitude higher than non-lasing lines. The laser and experimental conditions are tabulated below Fig. 1(b). This was a spectra where the $3d$-$3p$ transition at 45.1 nm was more intense than the $3p$-$3s$ transition at 46.9 nm [12].

A transient gain soft x-ray laser for neon-like argon has been demonstrated also at the APR using the same type of gas puff target irradiated with two picosecond laser pulses of a total energy of 9 J [16]. A higher gain coefficient of 18.7 cm^{-1} and a narrower beam divergence of less than 3.7 mrad was measured for the $3p$ – $3s$ transition at 46.9 nm with gas puff targets up to 0.45 cm long. No $3d$ – $3p$ line at 45.1 nm was observed. A possible explanation of the differences between the two experiments can be connected to the larger than expected deflection angle of ~ 26 mrad for the x-ray laser beams presumably due to refraction of the x-rays in a higher density plasma column [12, 15]. Indeed, when the x-ray output measurements for the 0.9 cm long plasma column are neglected, the estimated gain coefficient for the shorter column will be similar for both experiments. This subject has been disscussed in a previous paper [17], however, more precise measurements of the gas density and the plasma density profiles as well as detailed studies of laser gas puff coupling are required to give better insight into the amplification process. The density measurements are planned for the next series of joint experiments at LLNL using x-ray spectroscopy and soft x-ray laser interferometry, in collaboration with Colorado State University. In this experiment we also plan to perform the gain measurements using a gas puff target up to 2 cm long. It is expected to reach the saturated operation for the argon laser.

The problem of the refraction and the gain decreasing for longer plasma columns could be solved by using a gain medium in a form of a plasma channel for waveguiding of the x-ray laser beam during amplification [18]. The possibility of creating a plasma channel in a gas puff target irradiated with a nansosecond laser pulse focused with an axicon, that should be useful for x-ray amplification, was demonstrated a few years ago [19]. The experiments on a transient gain soft x-ray laser produced in a plasma channel formed with an axicon will be performed.

CONCLUSIONS

Experimental studies on transient gain soft x-ray lasers using gas puff targets irradiated with picosecond laser pulses were presented. The x-ray laser experiments were performed at the Lawrence Livermore National Laboratory using the COMET laser to irradiate argon, krypton or xenon gas puff targets formed by pulsed injection of gas through a slit nozzle using a gas puff valve developed at the Institute of Optoelectronics. Amplification in neon-like argon ions was demonstrated, however, without amplification in nickel-like krypton or xenon ions. A strong effect of refraction of x-rays on the gain process was observed.

ACKNOWLEDGMENTS

The authors would like to thank Albert L. Osterheld for his support. Work performed under the auspices of the US Department of Energy by the University of California Lawrence Livermore National Laboratory under Contract No. W-7405-Eng-48 and supported, in part, by the grant No 2 P03B 093 16 of the State Committee for Scientific Research of Poland.

REFERENCES

1. J. Nilsen, B. J. MacGowan, L. B. Da Silva, J. C. Moreno, *Phys. Rev. A* **48**, 4682 - 4685 (1993).
2. J. C. Moreno, J. Nilsen, L. B. Da Silva, *Opt. Commun.* **110**, 585 (1994)
3. H. Daido, Y. Kato, K. Murai, S. Ninomiya, R. Kodama, G. Yuan, Y. Oshikane, M. Takagi, H. Takabe, F. Koike, *Phys. Rev. Lett.* **75**, 1074 (1995).
4. J. Nilsen and J. C. Moreno, *Opt. Lett.* **20**, 1386 (1995).
5. J. Zhang, A. G. MacPhee, J. Nilsen, J. Lin, T. W. Barbee, Jr., C. Danson, M. H. Key, C. L. S. Lewis, D. Neely, R. M. N. O'Rourke, G. J. Pert, R. Smith, G. J. Tallents, J. S. Wark, E. Wolfrum, *Phys. Rev. Lett.* **78**, 3856 (1997).
6. P. Nickles, V. N. Shlyaptsev, M. Kalachnikov, M. Schnürer, I. Will, W. Sandner, *Phys. Rev. Lett.* **78**, 2748 (1997)
7. J. Dunn *et al.*, *Phys. Rev. Lett.* **80**, 2825 (1998)

8. M. P. Kalachnikov *et al.*, *Phys. Rev. A* **57**, 4778 (1998).
9. J. Dunn, Y. Li, A. L. Osterheld, J. Nilsen, J. R. Hunter, V. N. Shlyaptsev, *Phys. Rev. Lett.* **84**, 4834 (2000).
10. H. Fiedorowicz, A. Bartnik, Y. Li, P. Lu, E. E. Fill, *Phys. Rev. Lett.* **76**, 415 (1996).
11. H. Fiedorowicz, A. Bartnik, J. Kostecki, R. Rakowski, M. Szczurek, J. Dunn, R. Smith, J. Hunter, J. Nilsen, V. N. Shlyaptsev, A. L. Osterheld, in *Proc. SPIE vol.* **4505**, *Soft X-ray Lasers and Applications IV,* eds. E. E. Fill and J. J. Rocca (SPIE Press Bellingham, 2001) p. 179.
12. J. Dunn, R. F. Smith, J. Nilsen, H. Fiedorowicz, A. Bartnik, V. N. Shlyaptsev, "Picosecond laser-driven gas puff neonlike argon x-ray laser", submitted to *JOSA B* (2002).
13. H. Fiedorowicz, A. Bartnik, J. Kostecki, R. Rakowski, M. Szczurek, J. Dunn, J. Hunter, J. Nilsen, V.N. Shlyaptsev, A.L. Osterheld, "Investigations on transient gain soft X-ray lasers using a laser-irradiated gas puff target" presented at 7th *International Conference on X-ray Lasers,* Saint Malo, France, June 19-23, 2000.
14. P. Lu, T. Kawachi, M. Kishimoto, K. Sukegawa, M. Tanaka, N. Hasegawa, M. Suzuki, R. Tai, M. Kado, K. Nagashima, H. Daido, Y. Kato, A. Bartnik, H. Fiedorowicz, "Demonstration of a transient gain nickel-like xenon ion x-ray laser", submitted to *Opt. Lett.* (2002)
15. H. Fiedorowicz, A. Bartnik, J. Dunn, R.F. Smith, J. Hunter, J. Nilsen, A.L. Osterheld, V.N. Shlyaptsev, *Opt. Lett.* **26**, 1403 (2001).
16. P. Lu, T. Kawachi, M. Suzuki, K. Sukegawa, S. Namba, M. Tanaka, N. Hasegawa, K. Nagashima, H. Daido, T. Arisawa, Y. Kato, H. Fiedorowicz, *Jpn. J. Appl. Phys.* **41**, L133 (2002)
17. V.N. Shlyaptsev, J. Dunn, K.B. Fournier, S. Moon, A.L. Osterheld, J.J. Rocca, F. Detering, W. Rozmus, J.P. Matte, H. Fiedorowicz, A. Bartnik, and M. Kanouff, in *Proc. SPIE vol.* **4505**, *Soft X-ray Lasers and Applications IV,* eds. E. E. Fill and J. J. Rocca (SPIE Press Bellingham, 2001) p. 14.
18. H. M. Milchberg, C. G. Durfee III, J. Lynch, J. Opt. Soc. Am. B **12**, 731 (1995)
19. H. Fiedorowicz, A. Bartnik, J. Kostecki, M. Szczurek, A. Morozov, S. Suckewer, in Proc. SPIE, vol. **3156**, *Soft x-ray lasers and applications II,* eds. J. J. Rocca and L. B. DaSilva (SPIE Press, Bellingham, 1997) p. 296.

A table-top collisional Ni-like Ag X-ray laser at 13.9 nm pumped by single picosecond laser pulse

K.A. Janulewicz, A. Lucianetti, G. Priebe, W. Sandner, P.V. Nickles

Max Born Institute, Max-Born-Str. 2A, D-12489 Berlin, Germany

Abstract. The saturated operation of a soft X-ray laser in Ni-like Ag at 13.9 nm pumped by a single profiled picosecond laser pulse with the length of 6 ps has been demonstrated. The pump energy of 3 J used for lasing in this scheme is equal to half of the minimum energy applied in the previously reported experiments. Small signal gain coefficient as high as 33 cm^{-1} has been estimated from the data. The influence of the pump beam characteristics on the output is discussed as well.

INTRODUCTION

There has been much progress recently in development of table-top X-ray lasers pumped by optical (IR) laser drivers. The pumping scheme with a transient inversion demonstrated and commonly used for $3p - 3s$ transitions in Ne-like Ti [1] has been successfully applied to Ni-like ions ($4d - 4p$ transitions) [2, 3, 4]. The saturated gain with the gain coefficient as high as 63 cm^{-1} has been reported for Ni-lke Pd at 14.7 nm [3]. The total pump energy applied in this experiment was as low as 7 J. This was an important step towards the table-top X-ray lasers. The LLNL-group has also reported lasing for many elements in Ni-like isoelectronic sequence. The saturated lasing was observed for molybdenum (Mo) at 18.9 nm and silver (Ag) at 13.9 nm. Additionally, laser action was observed in cadmium (Cd) at 13.2 nm and tin (Sn) at 11.9 nm [3]. In all these experiments a double-pulse (long/short) pump arrangement, typical for the transient inversion scheme, has been applied. The best results has been obtained in the focusing setup accompanied by a controlled travelling wave target irradiation with the excitation velocity $v_{exc} \leq c$. In such an arrangement a long (nanosecond) pulse preformed the plasma and then, after an optimised delay, a short (picosecond) pulse heated rapidly the medium. Depending on the required plasma density, temperature and the ionization stage the delay between both pulses was regulated. There were also attempts to use a combination of subpicosecond pulses to produce and excite the X-ray laser active medium [4]. On the other hand the experiment conducted on the VULCAN facility at Rutherford-Appleton Laboratory (RAL) with the pump energy noticeably higher than 10 J delivered the shortest ever measured X-ray laser pulse from Ni-like Ag at 13.9 nm with the length of 2 ± 0.8 ps [5]. As the plasma density gradient is extremely important due to the strong refraction effect a double pulse irradiation scheme with different variants of the double-colour ($1\omega/2\omega$, $2\omega/1\omega$) irradiation (2ω is the second harmonics of the pumping wavelength) as well as triple-pulse structure were tested as the pump of Ni-like Ag XRL [4, 6]. This experiment showed for the first time lasing at 16.1 nm suggesting reduction in the density gradients as well as influence of the self-photopumping effect. The shortest wavelength

CP641, *X-Ray Lasers 2002: 8th International Conference on X-Ray Lasers*, edited by J. J. Rocca et al.
© 2002 American Institute of Physics 0-7354-0096-2/02/$19.00

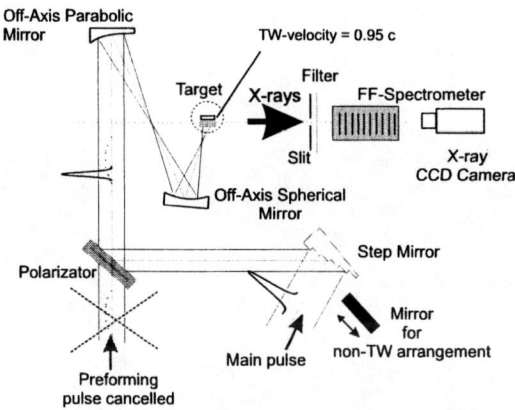

FIGURE 1. Experimental setup. The long preforming pulse marked by the dotted lines was switched off during the experiment but the scheme generally allows both the single and double-pulse irradiations.

obtained in the experiment with Ni-like ions in the transient regime was 7.3 nm from samarium [7]. However, the latter needed the energy which was far above that available from the table-top laser drivers. Lasing was also observed in Ni-like Nb, Zr and Y at the wavelengths between 20 and 24 nm with overall gain-length products of 11-12 [8]. The LLNL group observed for the first time the effect of the output signal increase for longer driving pulses with the length between 3 and 13 ps [9]. The driving pulse intensity was significantly reduced for such pulses. However, two pumping pulses were still required.

In this communication we are reporting the first demonstration of the saturated X-ray laser pumped by a single ultrashort laser pulse with the pump energy more than halved in comparison to the best achievements in the double-pulse scheme [3].

EXPERIMENTAL RESULTS

The experiment was performed at Max Born Institute (MBI) on the hybrid glass laser system operating at 1053 nm wavelength. This laser produces two beams with nominally 1 ps and 1.2 ns (FWHM) pulse durations and the repetition rate of 1 shot every 12 min. The short pulse was lengthened to ~6 ps and delivered energy between 2.5 and 3 J on the target. The line-formed focus had variable length in the range between 5 and 15 mm achieved by a combination of an off-axis paraboloid and off-axis spherical mirrors. A simple reflection echelon with 6 steps was applied for travelling wave irradiation arrangement. This echelon was aligned and positioned to give the speed of the excitation wave slightly lower (factor of 0.9-0.95) than the velocity of light c. The measured excitation speed for the intrinsic travelling wave produced by the focusing optics was $2.6c$. Both arrangements were used in the experiment on lasing in Ni-like Ag. The irradiation length during the experiment was constant but the target width was varied between 2 and 8 mm. The width of the focus line was 80-120 μm.

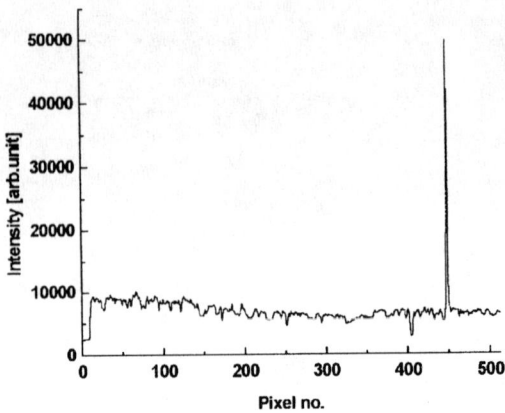

FIGURE 2. The emission spectrum showing the dominating 13.9 nm line (4d-4p J=0-1 transition in Ni-like Ag).

The spectral composition of the X-ray output was observed with a flat-field spectrometer and a 16-bit back-thinned 512x512 CCD camera. The spectrometer applied a 1200 l/mm variable-spaced grating of Harada type. Filter of different materials and thicknesses were used to control the radiation intensity reaching the camera's chip. The filter set included 0.2 μm Zr, 538 $\mu g/cm^{-3}$ (0.5 μm) Mo and combination of Zr/Si/Zr with the transmission of 0.03 at 13.9 nm. Some of the filters had a mesh support and this was used as a set of fiducial wires to estimate quantitatively the X-ray beam divergence and deflection. The Si absorption edge of the filter facilitated spectrometer calibration.

Figure 2 shows the cross section of the temporally integrated spectrum registered in the experiment. A polished, slab silver target with the thickness of 2 mm was irradiated normal to the surface. Different target lengths were used in the experiment and the X-ray output signal as a function of length is shown in Figure 3. The solid straight line determines the gain coefficient for the short targets using commonly applied Linford formula for the gaussian line broadening. This value for the non-travelling wave has been found to be equal to 33 ± 0.2 cm^{-1}. This value is comparable with that obtained in other experiments applying higher energies and controlled (matched) travelling wave excitation. The values put into the plot (Fig.3) were calculated from the raw data taking into account the theoretical (given) transmission of the filters. This could also contribute to the typical scatter of the measurement points, especially at the roll-off region. The roll-off effect, i.e. flattening of the output signal deviation from the exponential dependence on the target length, suggests that the saturation of the output signal has been achieved. However, the short-lived gain in the pumping schemes with the transient inversion does not allow to distinguish clearly between both effects. Travelling wave arrangement implemented increased the output signal noticeably and the same influence of the finite transit time on the gain could be extracted. Thus, the assumption that a significant contribution to the roll-off effect really comes from the saturation effect seems to be fairly justified.

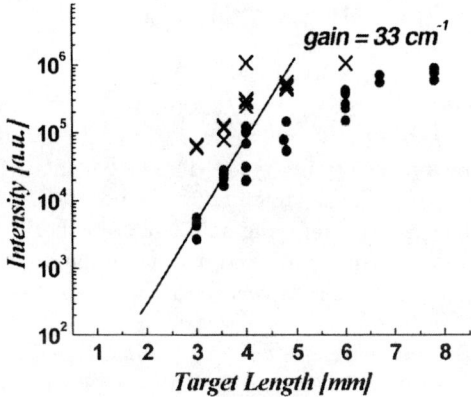

FIGURE 3. Dependence of the output signal on the target length. The dots correspond to the intrinsic travelling wave pumping scheme; the crosses correspond to pumping with the controlled speed of the travelling wave excitation.

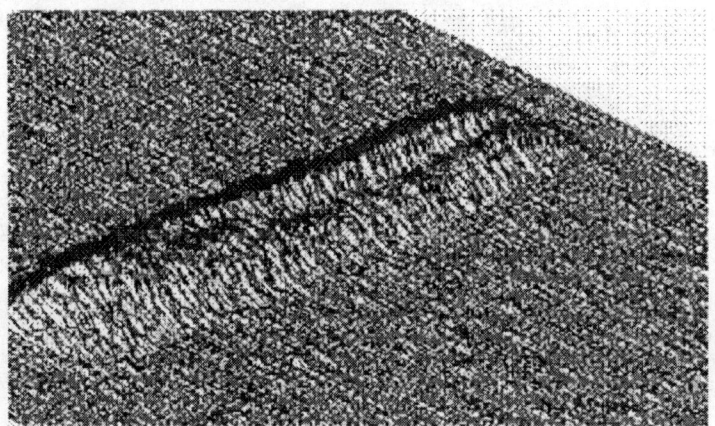

FIGURE 4. Intensity spatial distribution in the focus.

The focus quality is an another problem appearing in the experiment. A crossed-slit camera was used to monitor the intensity distribution in the focal line. A 3D-plot of the intensity distribution is shown in Fig.4. The distribution is strongly non-uniform in both axial and radial directions. It was suspected that the concave structure seen on the right side of the plot could improve lasing performance by the guiding effect. However, strong lasing was also present with the focal line structure without the concave density profile. Thus, this result actually excludes guiding as a possible source of the performance improvement.

DISCUSSION

The most important question which has to be answered here is the reason for such a strong lasing effect with very soft pumping. Some indications consistent with our observation could be concluded from the results reported in [3, 9]. It was observed there that lower energy of the preforming pulse is more beneficial for the X-ray laser output and it was assumed that lower temperature of the preformed plasma reduces population of the excited levels. The pumping pulses in our experiment were registered with an optical streak camera (HAMAMATSU) but due to the limited dynamics of the camera no noticeable shape deformations was registered. To overcome this, the input signal has been increased and the main part of the pumping pulse saturated the camera, while the low-level part of the signal has been strengthened and revealed. Neglecting the broadening effect of the main pulse due to the camera saturation, a long (a few hundred picosecond) leading edge (pedestal) was clearly seen. It is suggested that this low-level pedestal (with the duration in subnanosecond range) is responsible for the plasma preforming in our experiment. It was attempted to estimate quantitatively the pedestal level and the measurement of the total background energy gave 7 mJ what led to the corresponding energetic contrast of 3.2×10^3. Also other output characterisctics of the X-ray laser under investigation have been either directly measured or estimated. Divergence in the horizontal plane estimated from the spectroscopic images has shown minimum of 4 mrad at 4 mm target length which corresponds very well with the trends observed in [3]. Using theoretical values of the filter transmission and efficiency of the well electron generation by the photons the output energy between 2 and 5 μJ has been roughly estimated.

In conclusion we observed saturated emission of an X-ray laser in Ni-like Ag at 13.9 nm using as a pump a profiled single laser pulse with the duration (FWHM) of a few picoseconds. Gain as high as 33 cm^{-1} has been measured and the gain-length product of $\sim$ 16.5 has been achieved with the pumping energy of 2.5-3 J. We believe that this result opens new possibilities on the way towards efficient table-top soft X-ray lasers.

REFERENCES

1. Nickles, P.V., Schnuerer, M., Kalachnikov, M.P., Will, I., Sandner, W., Shlyaptsev, V.N., Phys. Rev. Lett. **78**, 2748 (1997).
2. Dunn, J., Osterheld, A.L., Shepherd, R., White, W.E., Shlyaptsev, V.N., Stewart, R.E., Phys. Rev. Lett. **80**, 2825 (1998).
3. Dunn, J., Li, Y., Osterheld, A.L.,Nilsen, J., Hunter, J.R., Shlyaptsev, V.N., Phys. Rev. Lett. **84**, 4834 (2000).
4. A. Klisnick *et al.*, J. Opt. Soc. of Am. B **17**, 1093 (2000).
5. A. Klisnick *et al.*, Phys. Rev. A **56**, 2493 (2002).
6. J. Kuba *et al.*, Phys. Rev. A **62** , 043808 (2000).
7. A.G. MacPhee *et al.*, Phys. Rev. A **64**, 053810 (2001).
8. Dunn, J., Nilsen, J., Osterheld, A.L., Li, Y., Shlyaptsev, V.N., Opt. Lett. **24**, 101 (1999).
9. Dunn J., *et al.*, "X-Ray Lasers 2000" Proceedings of the 7th IXRLC in St. Malo, France, June 19-23 2000; EDP Sciences, Les Ulis; Eds. G. Jamelot, C. Möller, A. Klisnick, J. Phys. IV France, **11**, (2000) Pr2-19.

Progress in X-Ray Laser Development at JAERI

Yoshiaki Kato, Hiroyuki Daido, Keisuke Nagashima, Tetsuya Kawachi, Noboru Hasegawa, Momoko Tanaka, Huajing Tang, Renzhong Tai, Peixiang Lu, Maki Kishimoto, Masataka Kado, Kouta Sukegawa, Masato Koike, Akira Sasaki, Kengo Moribayashi, Koichi Yamakawa, *Kazumichi Namikawa, **Henryk Fiedorowicz, **Andrzej Bartnik, and ***Betrich Rus

Advanced Photon Research Center, Kansai Research Establishment,
Japan Atomic Energy Research Institute, 8-1 Umemidai, Kizu, Souraku, Kyoto, 619-0215 Japan
*Department of Physics, Tokyo Gakugei University, 4-1-1 Nukui-Kita, Koganei, Tokyo, 184-8501
Japan
** Institute of Optoelectronics, ul. Kaliskiego 2, 00-908 Warsaw, Poland
*** PALS Research Center, Institute of Physics,18221 Prague 8, Czech Republic

Abstract. By traveling wave irradiation with a compact, 2-beam, 1-ps, 20-J CPA Nd-glass laser dedicated to x-ray laser research, saturated amplification has been demonstrated in Ni-like Ag at 13.9 nm and Ni-like Sn at 12.0 nm. High gain amplification has been also obtained in Ni-like La at 8.8 nm and with gaseous targets in Ne-like Ar at 46.9 nm and Ni-like Xe at 9.98 nm. Generation of highly coherent x-ray laser is planned by injection seeding of the plasma amplifier. The newly developed 340-TW Ti-sapphire laser may enable generating exceedingly high power x-ray radiation which will be useful for pumping inner-shell ionization x-ray lasers.

INTRODUCTION

Development of the transient-gain collisional excitation (TCE) x-ray laser [1] has enabled down-sizing of the collisional excitation x-ray lasers to the table-top [2] while preserving the advantage of high photon flux. With the traveling-wave (TW) irradiation by less than 10-J pumping energy, saturated amplification in Ne-like Cr and Fe and in Ni-like Mo, Pd, Ag, and unsaturated high-gain amplification in Ni-like

CP641, *X-Ray Lasers 2002: 8th International Conference on X-Ray Lasers,* edited by J. J. Rocca et al.
© 2002 American Institute of Physics 0-7354-0096-2/02/$19.00

Cd and Sn have been demonstrated at LLNL, with the shortest wavelength lasing in Sn at 12.0 nm [3]. Although the TCE x-ray lasers have been very successful, there still remain various issues, such as optimization of the lasing conditions for realization of more compact facilities, extending the lasing to shorter wavelengths, improvement of the coherence, and development of various x-ray laser applications. Research on other x-ray laser schemes is also important since the TCE x-ray laser has limitations to fulfilling all the requirements.

The Advanced Photon Research Center (APRC) has moved to the present new site in Kyoto Prefecture from JAERI Tokai Laboratory in 1999, and started new research activities on development and applications of high-intensity ultra-short pulse lasers [4, 5]. The x-ray laser research has been concentrated mostly on the TCE x-ray lasers using a 2-beam CPA Nd-glass laser which is designed for flexible operation to investigate various issues as described above.

In this paper we report progress in our x-ray laser research at APRC after the 7th International Conference on X-Ray Lasers in 2000 [6]. In these 2 years, we have implemented traveling wave pumping to the CPA Nd-glass laser [7], and succeeded in achieving saturated amplification in Ag at 13.9 nm and Sn at 11.9 nm [8, 9], and strong amplification in Sm at 8.8 nm [10]. We have also obtained strong amplification in Ar at 46.9 nm [11] and Xe at 9.98 nm with gas-puff targets, all based on the TCE scheme.

Injection seeding of the x-ray laser for generating high-coherence x-ray laser has been started [12]. Coherence of the Ag laser has been characterized, x-ray laser interferometer has been developed, and several x-ray laser application experiments have been started [13].

EXPERIMENTAL FACILITY

We have developed a high performance and fairly compact 2-beam CPA Nd-glass laser for x-ray laser experiments [7]. The 2-beam system, operating at 1.053 μm with 10-min shot interval, allows us to test various irradiation conditions, such as the oscillator-amplifier configuration and the 2-sided irradiation of single targets. Each beam of this laser is capable of generating a pre-pulse of variable pulse width and energy at a proper timing to the main pulse for optimum plasma formation and a main pulse of 1-ps duration and 20-J energy. The laser beam is focused with reflection optics without chromatic aberration to a line of 6-mm length and 15-μm width on target. A 6-step mirror has been installed into the line focusing optics for quasi-traveling wave pumping at the velocity of light with an optical delay of 4 ps between each step.

A tunable-wavelength and variable-bandwidth Ti-sapphire laser, which is synchronized to the Nd-glass laser, is being developed in order to generate high-power, coherent x-ray laser radiation by seeding the high-harmonics generated with the Ti-sapphire laser into the amplifier produced with the Nd-glass laser [12].

The spectrum of the x-ray laser is analyzed in the axial-direction with a grating spectrometer, which employs a newly-developed holographic laminar-groove diffraction grating of varied line-spacing for flat-field imaging with high diffraction efficiency to the first order, especially at shorter wavelengths [14]. The near field pattern of the x-ray laser beam is imaged with a 10 x optics on a CCD detector [15]. The ionization balance is diagnosed by the x-ray spectra of resonance lines with a crystal spectrograph.

EXTENDING X-RAY LASING TO SHORTER WAVELENGTHS

We have been investigating on extending the lasing of the TCE x-ray lasers to shorter wavelengths with our compact pumping laser. The major issue is to prepare a plasma with highly charged ions in a region with a modest density gradient, and pump the plasma for excitation of the Ni-like ions to the upper laser level. Direct extension of the long-pulse irradiation for pre-plasma formation adopted in previous experiments is not applicable, since generation of the Ni-like ions of higher-Z elements requires high-intensity irradiation, resulting in high-energy laser for pre-plasma formation.

We have irradiated solid targets with a pre-pulse of short-duration and high-intensity to generate highly charged ions, and with a main pulse after some delay from the pre-pulse to form a long scale-length plasma by thermal expansion. More specifically, a Ag slab target was irradiated with a 4-ps pre-pulse and, after 1.2 ns, with a 4-ps main pulse with the total laser energy of 12 J at the intensities of 3.0 x 10^{14} W/cm^2 and 2.3 x 10^{15} W/cm^2, respectively. Since these short laser pulses were accompanied with lower intensity ASE, the target was kept irradiated at an intensity of ~10^{12} W/cm^2. Similar irradiation condition was kept for Sn targets except the total laser energy was increased to 14 J.

Figure 1 shows spectra and near field patterns of the Ag x-ray laser taken at different amplifier lengths. As the amplification length increases up to 6 mm, the lasing line becomes very strong and the near field pattern becomes more confined to the central part, resulting in a narrower beam with the diameter of 20 μm by 40 μm and the beam divergence of ~5 mrad by ~10 mrad in the horizontal and vertical directions, respectively. The energy of the Ag x-ray laser increases exponentially with the plasma length up to 4.5 mm, and saturates approaching to linear increase with the length (Fig. 2). The measured small-signal gain is $g = 35$ cm^{-1} resulting in the gain-length product of $gl = 13.6$ at the saturation length $l_{sat} = 3.9$ mm. These values should be compared with the unsaturated amplification without the TW irradiation of $g = 24$ cm^{-1} and $gl = 10$ for $l = 4$ mm [7]. The measured energy of the x-ray laser is $E \sim 25$ μJ, corresponding to 1.8 x 10^{12} photons per pulse. The gain duration of the lasing transition has been estimated to be $\tau \sim 8$ ps from the measured ratio of the x-ray laser energies without and with TW

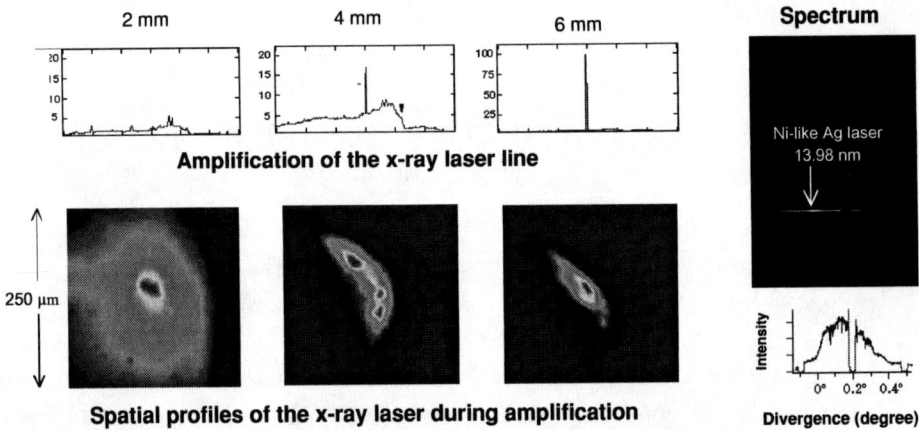

FIGURE 1. Spectra and near field patterns at different gain lengths (left) and horizontal beam divergence at 6mm (right) of the Ag x-ray laser.

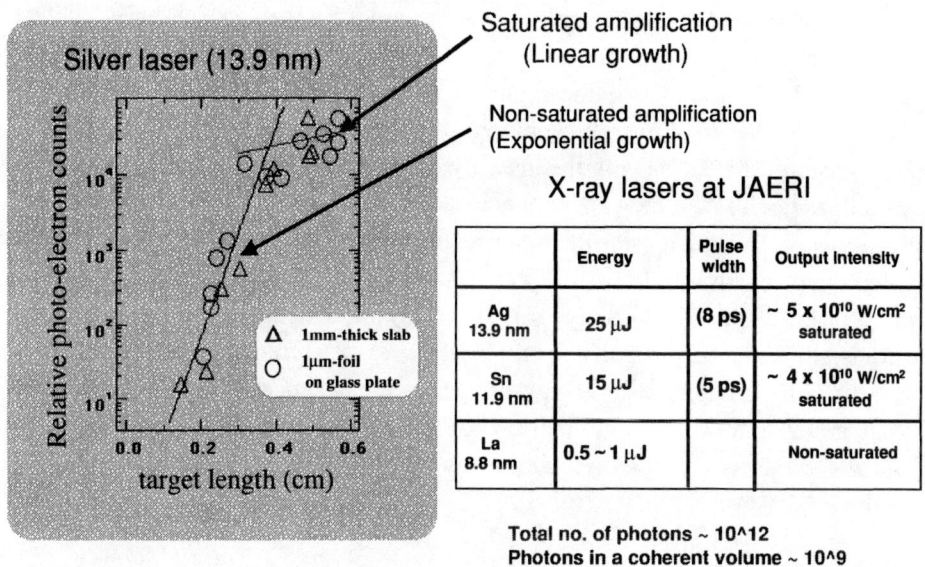

FIGURE 2. Energies of Ag x-ray laser vs. the gain length (left) and energies and estimated pulse widths and saturation intensities of the Ag, Sn and La x-ray lasers.

irradiation. The saturated output intensity is then estimated to be $I_{sat} \sim 4.5 \times 10^{10}$ W/cm^2 [9].

Saturated amplification has been obtained also with the Sn laser at 11.9 nm, with $g = 30$ cm^{-1} and $gl = 13.6$ for $l_{sat} = 4.4$ mm and the measured laser energy of $E \sim 17$ µJ. The estimated gain duration and the saturated intensity for the Sn laser are $\tau \sim 5$ ps and $I_{sat} \sim 5.0 \times 10^{10}$ W/cm^2, respectively [9].

Based on the hydrodynamic and atomic physics code simulations for the Ag laser, the electron temperature and the electron density of the gain region are estimated to be $T_e = 500$ eV and $n_e = 5 \times 10^{20}$ cm^{-3}, respectively. The analyses show the importance of the ASE irradiation of the plasma at the intensity of $\sim 10^{12}$ W/cm^2 for keeping the electron temperature to ~ 200 eV and thus sustaining the proper ionization degree, enabling rapid ionization to the Ni-like stage and excitation of the Ni-like ions to create the population inversion by the main laser pulse irradiation.

We have extended this study to test lasing at a shorter wavelength of 8.8 nm with Ni-like La [10]. With the total pumping laser energy of 18 J, lasing was obtained with $g = 14.5$ cm^{-1} and $gl = 7.7$ at the maximum length of 5.3 mm. It is expected for La that the density of the gain region is higher than the critical density of the pumping laser and it is heated by thermal conduction from the high-temperature critical density region.

We have also tested amplification in gaseous targets, which have inherent advantages of higher repetition rate with less debris formation. A high density Ar gas flow ejected from a slit of 6-mm length and 500-μm width was irradiated with a TW laser beam of 9-J total energy composed of a 1.5-ps pre-pulse and a 1.5-ps main pulse separated by 1.2 ns with the intensity ratio of 1:6. Strong amplification in the 3p-3s transition in Ne-like Ar at 46.9 nm was obtained with $g = 18.7$ cm^{-1} with small divergence of 3.7 mrad with interference–like structures [11]. This gain is higher than the previous result of $g = 10.6$ cm^{-1} obtained with a 0.6 ps pre-pulse and a 6 ps main pulse with 10 J total energy [16]. This study was extended to lasing at a shorter wavelength using Xe gas puff target. Under the similar irradiation condition as Ar, but with a higher total laser energy of 18 J, we have succeeded in observing the lasing at 9.98 nm in the 4d-4p transition of the Ni-like Xe ions with the gain of $g = 17.4$ cm^{-1}. This is the first definite demonstration of lasing in Xe with the TCE scheme.

COHERENCE IMPROVEMENT

Measurement of the coherence properties of the Ag laser with a multi-slit array shows that the spatial coherence at a far field is determined by the x-ray laser source as an almost incoherent disc. In order to generate an x-ray laser with higher spatial coherence, x-ray laser with a smaller beam size has to be produced by using a smaller area for amplification. This condition may be achieved by injecting a coherent beam into the plasma amplifier for controlled amplification. The intensity of the seeded radiation has to be higher than the spontaneous emission intensity to overcome the growth of the ASE in the amplifier. Another requirement is the homogeneity of the plasma amplifier so as not to disturb to quality of the injected beam during amplification.

Two schemes are being prepared for this objective. The first is the oscillator-amplifier configuration, where the amplifier is placed at a distance from the oscillator to fulfill the condition of small Fresnel number (Fig. 3). This approach has an advantage of perfect frequency matching between the oscillator and the amplifier. Another scheme is to inject a high order harmonics to a plasma amplifier [12]. The advantage of this approach is that the spatial as well as the temporal coherence of the x-ray laser will be controlled if we can control these properties of the high-order harmonics. The 57-th order harmonics of the 800 nm laser, which matches to the 13.9 nm lasing transition of the Ag laser, has been generated using a

100-fs, 1-TW Ti:sapphire laser in a preliminary experiment.

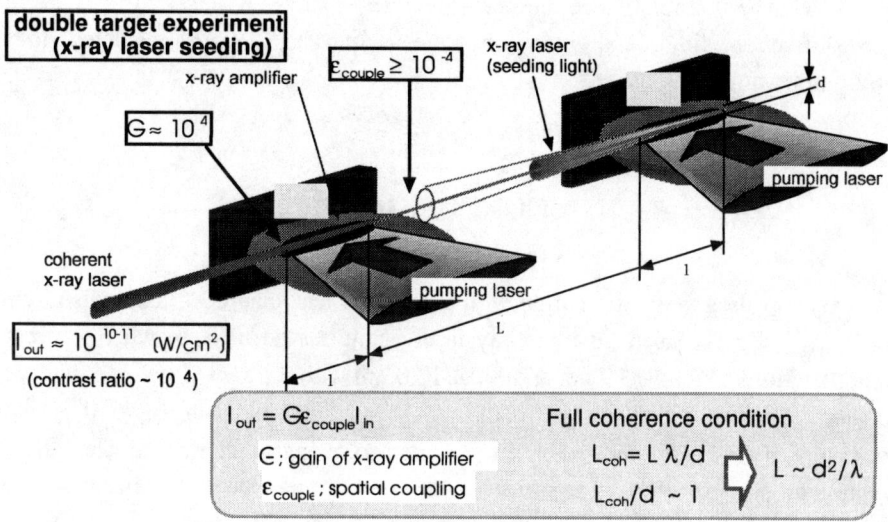

FIGURE 3. Concept for generation of highly coherent x-ray laser beam by injection seeding with the oscillator – amplifier configuration.

HIGH-POWER, ULTRASHORT-PULSE LASER

Investigation of various schemes other than the collisional excitation pumping is important for developing x-ray lasers with higher efficiency and shorter wavelength lasing. As an extension of the photo-pumped inner-shell ionization x-ray lasers [17], we have proposed the hollow atom x-ray laser, where lasing at shorter wavelengths with high gain over loner duration is expected when the inner shell electrons are completely ionized [18]. Especially for the hollow atom laser, intense and short duration x-ray sources are required for pumping the inner-shell ionization x-ray lasers.

Recent PIC simulation including the radiation damping predicts that very high x-ray conversion rate exceeding 30 % is expected as the irradiation intensity exceeds $\sim 10^{22}$ W/cm^2 [19]. This very intense x-ray radiation, when realized, will open vast new applications including the x-ray laser pumping. We have recently succeeded in producing a 18 J, 52 fs pulse, corresponding to 340 TW, by adding a large aperture amplifier to the 100-TW Ti-sapphire laser [20]. Achieving the irradiance of ~ 10^{22} W/cm^2 will be possible, since the beam divergence of the 340 TW laser is close to the diffraction limit and increasing the output to PW is expected to be straightforward.

CONCLUSIONS

By traveling wave irradiation with a 1-ps and less than 20-J laser pulse from a compact Nd-glass laser, intense x-ray laser beam has been generated by saturated amplification in Ag at 13.9 nm and Sn at 12.0 nm with the x-ray laser energies of 25 μJ and 17 μJ, respectively. The flux of the Ag laser corresponds to 1.8 x 10^{12} photons per pulse. This experiment has been extended to lasing at shorter wavelengths and high-gain amplification in Ni-like La at 8.8 nm has been obtained. In these experiments, a short-duration, high-intensity laser pulse was used to form a pre-plasma before the main pulse irradiation. This is in contrast to the use of long-duration, low-intensity laser pulse for pre-plasma formation, where high energy is required for extending the lasing to shorter wavelengths. With the similar irradiation conditions, high gain amplification has been also obtained with gas-puff targets in Ne-like Ar at 46.9 nm and for the first time in Ni-like Xe at 9.98 nm. More detailed studies are necessary to clarify the lasing conditions in these gaseous targets, such as the location, density and temperature of the gain region.

Several experiments have been started on application of the x-ray lasers, including x-ray interferometer for high-density plasma diagnostics, differential x-ray microscope for observing transparent small objects such as biological specimen, x-ray laser diffraction to study the phase transitions of ferro-electric crystals, and induced x-ray parametric scattering for mixing the x-ray laser and optical laser frequencies. The applications of the x-ray lasers will be significantly extended by the development of high-coherence and high-power x-ray lasers, which will enable single-shot, ps-resolution observation, manipulation, or fabrication of nano-scale structures.

A 340-TW, 50-fs laser pulse has been generated recently at APRC with the Ti-sapphire laser. The extremely high-intensity reaching 10^{22} W/cm^2 generated with the PW-class lasers will open various new phenomena, including x-ray generation of PW power. This new experimental regime may enable realizing x-ray lasers in shorter wavelengths.

REFERENCES

1. Nickles P. V., et al., Phys. Rev. Letters **78**, 2748 (1997).

2. Dunn J., et al., Phys. Rev. Letters **80**, 2825 (1998).

3. Dunn J., et al., in *X-Ray Lasers 2000*, ed. G. Jamelot, C. Moller and A. Klisnick, *J. Physique IV* **11**, Pr2-19 (2001), and references cited therein.

4. *Annual Report of Kansai Research Establishment 1999*, JAERI-Review 2001-003, Japan Atomic Energy Research Institute, March 2001.

5. *Annual Report of Kansai Research Establishment 2000*, JAERI-Review 2001-046, Japan Atomic Energy Research Institute, February 2002.

6. Kato Y., et al., in *X-Ray Lasers 2000*, ed. G. Jamelot, C. Moller and A. Klisnick, *J. Physique IV* **11**, Pr2-3 (2001).

7. Kawachi T., e al., Applied Optics, to be published.

8. Kado M., et al., Proc. SPIE **4505**, 54 (2001).

9. Kawachi T. et al., Phys. Rev. A, submitted.

10. Kawachi T, et al., in *X-Ray Lasers 2002*, these Proceedings.

11. Lu P., et al., Jpn. J. Appl. Phys. **41**, L133 (2002).

12. Hasegawa N., et al., Proc. SPIE **4505**, 204 (2001).

13. Daido H., et al., in *X-Ray Lasers 2002*, these Proceedings.

14. Koike M., et al., Proc. SPIE **4146**, 163 (2000).

15. Tanaka M., et al., Surface Rev. Letters **9**, No. 1 & 2 -VUV Special Issue- (2002), to be published.

16. Fiedorowicz H., et al., Opt. Letters **26**, 1403 (2001).

17. Moon S. J. and Eder D. C., Phys. Rev. A **57**, 1391 (1998).

18. Moribayashi K., Sasaki A., and Tajima T., Phys. Rev. A **58**, 2007 (1998).

19. Zhidkov A., Koga J., Sasaki A., and Uesaka M., Phys. Rev. Letters **88**, 185002-1 (2002).

20. Aoyama M., et al., presented at *CLEO 2002*, paper CMW5 (2002).

Observation of strong amplification at 8.8 nm in the TCE scheme by a table-top pumping system

Tetsuya Kawachi, Momoko Tanaka, Akira Sasaki, Maki Kishimoto,
Mamiko Nishiuchi, Kazuhito Yasuike, Noboru Hasegawa,
Alexander V. Kilpio[†], Peixiang Lu, Renzhong Tai, Huajin Tang[fl],
Masataka Kado, Keisuke Nagashima, Masato Koike,
Hiroyuki Daido and Yoshiaki Kato

Advanced Photon Research Center, Kansai-Research Establishment,

Japan Atomic Energy Research Institute, 8-1, Umemidai, Kizu, Kyoto, 616-0215, Japan

† General Physics Institute, Russian Academy of Science, 38 Vavilov Street, 117942, Moscow, Russia

fl LIXAM, Unversité Paris-Sud, CNRS UMR 8624, 91405, ORSAY CEDEX, France

Abstract. We observed strong amplification of the transition of $4d$ $4p$, $J = 0$ 1 (the transition from $(3d_{3/2}, 4d_{3/2})_0$ to $(3d_{3/2}, 4p_{1/2})_1$) of the Ni-like lanthanum (La) ions at a wavelength of 8.8 nm pumped by a compact CPA Nd:Glass laser light at a wavelength of 1.053 μm with a pumping energy of 18 J. The experimental gain coefficient and the achieved gain-length product was 14.5 cm^{-1} and 7.7, respectively. In this experiment, the pumping laser pulse consisted of a pre-pulse with a duration of 200 ps and a 7ps-duration main pulse, separated by 250 ps. A hydrodynamics simulation coupled with a collisional-radiative model showed that the present experimental condition generated a pre-formed plasma with small volume and made it possible by the main pulse to heat the high density region effectively.

INTRODUCTION

Recent advent of table-top compact soft x-ray lasers makes it possible to generate intense x-ray beam in a wavelength region upto $\sim$ 12.0 nm. Especially for the transient collisional excitation (TCE) scheme, gain-saturation behaviors have been observed with a pumping energy of less than 15 J [1-4]. In a shorter wavelength region than 10 nm, there has been only one report on substantial amplification of the Ni-like Sm ion laser at 7.3 nm with a pumping energy more than 60 J [5]. Demonstration of substantial amplification in this wavelength region using a small pumping system is valuable for the development of high efficient compact x-ray lasers.

CP641, *X-Ray Lasers 2002: 8th International Conference on X-Ray Lasers*, edited by J. J. Rocca et al.
© 2002 American Institute of Physics 0-7354-0096-2/02/$19.00

In the case of Ni-like x-ray lasers with the wavelengths of around 12.0 nm, the optimum electron density, n_e, and temperature, T_e, of the gain region is around ~ 5 10^{20} cm^{-3} and 400 eV, respectively. If we assume that the electron temperature, density for ions with an effective nuclear charge z_{eff} are approximately scaled as T_e z_{eff}^2, and n_e z_{eff}^7, respectively [6], the optimum gain condition for the shorter wavelength x-ray lasers is expected in high temperature and dense region, where z_{eff} is z - 28 in the Ni-like ion case. For an example, the Ni-like La (z =57) ion laser at a wavelength of 8.8 nm [7], the optimum plasma parameters of the gain region are inferred to be T_e ~ 900 eV and n_e ~ 8 10^{21} cm^{-3}, which is higher than the critical density of a pumping laser pulse at a wavelength of 1 μm.

In order to provide the pumping energy to high density region, frequency conversion of the pumping laser light, e.g., 2ω, has been often used. However the frequency conversion of a few ps laser pulse with an energy of ~ 15 J needs a large-size non-linear medium, and its low conversion efficiency (~ 50%) requires large energy for the fundamental laser light, which is not practical and contradict against the concept of table-top x-ray lasers. Rather, it is realistic for our objective to optimize the pre-formed plasma condition to achieve effective heating of the high density region by the fundamental laser light. Standing upon these concepts we conducted an experiment.

EXPERIMENT AND DISCUSSION

The details of our pumping laser system and experimental setup were described in elsewhere [4, 8]. A 2 μm-thick La slab target fabricated on a 1 mm-thick glass substrate was irradiated by a line-focused compact CPA Nd:glass laser at a wavelength of 1.053 μm. The length and the width of the line focus was 5.3 mm and 30 μm, respectively. The laser pulse consisted of two pulses; a pre-pulse with a duration of 200 ps and a main pulse with a duration of 7 ps, separated by 250 ps. The intensity on the target was 3.0 10^{12} W/cm^2 for the pre-pulse and 1.4 10^{15} W/cm^2 for the main pulse. We employed a quasi-traveling wave pumping by use of a step mirror with a step of 4 ps [4]. A laser-produced La plasma was observed by use of a grazing incidence spectrograph, GIS1, and a KAP crystal spectrograph (KAPCS). GIS1, which had a concave collecting mirror and uneven spacing holographic laminar grating (1200 grooves/mm) fabricated by a collaboration with Shimadzu [9], covered spectral range of 6 nm through 40 nm and was set in the propagation direction of x-ray laser beam. In order to attenuate the emission from the plasma, we used a Zr filter fabricated on a Ni-mesh with a transmittance of 10% at 8.8 nm. The KAPCS was set at the top port of the target chamber to observe the plasma. It covered 0.8 nm through 1.0 nm and the n = 3 - 5 transitions of the Ni-, Co- and Fe-like La ions were observed.

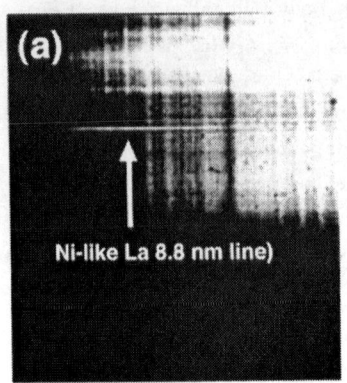

Ni-like La 8.8 nm line)

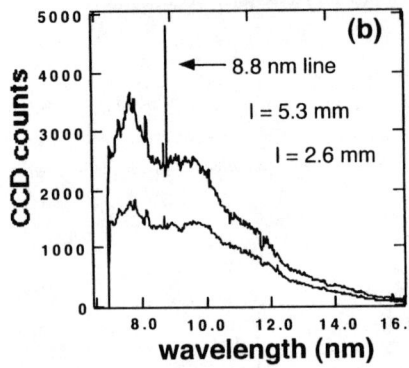

8.8 nm line
l = 5.3 mm
l = 2.6 mm

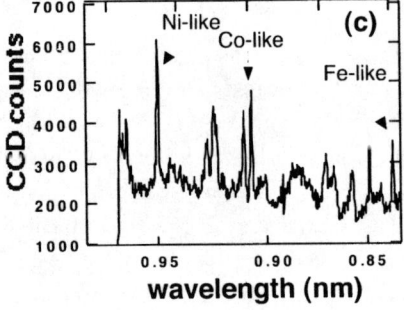

Ni-like
Co-like
Fe-like

Figure 1. (a); CCD image of the 8.8 nm line taken by the GIS1. Typical spectra of La plasma for the plasma length of 5.3 mm and 2.6 mm taken by GIS1; (b) and KAPCS; (c). The laser lines and typical 3d-5f resonance lines of the Ni-, Co- and Fe-like are marked by arrows.

The detectors of the two spectrographs were back-illuminated CCD (Charge coupled device, Princeton instrument Model SX1024). The wavelength calibration of GIS1 was done by use of the 2nd- and 3rd-order light of resonance series lines of the He-like C ions. For KAPCS, we used resonance series lines of the He-like Na ions, *i.e.*, He through He. By use of this calibration curve together with the HULLAC code [10], several lines of the Ni-, Co- and Fe-like La ions taken by KAPCS were identified, which were used as a monitor of the production of these ions.

Figure 1 (a) and (b) show the spectra taken by GIS1 for the target lengths of 5.3mm and 2.6mm, and the spectrum taken by KAPCS, respectively. The identified lasing line and the resonance lines of the Ni-, Co- and Fe-like ions are marked by arrows. Substantial amplification of the 8.8 nm line of the nickel-like La ions is obvious. In this target shot, the duration of the main pulse was set to be 7 ps, and the energy of the pre-pulse and main pulse was 1 J and 17 J, respectively. Figure 2 shows the output intensity of the 8.8 nm line versus various target length from 2 mm through 5.3 mm. The gain coefficient, g, and gain-length product, gL, were derived to be 14.5 ± 1.5 cm^{-1} and $gL = 7.7 \pm 0.8$, respectively, by use of the Linford fitting formula [11].

We observed the dependence of the output of the 8.8 nm line on the pre-pulse intensity. The main-pulse energy was fixed to be 17 J, and the pre-pulse intensity was changed from 5 $\times$ 10^{11} through 1.2 $\times$ 10^{13} W/cm². The 8.8 nm output has a peak at around pre-pulse intensity of 3 $\times$ 10^{12} W/cm².

In the pre-pulse intensity region of 1 $\times$ 10^{12} ~ 1.2 $\times$ 10^{13} W/cm², the KAPCS spectrograph showed that the ion abundance of the Co-like ions and Fe-like ions with respect to the Ni-like ions increase with decrease of the pre-pulse intensity. This indicates that by reducing the pre-pulse intensity, the volume of the pre-formed plasma is reduced, and the main pulse energy is absorbed in a localized area. This tendency is reproduced by a 1-D hydrodynamics code HYADES [12]. Figure 2 shows temporal evolution of the spatial profile of T_e under the laser irradiation condition similar to the present experiment. The time is measured from the peak of the pre-pulse. The pre-pulse intensity is (a); 1 $\times$ 10^{12} and (b); 4 $\times$ 10^{12} W/cm², and the main pulse intensity is fixed to be 5×10^{14} W/cm². After the main pulse irradiation (See the trace at 0 ps.), the size of the high temperature region in Fig. 2 (a) is smaller than that in Fig. 2 (b), and the peak value of T_e becomes higher with decreasing the pre-pulse intensity.

Under the condition of Fig. 2 (a), we calculated the gain coefficient of the 8.8 nm line by use of our collisional-radiative model (WHIAM code [13]), in which virtually all the rate coefficients of the atomic processes are calculated by the HULLAC code [10]. We assume Voigt profile for the lasing line, a convolution of a Gaussian corresponding to the ion temperature and a Lorentzian

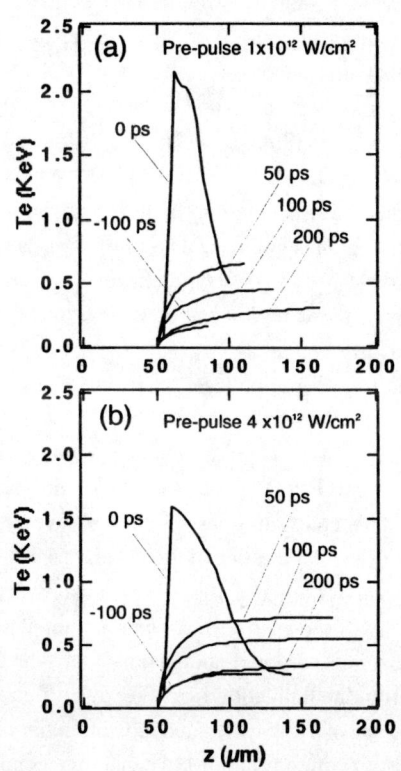

Figure 2. Temporal evolution of electron temperature calculated by the HYADES code under the pre-pulse level of 1×10^{12} W/cm²; (a) and 4×10^{12} W/cm²; (b). The main pulse intensity is fixed to be 5×10^{14} W/cm². Time is measured from the peak of the main pulse. High temperature region becomes smaller with the decrease of the pre-pulse intensity, which implies heating efficiency increases with decrease in the volume of the pre-plasma.

corresponding to the collision and natural broadening of the upper and lower lasing levels. In the typical gain region, $n_e = 3 \ 10^{21}$ cm^{-3}, $T_e \sim 1000$ eV and $T_i \sim 400$ eV, the Full Width at Half Maximum (FWHM) of the Gaussian is 11.7 mÅ, and that of the Lorentzian is 12.1 mÅ, which corresponds to the upper and lower collisional-radiative destruction rate of 1.60 10^{13} s^{-1} and 1.32 10^{13} s^{-1}, respectively. The peak gain is obtained in the over-dense region ($\sim 3 \ 10^{21}$ cm^{-3}) and the value of the calculated gain is larger by a factor of 2 than the experimental gain ($\sim$14.5). The discrepancy may be due to that the experimental gain is spatially averaged one.

The present result implies that the over-dense region can be heated effectively by electron thermal conductivity. In the HYADES code [12], under the plasma parameter of the present gain region ($n_e = 3 \ 10^{21}$ cm^{-3}, $T_e \sim 1000$ eV), electron thermal flux is calculated by use of Classical (Spitzer-Härm) method including the effect of heat flux limit. Although the present gain region is not in a category of classical (low density) plasma rather in a strongly coupled plasma, the present treatment is reasonable, because the electronic thermal conductivity in this region can be treated in the same fashion with that in the low density plasma [14].

In order to increase the peak value of the pumping laser pulse, we changed the duration of the main pulse into 1 ps under the pre-pulse intensity of 3 10^{12} W/cm^2. However, the output x-ray became smaller by a factor of 4 compared with the 7 ps-duration case. This is due to the followings. Firstly, the expected gain duration of the Ni-like La ions may be smaller than the step of our quasi-traveling wave system (= 4 ps) [4]. In this case, the absorption of the 8.8 nm line in the plasma takes place. Secondly, under the present pre-pulse intensity of order of 10^{12} W/cm^2, the observation of the pre-formed plasma by use of the KAPCS shows that the La ions are populated in the lower ionization stage than the Ni-like ions. Under such a condition, the main pulse has to sustain high temperature during the ionization from the lower ionization stages to the Ni-like ions. This is consistent with the calculated ionization time from La^{28+} to La^{29+} of $\sim$ 10 ps by use of hydrogenic approximation [15] under the condition of $n_e \sim 10^{21}$ cm^{-3} and $T_e \sim 1$ KeV.

In summary, we demonstrated substantial amplification of 8.8 nm line in the transient collisional-excitation scheme with a pumping energy of 18 J. This was the first demonstration of substantial amplification at this wavelength in the TCE scheme by use of a laboratory-size pumping system and was valuable in terms of development of high efficient compact x-ray laser.

ACKNOWLEDGMENT

This work was done under the auspices of JAERI Foreign Researcher Visiting Program and Russian Fundation for Basic Research (RFBR) No. 00-02-17060.

REFERENCES

1. M. P. Kalachnikov, P. V. Nickles, M. Schnürer, W. Sandner, V. N. Shlyatsev, C. Danson, D. Neely, E. Wolfrum, J. Zhang, A. Behjat, A. Demir, G. J. Tallents, P. J. Warwick, C. L. S. Lewis, Phys. Rev. A **57**, 4778-4783 (1998).

2. J. Dunn, A. L. Osterheld, J. Nilsen, J. R. Hunter and V. N. Shlyaptsev, Phys. Rev. Lett. **84**, 4834-4837 (2000).

3. A. Klisnick, P. Zeitoun, D. Ros, A. Carillon, P. Fourcade, S. Hubert, G. Jamelot, C. L. S. Lewis, A MacPhee, R. O'Rourcke, R. Keenan, P. Nickles, K. Janulewicz, M. Kalachnikov, J. Warwick, J. C. Chanteloup, A. Migus, E. Salmon, C. Sauteret, J. P. Zou, J. Opt. Soc. Am **B17**, 1093-1097 (2000).

4. T. Kawachi, M. Kado, M. Tanaka, A. Sasaki, N. Haegawa, A. V. Kilpio, S. Namba, K. Nagashima, P. Lu, K. Takahashi, H. Tang, R. Tai, M. Kishimoto, M. Koike, H. Daido and Y. Kato, submitted in Phys. Rev. A.

5. R.E. King, G. J. Pert, S. P. McCabe, P. A. Smith, A. G. MacPhee, C. L. S. Lewis, R. Keenan, R. M. N. Rourke, G. J. Tallents, S. J. Pestehe, F. Strati, D. Neely, and R. Allot, Phys. Rev. A **64**, 053810 (2001).

6. T. Fujimoto, J. Phys. Soc. Jpn, **54**, 2905 (1980).

7. J. Nilsen, Y. Li, J. Dunn and A. L. Osterheld, *in Proceedings of X-ray Lasers–1998*, Vol. 159 of IOP (Institute of Physics Publishing, Bristol and Philadelphia, 1998), p. 135.

8. T. Kawachi, M. Kado, M. Tanaka, N. Hasegawa, K. Nagashima, K. Sukegawa, P. Lu, K. Takahashi, S. Namba, M. Koike, A. Nagashima and Y. Kato, to be published in Appl. Opt. (in press)

9. M. Koike, T. Namioka, E. Gullikson, Y. Harada, S. Ishikawa, T. Imazono, S. Mrowka, N. Miyata, M. Yanagihara, J. H. Underwood, K. Sano, N. Ogiwara, O. Yoda and S. Nagai, Proc. SPIE **4146**, 163 (2000).

10. M. Klapisch, A. Bar-Shalom, J. Quant. Spectrosc. Radiat. Transf., **58**, 687 (1997)

11. G. J. Linford, E. P. Poressini, W. R. Sooy and M. L. Spaeth, Appl. Opt. **13**, 379-390 (1974).

12. J.T.Larsen and S.M.Lane. J. Quant. Spectrosc. Radiat. Transf. **51** ,179(1994).

13. A.Sasaki, T.Utsumi, K.Moribayashi, M.Kado, M.Tanaka, N.Hasegawa, T.Kawachi, H.Daido, J. Quant. Spectrosc. Radiat. Transf. **71**, 665 (2001).

14. R. M. More, *Handbook of Plasma Physics, Vol. 3, "Physics of Laser Plasma"* (North-Holland, Amsterdam, 1991), pp.63-110.

15. D. H. Sampson and H. L. Zhang, Astrophys. J. **335**, 516 (1988).

Analytical Ray-Tracing of a Transient X-ray Laser: Ni-like Ag Laser at 13.9 nm

Jaroslav Kuba[1,2*], Djamel Benredjem[2], Clary Möller[2], Ladislav Drska[1]

[1]*Faculty of Nuclear Sciences and Physical Engineering, Czech Technical University in Prague, Brehová 7, 115 19 Prague 1, Czech Republic*
[2]*LSAI/LIXAM, Bâtiment 350, Université Paris XI, 91405 Orsay, France*

Abstract. Numerical codes predict for transient x-ray lasers (XRL) very high picosecond duration gains [e.g. 1] that are restricted in space to several microns at FWHM. XRL beam propagation in plasma is vital for estimation of the effective gain at the plasma output. In this paper, beam propagation in transient plasmas is analytically studied. General 2-D formulae are developed for beams in electron density gradient media, including those with the exponential profile that describes the plasma created from a solid target. The gradient is predicted to potentially limit the amplification length within the maximum gain to <2.6 mm in standard experiments. The result given by the analytical model is confirmed by numerical ray tracing of XRL beams within an amplifying medium as it is defined by the full numerical simulation using the EHYBRID code.

INTRODUCTION

The transient XRL experiments, including those on Ni-like silver, were simulated numerically by several authors, e.g. [1]. These simulations are in qualitatively good agreement with experiments, but differ by one order of magnitude in actual values of the gain. The simulations show typically a high local gain of >500 cm^{-1}, which is, however, very restricted both in time (~picosecond duration) and in space (~5 µm FWHM). This high gain zone is located within 10 µm from the target surface, i.e. it coincides spatially with the region of steep electron density gradients near the critical surface. Therefore, it would be expected that the spatial sampling of this region would be limited by refraction.

Several effects might contribute to the explanation of the disagreement between the high local gains (hundreds of cm^{-1}) predicted by simulations and the experimentally observed gains (tens of cm^{-1}, e.g. [2]). Tommasini and Strati [3] analytically studied in detail a possible mismatch of the gain creation (i.e. traveling wave – TW) velocity and the group velocity of the XRL beam. The mismatch would occur in a high gain medium, where the XRL photons propagate more slowly than the speed of light and hence also more slowly than the gain driven by the TW. The thorough analysis for the

---

* Present address: Lawrence Livermore National Laboratory, 7000 East Ave, L-251, Livermore CA 94550, USA
E-mail: kuba1@llnl.gov

CP641, *X-Ray Lasers 2002: 8th International Conference on X-Ray Lasers,* edited by J. J. Rocca et al.
© 2002 American Institute of Physics 0-7354-0096-2/02/$19.00

Ni-like Ag 13.9-nm lasing line shows, however, that the optimum TW velocity would be >0.9 c instead of the actually used c. The mismatch effect remains therefore relatively small and would not fully explain the disagreement between simulations and experiments.

In the present paper we will concentrate on another important effect in the transient XRL context, which is not taken into account in [3], i.e. refraction of the XRL beam. The refraction of the beam in the plasma is determined by a gradient of electron density, which occurs inherently – due to the pump geometry currently used – in the direction of the pump laser beam.

The beam paths and ray-tracing in the quasi-steady-state laser context were studied both analytically and numerically by several authors, e.g. [4, 5]. Chilla and Rocca [6] found simplified analytical solutions for optimum amplification in capillary discharge laser plasmas. Ray trajectories were also calculated numerically by Shlyaptsev [7] for a transient X-ray laser. In analytical works [4, 5], the effects of refraction are demonstrated by a model where the electron density gradient appears in the direction of the pump laser beam, while it is supposed to be zero in the directions perpendicular to the pump laser beam, i.e. parallel with the target. In order to demonstrate refraction, a *linear profile* of the electron density is often used because of its simplicity. However, its real physical application is limited.

London [4] obtained an analytical solution for a *parabolic* electron density profile. The parabolic profile describes electron densities in XRL plasmas from exploding foil targets in qualitative agreement with both theoretically predicted and experimentally observed profiles. In this case, however, the gradient rises with distance from the target, which is contrary to the profiles simulated for a solid slab target in the present experiments.

For solid-target-based XRL plasma simulations, one usually employs an *exponential* electron density profile [8], which is predicted by analytical considerations on isothermic plasma expansion into vacuum. The profile is also in good agreement with the simulations (e.g. [1]).

In this paper, we will develop a general analytical solution for any 1-D profile of the electron density. We will apply the formula to an exponential profile. The analytical solution will enable us to explain the limitations of the XRL beam amplification by refraction. The analytical results will be compared with numerical ray tracing using the electron density profile obtained by a full numerical simulation by means of EHYBRID 1.5D hydrodynamic/atomic code.

BASIC EQUATIONS AND APPROXIMATIONS USED

We will consider the co-ordinates system where z is the direction of XRL beam propagation, x is parallel to the pump laser beam and y is perpendicular to x and z. A beam trajectory in a refractive index gradient, $\nabla n(x,y,z)$, medium is fully described by the well-known beam equation

$$\frac{d}{ds}\left(n\frac{dr}{ds}\right) = \nabla n(x,y,z) \tag{1}$$

47

where ds, $ds = \sqrt{dx^2 + dy^2 + dz^2}$, is the path element, $r = (x,y,z)$ stands for the position vector, and n is the refractive index. In the beam approximation, one can derive the following relation from Maxwell's equations: $n = \sqrt{1 - n_e/n_c} \approx 1 - n_e/2n_c$, where n_e stands for the electron density and n_c is the critical electron density for the beam's wavelength.

We will also introduce three other approximations: (I) Assuming that the refraction is small compared to the propagation length, we can apply the *paraxial approximation*, i.e., $ds \approx dz$. (II) If the refractive index changes only in the x-direction and remains constant in y and z co-ordinates, its gradient reduces to $\nabla n = (dn/dx, 0, 0)$. This approximation necessitates a homogeneous irradiation along the whole focal line. Finally, (III) in XRL plasmas, the refractive index is close to $n \approx 1$, hence we can omit n on the left-hand side of Eq. (1). The beam equation Eq. (1) hence simplifies to

$$\frac{d^2x}{dz^2} = \frac{dn}{dx} \tag{2}$$

This beam equation was solved in the literature [4, 5] for linear and parabolic profiles. We will develop a general solution for any $n_e = n_e(x)$ and will show its application on a realistic exponential plasma profile.

GENERAL SOLUTION FOR ANY PROFILE n$_e$(x)

The simplified beam equation (2) can be solved for any realistic density profile $n_e(x)$. The equation can be transformed by substituting s

$$s = \frac{dz}{dx} \quad i.e. \quad 2\frac{d^2x}{dz^2} = \frac{d(s^{-2})}{dx}$$

The transformed beam equation can be integrated to obtain

$$\frac{dz}{dx} = s = \pm\frac{1}{\sqrt{2n + C_1^*}} \quad \rightarrow \quad z = \pm\int\frac{dx}{\sqrt{2n + C_1^*}} + C_2 \tag{3}$$

where C_1^* and C_2 are integration constants. Note that this is the general solution of the beam equation (2) without assuming a specific form of the refractive index n. The complete solution in elementary functions can be obtained for any profile of the refractive index, $n=n(x)$, which would allow for an analytical solution of the integral on the right-hand side. For example, considering a $n \sim 1/(a+bx)^t$ profile (a, b, constants, $t \in \Re-\{0\}$), the integral leads to a well-known Tchebyschev's integral. One concludes that it is solvable in elementary functions if and only if (a) $1/t \in Z$ or (b) $1/t$

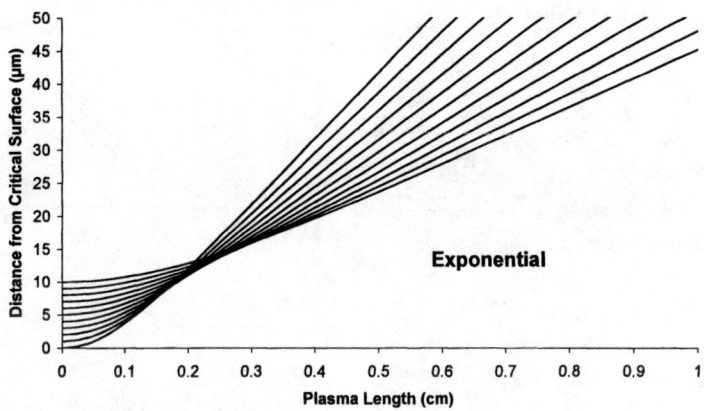

FIGURE 1. Beam trajectories in an exponential electron density profile. Beam parameters $\alpha_0 = 0$, $x_0 = 0 - 10 \ \mu m$; Plasma parameters $n_0 = 10^{21}$ cm^{-3} and a = -1670 (as inferred from plasma numerical simulations)

$-1/2 \in Z$, e.g. t = 1 or t = 2. However, we will show here only the solution in the realistic, exponential profile.

EXPONENTIAL PROFILE

Assuming an isothermal expansion of the plasma into the vacuum, one obtains an exponential profile of the electron density in the sub-critical region

$$n_e = n_0 e^{a(x-x_c)} \tag{4}$$

which agrees well with numerical simulations. Here $n_0 \approx 10^{21}$ cm^{-3}, is the critical electron density (for the pump laser wavelength) at $x = x_c$, x_c stands for the position of the critical surface with respect to the target.

In order to calculate the beam path through the exponential profile, we suppose that the plasma expansion is not significant on a time scale as small as the photon transit along the target. In this case, the refractive index does not depend on the z coordinate, and we can use the general solution with the above approximations.

Substituting $y = \sqrt{-ke^{a(x-x_c)} + C_1^2}$, ($k = n_0/n_c$, n_c being the critical electron density for the XRL beam, $n_c \approx 10^{25}$ cm^{-3}) into the right-hand side of the general solution, yields the beam path in the XRL plasma, described as (Fig. 1)

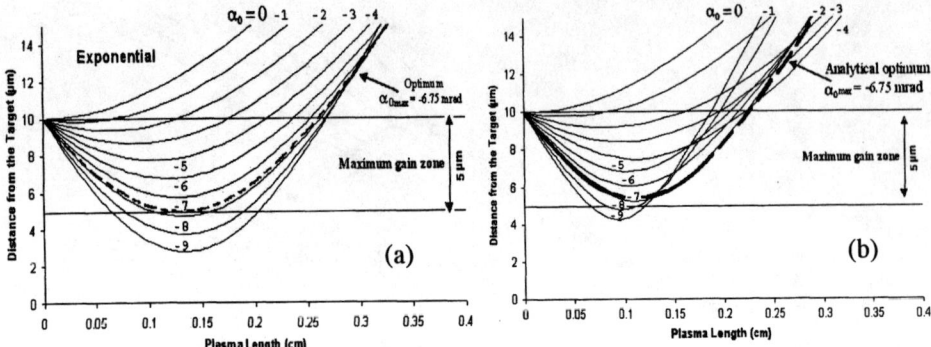

FIGURE 2. The refraction severely restricts the maximum achievable gain-length. (a) Analytical and (b) numerical ray tracings show that the beam leaves the maximum FWHM gain zone after <2.6mm travel.

$$x = \frac{1}{a}\ln\left[\frac{C_1^2}{k}\frac{\pm 4e^{aC_1(z-C_2)}}{\left(1\pm e^{aC_1(z-C_2)}\right)^2}\right] + x_c \tag{5}$$

Only the "+" solution is physical in most cases. For the boundary conditions $x = x_0$ and $dz/dx|_{x=x0} = 1/\alpha_0$ at $z = 0$, one finds for C_1 and C_2

$$C_1 = \sqrt{\alpha_0^2 + ke^{a(x_0 - x_c)}} \quad \text{and} \quad C_2 = -\frac{1}{aC_1}\ln\left|1 - \frac{2C_1}{\alpha_0 + C_1}\right| \tag{6}$$

GAIN-LENGTH LIMITATIONS BY REFRACTION

The analytical solution enables us to calculate also the maximal amplification length considering plasma from the simulation [1]. If we suppose that the peak gain (the spatial FWHM of the gain) is located within a distance of 5 to 10 μm from the target and that the TW is applied, the gain zone then forms a rectangular shape along the target (Fig. 2). We can hence calculate the starting point and the initial angle that would maximize the beam's path in this zone. Such a beam would be generated by spontaneous photons emitted at the extremity of the gain zone, farther from the target ($x_0 = 10$ μm at $z = 0$), 'touch' the extremity closer to the target, at $x = x_t$, and would leave – pulled by the high gradient – the gain zone at the farther extremity of the zone (Figure 4). In other words, we look for the beam's initial angle α_{0max}, for which the beam has its turning point at $x = x_t$ and $\alpha = 0$, i.e. $dx/dz = 0$. One can hence develop from Eq. (5) the formula

$$\alpha_{0\,max} = -\sqrt{k\left[e^{a(x_t - x_c)} - e^{a(x_0 - x_c)}\right]} \tag{7}$$

The beam exits the amplification zone at $z_{max} = 2C_2$ with the output angle $-\alpha_{0max}$.

For the values of $x_c = 3.7\ \mu$m (as inferred from [1]), $x_0 = 10\ \mu$m and $x_t = 5\ \mu$m, the initial angle of $\alpha_{0max} = -6.75$ mrad maximizes the path length to $z_{max} = 2.6$ mm (for the path length estimation the trajectory's curvature can be neglected). The amplification length for the other beams is calculated to be even shorter. Therefore, the model confirms that refraction plays an extremely important role in the beam amplification and that the amplification length of the XRL beam within the maximum gain zone would be limited to less than 2.6 mm by the refraction, in the present case, as confirmed also by numerical ray tracing code (Fig. 2b) taking the conditions given by the full EHYBRID code simulation [1].

CONCLUSIONS

In this paper we developed the general analytical 2D formula describing the beam path in any reasonable electron density profile plasma $n_e = n_e(x)$. The formula allows evaluate analytically also the realistic exponential profile as it was shown in this work. In turn, analytical solution enables us to optimize the beam path in plasma in terms of maximum gain-length product. The investigations shows that the refraction limits severely the maximum amplification length within the FWHM maximum gain zone to <2.6 mm, as it was confirmed also by numerical simulations.

ACKNOWLEDGMENT

The authors would like to thank to Jiří Limpouch from the Czech Technical University in Prague, and Vyacheslav N. Shlyaptsev from the Lawrence Livermore National Laboratory for the fruitful discussions on the analytical part of the paper. The work was supported by the grant No. CTU300111714 of the Czech Technical University in Prague.

REFERENCES

[1] J. Nilsen, and J. Dunn, Proc. of the SPIE Conf. **4505**, 100 (2001); J. L. Miquel et al., Proc. of the SPIE Conf. **3776**, 24 (1999); J. Kuba et al., Proc. of the 7th Intern. Conf. on X-Ray Lasers, J. Phys. IV France **11**, 35 (2001); J. Kuba et al., these proceedings

[2] J. Kuba *et al.*, Phys. Rev. **A 62**, 043808 (2000); J. Dunn et al., Phys. Rev. Lett. **84**, 4834 (2000).

[3] R. Tommasini and E. Fill, Phys. Rev. **A 62**, 034701 (2000); F. Strati and G. J.Tallents, Phys. Rev. **A 64**, 013807 (2001).

[4] Richard A. London, Phys. Fluids **31**, 184 (1988).

[5] B. Rus, Inst. Phys. Conf. Ser. No. 159, p. 119 (1999).

[6] J. L. A. Chilla and J. J. Rocca, JOSA B **13**, 2841 (1996).

[7] V. N. Shlyaptsev et al., Proc. of the SPIE Conf. Vol. 3156, p. 193 (1997).

[8] Ya. B. Zel'dovich and Yu. P. Raizer, Physics of Shock Waves and High-Temperature Hydrodynamic Phenomena (Academic, New York, 1966).

Modeling of Transient Ni-like Ag X-ray Laser

Jaroslav Kuba[1,2$♣], Raymond F. Smith[1$], Djamel Benredjem[1],
Clary Möller[1], Lee Upcraft[3], Robert King[3], Annie Klisnick[1],
Ladislav Drska[2], Geoff J. Pert[3], Jean-Claude Gauthier[4]

[1]LSAI/LIXAM, Bâtiment 350, Université Paris XI, 91405 Orsay, France
[2]Faculty of Nuclear Sciences and Physical Engineering, Czech Technical University in Prague,
Brehová 7, 115 19 Prague 1, Czech Republic
[3]Department of Physics, University of York, Heslington, YO10 6DD, York, UK
[4]LULI, École Polytechnique, 91128 Palaiseau, France

Abstract. Recent high temporal resolution Ni-like x-ray laser (XRL) experiments [1] have yielded important insights into the output characteristics of picosecond pumped XRL's and the shortest XRL pulse was demonstrated. However, important issues were raised that require to enhance our understanding of plasma and population dynamics, namely (a) short pulse duration, (b) XRL pulse occurring before the peak of continuum emission and (c) the role of (over-)ionization. A numerical study of the Ni-like transient silver XRL has therefore been undertaken to complement our experimental results. High gain coefficients existing with picosecond lifetimes and restricted in space (~5 μm FWHM) are predicted, which is consistent with short XRL durations experimentally observed. The simulations suggest that the gain is cut-off by fast over-ionization of Ni-like ions. The late onset of the continuum emission relative to the temporal peak of the XRL output (as observed experimentally) is explained as a signature of a thermal conductivity wave propagating toward the super-critical density region close to the target surface

1. INTRODUCTION

In this paper we present the results of our numerical modeling aimed to give insights into the gain dynamics of the transient Ni-like silver laser at 13.9 nm. Modeling is based on the EHYBRID numerical code developed by G. J. Pert *et al.* [2] that was customized to our silver atomic data file. The model case studied in the paper is the experiment carried out in 2000 at the Rutherford Appleton Laboratory (RAL, [1]).

2. TIME-RESOLVED EXPERIMENT

In a recent experiment at the RAL, a fast XUV Axis Photonique streak camera coupled to a flat-field spectrometer with a combined resolution as high as 1.9 ps was

---

♣ E-mail: kuba1@llnl.gov
$ Present address: Lawrence Livermore National Laboratory, 7000 East Ave, L-251, Livermore CA 94550, USA

CP641, *X-Ray Lasers 2002: 8th International Conference on X-Ray Lasers*, edited by J. J. Rocca et al.
© 2002 American Institute of Physics 0-7354-0096-2/02/$19.00

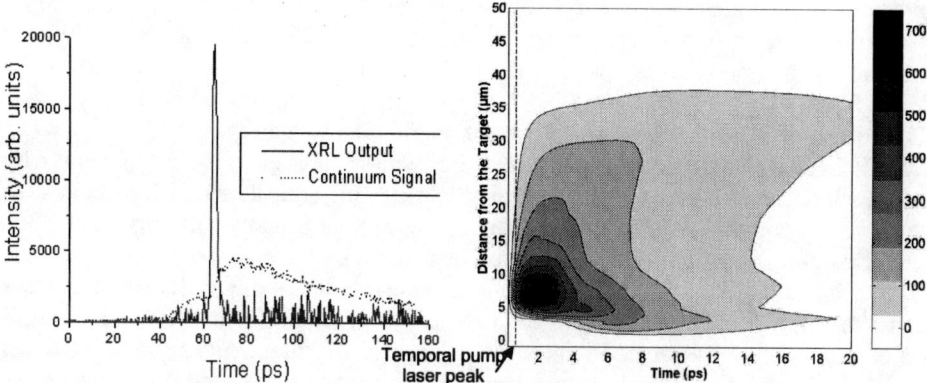

FIGURE 1. Temporally resolved transient Ni-like 4d $^1S_0 \rightarrow \,^1P_1$ 4p Ag laser at 13.9 nm. The pulse duration of 1.8 ± 0.7 ps is experimentally observed, which is the shortest x-ray laser pulse demonstrated thus far. The XRL appears in the rising edge of the continuum emission (dotted line).

FIGURE 2. Numerically simulated gain peak is very restricted both in time (3.1 ps at FWHM) and space (~5 μm at FWHM), which is consistent with the short XRL pulse duration.

used to diagnose the output of the Ni-like transient Ag J = 0→1 4d-4p lasing line at 13.9 nm [1]. In the first step the XRL output was maximized experimentally by varying several pump laser parameters. The optimum pump laser parameters were determined as follows: 4.3 J/cm in the 'long' 300-ps pre-forming pulse and 11.8 J/cm in the CPA 1.3-ps pulse with the temporal peak-to-peak delay of 200 ps between both pulses. 'Optimum' here refers to those conditions, which yield the maximum XRL output energy. The wavelength of the both pump laser pulses was 1.06 μm and the focal widths of the long and the short pulses were set at 120 μm and 80 μm, respectively.

The FWHM duration of the XRL pulse was measured (after a deconvolution taking into account the system response function) to be 1.8 ± 0.7 ps (Fig. 1). The XRL signal was extinguished at the rising edge of the continuum emission, which was repeatable for a number of different pump laser parameters. One possible interpretation of this result is that the rising edge of the continuum emission is a symptom of over-ionization within the plasma, which, in turn, signals the extinction of the population inversion on the lasing line. One may also postulate, however, that the relative time delay between the XRL output peak and the continuum emission peak is a result of a conductivity wave traveling towards the dense target surface. Such an effect would not necessarily impact on the population dynamics in the lasing ion.

3. SIMULATIONS

In order to clarify the issues raised by these experimental results, we modelled the experimental case by means of a full numerical simulation using the EHYBRID [2] hydrodynamic/atomic code. EHYBRID describes dynamics of laser plasma (including processes such as pump laser energy deposition, hydrodynamic motion, electronic thermal conduction, and ion-electron thermalization) coupled with the atomic physics of the lasant ions [3, 4]. The model is 1.5-D, designated to operate in planar geometry. The code uses a 296 Lagrangian cell matrix in the direction away from the target. The fluid is hence modeled numerically in the direction parallel to the driving laser by cells, which are assumed to be laterally isothermal, and, as such, the transverse expansion is considered to be self-similar [2], described by analytical formulae.

The atomic data input file for Ag was constructed by the authors of this paper. The Ni-like ion stage is modelled with 272 excited levels including all levels in the n=4 and 5 manifolds and with averaged contributions from the n = 6 to 8 levels. Ion stages other than Ni-like are treated with varying degrees of complexity using a screened hydrogenic model or a simpler two-level model based on the modified form of the Griem's model [5]. Electron-ion collision cross-sections were calculated at LULI (École Polytechnique) for all transitions within the n = 4 manifold using the code HULLAC [6]. For the calculation of the excitation/de-excitation rates, we did not use the d-coefficients introduced by van Wyngaarden et al [7]. In fact, it is more satisfactory to calculate all the rates for each electron density and temperature, with the help of a subroutine added into EHYBRID. Oscillator strengths for all transitions in the n = 4 to 8 manifolds were calculated with a multi-configurational Dirac-Fock code (MCDF) [8].

A. Study of the Gain

Figure 2 shows the gain on the Ni-like Ag, 4d $^1S_0 \rightarrow ^1P_1$ 4p lasing transition at 13.9 nm, as a function of time and distance away from the target surface. Two distinct regions are predicted. At earlier times when the 1.3 ps heating pulse is turned on there emerges a large gain (> 500 cm^{-1}) region with small dimensions both in space (< 5 μm within 15 μm of the target surface) and time (< 1 ps at FWHM). When the laser is turned off and the plasma expands, a larger gain plateau with gains < 400 cm^{-1} extending out to ~40 μm is observed.

The high gain peaks at electron densities close to 10^{21} cm^{-3} (Fig. 3). This is the critical density surface of the 1.06 μm laser driver and coincides spatially with a region of steep density gradients. It would be expected, therefore, that spatial sampling of this region would be limited by refraction. It is observed that the density gradients become more relaxed at distances away from the critical density surface.

The simulated maximum *local* gain is of about an order of magnitude higher than the *effective* gain observed experimentally. Considerations on XRL beam ray tracing can, however, help to explain the difference between the experimentally observed and simulated gains, cf. [9]. Also, a possible mismatch between gain creation velocity (in the traveling wave irradiation geometry) and the XRL photon propagation [10] would further reduce the effective gain.

The calculations described in this paragraph are similar to those predicted by other hydrodynamic codes for Ag [11, 12, 13].

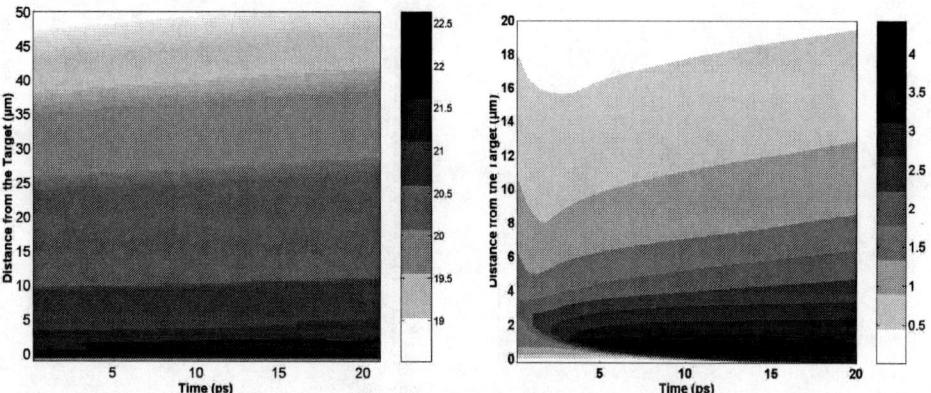

FIGURE 3. The peak gain appears in a region of a steep electron density gradient (cf. Fig. 2). Simulated electron density data also show that the Bremsstrahlung emission behavior is predominantly dictated by the heat wave propagating toward the target surface (N.B. log scale in electron density and Bremsstrahlung values).

FIGURE 4. Simulated Bremsstrahlung peaks well after the gain peak (at 1.6 ps) and is mainly emitted in the zone near the target surface (N. B. log scale in BS and electron density values)

B. Continuum Emission

In order to evaluate the continuum emission, we assume that it is dominated by Bremsstrahlung (BS) emission. The BS, $Q(\omega)$, emitted by plasma (in Maxwellian electron velocity distribution) per unit time and unit volume in a small frequency interval around ω is proportional to (e.g. [14])

$$Q(\omega)d\omega \propto Z^2 n_e n_i T_e^{-1/2} e^{-\frac{\hbar\omega}{kT_e}} d\omega$$

where n_e and n_i stand for electron and ion densities, Z is the average ionic charge, T_e electron temperature, ω the circular frequency, and k and $\hbar$ represent Boltzmann's and Planck's constants, respectively. In our computation we neglect the slowly varying Gaunt factor function as is usual in many practical applications [14]. Having calculated the plasma parameters n_e, n_i, Z and T_e, we can easily evaluate the radiation developed by BS in a small wavelength interval around the 13.9 nm Ni-like silver line, per unit volume and unit time, in each time-space cell in the plasma (Fig. 4). The evaluation clearly demonstrates that the continuum emission peak arrives *after* the XRL peak, following the behavior of the electron density (Fig. 3). The simulation hence shows that the BS peak in time is dictated by the heat wave propagating toward the target surface.

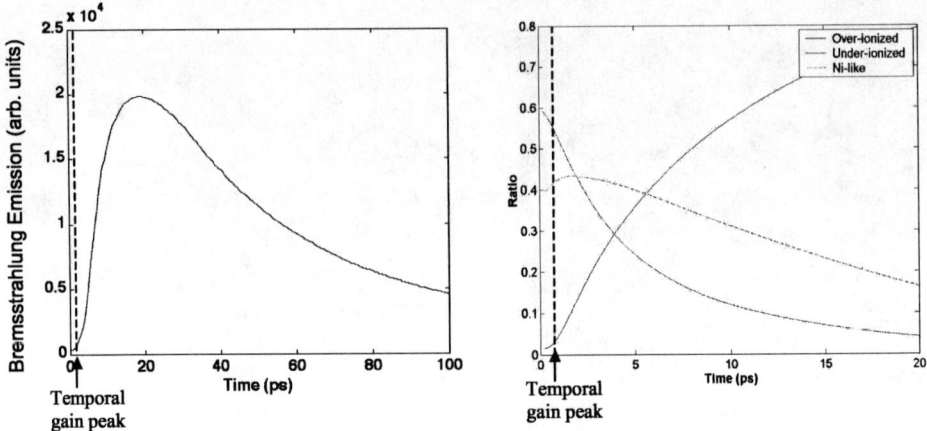

FIGURE 5. The simulated Bremsstrahlung emission integrated over the horizontal angle confirms the experimentally observed delay between the XRL and Bremsstrahlung peaks (cf. Fig. 1).

FIGURE 6. The ionization balance study shows that the gain is cut-off by fast over-ionization following the sub-ps heating laser pulse.

When calculating the actual continuum emission energy at the plasma exit, we should take into account the propagation of the radiation by different paths for each wavelength. In our estimations, we will show the time-resolved continuum emission (released from a small portion of plasma along the target) integrated over the horizontal angle only (which is – in this one-dimensional model – equivalent to the integration over the transverse distance from the target surface, Fig. 5). As observed during the experiment, the Bremsstrahlung emission rises slowly with its maximum well after the peak of XRL pulse (17.6 ps after the peak of the pump laser pulse).

C. Ionization Balance

The EHYBRID code allows the population ratio of each ion species to be evaluated in each time-spatial cell. The Ni-like ions form ~43% within the gain zone and their ratio even increases up to ~60% in the corona. The simulation also shows a rapid over-ionization within several picoseconds after the onset of the heating pulse. Studying carefully the ionization development of maximum gain cell (Fig. 6), we observe that the Ni-like ratio achieves its maximum after the gain peak, which could suggest that over-ionization does not play a role in the short XRL pulse duration. Looking however at over-ionization fraction (i.e. ratio of ions with ionization higher than Ni-like, id est higher than 19) we can see that it rises dramatically after the onset of the pump laser pulse, while the under-ionization fraction falls down at the same time scale. This is also the clue to the Ni-like ratio curve. The originally under-ionized ions have to 'transit' through the Ni-like stage (which is more stable than the surrounding ionization stages) before getting over-ionized. These 'new' Ni-like ions are not necessarily in an inversion of population and hence do not contribute to the gain. One concludes, therefore, that the over-ionization is the main factor in the short gain lifetime.

4. CONCLUSION

Full numerical modeling of a recent experiment was carried out. High gains were predicted near the critical surface with a broad lower gain (~400 cm^{-1}) plateau at sub-critical densities. The gain peaks 1.1 ps after the pump laser peak and lasts for 3.1 ps (FWHM) only and is also very restricted in space (~5μm FWHM). The simulation also shows that the maximum gain region appears in a zone of steep gradients of electron density resulting in high refraction index gradients, which would reduce dramatically the achievable effective gain.

The calculated parameters enabled us to explain the experimental observation that the XRL appears in the rising edge of the continuum emission from the plasma. This effect is caused by the fast variation in plasma electron density, which is a signature of a heat wave propagating toward the target surface. The simulation also shows that the short gain duration results from a fast over-ionization of ions in an inversion of population.

ACKNOWLEDGEMENTS

The authors would like to thank to Jiří Limpouch from the Czech Technical University in Prague for fruitful discussions on the subject. J. Kuba was in part supported by the grant No. CTU300111714 of the Czech Technical University in Prague.

REFERENCES

[1] J. Kuba, A. Klisnick et al., Proc. of the 7th Intern. Conf. on X-Ray Lasers, J. Phys. IV France **11**, 43 (2001); A. Klisnick, J. Kuba et al., Phys. Rev A **65**, 033810 (2002).
[2] G. J. Pert, J. Fluid. Mech. **131**, 401 (1983).
[3] S. B. Healy et al., IEEE Journal of Selected Topics in Quantum Electronics, Vol. 1, 3 949 (1995).
[4] P. B. Holden et al., J. Phys. B **27**, 341 (1994).
[5] J. H. Scofield, B. J. MacGowan, Physica Scripta **46**, 361 (1992).
[6] A. Bar-Shalom et al., J. of Quantit. Spectrosc. and Radiat. Transfer **71**, 169 (2001).
[7] W. L van Wyngaarden, K. Bhadra, and R. J. W. Henry, Phys. Rev. A **20**, 1409 (1979).
[8] I. P. Grant, Adv. Phys. **19**, 747 (1970).
[9] D. Benredjem, J. Kuba et al., these proceedings; J. Kuba, D. Benredjem et al., these proceedings.
[10] R. Tommasini and E. Fill, Phys. Rev. A **62**, 034701 (2000); F. Strati and G. J.Tallents, Phys. Rev. A **64**, 013807 (2001).
[11] J. Nilsen and J. Dunn, Proc. of the SPIE Conf. **4505**, 100 (2001).
[12] J. L. Miquel et al., Proc. of the SPIE Conf. **3776**, 24 (1999).
[13] J. Kuba et al., Proc. of the 7th Intern. Conf. on X-Ray Lasers, J. Phys. IV France **11**, 35 (2001).
[14] I. I. Sobelman: Atomic Spectra and Radiative Transitions, Springer 1992/1996, ISBN 3540545182.

X-ray laser program at MBI

P. V. Nickles[*], K. A. Janulewicz[*], A. Lucianetti[*], G. Priebe[*], A. Zigler[†], J.J. Rocca[**] and W. Sandner[*]

[*]Max Born Institute, Max-Born-Str. 2A, D-12489 Berlin, Germany
[†]Racah Institute of Physics, Hebrew University of Jerusalem, Israel
[**]Department of Electrical and Computer Engineering, Colorado State University, Fort Collins, CO 80523

Abstract. A survey of the Max Born Institute (MBI) activities in the field of X-ray lasers (XRLs) is presented. The main interest is focused on the transient soft X-ray lasers. Additionally, much work is put to look for new, efficient, compact (table-top) pumping schemes with a prospect to be applied in practice. The current state of the research and the plans for the future are described as well.

INTRODUCTION

Research on X-ray lasers began at Max Born Institute at the beginning of ninties. The first remarkable result achieved was the demonstration of the collisional scheme with the transient inversion in 1995 by applying combination of short (ps-) and long (ns-) laser pulses [1]. Since then, owing to direct access to both, CPA Nd:glass laser driver and 30 TW Ti:Sa laser system, MBI was still active and contributing noticeably to the research on the table-top soft X-ray lasers. The first hybrid x-ray laser combining capillary discharge technology with a picosecond laser pump pulse was demonstrated at MBI in 1999 [2]. Development of the capillary technology and general problems connected with longitudinally pumped X-ray lasers constitute significant part of the work at MBI-group in cooperation with the Colorado State University and Racah Institute of Physics from Jerusalem. However, the transient inversion scheme in the slab target geometry still dominates our research objectives. The latest XRL development has brought stronger interest of the X-ray laser community in possible applications of the coherent short-wavelength radiation. The applications area is determined by the output parameters of the XRLs. At the moment the XRLs with the brightness of 10^{24}-10^{26} photons/(s×mm^2×mrad^2×0.1%B$_w$) belong to the brightest sources available, synchrotron inclusive. The average number of 10^{12} photons/shot at the spectral range where relatively efficient XUV optics is available let us claim the XRLs are matured to be applied in practice. However, there is still need to look for new pumping schemes allowing development of more compact and efficient XRLs working in the short-wavelength range with a high repetition rate.

CP641, *X-Ray Lasers 2002: 8th International Conference on X-Ray Lasers*, edited by J. J. Rocca et al.
© 2002 American Institute of Physics 0-7354-0096-2/02/$19.00

MBI LASER DRIVERS

MBI has at its disposal two laser systems which are generally dedicated to the laser-matter interaction investigations. A Nd:glass laser system, which is a CPA upgrade of the former NIXE system (Neodymium:glass Induced X-ray Emitter) and consists of two subsystems delivering on the target a long nanosecond pulse and a short picosecond pulse with the controlled delay between them. The front-end of the system includes a Ti:Sa oscillator at 1053 nm (TSUNAMI) followed by a Ti:Sa regenerative amplifier with the output energy of 4 mJ in a single pulse and the repetition rate of 10 Hz. The CPA arm with a pulse compressor at the end delivers pulses with maximum energy of 6 J and the length of 1 ps. The length of the output pulse can be controlled by variation of the distance between the compressor gratings. The nanosecond pulse has length of 1.2-1.5 ns and energy of 10-12 J. One shot from this system is possible every 12 min. The multiterawatt Ti:Sa laser operating with the repetition rate of 10 Hz delivers 1 J energy in less than 50 fs and the peak intensity obtained in the focal spot is in excess of 10^{19} W/cm^2. While the glass laser system is dominantly used as an XRL driver both for the experiments in the slab target geometry and the longitudinally pumped active media, the titanium-sapphire laser was up to now applied exclusively for guiding proof in the capillary discharge plasma. Both systems, similarly to all existing facilities of this type, have a limited average power due to low repetition rate and as a result the average power of the XRLs pumped with these lasers is extremely low. Many of the future applications require increased average power and it means the laser drivers have to be developed in this direction. Now a new driver laser system with the repetition rate between 100 Hz and 1 kHz and energy of a few joules is under constraction at MBI. The first stage of the development characterised by the output energy of $\sim$1 J is planned to be commissioned in the year 2005.

TRANSIENT COLLISIONAL X-RAY LASERS

Transient inversion scheme using combination of two pumping pulses of different length and implementing collisional excitation mechanism opened a new way towards table-top X-ray lasers. It reduced the pump energy about two orders of magnitude and shortened significantly the output pulse. It is a very robust scheme with major advantages against the X-ray lasers using O(ptical)-F(ield)-I(onized) plasma as an active medium. The main advantage of this scheme is a high saturation parameter usually about two orders of magnitude higher than that of the OFI-lasers. This parameter decides about the output energy of the laser and the same about its possible applications. Further reduction in the pump energy and the following increase of the repetition rate could lead to X-ray lasers with remarkable average power according to the way delineated above.

Longitudinally pumped lasers

Longitudinal irradiation of the plasma column seems to be, even if being investigated very seldom, one of very promising routs in X-ray lasers down-sizing. The advantage is taken that in such an irradiation geometry it is possible to excite the medium efficiently using low-energetic but more tightly focused pump laser beam. The intrinsic travelling wave pump arrangement with the speed of the excitation wave $v_{exc} \leq c$ and the refraction problem eliminated by the cylindrical symmetry of both the pumping beam and the active medium, constitute major additional advantages of longitudinal pumping. However, high intensity radiation applied requests beam guiding to eliminate ionization induced divergence. It means that the plasma pipes are necessary for such lasers.

Hybrid X-ray laser

A hybrid X-ray laser (HXRL) pumped by a combination of the capillary discharge (plasma preforming) and a short (picosecond) laser pulse unifies two most popular pumping schemes in one. Lasing in such a laser working in Ne-like sulphur with the wavelength of the output signal at 60.8 nm has been demonstrated for the first time in 1999 [2]. The hybrid XRL is a transient inversion laser with increased efficiency due to applying capillary discharge for plasma preforming. The ratio of the gain-length product to the total pump energy estimated from the experimental data was equal to 3.1 J^{-1}, one of the highest figure of merits for the transient collisional XRLs, in spite of including the total electric energy which was dissipated in the medium [2]. There exist two versions of the capillary which could be used in the hybrid lasers. First one was designed to realise a fast electric discharge with the current rise time about 100 ns [3]. This version was used for the proof-of-principle experiment with the Ne-like sulphur laser at 60.8 nm. Major advantage of this kind of discharge is abundance of Ne-like ions in the discharge plasma being responsible for efficient lasing in the XUV range. On the other hand, the second variant - slow capillary discharge is characterised by lower and slower current and as a result a weak ionization of the medium with the average ionization stage $Z^* \leq$ 1 [4]. Hence it could be easily applied to the OFI-XRLs [5]. The capillary discharge plasma has a wall-sustained character, i.e. the physical processes on the capillary wall are deciding for the discharge plasma parameters. The plasma ablation process followed by its stagnation give a concave density profile with the minimum on the axis which is a prerequisite of successful guiding of the pump laser beam over the long distances. Sketch of such a capillary with the radial density distribution profile is shown in Fig.1

Laser guiding

In the experiment on laser guiding a capillary made of plastic with the dimensions: diameter of 0.5 mm and the length of 3 cm has been used. Guiding has been proved using beam from the Ti:Sa laser with the intensity of 10^{17} W/cm^2 in 50 fs pulse. Having used the capillary presented in Fig.1 under higher charging voltage of 8 kV the focusing

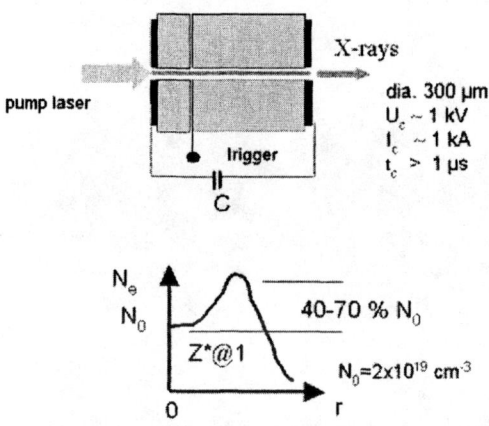

FIGURE 1. Work principle of the hybrid laser and the radial plasma density distribution in the channel.

effect of the discharge plasma channel was observed by focusing the transmitted beam on a CCD camera. The registered images are shown in Fig.2. Intensity distribution smoothening due to the presence of the plasma channel is clearly seen. This effect is a precondition of any successful scheme of the longitudinally pumped X-ray lasers. The guiding effect reduces significantly the refraction, scattering and absorption losses and the beam transmission estimated in the experiment was between 80 and 90%.

Excitation

The absorption process in the underdense plasma ($n_e = 2 \times 10^{19}$ cm^{-3}) characteristic for the capillary discharge is dominated by the inverse bremsstrahlung (IB) and the same depends strongly on the plasma density and its temperature. The intensity applied, connected directly with electric field of the light wave, changes the electron collisionality and hence influences the absorption process as well. The increased electron collisionality at low plasma temperature and low intensity boosts the absorption rate but at the same time it reduces the pump pulse energy and to some extent, assuming constant pulse shape, its intensity. Thus, the plasma temperature decreases and the absorption increases further with the propagation length. Pump energy depletion along the axis of the plasma channel, caused by the described effect is shown in Fig.3 for the case of a molybdenum capillary. As a consequence the pump pulse intensity has to be optimised to ensure uniform excitation and ionization levels along the plasma channel. It requires reduction in the beam absorption but at the same time preserving the necessary plasma temperature and ionization stage. The simplest way to achieve this is increase in the pump laser intensity to reduce the plasma collisionality and as a consequence the plasma absorption (Fig.3). For all of the elements analysed (Ti,Mo,S), the intensities above 10^{16} W/cm^2 are necessary to obtain a quasi-uniform excitation distribution in the axial direction. It is worth noting that this effect is present in any longitudinally pumped X-ray laser

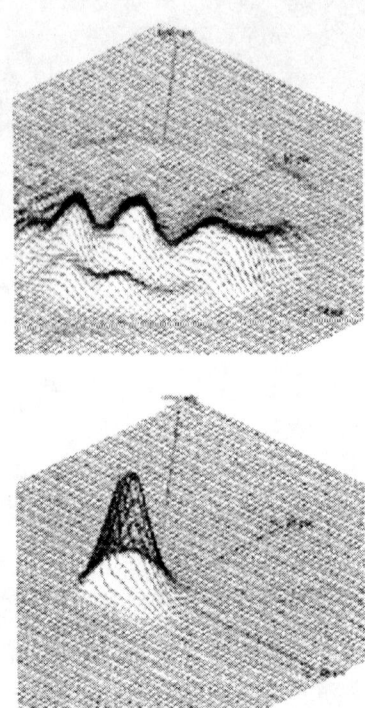

FIGURE 2. 3D-plots of the laser beam transmitted through the plasma channel and focused on the CCD camera: without discharge (upper) and with the discharge (lower).

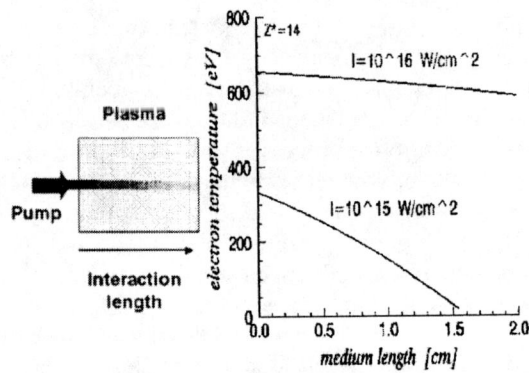

FIGURE 3. Axial pump energy depletion in the plasma channel of molybdenum plasma.

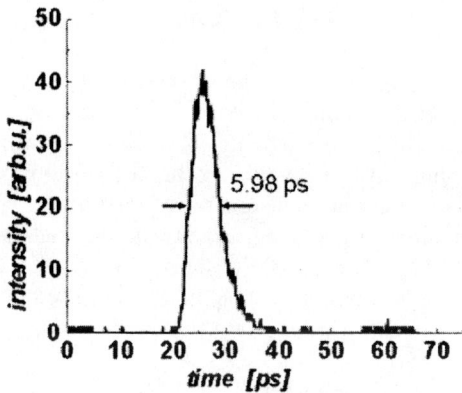

FIGURE 4. Shape of the pumping laser pulse used in the experiment. The pulse energy was equal to 1.5 J.

including the OFI-laser, laser ablated capillaries and gas-puff (gas-jet) targets. Control over it is a prerequisite of an efficient longitudinally pumped X-ray laser scheme.

Laser driven Ni-like Ag XRL

Current work on the X-ray lasers irradiated in the slab target geometry is concentrated on the transient inversion scheme applied to Ni-like ions emitting between 10 and 19 nm and possible applications of such a source. Specifically, Ni-like silver is investigated extensively due to its emission wavelength (13.9 nm) matching very well the spectral region of the efficient multilayer optics. While the most efficient Pd XRL (λ=14.7 nm) demonstrated at LLNL was pumped by a combiation of two lasers with the total energy as low as 7 J [6] we have recently demonstrated saturated Ni-like Ag XRL pumped by a single profiled picosecond laser pulse [7]. Pulse profiling means here that the level and length of the pedestal of the pulse generated and amplified in the pump laser system can be controlled. It has to be stressed that the total energy of the pumping pulse was about 3 J and the short pulse length (FWHM) was equal to 6 ps (Fig.4). These values make the final aim denoted as a high repetition rate XRL much more realistic. Gain as high as 33 cm^{-1} has been measured and the gain-length product of $\sim$ 16.5 has been estimated. The scheme has to be optimised but stable lasing at 13.9 nm with such a pump arrangement is irrefutable. We believe that this result opens new possibilisties on the way towards efficient tabe-top soft X-ray lasers.

OUTLOOK

We have reported the first demonstration of the saturated X-ray laser pumped by a single picosecond laser pulse with the pump energy more than halved in comparison to the best achievements in the double-pulse scheme. The reduction in the pump energy and lengthening of the pumping pulse are very beneficial for the prospects of the XRLs with high repetition rate and the same reasonable average power. The scheme required optimisation and test for other target arrangements as longitudinal irradiation both in gases and capillaries. The output energy of the short-wavelength radiation from the Ag-XRL estimated to be between 2 and 5 μJ let us think about its application in practice.

REFERENCES

1. Nickles, P.V., Schnuerer, M., Kalachnikov, M.P., Will, I., Sandner, W., Shlyaptsev, V.N., Phys. Rev. Lett. **78**, 2748 (1997).
2. Janulewicz, K.A., Rocca, J.J., Bortolotto, F.,Kalachnikov, M.P., Shlyaptsev, V.N., Sandner, W., Nickles, P.V., *et al.*, Phys. Rev. A, **63**, 033803(2001).
3. Moreno, C.H., Marconi, M.C., Shlyaptsev, V.N., Rocca, J.J., IEEE Trans. Plasma Sci., **27**,6(1999).
4. Ehrlich, Y., Cohen, C., Kaganovich, D., Zigler, A., Hubbard, R., Sprangle, P., Esarey, E., J. of Opt. Soc. Am. B,**13**, 68(1996).
5. Goltsov, A., *et al.*, "X-Ray Lasers 2000" Proceedings of the 7th IXRLC in St. Malo, France, June 19-23 2000; EDP Sciences, Les Ulis; Eds. G. Jamelot, C. Möller, A. Klisnick, J. Phys. IV France, **11**, Pr2-165 (2000).
6. Dunn, J., Li, Y., Osterheld, A.L.,Nilsen, J., Hunter, J.R., Shlyaptsev, V.N., Phys. Rev. Lett. **84**, 4834 (2000).
7. Janulewicz, K.A., Lucianetti, A., Priebe, G., Sandner, W., Nickles, P.V., this Proceedings.

Submilliradian Divergence Longitudinally Pumped Nickellike Molybdenum X-Ray Laser

T. Ozaki, H. Kuroda[a], R. A. Ganeev[a], A. Ishizawa[a] and T. Kanai[a]

NTT Basic Research Laboratories, NTT Corporation
3-1 Morinosato Wakamiya, Atsugi, Kanagawa 243-0198, Japan
[a] Institute for Solid State Physics, University of Tokyo
5-1-5 Kashiwanoha, Kashiwa, Chiba Japan

Abstract. We experimentally demonstrate an 18.9 nm nickellike molybdenum x-ray laser with a very small divergence of less than one milliradian, using the longitudinally pumping technique. Simulations show that the pedestal of the longitudinal pump beam produces a region within the preplasma with large electron density and gain gradients. It is shown that this can lead to the selective amplification of transverse modes with high spatial coherence.

INTRODUCTION

It has been shown that the longitudinally pumped transient-collisional-excitation (TCE) scheme is a promising candidate for realizing high repetition-rate x-ray lasers. Predictions [1] have shown that total pump energy of about 200 mJ is enough to saturate the 18.9 nm nickellike molybdenum x-ray laser. Strong amplification of this x-ray laser has been observed using a 240 mJ tabletop Ti:sapphire laser system coupled into molybdenum capillary targets [2]. These novel works opened up a new method for realizing high repetition rate x-ray lasers, which will have a large impact on various application experiments. In the present paper, we report on the observation of submilliradian divergence x-ray lasers using this scheme. Although saturated amplification is not obtained, simulations show that the unique electron density and gain profile generated by the pedestal of the longitudinal pump beam can selectively amplify x-ray laser transverse modes with high spatial coherence.

EXPERIMENTAL

Fig.1 is a schematic diagram of the experimental apparatus used in this work. 2-mm long slab molybdenum target (T) is placed within a vacuum chamber. First a 300 ps long pulse (LP) irradiates this target from the transverse direction, line-focused by a cylindrical lens (CL). After a delay, the 475 fs short pulse (SP) irradiates the target from the longitudinal direction, using a spherical lens (SL). The x-ray emitted in the on-axis direction is observed by a flat-field spectrograph equipped with a Hitachi grating (G), and the time-integrated spectrum is detected using a photocathode camera (PC). A cylindrical mirror (M) is used to spatially image a point (I) onto the

CP641, *X-Ray Lasers 2002: 8th International Conference on X-Ray Lasers,* edited by J. J. Rocca et al.

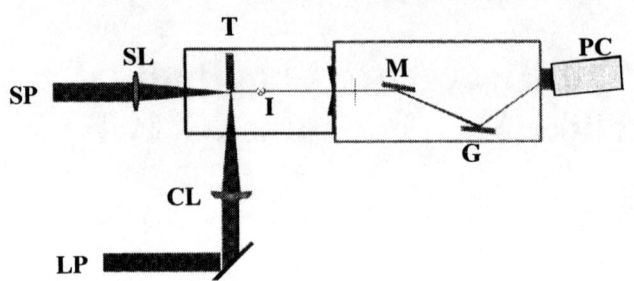

FIGURE 1. Schematic diagram of experimental layout.

detection plane. However, we deliberately defocus the imaging in the present work by moving the target away from this imaging position, in order to investigate the divergence of the x-ray laser. We can also direct the zero-order diffracted beam to a commercial Acton x-ray spectro-meter to calibrate the flat-field spectrograph system.

EXPERIMENTAL RESULTS

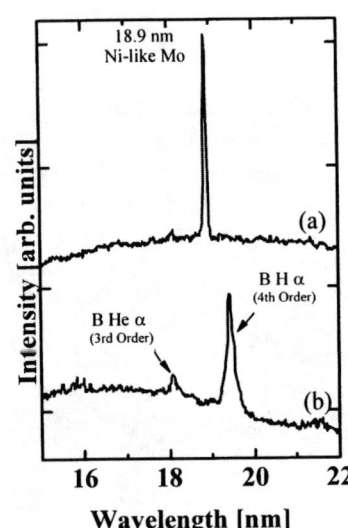

FIGURE 2. Trace of on-axis spectrum from (a) molybdenum plasma, and (b) boron plasma

In Fig.2 we show the trace of the on-axis spectrum that is observed using this apparatus. The pump energy used in this experiment is 30 mJ in the prepulse and 120 mJ in the main pulse, and the delay between the prepulse and main pulse is 4 ns. The main pulse has been measured to be accompanied by a pedestal with a contrast ratio of 1×10^{-5}. We calibrated the wavelength of our soft-x-ray spectrum using higher-order diffraction of several resonance lines from boron plasma. The results show that the molybdenum spectrum is completely dominated by a single 18.9 nm line, which can be assigned the 4d-4p lasing transition of nickellike molybdenum. The background in the molybdenum spectrum is due to the noise from the detector system, which can be inferred from the fact that there is no sign of an absorption edge at 17 nm, even though a 0.65 µm thick Al filter was used. What is also interesting about this spectrum is that the spectrograph slit is widely opened to 1.6 mm in this experiment, and still the observed laser line has a narrow spectral width.

Typical example of the raw image data obtained using this spectrograph system is shown in Fig.3. The 18.9 nm spectrum is a single spot, whose dimension is 90 µm wide and 80 µm high at the detector position. Judging from our defocused spectrograph configuration, we can estimate a maximum divergence of 1 mrad in the

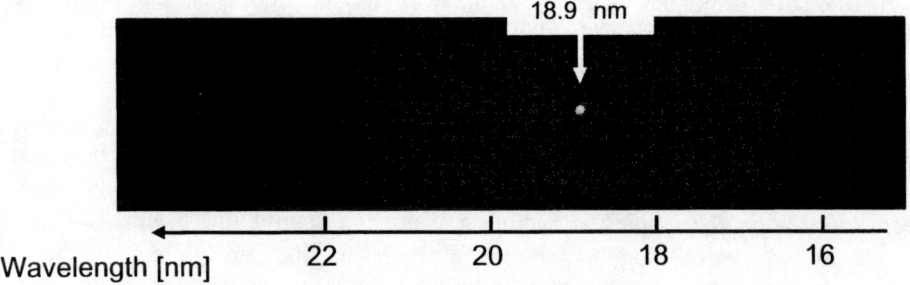

Wavelength [nm]   22   20   18   16

FIGURE 3. Image of on-axis spectrum near the 18.9 nm x-ray laser line.

vertical direction, and 0.3 mrad in the horizontal direction, which is, to the authors' knowledge, the smallest divergence ever observed for a TCE x-ray laser.

The 18.9 nm line is found to have several characteristics that are similar to conventional TCE x-ray lasers. One example is the variation of x-ray intensity with delay time between the prepulse and main pulse. This is shown in Fig.4, for a constant pump intensity of 3×10^{16} W cm^{-2}. The error bars in the figure are due to the shot-to-shot variation of the x-ray intensity. Maximum is observed for a delay of 4 ns, which drops by one order of magnitude for 2 ns delay. The 18.9 nm line is also observable up to a delay of 10 ns, but the intensity in this case drops by two orders of magnitude.

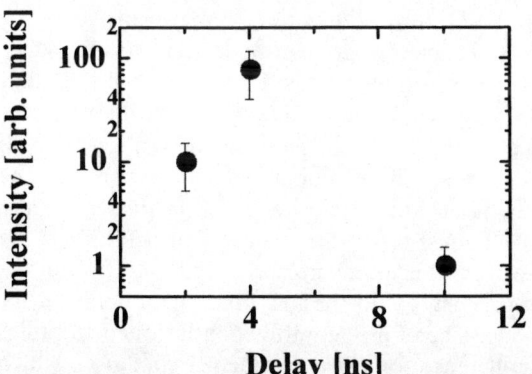

FIGURE 4. Dependence of 18.9 nm line intensity as a function of delay time between the prepulse and main pulse.

Gain evaluation in longitudinally pumped x-ray lasers is extremely difficult, since large nonuniformities in the gain coefficient are predicted along the x-ray laser axis. Therefore, we have chosen to measure the energy of our x-ray laser to estimate the magnitude of amplification. For this purpose, a commercial spectrograph system is used, equipped with a sodium salicylate scintillator and a photomultiplier tube. After careful calibration of the optics and filters used, the energy of the 18.9 nm line is evaluated to be 27 nJ.

DISCUSSIONS

In order to clarify what causes this small divergence x-ray laser, we performed some simulations. We combined several hydrodynamics and atomic physics codes [1], along with a ray tracing code to model the propagation of the longitudinal pump beam through the preplasma. The plasma was divided into 2000 meshes in the direction of the x-ray laser axis, and 200 meshes in the direction normal to the target surface. The code calculated the temporal gain for each of the 400,000 zones within the preplasma.

From the above simulation, we first learn that, due to its relatively low intensity, the prepulse alone is not enough to generate an abundance of nickellike molybdenum ions in the preplasma. Simulations also tell us that the pedestal in the longitudinal beam can deform the preplasma condition. Since the longitudinal beam is focused onto a 30 μm diameter region within the preplasma, the sub-TW cm^{-2} intensity pedestal produces a narrow region with electron temperatures high enough to generate an abundance of nickellike ions. Calculations also show that gain is restricted to regions with temperature sufficient for nickellike ion production, and only small or negative gain coefficients are observed in the colder region of the plasma.

The parabolic electron density and gain profile that are produced by the pedestal is found to limit the number of transverse modes that are amplified in the gain medium. In the work by London et al. [3], it was numerically shown that for gain medium with such unique profiles, the efficiently guided modes that receive large amplification could be restricted to the lower order modes. Using formulations from this paper, we can calculate the number of transverse modes that receive amplification greater than $0.18 \times gL_{max}$, where gL_{max} is the maximum gain-length product experienced by a single transverse mode. For a flat density profile, the number of effectively amplified one-dimensional Hermite-Gaussian modes is near 50, and in this case the geometrical dimensions of the gain region determines the divergence of the x-ray laser. However, this number is reduced to five for the parabolic profile predicted in our experiment, with the maximum gain obtained for the diffraction-limited transverse mode. Therefore, amplification is found to prefer transverse modes with high spatial coherence for our experimental condition, which explains the observed small divergence.

CONCLUSIONS

We have demonstrated longitudinal pumping of 18.9 nm nickellike molybdenum x-ray laser, resulting in very small divergence of less than one milliradian. Using simulations, we show that the pedestal of the longitudinal pump beam produces a region within the preplasma with large electron density and gain gradients, leading to the selective amplification of transverse modes with high spatial coherence.

REFERENCES

1. Li, R., Ozaki, T., Kanai, T., and Kuroda, H., *Phys. Rev E.*, **57**, 7093-7102 (1998).
2. Li, R., and Xu, Z. Z., *J.Phys. IV*, **11**, Pr2 27-34 (2001).
3. London, R. A., Strauss, M., and Rosen, M. D., *Phys. Rev. Lett.*, **65**, 563-566 (1990).

Recent Progress on the Understanding of the Transient Ni-like Ag X-ray Laser at 13.9 nm at LULI facilities

D. Ros[1], A. Klisnick[1], D. Joyeux[2], D. Phalippou[2], O. Guilbaud[1],
J. Kuba[1*♠], A. Carillon[1], G. Jamelot[1], R. Smith[1♠], M. Edwards[3], F. Strati[3],
G. J. Tallents[3], H. Daido[4], H. Tang[4], P. Neumeyer[5], D. Ursescu[5], T. Kühl[5],
J.-C. Chanteloup[6], K. Bouhouch[1].

[1]*Laboratoire d' Interaction du rayonnement X Avec la Matière, Université Paris XI, Bâtiment 350,
91405 Orsay, France*
[2]*Institut d'Optique Théorique et Appliquée, Laboratoire Charles Fabry, CNRS,
B. P. 147, 91403 Orsay, France*
[3]*Department of Physics, University of York, Heslington, YO10 6DD, York, UK*
[4]*Advanced Photon Research Center, Kansai Research Establishment, JAERI,
8-1 Umemidai, Kizu, Souraku, Kyoto 619-0215, Japan*
[5]*GSI Darmstadt, PHELIX Project,Germany*
[6]*LULI, UMR 3705 CNRS/CEA/Université Paris 6, Ecole Polytechnique,
91123 Palaiseau, France*

Abstract : This paper summarises our recent progress achieved in the characterisation and understanding of the Ni-like Ag transient X-ray laser pumped under traveling wave irradiation. We carried out two experiments at the LULI CPA laser facility. Several diagnostics of the plasma emission at the XRL wavelength or in the keV range indicate the presence of small-scale spatial structures in the emitting XRL source. Single-shot Fresnel interference patterns at 13.9 nm were successfully obtained with a good fringe visibility. For the first time we obtained plasma images with a high spatial resolution about 1 µm, showing the effects of pumping parameters on the X-ray laser far-field.

1. INTRODUCTION

Since the first experimental demonstration at MBI in 1995 [1] several experiments carried out in various laboratories [2-5] have confirmed that saturated X-ray lasers

---

♠ Also at Faculty of Nuclear Science Engineering, Czech Technical University, Prague, Czech Republic
♠ Now at LLNL, P. O. Box 808 mail Code L-251, 94550 Livermore, CA, USA

CP641, *X-Ray Lasers 2002: 8th International Conference on X-Ray Lasers*, edited by J. J. Rocca et al.
© 2002 American Institute of Physics 0-7354-0096-2/02/$19.00

(XRL's) pumped by Transient Collisional Excitation (TCE) may be obtained with moderate pump laser energy requirements. In this pumping scheme, low-intensity laser pulse preforms a plasma containing the lasant ions. A few hundreds of picoseconds later, a short (~ 1 ps) high-intensity laser pulse rapidly heats the plasma in a short timescale compared to ionisation typical times. The fast temperature rise induces transient population inversions in Ni-like or Ne-like ions through strong collisional excitation of ions in their ground state. Nevertheless due to the short duration of population inversions compared to the transit time of the XRL beam along the plasma column, the preformed plasma must be irradiated by the short heating pulse in a Traveling-Wave (TW) procedure. This is achieved by tilting the CPA pulse energy front [6] or by introducing step-delays across the CPA beam diameter [7]. At LIXAM, experimental investigations of transient X-ray lasers were initiated in 1998 [5, 6, 8]. We focused our studies on the Ni-like silver 4d-4p X-ray laser that emits at 13.9 nm, a wavelength that is particularly attractive for applications because it is close to the maximum reflectivity for X-UV optical devices based on Mo:Si multilayers. We have previously shown that the TW irradiation was essential to extract the X-ray laser photons from the plasma [5]. The method of detuning the compressor [6] that we have used to generate the TW allowed to control the direction of the X-ray laser emission [9]. In this paper we will summarise the main results obtained during our two most recent experimental campaigns, carried out with CPA high-power laser at LULI, Palaiseau (France). At LULI the 13.9 nm X-ray laser beam was characterised by several diagnostics, including a footprint monitor and a Fresnel interferometer. We observed during the first experiment that the wavefront of the XRL beam is not flat and that -for some shots- the XRL emitting plasma contains multiple coherent spatial structures that lead to interference patterns in the far field. We obtained interferograms with a good fringe visibility, showing that single-shot interferometry on a ps timescale is possible with transient XRL's. Finally strong lasing was observed on the Ni-like 4f-4d line emitted at 16 nm. Following this experiment [10], we decided to realise images of the emitting gain zone at the output of the plasma. Then, we will also present the first plasma images with a very high spatial resolution (~ 1 μm), showing large spatial variation in emission profiles depending on plasma production conditions.

2. A SATURATED X-RAY LASER AT 13.9 NM AT LULI

2.1 Experimental set-up

Figure 1 shows the geometry of the experimental set-up. The 450 ps, uncompressed beam was focused onto a silver slab target by a spherical and a cylindrical lens to produce a 2.1 cm x 150μm line focus. After compression by a, double-pass compressor under vacuum, the beam carrying the 0.7 ps compressed pulse was focussed by an off-axis parabola and a spherical mirror that produced a 3.2 cm x 70 μm line focus, which was superposed to the former.

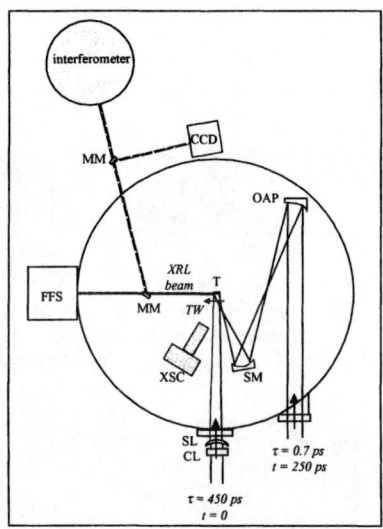

Figure 1: Experimental set-up inside the target chamber at LULI. Two retractable multilayer mirrors (MM) were used to direct the XRL beam at 13.9 nm towards either of the diagnostics shown. OAP: off-axis parabolic mirror; SM: spherical mirror; SL: spherical lens; CL: cylindrical lens; T: target; MM: multilayer mirror; XSC: cross-slit camera; FFS: flat-field spectrometer. An active pinhole camera, not shown in the figure, viewed the plasma column from above, with a viewing angle of 66° relative to the horizontal plane.

The peak to peak delay between the pump pulses was adjusted to 250 ± 50 ps. The pump laser irradiance at the target was respectively~$6\ 10^{11}$ W/cm^2 in the long pulse and ~$6\ 10^{14}$ W/cm^2 in the short pulse. A TW irradiation at the velocity of light, c, was generated in the direction towards a flat-field spectrometer. In our case, the duration of the pulse was calculated to vary from 700 fs at the centre of the line focus to 1.2 ps at the edge of a 20 mm target.

Several diagnostics were implemented to control the plasma size and homogeneity, as well as to analyze the X-ray laser emission. A cross-slit CCD camera monitored the width of the line foci as well as their superposition at the target surface. This diagnostic thus gave a magnified, time-integrated image ($M \approx 3$) of the plasma in the keV range, that allowed to control the size and the uniformity of the preformed plasma zone that is heated by the short pulse. Three diagnostics of the X-ray laser emission were implemented, with two retractable Mo:Si mirrors (see figure 1) allowing to direct the 13.9 nm laser beam towards one or the other diagnostics at each laser shot.

2.2 Time-integrated results

The on-axis emission of the silver plasma was analysed with a flat-field spectrometer equipped with a CCD detector. Fiducial wires were aligned close to the entrance slit of the spectrometer to provide angular calibration of the emission in the horizontal plane. Figure 2 shows one spectrum emitted by a 10 mm plasma pumped under similar irradiation conditions (the duration of the short pump pulse was of 1.3 ps in this case). Two lasing lines are visible: the "standard" Ni-like 4d-4p line at 13.9 nm and the 4f-4d line at 16 nm. The latter line has been previously observed in a silver plasma produced in a "two-colour" ($2\omega/1\omega$) irradiation configuration only [9]. Here it is observed with a $1\omega/1\omega$ irradiation, which is in contradiction with our previous observations [10, 11] The mean deflection angle of the 4f-4d line was observed to be slightly larger than for the 4d-4p line, namely 9 mrad instead of 7 mrad. One can see that the 4p-4d line in the bottom spectrum exhibits a double angular peak. Such angular structures were already observed occasionally in our previous experiments. They can be interpreted as the superposition of two laser beamlets having experienced two different paths in the active plasma or in two spatially distinct active zones.

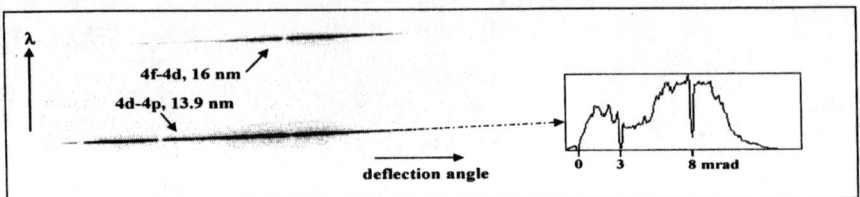

Figure 2: Time-integrated spectrum obtained from a 10 mm plasma under similar irradiation parameters (short pulse duration: 1.3 ps), and lineouts of the 4d-4p line along the deflection angle axis. Occasionally the 4d-4p line was observed to exhibit a double-peak angular structure. Strong lasing is also observed on the Ni-like 4f-4d line at 16 nm.

The control of the angular structure of the XRL beam will call for a better control of the deposition of the short pulse energy in the preformed plasma.

2.3 Time-resolved investigations

Using a sub-ps resolution streak we tried to study the temporal evolution of the two lasing lines. Due to triggering problems, we just obtain preliminary results any . Nevertheless, as shown in figure 3 we succeeded to obtain for the first time an image at the threshold of the sweep, including the two lasing lines.

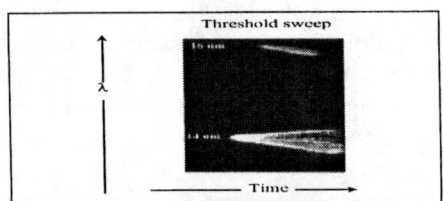

This should indicate that the two lasing lines are not emitted at the same time. But due to the difference of their intensity and the sensitivity of the photocathode, it would be necessary to use multi-thickness filters for further investigations.

Figure 3: CCD image from the Streak camera showing the time-resolved emission of the two lasing line.

More recently experiment performed at RAL gave important results about the duration of the X-ray lasing lines [12], completing our previous work [13].

3. FIELD CHARACTERISATION OF THE X-RAY LASER BEAM

In order to use the Ag X-ray laser for applications, we planed to characterise the optical properties of the beam (including far-field, interference patterns and near-field).

3.1 Far-field of the 13.9 nm laser

A footprint monitor consisting in the two multilayer mirrors and a back-illuminated CCD camera was used to characterize the 2-D spatial and angular structure of the 13.9 nm laser beam at a distance of 1.4 m from the source. Figure 4.a shows a typical image obtained from a 20 mm plasma showing the distribution of energy over the 13.9 nm laser beam section. A pointing angle of 7 mrad was inferred from a fiducial cross-wire, that was removed for this shot but can be seen on the image of figure 4.b. The horizontal divergence is around 5 mrad, while the vertical divergence is of the order of 10 mrad. For a limited number of shots however the footprint image exhibited fringe patterns, an example of which is shown in figure 4.b. Such interference fringes can be interpreted as being due to multiple coherent sources emitting XRL beamlets that interfer in the far field.

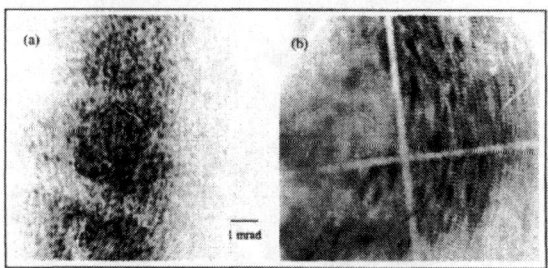

Figure 4: Examples of footprint images of the 13.9 nm laser obtained at a distance of 1.4 m from the source. For most of the shots, the distribution of energy was reasonably smooth, as shown in figure 4.a. For some shots fringe patterns, as shown in figure 4.b, were observed.

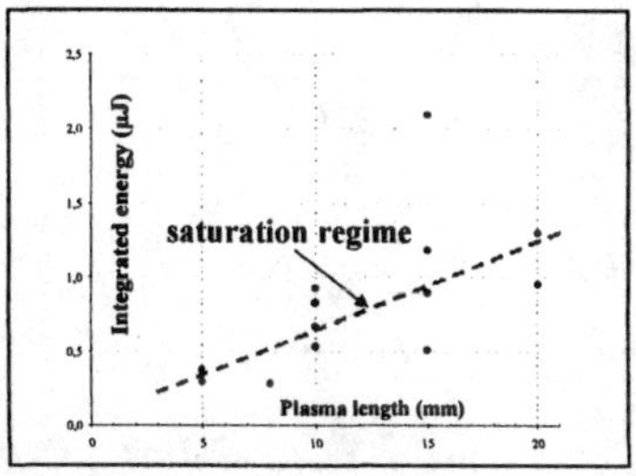

Figure 5: The energy of the 13.9 nm laser beam integrated over the whole section (from the footprint images) increases linearly with the length of the plasma, up to 20 mm.

A length-scan of the XRL energy, integrated over the beam section was carried out using the footprint monitor, as shown in figure 5. The CCD count number was converted into energy by taking into account the filter transmission and the reflectivity of the Mo:Si mirrors. One can see from figure 5 that, except for one data point at 15 mm, the XRL energy increases linearly with length. Since the measured level of energy is approximately twice below the calculated saturation energy E_{sat} ($E_{sat} = I_{sat} . s . \tau \approx 2.4$ µJ). It should be noted however that a reduction of a factor of two only in the XRL pulse duration (i.e. 1 ps instead of 2 ps) would lower E_{sat} to the experimentally observed energy.

3.2 Interferograms of the 13.9 nm laser

The 13.9 nm XRL beam was sent to a Fresnel bi-mirror interferometer, as shown in figure 6. This interferometer had been initially used to investigate the deformations of a niobium surface under high electric field [14, 15], using the QSS Ne-like zinc XRL emitted at 21.2 nm. The XRL beam is first reflected by a multilayer mirror to achieve spectral filtering of the radiation emitted by the plasma. It is then reflected by the bi-mirror, which forms an open book. Interference fringes are obtained in the region where the two XRL half-beams reflected by the bi-mirror overlap. Those fringes are recorded at the surface of a detector (optical CCD butt-coupled to a phosphor, positioned at a distance of 3.15 m from the source) which is set at a grazing angle of 9°, in order to increase the apparent fringe spacing. Figure 7 shows a typical interferogram obtained. Due to the low sensitivity of the detector used (optical CCD butt-coupled to a phosphor) the signal to noise ratio is poor and we evaluate the contrast to ~ 50%.

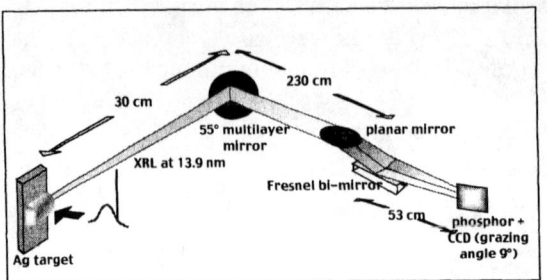

Figure 6: Schematic view of the interferometry set-up used with the 13.9 nm transient XRL.

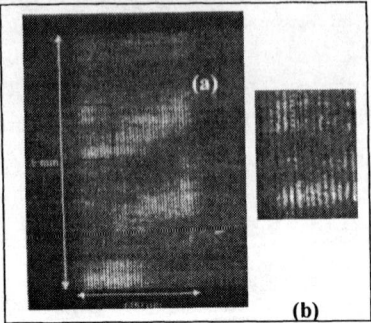

Figure 7: (a) example of interferogram of the 13.9 nm laser obtained at a distance of 3.15 m from the source. (b) detail of the interferogram, showing fringe shifts.

The intensity modulations appearing over the interference pattern with a larger spatial scale are connected to those, which were observed on the footprint images. Finally, on some regions of the interferogram (see detail in figure 7) we can observe fringe shifts of the order of λ, which correspond to aberrations of the XRL beam wavefront.

3.3 Near-field of the 13.9 nm laser

The previous results needed more detailed investigations to clarify the origin of the formation of these structures that could be closely related to the actual structure of the zone heated by the short, intense pulse. That's why in our last experiment at LULI, we used an aspherised microimaging mirror realised at LCFIO by ion erosion [16] to obtain images of the edge of the lasing Ag plasma. We present in figure 8 an artist's view of the system. The XRL beam is deviated by a Mo:Si plane mirror (incidence 35 deg, distance 32 cm from XRL) toward the imaging system box. The purpose of this mirror is twofold: as it is removable, the direct radiating axis of the XRL can be made free, to reach other diagnostics; moreover, it allows to place the imaging optics out of direct view of the plasma, which ensures a much better protection (the plane mirror, in direct view of the plasma, is much easier and cheaper to be replaced). The imaging block output is vertically shifted by 10.5 mm, so as to propagate above the relay mirror, towards the CCD camera, placed at 6 meters from the plasma source. An attenuating filter (2 µm CH + 0.5 µm Zr, T≈0.001 at 13.9 nm, diameter 15 mm) was inserted a few decimeter in front of the CCD, to block visible light and low energy radiation, and to attenuate the XRL beam. The optical magnification is 12.

Then a vertical knife edge, placed a well known distance (≈2 cm) from the desired object plane, was imaged with visible laser illumination, and using the full width of the original spherical mirror (30 mm), while masking the central corrected zone. Owing to this wide pupil, clean coherent illumination, and black field imaging, an accurate determination of the focussing position was possible, despite the use of visible light.

With the magnification used, the CCD pixel, back to object plane is 1.1 µm. Combined to the imaging resolution of 1 µm, and a possible small error in image focussing, the resulting overall resolution is about 1.5 to 2 µm, which could probably be improved by some image processing, at least to recover from the integration introduced by the square pixels. Among several images collected, under various XRL generation conditions, two of them are shown in figure 9.

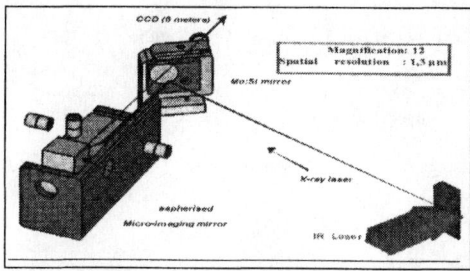

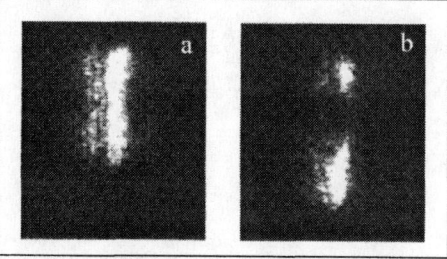

Figure 8: Schematic view of the interferometric microimaging sytem.

Figure 9: typical images obtained for two different values of the focus width of the long pulse beam : respectively 200 µm (a) and 150 µm (b).

Both shots were done using a 3 mm long target, a delay about 250 ps between the long and short pulse and a total pumping energy close to 20 joules (7 J for the short pulse and 12 J for the long). The length focus was about 21 mm, and the width of the short pulse about 70 µm, the long one varying between 200 µm (figure 9-a) and 150 µm (figure 9-b). We clearly observe the strong effect along the vertical direction of the width of the long pulse. This pulse, creating the preformed plasma, has a fundamental action on the vertical density gradient, leading to multiples coherent sources, as we suspected in our previous investigation about far-field [10]. The analysis of the results, depending on plasma production conditions, are under progress and will be the topic of a further more detailed paper.

4. CONCLUSIONS

Fooprints of the 13.9 nm laser indicate for some shots the existence of multiple coherent XRL sources emitted by the active plasma. Those structures are likely to be connected with the actual shape of the gain zone, and with the conditions of the ps pump laser deposition. Single-shot interferograms of the transient XRL at 13.9 nm were produced successfully. Applications to plasma probing are planned in a near future. Then we have observed fringe shifts in the interference patterns that indicate aberrations of the XRL beam wavefront and will require more detailed investigations.

Finally for the first time we obtained near field of the X-ray laser beam with a micrometer spatial resolution, giving us a large information about the effect of pumping conditions on the production of a single or multiple X-ray laser sources.

All these results demonstrate that the transient Ag XRL at 13.9 nm is suitable for applications, on condition to optimise the different pumping parameters. It will the main aim of our future work on the next facility : LASERIX [17].

ACKNOWLEDGEMENTS

The authors gratefully acknowledge the technical support of J-C. Lagron and L. Vanbostal (LIXAM). We would like also to thank all the support staff of LULI. The high quality of the imaging mirror realised by R. Mercier and his colleagues has been greatly appreciated. Finally, we wish to acknowledge the support of the TMR X-ray Laser Network contract N°ERBFMRXCT98-0185, the support of the HPRI-1999-CT00052 and the support of the "Région d'Île de France" (convention E1127).

REFERENCES

1. P.V. Nickles, V.N. Shlyaptsev, M. Kalachnikov, M. Schnürer, I. Will, and W. Sandner, Phys. Rev. Lett. **78**, 2748-2751 (1997)
2. A state of the art of the field can be found in Proc. of the 7th Int. Conf. on X-ray Lasers, Saint-Malo, France, X-ray Lasers 2000, ed. by Y. G. Jamelot, C. Möller and A. Klisnick, Journal de Physique IV France, **11**, Pr2 (2001)
3. M.P. Kalachnikov, P.V. Nickles, M. Schnürer, W. Sandner, V.N. Shlyaptsev, C. Danson, D. Neely, E. Wolfrum, J. Zhang, A. Behjat, A. Demir, G.J. Tallents, P.J. Warwick, and C.L.S. Lewis, "Satured operation of a transient collisional x-ray laser", Phys. Rev. A **57**, 4778-4783 (1998)
4. J. Dunn, A.L. Osterheld, R. Sheperd, W.E. White, V.N. Shylyaptsev, and R.E. Stewart." Demonstration of X-Ray Amplification in Transient Gain Nickel-like Palladium Scheme", Phys. Rev. Lett. **80**, 2825-2828 (1998)
5. A. Klisnick, P. Zeitoun, D. Ros, A. Carillon, P. Fourcade, S. Hubert, G. Jamelot, C.L.S. Lewis, A. Mac Phee, R. O'Rourcke, R. Keenan, P. Nickles, K. Janulewicz, M. Kalashnikov, J. Warwick, J.C. Chanteloup, A. Migus, E. Salmon, C. Sauteret, J.P. Zou, J. Opt. Soc. Am. B **17**, (2000) 1093
6. J. C. Chanteloup, E. Salmon, C. Sauteret, A. Migus, Ph. Zeitoun, A. Klisnick, A. Carillon, S. Hubert, D. Ros, P. V. Nickles, and M. Kalashnikov, J. Opt. Soc. Am. B **17** (2000) 151
7. J. Dunn Y. Li, A.L. Osterheld, J. Nilsen, J.R. Hunter, and V.N. Shlyaptsev, Phys. Rev. Lett. **84** (2000) 4834
8. D. Ros, "Extension du pompage des lasers X-UV aux ions nickelloïdes. Réalisation d'un laser à 13.9 nm. Modélisation de la cohérence spatiale des lasers X-UV", Thèse de doctorat en science de l'Université Paris-Sud, Orsay 1998. (in french)
9. J. Kuba, A. Klisnick, D. Ros, P. Fourcade, G. Jamelot, J.-L. Miquel, N. Blanchot, J.-F. Wyart, Phys. Rev. A **62** (2000) 43808
10. A. Klisnick, J. Kuba, D. Ros, A. Carillon, G. Jamelot, R. Smith, F. Strati, G. J. Tallents, R. Keenan, S. J. Topping, C. L. S. Lewis, P. Nickles, K. A. Janulewicz, F. Bortolotto, D. Neely, R. Clarke, J. Collier, A. G. MacPhee, C. Chenais-Popovics, J.-C. Chanteloup, D. Joyeux, D. Phalippou, H. Daido, H. Tang, SPIE Vol. **4505**, 75 (2001)
11. Miquel, J.-L.; Blanchot, N.; Bonnet L.; Jacquemot, S.; Klisnick A.; Kuba, J.; Ros, D.; Fourcade, P.; Jamelot, G.; Dorchies, F.; Chanteloup, J.-C., SPIE Vol. **3776**, 24 (1999)
12. M. Edwards, Y. Abou-Ali, J. Pestehe, F.Strati, G.Tallents, S. Hubert, R. Keenan, S. Topping, C.Lewis, O. Guilbaud, A. Klisnick, D. Ros, R.Clarke, D.Neely, M.Notley, "Time-resolved measurements of the transient X-ray laser emission", in this proceeding book
13. A. Klisnick, J. Kuba, D. Ros, R. Smith, G. Jamelot, C. Chenais-Popovics, R. Keenan, S. J. Topping, C. L. S. Lewis, Strati, G. J. Tallents, D. Neely, R. Clarke, J. Collier, A. G. MacPhee, F. Bortolotto, P. V. Nickles, and K. A. Janulewicz, Phys. Rev. A **65**, 033810 (2002)
14. F. Albert, D. Joyeux, P. Jaeglé, A. Carillon, J.P. Chauvineau, G. Jamelot, A. Klisnick, J.C. Lagron, D. Phalippou, D. Ros, S. Sebban, P. Zeitoun, Optics Communications **142** (1997) 184
15. F. Albert, P. Zeitoun, P. Jaeglé, D. Joyeux, M. Boussoukaya, A. Carillon, S. Hubert, G. Jamelot, A. Klisnick, D. Phalippou, D. Ros, A. Zeitoun-Fakiris, Phys. Rev. B **60** (1999)11089
16. R. Mercier, M. Mullot, M. Lamare, G. Tissot, "Ion beam milling fabrication of a small off-axis ellipsoidal mirror, diffraction limited to 1 μm resolution at 14 nm", 2001, Rev. of Sc. Instr., 72, 2, pp. 1559-1564
17. G. Jamelot, D. Ros, A. Klisnick, P. Zeitoun, J. Dubau, J.C. Lagron, L. Vanbostal, J-P. Chambaret, M. Pitman, "present state oand future of laser-pumped X-ray lasers in France: LASERIX", in this proceeding book.

Recent Progress of Research on X-ray Lasers at the Institute of Physics, CAS

Jie Zhang, Xin Lu and Ying-jun Li

Laboratory of Optical Physics, Institute of Physics, Chinese Academy of Sciences, Beijing 100080, China

Abstract. The transient nature of x-ray laser is studied. For a drive pulse even as short as 500fs, the x-ray laser gain still shows a quasi-steady state behavior. A real transient nature can only be seen when the x-ray laser is driven by a laser pulse with a duration as short as tens of femtoseconds. A femtosecond laser pulse driven collisional Ne-like Ti x-ray laser at 32.6 nm is numerically investigated. By using the optimized drive pulse configuration, a gain of 40 cm^{-1} can be generated from a 5 mm × 50 μm line focus using about 1 J pump energy. Effects of delay time on transient collisional excitation nickle-like x-ray lasers are investigated analytically using a simple model. The calculations show that the longer delay time can greatly relax the density gradient. Similarly, increasing the intensity of the short pulse or extending the duration can also raise the electron temperature, resulting in higher gain coefficient. Our results indicate that extending the short pulse duration is more efficient than that of increasing the intensity.

TRANSIENT CHARACTERISTICS OF A NEON-LIKE GE

X-RAY LASER AT 19.6 NM

1.1. Introduction

Since the first demonstration of soft x-ray laser in 1985 [1], quasi-steady state (QSS) collisional x-ray lasers have been investigated intensively [2-9]. By comparison, the transient collisional excitation (TCE) was proposed as an alternative scheme to significantly save drive energy [10-14]. In experiments the hydrodynamic conditions of TCE scheme can be realized by using a low intensity long prepulse to produce a preplasma with rich population of Ni-like or Ne-like ions in the ground

CP641, *X-Ray Lasers 2002: 8th International Conference on X-Ray Lasers*, edited by J. J. Rocca et al.
© 2002 American Institute of Physics 0-7354-0096-2/02/$19.00

state. After the prepulse, a high intensity short main pulse is used to make a jump of electron temperature. After 1996, a series experiments was performed by using a nanosecond prepulse and a picosecond main pulse [12-16]. The pump energy for a saturated laser output has been reduced to less than 10J [16]. Nilsen analyzed the population kinetics for a transient Ne-like Ti x-ray laser at wavelengths of 32.6 and 30.1nm. The results show that over picosecond scales, the gain of the Ne-like Ti laser has a quasi-steady state nature with regard to the equilibrium of the excited-state population [17]. In this chapter, we model the Ne-like Ge x-ray laser at 19.6nm numerically to investigate plasma conditions and the gain characteristics in QSS and TCE regimes, respectively. Comparison of gain characteristics and the plasma state for different drive pulses shows, that real transient nature of the Ne-like Ge x-ray laser at 19.6 nm can be generated by laser pulses with tens of femtoseconds duration.

1.2. Plasma Modeling

Simulations were carried out for 100μm thick germanium slab targets irradiated by a 1ns prepulse and a picosecond main pulse at 1.06 μm wavelength. Both pulses have Gaussian profile in time. The 1-D Lagrangian hydrodynamic code MED103 was used to simulate the time-evolution of laser-plasma interactions and to calculate the gain coefficients. The optimum peak intensity of 1ns prepulse was found to be 5×10^{12}W/cm^2. In order to see the difference between TCE and QSS excitation we did calculations with different main pulse duration but a fixed energy. The main pulse durations used in the simulations are 50ps, 5ps, 500fs and 50fs. The peak intensities of the main pulse are 10^{14}W/cm^2, 10^{15}W/cm^2, 10^{16}W/cm^2 and 10^{17}W/cm^2. All simulations performed with a 1ns prepulse at $5*10^{12}$W/cm^2. The peak to peak time delay between the prepulse and main pulse was set to be 0.5ns.

1.3. Results and Discussions

Fig. 1.1 shows the spatio-temporal behaviors of gain contours for four different drive pulse durations of 50fs, 500fs, 5ps and 50ps. The fraction of ground state F-like ions in plasma is presented in Fig. 1.2. This reflects the ionization balance.

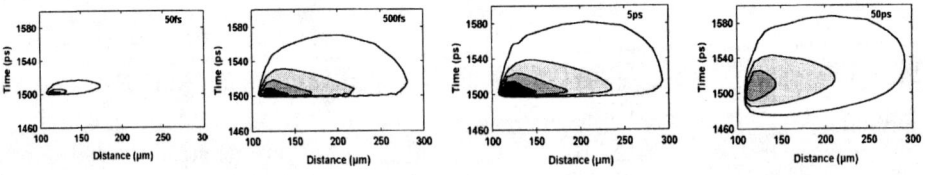

FIGURE 1.1. Gain Contours versus space and time for four values of the drive pulse durations 50fs, 500fs, 5ps and 50ps. The gain in plot is from 80cm^{-1} (blackest) reduces to 20cm^{-1} (white) by step of 20cm^{-1}.

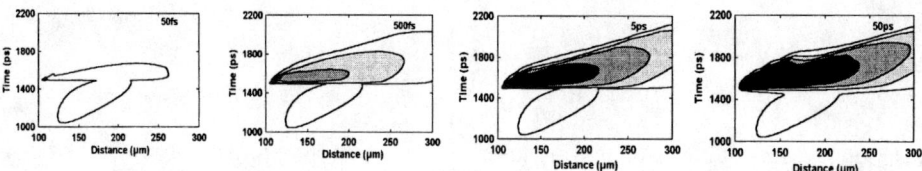

FIGURE 1.2. Contours of the fraction of ground state F-like ions versus space and time for four values of the drive pulse durations 50fs, 500fs, 5ps and 50ps. The contours from blackest to white represent fraction greater than 15%, 10%, 5% and 1%.

It is apparent that the characteristics of the x-ray laser driven by a 50ps main pulse are a typical QSS case. Detailed investigation shows that from 500fs to 50ps pulse duration, the maximum gain coefficients slowly decrease from about 100cm^{-1} to 83cm^{-1}. We should note that the peak gain coefficients and electron temperature for 500fs ~ 5ps drive pulse are located near the critical density and have narrow spatial width. In the corona region where the laser energy is mainly deposited, the spatial gain extent, electron temperature of plasmas are similar for 500fs, 5ps and 50ps drive pulse durations. So we believe that for Ne-like Ge x-ray lasers, the so-called "TCE" scheme which uses picoseconds pump pulse has no difference from traditional QSS excitation in the framework. The high gain of short pulse driven x-ray laser is only due to the relative high electron temperature.

For the shortest drive pulse of 50fs, the maximum gain is 91cm^{-1} and comes 1.2ps later after the drive pulse. The peak electron temperature near the critical density can be grater than 6keV. But this electron temperature is not suitable for collisional x-ray lasers. In the corona region, the gain and electron temperature are much lower than those dirven by the other excitation schemes with longer drive pulses as 500fs, 5ps, and 50ps durations. This is because the efficiency of absorption by the inverse bremsstralung for ultrashort laser pulses is much lower than long pulses. From Fig. 4 we can see, that for 50fs laser pulse pumping, there is no over-ionization.

The ionization balance is the most important criteria to determine QSS and TCE x-ray lasers. The pure TCE does not allow any change of ionization balance during pumping, but actually this condition can not be realized. The simulation results show that only 50fs drive pulse can basically avoid the over-ionization in the plasma. Thus it can be seen that the Ne-like Ge x-ray laser at 19.6nm pumped by a pulse as short as 50fs is the real transient collitional excitation.

2. SIMULATIONS STUDY OF A NE-LIKE TI X-RAY LASER AT 32.6 NM DRIVEN BY FEMTOSECOND LASER PULSES

2.1 Introduction

Up to now, most collisional x-ray lasers are pumped by Nb:glass lasers. However at university laboratories, the most popular CPA lasers are Ti:sapper lasers at 800 nm. In this paper we numerically investigated the possibility to generate a Ne-like Ti x-ray laser of 3p $\rightarrow$ 3s, J=0 $\rightarrow$ 1 transition at 32.6 nm driven by femtosecond Ti:sapper lasers. The pump geometry was set to be the standard line focus on a 100 µm thickness slab target. The 1D Lagrangian hydrodynamic code MED103 coupled with atomic physics code and atomic data package was used to predict the time evolution of laser-plasma interactions and to calculate the gain coefficient. Three different drive pulse configurations was studied: single prepulse, double prepulse with short interval (hundreds of picosecond) and long interval (nanoseconds). Detailed simulations were performed to optimize the drive pulse configurations.

2.2 Simulations for Single Prepulse Configuration

The plasma modeling aims to find the optimum intensity and temporal delay between drive pulses. The prepulse was assumed to be from the chirped pulse before final compression. Such chirped pulses usually have durations of a few hundreds of picosecond. The main pulse duration can be regulated from tens of femtosecond to a few picoseconds by changing the distance between the gratings in the compressor. In our simulations the duration of prepulse and main pulse were set to be 300 ps and 300 fs FWHM respectively. All prepulse and main pulse had Gaussian shape. The peak intensity of the main pulse at the line focus was fixed at 10^{15} W/cm^2.

The optimum peak intensity of a 300 ps prepulse was found to be 3×10^{11} W/cm^2. Fig. 1(a) shows the spatio-temporal profile of the Ne-like ion fraction in plasmas produced by the prepulse. The horizontal axis presents the spatial distance in the direction of plasma expansion. The initial position of target surface is located at 100 µm. The prepulse reaches its peak at 720 ps on the time scale used in Fig. 2.1(a). From Fig. 2.1(a) we can also get the best starting time of the main pulse. Obviously, the main pulse should turn on in time interval from 850 ps to 1 ns. The contours of the gain at 32.6 nm versus space and time for pulse delay of 130 ps are shown in Fig.

2.1(b). We can see the gain region has small space-time extent. That is because after the prepulse irradiation the preplasma is not well expanded. Fig. 2.1(c) shows the spatial profile of the locale gain coefficient and electron density at the moment when the highest gain exists. The maximum local gain is 44 cm^{-1}. The critical density surface is at 112 μm, where the electron density curve crosses with the top frame of figure. We can see that the gain region is located close to the critical density surface and have high gradient of electron density. So this case is not good for the amplification of x-ray lasers.

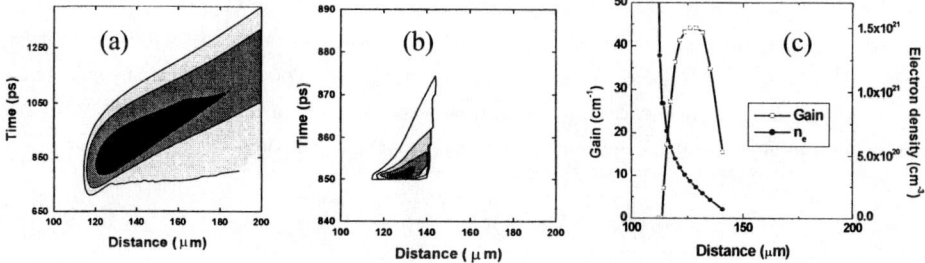

FIGURE 2.1. (a) - The Ne-like Ti ion fraction versus space and time in the plasma generated by a 300 ps prepulse with a peak intensity of 3×10^{11} W/cm^2. The fraction in plot is from 80% (blackest) to 40% (white) by a step of 20%. (b) - Contours of Ne-like Ti laser gain at 32.6 nm generated by a 300 fs main pulse. The gain in the plot is from 40 cm^{-1} (blackest) to 10 cm^{-1} (white) by a step of 10 cm^{-1}. (c) - Spatial profile of the local gain coefficient and electron density at the peak gain time.

2.3. Double Prepulse with a Short Interval

In order to improve the characteristics of preplasmas, a double prepulse configuration has been considered. The first prepulse was set at 3×10^{11} W/cm^2 peak intensity and 300 ps duration. We used two prepulse with a 360ps time interval (peak to peak). The first prepulse can generate a Ne-like preplasma. The second prepulse is used to hold the ionization stage of the preplasma and give it more time for expansion. This pulse configuration is an equivalent scheme of using a nanosecond prepulse followed by a ultra-short main pulse. The optimum peak intensity of the second prepulse was found to be 2×10^{11} W/cm^2. Fig. 2.2(a) shows much improved distribution of the Ne-like ion fraction in the preplasma generated by such prepulses. Fig. 2.2(b) shows gain generated by a 300fs, 10^{15} W/cm^2 main pulse, which reaches its peak intensity 120 ps later than the second prepulse. The spatial profiles of the local gain coefficients and electron density at the time for the highest gain are given in Fig. 2.2(c). The maximum local gain was 40 cm^{-1}. From Fig. 2.2(c) we can see that the gain extent is about two times wider than the case of single prepulse pumping, and the density gradient in gain region becomes much lower too.

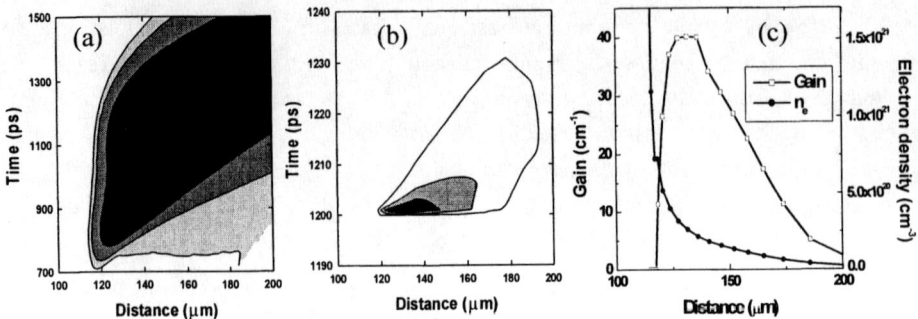

FIGURE 2.2. (a) - The Ne-like Ti ion fraction versus space and time in the plasma generated by a double 300 ps prepulse with an interval of 360 ps. The fraction in plot is from 80% (blackest) to 40% (white) by a step of 20%. (b) - Contours of Ne-like Ti laser gain at 32.6 nm generated by main pulse. The gain in the plot is from 30 cm^{-1} (blackest) to 10 cm^{-1} (white) by a step of 10 cm^{-1}. (c) - Spatial profile of the local gain coefficient and electron density at the peak gain time.

2.4. Double Prepulse with a Long Interval

In our simulations we also tried to use double prepulse with a long delay. The time interval between two prepulses was increased up to 3 ns. The similar pump scheme was successfully used in some earlier x-ray laser experiments [18, 19]. After 3 ns expansion, the preplasma becomes cold and the ionization degree will fall. The second prepulse with 5.3×10^{11} W/cm^2 intensity is used to heat the preplasma to the Ne-like state. The fraction of the Ne-like ions in the plasma produced by these two prepulses is shown in Fig. 2.3(a). By irradiating the 300 fs, 10^{15} W/cm^2 main pulse 300 ps later than the second prepulse on the plasma, we obtained the electron temperature and gain contours as shown in Fig. 2.3(b, c). The spatial profiles of the gain and the electron density at the peak gain time are shown in Fig. 2.3(d). For longer interval between prepulses, the maximum gain falls to 36 cm^{-1}, but the gain region has boarder spatial extent and lower density gradient.

In experiments if we use 5 mm × 50 μm line focus, then the total pump energy for the three pump schemes described above will be 1 J, 1.2 J and 1.4 J respectively. We have done also optimization for Nb:glass laser wavelength using the same pump schemes, the gain coefficients is about two times higher. This situation is caused by relatively long wavelength of the pump laser pulses, because the efficiency of the inverse bremsstralung absorption is inversely proportional to the square of the laser wavelength.

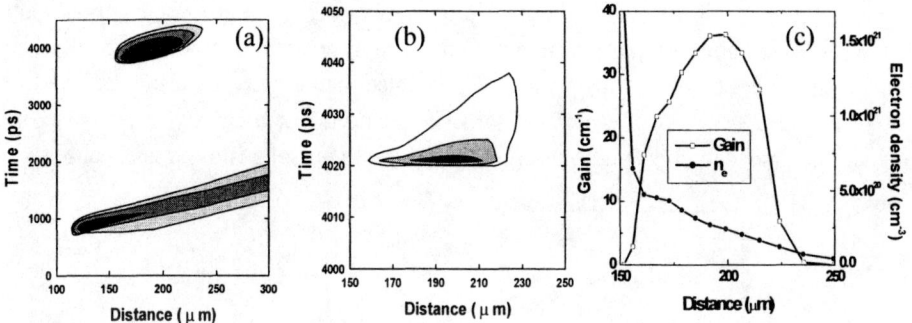

FIGURE. 2.3. (a) - The Ne-like Ti ion fraction versus space and time in the plasma generated by a double 300 ps prepulse with an interval of 3 ns. The fraction in the plot is from 80% (blackest) to 40% (white) by a step of 20%. (b) - - Contours of Ne-like Ti laser gain at 32.6 nm generated by main pulse. The gain in the plot is from 30 cm^{-1} (blackest) to 10 cm^{-1} (white) by a step of 10 cm^{-1}. (c) - Spatial profile of the local gain coefficient and electron density at the peak gain time.

2.5. Ray Tracing Calculations

It is difficult to compare the efficiency of pump schemes using double pre-pulse with short and long delays just by looking at the Fig. 2.2 and Fig. 2.3. In order to illustrate the role of refraction, we developed a 2-D ray tracing postprocessor to calculate the relative output intensity of x-ray laser for each pump scheme described above. The plasma length was set to be 5 mm. For simplify the time evolution of x-ray laser output was not included. The plasma expansion during short pulse pumping can be ignored. We assumed that the travelling wave excitation is used, so the spatial profiles of gain and electron density can keep their form unchanged along the line focus. In our tray tracing calculation the spatial profiles of gain and electron density are taken form Figures 2.1(d), 2.2(d), and 2.3(d). Fig. 2.4 shows the ray integrated output intensity of 32.6 nm laser line in the near (a) and far (b) fields. The intensity of each ray is weighted by the upper-level population of the laser transition at the beginning of the ray. The result shows that the output intensity of single prepulse pumping is very small. The double prepulse with short delay produces highest output intensity, and the double prepulse with long delay do not have advantages in comparison with the case of short delay beetween prepulses. However, in some experiments at Rutherford Appleton Laboratory with Ne-like Ge, the double prepulse with long delay time were found to be a substantial improvement on single nanosecond prepulse [18, 19]. We believe that for ultra-short pulse pumping x-ray laser, to use double prepulse with long delay is more effective for highly ionized ions like Ne-like Ge, Ni-like Sn et. al. But, for relatively lower ionized ions like Ne-like Ti or Ni-like Mo, the single nanosecond prepulse works best. The reason is becouse in x-ray laser generation of highly ionized ions, most part of pump energy spend on preparation of suitable pre-plasma, using double prepulse with long delay can increase the efficiency of the energy absorption of pre-pulse, and finally save the total pump energy. The x-ray laser of low ionized ions do not need much energy in prepulse, so using single nanosecond prepulse and picosecond main pulse is enough good choose.

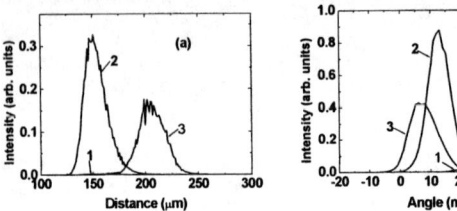

FIGURE. 2.4. The 32.6 nm x-ray laser output versus the source position (a) and the distribution angle (b) for single prepulse pumping (1), double prepulse with short interval (2) and long interval (3).

3. EFFECTS OF DELAY TIME ON TRANSIENT

NI-LIKE X-RAY LASERS

3.1. Introduction

A suitable delay time between the long-pulse and the short-pulse is beneficial for relaxing the plasma density gradient and thus is very critical for the propagation of x-ray lasers [17]. For the TCE scheme, a longer delay can make a longer scale length. But a too long delay will also reduce the temperature in the plasma [20]. Thus, if we would like to use the delay to relax the plasma density gradient, the delay and the pulse duration between the long pulse and the short pulse should be optimized.

In this chapter, we investigate the effects of the delay time on hydrodynamics of transient Ni-like Pd x-ray lasers using the formulas of Ref. [20]. In order to understand the optimization conditions of plasmas, we calculate the electron temperature, scale length, electron density for different delay time. The results show that extending the pulse duration is more efficient than that of increasing the intensity to generate transient x-ray lasers with high gain.

3.2. Analytical Results and Discussion

According to the formula of Ref. [20], useful scaling laws for plasma variables are used to describe the hydrodynamic process of TCE Ni-like x-ray lasers. The processes of laser pulses interacting with plasmas are divided into four distinct periods. They are $t \leq t_{1L}$, $t_{1L} \leq t \leq t_m$, $t_m \leq t \leq t_{2L}$ and $t_{2L} \leq t$, respectively, where $t_{1L} = \Delta t_{1L}$ is the long pulse duration, Δt_m is the delay time, $t_m = \Delta t_{1L} + \Delta t_m$ is the time when the short pulse arrives, Δt_{2L} is the short pulse duration, $t_{2L} = t_m + \Delta t_{2L}$ is the turning off time of the short pulse. In order to understand the effects of the delay time on hydrodynamics of the transient collisional x-ray laser under different conditions, we calculate the hydrodynamics of the transient collisional Ni-like Pd x-ray laser for different intensity and duration for the long pulse and the short pulse, respectively. First, we discuss why the transient x-ray laser is sensitive to the delay time. Second, we investigate the effects of the delay time by changing the intensity and the duration for long pulse. Finally, we discuss how to effectively enhance the electron temperature in the preplasma before the short pulse comes.

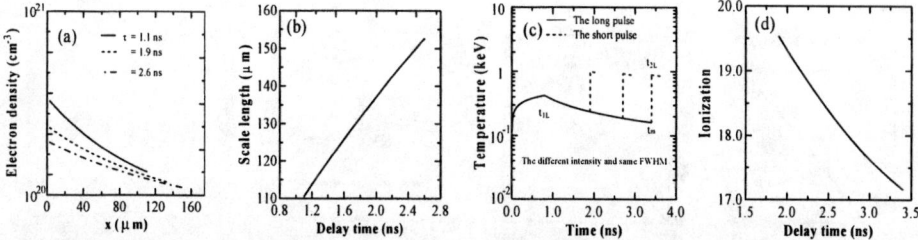

FIGURE 3.1. (a) electron density versus x with deferent delay, (b) temperature versus delay, (c) Temperature history, (d) ionization versus delay. The conditions are $I_1 = 0.7 \times 10^{12} \text{W/cm}^2$, $t_{1L} = 0.8 \text{ns}$, $\lambda = 1.053 \mu\text{m}$, $I_2 = 5.2 \times 10^{14} \text{W/cm}^2$, $t_{2L} = 1.1 \text{ps}$, the delay time τ is from 1.9 to 3.4ns.

We calculate hydrodynamics of the transient collisional Ni-like Pd x-ray laser with three different values of delay time of 1.1, 1.9 and 2.6ns. The conditions in the calculations are as almost the same as in Ref. [14], except the intensity of the long pulse of $I_1 = 3.0 \times 10^{12}$ W/cm^2. The calculations show that the densities are 5.79×10^{20}, 4.41×10^{20}, and 3.68×10^{20} cm^{-3} for the three different delay times, respectively, as shown in Fig. 3.1(a). The difference among the three densities will not much affect the gain of the x-ray lasers. In the calculations, 152.7μm is the longest scale length for the 2.6ns delay time, as shown in Fig. 3.1(b). For this delay time, the electron temperature is 160.9eV, as shown in Fig. 3.1(c). This results in an average ionization of only 17.16, as shown in Fig. 3.1(d). This is lower than the requirement for rich Ni-like Pd ionization population. For the 1.9ns of the delay time, the scale length is 134.3μm, which is middle among the three scale lengths. Of course, a longer delay time can make the scale length longer. However, a rich Ni-like (or Ne-like) ionization population is the most important condition. Thus in the design of the x-ray laser experiment, we not only need to optimize the scale length, but also need to optimize the ionization, which is more important than the scale length.

There are two ways to increase the temperature of preplasma. The first one is to enhance the intensity of the long pulse. The another one is to extend the duration of the long pulse. Here what we want to know is which one is more efficient. We calculate the change of the intensity, scale length, density and temperature while extending the delay time from 0 to 2.6ns and keeping the duration to be a constant of 0.8ns. The results show that the intensity has to be increased from 0.7×10^{12} to 4.0×10^{12} W/cm^2 so that the most Pd ions are kept in Ni-like ionization stage, as shown in Fig. 3.2(a). The scale lengths are prolonged from 40 to 169.8μm with the extension of the delay, as shown in Fig. 3.2(b). And the electron density range is from 3.90×10^{20} to 8.44×10^{20}cm^{-3}, as shown in Fig. 3.2(d). However, the highest temperature during the short pulse laser is dropped from 1.84keV to 800eV with the extension of the delay, as shown in Fig. 3.2(c).

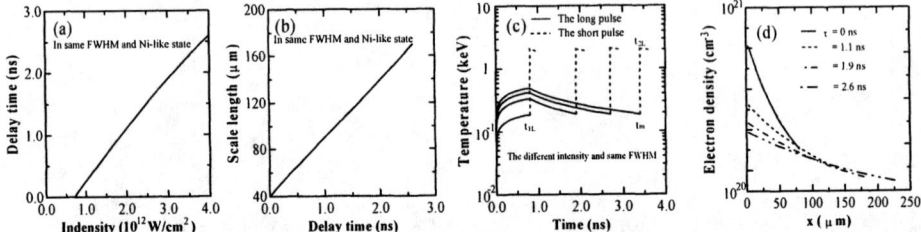

FIGURE 3.2. (a) Delay time versus long pulse intensity, (b) scale length vs delay time, (c) temperature history, (d) electron density vs x with deferent delay for the same long pulse duration and Ni-like ionization. The conditions are I_1 from 0.7×10^{12} to 4×10^{12}W/cm^7, $\lambda=1.053\mu$m, $t_{1L}=0.8$ns, $I_2=5.2 \times 10^{14}$W/cm^2, $t_{2L}=1.1$ps, the delay time τ is from 0 to 2.6ns.

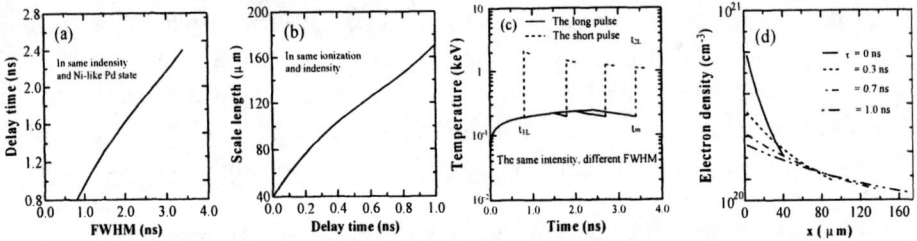

FIGURE 3.3. (a) Delay time versus FWHM (long pulse duration), (b) scale length versus delay time, (c) temperature history, (d) electron density vs x for deferent delay for the same intensity and Ni-like ionization. The condition are $I_1=0.7 \times 10^{12}$W/cm^2, the long pulse duration t_{1L} are from 0.8 to 2.4ns, $\lambda=1.053\mu$m, $I_2=5.2 \times 10^{14}$W/cm^2, $t_{2L}=1.1$ps, the delay time τ is from 0 to 1.0ns.

As a comparison with the case of changing the pulse intensity, we calculate the same parameters' change by extending the delay time from 0 to 1.0ns and keep the intensity as a constant of 0.7×10^{12} W/cm^2. The results show that the durations are increased from 0.8 to 2.4ns with the extension of the delay time while keeping the Pd ions as Ni-like ions, as shown in Fig. 3.3(a). The scale lengths are prolonged from 40 to 170.8μm with the extension of the delay, as shown in Fig. 3.3(b). And the highest temperature is only dropped from 1.84keV to 1.1keV during the short pulse with the extension of the delay time, as shown in Fig. 3.3(c). The electron density range is from 3.64×10^{20} to 8.44×10^{20} cm^{-3}, which is almost the same as the case above, as shown in Fig. 3.3(d). The efficiency of the heating pulse for prolonging the long pulse duration while keeping the constant intensity is greater than that for increasing the intensity while keeping the constant duration. It is significant because prolonging the pulse duration is much easier than increasing the pulse intensity in the x-ray laser experiment.

The ways to enhance the temperature are also to increase the intensity or the duration of the short pulse. We first calculate the case of changing short pulse intensity while keeping the pulses duration the same for different long pulse intensity or duration and delay time. The calculations show that the short pulse intensity needs to increase from 5.2×10^{14} to 1.6×10^{15} W/cm^2 with the delay extension from 0 to 2.6ns for a constant long pulse. And the intensity only needs to increase from 5.2×10^{14} to 1.2×10^{15} W/cm^2 with the delay extension from 0 to 1.0ns for a constant long pulse intensity. The efficiency for keeping the long pulse intensity constant is greater than that for keeping the long pulse duration constant.

Then we calculate the case of changing short pulse duration while keeping the pulses intensity the same for different long pulse intensity or duration and delay. The calculations show that the short pulse durations need to increase from 1.1 to 3.4ps with the delay extension from 0 to 2.6ns for a constant long pulse duration. And the duration only need to increase from 1.1 to 2.3ps with the delay extension from 0 to 1.0ns for a constant long pulse intensity. The efficiency for keeping the long pulse intensity constant is also greater than that for keeping the long pulse duration constant.

ACKNOWLEDGMENTS

This work was supported by the National Nature Science Foundation of China under Grant No. 19974074, 19825110, and 60108007, and National Hi-tech Program.

REFERENCES

1. Matthews D. L., Hagelstein P. L., Rosen M. D. et. al., *Phys. Rev. Letters* **54**, 110-113 (1985).
2. Nilsen J., MacGowan B. J., Da Silva L. B., and Moreno J. C., *Phys. Rev. A* **48**, 4682-4685 (1993).
3. Moreno J. C., Nilsen J., and Da Silva L. B., *Opt. Communications* **110**, 585-589 (1994).
4. Fill E. E., Li Y. L., Schlögl D., Steingruber J., and Nilsen J., *Opt. Letters* **20**, 374-376 (1995).
5. Zhang J., Warwick P. J., Wolfrum E. et. al., *Phys. Rev. A* **54**, R4653-R4656 (1996).
6. Zhang J., MacPhee A. G., Nilsen J. et. al., *Phys. Rev. Letters* **78**, 3856 (1997).
7. Zhang J., MacPhee A. G., Lin J. et. al., *Science* **276**, 1097-1100 (1997).
8. Smith R., Tallents G. J., Zhang J. et. al., *Phys. Rev. A*, **59**, R47-R50 (1999).
9. Löwenthal F., Tommasini R., and Balmer J. E. et. al., *Opt. Communications* **154**, 325-328 (1998).
10. Afanasiev Yu. V. and Shlyaptsev V. N., *Sov. J. Quant. Electron* **19**, 1606-1612 (1989).
11. Healy S. B., Janulewicz K. A., Plowes J. A., and Pert G. J., *Opt. Communications* **132**, 442-448 (1996)
12. Nickles P. V., Shlyaptsev V. N., Kalachnikov M. P., et. al., *Phys. Rev. Letters* **78**, 2748-2751 (1997).
13. Warwick P. J., Lewis C. L. S., Kalachnikov M. P., et. al., *J. Opt. Soc. Am. B* **15**, 1808-1814 (1998).
14. Dunn J., Osterheld A. L., Shepherd R. et. al., *Phys. Rev. Letters* **80**, 2825-2828 (1998).
15. Kalachnikov M. P., Nickles P. V., Schnürer M. et. al., *Phys. Rev. A* **57**, 4778-4783(1998).
16. Li Y. L., Dunn J., Nilsen J. et. al., *J. Opt. Soc. Am. B* **17**, 1098-1101 (2000).
17. Nilsen J., Li Y. L., and Dunn J., *J. Opt. Soc. Am. B* **17**, 1084-1092 (2000).
18. Lin J. Y., Tallents G. J., MacPhee A. G. et. al., *Opt. Communications* **166**, 211-218 (1999).
19. King R. E., Pert G. J., McCabe S. P. et. al., *Phys. Rev. A* **64**, 053810 (2001).
20. Li Y. J. and Zhang J., *Phys. Rev. E.* **63**, 036410(2001).

CAPILLARY DISCHARGE-PUMPED
X-RAY LASERS AND XUV SOURCES

EXCITATION AND RADIATION OF HIGHLY IONISED IONS IN THE FAST GAS-FILLED-CAPILLARY DISCHARGE

Karel Koláček[a], Jiří Schmidt[a], Vladislav Boháček[a], Milan Řípa[a],
Pavel Vrba[a], Olexandr Frolov[b], Alexander Jančárek[c],
and Miroslava Vrbová[c]

[a]*Institute of Plasma Physics, Academy of Sciences of the Czech Republic, Za Slovankou 3, P.O.Box 17,
182 21 Prague 8, Czech Republic*
[b]*Department of Electronics and Vacuum Physics, Faculty of Mathematics and Physics, Charles
University in Prague, Ke Karlovu 3, 121 16 Prague 2, Czech Republic*
[c]*Department of Physical Engineering, Faculty of Nuclear Sciences and Physical Engineering,
Břehová 7, 115 19 Prague 1, Czech Technical University, Czech Republic*

Abstract. The capillary experiment CAPEX was reconstructed to approach conditions suitable
for creation of population inversion in Ne-like Ar. The reconstruction consisted in substitution
of ceramics capillary for former plastic one, in remarkable reduction of pre-ionisation current,
and in change of Ar filling and pumping geometry. The soft X-ray spectroscopic measurements
were performed with survey flat field spectrograph. It is shown that under certain conditions
the strong spectral line at the wavelength of laser transition (46.9 nm) appears and dominates
the spectrum even at exposition 50 ns.

INTRODUCTION

Capillary discharges are studied as intense soft X-ray sources – a cheap alternative
to laser-plasma sources, free-electron sources, synchrotron sources, and sources based
on higher harmonics generation. One of the most attractive features of the capillary
discharges is their potential to lase. This was proved by J.J.Rocca (Colorado State
University), who not only demonstrated lasing on Ne-like Ar ($\lambda = 46.9$ nm) [1], Ne-
like S ($\lambda = 60.12$ nm) [2], and Ne-like Cl ($\lambda = 52.9$ nm) [3], but also achieved on Ne-
like Ar a saturation limit [4]. Since that time many laboratories tried and try to repeat
these results, but only in last two years the Technion – Israel Institute of Technology
(Haifa, Israel) [5], University of L'Aquila (L'Aquila, Italy) [6], and Tokyo Institute of
Technology (Yokohama, Japan) [7, 8] have been successful (see also the review
papers [9, 10]).

CP641, X-Ray Lasers 2002: 8th International Conference on X-Ray Lasers, edited by J. J. Rocca et al.
© 2002 American Institute of Physics 0-7354-0096-2/02/$19.00

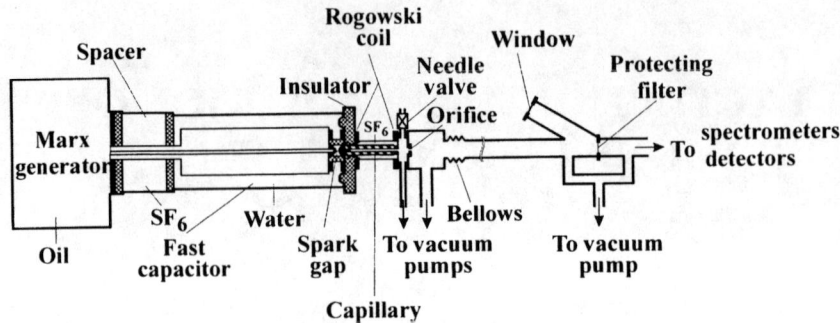

FIGURE 1. Apparatus

One of the ultimate conditions for achieving a population inversion and lasing is an efficient and fast "impurity-free"-plasma compression and "heating". Namely how much impurities the plasma contains and how efficient this heating is, could be inferred from the ionisation stage of ions the spectral lines of which appear in the soft-X-ray spectrum [5, 11].

Therefore, we performed gated spectroscopic measurements with a survey flat field spectrograph equipped with double-MCP/CCD registration and studied, how soft X-ray spectra change in different discharge conditions.

APPARATUS

The CAPEX apparatus (see Fig. 1) consists of a Marx generator, a coupling section (spacer), a fast capacitor (pulse forming line) with a closely coupled main spark gap and a capillary. The gas filling and pumping assembly is attached to the outer end of the capillary. The full description of the apparatus can be found in [12].

From our previous measurements we learned that plastic (polyamide or polyamide/polyimide) capillary ablates (gradually enlarges its inner diameter) and, hence, in uncontrollable way changes the boundary discharge conditions; moreover, such ablating capillary is a rich source of soot, which is extremely harmful to thin filters and all usually very expensive diagnostic equipment. Therefore, the plastic capillary was substituted by the ceramics (alumina) one. This issue was connected with a number of technological problems (reduction in volume (in dimensions) and change of shape (bending) of the machined alumina semi-finished product when it was conditioned (burned) in an oven, vacuum tightening of porous material, …).

The CAPEX main spark gap (coupling after breakdown the fast capacitor with the capillary) had large electrodes that extended beyond polyamide insulator casing (filled with pressurised SF_6 gas) to water region. Conductivity of water between electrodes as well as enlarged capacitance ($\sim$300 pF) supplied to the capillary the pre-pulse current >100 A. As a by-product this capacitance fed the capillary in the first moments after spark gap breakdown enhancing the discharge current rise-rate. This last advantage had to be abandoned: water from the space between electrodes was partly dislodged by polyethylene plug-in and pre-pulse current was reduced (following suggestions in [5]) to values <50 A. Unfortunately, even this arrangement is not ideal: a change of water conductivity influences the pre-pulse current and modifies discharge conditions; hence, in future fully independent pre-pulse is highly desirable.

In the original CAPEX device the filling gas was fed into the capillary around the outer surface of grounded electrode, while it was pumped through the axial orifice $\varnothing$2x20 mm in this electrode through which also the emitted radiation was brought out (see e.g. [13]). This turned to be unpractical and in new arrangement the pumping and radiation outputs are decoupled retaining in the axis a pinhole ($\varnothing$0.8x1 mm) for radiation exit only (as it is shown in Fig. 1).

EXPERIMENT

X-ray Measurements before CAPEX Reconstruction

The time curve of the axial soft X-ray radiation was measured by PIN diode covered by 0.75 μm Al filter (in the "long-wavelength region" its transmission range is from 17 nm to 70 nm). Already in autumn 1998 we found [14] that the time curve of the soft X-ray radiation is very similar to that published three years before us by Rocca [4] when presenting the achievement of the saturation limit. Therefore, we believed that short intense spike of soft X-ray radiation can be attributed to an amplified spontaneous emission [15]. However, further measurements with simple transmission grating spectrographs (50 lines, step 1.4 μm, slit 70 μm x 1.8 mm or 100 lines, step 1.0 μm, slit 100 μm x 1.8 mm), Seya-Namioka spectrograph (Rowland circle $\varnothing$500 mm, Au-coated grating 1200 gr/mm), and small grazing incidence spectrograph (Rowland circle $\varnothing$1000 mm, grazing angle $4°$, Au-coated grating 300 gr/mm, photographic (time integrated) detection) did not detected any intense spectral line around the wavelength 46.9 nm. Therefore, supported by extensive theoretical 1D MHD modelling (that included dissipative processes, ablation and ionisation of wall material, etc.) it was concluded [16] that this short intense spike of soft X-ray radiation detected by PIN diode could be explained by "efficient" pinching only.

X-ray Measurements after CAPEX Reconstruction

After CAPEX-device reconstruction described in the chapter Apparatus we repeated both the time resolved spectrally integrated measurements with PIN diode covered by 0.75 μm Al filter as before and the spectroscopic measurements with newly built survey flat field spectrograph (Jobin Yvon grating, average groove density 450 gr/mm, non-equidistantly spaced) for the spectral interval 10-110 nm and with double-MCP/CCD detection.

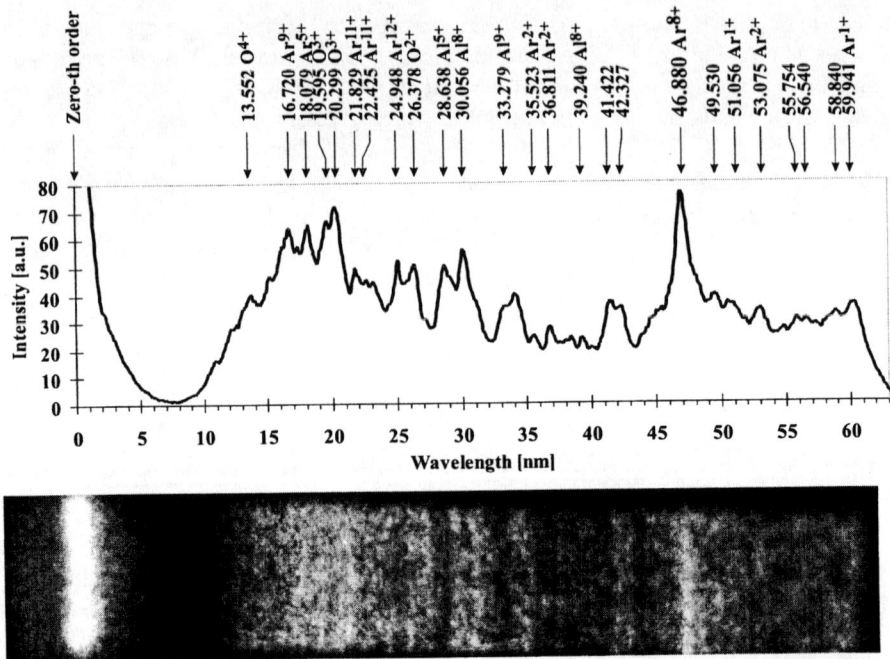

FIGURE 2. Profile of the soft X-ray spectrum of the radiation emitted from ⌀3.2x180 mm capillary filled with Ar to the pressure 55 Pa and excited by the discharge current pulse of the amplitude 44 kA (above) and the same spectrum as registered by double-MCP/CCD detector (below).

While PIN diode signal had the same shape (sharp narrow peak in the beginning of X-ray emission followed by long pulse of X-ray radiation) as before reconstruction, the spectra of the small-diameter capillary differed significantly.

In case of alumina capillary of the inner diameter 3.2 mm and of the length 18 cm the strong spectral line at the laser-transition-wavelength immediately appeared and dominated the spectrum (see Fig. 2) even at 50 ns gating of MCP (the shortest available at the moment); the discharge current was as high as 44 kA, and filling Ar pressure was 55 Pa. Due to long exposition, which covered even long start-up phase, many spectral lines (according to our preliminary identification) belonged to ions of low ionisation stages.

On the other hand the spectrum of alumina capillary of the inner diameter 4 mm and of the length 18 cm (see Fig. 3) resembled that measured before CAPEX reconstruction (without bright line at the laser-transition-wavelength). It is probably due to the fact that maximum pinching comes on after the discharge current maximum – in accord with scaling-law derived in [16]. In this case in the pressure range up to 50 Pa there is a group of bright spectral lines in the region 10-24 nm ("enhanced short-wavelength regime"). In the pressure range above 50 Pa the spectral lines in the above mentioned region 10-24 nm pale, while lines in the region 24-31 nm rise in intensity reaching the same values as shorter wavelength group ("enhanced broadened short-

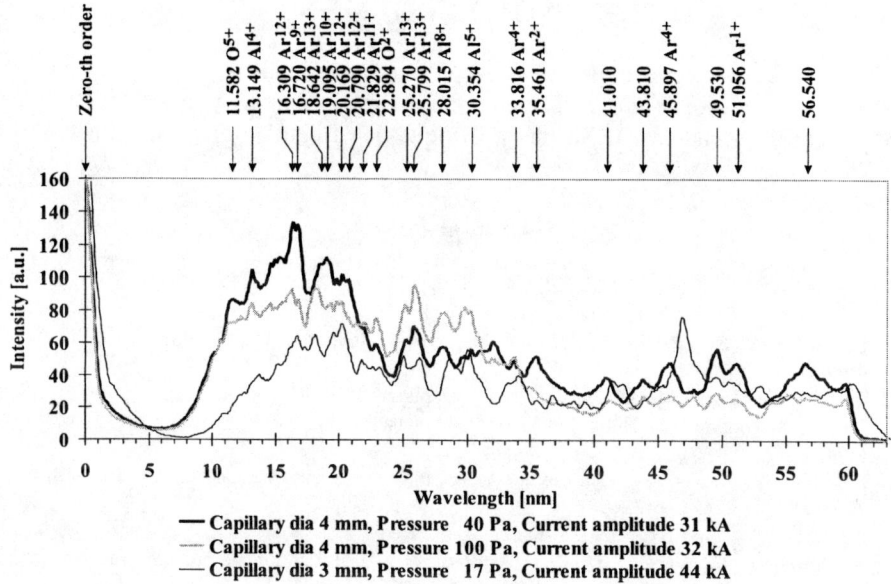

FIGURE 3. Profile of the spectrum of the capillary with inner diameter ∅4 mm; in this case the maximum discharge pinching comes on after discharge maximum –that is probably the reason, why intense line at the laser transition wavelength do not appears. For comparison the spectrum of the 3.2-mm capillary is shown.

wavelength regime"). An influence of the discharge current has not yet been systematically studied, but it could be generally said that a higher discharge current shifts 50 Pa border-line between two above mentioned regimes to higher pressure values.

CONCLUSION

The sharp narrow peak in the beginning of X-ray emission (detected by filtered PIN diode) indicates efficient pinching of plasma column and does not represent an evidence of lasing (condition necessary, but not sufficient).

The CAPEX apparatus was reconstructed to reduce the pre-pulse current, to substitute the ceramics capillary for former plastic one and to change the gas filling and pumping dynamics.

The spectroscopic measurements on reconstructed CAPEX apparatus with 3-mm-inner-diameter capillary detected the strong spectral line at the laser transition wavelength that dominates the spectrum (hopefully an amplified spontaneous emission). No such line was detected with 4-mm-inner-diameter capillary, probably because the pinching time is longer than the discharge-current-quarter-period.

ACKNOWLEDGEMENT

This work has been performed under auspices and with the support of both the Ministry of Education Youth and Sports of the Czech Republic under Contract INGO LA055 and the Academy of Sciences of the Czech Republic under Contract AVOZ 2043910.

REFERENCES

1. Rocca, J.J., Shlyaptsev, V., Tomasel, F.G., Cortázar, O.D., Hartshorn, D., and Chilla, J.L.A., *Phys. Rev. Letters* **73**, 2192-2195 (1994).
2. Tomasel, F.G., Rocca, J.J., Shlyaptsev, V.N., and Macchietto, C.D., *Phys. Rev. A.* **55**, 1437-1440 (1997).
3. Frati, M., Seminario, M., and Rocca, J.J., *Optics Letters* **25**, 1022-1024 (2000).
4. Rocca, J.J., Clark, D.P., Chilla, J.L.A., and Shlyaptsev, V.N., *Phys. Rev. Letters* **77**, 1476-1479 (1996).
5. Ben-Kish, A., Shuker, M., Nemirovsky, R.A., Fisher, A., Ron, A., and Schwob, J.L., *Phys. Rev. Letters* **87**, 015002-1 – 4 (2001).
6. Ritucci, A., Tomassetti, G., Palladino, L., Reale, L., Gaeta, G., Flora, F., Mezi, L., Kukhlevsky, S.V., Kaiser, J., Faenov, A., Pikuz, T., "Investigation of pulse characteristics of a 46.9 nm Ar capillary discharge soft X-ray laser", in this *AIP Conference Proceedings*, to be published.
7. Niimi, G., Hayashi, Y., Nakajima, M., Watanabe, M., Okino, A., Horioka, K.,Hotta, E., *J. Phys. D – Appl. Phys.* **34**, 2123-2126 (2001).
8. Niimi, G., Sakamoto, N., Nakajima, M., Hayashi, Y., Watanabe, M., Okino, A., Horioka, K., and Hotta, E., "Study of low current capillary discharge for compact soft X-ray laser", in this *AIP Conference Proceedings*, to be published.
9. Rocca, J.J., *Rev. Sci. Instrum.* **70**, 3799-3827 (1999).
10. Koláček, K., "Discharge-pumped soft X-ray lasers", in *Proc. National Laser Symp.*, Indore, India, Dec. 17-21, 2001, to be published
11. Rocca, J.J., Cortázar, O.D., Szapiro, B., Floyd, K., and Tomasel, F.G., *Phys. Rev. E.* **47**, 1299-1304 (1993).
12. Koláček, K., Schmidt, J., Boháček, V., Řípa, M., Vrba, P., Frolov, O., Jančárek, A., Vrbová, M., *Czech. J. Phys.* **52**, Suppl. D, D199-D204 (2002).
13. Kolacek, K., Schmidt, J., Bohacek, V., Ripa, M., Ctibor, P., Turek, K., Baranowski, J., Skladnik-Sadowska, E., Sadowski, M., Kishinets, A.S., Rupasov, A.A., Jancarek, A., "Axial Particle and Soft X-ray Emission from the Fast Capillary Discharge", in *Digest of Technical Papers of IEEE Pulsed Power Plasma Science Conf., PPPS-2001*, edited by R. Reinovsky and M. Newton, IEEE, Inc., IEEE Catalog #01CH37251, Piscataway, NJ, USA, 2001, pp.761-764
14. Koláček, K., Schmidt, J., Boháček, V., Řípa, M., Šunka, P., Piffl, V., and Ullschmied, J., *J. Techn. Phys.* **40**, 493-496 (1999)
15. Koláček, K., Schmidt, J., Boháček, V., Šunka, P., Ullschmied, J., Řípa, M., Piffl, V., Rupasov, A.A., Shikanov, A.S., Kubeš, P., Kravárik, J., "Spectroscopic Study of the Fast Capillary Discharge", in *Proc.13th Int. Conf. on High-Power Particle Beams, BEAMS'2000*, edited by K. Yatsui and W. Jiang, Nagaoka Univ. of Technology, Nagaoka, Japan, 2001, pp. 151-154
16. Vrba, P., Koláček, K., Schmidt, J., Řípa, M., Boháček, V., Bobrova, N.A., Sasorov, P., and M.Vrbová, in *Proc. Int. Symposium on Plasma Research and Application, PLASMA-2001*, edited by M. Sadowski, Inst. of Plasma Physics and Laser Microfusion, Warsaw, Poland, 2001, pp. 3.23.1-3.23.4

Z-scaling for the H-like recombination laser in the capillary discharge

K. Lee*, J. H. Kim¶, and D. Kim¶

*Laboratory for Quantum Optics, Korea Atomic Energy Research Institute,
P. O. Box 105, Deokjin-Dong, Yuseong-Ku, Daejeon, 305-600, Korea.
¶Physics Department, Pohang Univ. of Science and Technology,
San 31, Hyoja-Dong, Nam-Ku, Pohang, 790-784, Korea.

Abstract. Z-scaling formulae for adequate experimental conditions of a capillary discharge for e- collisional recombination X-ray laser using H-like ions were obtained analytically, which give a guide line for the design of an experimental set up. The dynamics of the capillary discharge were classified into three phases of pinch, expansion, and formation of gain. Dominant physical processes such as shock heating, adiabatic cooling, three-body recombination, and optical trapping of H-like Lyman-alpha line were taken into account. A series of one-dimensional MHD simulation with the parameters obtained by the z-scaling formulae was performed for the atomic species of Be (z=4) through Al (z=13) and showed significant gain values for all the species calculated. This verifies that our formulae well predict appropriate conditions for the amplification of lines in H-like ions by a capillary discharge. This analysis also shows the limit on the atomic number to which the capillary discharge may be applied for recombination X-ray lasers.

INTRODUCTION

The amplification of extreme ultraviolet radiation has been widely pursued and successfully demonstrated in various ways during the last decade [1]. However the usage of huge size, high power optical lasers to produce adequate gain media has made it difficult to build a laboratory scale, table-top x-ray laser.

Efforts to develop compact x-ray laser has been taken in parallel. Among them, Hagelstein proposed Ni-like Mo XV (19.1 nm) scheme [2] and Kim *et al.* demonstrated the amplification of C VI (18.2 nm), and Al XI (15.4 nm) lines using small-size lasers [3,4]. Recently, the development of femtosecond laser not only made it possible to construct a compact x-ray laser [5] but also encourage one to propose a femtosecond x-ray laser [6].

A capillary discharge has been also studied due to its compactness and efficient energy coupling. Rocca *et al.* demonstrated the amplification of the Ne-like Ar IX line at 46.9nm [7] and constructed a table-top scale x-ray laser [8], which uses the collisional excitation pumping scheme (CEPS). For collisional recombination pumping scheme (CRPS), Shin *et al.* [9] and Wagner *et al.* [10] amplified the H-like C VI line (18.2 nm) and the Li-like O VI lines (49.8, 52.0 nm) respectively but they

CP641, *X-Ray Lasers 2002: 8th International Conference on X-Ray Lasers*, edited by J. J. Rocca et al.
© 2002 American Institute of Physics 0-7354-0096-2/02/$19.00

are not satisfactory due to short gain length [9] and small gain [10]. Toward the successful
amplification in the water-window region (2.3 - 4.4 nm) using the H-like H_α line in a capillary discharge, the following questions need to be addressed:

- Scaling behaviors of experimental conditions on atomic numbers.
- The possibility of the amplification in the water window region.

To address such a question, z-scaling formulae of the experimental parameters for a recombination x-ray laser using H-like ions have been obtained. Even though the dynamics of fast capillary discharge is very complicated, with the snowplow model [11], the dominant physical processes such as shock implosion and adiabatic expansion well describe overall dynamics [12]. The results show that the capillary system can amplify up to the water-window region with aluminum but with a severe discharge condition.

Z-SCALING

The dynamics of capillary discharge plasmas is much complicated from atomic processes to hydrodynamics, which makes it impossible to describe all the details of dynamics analytically. However the authors simplified the overall dynamics considering the dominant mechanisms of heating and cooling processes in a pinch and an expansion phases respectively, and the conditions for amplification suggested by Elton [13] are adopted. More detailed derivations will be described in Ref. [14].

For the description of the pinch state, the snowplow model, which well estimates the pinch time, is adopted. Considering the dominant heating mechanism in a pinch state as a shock heating, i.e. thermalization of imploding kinetic energy, the plasma temperature at the pinch, $T_{e,p}$ can be estimated as

$$T_{e,p} \approx 1.2 \times 10^{11} \left(\frac{z}{N_o[cm^{-3}]} \right)^{1/2} \left(\frac{I_o[kA]}{\tau[ns]} \right) \quad [eV], \quad (1)$$

where z is the atomic number, N_o the initial density, I_o the peak current, and τ the period of a sinusoidal discharge current pulse. Taking this temperature to be high enough to make fully stripped ions as a half of the ionization energy of a H-like ion, the following condition for the initial parameters can be obtained.

$$\left(\frac{I_o[kA]}{\tau[ns]N_o^{1/2}[cm^{-3}]} \right) \approx 5.7 \times 10^{-11} z^{3/2}. \quad (2)$$

The pinch radius, R_p is estimated from a series of MHD simulations [15] as a tenth of an initial radius, R_o then the pinch density can be obtained by a mass conservation.

In an expansion phase, a fast adiabatic expansion has been found to be a dominant cooling mechanism from a MHD simulation [12], which can be described analytically

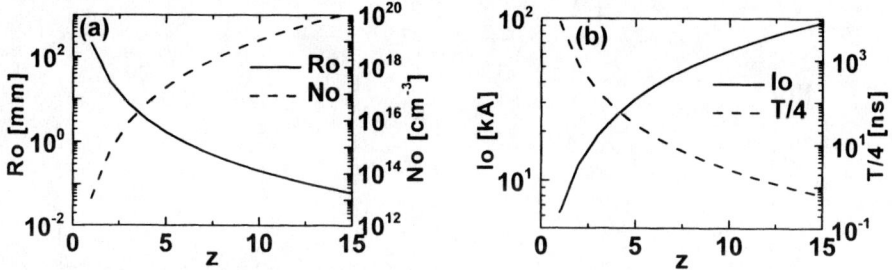

FIGURE 1. Experimental parameters obtained by the z-scaling formulae are plotted in (a) initial radius (R_o) and initial density (N_o) and (b) peak current (I_o) and its quarter period (T/4).

assuming uniform density and temperature profiles with the following characteristic time, t_o

$$t_o \approx \frac{\sqrt{3}}{5} \frac{R_p}{\sqrt{\frac{2}{m_i}T_{e,p}(z+1)}}, \tag{3}$$

The plasma quantities estimated in above analysis were compared with the Elton's conditions for the population inversion of the H-like H_α line and the condition of optically thin H-like L_α line was also imposed with an optical depth taking into account the velocity gradient effect [16]. These analyses lead to the following z-scaling formulae for the experimental parameters such as initial radius (R_o), an initial density (N_o), the period (τ) of a sinusoidal discharge current pulse, and its peak (I_o).

$$N_o = 1.1 \times 10^{13} z^6 [\text{cm}^{-3}], \tag{4}$$
$$R_o \leq 20.8 z^{-3} [\text{cm}], \tag{5}$$
$$I_o = 6.3 z [\text{kA}], \tag{6}$$
$$\tau = 3.3 \times 10^4 z^{-7/2} [\text{ns}]. \tag{7}$$

In obtaining above formulae, the assumption that no significant recombination process occurs in the expansion phase has been made because the cooling rate should be faster or comparable with the recombination process to build a population inversion,

$$\frac{1}{T_e}\left|\frac{dT_e}{dt}\right| > P^{3b},\tag{8}$$

where P^{3b} is the three-body recombination rate. Since the three-body recombination

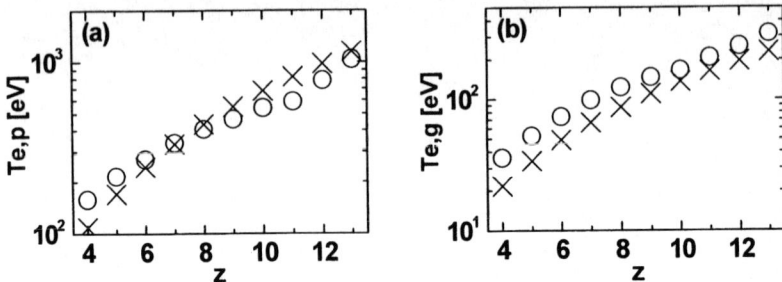

FIGURE 2. The electron temperatures (a) at the pinch and (b) at the time of maximum gain, estimated in the analysis (Cross) and obtained in the MHD simulations (open circle) are compared.

rate that is relevant in the CRPS increases much faster than the cooling rate as atomic number increases, above requirement limits the application of the capillary discharge to CRPS x-ray laser to aluminum (z=13). But this does not mean that heavier atom than aluminum cannot be applied to the recombination laser in the capillary discharge, instead it means that for atoms with z>13, the recombination from fully-stripped ions to H-like ions becomes significant even during the initial period of expansion.

The z-scaling formulae allow examining the experimental parameters for CRPS x-ray laser. The four experimental parameters obtained in above analysis are plotted in Fig. 1. The results show that as z increases or the lasing wavelength decreases, the experimental conditions become severer. For the amplification of the wavelength in the water-window region, which can be obtained with aluminum, the required conditions are $R_o \approx 95\,\mu m$, $N_o \sim 5.3 \times 10^{19}\,cm^{-3}$, $I_o = 82$ kA, and $\tau/4 = 1.0$ ns. The wire discharge may be employed to meet the initial radius and density. For a current pulse, the peak current looks manageable but the realization of such a short pulse may be hardly possible with the present high-power discharge technology.

For the parameters obtained by the z-scaling formulae, a series of MHD simulation has been performed to justify assumption made in the derivation of the z-scaling formulae and to confirm the results for atomic species of z=4 through 13. A complete description of the MHD code can be found in Ref. [12]. The temperatures at the pinch time and at the time of maximum gain are plotted in Fig. 2, which shows good agreement between the simple estimation and the MHD simulation.

The calculated maximum gains at axis are plotted in Fig. 3. For low-z elements, the gain increases rapidly as z increases. However the increase gets slower around z=9 and it begins to decrease from z=13, which is caused by the faster recombination process in higher-z elements. The gain values optimized by Shlyaptsev *et al.* [17], which have been calculated in a similar MHD simulation for CEPS, are also plotted in

Fig. 3 for comparative study. The CEPS has a higher gain than the CRPS in the long wavelength region but the CRPS has comparable or higher gains below 9 nm. It should be noticed that the experimental parameters suggested by the z-scaling formulae are not optimized but appropriate ones. An optimized condition can yield a higher gain in a shorter wavelength region

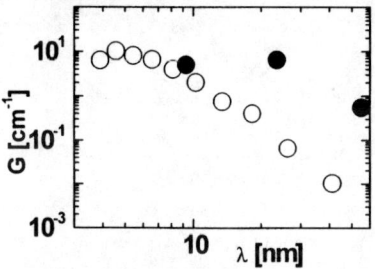

FIGURE 3. Gains obtained in the MHD simulation for the parameters suggested by the z-scaling formulae (open circles). For CEPS, optimized gains obtained by Shlyaptsev [17] are also plotted with solid circles.

The conditions of a discharge current between the CRPS and the CEPS are also compared in Fig. 4. Around 9 nm, the pulse length of a current is similar but the CRPS requires a lower current peak value, which suggests that for the amplification of shorter wavelength, the CRPS may be more efficient in power coupling.

CONCLUSION

Useful z-scaling formulae for the experimental parameters for CRPS x-ray lasers using H-like ions in a capillary discharge have been obtained through simplified assumptions in pinch, expansion, and gain phases. Such assumptions were justified by comparing with detailed MHD simulations. The formulae show that the capillary discharge can be used for the amplification of the wavelength in the water-window region with aluminum, however it requires rather severe electrical system, which seems difficult to be realized with the present technology.

The comparison with the gain and the experimental parameters obtained for the CEPS shows that the CRPS is a more efficient and promising lasing scheme for the amplification in the wavelength shorter than 10 nm.

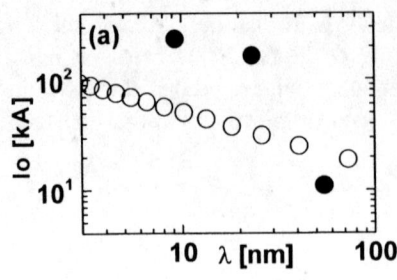 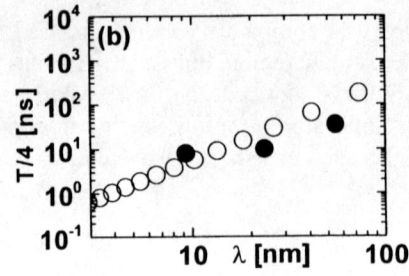

FIGURE 4. The required discharge conditions for CRPS (open circles) and CEPS (solid circles) are compared. The data for CEPS are taken from Ref. [17].

ACKNOWLEDGMENTS

This work has been supported in part by Korea Research Foundation Grant (KRF-2000-015-DP0175), by POSTECH research fund, by ADD and by electron Spin Science Center founded by Korea Science and Engineering Foundation.

REFERENCES

1. Y. Kato, H. Takuma, and H. Daido, eds. *X-ray laser, 1998*, Inst. of Physics Pub., Bristol, 1999 v. 159; C. H. Skinner, *Phys. Fluids B* **3**, 2420 (1991).
2. P. L. Hagelstein, *Ultrashort-Wavelength Lasers*, SPIE **1551**, 254 (1991).
3. D. Kim, C. H. Skinner, G. Umesh, and S. Suckewer, *Opt. Lett.* **14**, 665 (1989).
4. T. Hara, K. Ando, N. Kusakabe, H. Yashiro, and Y. Aoyagi, *Jpn. J. Appl. Phys.* **28**, L1010 (1989).
5. B. E. Lemoff, G. Y. Lin, C. L. Gordon, III, C. P. J. Barty, and S. E. Harris, *Phys. Rev. Lett.* **74**, 1574 (1995).
6. D. Kim, C. Toth, and C. P. J. Barty, *Phys. Rev. A Rap. Comm.* **59**, R4129 (1999).
7. J. J. Rocca, V. Shlyaptsev, F. G. Tomasel, O. D. Cortazer, D. Hartshorn, and J. L. A. Chilla, *Phys. Rev. Lett.* **73**, 2192 (1994).
8. B. R. Benware, C. H. Moreno, D. J. Burd, and J. J. Rocca, *Opt. Lett.* **22**, 796 (1997).
9. H. J. Shin, D. E. Kim, and T. N. Lee, *Phys. Rev. E* **50**, 1376 (1994).
10. T. Wagner, E. Eberl, K. Frank, W. Hartmann, D. H. H. Hoffmann, and R. Tkotz, *Phys. Rev. Lett.* **76**, 3124 (1996).
11. N. A. Krall and A. W. Trivelpiece, *Principles of Plasma Physics*, McGraw-Hill, New York, 1973, p. 123.
12. K. T. Lee, S. H. Kim, D. Kim, and T. N. Lee, *Phys. Plasmas* **3**, 1340 (1996).
13. R. C. Elton, *X-ray Lasers*, Academic, New York, 1990.
14. K. Lee, J. H. Kim, and D. Kim, *Phys. Plasmas* submitted.
15. K. Lee and D. Kim, *Appl. Phys. Lett.* **79**, 1968 (2001).
16. A. I. Shestakov and D. C. Eder, *J. Quant. Spectrosc. Radiat. Transfer* **42**, 483 (1989); Y. T. Lee, R. A. London, G. B. Zimmerman, and P. L. Hagelstein, *Phys. Fluids B* **2**, 2731 (1990).
17. V. N. Shlyaptsev, A. V. Gerusov, A. V. Vinogradov, J. J. Rocca, O. D. Lortazar, F. Tomasel, and B. Szapiro, *Ultrashort Wavelength Lasers II*, SPIE **2012**, 99 (1994).

Study of low current capillary discharge for compact Soft X-ray laser

G. Niimi, N. Sakamoto, M. Nakajima, Y. Hayashi, M. Watanabe,
A. Okino, K. Horioka and E. Hotta

*Department of Energy Sciences, Tokyo Institute of Technology,
Nagatsuta, Midori-ku, Yokohama 226-8502, Japan*

Abstract Capillary discharge experiments were carried out to get a lasing of Ne-like Ar. The relation between the appearance of the laser output and the instabilities of longitudinal modes has been studied by side-view observation. And observation of lasing at a low discharge current of 9 kA provides us a possibility of designing a compact and high-repetition-rate soft X-ray laser by use of a semiconductor switch instead of a gap switch.

1. INTRODUCTION

In 1994, J.J.Rocca et al. reported that a capillary discharge was found to produce a laser amplification between 3p - 3s levels of Ne-like Ar (J = 0 - 1, λ = 46.9 nm) [1]. This has led to a successful construction of a tabletop-sized soft X-ray laser [2], which is capable of generating saturated laser energy of 1mJ at 4 Hz with a discharge current of 26 kA (dI/dt = $1.5*10^{12}$A/s). But, it is difficult to increase repetition rate more than 10 Hz and to improve reproducibility, because of using a gap switch. To increase repetition rate of soft X-ray laser with improved reproducibility, it is necessary to decrease the input electric energy (and/or lasing current) and to use a semiconductor switch instead of a gap switch. For example, a static induction thyristor (NGK: RS1600PA40T1) can control a current amplitude more than 10 kA and attains dI/dt of $1.5*10^{11}$A/s. And lasing at low current provides us to construct a compact soft X-ray laser, because a lower discharge current permits a lower charging voltage and consequently a shorter insulation distance.

In our laboratory, research has also been performed on capillary discharge to achieve the Ne-like Ar soft X-ray lasing [3,4]. And, laser amplification has been observed. In the present paper, side-view framing photographs have been taken to observe the behavior of pre-discharge plasma column. And, we study the characteristics of the capillary

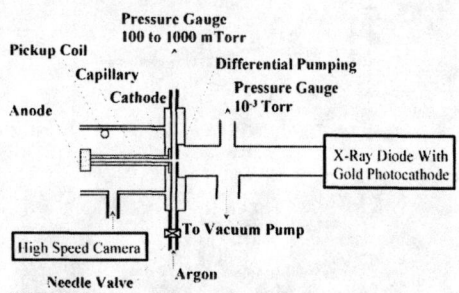

FIGURE 1. Schematic of the experimental apparatus.

CP641, *X-Ray Lasers 2002: 8th International Conference on X-Ray Lasers*, edited by J. J. Rocca et al.
© 2002 American Institute of Physics 0-7354-0096-2/02/$19.00

discharge soft X-ray laser by varying experimental conditions, such as a pre-discharge current, a main discharge current and a gas pressure. The results will be given in the following sections in detail.

2. EXPERIMENTAL SETUP

The capillary discharge current is driven by a pulsed power supply, which consists of a Marx generator, a transformer, a water capacitor, a gap switch and a capillary. With this power supply, it is able to obtain a discharge current pulse up to 35 kA with a rise time of 55 ns ($dI/dt = 8*10^{11}$ A/s). Two capillaries have been used in our experiments. One is made of Al_2O_3 ceramics with an inner diameter of 3 mm and a length of 150 mm, which is used in end-on X-ray diode (XRD) measurement. The other is used for side-view observation, which is made of Pyrex glass with an inner diameter of 3 mm and a length of 60 mm. During the experiments, the filling pressure of the argon gas is adjustable in a range from 100 to 1000 mTorr. The capillary is pre-discharged by a current of ~20 A, which is provided by a separate circuit.

Schematic diagram of the experimental apparatus is shown in Figure 1. The photographs of pre-discharge plasma were taken from the side direction, using a high-speed camera (Hadland, Imacon468). An XRD with gold photocathode is placed at a distance from the pinhole for the detection of the laser output.

3. RESULTS AND DISCUSSION

3.1 Observation of pre-discharge plasma column

A series of side-view observation has been performed on the pre-discharge plasmas column. Figure 2 shows the results observed with the framing camera at the gas filling pressure of 500 mTorr. It is found that within the period of approximately 40

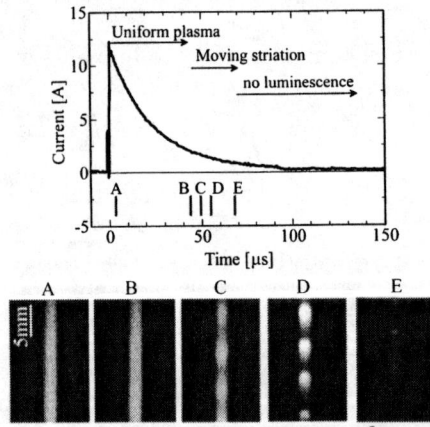

FIGURE 2. Pre-discharge current waveform and framing photographs of pre-discharge plasma.

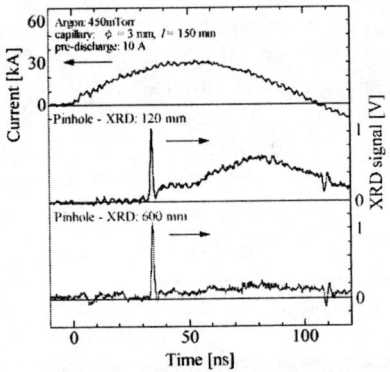

FIGURE 3. Waveforms of discharge current and the XRD signals measured at distance of 120 mm and 600 mm behind the pinhole.

μs after the start of the pre-discharge current, a uniform plasma column has been formed as shown in Figures 2A and 2B. During the period from 40 to 70 μs, the plasma becomes non-uniform and moving striations have been observed (see Figures 2C and 2D). As shown in Figure 2E, no luminescence of the plasma column has been observed after 70 μs. To obtain a stable output of the laser, it is considered that the main discharge should be initiated in uniformly pre-ionized plasma. The moving striations observed in Figures 2C and 2D may cause the axial modes of instabilities. These instabilities are considered to be responsible to the decreases of the amplitude and the reproducibility of the laser output, when starting the main discharge at relatively late timing.

3.2 Observation of lasing and optimum discharge parameters

Typical waveforms of the discharge current and the output signals of XRD are shown in Figure 3. The experimental conditions are as follows. The filling pressure of Ar gas pressure is 450 mTorr. The amplitudes of the pre-discharge and the discharge currents are 10 A and 30 kA, respectively. Main discharge current is initiated on 5 μs after the initiation of pre-discharge in the waveform of Figure 2. The middle and the lower trace in Figure 3 show the XRD output signals measured at a distance of 120 mm and 600 mm behind the pinhole, respectively. At about 34 ns after the initiation of main discharge, spike output has been found on the XRD signal. Compared to the middle trace, the ratio of the spike signal to the background signal in the lower trace has been improved. Therefore, the spike appears to have directivity, which is the characteristic of a laser.

In the present study, we have measured the laser output by changing the timing of the main discharge current with respect to the pre-discharge current, while maintaining the peak of pre-discharge current. The dependence of XRD signals on the pre-discharge current is shown in Figure 4. This experiment was conducted with a discharge current of 18 kA at a gas pressure of 350 mTorr. Pre-discharge current was varied from 0 to 16 A. When the pre-discharge current is higher than 10 A, the amplitude of the spike of the laser shows good reproducibility, while the amplitude of the spikes decreases and shows less reproducibility at pre-discharge current of 5 A.

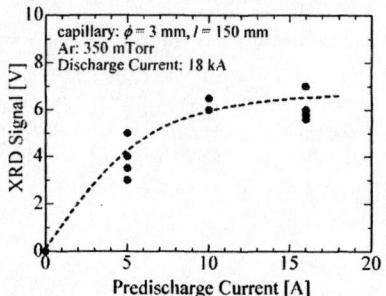

FIGURE 4. Variation of the laser signals as a function of pre-discharge current.

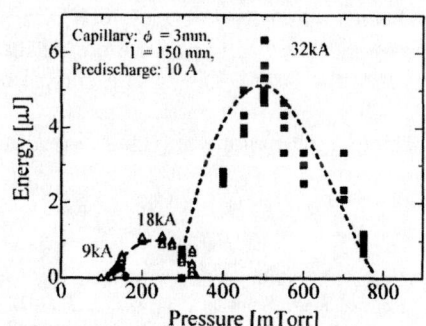

FIGURE 5. Amplitude of the laser energies as a function of discharge current and pressure with a pre-discharge current of 10 A.

105

Without pre-discharge, spike output was hardly obtained. The unstable spike output in the case of low pre-discharge current is quite possibly related to the development of the instabilities. This pre-discharge-dependency suggests that a sufficient pre-discharge current and early timing of main discharge are indispensable to avoid axial mode of instabilities and to obtaining laser outputs.

Next, in order to explore the region of optimum discharge parameters for laser outputs to occur, studies were conducted at different conditions of main discharge current and gas pressure. The dependence of the amplitude of the laser output on the argon gas pressure at 9 kA ($2*10^{11}$ A/s), 18 kA ($4*10^{11}$ A/s) and 32 kA ($7*10^{11}$ A/s) is shown in Figure 5. In the case of 9 kA, adequate plasma conditions for the laser output are obtained in a pressure range from 120 to 160 mTorr. In the case of 18 kA, adequate plasma conditions are obtained in a pressure range from 120 to 320 mTorr. In the case of 32 kA, adequate plasma conditions are obtained in the pressure range from 300 to 750 mTorr. With a higher discharge current, the optimum initial filling gas pressure increases and the pressure range becomes wider, which also yields a higher extracted energy of the laser output. It is shown that the discharge condition or the capacity of power source dominates the extracted energy of laser output.

We have not conducted the optimization of a capillary diameter, and have not confirmed whether a pinch plasma at 9 kA has already attained the optimum plasma condition or not with spectroscopic study. If it is impossible to attain the optimum plasma condition with changing discharge parameters, hybrid pumped soft X-ray laser with the transient gain [5] will offer a way towards the optimum plasma condition and efficient compact soft X-ray laser device. However the fact that laser signals can be even observed at a current below 9 kA with an adequate argon gas pressure provides us a possibility to construct a compact and high-repetition-rate soft X-ray laser by using static induction thyristors connected in series instead of a gap switch.

4. CONCLUSION

Using a capillary discharge, Ne-like Ar lasing has been confirmed. The relation between the appearance of the laser output and the instabilities of longitudinal modes has been studied by side-view observation. Adequate amplitude of the pre-discharge current and the timing of the main discharge have been obtained for achieving stable output of the laser. Lasing at a current of 9 kA was confirmed, which provides us a possibility to construct a compact and high-repetition-rate soft X-ray laser by replacing a gap switch with static induction thyristors.

REFERENCES

1. Rocca J.J. and Shlyaptsev V.N. et al., *Phys. Rev. Letters* **73**, 2192 (1994).
2. Rocca J.J. and Seminario M. et al., *J.Phys.IV France* **11**, 459 (2001).
3. Niimi G. and Hayashi Y. et al., *J. Phys. D* **34**, 2123 (2001).
4. Niimi G. and Hayashi Y. et al., to be published in *IEEE Trans. on Pla. Sci.* **30** (2002).
5. Nickles P.V. and Janulewicz K.A. et al., *J.Phys.IV France* **11**, 93 (2001).

Experimental and theoretical study of a capillary discharge dedicated to coherent XUV production

J. Pons, Ke Lan*, C. Cachoncinlle, E. Robert, R. Viladrosa, J.-M. Pouvesle, C. Fleurier

GREMI, Université d'Orléans & CNRS, BP 6744, 45067 Orléans Cedex 2 (France)
* Permanent address : Institute of Applied Physics and Computational Mathematics, P.O. Box 8009-12, Beijing 100088 (People's Republic of China)

Abstract. An argon plasma generated by a fast capillary discharge for the amplification of Ar IX 46.9 nm line has been studied experimentally by time-resolved pinhole imaging and XUV spectroscopy. Self breakdown by hollow cathode effect induces plasma ignition along the capillary axis. Numerical simulations of the emission radial profiles have been performed using a MHD and atomic kinetics code. Electron temperatures and densities have been calculated, and an interpretation of the plasma behavior is given. These studies have shown that no efficient compression is observed and that relevant conditions for the required lasing effect have not been presently reached using our device, but progress has been made in the understanding of the plasma emission dynamics.

INTRODUCTION

A novel capillary discharge device has been studied experimentally by time-resolved imaging and spectroscopy. The aim is to obtain the amplification of the Ar IX 46.9 nm line by collisional excitation scheme. The setup differs significantly from previous capillary discharge driven XUV lasers [1-5], especially concerning its triggering mode. Theoretical work has been performed for better understanding of the plasma processes. This work was motivated by prospecting for new conditions of lasing, in terms of current maximum and rise-time.

A fast discharge is generated in a 5 cm long, 1.5 mm diameter capillary filled with argon at a pressure of 1 mbar. The electrical energy is delivered by means of two flat 45 nF capacitors mounted in Blumlein configuration (Fig. 1a) with a surface-guided atmospheric spark-gap (described in [6]) as a switching element. The whole device is highly compact, and can be charged up to 25 kV.

CP641, *X-Ray Lasers 2002: 8th International Conference on X-Ray Lasers,* edited by J. J. Rocca et al.
© 2002 American Institute of Physics 0-7354-0096-2/02/$19.00

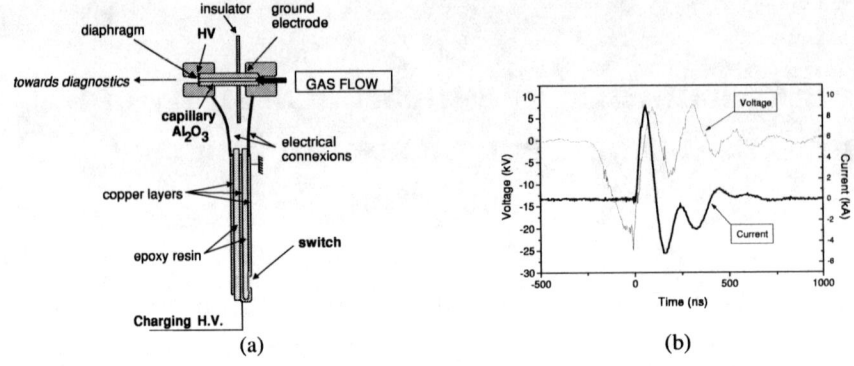

FIGURE 1. (a) Experimental setup: discharge driver (b) Discharge current and voltage

Alumina (Al_2O_3) capillaries are used to minimize wall ablation. Breakdown occurs without preionization by hollow cathode effect which generates a fast electron beam along the capillary axis due to electric field lines narrowing in the electrodes gap. A 9 kA current is obtained with a 0.2 kA/ns rise time when the circuit is charged at 20 kV (Fig. 1b).

Time-resolved observations of the plasma have been performed, by pinhole imaging and XUV spectroscopy. Radial emission profiles have been obtained using a 50 µm diameter pinhole camera having a 5.5 magnification, and spectra were taken using a transmission grating spectrograph called Spartuvix [7]. A gated MCP detector and a CCD camera have been used for pictures and spectra acquisition. The same kind of device has been used for the study of ablative carbon discharges [8]. All measurements have been done with a 20 kV charging voltage in argon at 1 mbar, as described before.

A one-dimensional magneto-hydrodynamics Lagrangian code coupled to an atomic kinetics code has been used for simulating the temporal evolution of the plasma emission, calculating the electron temperature and density as well. The basic equations of this code are described in [9], and some modifications have been brought to take new effects into account: the radial profile of the current density has been assumed to be Gaussian, and a correction factor has been considered for the ion excitation rates, to take the fast electrons contribution into account, according to [10].

EXPERIMENTAL RESULTS

Figure 2 shows the images obtained at different times of the discharge evolution, for the current displayed in Fig. 1b, in the detector sensitivity range (5-200 nm). Corresponding radial profiles of the emission intensity are given below. The plasma radius was deduced from the $1/e^2$ half width of a Gaussian fitting curve applied to each profile. Its evolution with time is plotted in Fig. 3a and 3b as well as the intensity profile on-axis value (further called peak intensity) and the two-dimensional integral of each profile (further called full intensity).

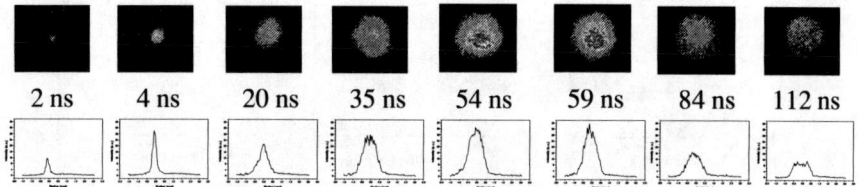

FIGURE 2. Pinhole images and corresponding radial intensity profiles for different times, in the spectral range 5-200 nm

The principal observation that should be made at first sight is that plasma initiates on the capillary axis, and then grows towards the walls; moreover, plasma emission is always maximum on the axis. This kind of plasma behavior differs significantly from preionized plasmas, in which the emission starts with an annular profile [11].

Five different periods can be noticed in the peak intensity evolution (Fig. 3a): an intense radiation is emitted on the axis a few nanoseconds after the discharge starts and rises rapidly (1), then decreases (2) as the radius increases. After the plasma has filled the whole capillary, the peak intensity increases again (3), and reaches its maximum (4) with the current. Finally both intensity and plasma radius decrease (5). In the meantime, full intensity evolution follows the one of the current (Fig. 3b).

Fig. 4 shows spectra observed at different times, in the wavelength range 40-60 nm. The 46.9 nm line has not been detected at any time, but the spectra are mostly constituted by Ar VII and Ar VIII lines. For instance, the Ar VIII 52 and 52.7 nm 3d-3p line evolutions are plotted in Fig. 3c. These curves show two different maximums as in the peak intensity evolution (Fig. 3a), the first one being higher than the second.

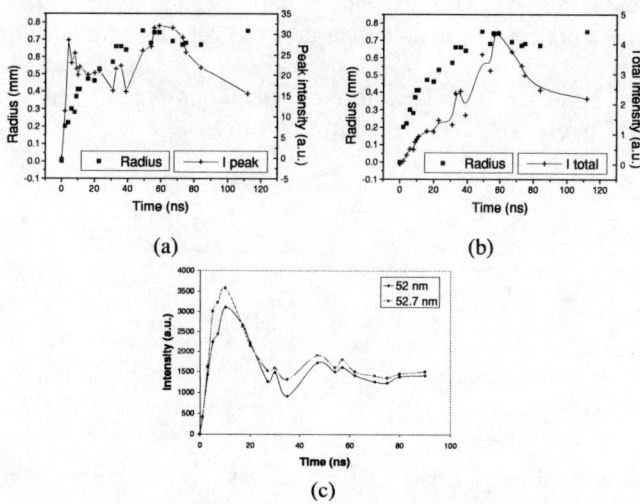

FIGURE 3. (a) Plasma radius and peak intensity evolution (b) Radius and full intensity evolution (c) Evolution of the 52 and 52.7 nm line intensities

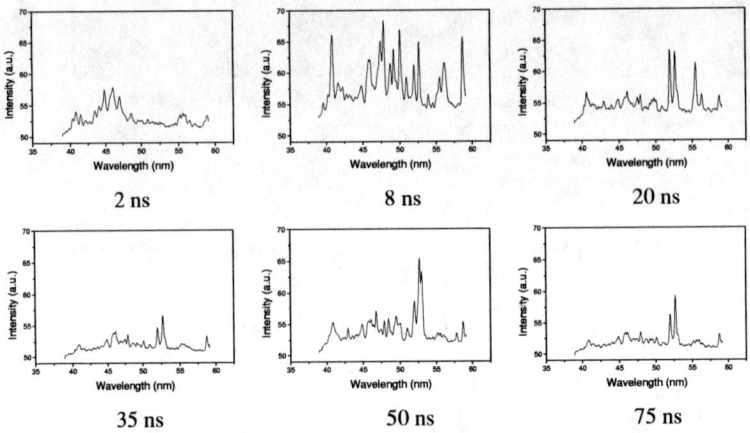

2 ns   8 ns   20 ns

35 ns   50 ns   75 ns

FIGURE 4. Spectra at different times in the range 40-60 nm

CALCULATIONS AND INTERPRETATION

The five periods described before can be explained as follows: in period 1, strong emission on the axis is due to Joule heating of the plasma and compression by Lorentz force. Since current is low during this period (less than 3 kA), magnetic pressure is not strong enough compared to kinetic pressure to maintain the compression. Therefore, the plasma relaxes and expands during period 2. This leads to a plasma cooling and emission decay. Then plasma is re-heated as the current increases (period 3) inducing emission increase. Period (5) corresponds to the plasma relaxation after current peak. Current density is higher on the axis than near the capillary walls, and consequently Lorentz force is too weak to compress the plasma situated in the wall vicinity towards the axis during periods 3 to 5. Therefore magnetic compression does not efficiently occur, in spite of a relatively high current.

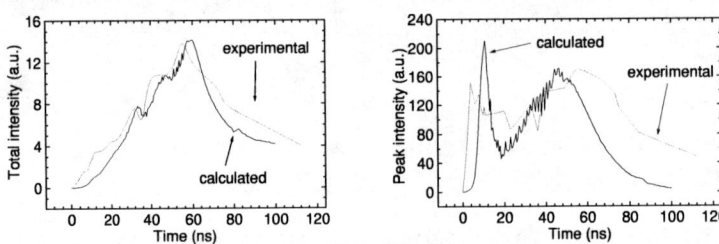

FIGURE 5. Calculated and experimental full (left) and peak (right) intensities in the range 5-200 nm

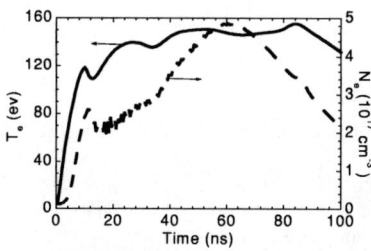

FIGURE 6. On-axis electron temperature and density yielded by the code

Figure 5 shows the emission peak and full intensities evolution in the observed spectral range 5-200 nm calculated by the code described in [9], adapted for the presence of an on-axis electron beam at the onset of the discharge, as described in the previous paragraph. This simulation qualitatively fits well with the experimental results. Calculation of the on-axis plasma electron temperature and density evolutions with time has been performed, and the results are shown in Fig. 6. For this purpose, the electron velocity distribution has been assumed to be non-Maxwellian. According to the code, T_e reaches 100 eV in less than 10 ns and is around 150 eV at maximum, and N_e is in the range $1\text{-}5.10^{17}$ cm^{-3}. After 60 ns, results of the code indicate a quite steady high temperature which does not seem completely coherent with the observation of the strong decrease of the emission.

The deduced temperature and electron densities are quite far from the optimal conditions for lasing at 46.9 nm as proposed in [11]. Excitation of highly ionized ions is very sensitive to the electron density which seems too low in our case. This can explain why no line from Ar IX or ions with a higher ionization level have been detected in the present experimental conditions.

CONCLUSION

Good efficiency for producing XUV radiation has been proved with our device, but lasing conditions at 46.9 nm have not been reached in our experimental conditions, mainly due to a relatively low electron density. Efficient compression and ionization have not obtained with this device. Pinch effect does not occur, because the current starts from the axis and has a Gaussian radial profile, with a low value near the capillary walls during the whole discharge.

The code has allowed to better understand the experimental results and knowledge of the observed phenomena in electron-beam self-triggered discharges has progressed, giving information for further modifications of our device. Higher densities can be obtained if a strong magnetic compression is generated, which means that current density must have an annular profile at the beginning of the discharge. Preionization seems to be one way for the achievement of such conditions, according to [11], and its application to our experimental setup will be subject to future investigations.

REFERENCES

1. Rocca, J.J., Shlyaptsev, V., Tomasel, F.G., Cortázar, O.D., Hartshorn, D., and Chilla, J.L.A., *Phys. Rev. Lett.* **73**, 2192-2195 (1994)

2. Ben-Kish, A., Shuker, M., Nemirovsky, R.A., Fisher, A., Ron, A., and Schwob, J.L., *Phys. Rev. Lett.* **87**, 015002 1-4 (2001)

3. Koshelev, K.N., Antsiferov, P.S., Dorokhin, L.A., Nazarenko, A.V., Sidelnikov, Yu.V., and Glushkov, D.A., "Observation of ASE effect for Ne-like Ar in a capillary discharge driven by inductive storage with plasma erosion opening switch", in *X-Ray Lasers 2000*, edited by G. Jamelot et al., EDP Sciences Journal de Physique IV 82, Les Ulis (France), 2001, pp. Pr2 119-122

4. Niimi, G., Hayashi, Y., Nakajima, M., Watanabe, M., Okino, A., Horioka, K., and Hotta, E., *J. Phys. D: Appl. Phys.* **34**, 2123-2126 (2001)

5. Tomassetti, G., Ritucci, A., Reale, A., Palladino, L., Reale, L., Kukhlevsky, S.V., Flora, F., Mezi, L., Kaiser, J., Faenov, A., and Pikuz, T., *Eur. Phys. J. D* **19**, 73-77 (2002)

6. Götze, S., Hong, D., Dussart, R., Cachoncinlle, C., Pons, J., Pouvesle, J.-M., Fleurier, C., and Viladrosa, R., "Development of a Blumlein generator dedicated to a fast capillary discharge XUV source", in *X-Ray Lasers 2000*, edited by G. Jamelot et al., EDP Sciences Journal de Physique IV 82, Les Ulis (France), 2001, pp. Pr2 609-612

7. Bourgade, J.L., Combis, P., Louis-Jacquet, M., LeBreton, J.P., de Mascureau, J., Naccache, D., Sauneuf, R., Thiell, G., Keane, C., MacGowan, B., and Matthews, D., *Rev. Sci. Instrum.* **59**, 1840-1842 (1988)

8. Dussart, R., Hong, D., Götze, S., Rosenfeld, W.E.S., Pons, J., Viladrosa, R., Cachoncinlle, C., Fleurier, C., and Pouvesle, J.-M., *J. Phys. D: Appl. Phys.* **33**, 1837-1842 (2000)

9. Lan, K., Zhang, Y., and Zheng, W., *Phys. Plasmas* **6**, 4343-4348 (1999)

10. Rutkevich, I., Mond, M., Kaufman, Y., Choi, P., and Favre, M., *Phys. Rev. E* **62**, 5603-5617 (2000)

11. Rocca, J.J, *Rev. Sci. Instrum.* **70**, 3799-3827 (1999)

Excitation of the 13.2 nm laser line of Nickel-like Cd in a capillary discharge plasma column

A. Rahman, E.C. Hammarsten, S. Sakadzic, J.J. Rocca

Colorado State University. Fort Collins, CO 80523,

V.N. Shlyaptsev [a,b] and A. Osterheld[a]

Lawrence Livermore National Laboratory [a]/University of California Davis [b]

J.-F. Wyart

Laboratoire Aime Cotton, CNRS (UPR 3321)
Centre universitaire 91405 - Orsay (France)

Abstract. Emission at 13.2 nm from the 4d $^1S_0$ -4p $^1P_1$ laser transition of Ni-like Cd was observed in a cadmium vapor plasma column excited by a high-current capillary discharge. The dynamics of the plasma were studied using time-resolved soft x-ray pinhole images and hydrodynamic simulations. The pinhole images show that the plasma column maintains a good symmetry up to the time of maximum compression. Fifty-five CdXXI lines were identified in the 12.7-18.4 nm region with the assistance of calculations performed using the Slater-Condon method with generalized least-squares fits of the energy parameters.

INTRODUCTION

Capillary discharge excitation has been demonstrated to be a very successful method for the generation of compact and efficient soft x-ray lasers [1-5]. Large amplification in capillary discharge-pumped ultra short wavelength lasers has been achieved in Ne-like Ar, Ne-like S and Ne-like Cl at wavelengths ranging from 46.9 nm to 60.8 nm [1-3]. Utilizing this excitation technique a high repetition rate tabletop laser operating at 46.9 nm has produced the highest average power reported to date for a soft x-ray laser, 3.5 mW [4], and it has been successfully utilized in several applications [5]. There is significant interest in the possibility of extending capillary discharge lasers to shorter wavelengths. Of particular interest is the possibility of obtaining lasing near 13.5 nm, a wavelength that is of interest for extreme ultraviolet lithography [6]. Scaling of the collisional laser scheme to this wavelength can be accomplished by the excitation of Nickel-like Cd ions (CdXXI), which can provide amplification at 13.2nm in the 4d $^1S_0$ -4p $^1P_1$ transition [7]. However, excitation of the laser upper level of Ni-like Cd requires a large increase in the discharge excitation power with respect to that used to produce lasing in the 46-61 nm region. We have developed a high power capillary discharge capable of generating the hotter and denser plasmas that are necessary for this task [8]. This pulsed power generator produces pulses of up to 200 kA with a 10-90% rise time of less than 15 ns. The initial study of the generation of hot dense plasmas with this high power capillary discharge involved the excitation of Ar gas, which resulted in electron temperatures of > 250 eV and electron densities of ~ 1×10^{20} cm^{-3} [8].

In this proceedings we discuss results of the generation of highly ionized cadmium plasma columns for amplification in Ni-like Cd, including the observation of strong line

CP641, *X-Ray Lasers 2002: 8th International Conference on X-Ray Lasers,* edited by J. J. Rocca et al.

emission from the 13.2 nm laser transition. The dynamics of the column was studied using time-resolved soft x-ray pinhole images and model simulations. Spectroscopy of the plasma column in the 12.7-18.4 nm region resulted in the classification of fifty-five $3d^9$ 4p-$3d^9$ 4d and $3d^9$ 4d- $3d^9$ 4f CdXXI lines.

GENERATION OF HIGHLY IONIZED CD PLASMA COLUMNS

The plasma columns were generated by exciting a capillary channel filled with Cd vapor with a fast high current pulse. Cd vapor was generated by a metal vapor gun, designed to generate vapor in a room temperature environment by rapidly heating a cadmium electrode with a capacitive discharge, and was injected into the capillary channel though an axial hole in one of the electrodes. The high current pulses used to excite the capillary plasma were produced by a pulse power generator consisting of a three-stage pulse compression scheme. The first stage, which produces a current pulse that is subsequently shorten by each of two following stages, consists of a conventional eight stage Marx generator that for the present experiment was operated at an erected voltage of ~ 650 kV. The Marx generator is used to charge the second pulse compression stage that consists of a 26 nF coaxial water-dielectric capacitor in about 1 µs. In turn this water capacitor is discharged through a self-breakdown spark gap pressurized with SF_6 gas to charge the third and final stage in about 75 ns. The third stage consists of two radial water dielectric transmission lines connected in a Blumlein configuration. The capillary channel is located at the axis of this radial Blumlein. The fast current pulse that excites the capillary plasma is produced by discharging the Blumlein transmission line through an array of seven synchronized triggered spark-gap switches distributed along the outer diameter of the water transmission line. The very rapid switching of the Blumlein transmission line produces current pulses with amplitudes of up to 200 kA and risetimes of less than 15 ns through the capillary load.

DYNAMICS OF THE CADMIUN PLASMA COLUMN

The evolution and stability of the Cd plasma columns was studied using a soft x-ray pinhole camera equipped with a microchannel plate MCP-CCD detector. The set up provided a magnification of 2.76 times. Gating of the MCP allowed for images of the plasma column with ~ 3 ns temporal resolution. A combination of a 1 µm thick carbon filter and a 0.2 µm thick aluminum filter was used to limit the response of the camera to photons with wavelengths between 45 Å and 100 Å and below 20 Å, allowing for the differentiation of the hotter regions of the plasma from the colder regions that emit longer wavelength radiation. Figure 1 consists of a sequence of end-on images of the plasma column, showing the evolution of the soft x-ray-emitting region of the Cd plasma in a 2.5 cm long polyacethal capillary of diameter 5 mm. The time of each image with respect to the initiation of the current pulse is indicated. The images show that the onset of significant soft x-ray emission occurs about 27 ns after the initiation of the current pulse, with rapidly increasing intensity in the few nanoseconds after that. The diameter of the soft x-ray emitting regions achieves a minimum diameter of 250-350 µm at 32-34 ns after the onset of the current pulse. Shortly afterwards the emitted soft x-ray intensity rapidly decreases. The plasma column is observed to have a good degree of symmetry, comparable to that observed in the plasma columns of the Ne-like Ar laser. This occurs up to the time of maximum compression. Later in time the plasma becomes asymmetrical. However, these late un-uniformities should not be of concern for soft x-ray laser development because they appear after the time of interest for laser amplification.

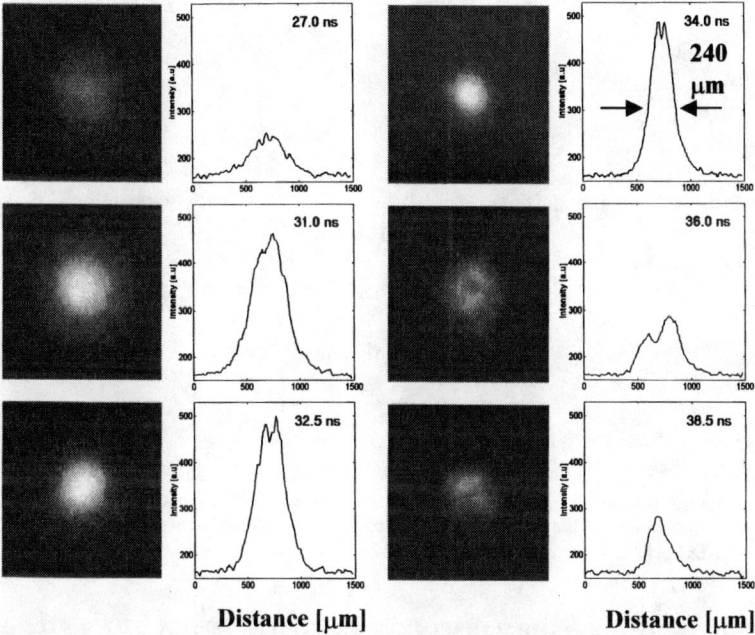

Distance [μm]   Distance [μm]

Figure 1. Sequence of end-on pinhole images of the soft x-ray emitting region of the Cd plasma column acquired using a 1 μm thick carbon filter and a 0.2 μm thick aluminum filter. The time of each image with respect to the initiation of the current pulse is indicated. The discharge current was ~190 kA and the capillary diameter was 5 mm.

These experimental observations are in good agreement with hydrodynamic model simulations conducted using RADEX. Results corresponding to a 190 kA cadmium discharge in a 4 mm diameter capillary at a Cd pressure of 0.9 Torr are shown in figure 2. The graphs of the evolution of the computed plasma parameters shows the dynamics of a cylindrical shock wave shell that is driven towards the axis of the discharge by the Lorentz force and thermal pressure in the skin layer near the walls. The electron temperature plot shows that the heat wave moves ahead of the mass, pre-heating the plasma in front of the shock wave. When this heat wave reaches the axis, the maximum of the current density switches to the center of the capillary, and the temperature rapidly increases. The main amount of the mass in the plasma shell is computed to reach the axis of the capillary several nanoseconds later (see electron density graph, Fig 2.). The electron density is calculated to reach a maximum value of $1-2 \times 10^{20}$ cm^{-3}. It is also interesting to notice that due to maximum joule heating on axis the computed electron density profile has a concave radial profile that could potentially guide the laser beam. Preliminary computations estimate the gain in the 13.2 nm line of Ni-like Cd at 1-2 cm^{-1}, and potentially > 3 cm^{-1} under optimized excitation conditions. The simulation results are further discussed in another paper in these proceedings [9].

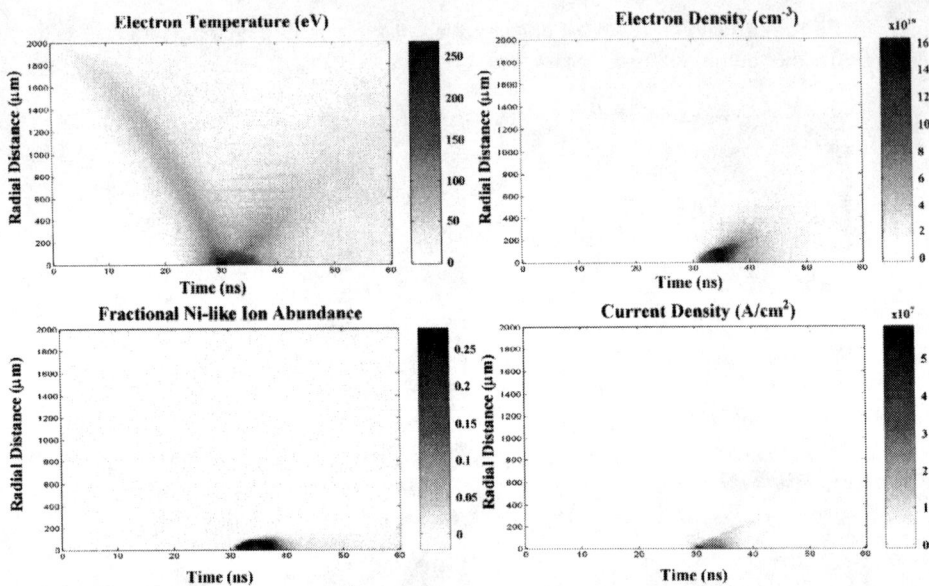

Figure 2. Computed evolution of the plasma parameters of the Cd capillary discharge plasma column. The graphs show the evolution of the electron temperature, electron density, density of Ni-like ions, and the current density. The peak discharge current is 190 kA, the capillary diameter is 4 mm and the initial cadmium pressure is 0.9 Torr.

SPECTROSCOPY: OBSERVATION OF EMISSION FROM THE 4d $^1S_0$- 4p $^1P_1$ LASER LINE OF Ni- LIKE Cd AND NEWLY IDENTIFIED CdXXI LINES

Time resolved spectroscopy of the Cd plasma column was carried out through selected intervals in the spectral range between 12.7 nm and 18.4 nm. The radiation axially emitted by the plasma was focused by a gold-coated grazing-incidence mirror into the slit of a 2.217 m grazing-incidence spectrograph with a 2400 lines per millimeter gold-coated diffraction grating. Figure 3 shows end-on spectra of the capillary discharge plasma for 15.5-16.6 and 17.2-18.4 nm regions. The spectra are observed to be dominated by lines from Ni-like (CdXXI) and Cu-like (CdXX) ions. From these and other spectra we identified fifty five lines corresponding to $3d^9$ 4p-$3d^9$ 4d and $3d^9$ 4d- $3d^9$ 4f CdXXI transitions with the assistance of calculations performed using the Slater-Condon method with generalized least-squares fits (GLS) of the energy parameters [10]. The average deviation between the measured and theoretical wavelengths is $<\lambda exp-\lambda th> = 0.0065$ nm. The classification of lines of Nickel-like CdXXI is shown in Figure 3. The measured transition wavelength (in Å) is followed by the level designation given as the J-value and index N_{th}, which numbers the levels from the lowest energy in the same J-values and configuration, as used in [10]. Figure 4 shows strong line emission observed at (13.162 ± 0.010) nm which is assigned to correspond to the 4d $^1S_0$ -4p $^1P_1$ laser line of Ni-like Cd (this value is more accurate than our previously reported wavelength of 13.172 nm). This measured wavelength agrees well with the previously reported wavelength (13.166 ± 0.015) nm of this line [7]. It also is in good agreement with the computed GLS theoretical value of 13.1719 nm.

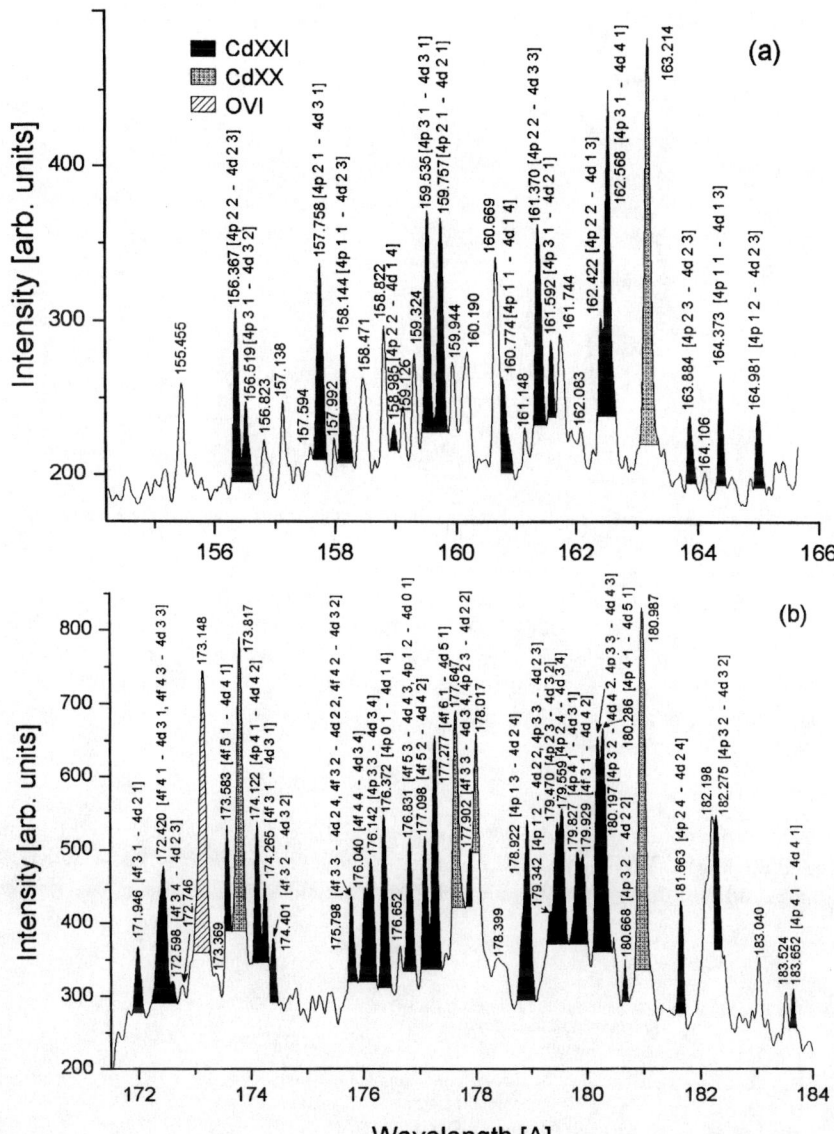

Figure 3. On axis spectra of a cadmium capillary discharge plasma column obtained at discharge currents of ~ 180 kA, and a capillary diameter of 5 mm. The measured transition wavelength (in Å), is followed by the level designation given as the J-value and the index level N_{th}. For some lines the wavelength values is the average of two spectra.

CONCLUSIONS

In summary, we have demonstrated that a fast capillary discharge is capable of generating radially symmetric plasma columns of small diameter in which Cd atoms are ionized to the

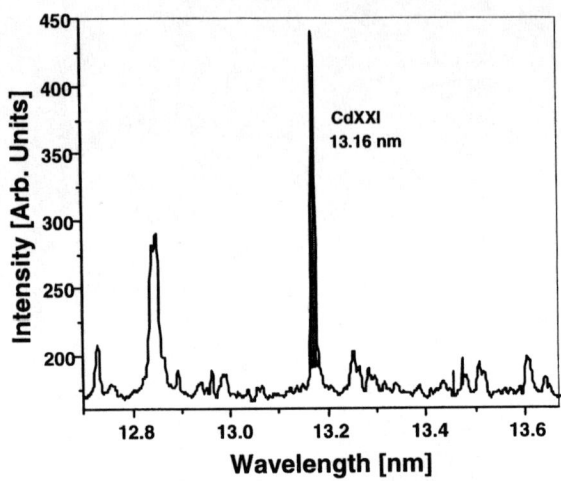

Figure 4. On axis spectra of the Cd capillary discharge plasma showing emission from the 13.2 nm line of Ni-like Cd.

Ni-like stage. Axial spectra of the discharge identified strong line emission at 13.2 nm, corresponding to the 4d $^1S_0$- 4p $^1P_1$ laser transition of Ni-like Cd. This identification is supported by the simultaneous observation of numerous Ni-like Cd lines. These results advance the possibility of the demonstration of a discharge-pumped laser in this transition, and in laser lines of other Nickel-like ions.

ACKNOWLEDGMENTS

This work was supported by DARPA grant DE-FG 03-00ER15084 and by the National Science Foundation. We also gratefully acknowledge the support of the W.M. Keck Foundation and the earlier contributions of F. Tomasel and M. Frati.

REFERENCES

1. J.J. Rocca, V. N. Shlyaptsev, F.G. Tomasel, O.D. Cortazar, D. Hartshorn, J.L.A. Chilla,, Phys. Rev. Lett., **73**, 2192, 1994; J.J. Rocca, D.P.Clark, J.L.A. Chilla, and V.N. Shlyaptsev, Phys. Rev. Lett. **77**, 1476-1479, (1996).
2. F.G. Tomasel, J.J.Rocca, V.N. Shlyaptsev, and C.D. Macchietto, Phys. Rev. A **55**, 1437-1440 (1997).
3. M. Frati, M. Seminario, and J.J. Rocca, Opt. Lett. **25**, 1022-1024 (2000).
4. C. Macchietto, B.R. Benware and J.J. Rocca, Opt. Lett. **24**, 1115-1117, (1999).
5. J.J. Rocca, et. al., Comptes Rendus De L' Academie Des Sciences Serie IV, Physique Astrophysique, **8**, 1065-1081 (2000).
6. D.T. Attwood, X-Ray Lasers, 2000 – J. Phys. IV **1**, Pr2-443-449 (2001).
7. Y. Li, J. Nilsen, J. Dunn, and A. L. Osterheld, Physical Review **A 58**, R2668 – 2672 (1998).
8. J.J. Gonzalez, M. Frati, Rocca J.J. and V.N. Shlyaptsev, and A.L. Osterheld, Phys. Rev E. **65**, 026404, (2002).
9. V.N. Shlyaptsev, et al., in these Proceedings
10. S.S. Churilov, A.N. Ryabtsev, and J.-F. Wyart, Physica Scripta **38**,326-35 (1988).

Investigation of the output pulse characteristics of a 46.9 nm Ar capillary discharge soft x-ray laser

A.Ritucci [1], G.Tomassetti [1], L.Palladino [1], A.Reale [1], G.Gaeta [1], L.Reale, T.Limongi[1] F.Flora [2], L.Mezi [2], S.V.Kukhlevsky [3], J.Kaiser [4], A.Faenov [5], T.Pikuz [5]

[1] Phys. Dept. Univ. of L'Aquila, gc LNGS of INFN, INFM, 67010 Coppito, L'Aquila, Italy
[2] ENEA Dip. Innov., Div. Fisica Applicata, C R E Frascati C.P. 65, 00044, Italy
[3] Department of Experimental Physics, University of Pécs, Ifjúság ú. 6, Pécs, 7624, Hungary
[4] Institute of Phys. Eng., Brno Univ. of Technology, Techn. 2, 616 69 Brno, Czech Republic
[5] MISDC of VNIIFTRI, Mendeleevo, Moscow Region, 141570, Russia

Abstract. In this paper, we report on the realization of a capillary discharge soft x–ray laser operating at 46.9 nm pumped by a 30 kA peak value, 150 ns half cycle duration current pulse (corresponding to a mean current slope of about $5 \ 10^{11}$ A/s). The slope of the pumping current is sufficiently high to produce the plasma compression and laser amplification on the 3p-3s, J=0-1 transition of Ne-like Ar, in 2.4-4 mm in diameter alumina capillary channels. We have analyzed the output pulse characteristics of the produced laser beam, such as the lasing time and the pulse duration, the saturation and the output pulse energy, the near field image as a function of different experimental parameters. Using the same current pulse, the lasing effect has not been observed in polyacetal capillaries, demonstrating the damning role of the wall capillary ablation in the heating and in the stability of the plasma column during the z-pinch compression.

INTRODUCTION

The demonstration of a capillary discharge soft x-ray laser operating at 46.9 nm in Ne-like Ar by Rocca et al. in 1994 [1], opened the possibility of the actual realization of compact and high repetition rate soft x-ray lasers. However, since that time, in spite of the fast developments of the capillary-discharge soft x-ray laser in Ne-like Ar carried on by that group [2], in other laboratories, it has been very difficult for a long time just to reproduce those original experiments. The reason resides on the fact that, in order to produce an optimal plasma column during the capillary discharge with the required temperature and density for the population inversion many experimental parameters should be determined. Consequently, only very recently, in 2000 and 2001 [3-4], some other group started to report some result.

Recently [5], we also reported on the demonstration of the soft x-ray laser amplification in Ne-like Ar using a capillary discharge device. In the present paper we

CP641, X-Ray Lasers 2002: 8th International Conference on X-Ray Lasers, edited by J. J. Rocca et al.
© 2002 American Institute of Physics 0-7354-0096-2/02/$19.00

will show further results concerning the optimization of the laser beam intensity as a function of some of the main experimental parameters of the capillary discharge: the current pulse amplitude, the initial gas pressure, the capillary diameter and material. Moreover, we have characterized the output pulse parameters of the laser beam such as the gain-length product and the saturation, the output pulse energy and duration, the divergence and the near field structure of the beam.

EXPERIMENTAL DEVICE

The electrical pumping device has been described in detail elsewhere [6]. The high current pulse through the capillary channel is produced by discharging a 10 nF water dielectric capacitor initially charged to high voltage by a six-stage Marx generator. This electrical device can generates current pulses with a peak value ranging between 26 and 36 kA. Due to the high value of the capacitor, the current has a relatively long half cycle duration, which changes with the capillary length between 135–150 ns. These parameters correspond to a current rise time ranging between 45 and 50 ns and to a mean current slope of ~ $0.5 \ 10^{12}$ A/s. A 20 A pre-ionization current, having a duration of 3–4 µm, precedes the main discharge and it is produced by an independent circuit. In the actual setup the capillary channel is separated from the detection line through a shutter valve, opening 1 ms before the discharge is fired. This system allows us to fill the capillary channel with Ar at the desired pressure, avoiding the gas outflow into the detection line. On the other hand, this arrangement actually limits the repetition rate of the capillary discharge device to 1 shot every 1 or 2 minutes.

Different detection lines have been used to analyze the radiation. A 1-m grazing angle spectrometer having a toroidal reflection grating with 550 g/mm has been used to spectrally analyze the soft x-rays with a spectral resolution of about 2 Å (with a 100-µm entrance slit). Another detection line has been used to acquire the near field images of the laser beam. It consists of a couple of two 75-cm focal length Sc/Si multi-layer mirrors, having a nominal peak reflectivity of 30% at 46.9 nm and a bandwidth of ±2 nm. The mirrors have been positioned in order to reproduce on the detector a 1:1 image of the laser beam at the capillary output. In both cases the detector, which is characterized by a spatial resolution of about 20 µm, consists of a MCP coupled to a CCD camera. At the present state, the MCP is gated with a 500 ns long voltage pulse so that it acquires time-integrated images. The time evolution and the laser pulse duration have been studied using a fast vacuum photodiode, positioned at 1 m far from the capillary output and coupled with a 1 GHz Tektronix digitizing oscilloscope. A calibrated silicon p-i-n diode detector has been utilized to estimate the laser output energy.

EXPERIMENTAL RESULTS

The measurements of the spectra, which have been obtained utilizing a 20-cm long, 3-mm in diameter alumina (Al_2O_3) capillary channel, gave the results shown in fig.1. In a very narrow region of both current and initial gas pressure the line of the Ar at 46.9

dominates the spectra in the spectral region analyzed (43– 49 nm). This clearly demonstrates the laser amplification on the 3p–3s (J=0–1) transition of the Ne-like Ar.

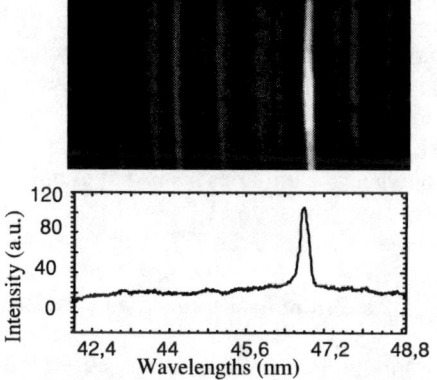

Fig.1. Measured spectra using a 3-mm in diameter 20-cm long alumina capillary channel initially filled with 320 mTorr of Ar and pumped with a 30 kA current pulse.

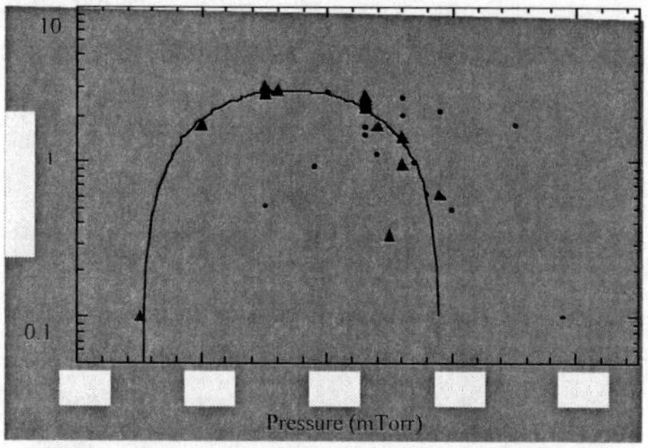

Fig.2. Dependence of the laser pulse intensity on the initial Ar gas pressure in the range of 100 – 500 mTorr. The two curves are related to two different alumina capillary channels.

The optimum current interval ranges between 27 and 32 kA, while the optimal pressure interval slightly changes from one capillary to another (see fig.2) in the interval between 200 and 450 mTorr.

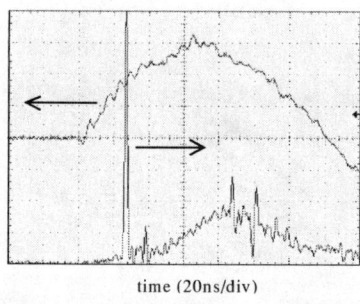

Fig.3. Time evolution of soft x-rays emitted by the discharge in a 3-mm in diameter, 20-cm long capillary channel initially filled with 0.32 Torr of Ar.

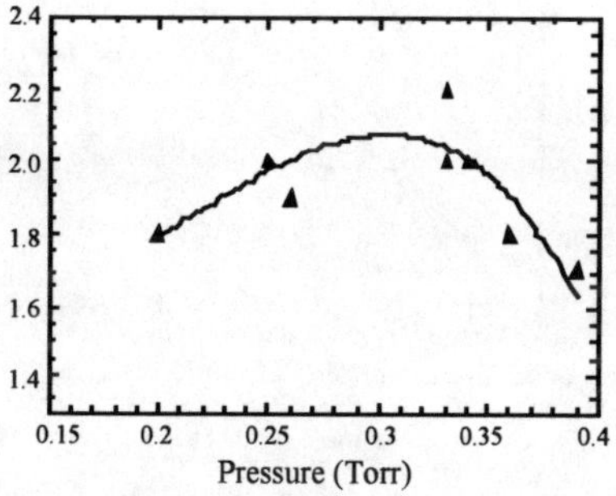

Fig.4. Duration of the laser pulse as a function of the initial gas pressure in a 20-cm long capillary.

The time evolution of the soft x-ray emission, measured with the vacuum photodiode and a 0.8 µm thick Al filter, is shown in fig.3. The laser amplification at 46.9 nm generates a short and high intensity peak appearing at 31 ns from the starting of the current pulse. This peak can be distinguished from the long lasting signal due to the spontaneous radiation emitted during the discharge by the high temperature plasma and having duration of about 100 ns. In the experimental conditions of fig.3 the laser emission has duration of about 2.2 ns at FWHM. However the pulse duration depends on the experimental discharge conditions. For example, in fig.4 the pulse duration has been analyzed as a function of the initial gas pressure. It can been seen that it reaches a maximum value of 2.2 ns at 330 mTorr and that it decreases up to 1.8 and 1.6 ns respectively at 250 and at 420 mTorr. The laser beam divergence has been roughly estimated by putting a slit on the photodiode and moving it orthogonal to the direction of propagation of the x-rays. By making several shots we have estimated a divergence of 4 mrad. This value is comparable to that also reported by other groups.

Using the same excitation current pulse and the same capillary diameter no indication of laser amplification has been observed using polyacetal capillaries instead of alumina in the pressure interval between 100 and 1000 mTorr. In spite of the fact that laser amplification has been previously reported in polyacetal capillaries [7], in our case we believe that the lack of amplification is an effect of the much slower current rise time, which allows a bigger amount of material to be ablated from the capillary walls before the laser emission.

In order to optimize the laser output intensity in alumina capillaries, the gain of the laser line has been studied as a function of the initial gas pressure, by changing the plasma column length from 10 up to 20 cm. As it can be seen in fig.5, the gain strongly changes from one pressure to another, and it jumps from 0.7 cm^{-1} at 250 mTorr up to 1 cm^{-1} at 330 mTorr.

In the same figure, it can be clearly observed the transition at 14 and 16 cm respectively at 330 and 250 mTorr between the region of the exponential increase of the laser intensity from that of saturation.

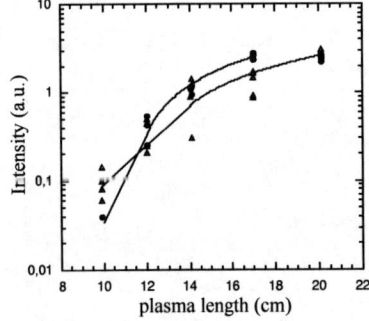

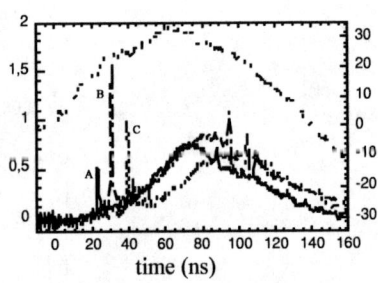

Fig.5. Intensity of the laser line at two different initial gas pressures: 330 mTorr (full circles) and 250 mTorr (triangles) as a function of the capillary length. The gain resulted to be respectively of 1 and 0.7 cm^{-1}.

Fig.6. Soft x-ray laser emission obtained in 20-cm long capillary channel initially filled with 0.3 Torr of Ar and the current pulse. The measurements were obtained in a (A) 2.4-mm (B) 3.2-mm and a (C) 4-mm in diameter alumina capillary channel.

However, in spite of the different values of the gain the laser intensity reaches quite comparable values at the two pressures in the saturation regime. With a 20-cm long plasma column the laser output energy has been estimated to be 5 µJ. This value is much smaller than that has been obtained (about few hundreds of micro-joules) in recent Rocca's experiments [2] conducted in quite similar experimental conditions. Also in this case we think that the reduced laser output energy can be an effect of the bigger current rise time. So, investigations of the laser output energy, as a function of the current slope, will be performed in the near future. The laser pulse energy doesn't change significantly with the capillary diameter, from 2.4 to 4.0 mm (fig.6) and it reaches its maximum intensity just at 3.2 mm. In fig.6, it can be observed the interesting effect of the scaling of the time of the laser emission with the capillary diameter. Laser line is emitted respectively at 20, 31 and 40 ns from the starting of the current pulse respectively with 2.4, 3.2 and 4-mm in diameter capillary channels. However it should be noted the short time necessary to achieve the population inversion inside the plasma column, which is well below the time of the maximum of the current pulse. This behaviour let us suppose that the laser amplification can be reached with much smaller current pulse amplitudes by properly matching the capillary diameter with the initial gas pressure.

The near field images of the laser beam in 20-cm long alumina capillary channels has been analyzed as a function of the initial Ar gas pressure. The results of the measurements are shown in fig.7. The shape of the laser beam changes with the pressure from an annular one at lower pressures to a single peaked one at higher pressures. Correspondingly the laser beam dimension scales from 600 µm up to 250 µm. These values are two times bigger respect to that reported in similar experiments by Rocca [8]. Measurements of the far field structure and dimension of the laser will clarify if the large near field image, obtained in our experiment, has to be attributed to a bigger gain region or to an increased refraction of the beam.

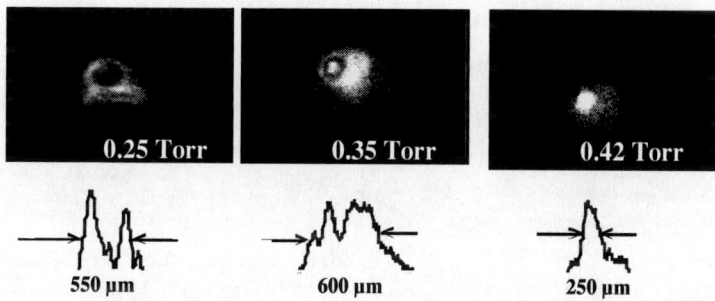

Fig.7. Near field images of the laser beam measured at different initial gas pressures. The measurements are obtained using a 30-kA current pulse through a 20-cm long, 3-mm in diameter alumina capillary channel.

CONCLUSIONS

In the present paper we have reported on the realization and the characterization of a capillary discharge soft x-ray laser operating at 46.9 nm in Ne-like Ar. Using a relatively long current pulse having a half cycle duration of about 150 ns and a peak value ranging from 27 up to 32 kA, we have measured a gain changing from 0.7 up to 1 cm^{-1} at the initial gas pressure respectively of 250 and 330 mTorr. We have clearly achieved the saturation regime, which leads us to an estimated energy of about 5 µJ in a 20-cm long capillary. All the measurements, which we have performed, show a clear evidence of the laser amplification in alumina capillary channels also using different capillary diameters. On the other side, using the same excitation current pulse no indication of lasing was observed in polyacetal capillaries.

Near field images of the laser beam has also been obtained. The shape of the laser beam changes from an anular to a single peaked shaped respectively at lower and at higher pressures. The laser beam dimension so ranges from 600 to 250 µm, giving a maximum peak spectral brigthness of 10^{21}ph/sec^{-1}mm^{-2}.

ACKNOWLEDGMENTS

We want to acknowledge the help of the technical staff of the Physics Department of University of L'Aquila for the development of the apparatus and particularly O. Consorte, A. Gaudieri, W. Galli and F. Del Grande.

This work is supported by the Italian National Institute of Nuclear Physics and in part by the Italian National Institute of Matter Physics.

S.V. Kukhlevsky thanks the Hungarian Scientific Research Foundation (OTKA, Contract T 026644). J. Kaiser acknowledges the Ministry of Education of Czech Republic (Grants J22/98:26L100002), the Czech Grant Agency (Grant GACR202/02/P113) and the Hungarian Ministry of Education.

REFERENCES

1. J.J. Rocca, V. Shlyaptsev, F.G. Tomasel, O.D. Cortàzar, D. Hartshorn, and J.L.A. Chilla, Phys. Rev. Lett., **73**, 2192 (1994).
2. B.R. Benware, C.D. Macchietto, C.H. Moreno, and J.J. Rocca, Phys. Rev. Lett. **81**, 1998 (1998).
3. A. Ben-Kish, M. Shuker, R.A. Nemirowsky, A. Fisher, A. Ron, and J.L. Schwob, Phys. Rev. Lett., **87**, 1 (2001).
4. Ghota Niimi, Yasushi Hayashi, Mitsuo Nakajima, Masato Watanabe, Akitoshi Okino, Kazuhiko Horioka and Eiki Hotta, J. of Phys. D: Appl. Phys. **34**, 1 (2001).
5. G.Tomassetti, A.Ritucci, A.Reale, L.Palladino, L.Reale, S.V.Kukhlevsky, F.Flora, L.Mezi, J.Kaiser, A.Faenov and T.Pikuz, Eur. Phys. J. D, **19**, 73-77 (2002).
6. G.Tomassetti, A.Ritucci, L.Palladino, L.Reale, O. Consorte, S.V.Kukhlevsky, I.Zs. Kozma, F.Flora, L.Mezi, J.Kaiser, O.Samek, M.Liska,Czech J. of Phys., **52**, 1 (2002).
7. C.H.Moreno, M.C. Marconi, V.N. Shlyaptev,B.R. Benware, C.D.Macchietto, J.L.A. Chilla, and J.J. Rocca, A.L. Osterheld, Phys. Rev.A, 58, 1509 (1998).

Extremely compact capillary discharge-based soft x-ray laser development and application to dense plasma diagnostics

J.J.Rocca, B.Luther, M.C.Marconi, T.Whiteaker, D.A.Braley,
J.Filevich, E.C.Hammarsten, A.Rahman, B.T. Szapiro, Y.Wang,
E.Jankowska, and M.Grisham

Department of Electrical and Computer Engineering. Colorado State University
Fort Collins, Colorado 80523

V.N. Shlyaptsev

University of California Davis

S. Moon

Lawrence Livermore National Laboratory

Abstract. We give an overview of recent capillary discharge-driven soft x-ray laser development experiments and applications at Colorado State University. We report the demonstration of the first desktop size soft x-ray laser, a capillary discharge Ne-like Ar soft x-ray laser that was measured to emit laser pulses with energy up to 10 μJ at 46.9 nm. In relation to the development of capillary discharge lasers at shorter wavelengths, spectra of the highly ionized cadmium plasmas identified strong emission from the $4d^1S_0$- $4p^1P_1$ laser transition of Ni-like Cd at 13.16 nm. We have also demonstrated optical guiding of intense laser pulses in the plasmas of an Ar capillary discharge, a result that is of interest for the development of efficient longitudinally pumped collisional soft x –ray lasers in a gaseous media. In term of applications, we summarize our continued progress in establishing capillary discharge lasers as a compact soft x-ray source of coherent radiation for dense plasma diagnostics. The combination of a tabletop 46.9 nm capillary discharge Ne-like Ar laser and a Mach-Zehnder interferometer based on diffraction gratings was used to study two-dimensional hydrodynamic effects in laser-created plasmas. The short wavelength and high brightness of the capillary discharge soft x-ray laser allowed us to map the density profile up to ~10^{21} cm^{-3}, close to the critical density. The interferometry of laser-created plasmas created at moderate irradiation intensity (0.1 -7 x 10^{12} W cm^{-2}) shows the development of a concave electron density profile that differ significantly from that expected for a classical expansion. Measurements involving line-focus and spot-focus laser created plasmas and hydrodynamic model simulations confirm this two-dimensional effect is essentially a universal effect that exists over a wide space of plasma parameters. In a separate experiments we have used the same soft x-ray laser interferometry tools to study the initial phase of the current-driven explosion of thin Al wires which are of current interest for the generation of x-ray radiation at large pulse power facilities. These measurements demonstrate the use of portable soft x-ray laser interferometry as a high-resolution tool for the study of high-density plasmas and for the validation of hydrodynamic codes.

DEMONSTRATION OF A DESKTOP SOFT X-RAY LASER

We have successfully operated a desktop/laptop version of the 46.9 nm Ne-like Ar capillary discharge laser that is the most compact soft x-ray laser demonstrated to date.

CP641, *X-Ray Lasers 2002: 8th International Conference on X-Ray Lasers,* edited by J. J. Rocca et al.
© 2002 American Institute of Physics 0-7354-0096-2/02/$19.00

The laser, which can easily fit on top of a very small desk, is illustrated in Fig.1. The body of the laser has a size of ~ 40x40x14 cm^3 excluding the turbo-molecular pump. For the tests reported herein the laser was operated at repetition rates of 0.5 to 1 Hz. Figure 1b shows a spectrum of the laser output in which the 46.9 nm laser line is observed to completely dominate all other lines. Figure 2c shows the temporal evolution of the radiation axially emitted by the discharge monitored with a vacuum photodiode. The laser beam was attenuated to avoid saturation of the photodiode. The intensity of the laser pulse is observed to overwhelm the spectrally integrated spontaneous emission of hundreds of lines emitted by the hot dense plasma column. A laser output pulse energy of ~ 10 μJ was measured for 21 cm long ceramic capillaries

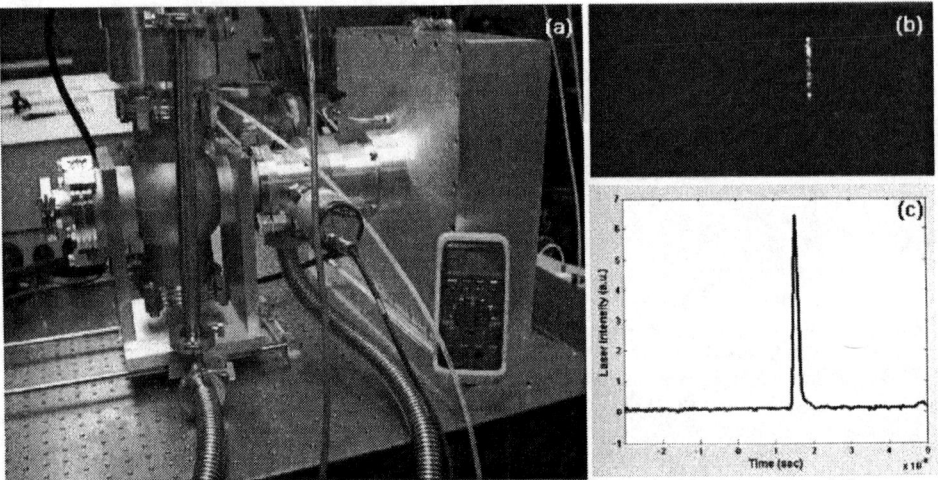

Fig 1. a). Size of desk-top Ne-like Ar laser compared with a hand-held multimeter. b) Axial spectra of the capillary discharge plasma in which only the amplified 46.9 nm laser line is observed. c) Temporal characteristics of the laser pulse. The laser pulsewidth is ~ 1.5ns.

filled with 250 mTorr of Ar excited by discharge pulses with 18.9 kA peak amplitude. Laser emission is observed even with excitation current pulses that reach less than 10 kA at the time of lasing. This very portable laser unit is likely to open new applications for soft x-ray lasers.

OBSERVATION OF THE 13.2 nm LINE OF NICKEL-LIKE Cd IN A CAPILLARY DISCHARGE PLASMA

We are also investigating the possibility of extending capillary discharge lasers to shorter wavelengths. In relation to this goal we have observed strong emission at 13.16 nm from the $4d^1S_0$ -$4p^1P_1$ laser transition of Ni-like Cd in a cadmium vapor plasma column excited by a high-current capillary discharge. The pulse generator used in these studies is capable of providing current pulses with < 15 ns risetime and peak currents up to 200 kA. The dynamics of the plasma was studied using time-resolved soft x-ray pinhole images and hydrodynamic simulations. The pinhole images show that the plasma column maintains a good symmetry up to the time of maximum compression. Spectroscopy of the plasma column in the 12.7-18.4 nm region resulted in the classification of fifty-five $3d^9 4p$-$3d^9 4d$ and $3d^9 4d$-$3d^9 4f$ CdXXI lines. These results are discussed in in the paper by Rahman et al. in these proceedings [1].

DEMONSTRATION OF A PLASMA WAVEGUIDE IN A FAST ARGON CAPILLARY DISCHARGE

Fast capillary discharges also have the potential to impact the development of efficient longitudinally excited laser-pumped collisional lasers by providing long plasma channels that can be heated with intense optical laser pulses. Ablative capillary discharges have been used to create plasma waveguides for laser-pumped recombination [2] and collisional excitation lasers [3]. In recent experiments we have demonstrated that fast capillary discharges of the type used for discharge-driven soft x-ray lasers can form plasma waveguides of long length (> 10 cm) and good axial uniformity. The guiding ability of an Ar plasma column excited by a fast 20 kA current pulse through a 3.2 mm capillary is illustrated in Fig. 2. The intensity of a cw laser beam (λ= 800 nm) propagating through the 11 cm long Ar capillary plasma is observed to increase significantly about 35 ns after the beginning of the current pulse. This is the result of guiding of the beam by the concave electron density profile that forms as a current-driven shockwave originating near the capillary wall rapidly moves towards the axis. Later in time, when the shockwave collides on axis, the transmission falls to nearly zero due to strong due to strong beam refraction and absorption. Similar results were obtained propagating 40 mJ laser pulses of 2.5 ps duration. Imaging of the output of the plasma column shows different waveguide modes whose structure

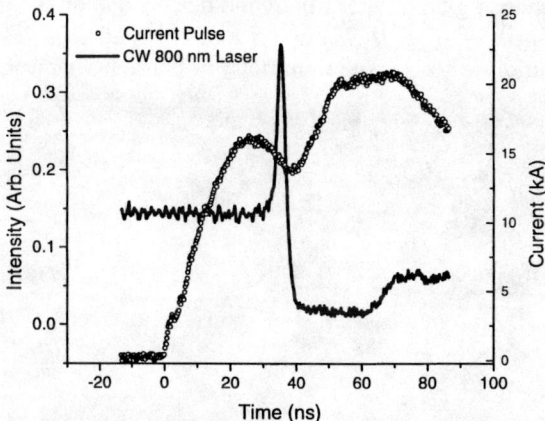

Figure 2. Measured variation of the transmission of cw optical laser probe beam through a 11 cm long Ar capillary discharge plasma column as a function of time relative to the discharge current pulse.

changes rapidly as the plasma column compresses sustaining guided beams of progressively smaller spot diameter. Guided propagation of beams with ~ 50 μm FWHM diameter corresponding to the lowest waveguide mode is observed shortly before the shockwave collides on axis. These results are discussed in more detail in the paper by Luther et al. in these proceedings.

DENSE PLASMA DIAGNOSTICS WITH A TABLE-TOP CAPILLARY DISCHARGE SOFT X-RAY LASER
Soft X-Ray Laser Interferometry Uncovers Two-Dimensional Effects In Laser-Created Plasmas

Soft x-ray lasers allow the probing of large-scale high-density plasmas beyond the limits that plasma refraction and absorption impose to visible and ultraviolet laser probes. Da Silva et al. conducted the first soft x-ray laser interferometry experiments using a laboratory-size 15.5 nm Ne-like Y laser pumped by the Nova laser in conjunction with a Mach-Zehnder interferometer based on thin-film beam splitters [4]. Capillary discharge pumped soft x-ray lasers, with an extremely high brightness that is similar or higher than that of their laboratory–size predecessors and a much higher repetition rate, offer the opportunity to develop portable soft x-ray tools for the diagnostics of a large variety of dense plasmas. Our group has realized the first demonstrations of soft x-ray interferometry using a table-top laser [5-7]. Experiments with the capillary discharge 46.9 nm Ne-like Ar laser were conducted using either a wavefront-division interferometer based on Lloyd's mirror [5-6], or an amplitude division interferometer in which diffraction gratings are used as beam splitters [7]. While simplicity is an advantage of the Lloyd's mirror interferometer, the diffraction grating interferometer (DGI) has the advantage of producing higher quality interferograms. The DGI can also be designed to operate at different soft x-ray wavelengths by choosing the proper ruling and blaze angle on the gratings. Recently a DGI was combined with a 14.7 nm Ni-like Pd transient laser to demonstrate picosecond-resolution soft x-ray interferometry of dense laser-plasmas [8].

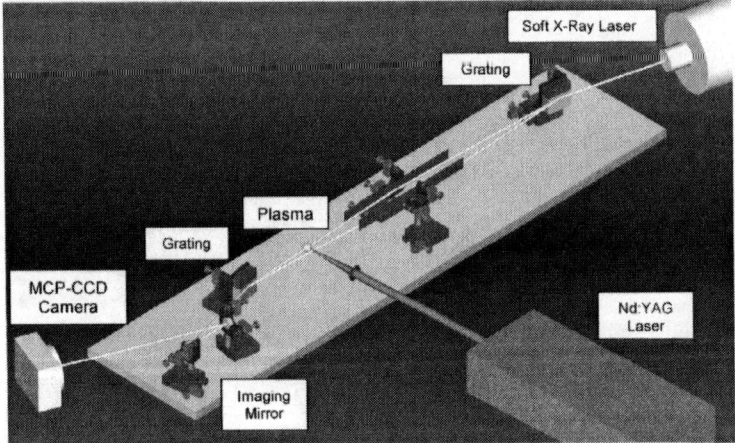

Figure 3. Schematic representation of the experimental set up. The interferometer was at ~2m from the exit of the soft x-ray laser. The detector was placed at ~7m from the interferometer.

The dynamics of laser-created plasmas has been extensively investigated and remains of significant current interest. Measurements of laser-plasmas created by irradiated solid targets at moderate irradiation intensities normally show plasma profiles that are well described by adjusting the angle of expansion in a one-dimensional (1D) simulations, often described as 1½ D simulations. At higher irradiation intensities (> 10^{14} W/cm^2) the radiation pressure can be sufficiently large respect to the plasma pressure to significantly alter the electron density profile by excluding the plasma from regions of otherwise high density [9]. In recent soft x-ray laser interferometry studies of seemingly typical laser-plasmas created at relatively low irradiation intensity (0.1 – 7 x 10^{12} W cm^{-2}), where the ponderomotive force and other high intensity effects are negligible, we clearly observed plasma density distributions that differ significantly from the expected classical conical expansion. To study these unexpected phenomena we conducted extensive series of experiments in line-focus and spot focus plasmas. In this paper we focus our attention on the spot-focus experiments conducted at irradiation intensities of ~ 7×10^{12} W cm^{-2}. The line-focus results at lower intensity (1×10^{11} W cm^{-2}) are reviewed in the paper by Filevich et al. in these proceedings [10].

The soft x-ray laser interferometry system used in the experiments is schematically shown in Fig.3. It consists of a tabletop capillary discharge laser emitting ~ 0.1 mJ pulses of ~ 1.2 ns duration at 46.9 nm, the DGI, and a Nd-YAG laser that is focused onto a polished Cu target to generate the plasmas that are the subject of this study. The interferometer consists of a modified Mach-Zehnder configuration of rhomboidal shape that uses diffraction gratings as beam splitters and elongated grazing incidence mirrors. The zero and first diffracted orders from the first grating are used to form the two arms of the interferometer. To ensure a similar intensity in both arms, the gratings were ruled with a blaze angle that produces an equal distribution of energy between the zero and first order for an angle of incidence of 79 degrees. The elongated gold coated mirrors redirect the beam towards the second grating where they are recombined and the interference pattern is generated. Advantages of the DGI scheme over other amplitude-division soft x-ray interferometers based on thin film beam splitter [4,11] include a higher throughput (~ 6% percent per arm) and a significantly increased resistance of the beam splitters to plasma debris. A flat relay mirror and a spherical imaging mirror of 15 cm focal length, both coated with Si/Sc multilayers with reflectivity of ~40% at 46.9 nm [12] , were used to image the plasma onto a gated CCD/MCP detector setup with a magnification of 51.2X. To make possible the alignment of the interferometer with a 824 nm semiconductor laser, the gratings were designed to have two sets of ruled lines with densities of 300 lines/mm for the soft x-ray laser beam and 17.06 lines/mm for the IR laser beam respectively. The alignment diode was selected to have a coherence length similar to that of the soft x-ray laser (~ 200 μm). This concurrent checking of both the alignment and the optical path length results in the setup only needing minor additional adjustments under vacuum. In the elongated plasma columns generated by the capillary discharge, gain guiding and strong refractive anti-guiding results in the rapid buildup of spatial coherence [13]. This intrinsic mode selection mechanism makes it possible to achieve a very high degree of spatial coherence, which in turn allows for the generation of high visibility interferograms. The interferograms obtained have a fringe visibility of ~ 0.5 over the entire 400 x 500 μm^2 field of view.

In the present experiments the plasmas were formed from a copper target using 0.62 J pulses of 13 ns FWHM duration from a Nd:YAG laser operating at its fundamental wavelength of 1.06 μm. The laser was focused onto a ~ 30 μm diameter spot with a 15 cm focal length aspheric lens. Detailed series of interferograms were obtained for plasmas created by firing the heating laser either onto a fresh polished copper target or into a crater formed by previous shots. In the latter case the plasmas display a significant density increase at large distances from the target surface. This results from the fact that the crater constrains the plasma's lateral expansion and guides it's motion into the direction normal to the target. In all cases the measurements unveiled the formation of an "inverted center" electron density profile with a density minimum on axis , as we also observed in the line-focus experiments at lower irradiation intensities [10]. The plasma evolution was mapped by taking interferograms at different times relative to the initiation of the heating laser pulse (Fig. 4).

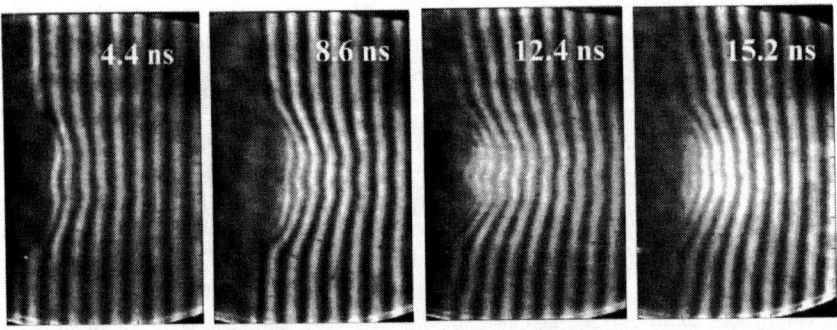

Figure 4. Interferograms of plasmas generated focusing a Nd-YAG laser into a spot ~ 30 μm, and the irradiation intensity was 7 10^{12} W cm^{-2}. The interferograms correspond to plasmas generated firing the laser for a second time in the same target location. The times delays are measured respect to the beginning of the laser pulse. The heating laser is incident from the left.

In the earliest interferogram obtained at 4.4 ns after the beginning of the laser pulse, a slightly concave shape is already visible in the fringes closest to the target. In this axisymmetric geometry such fringe pattern is indicative of a concave electron density profile with a significant minimum on axis. This density cavity becomes more pronounced as time progresses. At the time of the maximum laser intensity this central minimum in the density profile is observed to extend through practically the entire sub-critical region of the plasma. Figure 5 shows the evolution of the electron density distributions derived from Abel inversion of the interferograms of Fig.4. The formation of a concave electron density profile with pronounced plasma sidelobes and a density cavity on the irradiation axis is observed. At 10.5 ns the density of the sidelobes is measured to reach electron ~ 9 x 10^{20} cm^{-3} at a distance of 27 μm from the target. This density corresponds to 90% of the critical density of this λ=1.06 μm pump laser.

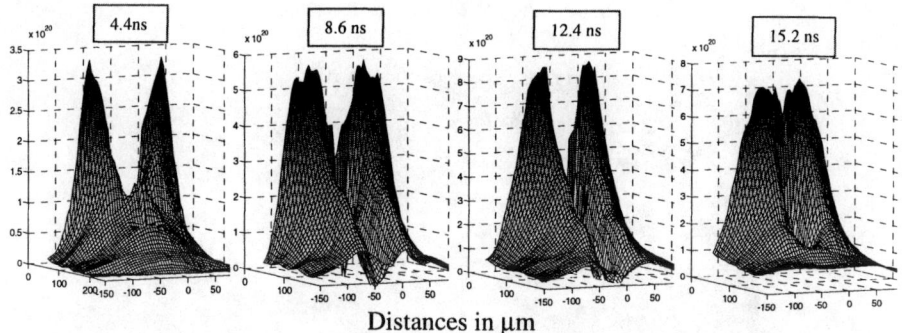

Distances in μm

Figure 5. Computed electron density corresponding to the interferograms of Fig. 4

This Observed "inverted" density profile is neither a result of the ponderomotive force nor the effects of laser radiation refraction or filamentation. Hydrodynamic simulations show that the observed two-dimensional profiles are caused instead by the generation of cold plasma sidelobes outside the laser-irradiated target area and by the subsequent establishment of pressure balance between the sidelobes and the hotter central plasma. The sidelobes result from the build-up of cold material generated by an increased ablated area caused mainly by XUV plasma radiation. Moreover, the simulations indicated that this is essentially a universal effect that should be observed over a wide range of plasma parameters. The simulations and the line-focus experiment results are discussed in detail in another paper in these proceedings [10].

Soft X-Ray Laser Interferometry/Shadowgraphy of Current-Driven Exploding Wires

In another application of capillary discharge soft x-ray laser to dense plasma diagnostics we have used the soft x-ray interferometry setup discussed in the previous section to measure the evolution of thin Al exploding wires. The understanding of the initial stage of plasma formation in current-driven exploding wires is of significant current interest in relation to the efficient generation of large pulses of incoherent x-rays by the implosion of arrays of thin wires driven by large Z-pinch machines (nearly 2 MJ of x-rays have been generated in pulses of a few ns duration with Sandia's Z generator) [14]. As an example Fig. 6 shows a soft x-ray laser interferogram corresponding to the explosion of a 25 μm diameter Al wire, acquired 83 ns after the initiation of the current pulse that heats the wire. Deflection of the interference fringes are observed very close to the wire core. Photoionization of the large density of atoms in the cold wire core absorbs the x-ray laser probe, effectively creating a shadowgram in that region. The acquisition of sequences of shots yields the velocity of expansion of the wire core, and information on the development of instabilities. The measurement of the evolution of the electron density and vapor density distributions require the simultaneous analysis of the fringe deflection and absorption data in each interferogram. A detailed discussion of this experiment can be found in the paper by Jankowska et al [15].

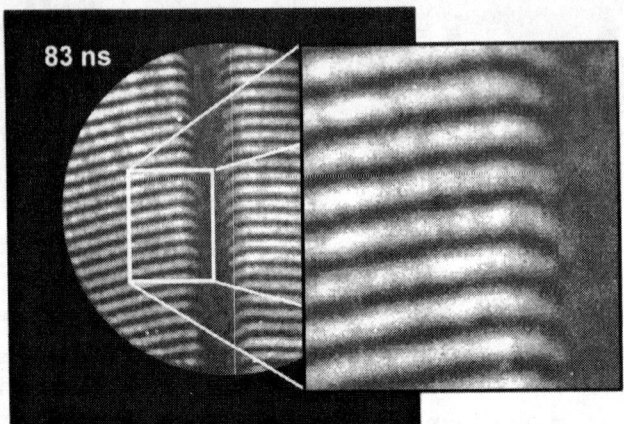

Figure 6. Soft x-ray laser interferogram/shadowgram of the explosion of a 25μm diameter Al wire acquired 83 ns after the initiation of the heating current pulse (4.7 kA peak amplitude, 30A/ns current increase rate).

ACKNOWLEDGMENTS

The soft x-ray laser development work was supported by the National Science Foundation and DARPA The dense plasma interferometry was supported by U.S. Department of Energy grant No. DE-FG03-98DP00208. We also gratefully acknowledge the support of the W.M. Keck Foundation

REFERENCES

1. A. Rahman et al. Proceedings on 8th Int. Conf. on X-Ray Lasers, May 2002, Aspen, Colorado, USA.
2. A. Janulewicz, J. J. Rocca, F. Bortolotto, M. P. Kalachnikov, V. N. Shlyaptsev, W. Sandner, and P. V.Nickles, Physical Review A 63, 033803 (2001).
3. Avitzour, I. Gelzner, Y. Ping and S. Suckewer, Proceedings of the 8th Int. Conf. on X-Ray Lasers, May 2002, Aspen, CO, USA.
4. L.B. Da Silva. et al. Phys. Rev. Lett. **74**, 20, 3991 (1995).
5. J.J. Rocca, C.H. Moreno, M.C. Marconi, and K. Kanizay- Optics Letters **24** 420 (1999).
6. C.H. Moreno, M.C. Marconi, K. Kanizay, and J.J. Rocca Yu. A. Uspenskii and A. V. Vinogradov Yu. A. Pershin Phys. Rev E. **60,** 1, 911 (1999).
7. J. Filevich, K. Kanizay, M.C. Marconi, J.L.A. Chilla, and J.J. Rocca. Optics Lett. **25,** 356, (2000).
8. R. F. Smith, J. Dunn, J. Nilsen, V. N. Shlyaptsev, S. Moon, J. Filevich, J. J. Rocca, M. C. Marconi, J. R. Hunter, and T. W. Barbee, Jr **89**, Phys. Rev. Lett. **89** 065004-02 (2002).
9. D.T.Attwood, D.W.Sweeney, J.M. Auerbach and P.H.Y. Lee, Phys. Rev. Lett. **40**, 184, (1978).
10. J. Filevich et al. Proceedings of the 8th Int. Conf. on X-Ray Lasers, May 2002, Aspen, CO,USA.
11. A.S. Wan et. al, Phys. Rev. E, **55**, 6293, (1997).
12. Y.Liu, M.Seminario, F.Tomasel, J.J.Rocca and D.T.Attwood, Phys. Rev. A **63**, 033802 (2001).
13. Yu. A. Uspenskii, V. E. Lebashov, A. V. Vinogradov et al. Opt. Lett. **23**, 771, (1998).
14. C. Deeney, M.R.Douglas, R.B. Spielman, T.J.Nash,D.L. Peterson, P.L. Eplattenier, G.A. Chandler, J.F. Seamen, and K.W. Struve. Phys. Rev. Lett. **81**, 22, 4883 (1998).
15. E.Jankowska et al. Proceedings on 8th Int. Conf. on X-Ray Lasers, May 2002, Aspen, CO, USA.

Dynamics and Emission Characteristics of Xenon Capillary Discharge

P. Vrba[+], M. Vrbova[¶], N. A. Bobrova[&], P. V. Sasorov[&], C. Cachoncinlle[*],
J.M. Pouvesle[*], E. Robert[*], O.Sarroukh[*], T. Gonthiez[*], R. Viladrosa[*],
C. Fleurier[*]

[+]Institute of Plasma Physics, AS CR, CZ-18221 Prague 8, Czech Republic
[¶]Czech Technical University, Brehova, 7, CZ-115 19 Prague 1, Czech Republic
[&]Institute of Theoretical and Experimental Physics, Moscow, Russia
[*]GREMI-ESPEO, Université d'Orléans, BP 6744 Orleans Cedex 2, France

Abstract. Xenon filled fast capillary discharge is studied as a source of intense EUV radiation. Experiments are performed with GREMI apparatus. One-dimensional two temperatures MHD code is used to evaluate time and radial dependences of capillary plasma properties. Plasma thermodynamic and radiative characteristics are evaluated by means of steady state IONMIX code. Measured and estimated emission spectra are compared.

EXPERIMENTAL SET UP AND MEASUREMENTS

The performance of xenon filled capillary discharge system (Fig.1) has been studied to produce intense source of soft X-rays radiation [1]. Few kA currents having fast rise time (50 ns) was applied across the xenon filled alumina capillary to produce radiation mostly in EUV region from 10 to 16 nanometers. The electrical energy (2J) to initiate the discharge inside the capillary was provided by two capacitor banks configured in a Blumlein fashion pulse forming line. The dynamics of the plasma and emitted EUV radiation are characterized by employing filtered photodiode EUV pinhole camera and EUV spectrometers. A fast pinching effect is observed from the pinhole images even when the current is kept below 10 kA.

Variations of charging voltage have a strong influence on the current waveforms (see Fig. 2a, b) and time

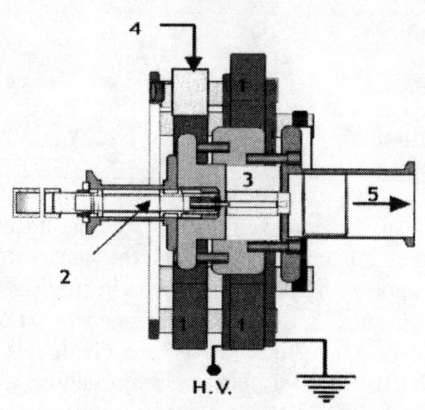

FIGURE 1. Experimental set up;
1 – Knob capacitors, 2 – Gas inlet, 3- Capillary,
4 – Fast switch, 5 - To Detection chamber

CP641, X-Ray Lasers 2002: 8th International Conference on X-Ray Lasers, edited by J. J. Rocca et al.
© 2002 American Institute of Physics 0-7354-0096-2/02/$19.00

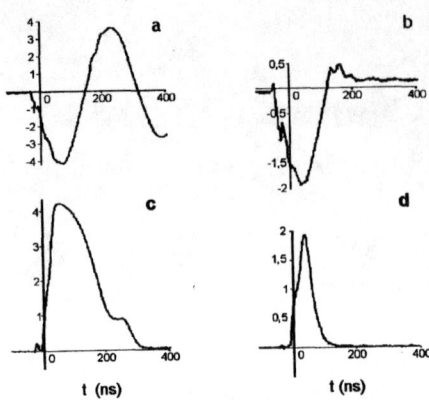

FIGURE 2. Measured electric currents (a,b) and EUV radiation intensities (c,d); a, c correspond to charging voltage $U_0 = 28$ kV and b, d to 12 kV

dependences of EUV radiation intensity (Fig. 2c, d). Also the emission spectra are remarkably changing during every current pulse (see Fig. 7) and are strongly influenced by the variation of filling pressure and initial charging voltage.

MAGNETOHYDRODYNAMIC SIMULATIONS

One dimensional two temperatures MHD code NPINCH [2] is used to evaluate time and radial dependences of capillary plasma characteristics. Capillary radius $R_0 = 0.5$ mm and non-ablating material (alumina) of the wall are judged. The gas flow through the small diameter capillary imposes a strong pressure gradient. The pressure inside the channel varies from 0 to 1 mbar. We present the modelling for the two limit pressures $p_0 = 0.2$ and 1 mbar (corresponding initial mass density $\rho_0 = (1.1$ and $5.3). \, 10^{-6} \, \text{g.cm}^{-3}$) and two charging voltages $U_0 = 12$ and 28 kV of the 16 nF capacitor. The corresponding measured waveforms of electrical current $I(t)$ are shown as Figs. 2a, 2b. For the purpose of the NPINCH code the observed waveforms are approximated by:

$$I(t) = -I_0 \sin(\frac{\pi t}{2t_0})\exp(\frac{-t}{t_1}), \tag{1}$$

where I_0, t_0 and t_1 are the fitted parameters. We present here the results for four representative runs (cases A,B,C and D) with input parameters summarized in Table 1.

TABLE 1. Input Parameters for NPINCH

Case	Voltage kV	Current kA	Pressure mbar	Density g/cm³	N_e cm⁻³	I_0 kA	t_0 ns	t_1 ns
A	28	6	1.0	5.2630e-6	2.4 10¹⁶	6.53	85	986.4
B	28	6	0.2	1.0526e-6	4.8 10¹⁵	6.53	85	986.4
C	12	2.6	1.0	5.2630e-6	2.4 10¹⁶	4.80	85	123.1
D	12	2.6	0.2	1.0526e-6	4.8 10¹⁵	4.80	85	123.1

The simulations show that feeble magnetic plasma compression takes place in all these cases (see Figs. 3). Plasma is initially detached from the wall and the pinch time varies from 37 to 49 ns. The highest compression ratio 12.8 is achieved in the case B (see Table 2). In the period of pinch decay, plasma is expanded and impacts on the wall. Being reflected back converging and diverging shock waves are created. The highest peak of the plasma electron density due to first fast plasma compression is also achieved in the case B. The increase of the local plasma electron temperature is caused by the current density increase (e.i. by plasma contraction and/or due to total current increase). Space-time profiles of electron temperature in all the cases are rather smooth and the peak values are achieved in the region of the current maximum (Fig.4).

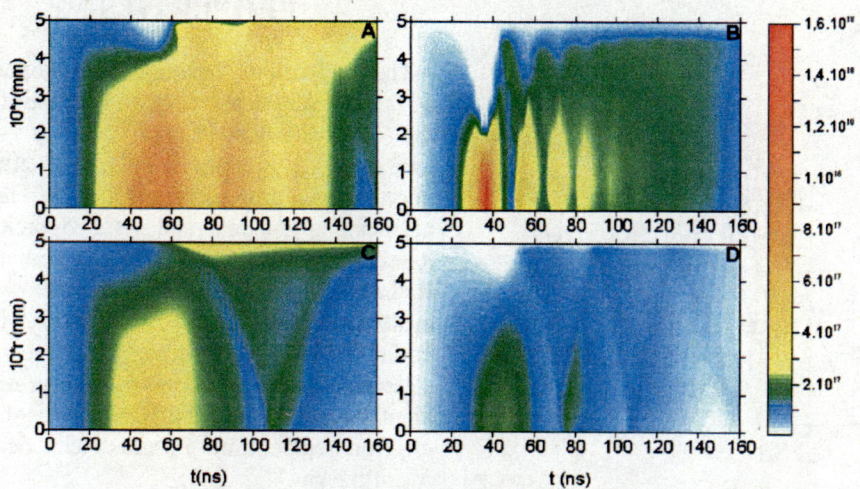

FIGURE 3. Radial-time dependences of plasma electron density N_e (in cm^{-3})
Comparison of the results for cases A, B, C, D, see Table 1

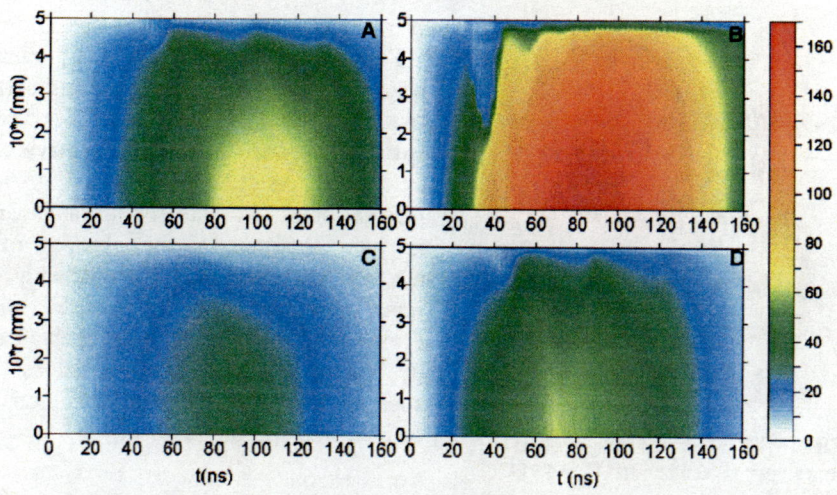

FIGURE 4. Radial-time dependences of plasma electron temperature T_e (in eV)
Comparison of the results for cases A, B, C, D, see Table 1

TABLE 2. Summary of Representative Calculated Values

Case	Run #	Pinch time t_p ns	Compression ratio ρ/ρ_0	Pinch T_e eV	Max. T_e eV	Pinch N_e cm^{-3}	Max. N_e cm^{-3}
A	1036	44	2.35	33.4	66.0	$8.10 \cdot 10^{17}$	$8.90 \cdot 10^{17}$
B	1037	37	12.8	92.7	167.0	$1.82 \cdot 10^{18}$	-
C	1038	49	1.57	22.7	36.5	$4.16 \cdot 10^{17}$	$4.22 \cdot 10^{17}$
D	1039	37	2.40	39.2	61.8	$2.03 \cdot 10^{17}$	-

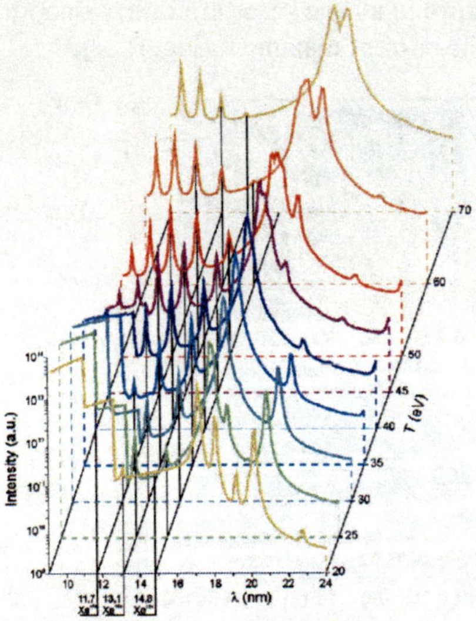

THERMODYNAMIC AND RADIATIVE PLASMA PROPERTIES

Thermodynamic and radiative properties of the hot xenon plasma are evaluated using the IONMIX code [3]. Steady state ionization fractions and excitation populations are determined by detail balancing arguments and rate coefficients, based on hydrogenic ion approximation. Radiative emission coefficients are evaluated considering contributions from bound-bound, bound-free, free-free and electron scattering processes. The principle input parameters for the IONMIX code are electron temperature T_e and nuclei density N_{at} of the gas.

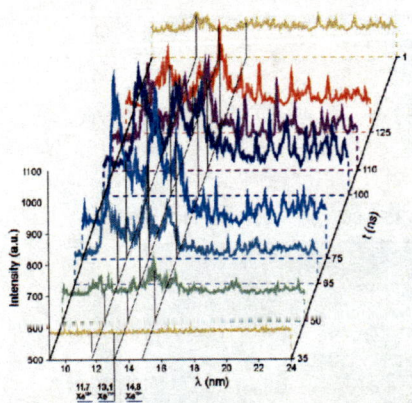

FIGURE 6. Evaluated spectral emission intensity; for $N_{at} = 3 \cdot 10^{17} \text{cm}^{-3}$ and various T_e

Ionization State

To get dependence of the ionization fraction on the plasma temperature repetitive runs of the IONMIX code for variable plasma temperature have been performed. As it is seen from Fig. 5, the plasma ionization state (or ion abundances) is very sensitive to the changes of the electron temperature. Consequently, the remarkable variations of emitted spectra during the shot (Fig. 7) are probably caused by plasma temperature variations. The spectral sensitivity to initial conditions may be interpreted in the same way.

FIGURE 7. Measured time-resolved spectra; $p_0 = 0.66$ mbar and $U_0 = 15$ kV

TABLE 3. Selected Spectral Lines of Xe Ions (Bohr-like Model)

Ion		Xe^{7+}	Xe^{8+}	Xe^{9+}	Xe^{10+}	Xe^{11+}	Xe^{12+}	Xe^{13+}	Xe^{14+}	Xe^{15+}	Xe^{16+}	Xe^{17+}
Φ_j	[eV]	112.3	170.8	201.7	232.6	263.5	294.4	325.3	358.3	389.6	420.9	452.2
n_0		5	4	4	4	4	4	4	4	4	4	4
λ_L	[nm]	36.1	20.2	17.1	14.9	13.1	11.7	10.6	9.6	8.8	8.1	7.6
λ_B	[nm]	59.9	37.1	31.4	27.3	24.1	21.5	19.5	17.7	16.3	15.1	14.0
λ_{Edge}	[nm]	11.04	7.25									

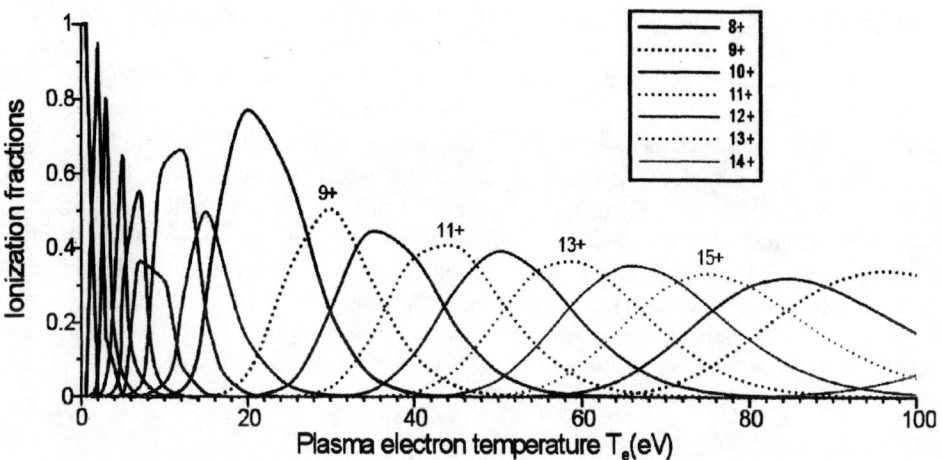

FIGURE 5. Xenon ionization fractions for initial atom concentration $N_{at} = 3.2 \ 10^{16} \ cm^{-3}$

Bohr-Like Model for Xenon Ions

The estimations of the radiative emission characteristics are based on the "Bohr-like model" of the energy levels for every xenon ion. The energy of ion with outermost electron residing in the shell n is given by

$$E_{n,j} = - \, \Phi_j \, (n_0/n)^2 \; ; \quad n > n_0 \, , \tag{2}$$

where n_0 represents the principal quantum number of the ground state and Φ_j is the ionization potential of the jth ion. The spectral lines are determined by energy level structures of the ions. The radiation wavelength λ_L corresponds to the transition between ground and first excited state (Lymann-like α) whereas λ_B is the Balmer-like α line. As the ionization potentials are only moderately increasing with the ionization state j, the wavelength corresponding to the same quantum transitions of the adjacent ions are only moderately decreasing with increasing j. The studied spectral lines that coincide with the investigated wavelength region are marked (see Table 3).

Spectral Emissivity

The spectrum of the radiation emitted by plasma consists of continuous and line parts. Our estimation of the plasma spectral emissivity $\eta(\lambda)$ is based on the Kirchhoff-Planck law: $\eta(\lambda) = k(\lambda) . w(\lambda)$, where $k(\lambda)$ is the spectral emission coefficient calculated by the IONMIX code and $w(\lambda)$ is the Planck formula describing the spectral energy density of the black-body radiation.

The spectral emissivities $\eta(\lambda)$, calculated for various plasma temperatures $T_e = 20 - 70$ eV and initial atom density $N_{at} = 3 \cdot 10^{17}$ cm^{-3} are summarized as Fig. 6. The range of the temperatures and the initial atom density are chosen according to the experiments and results of the NPINCH code. Lyman-like transitions (at wavelengths λ_L) are easily identified there. The three emission peaks at 11.7, 13.5 and 14.7 nm corresponds to Xe^{12+}, Xe^{11+} and Xe^{10+} ions. We suggest the three peaks observed in time resolved spectra (Fig. 7) to be interpreted as these three Lyman-like α transitions and the changes of their relative amplitudes to interpret as the changes of the concentrations of corresponding ions during the shot. Consequently, the highest concentration of Xe^{12+}, corresponding to the highest peak at 11.7 nm, is detected at $t_{exp} = 75$ ns. At this time the calculated plasma electron temperature (by means of the NPINCH code) is approximately equal to 50 eV.

CONCLUSIONS

MHD computer simulations proved that the plasma parameters are remarkably varied if the initial pressure and electrical current are modestly varied. If the peak of the current is changed in the range 2.6 – 6.3 kA and initial xenon pressure in the range 0.2 – 1 mbar, the pinch effect is relatively feeble and namely with higher initial pressures causes very limited heating. The plasma ionization state, as evaluated according to the IONMIX code, is very sensitive to plasma temperature variations. Time dependences of plasma temperature, estimated on the bases of spectral measurements and IONMIX emissivity calculations are in a good agreement with plasma temperature profile calculated according to NPINCH code.

ACKNOWLEDGMENTS

This research was done in the frame of collaboration of four research institutions being partially supported by IPP AV CR AVK 2043105/2752 *"X-ray Source Optimization"* and by MSM210000022 *"Laser System and Applications"*.

REFERENCES

1. Cachoncinlle C. et al.: *"Capillary Discharge Sources of Hard UV Radiation"*, Proc. of XXV ICPIG Nagoya, Japan 2001, vol. 4, 345.
2. Bobrova N.A., Bulanov S.V., Razinkova T.L., Sasorov P.V., *Plasma Physics Reports* **22** (1996), 387-402.
3. MacFarlane J.J., *Comput. Phys. Commun.* **56** (1989) 259-278.

EUV Emission Spectra and Gain in Polyacetal Capillary Discharge

M. Vrbova[*], P. Vrba[¶], A. Jančárek[*], N. A. Bobrova[&], P. V. Sasorov[&], J. Limpouch[*], L. Pina[*], L. Nadvornikova[*], A. Fojtik[*]

[*] Czech Technical University, Brehova 7, CZ-115 19 Prague 1, Czech Republic,
[¶] Institute of Plasma Physics, AS CR, CZ-18221 Prague 8, Czech Republic,
[&] Institute of Theoretical and Experimental Physics, Moscow, Russia

Abstract. Experimental and computer studies of polyacetal capillary discharge are reported. Time resolved spectra in the wavelength region 3 – 25 nm are measured. Space-time dependences of plasma electron density and temperature are calculated by means of MHD code. Time profiles of selected lithium-, helium- and hydrogen-like carbon and oxygen ion populations and time resolved spectra are evaluated by means of the FLY code. Gain factors for a capillary initially either evacuated or filled by polyacetal vapors are calculated.

EXPERIMENTAL SET UP

Our experimental set up is composed of two parts: the capillary discharge system and the diagnostic one (Fig.1). The energy is stored in 6 knob capacitors of 2.55 nF each

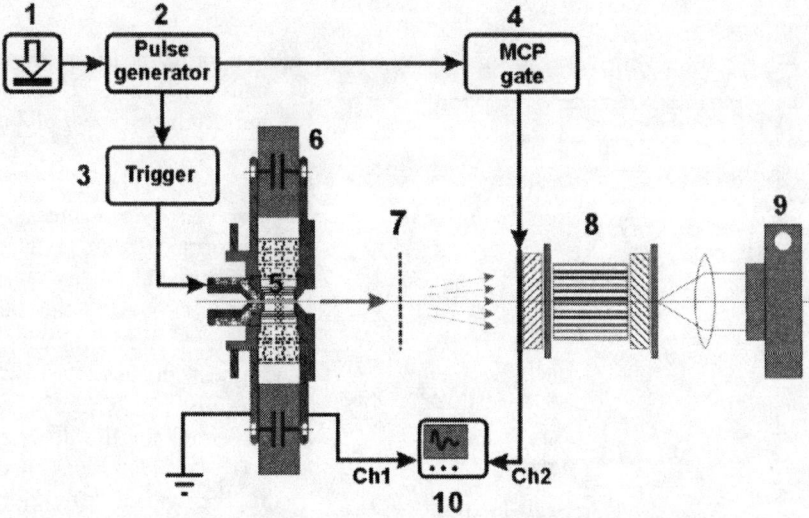

FIGURE 1. Experimental set up; 1- Manual switch, 2- Two channel generator, 3- HV trigger, 4- MCP electronics, 5- Capillary, 6- Capacitors, 7- Transmission grating, 8- MCP, 9 – CCD camera, 10 – Scope

CP641, *X-Ray Lasers 2002: 8th International Conference on X-Ray Lasers,* edited by J. J. Rocca et al.
© 2002 American Institute of Physics 0-7354-0096-2/02/$19.00

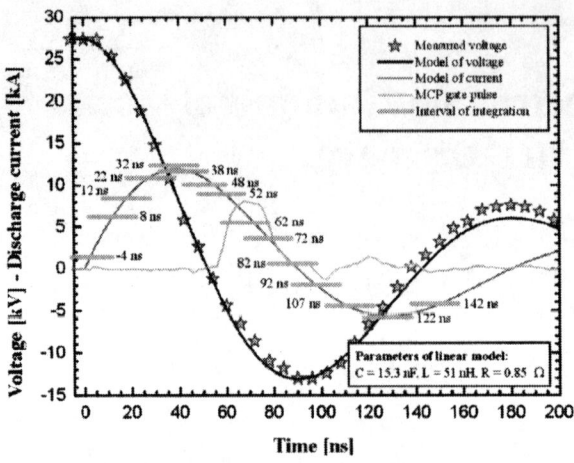

FIGURE 2. Measured voltage, evaluated current and measured gating pulse

mounted in parallel. Maximum applied voltage is 40 kV. The polyacetal capillaries with 1.1 mm diameters and the lengths 2.5 cm are used [1]. The measured voltage waveforms (Fig. 2) have character appropriate to a serial RLC circuit with quasi-period about 180 ns. The XUV spectrometer was built on the basis of a freestanding 1.4 μm period gold grating in normal incidence geometry without imaging optics. The distance between the end of the capillary and grating was 910 mm and between the grating and gated open MCP framing camera 1330 mm. Resulted spectrometer resolution is about 1 nm. The trigger pulse pair with variable delay initiates the main discharge and the opening of the MCP framing camera X-SHOT for time interval equal to 20 ns. The output visible signal is recorded by BI CCD camera with 16-bit resolution and high dynamic range (20 000:1), stored in the PC and further processed using 16-bit image data algorithm. The 0.75 μm thick Al foil is used for absolute spectrometer calibration.

Time Resolved Spectra

The MCP gating with variable time delay trigger allows us to measure the framed emission spectra (Fig. 3). The most intense spectral lines are ascribed to C^{5+}, C^{4+}, O^{6+} and O^{5+} (see Table 1). The time resolved spectra measurements performed with polyethylene capillary have been almost the same. However, the peaks at 15 nm were missing. The highest intensity of Lyman α line of C^{5+} (3.37 nm) is observed for the delay of 32 ns. The identification of lines in the wavelength range 6-10 nm is rather complicated due to our small spectral resolution.

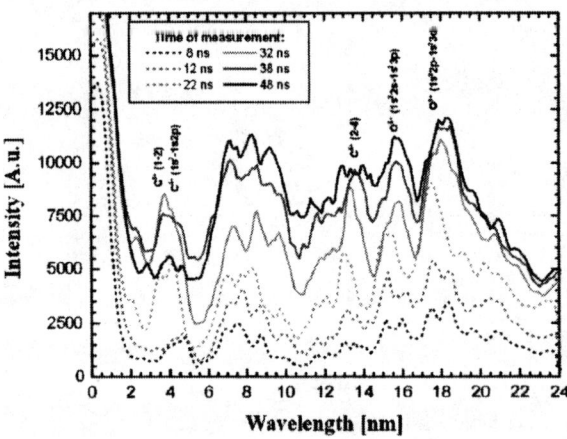

FIGURE 3. Instantaneous emission spectra; $U_0 = 27$ kV

140

TABLE 1. Identified spectral lines

Wavelength [nm]	Ion	Configuration	g_i	g_k	A_{ik}	Center of gate time interval [ns]
3,37	C^{5+}	1*-2*	2	8	$6,09.10^{11}$	12, 32
4,03	C^{4+}	$1s^2 - 1s2p$	1	3	$8,89.10^{11}$	12, 32, 42
13,50	C^{5+}	2*-4*	8	32	$1,09.10^{10}$	-3, 12, 32, 42
15,01	O^{5+}	$1s^2 2s - 1s^2 3p$	18	98	$2,62.10^{10}$	-3, 12, 32, 42
17,29	O^{5+}	$1s^2 2p - 1s^2 3d$	2	10	$8,78.10^{10}$	-3, 12, 32

MHD SIMULATIONS

Spatial and time dependences of plasma electron density and temperature in capillary discharge have been evaluated by means of NPINCH code under the one-dimensional two-temperature (ion and electron) one-fluid MHD approximation [2]. The model considers the material of the wall as dense cold neutral gas. For our polyacetal capillary atomic number $Z=7$, average atomic weight $A=14$, and initial density $\rho_0=1$ g/cm^3 for the wall material are used. The current pulse is considered in the form:

$$I(t) = I_0 \sin\left(\frac{\pi t}{2t_0}\right)\exp\left(-\frac{t}{t_1}\right), \quad (1)$$

where $I_0 = 17$ kA, $t_0 = 45$ ns, $t_1 = 120$ ns correspond to the current waveform measured with $I_{max} = 12$ kA, $t_{max} = 38.3$ ns (according to experimental results, Fig. 2).

The capillary is gradually filled by material ablated from the wall. The first peak of electron density $N_e = 2.10^{18}$ cm^{-3} on the axis at 20 ns (Fig. 4) reflects a quick compression of are plasma. The plasma electron temperature T_e is increasing due to increasing local current density. The peak value about $T_e = 130$ eV is reached at on the axis at $t_{max} = 40$ ns.

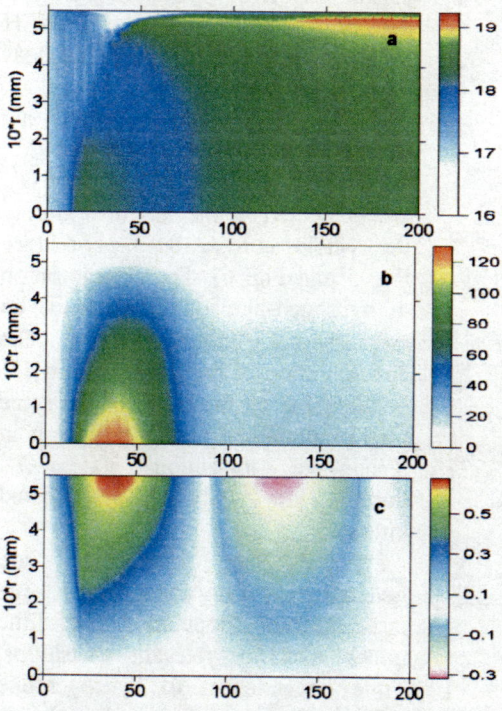

FIGURE 4. Results of computer simulation:
a - decimal logarithm of electron density in cm^{-3},
b - the electron temperature in eV,
c - electric current inside the cylinder of radius r in a.u.

Relatively quick cooling is found at the end of the first half period. In the second half period the plasma is relatively cold $T_e \leq 20$ eV and electron density $N_e > 2.10^{18}$ cm^{-3} on the axis grows due to wall ablation. The ion and electron temperatures are almost equal all the time and everywhere.

POPULATION DENSITIES AND TIME RESOLVED SPECTRA EVALUATION

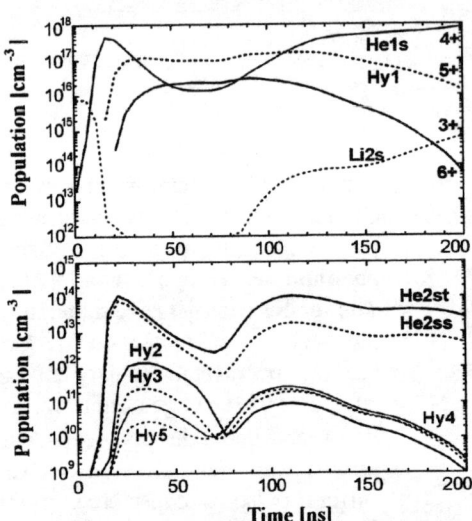

FIGURE 5. Population densities of the ground and excited states for carbon ions

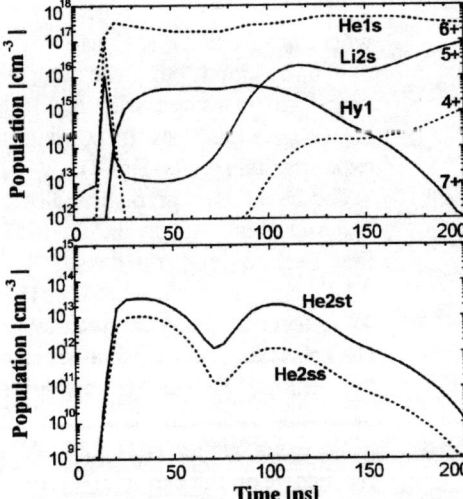

FIGURE 6. Population densities of the ground and excited states for oxygen ions

Time dependencies of ionization fractions and energy level populations for lithium-, helium- and hydrogen-like carbon, oxygen and aluminum ions have been evaluated by means of the computer code FLY [3]. The file containing temporal history of axial plasma electron temperature and density, resulting from NPINCH MHD simulations, has been used as an input for the FLY code.

The results of evaluations show that predominantly the of carbon ions C^{4+}, C^{5+} and oxygen ions O^{5+}, O^{6+} and O^{7+} are presented in the mixture during the whole current quasi-period.(see Fig. 5 and Fig. 6). The concentration of hydrogen-like ion C^{5+} prevails on the interval $25 \div 90$ ns. The concentration of fully stripped ion C^{6+} is negligible all the time. In the same time predominant part of oxygen is in helium-like ionization state O^{6+} during the whole investigated interval.

Spectral line intensities are generally proportional to the instantaneous populations of the upper levels of relevant transitions. Time dependences of hydrogen-like C^{5+} populations of first five levels (see Fig. 5) indicate double peaks and the inversion population (including Balmer α-transition) during a long time interval started from 81 ns. However, the absolute values of populations of the excited hydrogen

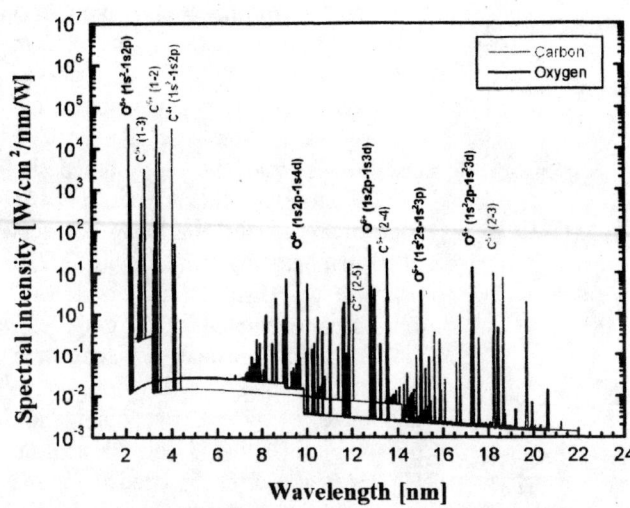

FIGURE 7. Instantaneous spectra at 31 ns emitted by hydrogen-, helium-, lithium-like Carbon and Oxygen ions

like-ions of carbon are two orders lower than that for helium-like carbon ions. The time resolved emission spectra may be expressed by means of subprogram FLYSPEC for any chosen time. On the leading edge of the current pulse (at 31ns) the most intensive spectral lines belong to C^{5+}, C^{4+} and O^{5+}, O^{6+} (Fig. 7).

In second half period of the current the evaluated spectra contain also strong lithium-like oxygen lines at 11.5 ÷ 12 ns and at 15 nm, 17.3 nm. Unfortunately, due to the principle limitations of the FLY code, no spectral lines corresponding to ions with lower than lithium-like ionization stages could be evaluated.

Gain Factor Evaluation

Time dependence of hydrogen-like populations of the carbon ion C^{5+} excited levels (Fig. 5) indicate the inversion population on a long time interval started from 81 ns. The gain g for the transition $3\rightarrow2$ (Balmer α-transition) of H-like ion C^{5+} may be evaluated according to well known formula [4]

$$g = 4,3.10^{-33} \lambda^3 A_{32} N_3 F_{inv} \quad [\text{cm}^{-1}], \tag{2}$$

where $F_{inv} = 1 - \dfrac{g_3 N_2}{g_2 N_3}$ is inversion factor, λ is resonant radiation wavelength in Angstroms, Einstein coefficient $A_{32} = 5.7 \ 10^{10} \ \text{s}^{-1}$, degeneracy factors $g_2 = 8$, $g_3 = 18$ and N_2, N_3 are lower and upper level populations in cm^{-3} calculated by the FLY code. The resulting peak value of the gain under current experimental situation ($R_0 = 0.55$ mm, $I_{max} = 15$ kA, $T_{1/4} = 45$ ns) $g = 0.0109 \ \text{cm}^{-1}$ and is achieved at the time about 111 ns, i.e. in the third quarter period of current pulse (Fig. 8). The calculated peak temperature in this case is 160 eV at 35 ns. At this period the plasma temperature decreases very quickly due to incoming cold material ablated from capillary wall. At the time of gain maximum the electron temperature has the value 20 eV and the electron density is about $N_e = 3. \ 10^{18} \ \text{cm}^{-3}$.

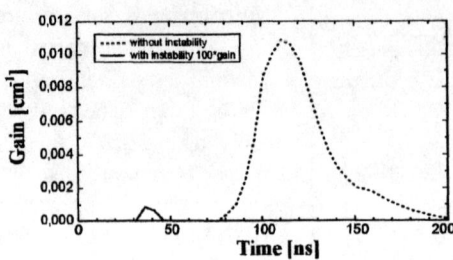

FIGURE 8. Time dependence of gain factor in evacuated polyacetal capillary discharge

TABLE 2: Evaluated peak values of the gain factor in cm⁻¹

I \ p_0	0.5 torr	1.0 torr	5.0 torr
15 kA	0.018	0.032	0.003
30 kA	0.265	0.205	0.135
45 kA	0.305	0.620	0.255

If MHD turbulence is taken into account the evaluated peak plasma temperature would be 65 eV at maximum and the achievable gain factor is even smaller. The instabilities may be suppressed by initial capillary filling by gas or vapors. When the modeling for gas filled capillary is done (with initial pressure of polyacetal vapors 1 torr) gain factor evaluated is 0.27 cm⁻¹.

Generally, the peak concentration of fully stripped ions is too low to guarantee efficient recombination pumping of the laser transition. Higher peak current is needed for the reasonable gain increase with initial filled capillaries (see Table 2).

CONCLUSIONS

Dynamics of capillary discharge and its emission characteristics are analyzed on the bases of MHD and FLY codes simulation. Time resolved spectra evaluated and measured have been compared and found that correspond each other. Further increase of the current passing through the capillary and initial gas filling have been modeled and have shown the way for further system optimization.

ACKNOWLEDGMENTS

This work was supported by a grant of the Ministry of Education Youth and Sports of the Czech Republic on the project LN00A100 *Laser Plasma Research Center*.

REFERENCES

1. Vrbová M., Jančárek A., Pína L., Vrba P., Bobrova N.A., Sasorov P.V., Kalal M., Nádvornikova L.: *J. Phys. IV France* **11** (2001), Pr2, 575-578.
2. Bobrova N.A., Bulanov S.V., Razinkova T.L., Sasorov P.V.: *Plasma Physics Reports* **22** (1996) 387-402.
3. Lee R.W., Larsen J.T.: J.Q.S.R.T. **56** (1996) 535-556.
4. Elton C.: *X – Ray Lasers*, Publishers, New York, London, 1990, pp. 21 – 28.

OTHER COLLISIONAL X-RAY LASERS

Characterization of saturated Ni-like x-ray lasers

J.E. Balmer, M. Braud, C. Siegel, and J. Nilsen[1]

Institute of Applied Physics, University of Berne
Sidlerstr. 5, CH-3012 Berne, Switzerland
[1]*Lawrence Livermore National Laboratory, Livermore, CA 94551*

Abstract. We report on recent experiments aimed at a full characterization of the x-ray laser output with respect to output energy, pulse duration, near- and far-field intensity distributions, gain-length product required for saturation, and coherence properties. Most of these measurements were performed on the 14.7-nm Pd and the 12.0-nm Sn lasers at a drive irradiance of 10 TW/cm^2, using a 1054-nm drive energy of 30 J in 100-ps pulses.

1. INTRODUCTION

Since the first demonstration of substantial x-ray laser gain in neon-like selenium [1], the electron-collisional excitation scheme has proven to be the most successful method on the route towards high output power and saturated gain. Saturated operation of soft-x-ray lasers is important because it assures the maximum stimulated-emission power extraction from a given volume of excited plasma. Values of the gain-length product gL $\geq$ 14 are typically required to reach saturation in laser-driven soft-x-ray lasers [2]. Clear evidence of saturation has been reported in the last few years by a number of authors for the 4d $\rightarrow$ 4p, J = 0–1 transition in nickel-like elements at wavelengths between 14.7 and 7.3 nm [3-6]. Almost exclusively, these experiments have exploited the prepulse and multiple-pulse techniques [7]. This has resulted in considerable reduction of the drive irradiance required for high gain. Characterization of these lasers has included measurements of the output energy, pulse duration, near-field and far-field intensity distributions, and divergence [3,4,8,9].

Here, we review our experimental investigations aimed at the full characterization of gain-saturated nickel-like soft-x-ray lasers at wavelengths between 14.7 and 12.0 nm. In earlier work, we have demonstrated the achievement of gain saturation on the 4d $\rightarrow$ 4p, J = 0-1 transition in nickel-like Pd, Ag and Sn at 14.7, 14.0 and 12.0 nm, respectively, as a result of a systematic optimization of the prepulse configuration, target curvature, and target surface roughness [5,6,10]. The required drive energy used was only 30 J in a 100-ps pulse, corresponding to an irradiance of 12 TW/cm^2. The pumping configuration used was a 0.5% (or "twin-0.25%", see below) prepulse preceding the 100-ps duration main pulse by 6 ns. The measured gain coefficients were in the range of 8-10 cm^{-1}, with gain-length products of ~16 at saturation. More recent

CP641, *X-Ray Lasers 2002: 8th International Conference on X-Ray Lasers*, edited by J. J. Rocca et al.
© 2002 American Institute of Physics 0-7354-0096-2/02/$19.00

experiments have included the measurement of the near-field and far-field intensity distributions, the coherence properties, and the pulse duration.

2. EXPERIMENTAL SETUP

The experiments were performed using the 1054-nm Nd:glass laser at the Institute of Applied Physics (IAP) of the University of Berne. The system, having a final amplifier of 90-mm diameter, is capable of delivering up to 40 J at a pulse duration of 100 ps (FWHM). The system can be fired at intervals of 30 minutes, given by the cooling time of the final amplifier. The maximum irradiance on target of ~15 TW/cm^2 is achieved in a line focus of 2.5-cm length and approximately 100-μm width. Defined prepulses at variable delays are obtained by inserting beamsplitters into an optical delay line in the double-passed final amplifier stage. The beamsplitter used in the recent experiments was a doubly antireflection-coated plate of BK7 glass with a residual reflectivity of 0.25% at 1.054 μm on each surface. The beamsplitter thus generated two prepulses of equal amplitude, separated in time by 200 ps. For the remainder of this article, this configuration will be termed a 0.5% prepulse. The spatial overlap of the prepulse line focus and the main pulse line focus was adjusted to an accuracy of ± 8 μm in the focal plane. All the targets had a length of 2.5 cm and the line focus was aligned to slightly overfill the target on the spectrometer side.

The standard diagnostic in these experiments was an on-axis, time-integrating spectrometer configured in an angle-resolving arrangement. It has been described previously in more detail [5,6,10]. For the more recent experiments, additional x-ray diagnostics were installed on the opposite side of the target. In the case of the spatial coherence experiments, this consisted of a 40-mm diameter P20 phosphor screen imaged to a cooled CCD camera having a pixel size of 9 μm square. For the measurements of the near-field and far-field intensity distributions, the P20 phosphor was replaced by a P43 phosphor resulting in an increase in overall sensitivity of approximately 3×.

2. RESULTS

Here, we will restrict ourselves to a summary of the recent results on the spatial coherence properties, the near-field and far-field intensity distributions, and the pulse duration. More details on these experiments can be found in Refs. [11-13].

Spatial coherence

The spatial degree of coherence of the 12-nm Sn x-ray laser was measured with a diagnostic based on a Young's double-slit experiment. The double slits used in this experiment were laser cut in-house with a Nd-YAG laser through 10-μm thick steel foils. The slits are 4 mm long and approximately 4 μm wide. The double slits were placed horizontally in the center of the x-ray laser beam, so that the interference patterns were observed in the vertical direction. The distance between the target and the double slits was approximately 21 cm. The diffracted intensity distribution was recorded by the P20 phosphor screen situated 45 cm behind the double slits and imaged to the CCD camera with a 1:1 magnification.

For the measurements, we used six different double-slit masks with slit spacings ranging from 10 μm to 51 μm. A typical interference pattern is shown in Figure 3a) for a slit spacing of 17 μm. It clearly shows interference at the location of the x-ray laser, whereas no evidence for interference is seen for the spontaneous emission. However, as can be seen in the figure, there is a non-negligible contribution of spontaneous emission of VUV and soft x-rays along the slits that has to be accounted for in the analysis. The approximate value of the spontaneous emission at the location of the x-ray laser is obtained by taking a horizontal scan and interpolating. The solid line in Figure 3b) is the intensity scan of the interference pattern of Fig. 3a). The dashed line is the result of an analytical calculation obtained by varying the width of the two slits and taking into account the visibility of the fringes.

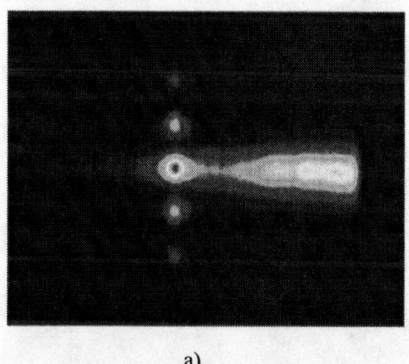 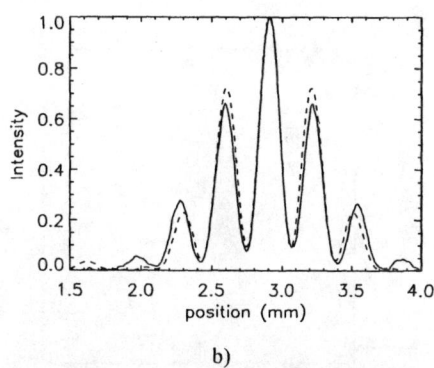

a)   b)

FIGURE 1. a) Interference pattern and b) intensity scan (solid line) of a Young's double slit experiment with 4-μm wide slits separated by 17 μm. The dashed line represents the best fit to the experimental curve with the widths of the slits and the visibility as the fit parameters.

The interference patterns of the different slit separations were analyzed with respect to their visibility, usually defined as $V = (I_{max} - I_{min})/(I_{max} + I_{min})$ where I_{max} refers to the maximum intensity at the center of the diffraction pattern, and I_{min} to the average of the intensities at the two adjacent minima. Assuming a Gaussian Schell-model [14] for the source shape, this allows us to determine the diameter of the emitting region and the transverse coherence length at the diffraction (slit) plane. As described in [12], the value obtained for the total transverse coherence length (the sum of the coherence length at the source plane, ρ_s, and the coherence length gained through propagation from the source to the diffraction plane, ρ_p) is $\rho_y = 35$ μm. The value obtained for the coherence length at the source plane in the vertical direction is $\rho_s = 2.5$ μm and the value for the diameter of the emitting region is $\sigma_s = 46$ μm.

Far-field measurements

Additional information on the spatial coherence properties of the x-ray laser beam has been obtained by analyzing the far-field intensity distribution. Fig. 2 shows the far-field images of the 14.7-nm Pd x-ray laser after 2.0 m of free space propagation for two different prepulse amplitudes. The divergence angles Θ_x and Θ_y (FWHM) obtained from the distributions of Fig. 2a) and b) are 2.2 x 6.5 and 1.4 x 3.7 mrad^2, respectively. It is seen that for the larger prepulse amplitudes (and thus shorter pre-pulse delays) the divergence angle decreases – mainly in the vertical direction. Again, making use of the theory of partially coherent beams (Gaussian Schell-model beams), the knowledge of the far-field divergence angle allows us to calculate the coherence length at the laser output aperture [13]. The results are 6.7 µm x 2.5 µm for the 8% prepulse and 4.3 µm x 1.4 µm for the 0.5 % prepulse for the x- and y-directions, respectively. These results are consistent with the coherence measurements based on the double-slit experiment and they agree with the dimensions of the observed small-scale structures in the near-field images described below.

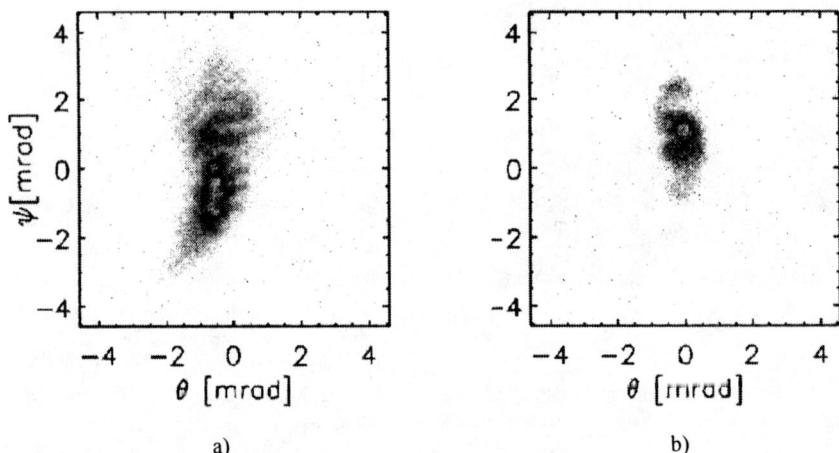

a)   b)

FIGURE 2. Far-field intensity distribution of the 14.7-nm laser emission after 2.0 m of free space propagation with the 0.5% prepulse of 6-ns delay (a) and 8% prepulse of 1-ns delay (b). θ and ψ denote the angular coordinates in the x- and y-directions, respectively.

Near-field measurements

The setup of the 2D spatial imaging diagnostic consisted of a 50-cm concave, multilayer Mo/Si x-ray mirror that images the output aperture of the Pd 14.7-nm laser line onto the P43 phosphor/CCD camera detection system via a flat normal-incidence Mo/Si mirror and a 45 degree Mo/Si turning mirror. Distances were chosen such as to give a magnification of 10. The spatial resolution at the output plane of the Pd laser was measured to be ~3 µm. This value is consistent with the smallest structures that could be resolved in the near-field images. In these experiments, the targets had a 25-μm diameter steel wire placed ~180 μm in front of the target surface at the output end of the x-ray-laser that served as an absolute spatial fiducial. The setup is described in more detail in Ref. [13].

Figure 3 shows typical images of the Pd J = 0–1 line at 14.7 nm at the output aperture of the lasing plasma. The plasma blow-off direction is the horizontal axis in the figures with zero corresponding to the original target surface. The 25-μm thick wire fiducial is best visible in the contour plot and clearly evident at ~180 μm from the target surface. Differences in the near field intensity distribution are visible for different prepulse amplitudes and prepulse delays. There are small-scale inhomogeneities that may be caused by the different gain lengths experienced by the individual rays. The higher prepulse amplitudes (2.8 % and 8%) gave more reproducible structures than the 0.5% prepulse which showed large pulse-to-pulse variations even for identical settings. An interesting detail are the jet-like structures in the direction of expansion of the plasma, e.g. in Fig. 4c) and d).

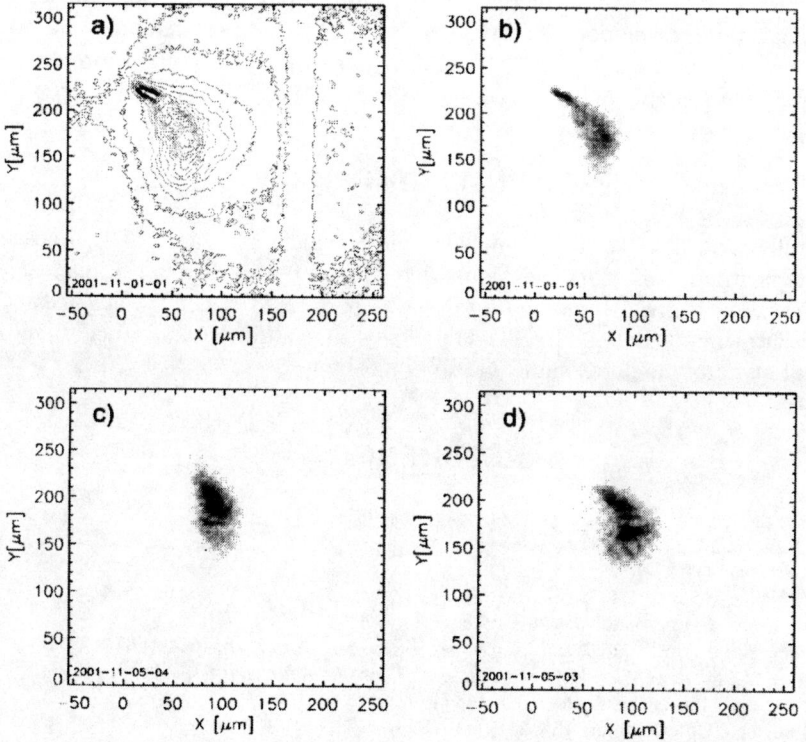

FIGURE 3. 2D images of the 14.7-nm laser emission at the output aperture of a 2.5-cm long flat slab target with a 0.5% prepulse and 2-ns delay (a) and (b), a 2.8% prepulse and 2-ns delay (c), and a 8% prepulse and 1-ns delay (d). The plasma blow-off direction is the horizontal axis in the figures with zero corresponding to the original target surface. The 25-μm thick fiducial wire is clearly evident at ~180 μm from the target surface.

The position of the laser peak intensity relative to the target surface was measured as a function of main pulse-to-prepulse delay for the three prepulse amplitudes (0.5%, 2.8% and 8%). The curve of the 0.5% prepulse shows a flat maximum at ~60 μm from the target at a prepulse delay of ~6 ns. For the higher prepulse amplitudes the emission zone is seen to move away from the target nearly linearly with the delay time. In contrast to the 0.5% prepulse, optimum lasing conditions are limited to a narrow delay-time window and correspond to peak positions of ~80 μm and ~90 μm in front of the target surface for the 2.8% and the 8% prepulses, respectively.

The experimental data were compared to simulations using the LASNEX [15] and CRETIN [16] codes. The calculations are found to be in very good agreement taking into account the pulse-to-pulse variation of the experimental data and the residual uncertainty in position measurement with the 25 μm fiducial wire [13].

Time-resolved measurements

Time-resolved measurements were performed with the aid of a Kentech 2× magnification x-ray streak camera equipped with a 140-μm wide CsI slit cathode. The instrumental time resolution is estimated as ~12 ps. Pulse durations of ~40 ps (FWHM) were measured for the Ni-like Pd, Ag and Sn x-ray lasers. More details of these experiments are given in an accompanying paper [17].

ACKNOWLEDGMENTS

The authors would like to acknowledge the technical assistance of B. Locher for target preparation. This work was supported in part by the Swiss National Science Foundation. The work of one author (JN) was performed under the auspices of the U.S. Department of Energy by the University of California Lawrence Livermore National Laboratory under Contract No. W-7405-Eng-48.

REFERENCES

1. D.L. Matthews et al., Phys. Rev. Lett. **54,** 110-113 (1985).
2. M.H. Key, Nature **316,** 314-319 (1985).
3. H. Daido et al., Opt. Lett. **21,** 958-960 (1996).
4. J. Zhang et al., Science **276,** 1097-1100 (1997).
5. R. Tommasini, F. Loewenthal, and J.E. Balmer, Phys. Rev. A **59,** 1577-1581 (1999).
6. R. Tommasini, F. Loewenthal, and J.E. Balmer, J. Opt. Soc. Am. B **16,** 1664-1667 (1999).
7. J. Nilsen, J.C. Moreno, B.J. MacGowan, and J.A. Koch, Appl. Phys. B **57,** 309-311 (1993).
8. J. Nilsen, et al., Phys. Rev. A **56,** 3161-3165 (1998).
9. J.Y. Lin, et al., Opt. Commun. **158,** 55-60 (1998).
10. J.E. Balmer, R. Tommasini, and F. Loewenthal, IEEE J. Sel. Topics Quantum Electron. **5,** 1435-1440 (1999).
11. J.E. Balmer, M. Braud, and Ch. Siegel, "Towards full characterization of nickel-like x-ray lasers", in *Soft X-Ray Lasers and ApplicationsIV*, edited by E.E. Fill and J. Rocca, Proceedings of SPIE Vol. 4505, 93-99 (2001).
12. M. Braud, Ch. Siegel, F. Loewenthal, and J.E. Balmer, *Opt. Commun.*, to be published.

13. Ch. Siegel, M. Braud, J.E. Balmer, and J. Nilsen, *Opt. Commun.*, to be published.
14. L. Mandel and E. Wolf, *Optical coherence and quantum optics*, Chap. 4-5, Cambridge University Press, 1995.
15. G. B. Zimmerman and W. L. Kruer, Comm. Plasma Phys. Contr. Fusion **2**, 51 – 61 (1975).
16. H. A. Scott, JQSRT **71**, 689 – 701 (2001).
17. M. Braud, C. Siegel, J.E Balmer and J. Nilsen, "Near-Field Spatial Imaging and Time Resolved Measurements of the Ni-like Palladium Soft X-Ray Laser", *These Proceedings*

Near-Field Spatial Imaging and Time-Resolved Measurements of the Ni-like Palladium Soft X-Ray Laser

M. Braud, C. Siegel, J.E Balmer and J. Nilsen*

Institute of Applied Physics, University of Bern, Sidlerstrasse 5, 3012 Bern
Lawrence Livermore National Laboratory

Abstract. We have performed a series of near-field imaging experiments for the nickel-like 14.7-nm Pd x-ray laser with the aim to characterize the two-dimensional source intensity distribution and its position relative to the target surface. The effect of different prepulse amplitudes at variable main pulse-to-prepulse delays is investigated. To further characterize the nickel-like 14.7-nm Pd x-ray laser, we then performed measurements of its pulse duration with the help of a streak camera.

INTRODUCTION

Since the first demonstration of substantial x-ray laser gain in neon-like selenium [1], the electron-collisional excitation scheme has proven to be the most successful method on the route towards high output power and saturated gain. Clear evidence of saturation has been reported in the last few years by a number of authors for the 4d → 4p, J = 0-1 transition in nickel-like elements at wavelengths between 14.7 and 7.3 nm [2-3]. Considerable reduction of the drive irradiance required for high gain was obtained in these experiments by exploiting the prepulse and multiple-pulse techniques [4-5]. Reports on the properties of these lasers have included measurements of the output energy, pulse duration, near- and far-field intensity distributions, and divergence [2,6].

With the availability of highly reflective multilayer imaging mirrors at several wavelengths of neon- and nickel-like soft-x-ray lasers, high-resolution two-dimensional imaging of the laser source became possible [6-7]. Experiments were performed to measure the near-field spatial dependence of the laser output under various illumination conditions (e.g. different prepulse techniques) and for different target configurations. Further valuable information could be obtained by the measurement of the lasing position relative to the target surface, and of the temporal profile, with the aim to understand the plasma conditions under which lasing occurs and how to improve these laser systems [6].

In this paper we report on a series of near-field imaging experiments for the Pd 14.7-nm x-ray-laser, together with time-resolved measurements.

CP641, *X-Ray Lasers 2002: 8ᵗʰ International Conference on X-Ray Lasers*, edited by J. J. Rocca et al.
© 2002 American Institute of Physics 0-7354-0096-2/02/$19.00

EXPERIMENTAL SETUP

The experiments were performed using the 1054-nm Nd:glass laser at the Institute of Applied Physics (IAP) of the University of Berne. The system, having a final amplifier of 90-mm diameter, is capable of delivering up to 40 J at a pulse duration of 100 ps (FWHM). A line focus of 2.5-cm length and approximately 100-μm width allows a maximum irradiance on target of ~15 TW/cm^2. The targets were 25-mm long slabs of diamond-machined Pd, Sn and Ag. Defined prepulses at variable delays were obtained by inserting different beamsplitters into an optical delay line in the double-passed final amplifier stage. For these experiments three different prepulse schemes are investigated: a 2 x 0.25% twin prepulse (called the 0.5% prepulse for simplicity), a 2.8% single prepulse, and finally a 8% single prepulse

The setup of the first diagnostic, the 2D spatial imaging diagnostic, is shown in the left part of Fig. 1. It consists of a concave, multilayer, normal-incidence Mo/Si x-ray mirror that images the output aperture of the Pd 14.7-nm laser line onto the P43 phosphor/CCD camera detection system via a flat normal-incidence Mo/Si mirror and a 45-degree Mo/Si turning mirror. The concave mirror of 2.5-cm diameter has a radius of curvature of 50 cm and is placed at a distance of 27.5 cm from the output aperture of the Pd laser, while the phosphor screen is placed 275 cm from the mirror to give a magnification of 10. The spatial resolution at the output plane of the Pd laser is ~3 μm. The flat slab targets had a 25-μm diameter steel wire placed ~180 μm in front of the target surface at the output end of the x-ray-laser that served as an absolute spatial fiducial in the measurements.

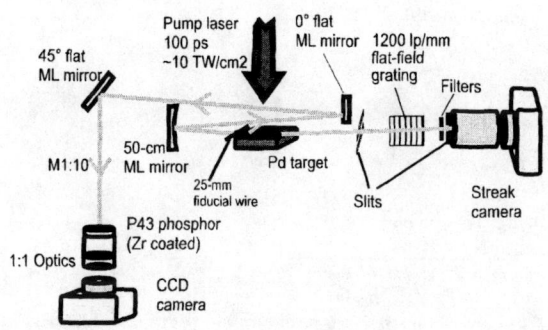

FIGURE 1. Setup of the experiment

The setup of the second diagnostic, the time resolved diagnostic is shown in the right part of Fig. 1. It consists of a 1200 lp/mm, aberration-corrected, concave Hitachi grating (radius of curvature: 5649 mm), which disperses the incident radiation perpendicularly to the direction of the angular resolution on the streak camera. The streak camera is equipped with a 140-μm wide CsI cathode, and a P11 phosphor screen. The emitted light is then amplified by an image intensifier (VARO) and imaged onto a cooled CCD camera, having a 23-μm pixel size. To avoid saturation of the streak camera, a 5-μm wide slit and several Al/formvar filters were used. The time resolution of the streak camera is ~12-15 ps.

NEAR-FIELD SPATIAL IMAGING

In an earlier experiment, we optimized the prepulse conditions for the three prepulse amplitudes mentioned above and found an optimum with the 0.5% prepulse, providing a large prepulse-delay range, where efficient laser emission occurs. A flat maximum was estimated at a prepulse delay of ~6 ns, resulting in the use of this scheme as the standard setting [8]. Typical spatial images of the Pd J = 0–1 line at 14.7 nm at the output aperture of the flat slab target for the 0.5%-6 ns prepulse scheme are shown in Fig. 2. The plasma blow-off direction is horizontal in the figures with zero corresponding to the original target surface, in reference to the wire fiducial (best visible in the contour plot).

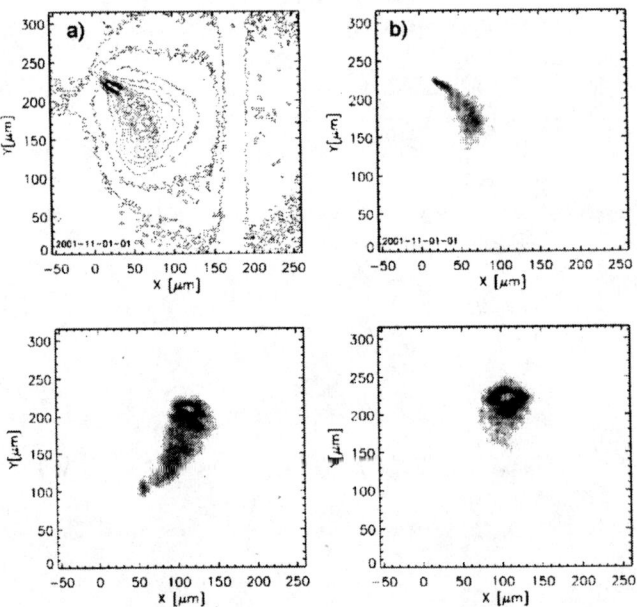

FIGURE 2. Spatial images of the 14.7-nm laser emission at the output aperture of the flat slab target

Interesting are the jet-like structures in the direction of expansion of the plasma. The higher prepulse amplitudes (2.8% and 8%) gave more reproducible structures than the 0.5% prepulse, which showed large pulse-to-pulse variations even for identical settings (see Fig. 2).

We obtained more reproducible results for the positions of the laser peak intensity as a function of main pulse-to-prepulse delay. Fig. 3 shows the position of the laser peak intensity relative to the target surface as measured for the three prepulse amplitudes (0.5%, 2.8% and 8%) with different delays, as far as laser emission occurs. The curve of the 0.5% prepulse shows a flat maximum at ~60 μm from the target for a prepulse delay of ~6 ns. For the higher prepulse amplitudes the emission zone is seen to move away from the target nearly linearly with the delay time. In contrast to the 0.5% prepulse, optimum lasing conditions are limited to a narrow delay-time window corresponding to a peak position of ~80 μm and ~90 μm in front of the target surface for the 2.8% and the 8% prepulse, respectively.

Additional information on the beam quality, especially on the spatial coherence properties of the x-ray laser beam can be obtained by analyzing the far-field distribution. With little modification of the setup shown in Fig. 1 (the imaging mirror and the 0°-incidence relay mirror are removed) we were able to obtain highly resolved far-field images, and thus to measure the divergence angles for the horizontal and the vertical axes.

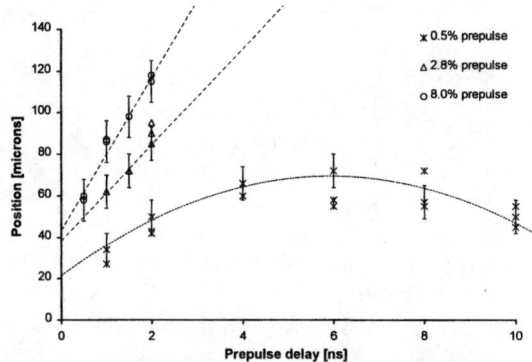

FIGURE 3. Position of the peak intensity relative to the target surface at the output aperture of the flat slab target for the three prepulse amplitudes.

Making use of the theory of partially coherent beams (Gaussian Schell-model beams), the knowledge of the far-field divergence angle allows us to calculate the coherence length at the laser output aperture. We obtained for the x- and y-direction respectively 6.7 μm x 2.5 μm for the 8% prepulse, 6.7 μm x 2.2 μm for the 2.8% prepulse and 4.3 μm x 1.4 μm for the 0.5% prepulse.

These results agree with the dimensions of the small-scale structures observed in the near-field images – that may be caused by different gain lengths experienced by the individual rays – and are consistent with the coherence measurements based on a Young's double slit experiment of the nickel-like 12-nm Sn x-ray laser [9].

TIME RESOLVED MEASUREMENTS

With help of the setup shown in Fig. 1, time-resolved measurements of the 14.7-nm laser line of the Pd x-ray laser were performed with the aid of a Kentech 2× magnification x-ray streak camera equipped with a 140-μm wide CsI slit cathode. The drive irradiance for these shots was ~10 ± 2 TW/cm^2, and the prepulse scheme used was the 0.5% prepulse at 6-ns delay.

Due to effects of saturation of the streak camera, we measured its dynamic range. Fig. 4 shows the results obtained by varying the intensity of the Pd x-ray laser. For this measurement, we used a streak rate of 23 ps/mm and a gain of 2000 for the image intensifier.

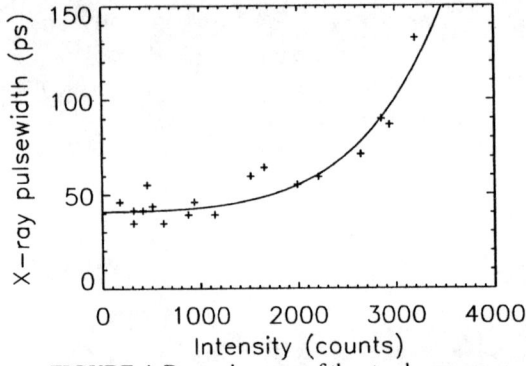

FIGURE 4. Dynamic range of the streak camera

The upper limit of the dynamic range, defined as the intensity at which 20% broadening of the pulse occurred, is about 1500. Assuming these results, the use of a 5 μm wide slit and two Formvar/Al filters (T=10%) were necessary to operate the streak camera within its dynamic range and thus avoid broadening of the pulse due to saturation.

The time resolved pulse profile of the Pd 14.7-nm laser line is shown in Fig. 5. The FWHM of the pulse duration is measured to be 40 ps, whereas the spontaneous emission lasts for about 160 ps. The intensity of the spontaneous emission is multiplied by a factor of 50 to be more visible on the plot.

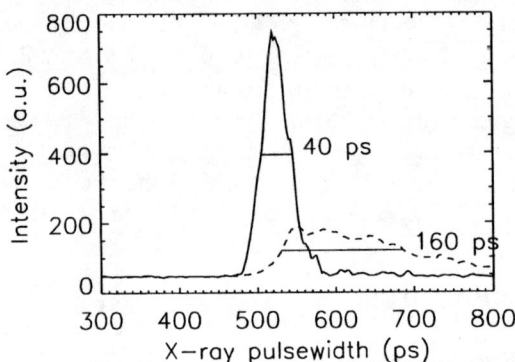

FIGURE 5. Time-resolved profile of the 14.7-nm Pd x-ray laser and of the spontaneous emission

As is evident in Fig. 5, the laser emission occurs on the rising edge of the spontaneous emission. Nevertheless, this is not an absolute timing measurement, but this is a subject of ongoing experiments.

Having optimized the diagnostic setup, we kept the same prepulse conditions, and measured the pulse duration of the Ni-like Ag and Sn x-ray laser at 14 and 12 nm, respectively. The results are shown in Fig. 6.

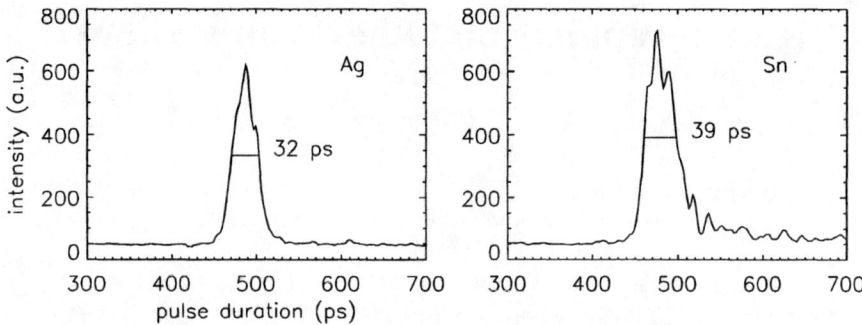

FIGURE 6. Time-resolved profiles of the Ni-like Ag and Sn x-ray laser

Both the Ag and the Sn x-ray laser show a pulse duration of the same order as the Pd x-ray laser (40 ps).

CONCLUSION

In conclusion we have reported on a series of near-field imaging experiments for the nickel-like 14.7-nm Pd x-ray laser that were conducted with the aim to characterize the two-dimensional source intensity distribution and its position relative to the target surface. The effect of different prepulse amplitudes at variable main pulse-to-prepulse delays was investigated. We observed a difference in the spatial dependence for the larger prepulses of 2.8% and 8% that moved the peak emission farther away from target surface with increasing delay whereas for the weak prepulse of 0.5% the position of peak emission remained rather stable over a delay-range of several nanoseconds. The size of the smallest structures detected is consistent with the coherence length – obtained from the correspondent far-field date – at the output aperture.

As for the time-resolved measurements, consistent results were obtained for the Ni-like Pd, Ag and Sn x-ray lasers, and a pulse duration (FWHM) of 40 ps was obtained.

REFERENCES

1. D. L. Matthews et al, *Phys. Rev. Lett.* **54**, 110 - 113 (1985).
2. Zhang et al, *Phys. Rev. Lett.* **78**, 3856-3859 (1997).
3. R. Tommasini, F. Loewenthal, and J. E. Balmer, *J. Opt. Soc. Am. B* **16**, 1664 - 1667 (1999).
4. T. Boehly et al, *Phys. Rev. A* **42**, 6962-6965 (1990).
5. H. Daido et al., *Phys. Rev. Lett.* **75**, 1074-1077 (1995).
6. J. Nilsen, J. Zhang, and A. G. MacPhee, *Phys. Rev. A* **56**, 3161-3165 (1997).
7. J. C. Moreno et al., *SPIE* **3156**, 71 - 77 (1997).
8. M. Braud, F. Loewenthal, and J. Balmer, *C.R. Acad. Sci. Paris 1*, 1019-1024 (2000).
9. M. Braud, C. Siegel, F. Loewenthal, and J. E. Balmer, *Opt. Commun.* submitted for publication, (2002).

Development Of X-ray Lasers For Radiographic And Other Applications

R Keenan*, C L S Lewis*, S J Topping*, J S Wark[†], E Wolfrum[†]

* Department of Pure and Applied Physics, Queen's University of Belfast, Belfast, BT7 1NN
[†]Department of Physics, University of Oxford, Oxford, OX1 3PU

Abstract. Saturated outputs for the Ne-like Ge and Ni X-ray Lasers (XRLs) at 196 and 231 Å respectively are now routinely produced at RAL from long single targets. The pumping configuration of ~100 ps pulses and a new simpler pump beam arrangement has been optimised with a typical XRL output of 3 mJ. Optimum coupling of the drive beams has been demonstrated through imaging of the XRL showing that target displacement of ~100 μm is needed. The output of the Ni XRL has been characterised and maximised with a view to applications of X-ray lasers such as Non-linear Optics and Thomson scattering. The Ge XRL was optimised to carry out radiography of Aluminium foils driven by optical laser radiation

INTRODUCTION

Development of the five beam configuration began with the Sm XRL at 73 Å with a view to doing XRL radiography of CH foils [1] and the five beam technique has now been extended and perfected for the Ne-like Ge and Ni XRLs at 196 and 231 Å. This is now the routine configuration for an XRL pumped by the 'long', ~100 ps, beams of Vulcan. We have fully characterised the output of these XRLs to produce an XRL suitable for applications. The target retraction has become and important factor and the initial work on a flat-field spectrometer has been backed up by imaging the XRL.

EXPERIMENTAL SETUP

Five Beam Setup

Targets were irradiated by 5 beams of the Vulcan Nd:glass laser at 1.05 μm. Line foci 100 μm wide and 20 mm long were generated using a lens and off axis spherical mirror. The beams, 80 ps duration with 40 J per beam, were distributed along a 25 mm x 100 μm line focus giving an on target intensity of up to 5 x10^{13} W/cm^2. A prepulse with variable delay was generated in the laser area by splitting the oscillator pulse into two before the preamplifiers of Vulcan. The beams were displaced axially to produce uniform illumination and timed along the target so that each beam was

CP641, *X-Ray Lasers 2002: 8ᵗʰ International Conference on X-Ray Lasers,* edited by J. J. Rocca et al.
© 2002 American Institute of Physics 0-7354-0096-2/02/$19.00

incident on target centre at the same time. The beams were configured as shown in Fig. 1 a) with one beam on axis, two at ±30° and two at ±60°.

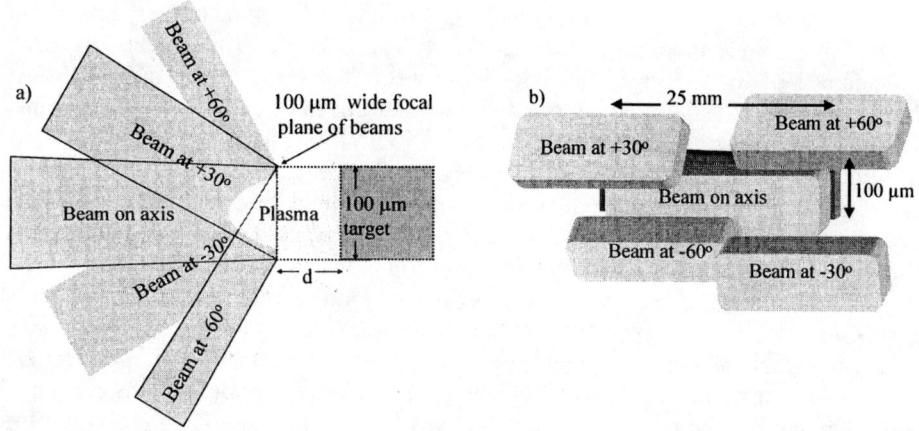

FIGURE 1. Sketch of five beam irradiation of a single target 100 μm wide x 25 mm long a) side view showing how target retraction, d, allows the 5 beams to heat the same region of preplasma and b) front view showing displacement of beams along the axis

The beams were displaced along the axis as shown in Fig. 1 b) with each beam delayed according to its position and positioned to provide maximum symmetry about the axis. The target base was mounted on a Sony magnascale so the target could be accurately retracted a distance d after focussing on the 100 μm wide target (see Fig. 1a)). This target retraction of a few 10s of μms allowed the beams arriving at an angle to heat the same region of the preformed plasma. This should allow all beams creating the conditions for gain to arrive at the same small area of pre-plasma on axis, without the retraction the beams at +30° and +60° (and conversely those at -30° and -60°) could create a separate gain region above (and below) the axis.

Measurement of the XRL

As Ge XRL was used as the probe for radiography experiments, targets 25 mm long were used as they gave a bright saturated output but also allowed the central, on-axis beam, to be extended so that it could illuminate the entire target and be used as the prepulse energy source. The energy balance between the prepulse beam and the other four set the prepulse ratio to 20% for this particular setup. The prepulse beam was timed so that it arrived 2 ns before the main heating beams. Initial optimisation of the target retraction was carried out using a time integrated, angle resolving, axial grazing incidence flat-field spectrometer (FFS) with a 1200 lines mm^{-1} aperiodic ruled grating, coupled to a back-thinned CCD with 24 μm pixels as the main diagnostic. The scan demonstrated the positive effect of retracting the target, with a retraction of between 70-100 μm producing the brightest output, however no knowledge of how the vertical divergence is changing could be gained from the FFS.

For the absolute numbers in Figures 3 and 5, knowledge and best estimates of the detection parameters is required. The FFS measures the angular divergence but makes no measure of the vertical divergence as this is in the dispersion direction. This is taken account of by estimating the vertical as 3 times the horizontal divergence, from imaging shots, and combining this with the collection angle of the spectrometer. A best estimate of the grating reflectivity is made at 5%. The transmission of the Al filter is taken from Henke[2] and the importance of this is discussed below. The final assumption is the conversion efficiency of the CCD from photons to counts, this is taken to be 2 photons/count for a 196 Å photon for the specific detector used here. The counts on the CCD are integrated in the horizontal direction and across the spectral width of the laser line, a background is subtracted, and the assumptions described above used to get the total number of photons and hence the total output at the XRL wavelength. For imaging shots the calibrated reflectivity of the X-ray mirrors was used to obtain absolute photon numbers.

Recently published experimental measurements of the mass absorption coefficient of Al show a minimum at 63 eV [3]. This gives a new value for its transmission at 196Å compared to the previously available data [2]. As Al is used as the filter in the flat-field spectrometer (FFS) for XRL measurements, above the Al L-edge at 170 Å, this value is required to make an accurate estimate of the XRL output. For a 25 mm long XRL target the filter used in the FFS is typically 4 μm, at this thickness the new data gives a transmission of 0.46 % compared to 0.013% with the old data, leading to a difference of ~x40 in any estimate of the measured XRL beam energy. A measurement of the Al transmission at 196Å was made with the Ge XRL in an attempt to resolve the discrepancy. The initial data was limited but the experiment has now been repeated for Ge and extended to include the Ni XRL at 231 Å. Transmission data at both 196 and 231 Å has confirmed the validity of the original Henke data: details of these measurements will be presented in a forth-coming publication.

The target retraction for Ge was further optimised with the imaging setup for radiography [4]. The output of the XRL 3 cm from the end of the target was imaged using Mo/Si multilayer X-ray mirrors (XRM), at x18 magnification, onto a CCD in a set up similar to Fig. 2 but with a 45° mirror relaying the beam from XRM1 to the CCD. The imaging mirror, 25 mm in diameter and with a focal length of 50 cm, had a peak reflectivity of 45% at 196 Å and a bandpass of 20 Å and was positioned using a wire grid to give a resolution better than 5 μm. Through imaging the retraction was truly optimised to produce a quasi-homogeneous beam in both the horizontal and vertical directions.

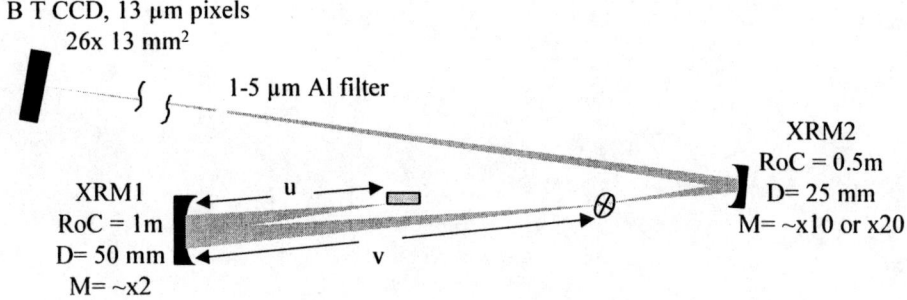

FIGURE 2. Setup for imaging the Ni XRL

The Ne-like Ni XRL at 231 Å is used as a mixing beam for non-linear optics and for these experiments the pump conditions were fully characterised in the five beam configuration [5]. A length scan was carried out on a FFS and the uniformity and ionisation balance were monitored as usual with a space resolving time integrated keV spectrometer and a cross-slit camera. An optimised output of ~3 mJ was obtained with a FWHM and off axis refraction of ~ 8 mrad. Full characterisation of the source has now been carried out using the imaging setup in Fig. 2. The first XRM images the beam at ~x2 magnification to an interaction plane where four wave mixing or Thomson scattering could occur. This plane is then imaged with a second XRM onto a CCD with 13 μm pixel at high, x10 or x20, magnification. The total output and exit point shape were measured for various target retractions.

RESULTS

For Ge initial optimisation of the target retraction was carried out using a FFS as the main diagnostic with Fig. 3 showing the output of a 25 mm long XRL target varying as we scan through target retraction. This demonstrates the positive effect of retracting the target with a retraction of between 70-100 μm producing the brightest output. The absolute output was calculated as described above.

Further optimisation was carried out through imaging with the images in Fig. 4 showing a double lobed structure for a target retraction of 70 μm and the output optimised at 90 μm. It can clearly be seen that target retraction is important, not only to achieve maximum brightness, but also the large, relatively uniform area needed particularly for radiography. For a retraction of 70 μm an obvious double lobed structure is present, see all the images in Fig. 4a). This is from the beams above and below axis creating separate gain regions. Only when the target is retracted 90 μm do all the beams heat the same region, though still creating an XRL much larger in the vertical direction. The optimised target retraction depends mainly on the configuration of the prepulse, where the objective is creation of a single quasi-homogeneous gain region. The beam shape with a retraction of 90 μm was highly reproducible as the shots in Fig. 4c) show.

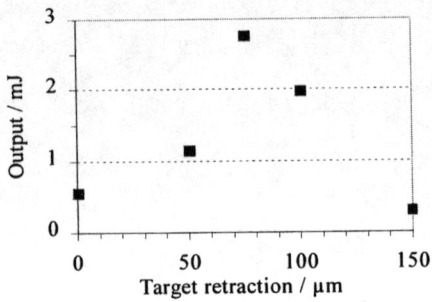

FIGURE 3. Output measured on a FFS of a 25 mm Ge XRL for various target retractions d.

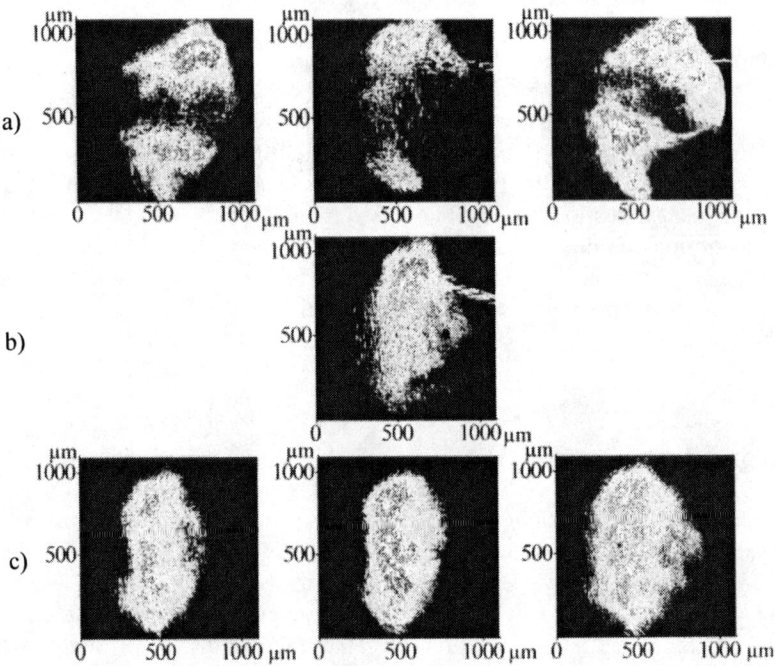

FIGURE 4. Optimisation of target retraction through imaging the Ge XRL 3 cm from the exit point showing a) 70 b) 80 and c) 90 μm retraction

For the Ne-like Ni images were taken at high magnification of the exit point for various target retractions. The graph in Fig. 5 below shows the results of this optimisation with the image in Fig. 5 showing an optimum output of 4 mJ at 100 μm retraction having a FWHM of 180 μm^2. The reproducibility of the source was also verified at various retractions and Fig. 5 has two shots taken at 75 μm retraction with an on target energy of 180 ± 2 J both producing a 3 mJ output in an exit point FWHM of ~150 μm^2.

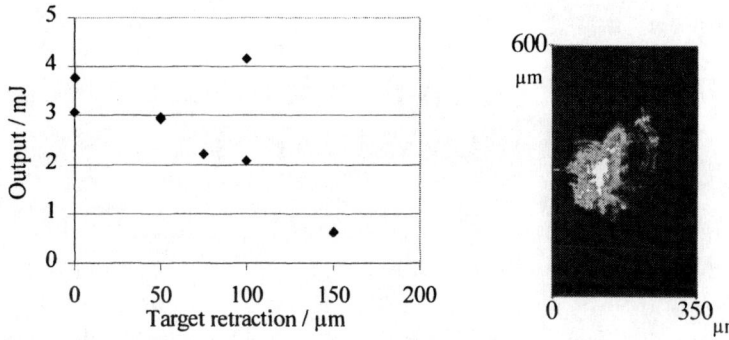

FIGURE 5. Target retraction optimisation through imaging the exit point of Ni XRL with image of exit point for 100 μm retraction.

CONCLUSIONS

The prepulse technique has allowed the use of long single targets up to 25 mm and the output from these targets has been found to improve when the target is retracted from optimum focus. The five beam pumping of single targets is now the routine configuration for an XRL produced by long, ~100 ps, pulses of Vulcan. The five beam setup has now been optimised such that a highly reproducible saturated output of ~3 mJ is obtained for both the Ne-like Ge and Ni XRLs at 196 and 231 Å respectively. Target retraction has been optimised through imaging allowing production of a well-defined source suitable for various applications.

REFERENCES

1. Keenan, R., et al., " Five beam irradiance of a single Sm X-ray laser target", in *CLF Annual report 1998-1999*, RAL-TR-1999-062, 1999, pp. 56-57.
2. Henke, B.L., Gullickson, E.M., Davis, J.C., *At. Data Nucl. Data Tables* **54**,181 (1993)
3. Gullickson, E.M., Denham, P., Mrowka, S., Underwood, J.H. , *Phys. Rev. B* **49**,16283 (1994)
4. Wolfrum, E., et al., "X-ray laser radiography of hydrodynamic perturbations due to laser imprint", in *X-ray lasers 1998*, edited by Y. Kato et al., IOP conference series 159, 1999, pp. 657-663
5. Topping, S.J., et al., "Progress in FWSD non-linear effects with XRLs", in *X-ray lasers 2000*, edited by G. Jamelot et al., Journal de Physique IV, 2001, pp. 487-490.

Development and applications of X-ray lasers at LSAI/LIXAM

Annie Klisnick[1], Gérard Jamelot[1], David Ros[1], Antoine Carillon[1], Pierre Jaeglé[1], Mustapha Boussoukaya[1], Olivier Guilbaud[1], Jaroslav Kuba[1§#], Raymond Smith[1#], Jean-Claude Lagron[1], Laurent Vanbostal[1], Denis Joyeux[2], Daniel Phalippou[2], Stéphane Sebban[3], Alain Touati[4], Marie Anne Hervé du Penhoat[4], F. Ballester[5], E.-J. Petit[5], Bedrich Rus[6], Tomas Mocek[6], Federico Strati[7], Matthew Edwards[7], Gregory J. Tallents[7], Roisin Keenan[8], Simon Topping[8], Ciaran L.S. Lewis[8], Paul Neumeyer[9], Daniel Ursescu[9], Thomas Kühl[9], Huajing Tang[10], Hiroyuki Daido[10]

[1] *Laboratoire d'Interaction du rayonnement X Avec la Matière (former LSAI), bâtiment 350, Université Paris-Sud, 91405 Orsay Cedex, France*
[2] *Laboratoire Charles Fabry-IOTA, Université Paris-Sud, CNRS, BP 147, 91403 Orsay, France*
[3] *Laboratoire d'Optique Appliquée, ENSTA, CNRS UMR 7639, 91761 Palaiseau cedex, France*
[4] *Groupe de Physique des Solides, Tour 23, 2 Place Jussieu 75251 Paris cedex 05, France*
[5] *CEA Saclay, DAPNIA-SEA, 91690 GIF-sur-Yvette, France*
[6] *Department of Gas Lasers/PALS, Institute of Physics, 18221 Prague 8, Czech Republic*
[7] *Department of Physics, University of York, Heslington, YO10 6DD, York, UK*
[8] *School of Mathematics and Physics Physics, The Queen's University of Belfast, Belfast BT7 1NN, UK*
[9] *Gesellschaft für Schwerionenforschung, PHELIX Project, Planckstraße 1, 64291 Darmstadt, Germany*
[10] *Advanced Photon Research Center, J.A.E.R. I., Kizu, Kyoto 619-0215, Japan*

Abstract. We present an overview of our research activity achieved since the last X-ray laser Conference in Saint-Malo. Our research program involves the development of laser-pumped collisional X-ray lasers under different regimes of irradiation, and the use of these sources for applications. The work presented involves a number of French and international collaborations and was carried out at different pump laser facilities: LULI (Ecole Polytechnique), LOA (ESNTA) in France; Rutherford Laboratory in U.K; PALS in Czech Republic.

1. INTRODUCTION

There has been considerable progress in the recent years on pumping plasmas through collisional excitation under different regimes of laser irradiation. These different regimes lead to a range of X-ray lasers operated at saturation and having distinct specific requirements and advantages. More specifically three types of laser-pumped, collisional X-ray lasers are now routinely generated. Optical-field ionization

§ Also at Faculty of Nuclear Science Engineering, Czech Technical University, Prague, Czech Republic
# Now at Lawrence Livermore National Laboratory, P. O. Box 808 mail Code L-251, 94550 Livermore, CA, USA

(OFI) of a gas by a circularly polarized, 30 fs, ~ 1 J laser pulse has led to saturated X-ray lasers at 41.2 nm and 32.8 nm [1,2]. These OFI X-ray lasers deliver a low energy (typically 10 nJ) pulse, but they operate at a relatively high repetition rate, of 1 Hz or above. On the other hand the fast heating of a preformed plasma with a ~ 1 ps, 10-50 J laser pulse was used to generate the so-called "transient" X-ray lasers [3,4,5]. The repetition rate of these XRL's is currently limited to 1 shot every ~10 min, but the output energy is higher than for the OFI systems, typically 10 μJ in a short, picosecond pulse. Finally the so-called "quasi-steady-state" (QSS) XRL's are pumped by longer (100-500 ps), higher energy (100 - 500 J) pulses [4, 6,7]. They thus rely on larger-scale laser facilities operating at low repetition rate, typically 2-3 shots/hour. Hence one could consider them as "old-fashioned", or even obsolete systems. However it should be emphasized that QSS XRL's are today the only XUV sources that can deliver up to a few mJ in a ~ 100 ps pulse. We thus believe that all three types of X-ray lasers above mentioned are worth being developed in the future, because they have complementary properties that are able to cover a large variety of applications.

Our research activity over the last two years actually covered the study of the three types of X-ray laser pumping, as will be shown in sections 2 and 4. An important part of our research program is devoted to the demonstration of the potential of X-ray lasers as a tool for investigations in several research areas. In 2001 we have been involved in two application experiments, carried out respectively at LULI, Palaiseau and at PALS (Prague, Czech Republic). The motivations underlying these projects and the first results obtained will be presented in section 3. Conclusion and future prospects of our work will be discussed in section 5. In particular we will briefly outline the status of our LASERIX project which aims at the construction of a dedicated X-ray laser facility.

2. DEVELOPMENT OF COLLISIONAL X-RAY LASERS

As mentioned in the introduction, one of the main interest in developing OFI soft X-ray lasers is that they require a low energy, high repetition rate pump laser. After the first demonstration by Lemoff et al. in 1995[8], we demonstrated saturated amplification on the Pd-like 5d-5p line at 41.8 nm in Xe^{8+} at the LOA, Palaiseau [1]. A 10 Hz Ti-Sa laser beam, yielding 330 mJ in 35 fs, was focused in a cell filled with xenon. More recently we have been involved in the extension of the OFI pumping scheme to Ni-like krypton, yielding strong lasing on the 4d-4p J 0-1 line, emitted at 32.8 nm. This work, led by S. Sebban, is described in detail in these Proceedings [2].

The target chamber of the new PALS laser facility in Prague has been designed, in collaboration with LIXAM, to accommodate several types of experiments, including the generation of X-ray lasers and the applications of theses sources. A first collaborative experiment was carried out in May 2001 in which the QSS Ne-like zinc laser, emitted at 21.2 nm was successfully implemented. This laser, operated in a double-pass, half-cavity geometry, was initially developed by the LIXAM/LSAI group at the 6-beam LULI laser facility [6]. By optimizing the conditions of irradiation of the zinc target, some of the characteristics of the zinc laser could be improved at PALS.

In particular the output energy could be brought to ~4 mJ, instead of 1 mJ previously obtained at LULI. More details are given by B. Rus in his Proceeding [9].

Finally regarding transient collisional pumping, we have been continuing our work on the temporal and spatial characterization of the Ni-like silver beam, emitted at 13.9 nm. In 2001 two experimental campaigns were carried out in collaboration at LULI and at the RAL facilities. The main results obtained will be summarized below in section 3. Preliminary results obtained more recently at LULI, in May 2002, will be also presented.

3. APPLICATIONS OF QSS X-RAY LASERS

Among the X-ray lasers that were discussed in the introduction, only QSS X-ray lasers were used in the application experiments that we have carried out until now. This is mainly because we have a longer experience of these systems which are now well controlled and characterized. Another reason is that we have had until now an easier access to the long pulse laser drivers that are used to generate the QSS X-ray lasers. In 2001 we have been involved in two application projects in collaboration with other groups. At LULI, the Ni-like Ag laser, pumped in the QSS regime, was used to irradiate biological samples. This project was started recently and is still at a preliminary stage, as will be explained below. For the second project, the Ne-like zinc laser, recently implemented at the PALS laser facility, was used to investigate niobium surfaces perturbed by a DC electric field with XUV interferometry. This project is the continuation of experiments initiated at LULI in 1999 [10]. New results obtained at PALS are presented below, and also discussed elsewhere [11] in these Proceedings.

3.1. Irradiation of biological samples

In collaboration with A. Touati and col. from GPS, Paris, we have used the 13.9 nm (~ 89 eV) Ag laser to irradiate dry plasmid DNA samples. The general context of this work is to improve the understanding, at a molecular level, of the damages induced to biological cells by XUV radiation. More specifically, the GPS group investigates the influence of inner-shell ionizations of C, O, and P atoms in the DNA molecule. Those processes, although of low occurrence, are believed to play a major role in inducing lethal damages to the cells [12]. The aim of our collaborative experiment was to complement measurements made by the GPS group at higher energies, with synchrotron light, by irradiating samples at 89 eV, that is below the inner-shell absorption edges of the DNA atoms. The sample used in the experiment was dry plasmid DNA pSp, which is able to survive under vacuum for about one hour. After irradiation the damage induced to DNA is quantified through the occurrence of single strand breaks (SSB) and double strand breaks (DSB, lethal) to the DNA helix. Those breaks change the morphology of the plasmids from the undamaged form, which can then be quantified by electrophoresis.

Figure 1 shows the experimental set-up used at the LULI 6-beam laser facility. The 13.9 nm Ag laser was generated at both ends of a 20 mm Ag plasma. A silver slab was irradiated by 6 focused and superposed laser beams, carrying a 80 J, 130 ps pulse preceded by a 1% prepulse. One of the X-ray laser beamlet was sent to a calibrated XUV CCD detector after reflection onto a 45° flat Mo:Si multilayer mirror. This was used to monitor from shot to shot the number of emitted XRL photons. The XRL beamlet emitted in the other direction was also reflected onto a 45° multilayer mirror for spectral selection and then sent to the biological sample, which was placed at 1.12 m from the XRL output end. In order to allow a reliable control of the potential damages induced by the X-ray laser radiation, another sample ("test-sample") was also placed under vacuum, at the back of the irradiated sample holder, where it did not see the X-ray laser photons. Both samples experienced the same time under vacuum and were later processed by electrophoresis with the same procedure.

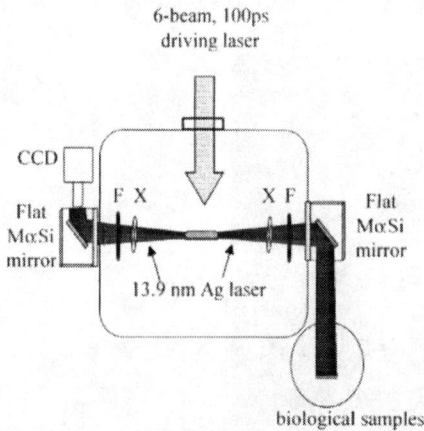

Figure 1. Experimental set-up: the biological sample is placed under vacuum at 1.12 m from the XRL source. The 45° Mo:Si mirror is used to select the XRL wavelength. The XRL beam emitted in the other direction is used to monitor the emitted number of XRL photons at each shot, by usin a calibrated XUV CCD. (F: filter; X: cross-wire)

Unfortunately during this experiment the output energy of the Ag X-ray laser was about 10 times lower than expected and we could only irradiate one biological sample. In order to reach the desired number of photons on the sample, we irradiated it with three successive laser shots. We estimated that the total number of photons onto the sample was of ~ 2. 10^{11}, corresponding to an absorbed dose of 1.6 kGy (at 89 eV, the absorption length in the biological material is ~100 nm). After electrophoresis, the irradiated sample did not show any evidence of damage induced by XRL irradiation, as the number of SSB's and DSB's was very similar to those found with the test (non-irradiated) sample. Thus a new experiment should be performed with an enhanced number of irradiating photons. It is considered that a dose of about 10 kGy is necessary at this photon energy to induce one SSB per plasmid in the irradiated sample [13].

3.2. Interferometry of Nb surfaces under electric field

Since 1999, in collaboration with the SEA Laboratory in CEA-Saclay and with the LCFIO, Orsay, we have been involved in the investigation of the deformations induced by strong electric field to niobium surfaces. The main motivation of this project is to get a better understanding of the phenomena, such as field electron emission, or breakdowns, that limit the performance of current superconductive cavities of accelerators. In an initial experiment, carried out at LULI, we have demonstrated that X-ray laser interferometry is an ideal tool for *in-situ* mapping of any nanometric change of the surface relief, during the very action of the electric field.

A new experiment was performed at PALS, using the newly implemented Ne-like Zn XRL. A description of the experimental set-up is given in [11]. During the experiment, DC electric fields of up to 100 MV/m were applied to the sample, which was a 2 µm niobium layer deposited on a glass substrate. Several interferograms of the perturbed surface were obtained for different values of the electric field, with either negative or positive polarisation.

Figure 2 shows a sequence of interferograms of the Nb suface under different values of the E-field, along with the reconstructed phase maps. Each map can be used to infer the elevation of the surface at each point with respect to a reference plane.

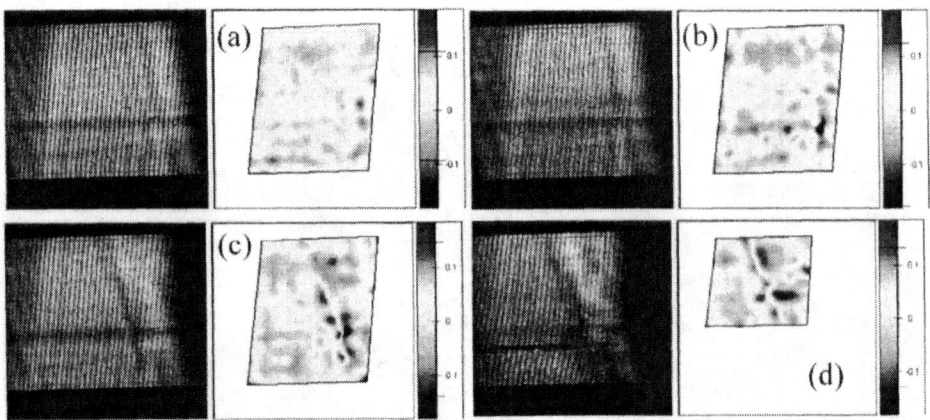

Figure 2. Examples of obtained interferograms and reconstructed phase maps for different applied DC E-fields: (a) no field; (b) 50 MV/m (negative polarity); (c) 37.5 MV/m (positive polarity); 50 MV/m (positive polarity).

It was observed that the negative polarity of the E-field did not induce any significant alterations to the surface. By contrast, the interferograms obtained for the positive polarity exhibited strong perturbations, and breakdowns were detected occasionally. In Figs. 3 (c) and (d) one can see that an elongated structure appears in dark in the reconstructed map. This can be interpreted as the formation of a ~30 µm high "step" on the surface, the position of which corresponds to the polarized knife edge anode which is set close to the niobium surface.

Detailed analysis and interpretation of the collected data is in progress. A new experiment is scheduled in September 2002, at the LULI, 6-beam laser facility, in which massive polished Nb samples will be investigated.

4. CHARACTERIZATION OF TRANSIENT X-RAY LASERS

In the transient pumping regime, a first low-intensity laser pulse preforms a plasma which is then rapidly heated by a short (~ 1 ps) high-intensity laser pulse. Because of the short (a few ps) duration of the induced population inversions a traveling-wave (TW) is achieved on the short heating pulse. At LIXAM, experimental investigations of transient X-ray lasers were initiated in 1998 [14]. We have focused our studies on the Ni-like silver 4d-4p X-ray laser line, emitted at 13.9 nm. This wavelength is particularly attractive for applications because it is close to the maximum reflectivity for X-UV optics based on Mo:Si multilayers.

In 2001 we have been continuing our work on the characterisation of the X-ray laser beam in time and space. Following the first demonstration of a very short, 2 ps XRL pulse at the RAL CPA-laser facility (UK) [15], numerical simulations were performed (see [16] in these Proceedings). A new collaborative experiment was performed at the same facility. The short duration of the silver XRL was confirmed, with a measured pulse duration of 3.5 ±0.5 ps, as shown in figure 3. A longer duration, of 13 ±2 ps, was measured the 3p-3s lasing line in Ne-like nickel, emitted at 23.1 nm. A more detailed discussion of the results obtained during this experiment is given in [17,18] in these Proceedings.

Figure 3. Streaked image of the 4d-4p Ni-like Ag lasing line. The measured pulse duration is 3.5 ±0.5 ps. More details are given in [23].

The spatial distribution of energy in the 13.9 nm XRL beam was investigated at the CPA LULI 100 TW facility. Figure 4.a shows the typical distribution of energy observed over the 13.9 nm laser beam section, at 1.4 m from the source. For a limited number of shots however this distribution exhibited very clear fringe patterns, an example of which is shown in figure 4.b. Such interference fringes can be interpreted as being due to multiple coherent sources emitting XRL beamlets that interfere in the far field. The origin of these multiple sources is not explained yet.

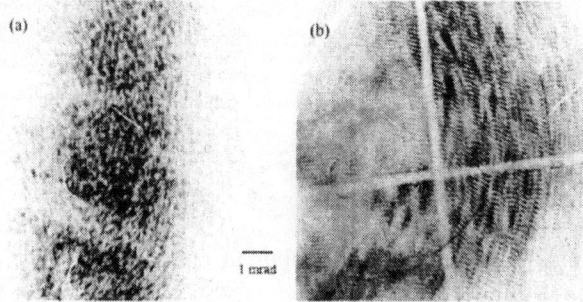

Figure 4. Examples of footprint images of the 13.9 nm laser obtained at a distance of 1.4 m from the source on a XUV CCD detector.
(a) typical image; (b) for a limited number of shots fringe patterns were observed.

However controlling and improving the quality of the beam is an important challenge for the future development of transient XRL's as a source for applications. A new experiment was performed recently in May 2002 in which images of the XRL emitting zone at 13.9 nm, at the output plane of the plasma filament (ie "near-field" images) were obtained with a high ~ 1 μm resolution [19]. The set-up of this experiment, performed in collaboration with LCFIO, is described elsewhere in these Proceedings [20].

Figure 5 shows two images obtained from a 3 mm long target. The difference between the two images is the width of the long pulse line focus, which was of 200 µm in figure 5a and 150 µm in figure 5b. The splitting of the XRL emitting pupil in two parts is probably due to "vertical" refraction (i.e. in the direction parallel to the target surface), which is stronger for the 150µm width. This interpretation has to be supported by 2D ray-tracing simulations, which are developed at LIXAM [21]. Other images obtained during this experiment, in various conditions of irradiation, show the presence of very small (a few µm) structures in the emitting pupil of the XRL. The detailed analysis and interpretation of these images are in progress.

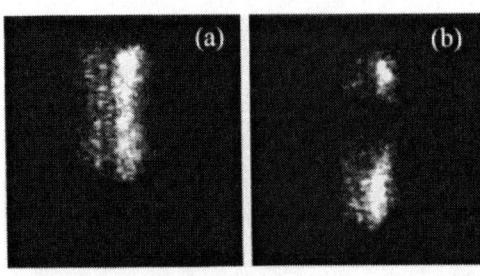

Figure 5. Near-field image of the transient Ni-like Ag XRL.

The dimensions of the frame containing each image is 122x144µm in the object plane. Laser: 450 ps, ~2 J, followed at t=250 ps by 450 fs, ~1 J (with TW on).

Width of the long pulse line focus:

(a) 200µm; (b) 150 µm.

5. CONCLUSION AND PROSPECT: THE LASERIX PROJECT

In this paper when have reviewed the studies of several variants of saturated, collisional X-ray lasers: OFI, QSS and transient systems. They have different pump laser requirements and yield sources with complementary properties for applications.

A new project of application was started recently in which the 13.9 nm laser is used to irradiate dry biological samples placed under vacuum. A first experiment was performed but the XUV dose sent to the sample did not allow to observe any induced effect. Further experiment, with a enhanced number of XRL photons will be scheduled.

A new experimental investigation of surfaces perturbed by an intense electric field was carried out at PALS. It confirms that the technique of Fresnel interferometry with an XRL is a powerful tool to investigate transient structures on perturbed surfaces with a high sensitivity. This method could thus be applied to other types of surface perturbations (laser irradiation at damage threshold, for example).

The pulse duration of the transient Ni-like Ag X-ray laser was confirmed to be very short, of a few ~ ps. Further experiments exploring the effect of varying the irradiation parameters are still required. The observed far-field distribution of energy in the 13.9 nm beam suggests that the emitting pupil of the XRL is composed of several mutually coherent sources that interfere at some distance from the source. The spatial structure of the emitting pupil of the XRL was investigated recently with a micronic resolution, using an aberration corrected, imaging, multilayer mirror. Preliminary results obtained were shown; the detailed analysis of the near-field images is in progress.

Finally the LIXAM laboratory in now involved in an ambitious and exciting project: the construction of a pump laser facility, dedicated to the development of applications of XRL's to a wide range of research areas. This facility, named LASERIX, will be based on transient XRL's, pumped by two ~ 10 J pulses with a duration of ~ 600 ps and ~1 ps respectively. The aim of LASERIX will be also the development of reliable transient XRL's, with controlled and stable characteristics, as well as the exploration of alternative pumping schemes, to extend XRL's to shorter wavelengths or higher repetition rate. LASERIX should be available to users in 2005. The study of the design of the experimental facility, including the pump laser, is now in progress. In particular we are now considering two different options for building a CPA laser delivering a total energy of ~ 30 J. These options involve two different amplifier technologies, based either on Nd:glass rods or Ti:Sa crystals. More details on the motivations and the status of LASERIX are given by G. Jamelot, leader of this project, in these Proceedings [22].

ACKNOWLEDGMENTS

The authors would like to acknowledge the technical support of P. Naucelles and G. Nicolas at LIXAM, as well as of the laser staff at the three laser facilities used for the experiments described in this paper: LULI, Palaiseau; RAL (UK) and PALS (Czech Republic). For these experiments we gratefully acknowledge the financial support from LEA (Laboratoire Européen Associé), from the European TMR Program (contracts FMGECT95-0053 -RAL facility, and HPRI-1999-CT00052 -LULI facility), and from the TMR X-ray Laser Network (contract N°ERBFMRXCT98-0185).

REFERENCES

1 . Sebban S. et al., *Phys. Rev. Lett.* **86,** 3004 (2001)
2 . Sebban S. et al., *these Proceedings*
3 . Kuba A. et al, *Phys. Rev. A* **62** (2000) 43808
4 . Lin J.Y. et al, *Opt. Comm.* **166** (1999) 211
5 . Dunn J. et al, *Phys. Rev. Lett.* **80** (1998) 2825
6 . Rus B. et al *Phys. Rev. A*. **55**, 3858 (1997)
7 . Sebban S. et al., *Phys. Rev. A* **61** (2000) 3810
8 . B.E Lemoff, G.Y. lin, C.L. Gordon III,C.P.J. Barty, and S.E. Harris, *Phys. Rev. Lett.74*, 1574 (1995)
9 . Rus, B. et al, *these Proceedings*
10 . Zeitoun, P. et al., Nucl. Instr. and Meth. in Phys. Res. A. **416** (1998) 189; Albert F. et al., *Phys. Rev. B* **60** (1999)11089
11 . Mocek, T. et al., *these Proceedings*
12 . Hervé du Penhoat M.A. et al., *Rad. Res.* **151** (1999) 649-658
13 . Hervé du Penhoat M.A., *private communication*
14 . Klisnick A. et al, *J.O.S.A. B* **17,** 1093 (2000)
15 . Klisnick A. et al, *Phys. Rev. A* **65** (2002) 033810
16 . Kuba J. et al., *these Proceedings*
17 . Edwards M., Guilbaud O. et al., *these Proceedings*
18 . Abou Ali Y. et al., *these Proceedings*
19 . R. Mercier, M. Mullot, M. Lamare, G. Tissot, *Rev. of Sc. Instr.*, **72**, 2 (2001) 1559-64
20 . Ros D. et al., *these Proceedings*
21 . Le Pape S. et al., *these Proceedings*
22 . Jamelot, G.. et al., *these Proceedings*

On the stability of the zinc x-ray laser beam quality using a half cavity

A. R. Präg*, T. Mocek, M. Kozlová and B. Rus

Inst. of Physics, Academy of Sciences of the Czech Republic, Na Slovance 2, 18221 Prague 8, Czech Republic
* corresponding e-mail: praeg@fzu.cz

Abstract. At the Prague Asterix Laser System Center (PALS) the Asterix laser delivering up to 700 J in 0.5 ns is used as a pump source for x-ray laser experiments and applications. The prepulse technique was applied which is known to improve the neon-like x-ray laser at the $J = 0 - 1$ transition dramatically. Since Zn slab targets were used the output wavelength was 21.2 nm. A prepulse having up to 20 J precedes the main pulse by 10 ns. The main beam and the prepulse beam are focussed by two different optical systems separately and their foci are superimposed at the target surface. By implementing a half-cavity for double-pass amplification using a Mo/Si multilayer mirror – which can be used for 100 shots – the x-ray laser output was more than 10 times stronger than at the single pass in a 3 cm long plasma. Double-pass amplification was observed to be most efficient when the pump pulse was at least 150 ps longer than the round trip time (≈ 260 ps) in the half-cavity. Under this fundamental condition the x-ray laser reached saturation in the double-pass regime containing ~4mJ energy which was proved to be enough for applications. In this contribution, the x-ray laser features like divergence in two dimensions, the beam quality (symmetry), and the pointing angle are investigated over 110 shots. To characterize the stability of the x-ray laser the shot distribution, the mean value and the standard deviation for these parameters are evaluated. For 18 shots of a one-day-series these values are given, and a statistical analysis carrying out a chi-squared test characterize the Zn x-ray laser as a robust tool suitable for future applications.

INTRODUCTION

Using different approaches X-ray lasers (XRL) were developed in a couple of laboratories [1]. The first approach used collisional excitation (CE) for that several intense XRL over a broad wavelength range were realised in Ne-like [2-6] and Ni-like ions [7,8]. Output energies of 7 mJ in Ne-like Y [2] were reported. Applying the modified pumping method of transient CE, XRL were demonstrated on Ne-like [9,10], and on the Ni-like scheme [11-14]. A second approach is based on capillary-discharges where lasing takes place in a gas-filled capillary driven by a high-current pulse. Strong lasing at 46.9 nm in Ne-like Ar was attained with this method [15]. In a third approach ultra-short pulses induce optical-field ionisation in a noble gas. Using this method lasing at 41.8 nm in Pd-like Xe [16] was realised. In this work we use the prepulse technique which improves the CE $J = 0 - 1$ XRL dramatically [17]. We examine the double-pass amplified Ne-like Zn at 21.2 nm, using 3-cm long plasma and a half-cavity set-up. Saturation of the half-cavity Zn XRL has been demonstrated earlier in [5].

The diversified application of XRL is coupled to the condition of an appropriate stability of the pulsed XRL beam. The stability of XRL beam parameters (beam size, divergence, symmetry, pointing angle, energy, coherence and pulse duration) was to

CP641, *X-Ray Lasers 2002: 8th International Conference on X-Ray Lasers,* edited by J. J. Rocca et al.
© 2002 American Institute of Physics 0-7354-0096-2/02/$19.00

our knowledge not yet investigated over a long series of shots. For practical applications of XRL, it is necessary to have not only a high-rep rate but also a high shot-to-shot reproducibility. Mostly, single-shot results are published, i.e. when long-pulse systems at low rep-rate are used. However, a considerable shot-to-shot fluctuation was reported [18], but without analysing the reasons in detail. Despite possible reasons for the fluctuations (pump laser instabilities, scattering of pulse duration, degradation of optical components due to damage) in this contribution we study the stability over 110 shots, at a low rep-rate of 20 minutes between two shots. Nevertheless, the low rep-rate constitutes a considerable disadvantage for the statistical analysis, because the confidence into statistics is increasing with the size of the random sample gained in a rational time. Typical values for the XRL beam divergences reported recently are between 2.2 and 3 mrad for Ni-like Ag in horizontal direction and between 5 and 6 mrad in vertical direction [19]. In both directions a slight dependence on the focus width was found [19]. Generally for the vertical divergence a higher value was measured than for the horizontal divergence [1]. Defining a dimensionless parameter, called the *beam symmetry* as the ratio of horizontal to vertical divergence, a value clearly < 1 was published. This result is due to the asymmetric plasma expansion in either direction (parallel or perpendicular to the target surface). Due to the same context from two-dimensional (2D) near-field images of the Ni-like Mo XRL can be extracted that the gain region in the plasma is in the direction parallel to the target larger than perpendicularly [14]. 2D far-field patterns are reported for Ni-like Ag [19], and for the capillary-discharge Ar laser [20]. The XRL pointing angle was investigated in [18]. Double-pass amplification applying a two-target geometry [21] or using a half-cavity [5] was realised. Time-resolved measurements exhibit XRL pulses of 100 ps for the Ne-like Zn, however, without analysing the stability over numerous consecutive shots [5]. This contribution demonstrates that a half-cavity Zn XRL, pumped by a 600 J/450 ps laser, has an excellent stability in the beam quality in a series of 100 shots thus representing a robust tool suitable for applications. It is shown that one multi-layer mirror is sufficient for >100 shots without re-aligning the half-cavity. The beam stability of a double-pass XRL is investigated under the condition that the prepulse and the half-cavity geometry are fixed. The shot-to-shot fluctuations of the beam divergence, symmetry, and pointing angle are analysed. The obtained results are discussed using χ^2-tests [22], preferably assuming Gaussian distributions.

EXPERIMENTAL SET-UP

The experiments were performed at PALS [23], where the Asterix laser, developed at the Max-Planck-Institut für Quantenoptik (Garching) was installed [24]. This laser is capable of delivering 700 J in 0.5 ns and is hence suitable to pump XRL efficiently. The laser operates in the infrared (IR) at 1315 nm. The experimental set-up including the mainly used components is shown in [25]. Two focussing optics assembled in front of the target chamber [23] produce a narrow line focus. The main beam is focussed by an asphere combined with a cylindrical lens matrix consisting of 10 cylindrical lenses. The line focus is 3 cm long and ~130 μm wide. The cylindrical lens matrix [25] is arranged as two arrays of five lenses (5x2 matrix). The prepulse guided separately to the target chamber is focused by a second optics consisting of a spherical and a cylindrical lens. The independent focussing has the unique advantage that both line foci can be modulated separately. The 35-mm long prepulse line was slightly defocused to a width of ~700 μm, so that the centre of the pre-plasma is as uniform as possible. The superposition of both line foci was controlled at each shot by an off-axis x-ray pinhole camera. This camera using an off-shelf CCD combined with a 3-μm Al filter images the incoherent x-rays emitted from the plasma and hence monitors the width and uniformity of the line focus. An electronic device constructed recently uses two fast IR-sensitive InGaAs photo-diodes and controls the pepulse delay at every shot. The response time-gap between both diodes was measured on a 3-GHz bandwidth Tektronix scope. This prepulse monitor is able to detect parasitic prepulses of the pump laser at a contrast of 10^{-4}, however, in fact parasites could be excluded during our experiments. The flat 3-cm wide targets were polished Zn slabs having a roughness of 5 μm and a purity of 99.9%. One target was used for more than 100 shots, any at fresh target surface. The half-cavity used flat Mo/Si multi-layer mirrors of 25-mm diameter. At 21.2 nm a reflectivity of 30% was measured [25]. The distance between multi-layer mirror and target edge was 8.5 mm. The mirror angle can be varied between −10 and +10 mrad. The unused mirror surface was protected by a 1-mm thick shield having a 1-mm pinhole as a gap for the XRL beam. Since after each shot the mirror surface was damaged in a 1-mm spot the mirror was moved behind the shield by 2 mm without changing the half-cavity geometry. This procedure enables 100 shots with one mirror. The mirror motion system assembled in the vacuum chamber was PC-controlled thus a breaking of vacuum for aligning was not necessary. The XRL beam was analysed by a spectrometer or by a footprint detector. The beam was switched between both diagnostics by the means of a flat retractable Mo/Si multi-layer mirror working at 45°. As primary diagnostics served a Wadsworth spectrometer consisting of a curved blazed grating (900 l/mm) and a phosphor coupled to a cooled CCD [25]. The phosphor is mounted tangentially on the Wadsworth circle of 1.2-m radius. To avoid detector saturation at half-cavity shots we used a 1.5-μm Al filter transmitting 11% at 21.2 nm. A second diagnostics detected a 2D far-field pattern (footprint) of the XRL giving the beam divergence parallel and vertical to the target surface. A Mo/Si multi-layer mirror reflects the x-ray beam onto a 38-mm diameter phosphor. The path length from the target to the phosphor via the mirror is 1.1 m. The phosphor (grain-size<3μm) was a Gd_2O_2S:Tb aluminised with a 50-nm thick layer, hence insensitive in the visible and IR. The phosphor is protected against scattered light by a 20-cm long metal tube thus only the radiation reflected by the multi-layer mirror is illuminating the phosphor. An objective and a CCD working in the visible record the footprint. A hair-cross of 120-μm-wires served as a fiducial 22 cm in front of the phosphor. The experiment had the following order. The prepulse was optimised for single pass using the Wadsworth spectrometer. Then the half-cavity geometry was optimised. Finally the beam quality was studied using the footprint diagnostics. Additionally, at any shot the pump energy and the pulse duration were measured.

EXPERIMENTAL RESULTS

The results were obtained using a 1.6 J prepulse, followed after 10 ns by the main pulse at an irradiance of ~2.8×10¹³ Wcm⁻². Due to the ~700 μm-wide prepulse focus the pre-irradiance was ~1.6×10¹⁰ Wcm⁻² which corresponds to a prepulse ratio of ~6×10⁻⁴. Former results confirm that this weak prepulse optimises the Zn XRL [3]. The laser at 21.2 nm dominated the spectrum, thus the negligible background enabled us to record footprints. A single-pass XRL footprint (3-cm-target) [26] showed an ellipsoidal beam having a horizontal divergence of 3 mrad and a vertical divergence of 5 mrad and thus a symmetry of 0.6. The pointing angle of 5.5 mrad was identical in all single-pass shots. Activating the half-cavity mirror, the XRL output was enhanced by an order of magnitude, thus the objective aperture of the footprint CCD was reduced without affecting the beam divergence measurement. Considering the round trip time of 257 ps it is emphasised, that the half-cavity is working efficiently only under the condition, that the pump pulse is at least 400 ps long [25]. For shorter pulses it is not guaranteed, that the inversion is lasting long enough for a whole round trip. For that reason the half-cavity length was put to the shortest feasible of 38.5 mm.

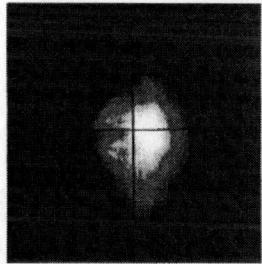

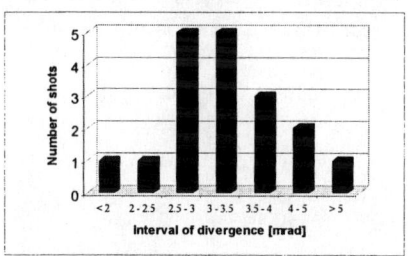

FIGURE 1. Footprint of the double-pass Zn XRL beam. The target is left and the shadows are due to the fiducial wires.

FIGURE 2. Distribution of the half-cavity XRL beam divergence in horizontal direction.

Figure 1 displays a footprint of the half-cavity XRL beam at the re-injection angle of 3 mrad. The pointing angle was ~5 mrad and the horizontal and vertical beam divergence are 3.8 and 5.8 mrad, respectively, and thus the beam symmetry of 0.65 is close to that of the single pass. The pointing angle was identical for all half-cavity shots. The statistical analysis carried out in this work uses the half-cavity shots

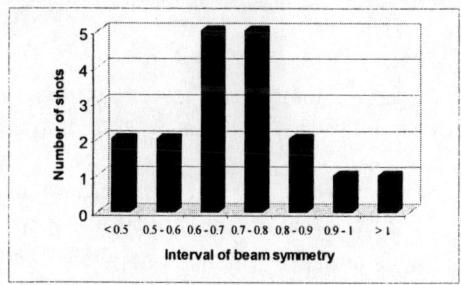

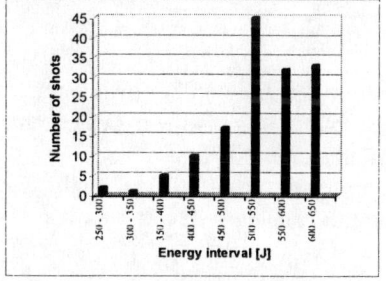

FIGURE 3. The XRL beam symmetry distribution.

FIGURE 4. The pump laser energy distribution observed at more than 100 XRL shots.

exclusively. Figure 2 shows the distribution of the horizontal beam divergence observed at 18 subsequent shots. The frequency at each interval is given on the

vertical axis. The vertical divergence can be found in [26]. The dimensionless ratio of horizontal to vertical divergence qualifies the beam symmetry. The distribution of observed beam symmetries is plotted in figure 3. For comparison, the distribution of the pump energies for more than 100 XRL shots shows obviously an asymmetric distribution (figure 4). The remarkable difference of both distributions is analysed by the means of a χ^2-test.

STATISTICAL ANALYSIS OF THE XRL BEAM QUALITY

The double-pass XRL output was observed to be robust and reproducible in numerous shots. We execute a statistical analysis including a χ^2-test that is used as a common tool to control the quality and reliability of products or processes. The procedure of a χ^2-test is described elsewhere [22] and is applied here onto the XRL beam quality. At the beginning stands a hypothesis of the underlying distribution, in fact, we assume a Gaussian distribution. This assumption makes sense in all experiments where the errors are *randomly* distributed and independent of controllable parameters. The observed fluctuations are considered to have statistical reasons. Possible systematical errors (target quality, alignment, temperature fluctuations etc.) are not investigated here. The χ^2-test is performed in three steps: (a) We define seven intervals and assign the experimental results to these intervals. The number of intervals leads to the degrees of freedom f of the test. (b) By assuming a Gaussian distribution we calculate the χ^2 using the formula (1). (c) We compare the result of χ^2 with a probability table [27] and extract a probability of making no error by accepting or rejecting the hypothesis. The probabilities printed in tables [27] are calculated by using equation (2). - It is emphasised that the test is not telling whether the hypothesis is true or not, it is only giving a probability to accept or to reject the hypothesis. In the case of rejecting the hypothesis the test is not explaining the reasons for doing that. In the case of acceptance the test is delivering a probability for the chance of making an error, anyway. The χ^2 measures the deviation between the observed (*figures 2-4*) and the expected distribution.

$$\chi^2 = \sum_{i=1}^{k} \left(\frac{(m_i - Np_i)^2}{Np_i} \right) \tag{1},$$

where k is the number of intervals, m_i is the observed number of shots (*frequency)* in the interval i, p_i is the expected probability of the interval i, and N is the total number of all shots.

The result of a statistical analysis determining the average a of all measured values, and the standard deviation s is given in table 1 for horizontal and vertical divergence and beam symmetry. For the χ^2-test we use a and s as input parameters for a normalised Gaussian distribution. Hence, the probability p_i to find a result in the interval i is fixed through the expected distribution. The detailed equation is written out in [26]. The expected frequency at the interval i is Np_i. The division of the spectrum into intervals is so far free if there is no overlap and additionally the full observable spectrum is covered. Organising seven intervals (s. figures 2,3), and counting the number of experimental hits (*frequency*) in each interval, we receive the f of the χ^2-test defined as $f = k - 1 -$ (number of estimated parameters). Hence with 7 intervals and two parameters a and s we get $f = 4$. The χ^2-distribution described by a fundamental density function $d(y,f)$ is given in [22]. Integration of the density function $d(y,f)$ leads to a probability $\beta(\chi^2,f)$ for rejection of the hypothesis given in equation (2):

$$\beta(\chi^2,f) = \left(\frac{1}{2^{f/2}\,\Gamma(f/2)}\right) \int_0^{\chi^2} \left(\exp(-y/2)\,y^{\frac{f}{2}-1}\right) dy = \int_0^{\chi^2} d(y,f)\,dy \qquad (2),$$

with the upper integration limit χ^2 given in equation (1). The substitution $y := \chi^2$ is used for simplification and Γ denotes the gamma-function. At fixed f with higher χ^2 the rejection probability β is enlarged. For $f = 4$ the integration can be solved analytically and we receive

$$\beta = 1 - (2 + y)\exp(-y/2)/2 \qquad (3).$$

The acceptance probability α of the hypothesis is $\alpha = 1 - \beta$. With extended χ^2 the α will be smaller and v.v. The density function $d(\chi^2,4)$ and the probability function $\beta(\chi^2,4)$ are plotted in [26] for $f = 4$. For higher f and several χ^2 the calculated $\beta(\chi^2,f)$ can be found in tables [27].

If χ^2 is above threshold the hypothesis should be rejected. Conversely, the hypothesis can be accepted since the χ^2-test is not in contradiction. Customarily there is set a critical probability level above that the hypothesis will be rejected, e.g., $\beta=0.50$ or $\beta=0.95$. In the latter case remains a 5%-error when the hypothesis is rejected. For $f = 4$ the thresholds are $\chi^2 = 3.36$ ($\beta=0.50$) and $\chi^2 = 9.49$ ($\beta=0.95$), respectively. Particularly, if the sample is small (<100) or the frequency in some intervals is <3, this can lead to an over-estimation of χ^2 and thus to an under-estimation of the acceptance probability. Thus we will not accept the hypothesis careless. Clearly the statistics will be improved with increasing number of shots. Examples for the execution of the χ^2-test are given in detail in [26] for the beam symmetry and the 2D-divergences of the half-cavity XRL observed in a series of shots. The result for the beam symmetry is $\chi^2 = 1.4$. Using (3) we calculate $\beta = 0.16$ and $\alpha = 0.84$. For the divergences similar tests are performed [26] and the results are listed in table 1. Concerning the beam

TABLE 1. Statistical parameters and results of chi-squared tests for the Zn XRL beam properties.

	Horizontal divergence	Vertical divergence	beam symmetry
Number of shots	18	18	18
Average a	3.43 mrad	4.77 mrad	0.75
Standard deviation s	0.92 mrad	1.31 mrad	0.25
Sum of χ^2	3.03	2.03	1.40
Probability $\alpha = 1 - \beta$	0.55	0.73	0.84

symmetry, as well as the horizontal and vertical divergences, β is less than the critical level: The hypothesis can be accepted, because there is no significance for rejection. In the case of rejection remains an error probability of more than 55% for these beam parameters as given in table 1. This is the only argument to accept the hypothesis. It is not clear, whether the hypothesis is really true. But from the statistical point of view the Gaussian distribution seems to be authentic. In the opposite case, if β would be higher than the critical level the hypothesis would be rejected doing an error of $< \alpha\%$. This latter case is not fulfilled here. Considering the both beam divergences and the beam symmetry there is no argument to reject the hypothesis. An objection that the χ^2-test is favouring the acceptance of hypothesis can be enfeebled when the pump energy showing an asymmetric-shaped distribution is analysed (figure 4). Here the statistics is improved using a larger random sample. A mean energy of 523 J and a standard deviation of 97 J are extracted. A χ^2-test gives $\chi^2 = 22.5$, $\beta = 0.999$, and hence $\alpha < 0.1\%$. The Gaussian-hypothesis must be rejected obviously. Definitely, the pump energy is not random-distributed. That means that the XRL demonstrates a higher stability than the drive laser, confirming additionally that the double-pass Zn XRL is saturated [25].

SUMMARY

The beam quality of a half-cavity Zn XRL (21.2 nm) delivering stable output over many shots is studied in a statistical analysis including χ^2-tests. At fixed half-cavity geometry the hypothesis that the divergence is Gaussian distributed is confirmed. The 450-ps long pump pulse was adequate to maintain the inversion over the whole roundtrip in the 38.5-mm short half-cavity usable routinely without exchange of any component. Protecting the unused mirror surface we performed 110 consecutive shots. High reproducibility recommends this XRL as a robust tool for applications. The Zn XRL had a nearly symmetrical beam at a stable pointing angle. This enables applications applying 3 reproducible shots per hour. It is planned to allocate XRL beam time to external scientists. Clearly the statistics will be improved with increasing number of shots, thus the beam diagnostics will be continued further. Concluding, the narrow-scattered XRL beam quality compared with the rather wide-scattered input pump energy illustrates an impressive stability of the saturated half-cavity zinc XRL.

ACKNOWLEDGEMENTS

Fundamental and technical support of J. Moravec, J. Sobota, K. Rohlena, J. Ullschmied, and K. Jungwirth is greatly appreciated. This work was financed from the EU Access to Research Infrastructures Grant HPRI-00108, from the Research Centres Project LN00A100, and from the Czech Academy of Sciences Grant A1010014.

REFERENCES

1. Rocca, J.J., *Rev. Sci. Instr.* 70, 3799 (1999)
2. DaSilva, L.B., MacGowan, B.J., Mrowka, S., Koch, J.A. et al., *Opt. Lett.* 18, 1174 (1993)
3. Präg, A.R., Löwenthal, F., Tommasini, R., and Balmer, J.E., *Appl. Phys. B* 66, 562 (1998)
4. Li, Y., Pretzler, G., and Fill, E.E., *Phys. Rev.* A 51, R4341 (1995)
5. Rus, B., Carillon, A., Dhez, P., Jaeglé, P., Jamelot, G. et al., *Phys. Rev. A* 55, 3858 (1997)
6. Präg, A.R., Löwenthal, F., and Balmer, J.E., *Phys. Rev. A* 54, 4585 (1996)
7. Zhang, J., MacPhee, A.G., Lin, J., Wolfrum, E., Smith, R., Danson, C. et al., *Science* 276, 1097 (1997)
8. Sebban, S., Daido, H., Sakaya, N., Kato, Y., Murai, K. et al., *Phys. Rev. A* 61, 043810 (2000)
9. Nickles, P.V., Shlyaptsev, V.N., Kalashnikov, M., Schnürer, M. et al., *Phys. Rev. Lett.* 78, 2748 (1997)
10. Warwick, P.J., Lewis, C.L.S., Kalashnikov, M.P., Nickles, P.V. et al, *J. Opt. Soc. Am. B* 15, 1808 (1998)
11. King, R.E., Pert, G.J., McCabe, S.P., Simms, P.A., Lewis, C.L.S.,et al., *Phys. Rev. A* 64, 053810 (2001)
12. Klisnick, A., Kuba, J., Ros, D., Smith, R., Jamelot, G. et al., *Phys. Rev. A* 65, 033810 (2002)
13. Dunn, J., Li, Y., Osterheld, A.L., Nilsen, J. et al., *Phys. Rev. Lett.* 84, 4834 (2000)
14. Li, Y., Dunn, J., Nilsen, J., Barbee, T.W. et al., *J. Opt. Soc. Am. B* 17, 1098 (2000)
15. Rocca, J.J., Clark, D.P., Chilla, J.L.A., and Shlyaptsev, V.N., *Phys. Rev. Lett.* 77, 1476 (1996)
16. Lemoff, B.E., Yin, G.Y., Gordon, C.L., Barty, C.P.J., and Harris, S.E., *Phys. Rev. Lett.* 74, 1574 (1995)
17. Nilsen, J., MacGowan, B.J., DaSilva, L.B., and Moreno, J.C., *Phys. Rev.A* 48, 4682 (1993)
18. Kuba, J., Klisnick, A., Ros, D., Jamelot, G. et al., *J. de Physique IV (France)*11, Pr2-35 (2001)
19. Tang, H.J., Daido, H., Suzuki, M., Yamagami, S. et al., *J. de Physique IV (France)*11, Pr2-129 (2001)
20. Moreno, C.H., Marconi, M.C., Shlyaptsev, V.N., Benware, B.R. et al., *Phys. Rev. A* 58, 1509 (1998)
21. Daido, H., Sebban, S., Sakaya, N., Tohyama, Y. et al., *J. Opt. Soc. Am. B* 16, 2295 (1999)
22. Kenney, J.F., and Keeping, E., *Mathematics of Statistics*, Princeton, Van Nostrand (1951)
23. Rus, B., Mocek, T., Präg, A.R., Lagron, J.C. et al., *J. de Physique (France) IV* 11, Pr2-589 (2001)
24. Jungwirth,K., Cejnarova, A., Juha, L., Kralikova, B. et al., *Phys.Plasmas* 8, 2495 (2001)
25. Rus, B., Mocek, T., Präg, A.R., Kozlová, M., Jamelot, G. et al., *submitted to Phys. Rev. A* (2002)
26. Präg, A.R., Mocek, T., Kozlová, M., Rus, B., Jamelot, G., and Ros, D., *submitted to Eur. Phys. J. D* (2002)
27. Owen, D.B., *Handbook of statistical tables*, Reading, Palo Alto, London, Addison-Wesley (1962)

Multi-millijoule, highly coherent X-ray laser at 21 nm as a routine tool for applications

B. Rus[1], T. Mocek[1], A.R. Präg[1], M. Kozlová[1], M. Hudeček[1], G. Jamelot[2], J-C. Lagron[2], A. Carillon[2], D. Ros[2], D. Joyeux[3], and D. Phalippou[3]

[1] Gas Lasers Department / PALS Centre, Institute of Physics, Academy of Sciences of the Czech Republic, Na Slovance 2, 18221 Prague 8, Czech Republic
[2] Laboratoire d'Interaction du rayonnement X Avec la Matière, Bâtiment 350, Université Paris-Sud, 91405 Orsay cedex, France
[3] Institut d'Optique Théorique et Appliquée, Laboratoire Charles Fabry, Bâtiment 503, Université Paris-Sud, 91403 Orsay cedex, France

Abstract. A recently developed double-pass, deeply saturated Ne-like Zn X-ray laser at 21.2 nm, is presented. The system consists of a 3-cm long plasma and a half cavity constituted by a flat multilayer mirror located at 8.5 mm. The pump sequence includes a weak prepulse and a main heating pulse, delivering ~500 J of net energy at 1.315 μm, in pulses with a length of ~450 ps. A novel prepulse illumination geometry is employed, in which the prepulse beam generates a wide preplasma buffer into which subsequently a much more tightly focused main beam is coupled. The emitted X-ray beam is narrowly collimated and possesses a remarkably high spatial quality. In the half-cavity regime, the system provides an output energy of about 4 mJ, corresponding to peak power of 40 MW. A fully routine operation of the half-cavity has been achieved, making it possible to perform, without any realignment or manual intervention, about 100 shots with one multilayer mirror.

INTRODUCTION

Despite tremendous progress made over the past few years in the development of transiently pumped and capillary-discharge based collisional X-ray lasers (XRL) [1,2], laser-plasma systems driven in the regime of a few hundred ps down to sub-100-ps offer the possibility to produce the largest pulse energy and peak power. If such XRLs are able to generate an adequately coherent beam with high spatial quality, they would find a number of unique specific applications.

The objective of the described experiments was to implement a robust XRL based on Ne-like Zn [3,4,5], delivering several mJ in a narrowly peaked beam, and employing an automated half cavity in a fashion compatible with the "adjust and forget" approach. The experiments, carried out with newly developed XRL hardware [6], were conducted at the recently established PALS (Prague Asterix Laser System) Centre [7] hosting the former Asterix IV iodine laser and a configurable tandem of interaction chambers.

CP641, *X-Ray Lasers 2002: 8th International Conference on X-Ray Lasers*, edited by J. J. Rocca et al.
© 2002 American Institute of Physics 0-7354-0096-2/02/$19.00

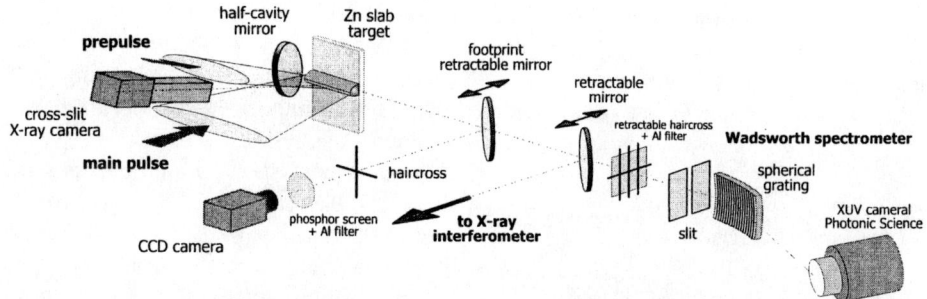

FIGURE 1. Experimental setup (not to scale). The X-ray laser consists of a 3-cm long Zn slab target and a flat multilayer mirror distant 8.5 mm. The pumping sequence consists of a weak prepulse beam injected at 25° from above, followed after 10 ns by the main pump pulse injected horizontally and focused by the composite optics represented in Fig.2a. The XRL beam is switched between the individual diagnostics by the respective 45° retractable multilayer mirrors.

EXPERIMENTAL ARRANGEMENT

The employed experimental setup is shown in Figure 1. The target consists of an optically polished zinc slab 30×50 mm, which may be used for about 100 XRL shots. It is irradiated with a joule-level prepulse, the line focus of which is deliberately placed out of the target plane, and by the main heating pulse delivering ~500 J of net energy focused tightly to a narrow line. Both pulses have the same duration, which is nominally 450 ps FWHM. While the prepulse beam of 15 cm in diameter is focused by a simple combination of a cylindrical and a spherical lens, the main beam of 29 cm in diameter is focused by a novel composite optics depicted in Figure 2a. The system provides a narrow, highly uniform 3-cm long line focus, owing to the precision (parallelism of the face surfaces, parallelism of the generating axis to the side edges) at which shorter cylindrical lens segments may be fabricated cost-effectively. As a consequence, the system outperforms the arrangements employing single-piece cylindrical segments extending over the full beam height, the sole penalty being almost a negligible screening effect of the central strut.

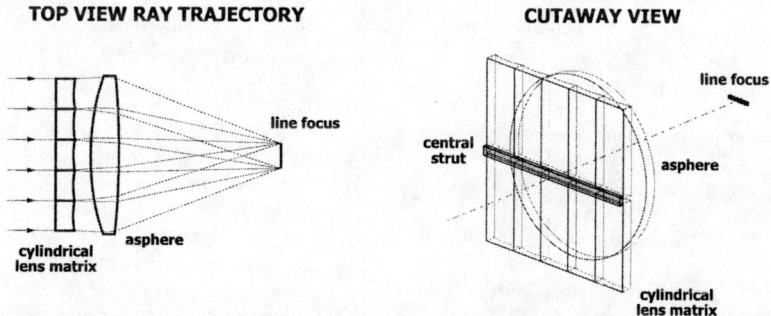

FIGURE 2a. Generic scheme of the composite focusing optics with 30-cm clear aperture. A matrix of 10 cylindrical lens segments 150×60 mm and a single $f/3$ aspherical lens are employed. The central 10-mm wide central support strut is covered by a volumetrically absorbing glass.

The half cavity, shown in Figure 2b, consists of a 25-mm-diam flat multilayer mirror, with a reflectivity near 21 nm of ~30%. The mirror is positioned 8.5 mm from the plasma end, which is the minimum distance possible due to the experimental configuration. The mirror is mounted on a compact 5-axis motorized support allowing quick re-positioning between subsequent shots, which consists in a displacement by 2 mm to expose a fresh multilayer surface to the XRL beam in the next shot. The re-positioning effectively takes less than a second. One single mirror can be used for about 100 shots, without any further realignment or manual intervention to the vacuum chamber.

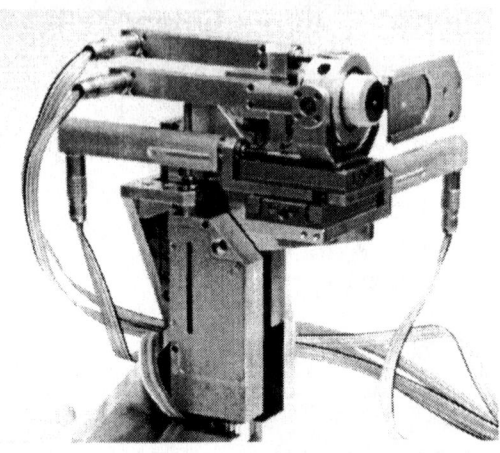

FIGURE 2b. The automated half cavity pad; the 25-mm diam flat multilayer Mo:Si mirror is protected from the plasma by a thin stainless shield having a 1-mm hole to pass the XRL beam.

The XRL emission may either be detected as a footprint by a monitor located at a distance of 110 cm downstream from the plasma exit plane, or spectrally analyzed by a Wadsworth-type spectrometer. The XRL beam may further be sent to the "application" channel, which was here occupied by an X-ray interferometer (see below).

RESULTS AND DISCUSSION

The geometry of the XRL active medium is apparent from Figure 3. The preplasma, produced by the deliberately de-focused prepulse beam with a nominal energy of 1.6 J, extends laterally to ≈700 µm and longitudinally overfills the target. The net irradiance created on the target by the prepulse is ≈2×10^{10} Wcm^{-2}. The main beam, applied to thus created plasma "buffer" after 10 ns following the prepulse, generates a highly uniform 130-µm wide narrow line, corresponding to net irradiance of ~2.8×10^{13} Wcm^{-2}.

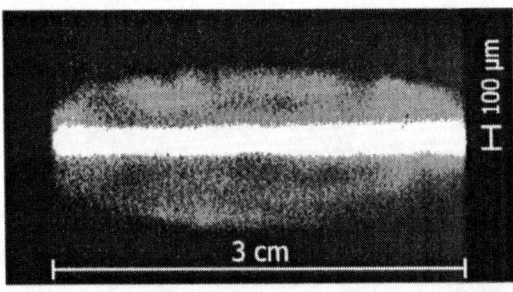

FIGURE 3. The amplifying plasma column viewed in the multi-keV X-ray spectrum by the cross-slit X-ray camera. The image is synthesized from two separate records, the first taken with the prepulse beam alone deliberately shot with high energy (~60 J) to visualize the preplasma, and the second one involving the main beam fired with nominal energy. Note different horizontal and vertical scales.

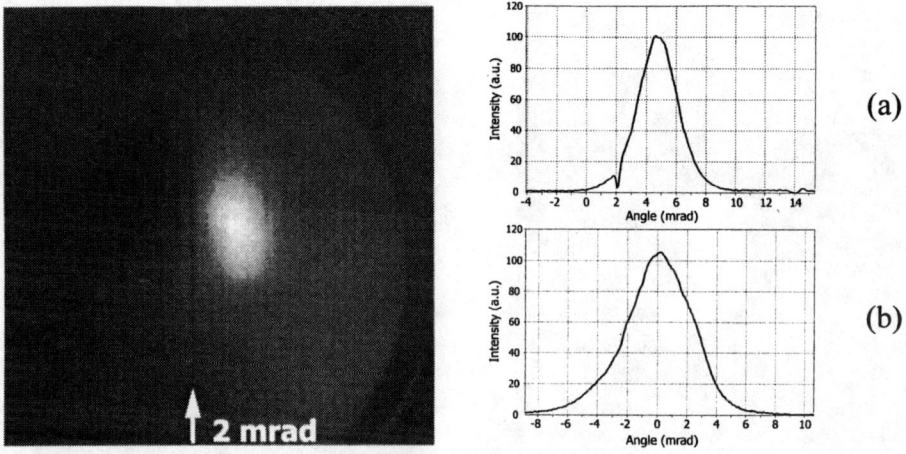

FIGURE 4. Footprint of the ASE single-pass XRL beam emitted by the plasma of a length of 3 cm, and its (a) horizontal and (b) vertical lineout. The target position is on the left from the beam.

A typical footprint of the single-pass ASE beam is shown in Figure 4. The footprint corresponds to the optimum prepulse energy of 1.6 J. A smooth ellipsoidal beam possessing a high degree of both horizontal and vertical symmetry and constituting a true "beam" in the usual sense of laser physics, is produced. Its horizontal and vertical divergence is 3(±0.5) mrad and 5(±0.5) mrad, respectively; the beam emerges from the plasma at an angle of ~5 mrad.

The dependence of the XRL output on the prepulse energy is displayed in Figure 5. The data were obtained both by keeping the prepulse focus conditions and the energy of the main pulse constant. They show that the laser action at 21.2 nm is optimized, under the given experimental conditions, for a prepulse energy close to 1.6 J, which corresponds to the prepulse-to-main pulse irradiance ratio of $~6\times10^{-4}$. We note that this observation of a specific weak prepulse optimizing the J=0-1 zinc laser corroborates with the former results obtained using a significantly different target illumination arrangement [8].

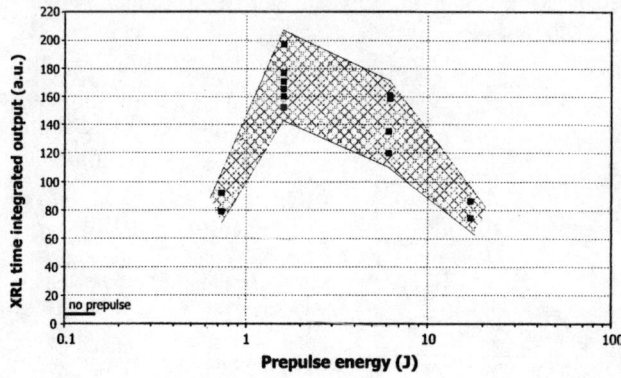

FIGURE 5. The 21.2-nm XRL single-pass output as a function of the prepulse energy; the error bar space is represented by a shaded region. The optimum prepulse energy 1.6 J corresponds to a net target irradiance of $\approx2\times10^{10}$ Wcm^{-2}.

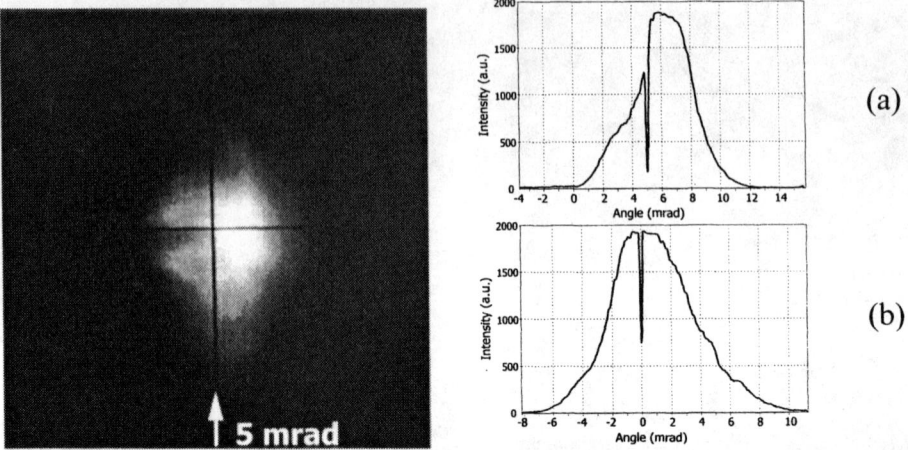

FIGURE 6. Footprint of the half-cavity XRL beam emitted by double-passing a 3-cm plasma, and its (a) horizontal and (b) vertical lineout. The target position is on the left from the beam.

In the double-pass regime, the system produces typically 11-fold enhancement of the XRL output compared with the single-pass ASE signal, depending on the re-injection conditions provided by the half cavity [9]. Figure 6 displays a footprint of the XRL beam corresponding to the optimum re-injection conditions. The beam emerges from the plasma under an angle of ~5 mrad and its horizontal and vertical divergence of 3.8(±0.5) and 5.8(±0.5) mrad, respectively, is close to those of the single pass. It is to be emphasized that both the XRL beam intensity and its profile produced by the half cavity were largely robust with respect to the pumping conditions [10].

Considering the half-cavity mirror reflectivity of ~30% and the geometry of the beam re-injection back to the gain region, the 11-fold enhancement of the XRL output implies the net amplification available in the return pass of ~200 (for details see [9]). To verify that this value corresponds to a saturated system and not to a gain fading (the radiation transit time of 256 ps in the employed half cavity is not negligible to the 450-ps pump pulse), shots with different lengths of the pump pulse were fired. The obtained dependence, displayed in Figure 7, exhibiting an increasing tendency with the pulse shortening down to ~380 ps below which it strongly diminishes, clearly indicates that under the nominal pump conditions the gain lifetime fully sustains the return amplification.

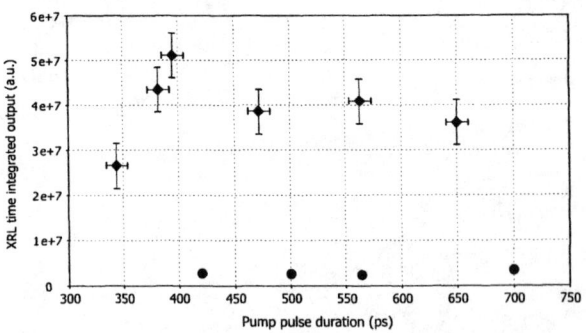

FIGURE 7. Dependence of the single-pass (●) and half-cavity (◆) XRL output on the pump pulse duration. The double-pass data show that pump pulses longer than ≈380 ps produce sufficient gain lifetime to support return amplification in the system.

The double-pass amplification under the given experimental conditions was also investigated numerically, employing a bi-directional modelling of radiation transfer in an XRL medium [9,11]. The model calculates the intensities I^+ and I^- of the emission propagating along the positive and negative direction of the amplifier axis z, which depend on the small-signal emissivity j_0 and gain g_0 through

$$\frac{\partial I^\pm(z,t,v)}{c\partial t} \pm \frac{\partial I^\pm(z,t,v)}{\partial z} = \frac{j_0(t,v)}{1+\dfrac{I_{tot}(z,t)}{I_{sat}(t)}} + \frac{g_0(t,v)I^\pm(z,t,v)}{1+\dfrac{I_{tot}(z,t)}{I_{sat}(t)}} \tag{1},$$

where I_{sat} is the saturation intensity and where

$$I_{tot}(z,t) = \int_{-\infty}^{+\infty} \left[I^+(z,t,v') + I^-(z,t,v') \right] dv' \tag{2}.$$

The feedback provided by the half-cavity mirror, located at a distance D from the plasma and re-injecting the XRL radiation back to the active medium with an efficiency F, enters to the radiation transfer equation (1) as a boundary condition

$$I^+(z=0,t,v) = F I^-(z=0,t-2D/c,v) \tag{3}.$$

The simulations reproduce the experimentally observed behaviour of the XRL system for a small-signal gain of 7.0($\pm$0.5) cm^{-1}, as shown in Figure 8. They illustrate that the half-cavity laser is deeply saturated, with the equivalent small-signal gain-length product $2g_0L+\ln(F)$ amounting to $\approx$39, or with the saturated gain-length product ($\int gl$ involving the actual gain progressively reduced by saturation) of ~21.

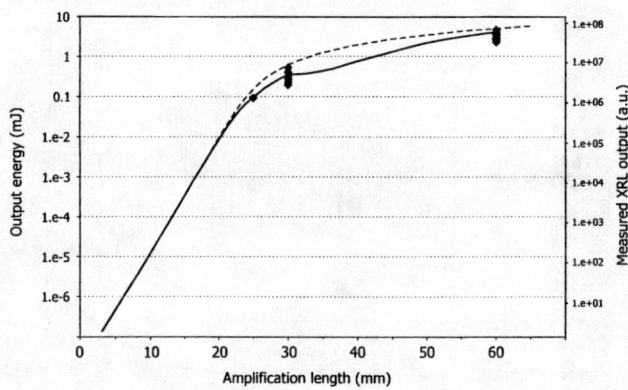

FIGURE 8. The XRL output as a function of the amplification length. Only a few experimental points of those collected are represented, but include the extreme values obtained for each length. The solid line is the result of the full bi-directional simulation corresponding to a small-signal gain of 7.0 cm^{-1}. The dashed line represents a simple, continuous ASE amplifier with unidirectional radiation propagation.

The performed experiments included also coherence measurements of the half-cavity XRL beam, combined with an early application in interferometry of solid surfaces subjected to large electrical fields [12]. The device employed for these measurements was a Fresnel double-mirror interferometer [13], located 3 m downstream from the plasma via a reflection off a 45° multilayer mirror (cf. Fig.1). Its arrangement is schematically shown in Figure 9. The resulting fringes are detected by phosphor-coated CCD camera located 70 cm from the double mirror center.

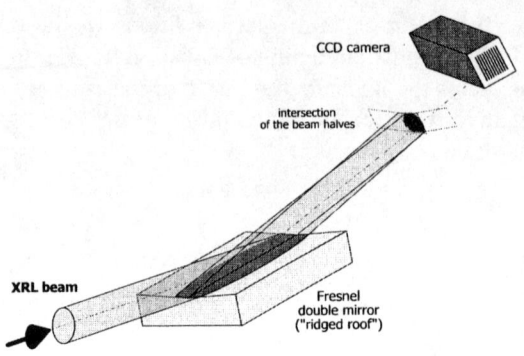

FIGURE 9. Fresnel double mirror wavefront-splitting interferometer. Two 15×60 mm adjacent mirrors, containing an angle of 5.2 mrad, reflect the XRL beam under an angle of 6°. The CCD camera is inclined with respect to the reflected beam in order to augment the period of the detected fringes.

A typical interferogram produced by the half-cavity XRL beam is shown in Figure 10. The detected fringes are straight along the whole record, indicating a high quality of the XRL beam wavefront. The fringe visibility $(S_{max}-S_{min})/(S_{max}+S_{min})$, where the signals S_{max} and S_{min} correspond to the maximum and the adjacent minimum in the fringe system, was inferred after subtracting from the raw interferograms a carefully determined background signal. A typical visibility for the nominal pump conditions amounts to ≈0.45, while exceeding a value 0.5 for the "best" shots. The collected visibility data, corresponding to a given pump energy, exhibit a non-negligible scatter [9], in stark contrast to the excellent reproducibility of the other measured XRL beam parameters. While the reasons of these shot-to-shot visibility fluctuations are not understood, it should be emphasized that their magnitude does not present any hindrance for interferometric applications.

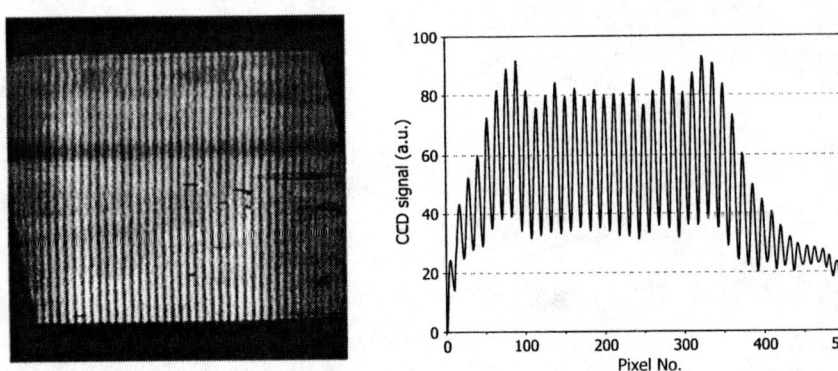

FIGURE 10. Interference fringe pattern produced by the half-cavity XRL beam and its averaged lineout. The small defects in this raw CCD record are due to imperfections in the converting phosphor. The fringe visibility in the central part of the XRL beam amounts to 0.43(±0.05).

SUMMARY

The parameters of the implemented XRL are summarized in the below table; its peak power of 40 MW is currently the highest value achieved by a device of this type. The engineering arrangement allows to perform about 100 shots in a fully automated regime.

Output pulse energy	4 mJ
Number of photons per pulse	4×10^{14}
Output pulse duration (*calculated*)	≈ 90 ps
Peak power	40 MW
Beam divergence (h × v)	3.8×5.8 mrad
No. of shots with one target / mirror	~100

ACKNOWLEDGMENTS

The authors are pleased to acknowledge the work of the PALS laser facility staff, as well as the work of the Institute of Physics technical support staff. We further thank J.Ullschmied for his administration efforts, J.Moravec from Foton s.r.o for his expertise in implementing the X-ray CCD, and J.Sobota of the Institute of Scientific Instruments for making the multilayer mirrors. The support of K.Rohlena, K.Jungwirth, and P.Jaeglé is also appreciated. This work benefited from the EU Transnational Access to Research Infrastructures grant HPRI-00108, from the National Research Centres project LN00A100, and from the Czech Academy of Sciences grant A1010014.

REFERENCES

1. X-Ray Lasers 1998, *Proceedings of the 6th International Conference on X-Ray Lasers*, ed. by Y.Kato, H.Takuma, and H.Daido, Institute of Physics Conference Series No.159, IOP, Bristol, 1999
2. X-Ray Lasers 2000, *Proceedings of the 7th International Conference on X-Ray Lasers*, ed. by G.Jamelot, C.Möller, and A.Klisnick, Journal de Physique IV 11, Pr2, EDP Sciences, Les Ulis, 2001
3. Rus, B., Carillon, A., Gauthé, B., Goedtkindt, P., Jaeglé, P., Jamelot, G., Klisnick, A., Sureau, A., and Zeitoun, P., *J.Opt.Soc.Am.B* **11**, 564-573 (1994)
4. Fill, E.E., Li, Y., Schlögl, D., Steingruber, J., and Nilsen, J., *Opt. Lett.* **20**, 374-376 (1995)
5. Rus, B., Carillon, A., Dhez, P., Jaeglé, P., Jamelot, G., Klisnick, A., Nantel, M., and Zeitoun, P., *Phys.Rev.A* **55**, 3858-3873 (1997)
6. Rus, B., Mocek, T., Präg, A.R., Lagron, J.C., Hudeček, M., Jamelot, G., and Rohlena, K., *J.Phys.IV* **11**, Pr2-589-596 (2001)
7. Rus, B., Rohlena, K., Skála, J., Králiková, B., Jungwirth, K., Ullschmied, J., Witte, K.J., and Baumhacker, H., *Laser Part.Beams* **17**, 179-194 (1999)
8. MacPhee, A.G., Lewis, C.L.S., Warwick, P.J., Weaver, I., Jaeglé, P., Carillon, A., Jamelot, G., Klisnick, A., Rus, B., Zeitoun, P., Nantel, M., Goedtkindt, P., Sebban, S., Tallents, G.J., Demir, A., Holden, M., and Krishnan, J., *Opt.Commun.* **133**, 525-533 (1997)
9. Rus, B., Mocek, T., Präg, A.R., Kozlová, M., Jamelot, G., Carillon, A., Ros, D., Joyeux, D., and Phalippou, D., accepted for publication in *Phys.Rev. A*
10. Präg, A.R., Mocek, T., Kozlová, M., and Rus, B., these Proceedings
11. Rus B., and Mocek, T., these Proceedings
12. Mocek, T., Ros, D., Rus, B., Joyeux, D., Präg, A.R., Kozlová, M., Carillon, A., Phalippou, D., Ballester, F., Jacques, E., Boussoukaya, M., and Jamelot, G., these Proceedings
13. Svatos, J., Joyeux, D., Phalippou, D., and Polack, F., *Opt.Lett.* **18**, 1367-1369 (1993)

OPTICAL-FIELD-IONIZATION LASERS

Theoretical Modeling of Recombination Gain in LiIII Transition to Ground State

Yoav Avitzour*, Stephan Brunner†, Ernest Valeo† and Szymon Suckewer*

*Princeton University, Princeton, NJ 08540
†Princeton Plasma Physics Laboratory, Princeton, NJ 08540

Abstract. We present numerical calculation of recombination gain in LiIII transition to ground state $(2 \rightarrow 1)$. The model includes the initial ionization of the plasma by an intense fs laser pulse, and continues through the expansion and cooling of the plasma simultaneously with the recombination process. We show that although initial estimations of the energy absorption by the plasma from the ionization laser does not allow for recombination gain, the expansion and cooling processes that take place immediately after ionization, in addition to the non-Maxwellian distribution of electrons in the plasma, give rise to high gain under feasible experimental conditions.

INTRODUCTION

Recombination gain in the $2 \rightarrow 1$ transition of LiIII ions at 13.5nm using optical field ionization (OFI) by high power femtosecond lasers has already been generated [1, 2] and lasing action was demonstrated [3, 4]. In addition, extensive theoretical work has been conducted to model the different processes involved in generating recombination gain (see e.g. [5, 6, 7]), specifically, the required initial conditions and the feasibility of creating them. Two parameters play a crucial role in the recombination process - the electron temperature and the electron (and ion) density. When calculating the gain by integrating the rate equations using simple temperature dependent rate coefficients, it is easy to show that in order to achieve gain in LiIII 2->1 transition, the electron density should be roughly between $10^{18} - 10^{20} cm^{-3}$ and the electron temperature should be under 10eV. The means of producing such cold, high density plasma are applying an ultra-short, high power laser pulse (pulse duration $\sim 100 - 500fs$, power $\sim 0.3 - 1TW$) to neutral gas or singly ionized plasma. The high electric field during the laser pulse completely strips the electrons from the nuclei, and the short pulse duration prevents substantial heating. In addition, Since the duration of the recombination gain is at the most up to $10 - 15psec$ after the ionization, the ionization time scale must be much shorter. However, using simple models to estimate the average energy of the ionized electrons, yields an electron temperature that is too high to allow for recombination gain. Even more sophisticated models, that take into account additional effects, predict only very small gain, if any (e.g . [8, 7]). We have shown that by taking into account two important properties of the recombining OFI plasma - it's non-Maxwellian nature and it's spatial distribution, which lead to rapid expansion and cooling - high gain can be achieved, notwithstanding the high average energy after ionization.

CP641, *X-Ray Lasers 2002: 8th International Conference on X-Ray Lasers*, edited by J. J. Rocca et al.

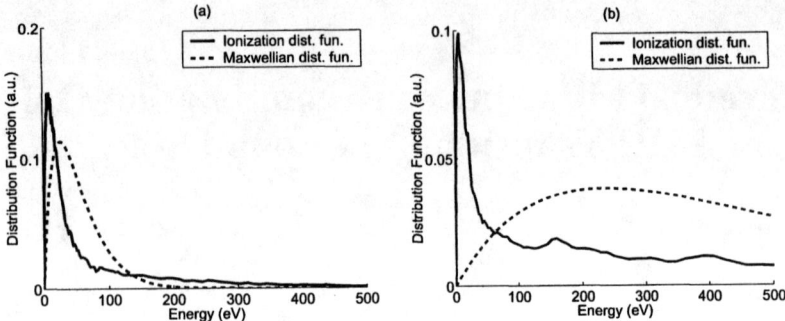

FIGURE 1. Calculated vs. Maxwellian energy distribution functions with the same average energy. (a) Ionization wavelength=248nm. (b) Ionization wavelength=800nm.

PHYSICAL SYSTEM

The main source for heating under the conditions of OFI by a short pulse is the so-called above threshold ionization (ATI) heating, or residual heating. Explained first in [5], ATI heating is due to the phase mismatch between the oscillating field and the ionized electron released at rest at some arbitrary time during the oscillation. The distribution function of the ionized electrons after the ionization is obtained by determining the probability for the electron to be ionized at each phase of the laser, using static field tunneling ionization rates [9]. The average residual energy is proportional to the quiver energy of the electrons in the laser field ($E_q = e^2 E^2 / 4m\omega^2$, where E is the laser electric field amplitude, ω is the laser angular frequency, and e and m are the electron charge and mass), hence scales quadratically with the wavelength ($\lambda = 2\pi c / \omega$) and close to linearly with the ionizing intensity (the actual dependence on the intensity is also affected by the tunneling rates). The functional form of this distribution function is given by [5]:

$$f_e = g(\eta; E, I_p, \omega) \equiv \frac{u}{\sqrt{\eta}(1-\eta)} \exp\left[\frac{-2/3 \left(I_p/I_h\right)^{3/2} E_0/E}{\sqrt{1-\eta}}\right] \tag{1}$$

where $\eta = \varepsilon/2E_q$ is the normalized energy, I_p is the ionization potential, normalized to the ionization potential of hydrogen atom I_h, and the electric field is normalized to the atomic field strength $E_0 = 5.1 \times 10^9 V/cm$. Since very short laser pulses are discussed, ATI heating is the main source for plasma heating during the pulse, and collisions (i.e. inverse Bremsstrahlung heating) have a secondary effect. The energy distribution function in (1) diverges like $\varepsilon^{-1/2}$ as $\varepsilon \to 0$. This divergence disappears once collisions are introduced, yet a large number of electrons are still concentrated near $\varepsilon = 0$. On the other hand, the distribution function extends to much higher energies than the Maxwellian distribution function with the same average energy. Fig 1 shows the difference between a Maxwellian distribution function and a calculated distribution function with the same average electron energy for two different ionizing beam wavelengths. Since the high energy electrons do not participate in the collisional processes, yet contribute substantially to the average energy, using the average energy as a parameter can be misleading. By looking at Fig. 1, it is clear that the difference between the calculated distribution function and the Maxwellian distribution function in the longer wavelength case is much more substantial, and hence gain generation is possible even though the average energy in this case is very high ($\sim 400 eV$).

Another important feature that should be taken into account is the spatial properties of the experiment. The high power beam was focused to a tight spot that propagated in the plasma. The typical dimensions in experiments are a microcapillary with diameter $\sim 350 \mu m$ while the laser is focused to a spot size of $\sim 10 \mu m$ in diameter. This means that the "hot" fully ionized plasma is formed in a very narrow channel embedded in a relatively cold ($\sim 1 eV$) plasma. This feature may lead to very rapid cooling by expansion of the very energetic electrons before they can participate in any Maxwellization process.

THEORETICAL MODEL

Ionization

In order to consider collisional heating and Maxwellization processes during the ionization, the ionization distribution function was calculated using a 3D velocity-space PIC code similar to the one described in [10, 11]. At each time step, electrons were added using the tunneling ionization rates that corresponded to the frequency resolved laser electric field at that time step. The ionizing laser pulse was assumed to have Gaussian profile in space and time, i.e. the electric field was given by: $\vec{E} = E_0 \times \exp\left(-(2r/d)^2\right) \times \exp\left(-(2t/\tau)^2\right) \times \sin(\omega t)\hat{z}$, where r is the transverse radius, d is the beam diameter and τ is the pulse duration. The initial plasma conditions were taken as He-like Li plasma with relatively low ($\sim 1 eV$) electron temperature. This value is taken from experimental estimates and agrees with the calculations in [6].

The code takes into account e-e and e-i collisions. In contrast to the arguments in [12, 6], e-e collision play an important role in our model. The e-e collision rate is much higher than the e-i collision rate since the relative velocities between electrons moving together in the electric field are on the order of the average energy of the final distribution function, whereas the relative velocities between the electrons and ions are on the order of the quiver energy. Although e-e collisions do not contribute to overall heating of the plasma, they do contribute to the Maxwellization process of the distribution function, and cause an "effective heating". Several effects are not taken into account. First, the effects of the ponderomotive force during ionization. Clearly, the ponderomotive force pulls out electrons from the plasma, and effectively reduces the number of electrons participating in the recombination. In addition, since an electric field is created in the core of the plasma, those energetic electrons will return to the plasma and provide additional heating. However, since the force peaks at $r_0/\sqrt{2}$ (where r_0 is the ionizing beam radius), and goes to 0 at $r = 0$, it is effectively narrowing the high density plasma channel, which may enable higher cooling efficiency. What can be said at this stage is that for the larger diameter calculations the ponderomotive force will have negligible effect, but we might expect a stronger effect as the ionizing beam diameter decreases. Therefore, the inclusion of the ponderomotive force into the calculations is in progress.

Expansion and cooling

As discussed above, the tightly focused laser ionizes a narrow channel in a relatively cold plasma. As a result, the system undergoes rapid expansion and cooling that may affect the distribution function and the temperature of the electrons in times relevant to gain generation, i.e. in the first $10 - 15$psec after ionization. The time evolution of the distribution function was calculated using a 1-D cylindrically symmetric Fokker-Planck code. The code is described in [13] and is essentially an implementation of the SPARK code described extensively in [14]. The importance of the expansion, apart from direct temperature diffusion, is in the Maxwellization process. Since the energetic electrons have high probability to escape the "gain region" before any collision has occurred, they will not participate in the Maxwellization in that region, resulting in an effectively larger cooling rate. The assumption of cylindrical symmetry is somewhat inaccurate, yet relaxing this assumption can only lead to faster cooling and higher gain. Since the velocities of the electrons ionized by a linearly polarized laser are directed straight out of the "gain region", and not averaged on all directions as in our isotropic model, the energetic electrons will have higher probability to escape the "gain region" before even one collision has occurred. Another effect that is neglected is the heating from the recombination, which physically occurs simultaneously with the expansion, but is calculated separately with no feedback. We estimated the overall heating from recombination and it is clear that it should not have a significant effect on the gain. The estimate is discussed below. The parameters that were taken for modeling resemble the actual experimental conditions. Since the experiment was done in a $350\mu m$ diameter microcapillary, and the "gain region" had diameter less than $30\mu m$, the microcapillary walls were assumed to be at infinity. This assumption is justified as we are only interested in phenomena in the first $10 - 15$psec after the ionization, and any reflection from the walls would contribute in later times.

Recombination and gain

Once we had the time dependent distribution function, the atomic processes' cross sections were integrated over it, to give time dependent rate coefficients for integrating the rate equations and calculating recombination process. In order to simplify the code and make it more robust, simple analytic forms of the ionization and excitation cross sections were used [15, 16]. The rest of the cross sections were obtained from the detailed balance relations between excitation (ionization) and de-excitation (recombination).

In order to estimate the heating of the plasma from the recombination and de-excitation processes [17], we took the population of the levels after solving the rate equations, and calculated the energy addition from all the recombined electrons up to the time when the maximum gain is achieved, using $< \varepsilon > (t) = \frac{1}{N_e(t)} \sum_n N_n^{(+2)}(t) I_{p,n}$ where $N_n^{(+2)}(t)$ is the population of level n of the LiIII ions at time t, and $I_{p,n}$ is the ionization potential from level n. The heating estimated did not exceed $0.5eV$ in any of the parameters checked, and was under $0.1eV$ in most of the cases. We must note that by dividing by N_e, we assumed the added energy is distributed equally among all the

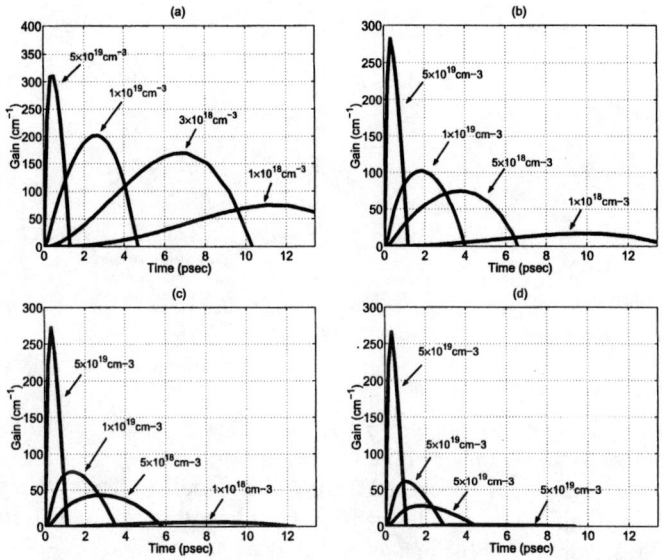

FIGURE 2. Gain vs. Time for different initial ion densities (Ionization intensity=$1.3 \times 10^{17} W/cm^2$, wavelength=$248nm$). (a) ionizing beam diameter=6μ, (b) 10μ, (c) 15μ and (d) 30μ.

electrons. In fact, since the electrons contributing to the recombination are in essence the cold electrons, they are the ones that will absorb the added energy, and the results above might be higher by a factor of 2-3. Even taking this into account, the amount of recombination heating is very small, in comparison to ATI heating.

RESULTS

The model above was used to characterize the behavior of the transient gain in LiIII $2 \rightarrow 1$ transition with varying initial conditions. We were able to show that high gain ($> 200cm^{-1}$) can be obtained under feasible experimental conditions. The following parameters were scanned in order to establish the conditions for optimum gain:

- Ion density: $10^{18} - 5 \times 10^{19} cm^{-3}$
- Ionizing beam diameter: $6 - 30 \mu m$
- Ionizing beam wavelength: $248nm$, $800nm$
- Ionizing beam peak intensity: $1.3 - 2.6 \times 10^{17} W/cm^2$

It is clear from the results presented in Fig. 2 that as the ionizing beam diameter is smaller, higher gain can be achieved due to the more effective expansion and plasma cooling. Higher ion densities also contribute to higher gain. However, higher densities tend to shorten the gain time due to faster collisional depopulation of the lasing level, while smaller diameters increase the gain time. In addition, collisional broadening of the lasing line is ignored in the gain calculation, and will become more important at high

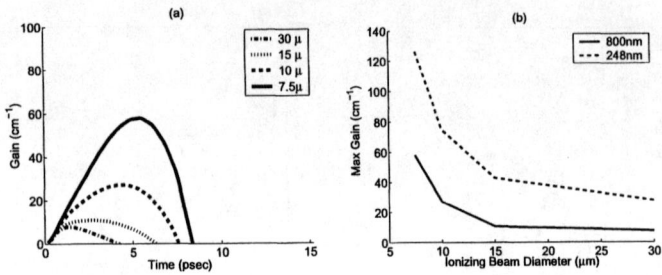

FIGURE 3. (a) Gain vs. Beam Diameter for 800nm (Intensity=$1.3 \times 10^{17} W/cm^2$, ion density=$5 \times 10^{18} cm^{-3}$) (b) Comparison with the 248nm results.

densities, resulting in decreased gain. We also notice that at higher densities, the gain is much less sensitive to the diameter of the ionizing beam.

In Fig. 3 we present the results obtained with longer ionization wavelength (800nm); while the gain is significantly smaller, it is still sufficiently high, which indicates that generation of high gain is feasible in a wider range of parameters.

ACKNOWLEDGMENTS

The authors would like to thank Prof. G. Pert for valuable discussions and comments about recombination heating and the effect of the ponderomotive force. This work was supported by INTEL Inc. and NSF(PHYS) grants.

REFERENCES

1. Nagata, Y., Midorikawa, K., Kubodera, S., Obara, M., Tashiro, H., and Toyoda, K., *Phys. Rev. Lett.*, *71*, 3774–3777 (1993).
2. Krushelnick, K. M., Tighe, W., and Suckewer, S., *J. Opt. Soc. Am. B-Opt. Phys.*, **13**, 306–311 (1996).
3. Korobkin, D. V., Nam, C. H., Suckewer, S., and Goltsov, A., *Phys. Rev. Lett.*, **77**, 5206–5209 (1996).
4. Goltsov, A., and et. al., *IEEE J. Sel. Top. Quantum Electron.*, **5**, 1453–1459 (1999).
5. Burnett, N. H., and Corkum, P. B., *J. Opt. Soc. Am. B-Opt. Phys.*, **6**, 1195–1199 (1989).
6. Janulewicz, K. A., Grout, M. J., and Pert, G. J., *J. Phys. B-At. Mol. Opt. Phys.*, **29**, 901–914 (1996).
7. Pert, G. J., *J. Phys. IV*, **11**, 181–187 (2001).
8. Janulewicz, K. A., Healy, S. B., and Pert, G. J., *Opt. Commun.*, **140**, 165–178 (1997).
9. Landau, L. D., and Lifshitz, E. M., *Quantum Mechanics*, Butterworth-Heinemann, 1999, 3 edn.
10. Takizuka, T., and Abe, H., *J. Comput. Phys.*, **25**, 205–219 (1977).
11. Ma, S., Sydora, R. D., and Dawson, J. M., *Comput. Phys. Commun.*, **77**, 190–206 (1993).
12. Ditmire, T., *Phys. Rev. E*, **54**, 6735–6740 (1996).
13. Brunner, S., and Valeo, E., *Phys. Plasmas*, **9**, 923–936 (2002).
14. Epperlein, E. M., *Laser Part. Beams*, **12**, 257–272 (1994).
15. Lotz, W., *Z Phys*, **216**, 241–& (1968).
16. Fisher, V. I., Ralchenko, Y. V., Bernshtam, V. A., Goldgirsh, A., Maron, Y., Vainshtein, L. A., Bray, I., and Golten, H., *Phys. Rev. A*, **55**, 329–334 (1997).
17. Pert, G. J., *Private Communication*.

Search For Optimum Conditions For Gain To Ground State In LiIII

Iddo Geltner, Yuan Ping, Anatoli Morozov, and Szymon Suckewer

School of Engineering and Applied Science, Princeton University
Princeton NJ, 08544

Abstract. The optimum conditions for gain generation for the transition to the ground state at 13.5 nm in LiIII ions were investigated in discharge created pre-plasma. The spatial and temporal evolution of plasma density was measured and a hollow density profile was observed. Waveguiding of high power laser pulses has been demonstrated in discharge created plasma channels, which can be helpful in increasing the interaction length, thus improving the gain-length product.

INTRODUCTION

Gain to ground state has been demonstrated in LiIII ions [1] and later also lasing was generated with a gain-length product (GL) of $\approx$6-7 [2]. However, in order to reach saturation a product of ~10 must be reached [3]. Much work has been done to theoretically estimate the gain in "recombination pumped" schemes [4,5]. Recent calculations suggest that gain can be improved by tightly focusing the ultrashort pumping beam and creating a narrow channel (few microns in diameter) of totally stripped Li ions in a cold background plasma [6].

In the search for optimum plasma conditions for lasing (large GL), the increase in propagation length of the pumping laser beam must be considered vs. shorter length and higher gain. We present recent work concentrated on improving the guiding [7] of the pumping beam, with the smallest diameter possible, in pre-plasma prepared in the conditions suggested by our calculations. The plasma was created using electric discharge ablation of a LiF microcapillary in vacuum. The current through the plasma heats the plasma on axis and creates a hollow density profile suited for waveguiding. The guiding efficiency and beam diameter were investigated as a function of focusing geometry and input intensity. Guiding of 25 µm beam diameters (FWHM) was demonstrated and intensities of up to 3×10^{15} W/cm^2 were achieved at the output of the waveguide.

The plasma conditions needed for gain in LiIII can be estimated using the dependence of the various rates of processes, involved in the creation of the population inversion, on the plasma density and temperature. In the recombination scheme the important processes are the radiative de-excitation (A_{21}, the Einstein A coefficient), collisional de-excitation ($\propto n_e T_e^{-1/2}$) and three-body recombination ($\propto n_e^2 T_e^{-4.5}$). The temperature should be as low as possible to allow for maximum

CP641, *X-Ray Lasers 2002: 8th International Conference on X-Ray Lasers,* edited by J. J. Rocca et al.

recombination and de-excitation. The lower limit to the density is determined by the need for the recombination rate to be faster than the depletion of level 2 and the upper limit is set by the need for the lifetime of level 2 to be longer than the pumping pulse duration. The plasma density fit for our experimental parameters (namely, Li ions and a 100 fs pumping beam) is $n_e \approx 10^{18}$-10^{20} cm^{-3}. The discharge created pre-plasma has a temperature of 3-5 eV [7], which, according to our calculations, is low enough.

EXPERIMENTAL SETUP

In our experiments we used a Ti:sapphire 3.5 TW laser system at a central wavelength of 800 nm. The system consists of four parts: a master oscillator producing 90 fs pulses at 82 MHz, a regenerative amplifier, a final 4-pass amplifier that amplifies the pulses up to >700 mJ at a repetition rate of 5 Hz, and a vacuum compressor, which compresses the beam back to 100 fs, with energies reaching 350 mJ per pulse. The beam propagates in vacuum to the experimental chamber and is focused by an f=19 cm off-axis parabolic mirror. An aperture is placed before the mirror to allow the control of the f-number.

The microcapillaries were 8-20 mm long and 350-500 μm in diameter. A capacitor (10-100 nF) was placed across the microcapillary and it was charged up to 7.5 kV. The discharge was initiated by a low power laser pulse (Nd:YAG at λ=1.06 μm, 20 mJ, 10 ns), which ablated the microcapillary wall. The fs beam was coupled into the microcapillary at delays varying between 0-500 ns (approximately the duration of the discharge current pulse). An additional Nd:YAG laser at the second harmonic was used to measure the electron density profile at the exit of the microcapillary in a Mach-Zehnder interferometer configuration [Fig. 1(a)]. Lenses L_2 and L_3 imaged the output of the microcapillary with large magnification onto a charge-coupled device camera [Fig. 1(b)] and the images of the exiting beam were taken at different delays and different discharge configurations.

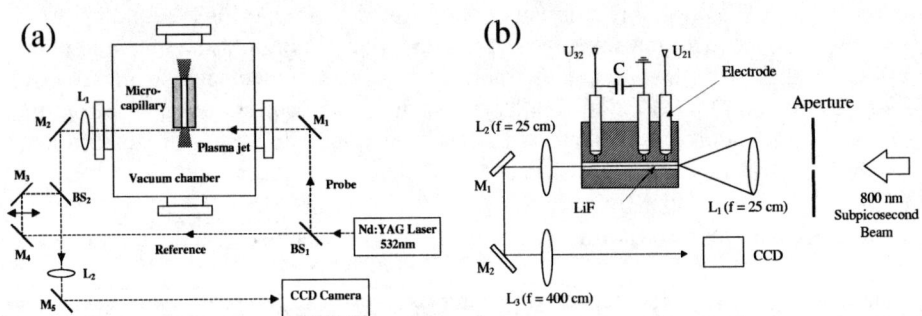

FIGURE 1. Experimental setup: (a) interferometric measurements of the plasma density at the microcapillary exit. (b) Imaging of beam at microcapillary output.

RESULTS

The interferometric measurements were done to verify that the plasma density on axis was up to ~1.5×10^{19} cm^{-3} near the microcapillary end, and that a hollow waveguiding profile is created. Figure 2 shows cross sections of the plasma profile across the center of the exit plane of the microcapillary at different delays and at the optimum discharge parameters for guiding. During the rising stage of the density a hollow profile appears (τ=320 ns). This delay corresponds to the delay at which guiding occurs. Using the profile shape the matching focal size of the beam [7] can be calculated and there is excellent agreement with our measurements [8]. By changing the aperture between 7 and 40 mm (fully open) the f-number is changed and the focal size can be adjusted to fit the matching diameter and there is a clear transmission peak at the matching geometry [8]. Figure 3 shows an example of a guided beam at the output of the microcapillary. The guided beam diameter is 24 μm (FWHM) with a Gaussian-like profile and a low intensity structure in the wings.

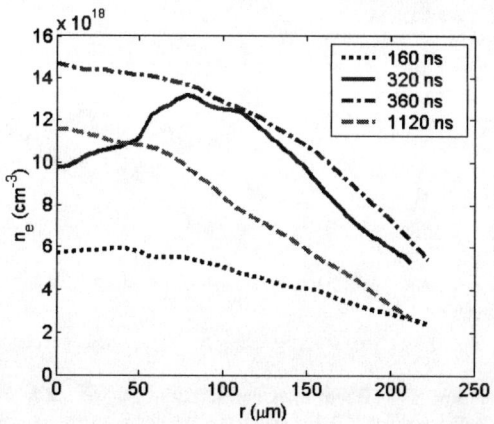

FIGURE 2. Plasma density at microcapillary exit at different delays after the discharge initiation. A hollow waveguiding profile is clearly seen at τ=320 ns. Discharge parameters: C=50 nF, V=5.5 kV.

The guided beam pulse energy was varied between 15 to 250 mJ and the dependence of the transmission on the input intensity at two different f-numbers was investigated. The results for 14 and 40 mm apertures are presented in Fig. 4. There is a clear decrease in transmission as the intensity increases and an average output intensity of 3×10^{15} W/cm^2 cannot be surpassed [Fig. 4(b)]. These results must be improved in order for this scheme to be viable for soft x-ray lasing in discharge created plasma in LiF microcapillaries. We should emphasize that the transmission in Fig. 4(a) is for the guided mode only. Furthermore, it is evident that for the same intensity the transmission for the full aperture case is larger. This can be explained by the fact that in this case 80% of the energy is contained in a 30 μm diameter, while the transmission is calculated relative to the 25% in the 8 μm central spot. Total transmission, including the wide wings of the beam reached more than 40% even for the highest input intensity. The loss of total energy can be explained mainly by collisional absorption and various other scattering mechanisms. Ionization-induced defocusing can explain why most of the transmitted energy is not confined to the guiding channel. Using the rates calculated by Ammosov *et al.* [9] for tunneling ionization, at 1×10^{16} W/cm^2 the beam cannot ionize LiII or FIV, but an order of magnitude increase in intensity increases the plasma density by almost a factor of 2 since LiII, FIV, FV and LiIII are ionized. The tunneling time for FV and LiIII is 10 fs so the front of the pulse sees a different density than the back; a density peak forms at the center of the hollow profile. As the intensity increases only the front of the pulse sees a hollow profile, while the rest of the beam is strongly defocused.

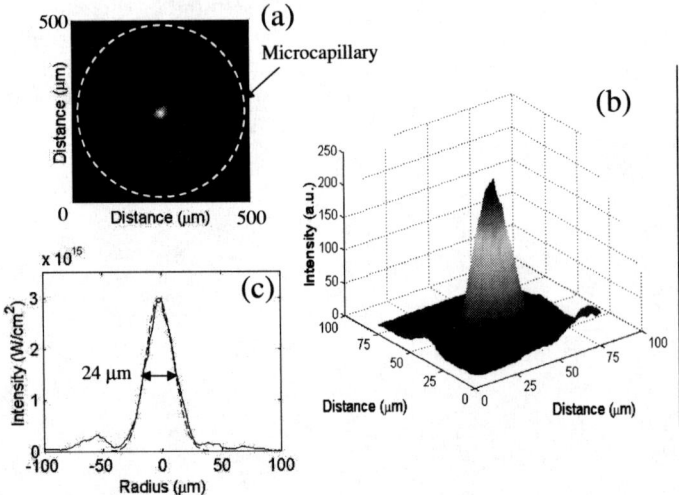

FIGURE 3. Guided beam profile. (a) Beam image at output of microcapillary. Dashed line represents microcapillary wall. (b) Beam profile (c) Gaussian fit to beam profile - cross-section showing a diameter of 24 μm FWHM.

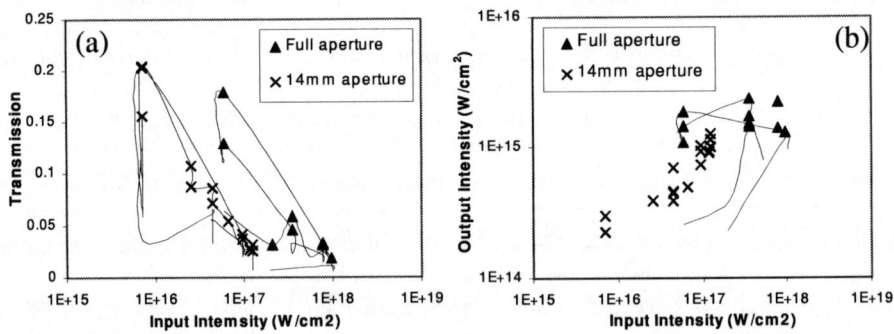

FIGURE 4. Transmission (a) and output intensity (b) vs. input intensity for two different apertures.

Simulations by Ehrlich *et al.* [10] for a 0.1 TW, 800 nm beam through a plasma channel 6.6 cm long with an initial density of 5×10^{18} cm^{-3} show a loss of ~50% to the total transmitted energy (due mostly to ionization-induced defocusing). While in our case the microcapillary length is only 2 cm, the change to the hollow density profile is much more severe since the ions created by the front of the pulse are Li^{+3} and F^{+5} for an intensity of 1×10^{17} W/cm^2 and F^{+7} for 1×10^{18} W/cm^2. In Ref. 10 the guided beam creates H^+ and C^{+4} ions from the singly ionized pre-plasma.

CONCLUSION

Discharge created pre-plasma in LiF microcapillaries is suitable for SXR lasing in terms of plasma density and temperature. At proper discharge conditions we are also able to guide ultrashort powerful beams through the 2 cm long microcapillaries. However, the output intensities are too low to create population inversion at the end of the microcapillary, mainly due to collisional absorption and ionization-induced defocusing.

ACKNOWLEDGMENTS

The authors would like to thank Dr. D. Kaganovich for helpful suggestions and for lending us his microcapillaries, Y. Avitzour for his important contribution to this work, and N. Tkach for excellent technical assistance. This work was supported by DARPA, INTEL and NSF (PHYS) grants.

REFERENCES

1. Nagata, Y., Midorikawa, K., Kubodera, S., Obara, M., and Toyoda, K., *Phys. Rev. Lett.* **71**, 3774-3777 (1993); Krushelnick, K., Tighe, W., and Suckewer, S., *J. Opt. Soc. Am. B* **13**, 306-311 (1996).
2. Korobkin, D. V., Nam, C. H., and Suckewer, S., *Phys. Rev.Lett.* **77**, 5206-5209 (1996); Goltsov, A. et al., *IEEE J. Sel. Top. Quant.* **5**, 1453-1459 (1999).
3. Casperson, L. W., *J. Appl. Phys.* **48**, 256-262 (1977).
4. Peyraud, J., and Peyraud, N., *J. Appl. Phys.* **43**, 2993-2996 (1972).
5. Janulewicz, K. A., Healy, S. B., and Pert, G. J., *Opt. Com.* **140**, 165-178 (1997).
6. Avitzour, Y., Brunner, S., Valeo, E., and Suckewer, S., "Theoretical Modeling of Recombination Gain in LiIII Transition to Ground State" in this volume; Avitzour, Y., Brunner, S., Valeo, E., and Suckewer, S., (to be published).
7. Ehrlich, Y., Cohen, C., Zigler, A., Krall, J., Sprangle, P., and Esarey, E., *Phys. Rev. Lett.* **77**, 4186-4189 (1996).
8. Geltner, I., Ping, Y., and Suckewer, S., (to be published).
9. Ammosov, M. V., Delone, N. B., and Krainov, V. P., *Sov. Phys. JETP* **64**, 1191-1194 (1986).
10. Ehrlich, Y., Cohen, C., Kaganovich, D., Zigler, A., Hubbard, R. F., Sprangle, P., and Esarey, E., *J. Opt. Soc. Am. B* **15**, 2416-2423 (1998).

Collisionally pumped Ni-like and Pd-like optical field ionization soft x-ray lasers

S. Sebban[1], T. Mocek[2], D. Ros[3], L. Upcraft[1], Ph. Balcou[1], R. Haroutunian[1], G. Grillon[1], B. Rus[2], A. Klisnick[3], A. Carillon[3], G. Jamelot[3], C. Valentin[1], A. Rousse[1], J. P. Rousseau[1], L. Notebaert[1], M. Pittman[1], and D. Hulin[1]

1- *Laboratoire d'Optique Appliquée, CNRS UMR 7639, F-91761 Palaiseau cedex, France*
2- *Institute of Physics, Academy of Sciences of the Czech Republic, Prague 8, Czech Republic*
3- *Laboratoire de Spectroscopie Atomique et Ionique, F-91405 Orsay cedex, France*

Abstract: We report recent investigations of Optical-Field Ionization soft-x ray lasers. A gain of 67 cm⁻¹ on the $4d^95p$-$4d^95d$ transition at 41.8 nm in Pd-like xenon and a gain-length product of 15 have been inferred at saturation. This source delivers about 5×10^9 photons per pulse. More recently we have demonstrated lasing at 32.8 nm in Ni-like krypton. The influence of the pumping energy and the laser polarization on the lasing output as well as the far-field pattern of the x-ray laser beam are presented and discussed.

INTRODUCTION

The first OFI soft x-ray laser was first demonstrated in 1993 (1) with a recombination scheme on the Lyman-α transition of H-like Li. In spite of subsequent improvements (2) and numerous other investigations (3-5) the gain-length product in that scheme has never exceeded 10. The main difficulty for OFI recombination XRLs is obtaining a low enough temperature plasma to allow significant recombination to occur, while keeping the laser intensity sufficiently high for ionization. A breakthrough in the collisionally pumped scheme was the demonstration of a gain-length product of 11 of the 5d-5p emission line at 41.8 nm of Xe^{8+}, by focussing a 70 mJ, 40 fs circularly polarized laser pulse into a gas cell filled with xenon (6). These results followed previous modeling predicting lasing action for three specific laser schemes (7) in the 30-50 nm range that can be driven by an ultra-short pulse focussed to an intensity of 10^{16} to 10^{17} W·cm⁻². These systems are based on electron collisional excitation of eight-time-ionized gases created by OFI: Ne-like Ar lasing at 48 nm, Ni-like Kr lasing at 32 nm, and Pd-like Xe at 41 nm. This promising work offered the prospect for high-repetition rate, table-top soft X-ray lasers which would be suitable for widespread applications. However it has taken five years to confirm the original results, showing that fundamental aspects of the process were not yet completely understood. In 1999, our group succeeded in demonstrating lasing on the $4d^95d\,^1S_0$-$4d^95p\,^1P_1$ transition in Xe^{8+} at 41.8 nm by focussing a 330 mJ, 35 fs circularly polarized laser pulse into a gas cell filled with 15 Torr of xenon (8);

CP641, *X-Ray Lasers 2002: 8th International Conference on X-Ray Lasers*, edited by J. J. Rocca et al.
© 2002 American Institute of Physics 0-7354-0096-2/02/$19.00

saturated amplification was been demonstrated for the first time using the collisional OFI x-ray laser technique. More recently, we have extended the OFI x-ray laser technique to the use of Ni-like ions. Strong amplification at 32.8 nm on the $3d^9 4d$ $^1S_0$-$3d^9 4p^1 P_1$ transition in Kr^{8+} has been observed. This source still represents the shortest wavelength high repetition rate OFI collisional x-ray laser produced (9) to date.

In this article we report the recent investigation on collisional OFI soft-X ray lasers using Pd-like and Ni-like ions performed at the Laboratoire d'Optique Appliquée (Palaiseau-France). We have studied how controlling the pressure, the position of the spot focus, the length of the gas cell, the driving energy and the ellipticity of the polarization, has optimized the lasing output. We have also characterized the far-field pattern of the 41.8 nm x-ray laser beam.

1. EXPERIMENTAL SET-UP

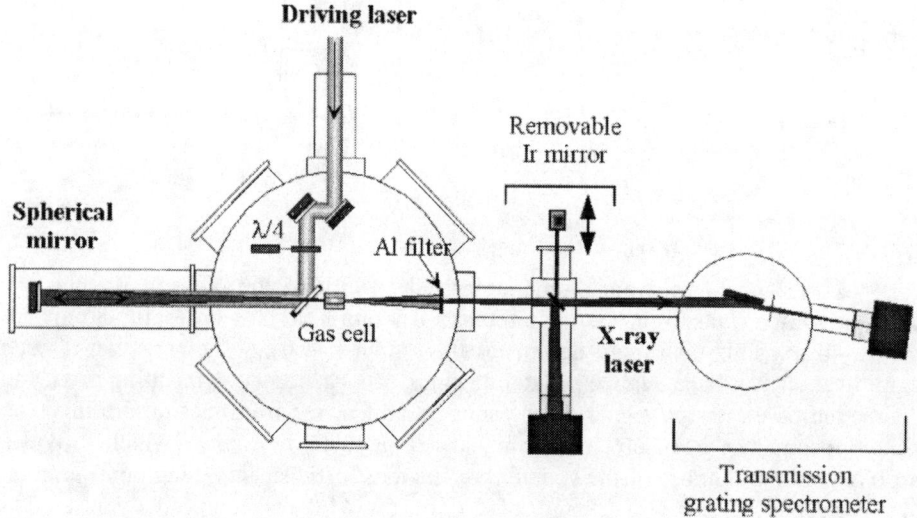

FIGURE 1. Schematic description of the experimental set-up

These experiments have been performed at the Laboratoire d'Optique Appliquée, Palaiseau (France). A schematic experimental set-up is shown in Figure 1. We used a 10 Hz multiterawatt Titanium-Sapphire laser system, yielding up to 750 mJ on target with a pulse duration of 30 fs (10). The resulting average laser intensity on target was up to ~5×10^{17} W·cm^{-2} in vacuum. The polarization of the driver can be varied from linear to circular using a 50 mm diameter motorized quarter-wave plate. The interaction medium is a gas cell filled with xenon held at a uniform pressure. Two 500 μm -diameter replaceable pinholes drilled by the laser were implemented on both sides of the gas cell to provide a differential pumping and isolate the gas-filled

205

and vacuum regions. The entrance of the gas cell can be moved under vacuum using a motorized translation stage making it possible to vary the length of the gas cell from 0 up to 5 mm. The main diagnostic is an on-axis soft x-ray spectrometer composed of a grazing incidence gold coated spherical mirror and a 2000 lines/mm transmission grating. The spectra were recorded with a back-illumination type soft x-ray CCD camera. A 2500 Å -thick, pinhole-free aluminum filter isolates the spectrometer from the incoming visible and laser light. Moreover, we have used 45° flat iridium mirror to get the far-field pattern of the 41.8 nm XRL beam.

2. RESULTS AND DISCUSSION

1. Time integrated spectra

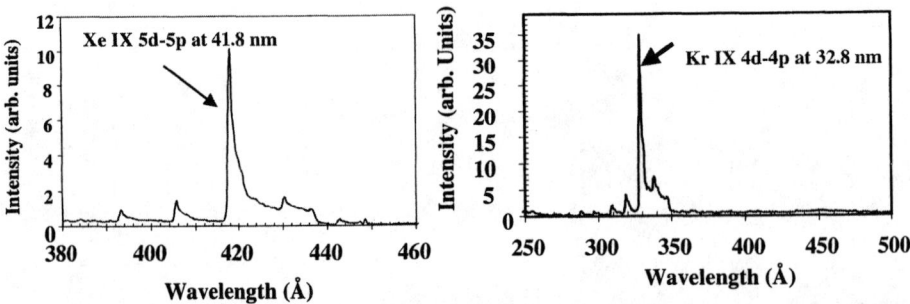

FIGURE 2. Time-integrated spectra of the Xe (a) and Kr (b) plasmas

Figure 2 shows typical time-integrated spectra of the Xe and Kr plasmas obtained for ten shots in the case of circular polarization. In both cases, the 5d-5p and the 4d-4p transitions strongly dominates the output spectrum showing that a large amplification has been realized. Satellite lines can be observed resulting from the diffraction of the lasing signal on the support grid of the transmission grating. The xenon spectra has been obtained using a calibrated CCD camera which provides absolute photon number of the source. For the most efficient pumping configuration (15 Torr, 330 mJ and a cell length of 4 mm), the 41.8 nm source delivers around 5×10^9 photons per pulse. The krypton laser has been measured using a different detector but the output intensity was only 2-3 time lower than the xenon laser. Theoretical estimates has shown that saturation intensities of these OFI soft XRL were computed (11,12) to around 10^7 W·cm^{-2}. Assuming a source size of 40 μm (our focal spot) and pulse duration from 5 ps to 15 ps (12), one respectively expects from $\sim 10^8$ to 10^9 photons at the saturation intensity of 10^7 W·cm^{-2} in the case of the xenon 41.8 nm laser. From this comparison, it appears that the laser has reached the saturation regime. This point will be later confirmed by the gain measurement. An important point to be noticed is that the lasing action was very sensitive to ASE level of the laser driver. The presence of a preformed plasma will result in decreasing the on-axis density and lower gain (13). Here we adjust the ASE level to be minimum by generating high order harmonics of a high energy laser beam (about 300 mJ/pulse).

These high order harmonics, which are probably produced on the rising part of the laser pulse could only be emitted if the plasma is not pre-ionized, i.e. low ASE level.

2. Pressure dependence of the 41.8 and 32.8 nm lasing signal

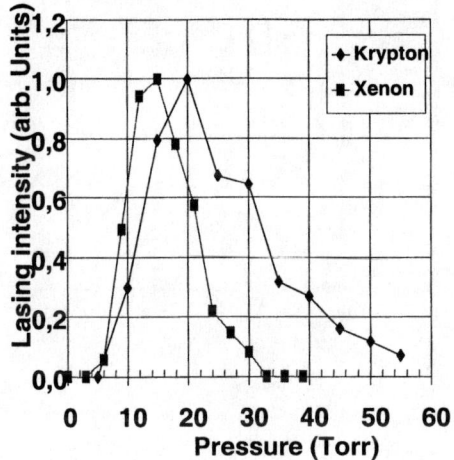

FIGURE 3. Normalized output intensities of the 41.8 and 32.8 nm lasing lines shown as a function of the gas pressure.

In figure 3 the normalized output intensities of the 41.8 and 32.8 nm lasing lines is shown as a function of the gas pressure. Lasing appears to be efficient in the range 15-25 Torr with a maximum at 15 torr for Xe and 20 Torr for krypton. The xenon laser is amplified over a narrow range of gas densities whilst the efficient pressure range is broader for Ni-like krypton. This finding is consistent with the fact that Ni-like ions are more stable against further collisional ionization than Pd-like ions. Nevertheless, the pressure dependencies of both laser sources are very similar in the sense that the lasing output first increases with the gas density being maximized for quite low gas pressure. This can be qualitatively understood since for both Ni-like krypton and Pd-like xenon plasmas, eight electrons have to be removed from the neutral atoms. As a consequence, propagation limitations due to the ionization-induced-refraction (21) of the driving laser beam are expected to be very similar for both. Our experimental results suggest that the measured optimum pressure is the best compromise between high plasma density and laser propagation obtainable in this geometry.

3. Laser energy dependence of the 41.8 and 32.8 nm lasing signal

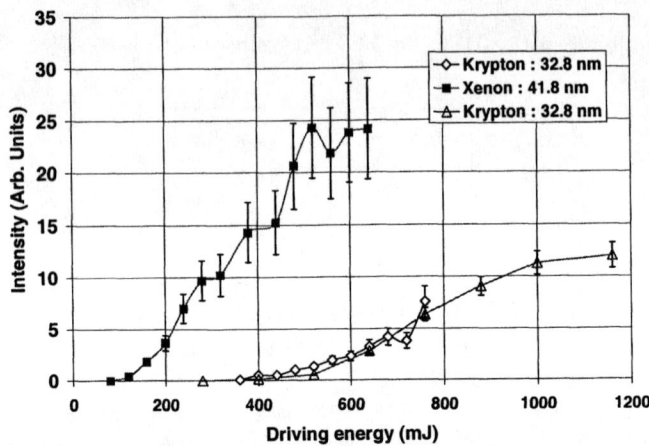

FIGURE 4. Lasing output of the 32.8 and 41.8 nm lasers as a function of the driving laser energy

Figure 4 shows the lasing output of the 32.8 and 41.8 nm lasers as a function of the driving laser energy ranging from 120 ($\sim 10^{10}$ W·cm^{-2}) up to more than one Joule ($\sim 8 \times 0^{17}$ W·cm^{-2}). The error bars arise from the uncertainty in the filter attenuation and from the driving laser energy fluctuations. For the xenon and the krypton sources, the driving energy thresholds are 120 and 400 mJ respectively. For both lasing lines the output signal monotonically increases while a saturation-like behavior can be observed for higher driving energies. We should point out that these energies are high enough to create significant regions of overionized plasma. In this experiment it seems that the large amount of driving energy was probably sufficient to counterbalance the ionization induced defocusing and to generate a sufficiently long high gain medium.

4. Gain measurement

Figure 5 shows the intensity in a single x-ray laser pulse as a function of the plasma length ranging from 1.5 up to 5 mm. For xenon, the lasing intensity is shown to increase exponentially up to the plasma length of 2.2 mm, where it begins to saturate.

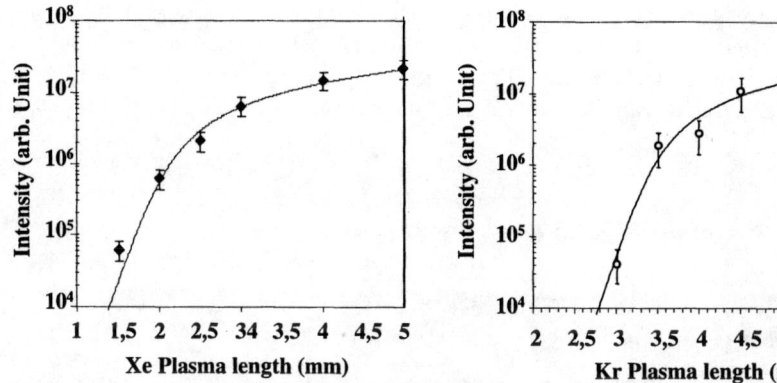

FIGURE 5. Output intensities of the 41.8 (left) and 32.8 nm (right) lasing lines shown as a function of gas cell length.

For krypton, no lasing has been observed for length shorter than 3 mm. By increasing the gas length from 3 to 3.5 mm, the lasing signal increased by a factor of ~40. Further increase of the cell length resulted in an enhancement of the output signal by one more order of magnitude before a saturation-like behavior has been observed. This disagrees with the previous calculations which predicts a similar behavior to Xe. It should be stated that maintenance work was performed on the laser between the Xe and Kr experiments which have probably modified its characteristics which therefore makes a comparison with the computational results difficult.

To estimate the gain coefficient, the output intensity $I_\nu(z, t)$ amplified along the z-axis a wavelength of λ_0 has been calculated using the following set of equations :

$$I(z+dz) = I(z)\exp\{g\,dz\} + \frac{J_0\left[\exp\{g.dz\}-1\right]}{g} \qquad (1)$$

$$\text{with } g(z) = \frac{G_0}{1 + I(z)/I_{sat}} \qquad (2)$$

where G_0 and g are the small-signal and the large gain coefficient respectively, I_{sat} is the saturation intensity. For simplicity the emissivity J_0 is assumed to be constant along the propagation axis.

In the case of the xenon x-ray laser, a good agreement is obtained for $G_0 = 67 \pm 3$ cm^{-1} and I_{sat} has been reached for a plasma length of 2.2 mm resulting in a gain-length product of 15 ± 0.7 at saturation. This confirms that the amplification of the 5d-5p line is saturated.

For the krypton laser, a good agreement between experimental results and calculations has been obtained for a gain coefficient of 78 ± 2 cm^{-1} and a saturation intensity being reached for a length of 3.5 mm. This corresponds to a gain-length product at saturation $(G.l)_s$ of about 27. The above gain coefficient is similar to the gain measured for the 5d-5p Pd-like xenon line. However, the gain-length product seems to be overestimated because it is not consistent with an expected output from so highly amplified x-ray laser source. The last point clearly indicates that only an unknown fraction of the gas cell length participates in the process of amplification of the 4d-4p line and that a realistic estimation of the gain-length product is not possible at the moment.

5. Far-field pattern of the 41.8 nm Pd-like x-ray laser

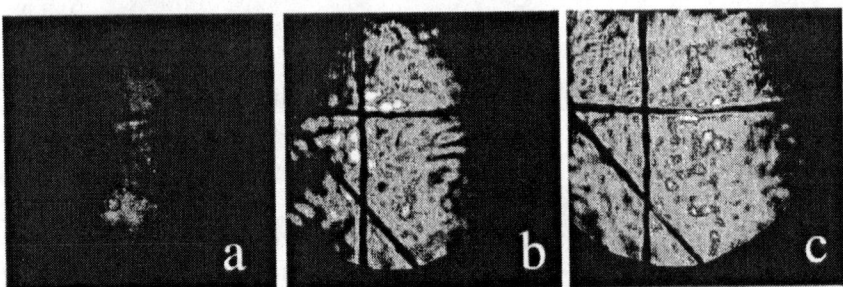

FIGURE 6. Far-field pattern for the Xe laser at three different gas pressure 5 (a), 10 (b) and 15 torr (c).

The figure 6 shows three imprints of the 41.8 nm laser beam obtained for gas pressures of 5, 10 and 15 torr. Each image corresponds to a single shot exposure. To obtain these data, a 45° flat iridium mirror has been used to send the x-ray laser beam onto a X-UV CCD. Two 0.25 μm thick aluminum filters were placed before the CCD camera in order to cut the driving laser beam and the visible emission from the plasma. The mirror has not been characterized but the calculated reflectivity is 20%. A cross wire which is placed in the path of the x-ray laser beam at a distance of 30 cm to the plasma edge gives the horizontal and vertical angle of the x-ray laser emission.

As anticipated from figure 3, the x-ray laser output increases with gas pressure up to 15 Torr. The imprints of the beam show some inhomogeneities for all pressures; some hot spots can be observed showing that all the rays are not amplified the same way when they propagate the plasma column. Due to some propagation problems of the driving laser beam, the lasing media is probably not as homogenous as is desirable. It is interesting to note that the spatial profile of the beam dramatically changes when increasing the pressure. The x-ray laser divergence increases in both directions showing that the lasing volume has been enlarged by increasing the gas pressure. For optimal lasing conditions (P=15 torr), the divergence of the beam is about 10 mrad in the horizontal direction and slightly more in the vertical direction.

CONCLUSION

In conclusion, we have presented recent investigations on collisionally pumped OFI soft x-ray laser. We have demonstrated the first saturated amplification of a such a x-ray laser systems in Pd-like Xe plasma at 41.8 nm. Moreover we have experimentally observed a large amplification of the 4d-4p line of Ni-like krypton at 32.8 nm which is the first extension of the OFI collisional x-ray laser technique to shorter wavelengths using Ni-like ions. These monochromatic high repetition rate soft x-ray sources were reproducible on a day to day basis and are new candidates for application experiments requiring narrow bandwidth high repetition rate collimated soft x-ray source. These x-ray lasers provide up to 5×10^9 photons per pulse in a ~10 to 15 mrad of divergence. Recent investigation which has not been presented in this paper has given some preliminary results on the measurement of the XRL pulse duration. Our data seems to show that the duration of the 41.8 nm pulse is shorter to 3 ps but more complete set of data are needed to conclude. An other important issue is the control of the propagation of the driving laser in order to increase the gas density and then boost the lasing output. This could be done by using more energetic laser drivers or waveguiding methods such as hollow and discharge-ablated capillaries. These issues will be addressed in future work.

REFERENCES

1. Y. Nagata, K. Midorikawa, S. Kubodera, M. Obara, H. Tashiro and K. Totada, Phys. Rev. Lett. **71**, 3774-3777 (1993)
2. D. V. Korobkin, C. H. Nam, Suckewer S and A. Goltsov, Phys. Rev. Lett. **77**, 5206-5209 (1996)
3. Y. Nagata, K. Midorikawa, S. Kubodera, M. Obara, H. Tashiro, K. Toyoda and Y. Kato, Phys. Rev. A **51**, 1415-1418 (1995)
4. A. Egbert, D. M. Simanovskii, B. N. Chichkov, and B.Wellegehausen, Phys. Rev. E **57**, 6761-6772 (1998)
5. E. Fill, S. Borgström, J. Larsson, T. Starczewski, C. G. Wahlström, and S. Svanberg, Phys. Rev. E **51**, 6016-6027 (1995)
6. B. E. Lemoff, C. L. Gordon III, C. P. J. Barty and S. E. Harris, Phys. Rev. Lett. **74**, 1574-1577 (1995)
7. B. E. Lemoff, C. P, J. Barty and S. E. Harris, Optics Letters **19**: 569-571 (1994)
8. S. Sebban, R. Haroutunian, Ph. Balcou, G. Grillon, A. Rousse, S. Kazamias, T. Marin, J. P. Rousseau, L. Notebaert, M. Pittman, J. P. Chambaret, A. Antonetti, D. Hulin, D. Ros, A. Klisnick, A. Carillon, P. Jaeglé, G. Jamelot and J. F. Wyart. Phys. Rev. Lett. **86**: 3004-3007 (2001)
9. S. Sebban, T. Mocek, D. Ros, L. M. Upcraft, Ph. Balcou, R. Haroutunian, G. Grillon, B. Rus, A. Klisnick, A. Carillon, G. Jamelot, C. Valentin, A. Rousse, J. P. Rousseau, L. Notebaert, M. Pittman and D.Hulin, submitted for publication (2002)
10. M. Pittman, S. Ferré, J. P. Rousseau, L. Notebaert, J. P. Chambaret, and G. Chériaux, App. Phys. B, in press (2002)
11. G. L Strobel, and P. Amendt,, App. Phys. B **58**, 45-51 (1994)
12. G. J. Pert, S. MacCabe, and P. A. Simms. in *Proceedings of the t 6th International Conference on x-ray lasers*, (Kyoto, 1998)
13. S. M. Hooker, P. T. Epp, and G. Y. Yin, J. Opt. Soc. Am. B **14**, 2735-2741 (1997)
14. C. Decker, D. C. Eder, and R. A. London,, Phys. Plasmas **3**, 414-419 (1996)

Experimental and Theoretical Simulations for Conditions for Lasing at 13.5 nm in LiIII

Y. Avitzour, I. Geltner, A. Morozov, Y. Ping, and S. Suckewer

School of Eng. & Appl. Sciences, MAE Dept., Princeton University
Princeton, NJ 08544, USA

Abstract. We present results related to the search for optimum conditions for lasing to ground state of H-like LiIII ions at 13.5 nm. These conditions are being considered from the point of view of the development of a prototype of a very compact 13.5 nm laser for metrology of soft x-ray (EUV) lithography. Theoretical simulations are discussed in relation to experimental data. Experiments on channeling of ultrashort high intensity pumping laser beam in microcapillary plasma are presented for conditions appropriate for lasing at 13.5 nm. We are also discussing Raman amplification of ultrashort pulses in microcapillary plasma as a possibility for future use in X-ray lasers.

INTRODUCTION

For a number of years we have recognized the importance of developing compact soft X-ray lasers (SXLs). More recently it became quite clear that such lasers should not only be compact but should also have a good repetition rate (1 – 10 Hz and higher), good beam quality (coherence), and energy in the - 1   10 μJ/pulse range. These issues were well articulated at this conference by P. Nickles [1]. At 13 – 14 nm range such lasers would be very valuable for metrology of soft X-ray lithography (also called EUV lithography) (see e.g. [2]). Ideal laser for this purpose would be discharge pumped collisional SXL ("Rocca Type" SXL) in Ni-like Cd at 13.2 nm, when high gain-length product will be demonstrated [3].

Our approach to SXL development for metrology of EUV lithography is based on the demonstration of a high gain-length product (GL) at 13.5 nm in H-like LiIII ions (in transition to ground state) using laser created preplasma (GL $\approx$ 6.5) [4] and discharge created preplasma, GL $\geq$ 7 [5] in LiF microcapillaries. The longitudinal "pumping" of Li plasma by means of Optical Field Ionization (OFI-tunneling ionization) using a high intensity, subpicosecond laser provides totally stripped Li^{3+} ions. They recombine rapidly in relatively cold plasma and create population inversion and high gain in LiIII ions between excited and ground levels. Such a 13.5 nm laser is quite efficient, therefore it can be compact and can operate at

CP641, *X-Ray Lasers 2002: 8th International Conference on X-Ray Lasers,* edited by J. J. Rocca et al.
© 2002 American Institute of Physics 0-7354-0096-2/02/$19.00

relatively high repetition rates. In order to satisfy the requirements for metrology of EUV lithography its parameters (rep. rate ~ 10 Hz, energy per pulse ~ 10 μJ, beam divergence ≤ 5 mrad, and transverse coherence ~ 3 μm) should be quite reproducible for a long running time [6].

In this paper we present the analysis of experiments and theoretical modeling in search for optimum pumping pulse intensity, its wavelength and duration as well as the issue of pumping beam channeling in plasma with optimal density for soft X-ray lasing. We also present the idea of using Raman amplification of ultrashort laser pulses in plasma by a low power (but higher frequency) counter propagating laser beam as a potentially efficient method for pumping SXLs.

EXPERIMENTAL CONSIDERATIONS

Population inversion for the 2 – 1 transition in H-like ions requires that recombination rates for totally stripped ions to H-like ions have to be larger than the decay rate of level n = 2 by collisional and radiative transitions. This means that the three-body recombination rate: $\beta_3 n_e^2 T_e^{-4.5}$ (in $\sec^{-1}$), where three-body recombination coefficient $\beta_3 \sim n^4$ and n is principle quantum number of H-like ions, n_e and T_e are electron density and temperature, respectively, has to be larger than the sum of the radiative de-excitation rate (A_{21}) and collisional de-excitation rate ($\alpha_{21} n_e T_e^{-1/2}$ in $\sec^{-1}$; α_{21} is the collisional de-excitation coefficient).

The decay rate of level 2 provides an upper limit for the pumping pulse duration, τ_L, namely $\tau_L <$(decay rate)$^{-1}$. Hence, the ratio of the three body recombination rate to the decay rates of level 2 is proportional to $T_e^{-4.5}$ for a radiative dominated decay and to $T_e^{-3.5}$ for a collisional dominated decay. Therefore, for fast recombination and high gain generation the T_e should be as low as possible. For LiIII ions it means that $T_e \leq 3 - 5$ eV for the optimal n_e in the range $10^{18} - 10^{19}$ cm^{-3}.

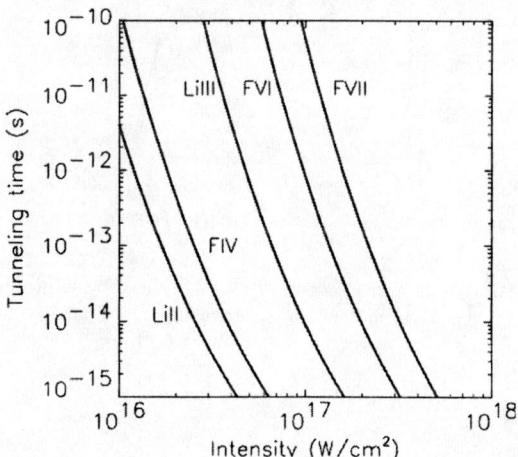

FIGURE 1. Tunneling ionization times of Li and F ions.

Optical Field Ionization. Fig.1 presents tunneling times for different Li and F ions (from wall ablation of the LiF microcapillary) as a function of laser intensity. One may see that for ionization of LiIII ions on a time scales of a few fsec the laser beam intensity has to be on the order of 10^{17} W/cm^2. For such an intensity the Keldysh parameter γ [7] is much below 1, hence tunneling ionization dominates mulitphoton ionization in OFI processes [8]. It is clear that just the front of such a laser pulse has enough intensity to ionize Li II and F IV, but even its peak intensity is not sufficient to ionize F VI.

Plasma Channeling experiments were conducted using our new, compact Ti/Sapphire laser (100 fsec, up to 400 mJ, 5 – 10 Hz, located on two optical tables). This laser and the experimental arrangement for the waveguiding experiment in discharge created plasma in microcapillaries are described in [9]. Interferometric measurements of plasma density at the ends of microcapillary have shown that the best conditions for waveguiding, i.e. a hollow density profile occurred about 70 – 120 nsec after the first peak (at 200 ns) of the discharge current. In Fig. 2 is shown a cross section of 100 fsec beam profile (solid line) in comparison with a Gaussian profile (dashed line) at the exit of a 20 mm long and 0.5 mm diameter microcapillary where plasma was created by discharge of a capacitor C = 50 nF charged to 5.5 kV. Although the beam profile was very good, so far the absorbed beam energy in such a long microcapillary was very high and absorption increased with input beam intensity.

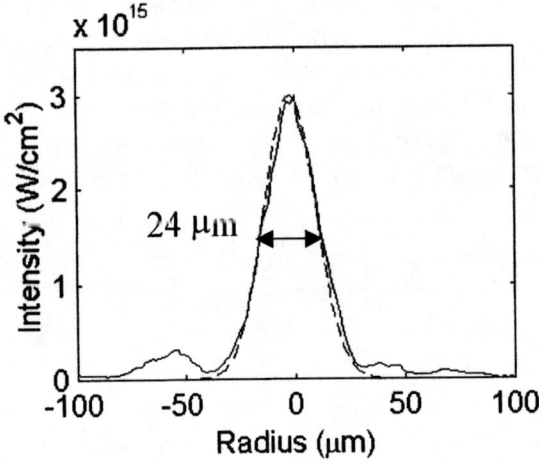

FIGURE 2. Cross section of ultrashort laser beam profile (solid line) at output of microcapillary. Dashed line represents Gaussian fit with FWHM ≈ 24 μm.

THEORETICAL MODELING

In the theoretical modeling of the gain generation in LiIII ions on 2 – 1 transition at 13.5 nm two important features were taken into account:

(a) The electron distribution function is highly non-Maxwellian after the ionization process (OFI)

(b) The pumping beam forms a narrow, fully ionized plasma, embedded in a cold background plasma ("preplasma").

These features have a crucial effect on gain generation in LiIII ions in recombining Li-plasma. The model, which is described in more detail in [10], consists of 3 parts:

1. Ionization and heating of the plasma channel.
2. Plasma expansion and cooling.
3. Recombination of Li^{3+} into LiIII ions and gain generation on 2 – 1 transition.

Here the tunneling ionization is the main ionizing process of LiII and LiIII ions whereas Above Threshold Ionization (ATI) [11] provides electron heating (very detrimental for recombining schemes) during pumping pulses. The ionization distribution function was calculated using a 3D velocity-space PIC code similar to the one described in [12,13]. In the code also e – e and e – i collisions were taken into account [10]. The initial plasma (preplasma) was assumed to have $T_e \approx 1 eV$ and consisted of LiII ions.

In order to calculate the expansion and cooling of a narrow hot channel in a relatively cold plasma the time evolution of the electron distribution function was calculated using a 1-D cylindrically symmetric Fokker-Planck code [14,15]. The expansion has an important effect on the lowering of the electrons average energy and the Maxwellization process. The obtained electron distribution function was used to calculate time and space dependent rate coefficients for the atomic processes important for the population inversion and gain creation on 2 – 1 transition in LiIII ions.

Results of the calculation of the gain as a function of time for two pumping beam wavelengths (248 nm and 800 nm) and for different channeling diameters (6 – 30 μm) are shown in Figs. 3 and 4. These results were obtained for an optimum electron density ($n_e \approx 5 \times 10^{18}$ cm^{-3}) and beam intensity $I = 1.3 \times 10^{17}$ W/cm^2 (this is optimum intensity for maximum gain generation with a 248 nm laser, but not necessarily for a 800 nm laser). From these figures one may see a strong effect of the pumping channel diameter on the gain. For a 248 nm pumping beam and 6μm channel diameter gain can reach $G \approx 200 cm^{-1}$. Gain is lower for the 800 nm beam due to larger ATI heating, but still reaches $G \approx 60$ cm^{-1} for a 7.5 μm diameter channel. These results suggest that using a very tightly focused pumping beam it should be possible to reach gain saturation for 13.5 nm laser in 1 – 2 mm long plasma. Hence, it may be possible to generate such gain in a plasma jet rather than in a microcapillary, which would be very advantageous for EUV lithography applications.

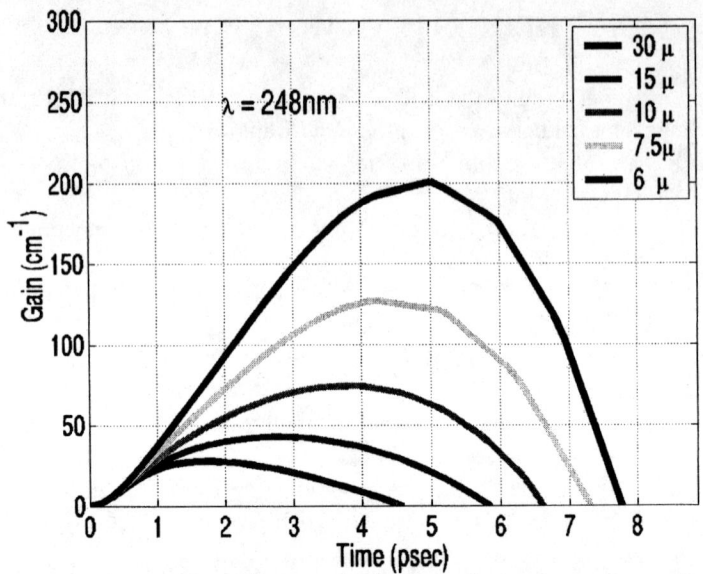

FIGURE 3. Time evolution of the gain for optimum electron density (n $_e \approx$ 5x10^{18}cm^{-3}) and for 6 – 30 μm channeling diameters (pumping beam intensity 1.3x10^{17} W/cm^2 at 248 nm).

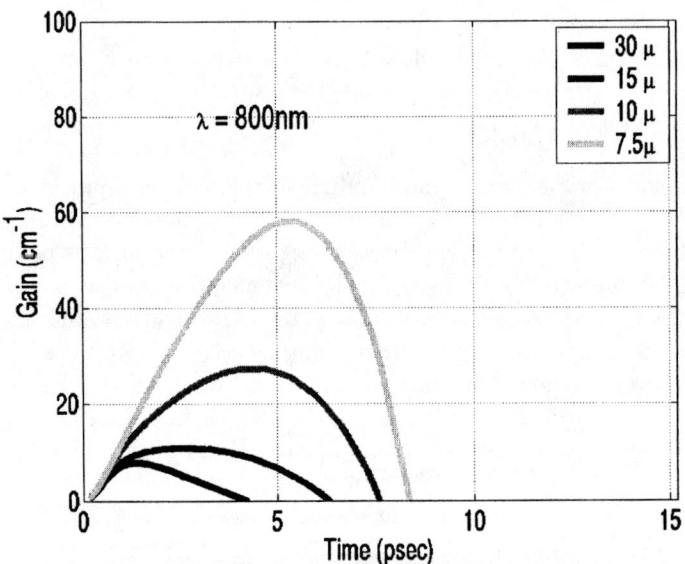

FIGURE 4. Time evolution of the gain for the same conditions as in Fig. 3, but for 800 nm pumping beam.

RAMAN AMPLIFICATION OF ULTRASHORT PULSES FOR PUMPING X-RAY LASERS (?)

Raman amplification of ultrashort pulses in plasma by a counter-propagating pumping pulse may lead in the future to a new method for generation ultrahigh intensity, ultrashort pulses [16,17]. Such Raman amplification is based on the three-wave interaction:

$$\omega_1 = \omega_2 + \omega_p, \qquad \vec{k}_1 = \vec{k}_2 + \vec{k}_p \qquad (1)$$

where $\omega_1, \omega_2, \omega_p$ are the frequencies of the pumping pulse, ultrashort pulse ("seed pulse"), and plasma waves, respectively, whereas $\vec{k}_1, \vec{k}_2, \vec{k}_p$ are the corresponding wave vectors. The basic idea of Raman amplification of ultrashort pulse is illustrated by Fig. 5.

The first experimental demonstration of Raman amplification of 200 fsec, 745 nm pulses was performed in LiF microcapillary plasma [18] in a set up very similar to our earlier experiment on demonstration of lasing in LiIII ions at 13.5 nm [4]. Recently we have shown that resonant conditions existed in a very short plasma with effective length $\ell_{\text{eff}} \sim 0.2$ mm, where amplification ($\sim 6 - 8$) took place and these results are presented in these proceedings [19].

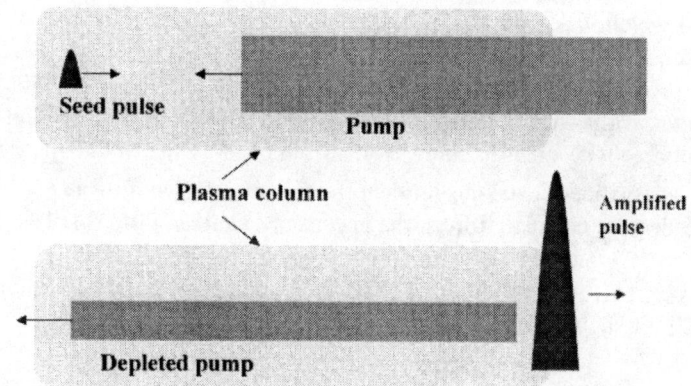

Initial experiments: $\lambda_{\text{seed}} = 745$ nm, $t_{\text{seed}} = 200$ fs
$\lambda_{\text{Pump}} = 532$ nm, $t_{\text{pump}} = 5$ ns
$\lambda_{\text{plasma}} = 1861$ nm, $n_e = 3 \times 10^{20}$ cm^{-3}

FIGURE 5. Illustration of principle of Raman amplification of ultrashort pulses in plasma and example of parameters from [18]

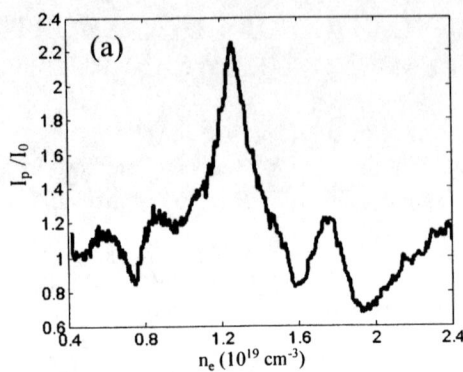

FIGURE 6. The ratio of amplified spectral intensity I_p to non-amplified intensity I_0 as a function of electron density in plasma jet. Resonance density is in range $n_e \approx 1.1 - 1.5 \times 10^{19} cm^{-3}$ which corresponds to 5 nm spectral broadening of ultrashort "seed" pulse.

In this paper we discuss only the results on Raman amplification in a high pressure jet plasma which was not presented in [19] and may be implemented in a 13.5 nm laser for EUV lithography.

In the experiment with a plasma jet the 800 nm, 170 psec laser beam (before compressor) was split into two beams: "energetic" one (95% of energy) which was compressed to 10 psec in order to serve as the pumping pulse, and second beam (5% of energy) which was compressed to 500 fsec with second compressor. This second beam is frequency-doubled by a SHG crystal to pump an Optical Parametric Oscillator (OPO) to provide a wavelength tunable seed pulse in the range of 850 – 920 nm. A spectrometer with a CCD detector was used to measure amplification of ultrashort "seed" pulses. The electron density distribution was measured interferometrically. With a 1 mm orifice for the gas jet the effective plasma length was $\ell_{eff} \approx 0.5$ mm. The resonance density calculated from the spectrum of the amplified pulse was $n_e \approx 1.1 - 1.5 \times 10^{19}$ cm^{-3}, which was confirmed by interferometric measurements. The observed amplification was slightly above 2 (Fig. 6), which is in good agreement with linear theory [17] for such short and not very dense plasma.

CONCLUSION

We have demonstrated guiding of ultrashort pulses in discharge microcapillary plasmas up to 2 cm long, which is important for a high and reproducible gain-length product, GL. However, in such long plasma large absorption of the pumping beam has a very negative effect on gain generation after a certain propagation distance. Therefore, we have analyzed, using theoretical modeling, the possibility of very high gain generation in short plasma column in order to obtain GL close to saturation. Calculations were performed for two different wavelengths of ultrashort (100 fsec) pumping beams: $\sim$ 0.25 μm and $\sim$ 0.8 μm. It was shown that the gain strongly depends on the plasma channel diameter. For a channel diameter d = 6 μm gain as high as G $\approx$ 200 cm^{-1} was predicted for 0.25 μm femtosecond laser beam. Smaller, but still very high gain ($\sim$ 60 cm^{-1}) was predicted for 0.8 μm pumping beam.

We have also discussed the possibilities of Raman amplification of ultrashort pulses in plasma and the potential of using such pulses for more efficient pumping of soft x-ray lasers, particularly for a 13.5 nm LiIII laser for metrology of EUV lithography.

ACKNOWLEDGEMENTS

We would like to express our appreciation to N. Tkach for his dedicated technical support. This work was supported by grants from NSF (PHYS) and INTEL Inc.

REFERENCES

1. Nickles, P., et al., Invited Talk at XRL – 2002 Conference (Aspen, May 2002); and in these Proceedings.
2. Attwood, D., ibid
3. Rocca, J., et al., ibid
4. Korobkin, D.V., Nam, C.H., Suckewer, S., and Goltsov, A., *Phys. Rev. Lett.* 77, 5206-5209 (1996).
5. Goltsov, A., et al., *IEEE J. Sel. Top. Quant. Elect.* 5, 1453-1459 (1999).
6. Bjorkholm J., private communications (2001).
7. Keldysh, L, *Sov. Phys. JEPT* 20, 1307-1314 (1965).
8. Ammosov, M.V., Delone, N.B., and Krainov, V.P., *Sov. Phys. JETP* 64, 1191-1194 (1986).
9. Geltner, I., Ping, Y., Morozov, A., and Suckewer, S., these Proceedings.
10. Avitzour, Y., Brunner, S., Valeo, E., and Suckewer, S., these Proceedings.
11. Burnett, N.H., and Corkum, P.B., *J. Opt. Soc., Am. B-Opt. Phys.* 6, 1195-1199 (1989).
12. Takizuka, T., and Abe, H., *J. Comput. Phys.* 25, 205-219, (1977).
13. Ma, S., Sydora, R.D., and Dawson, J.M., *Comput. Phys. Commun.* 77, 190-206 (1993).
14. Brunner, S., and Valeo, E., *Phys. Plasmas* 9, 923-936 (2002).
15. Epperlein, E.M., *Laser Particles & Beams* 12, 257-272 (1994).
16. Shvets, G., Fisch, N.J., Pukhov, A., and Meyer-ter-Vehn, J., *Phys. Rev. Lett.* 81, 4879 (1998).
17. Malkin, V.M., and Fisch, N.J., *Phys. Plasmas* 8, 4698 (2001).
18. Ping, Y., Geltner, I, Fisch, N.J., Shvets, G., and Suckewer, S., *Phys. Rev. E* 62, R4532 (2000).
19. Ping, Y., Geltner, I, Morozov, A., Fisch, N.J., and Suckewer, S., these Proceedings; also *Phys. Rev. E* (submitted).

Numerical Simulations of Collisionally Pumped OFI Soft X-Ray Lasers

L. M. Upcraft[1], G. J. Pert[2] and S. Sebban[1]

1 - Laboratoire d'Optique Appliquée, ENSTA,Chemin de la Huniere, 91761 Palaiseau, FRANCE
2 – Department of Physics, University of York, Heslington, York, YO10 5DD, UK

Abstract. Numerical simulations have been performed for the collisionally pumped soft X-Ray lasers in eight times ionized Kr and Xe which are driven by optical field ionisation of the gas by a high intensity femtosecond laser. Results are presented covering the expected state of the plasma formed, time dependant gain calculations and also a brief discussion on the nature of the electron distribution function.

INTRODUCTION

The use of high power, sub picosecond lasers as means of achieving soft X-Ray lasing through optical field ionisation (OFI) has been examined theoretically within the context of both recombination [1] and electron collisional excitation schemes [2, 3]. However, while the recombination schemes have yet to be demonstrated convincingly, the collisional systems are now routinely produced here at the Laboratoire d'Optique Appliquée (LOA) although further characterization work is needed.

The basis of the scheme, as first proposed by Lemoff *et al* [2] is that a high power, short pulse optical laser of circular polarisation, if focused into a gas cell, will produce both the lasant ion stage and the hot electrons necessary for collisional excitation of the upper laser level. The Ti:Saph laser at LOA is capable of providing around 1 J with pulse durations as short as 30 fs and has been used to generate soft X-Ray lasing in Ni-like Kr at 32.8 nm and in Pd-like Xe at 41.8 nm [4]. Numerical simulations on aspects of these lasers are presented below.

RESULTS

The main problem with the OFI scheme is in the formation of a suitable length of plasma for amplification of the X-ray photons to occur. Longitudinally driving the optical laser into the gas means that ionisation induced refraction of the beam can rapidly reduce the intensity below that needed to form the required ionisation stage. This problem is more apparent with collisional schemes than recombination since the formation of hot pumping electrons through the OFI process needs a longer wavelength. The use of 800 nm radiation as opposed to 248 nm for investigated

CP641, *X-Ray Lasers 2002: 8th International Conference on X-Ray Lasers,* edited by J. J. Rocca et al.

recombination schemes means that the degree of beam dispersion (as calculated through the refractive index) is an order of magnitude larger in this case.

Controlling the natural spread of the beam, as defined through the Rayleigh length of the focused beam radius is theoretically possible given that with >1 J of laser energy available, large focal radii of ~ 100 μm would still result in the required intensity to form the eight stage ion in both Kr and Xe. However, practically, the availability of focusing optics and the geometry of the target system have restricted the current experiments to working at spot radii of 30 μm.

Figure 1 shows the expected ionisation state of the plasma in an azimuthally symmetric axial / radial plane within the gas cell for the experimental conditions [4] in which the best lasing in Kr and Xe has been observed. For Kr this is at a cell gas pressure of 20 Torr with a laser energy of 700 mJ, for Xe the pressure is 15 Torr and the laser energy is 330 mJ.

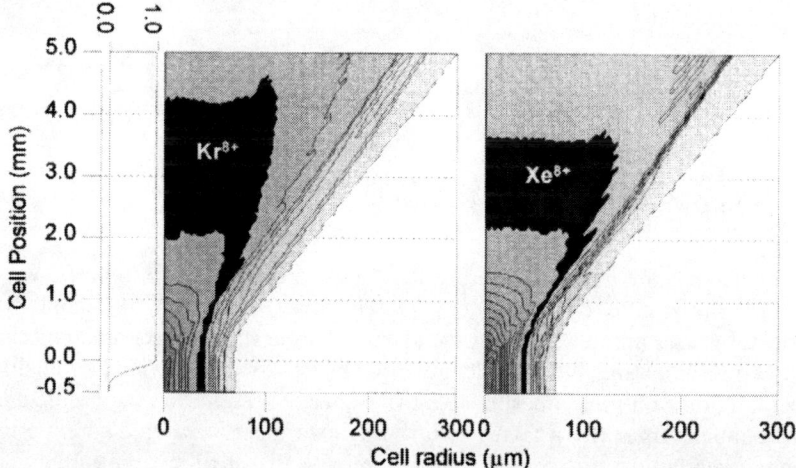

FIGURE 1. Ionisation contour plots for Kr at 20 Torr (left) and Xe at 15 Torr (right). The laser is assumed to be focused at the cell entrance at 0 mm and propagates the axial length of the cell. The region of eight times ionised lasant ion is shown with subsequent regions showing increasing ionisation stages towards the center of the plots. The gas pressure is assumed constant within the gas cell, but extends from the entrance along the axis with a drop off such that vacuum occurs after 500 μm. The axial gas density profile is shown on the left for both cases, normalised to the maximum of 20 and 15 Torr for Kr and Xe respectively.

Although figure 1 shows that we expect reasonably uniform columns which should be suitable for amplification of Kr^{8+} of about 2 mm in length, and around 1.5 mm for Xe^{8+}, an important feature is that large regions of strongly over-ionised plasma are predicted in both cases. This region will be a source of high energy electrons (up to 10 keV) which may well act to limit the action of the X-Ray laser through collisional ionisation of the lasant ion stage.

Simulations of the atomic processes have been performed using a time dependent collisional / radiative model [5] in order to calculate the gain coefficients for comparison with experiments. The atomic data necessary (atomic energy levels, collisional rates and radiative rates) has been obtained from the atomic code of Cowan [6]. The temporal evolution of the gain is shown in figure 2 for both Kr and Xe. The plasma conditions are taken as that predicted from the propagation results (figure 1) for an axial point in the center of the main region of the lasant ion stage.

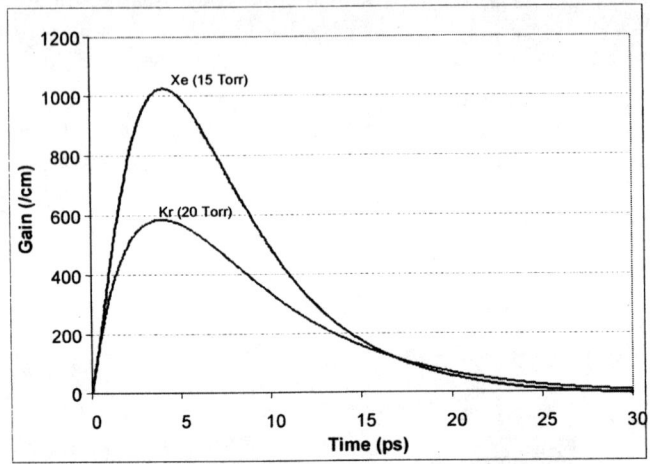

FIGURE 2. Calculated gain coefficients for Kr at 20 Torr and Xe at 15 Torr for the experimental conditions discussed in the text.

The predicted gains are much larger than the experimentally estimated values of 78 cm^{-1} for Kr and 68 cm^{-1} for Xe [4]. However, it should be stated that the experimental values are necessarily time averaged values. This fact and experimental difficulties in estimating the gain length product make direct comparison difficult. However, a point of a more questionable difference is seen in that the predicted gains for Xe are much larger (by a factor of up to 2) than those for Kr. The experimental values are essentially similar in value and moreover, that for Kr is measured as larger than for Kr. A significant possible reason for this difference is that although the Kr system has been observed to lase unambiguously, there has been insufficient data from the experiments to confirm the effect of saturated behaviour which will affect any measurement of the gain coefficient.

A major source of concern with the calculated values is that they are based on a model which assumes the pumping electrons are thermal, i.e. have a Maxwell-Boltzmann, distribution. This is certainly not the case given the nature of the OFI process. The nature of the distribution function can be examined using the method of Pert [7] which calculates OFI, inverse bremsstrahlung and collisional relaxation of the electrons. The calculated distributions for Kr and Xe are shown in figure 3. Note that the lack of eight distinct peaks for each ionisation stage is due to the combined effects of inverse bremsstrahlung and collisional relaxation. This shows that the distributions are clearly non-thermal in nature, and more importantly calculations show that at these relatively low densities they remain so for many tens of picoseconds. Given that the peak gain value, albeit on predictions with the thermal distribution, occurs at around 5ps, it is clear that this is a significant factor in the understanding of the system.

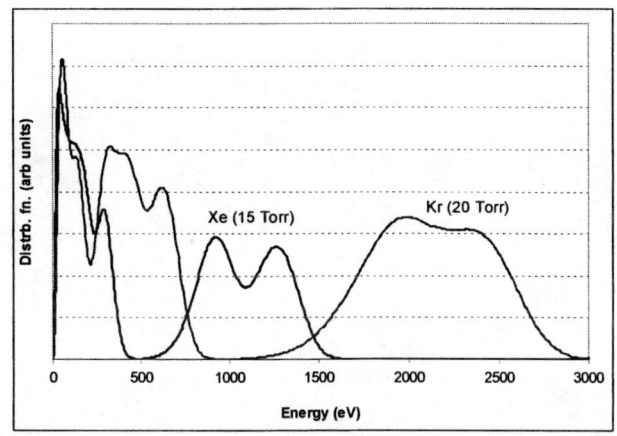

FIGURE 3. Calculated electron distribution functions for Kr and Xe due to OFI at the end of a 30 fs pulse. The intensity in both cases is taken to be that which results from the axial position in the center of the uniform regions of lasant ion as predicted from the propagation calculations of figure 1.

CONCLUSIONS

Driving laser pulse propagation simulations for the observed experimental conditions show that uniform regions of lasant ion are produced, but so too are significant regions of over-ionised plasma which may limit the lasing action through further collisional ionisation. Atomic calculations show discrepancy with the measured values but the use of thermal electron distributions is expected to be a major factor when strong non-thermal characteristics persist for longer than the estimated gain lifetime. Work is currently ongoing to examine these three aspects within a more coherent single numerical model.

ACKNOWLEDGMENTS

This work was supported by the European X-Ray Laser Network programme. Much of the numerical work is based upon codes developed and maintained by Prof. G. J. Pert at the University of York, UK.

REFERENCES

1. Grout, M. J., *PhD Thesis,* University of York (1997).
2. Lemoff, B. E., Barty, C. P. J. and Harris, S. E., *Opt. Lett.*, **19** 569 (1994).
3. Upcraft, L. M., *PhD Thesis,* University of York (2002).
4. Sebban, S., "Investigations on collisional optical field ionization soft x-ray lasers", *elsewhere in these proceedings.*
5. Pert, G. J., *J. Phys. B.,* **23** 619 (1990)
6. Cowan, R. D., *Theory of Atomic Structure and Spectra*, University of California Press, 1981
7. Pert, G. J., *J. Phys. B.*, **34** 881 (2001)

THEORY OF X-RAY LASERS
AND AMPLIFIED BEAMS

Atomic Data for F-like Molybdenum

K. M. Aggarwal and F. P. Keenan

Dept. of Pure and Applied Physics, Queen's University, Belfast BT7 1NN, UK

Abstract. Energy levels, radiative rates and collision strengths for transitions among the lowest 60 fine-structure levels belonging to the $(1s^2)$ $2s^22p^5$, $2s2p^6$, and $2s^22p^43\ell$ configurations of F-like Mo XXXIV have been calculated using the fully relativistic GRASP and DARC codes. Comparisons with other available results are made, and the accuracy of the present data is assessed. Energy levels are expected to be accurate to within 1%, while oscillator strengths and collision strengths for all transitions are probably accurate to better than 20%.

INTRODUCTION

Electron impact excitation of F-like ions is of considerable interest, particularly in x-ray lasers and tokamak plasmas. Therefore, to understand plasma characteristics and to model lasers, atomic data for energy levels, radiative rates, collision strengths, rate coefficients, etc. are required, as the corresponding experimental data are often not available, except for energy levels. Hence, in this paper we report our results for energy levels, radiative rates and collision strengths (Ω). The much desired results for excitation rate coefficients are not yet available, because the calculations for Ω have so far been performed only at energies above thresholds.

Earlier available results for this ion are of Sampson et al. [4], who have adopted their relativistic Dirac-Fock-Slater (DFS) program for generation of wavefunctions, and the Distorted-Wave (DW) code for the computation of Ω. However, they have computed results for transitions among the lowest three levels, and from these to higher excited levels, and have reported values of Ω at only 6 energies above thresholds. The omitted contribution of resonances in the threshold region may enhance the rates up to an order of magnitude, even at the high temperatures at which data are required in laser applications. To take care of these deficiencies, our calculations for Ω in the threshold region are in active progress.

CALCULATION DETAILS

The $(1s^2)$ $2s^22p^5$, $2s2p^6$ and $2s^22p^43\ell$ configurations of Mo XXXIV give rise to 60 fine-structure levels. To generate the wavefunctions, we have employed the GRASP code [2] with the option of extended average level, and configuration interaction (CI) has been included among the above stated configurations. For the computations of Ω, we have adopted the Dirac Atomic R-matrix Code (DARC) of Norrington and Grant [3]. This program includes relativistic effects in a systematic

CP641, *X-Ray Lasers 2002: 8th International Conference on X-Ray Lasers,* edited by J. J. Rocca et al.

way, in both the target description and the scattering model. It is based on the jj coupling scheme, and uses the Dirac-Coulomb Hamiltonian in the R- matrix approach. The R-matrix radius has been adopted to be 1.33 au, and 35 continuum orbitals have been included for each channel angular momentum for the expansion of the wavefunction. This allows us to compute Ω up to an energy of 3200 Ry, which will easily permit us to compute the excitation rate coefficients up to a temperature of 10^8 K. In order to obtain convergence of Ω for all transitions and at all energies, we have included all partial waves with angular momentum $J \leq$ 40, and to account for the higher neglected partial waves, a top-up based on the sum rules has also been included.

RESULTS AND DISCUSSION

In Table 1 we list our computed energies, with and without the Breit and QED corrections, for the lowest 40 levels only. The net effect of the Breit and QED corrections has lowered the corresponding Coulomb energies by up to about 0.5 Ry. Additionally, these corrections have altered the energy order of two levels, namely 16 and 17. The only other values available in the literature are those of Sampson et al [4], who have adopted their DFS code. Therefore, their energy levels are also included in Table 1 for comparison. The two sets of energy agree within 0.2 Ry (less than 0.1%), which is highly satisfying. However, the energy order is slightly different in a few instances: for example, levels 22/23, 25/26, and 55/56. These minor differences arise for two reasons. Firstly, the potential used by Sampson et al and the procedure adopted for the calculation of radial functions are comparatively less accurate (see [4] for details), and secondly, the DFS calculations include more configurations and hence more CI.

In Table 2 we compare the length form of our oscillator strengths with those of Sampson et al [4]. Also included in this table are the ratio of the velocity and length forms of the f- values. This ratio gives an indication of the accuracy of the results, but can often be quite deceptive. Additionally, for weaker transitions, with f- values < 0.10, it is difficult to obtain a near unity ratio, as such transitions are highly sensitive to mixing coefficients. Since such transitions often have f- values ≤ 0.001, they are not of much practical use and do not affect the diagnostic or modelling calculations.

The agreement between our GRASP and the Sampson et al's DFS f- values [4] is better than 20%, especially for the strong transitions. For weaker transitions, the differences between the two sets are up to two orders of magnitude. This is not surprising and such large differences, for weaker transitions, are often observed between any two calculations, especially when there are differences in the number of configurations included. Therefore, accuracy estimates for the weaker transitions are always insecure, but the length form of our oscillator strengths for the strong transitions are expected to be accurate to better than 20%. In fact, the only way to improve these results is to include more CI. This may or may not affect the present results, as molybdenum is comparatively a heavy element, for which the

Table 1. State definitions and energy levels (in Ry) for F-like Mo XXXIV.

Index	State Configuration	Level	J^π	GRASP^a	GRASP^b	DFS^c
1	$2s^2 2p_{1/2}^2 2p_{3/2}^3$	$^2P^o$	$(3/2)^o$	0.0000	0.0000	0.0000
2	$2s^2 2p_{1/2} 2p_{3/2}^4$	$^2P^o$	$(1/2)^o$	8.1860	8.0586	8.0704
3	$2s 2p_{1/2}^2 2p_{3/2}^4$	^2S	$(1/2)^e$	24.5954	24.4953	24.3654
4	$2s^2 2p_{1/2}^2 2p_{3/2}^2 3s$	^4P	$(5/2)^e$	180.926	180.715	180.767
5	$2s^2 2p_{1/2}^2 2p_{3/2}^2 3s$	^2P	$(3/2)^e$	181.331	181.127	181.193
6	$2s^2 2p_{1/2}^2 2p_{3/2}^2 3s$	^2S	$(1/2)^e$	183.070	182.960	182.796
7	$2s^2 2p_{1/2}^2 2p_{3/2}^2 3p_{1/2}$	$^4P^o$	$(3/2)^o$	185.350	185.166	185.192
8	$2s^2 2p_{1/2}^2 2p_{3/2}^2 3p_{1/2}$	$^2D^o$	$(5/2)^o$	185.521	185.307	185.368
9	$2s^2 2p_{1/2}^2 2p_{3/2}^2 3p_{1/2}$	$^2P^o$	$(1/2)^o$	187.189	187.024	186.890
10	$2s^2 2p_{1/2}^2 2p_{3/2}^2 3p_{3/2}$	$^4P^o$	$(5/2)^o$	187.556	187.306	187.345
11	$2s^2 2p_{1/2}^2 2p_{3/2}^2 3p_{3/2}$	$^4D^o$	$(7/2)^o$	187.586	187.331	187.404
12	$2s^2 2p_{1/2}^2 2p_{3/2}^2 3p_{3/2}$	$^2S^o$	$(1/2)^o$	187.884	187.706	187.691
13	$2s^2 2p_{1/2} 2p_{3/2}^3 3s$	^4P	$(3/2)^e$	188.877	188.588	188.639
14	$2s^2 2p_{1/2}^2 2p_{3/2}^2 3p_{3/2}$	$^4S^o$	$(3/2)^o$	189.124	188.918	188.977
15	$2s^2 2p_{1/2} 2p_{3/2}^3 3s$	^2P	$(1/2)^e$	189.217	188.928	188.999
16	$2s^2 2p_{1/2}^2 2p_{3/2}^2 3p_{3/2}$	$^4D^o$	$(3/2)^o$	189.968	189.810	189.675
17	$2s^2 2p_{1/2} 2p_{3/2}^3 3s$	^2D	$(5/2)^e$	190.062	189.680	189.734
18	$2s^2 2p_{1/2} 2p_{3/2}^3 3s$	^2D	$(3/2)^e$	190.226	189.854	189.911
19	$2s^2 2p_{1/2}^2 2p_{3/2}^2 3d_{3/2}$	^4P	$(3/2)^e$	192.906	192.683	192.740
20	$2s^2 2p_{1/2}^2 2p_{3/2}^2 3d_{3/2}$	^4D	$(5/2)^e$	192.955	192.694	192.762
21	$2s^2 2p_{1/2}^2 2p_{3/2}^2 3d_{3/2}$	^4P	$(1/2)^e$	193.087	192.848	192.909
22	$2s^2 2p_{1/2}^2 2p_{3/2}^2 3d_{3/2}$	^4D	$(7/2)^e$	193.156	192.870	192.961
23	$2s^2 2p_{1/2} 2p_{3/2}^3 3p_{1/2}$	$^4P^o$	$(1/2)^o$	193.194	192.912	192.946
24	$2s^2 2p_{1/2}^2 2p_{3/2}^2 3d_{5/2}$	^2F	$(7/2)^e$	193.500	193.208	193.299
25	$2s^2 2p_{1/2}^2 2p_{3/2}^2 3d_{5/2}$	^4F	$(9/2)^e$	193.543	193.239	193.343
26	$2s^2 2p_{1/2} 2p_{3/2}^3 3p_{1/2}$	$^2P^o$	$(3/2)^o$	193.594	193.283	193.336
27	$2s^2 2p_{1/2}^2 2p_{3/2}^2 3d_{5/2}$	^2P	$(1/2)^e$	194.091	193.821	193.880
28	$2s^2 2p_{1/2} 2p_{3/2}^3 3p_{1/2}$	$^2F^o$	$(5/2)^o$	194.482	194.092	194.159
29	$2s^2 2p_{1/2}^2 2p_{3/2}^2 3d_{5/2}$	^2P	$(3/2)^e$	194.520	194.278	194.291
30	$2s^2 2p_{1/2}^2 2p_{3/2}^2 3d_{5/2}$	^4P	$(5/2)^e$	194.894	194.617	194.688
31	$2s^2 2p_{1/2} 2p_{3/2}^3 3p_{3/2}$	$^4D^o$	$(5/2)^o$	195.438	195.111	195.144
32	$2s^2 2p_{1/2}^2 2p_{3/2}^2 3d_{3/2}$	^4F	$(3/2)^e$	195.459	195.262	195.195
33	$2s^2 2p_{1/2} 2p_{3/2}^3 3p_{3/2}$	$^2D^o$	$(3/2)^o$	195.694	195.373	195.408
34	$2s^2 2p_{1/2} 2p_{3/2}^3 3p_{3/2}$	$^4D^o$	$(1/2)^o$	195.741	195.400	195.423
35	$2s^2 2p_{1/2} 2p_{3/2}^3 3p_{1/2}$	$^2P^o$	$(3/2)^o$	195.946	195.629	195.673
36	$2s^2 2p_{1/2}^2 2p_{3/2}^2 3d_{5/2}$	^4F	$(5/2)^e$	196.018	195.813	195.695

Index	State Configuration	Level	J^π	GRASP[a]	GRASP[b]	DFS[c]
37	$2s^2 2p_{1/2} 2p^3_{3/2} 3p_{3/2}$	$^2F^o$	$(7/2)^o$	196.604	196.173	196.246
38	$2s^2 2p_{1/2} 2p^3_{3/2} 3p_{3/2}$	$^2D^o$	$(3/2)^o$	196.673	196.278	196.305
39	$2s^2 2p_{1/2} 2p^3_{3/2} 3p_{3/2}$	$^2D^o$	$(5/2)^o$	196.976	196.575	196.643
40	$2s^2 2p_{1/2} 2p^3_{3/2} 3p_{3/2}$	$^2P^o$	$(1/2)^o$	198.612	198.259	198.268

a: Coulomb energies
b: Breit and QED corrected energies
c: Sampson et al [4]

Table 2. Comparison between the present GRASP and earlier Dirac-Fock-Slater [DFS: 4] oscillator strengths for some transitions of Mo XXXIV.

Transition		GRASP		DFS	Transition		GRASP		DFS
i	j	f_L	f_V/f_L	f_L	i	j	f_L	f_V/f_L	f_L
1	3	4.7255-2	5.8-1	0.0492	2	6	2.6940-4	5.1-1	0.0001
1	4	1.2697-2	9.1-1	0.0121	2	13	2.0056-3	8.7-1	0.0017
1	5	6.5786-2	8.8-1	0.0611	2	15	4.2731-2	8.8-1	0.0399
1	6	1.4681-2	9.4-1	0.0147	2	18	8.7091-2	9.0-1	0.0824
1	13	1.4999-2	8.8-1	0.0143	2	19	1.4920-3	1.0-0	0.0017
1	15	1.1804-2	8.8-1	0.0110	2	21	4.5561-4	8.6-1	0.0003
1	17	4.9140-2	9.1-1	0.0474	2	27	2.8577-4	5.5-1	0.0002
1	18	7.6317-3	8.6-1	0.0074	2	29	4.4220-3	9.9-1	0.0037
1	19	7.8820-4	9.5-1	0.0009	2	32	9.6384-5	7.2-1	0.0005
1	20	2.2340-4	1.0-0	0.0002	3	7	4.5481-5	2.4-3	0.0000
1	21	8.0005-3	9.4-1	0.0087	3	12	9.3534-5	6.8-5	0.0001
1	27	1.0938-1	9.5-1	0.1070	3	14	2.6936-4	3.9-3	0.0004
1	29	2.2464-1	9.7-1	0.1811	3	26	1.2557-5	7.8-4	0.0000
1	30	4.9692-1	9.7-1	0.4578	3	33	1.7460-4	6.2-3	0.0003
1	32	1.9285-1	9.7-1	0.2379	3	34	1.0529-5	6.4-4	0.0000
1	36	3.8176-1	9.6-1	0.4304	3	35	4.7800-4	6.5-3	0.0011
2	3	3.0348-2	3.9-1	0.0315	3	38	6.4087-5	2.9-4	0.0001
2	5	2.4017-4	6.7-1	0.0002	3	40	4.2673-4	1.7-3	0.0016

relativistic effects are more important than CI. Furthermore, limited inclusion of CI is generally neither helpful nor decisive, and inclusion of larger CI for Mo XXXIV

Table 3. Comparison between the present R-matrix [RM] and earlier Distorted Wave [DW: 4] collision strengths for some transitions of Mo XXXIV.

i	j	RM		DW	RM	DW	RM	DW
E (Ry)		300	320	300	700	700	1100	1100
1	2	7.973-3	7.843-3	8.13-3	6.849-3	7.06-3	6.941-3	6.62-3
1	3	1.440-1	1.474-1	1.39-1	1.805-1	1.65-1	2.085-1	1.86-1
1	4	1.169-3	1.150-3	1.09-3	1.198-3	1.16-3	1.490-3	1.45-3
1	5	2.126-3	2.319-3	2.05-3	4.889-3	4.65-3	6.984-3	6.75-3
1	6	6.176-4	6.484-4	6.16-4	1.125-3	1.14-3	1.567-3	1.61-3
1	7	1.417-3	1.337-3	1.42-3	7.622-4	8.10-4	6.581-4	6.97-4
1	8	2.494-3	2.442-3	2.57-3	2.317-3	2.43-3	2.488-3	2.60-3
1	9	1.049-3	1.021-3	1.11-3	8.738-4	9.42-4	9.013-4	9.68-4
1	10	2.093-3	1.992-3	2.10-3	1.319-3	1.43-3	1.243-3	1.36-3
1	11	2.197-3	2.061-3	2.07-3	1.100-3	1.12-3	9.370-4	9.65-4
1	12	4.292-4	3.936-4	3.83-4	1.221-4	1.13-4	5.972-5	5.13-5
1	13	6.067-4	6.334-4	6.00-4	1.082-3	1.08-3	1.506-3	1.50-3
1	14	1.411-2	1.422-2	1.26-2	1.581-2	1.39-2	1.671-2	1.46-2
1	15	3.799-4	4.066-4	3.78-4	8.041-4	7.89-4	1.147-3	1.13-3
1	16	1.081-2	1.083-2	1.14-2	1.138-2	1.19-2	1.180-2	1.23-2
1	17	1.581-3	1.696-3	1.61-3	3.444-3	3.47-3	4.900-3	4.93-3
1	18	4.497-4	4.527-4	4.32-4	5.949-4	5.95-4	7.903-4	7.88-4
1	19	2.051-3	1.890-3	1.88-3	7.638-4	7.72-4	5.643-4	5.77-4
1	20	2.743-3	2.523-3	2.50-3	9.724-4	9.71-4	6.825-4	6.85-4
1	21	1.516-3	1.460-3	1.46-3	1.225-3	1.26-3	1.377-3	1.43-3
1	22	2.964-3	2.760-3	2.70-3	1.420-3	1.33-3	1.230-3	1.13-3
1	23	2.589-4	2.386-4	2.28-4	7.817-5	7.47-5	3.966-5	3.80-5
1	24	2.974-3	2.803-3	2.73-3	1.747-3	1.80-3	1.648-3	1.73-3
1	25	2.846-3	2.615-3	2.51-3	1.078-3	1.05-3	8.372-4	8.21-4
1	26	2.433-3	2.408-3	1.92-3	2.302-3	1.75-3	2.331-3	1.74-3
1	27	8.682-3	8.998-3	8.79-3	1.366-2	1.35-2	1.731-2	1.72-2
1	28	8.051-4	7.417-4	7.07-4	2.368-4	2.21-4	1.144-4	1.02-4
1	29	1.631-2	1.696-2	1.55-2	2.592-2	2.36-2	3.285-2	3.00-2
1	30	3.468-2	3.642-2	3.58-2	5.840-2	5.77-2	7.460-2	7.40-2
1	31	1.020-3	9.663-4	1.02-3	5.848-4	6.16-4	5.281-4	5.51-4
1	32	1.591-2	1.660-2	1.84-2	2.609-2	2.94-2	3.330-2	3.76-2
1	33	9.825-3	9.902-3	1.22-2	1.090-2	1.34-2	1.145-2	1.39-2
1	34	3.579-4	3.437-4	3.55-4	2.512-4	2.57-4	2.480-4	2.50-4
1	35	2.668-2	2.694-2	2.14-2	2.983-2	2.34-2	3.130-2	2.44-2
1	36	2.991-2	3.125-2	3.31-2	4.916-2	5.27-2	6.279-2	6.76-2
1	37	1.565-3	1.547-3	1.66-3	1.613-3	1.78-3	1.794-3	1.98-3
1	38	6.878-4	6.441-4	1.08-3	3.049-4	7.81-4	2.271-4	7.20-4
1	39	9.990-4	9.631-4	1.03-3	7.638-4	8.63-4	7.841-4	8.96-4
1	40	3.591-4	3.433-4	3.55-4	2.352-4	2.68-4	2.272-4	2.66-4

will make the calculations intractable, as the $2s2p^5 3\ell$ configurations alone result in an additional 53 fine-structure levels.

In Table 3 we compare our values of Ω with the DW calculations [4] for resonance transitions at three energies of ~ 300, 700 and 1100 Ry. The agreement between the two sets of Ω is generally highly satisfactory. However, there are a few transitions for which the differences are particularly noticeable. For example, the present results are about 30% higher for the 1-26 and 1-35 transitions, and are lower by about the same amount for the 1-51 transition, which is a direct consequence of the difference in the f- values. The other transitions for which the differences in the two sets of Ω are up to an order of magnitude are particularly those from the third excited level i.e. $2s2p^6\ {}^2S_{1/2}$. However, the magnitude of Ω for these transitions is invariably small ($\leq 10^{-5}$). Hence such transitions are often very sensitive to any perturbation, as a small error in any of the Ω_J value may significantly affect the result for $\sum \Omega_J$. Nevertheless, due to their small magnitude, such transitions do not affect the modelling calculations or diagnostic results. In general, we can say that the agreement between our present R- matrix and earlier DW values of collision strengths for transitions in Mo XXXIV is highly satisfactory.

To conclude, our calculated energy levels are expected to be accurate to better than 1%, whereas the radiative rates, especially for the strong transitions with f- values ≥ 0.01, are accurate to better than 20%. Similarly, the presently listed values of Ω should be accurate to better than 20% for a majority of transitions, and at energies below 2000 Ry. For weaker transitions, whose Ω magnitudes are very small ($\sim 10^{-4}$ or less), any accuracy estimates are insecure. Calculations for Ω in the thresholds region are still in progress, and therefore the results for rate coefficients are not yet available. But a complete set of results for energy levels, radiative rates and collision strengths for *all* transitions among the lowest 60 fine-structure levels of Mo XXXIV, along with detailed comparisons will be presented in our future publication [1].

ACKNOWLEDGMENT

This work has been financed by EPSRC and PPARC of UK, and we wish to thank Dr. Patrick Norrington for providing his code available to us prior to publication.

REFERENCES

1. Aggarwal K. M., and Keenan, F. P., *At. Data Nucl. Data Tables* - submitted.

2. Dyall, K. G., Grant, I. P., Johnson, C. T., Parpia, F. A., and Plummer, E. P., *Comput. Phys. Commun.* **55**, 425 (1989).

3. Norrington, P. H. and Grant, I. P., *Comput. Phys. Commun.* - in preparation

4. Sampson, D. H., Zhang, H. L., and Fontes, C. J., *At. Data Nucl. Data Tables* **48**, 25 (1991).

Atomic Data for Ne-like Iron

K. M. Aggarwal[1], F. P. Keenan[1] and A. Z. Msezane[2]

[1]*Dept. of Pure and Applied Physics, Queen's University, Belfast BT7 1NN, UK*
[2]*CTSPS, Clark Atlanta University, Atlanta, GA 30304, USA*

Abstract. Energy levels and radiative rates for fine-structure transitions among the lowest 89 levels of the $(1s^2)$ $2s^2 2p^6$, $2s^2 2p^5 3\ell$, $2s^2 2p^5 4\ell$, $2s 2p^6 3\ell$ and $2s 2p^6 4\ell$ configurations of Fe XVII, are calculated using the GRASP code [4]. Collision strengths are also calculated using the DARC program [6], but for transitions among the lowest 55 levels only. The results are compared with those available in the literature, and the accuracy of the data is assessed.

INTRODUCTION

Neon-like ions occur in a variety of plasmas including astrophysical, laser produced, magnetically confined, and Z-pinch plasmas, and have particularly been successful in laser applications. Iron is one of the most abundant and important element in the solar corona and chromosphere, and in fusion reactors. Emission lines of Fe XVII have particularly been observed in the X-ray and EUV (190-350 Å) ranges. Therefore, atomic parameters such as energy levels, radiative rates, collision strengths, rate coefficients, etc. are required for understanding the plasma properties, apart from modelling of lasers. Hence, in this paper we report our results for energy levels, radiative rates and collision strengths (Ω). The values for excitation rates are not yet available, as results for Ω have so far been computed at energies above thresholds only, whereas resonances in the threshold region need to be accounted in a fine enery mesh.

We have adopted the GRASP code [4] for generating wavefunctions and the Dirac Atomic R-matrix Code (DARC: [6]) for computing Ω. Energy levels and radiative rates have been computed for transitions among the lowest 89 fine-structure levels belonging to the $(1s^2)$ $2s^2 2p^6$, $2s^2 2p^5 3\ell$, $2s^2 2p^5 4\ell$, $2s 2p^6 3\ell$, and $2s 2p^6 4\ell$ configurations of Fe XVII. But calculations for Ω have been performed for transitions among the lowest 55 levels only. This is because of our present computational limitations.

Earlier calculations for Fe XVII have been performed by many workers, but have mostly been confined to the lowest 37 fine-structure levels only (see [1] for references). Recently, Bhatia and Doschek [2] have reported results for transitions among the lowest 73 fine-structure levels. They have adopted the SuperStructure (SS) and Distorted-Wave (DW) programs, for the generation of wavefunctions and computations of Ω, respectively. Their calculations are semi-relativistic in the LSJ coupling scheme, whereas we have adopted the fully relativistic jj coupling

CP641, X-Ray Lasers 2002: 8th International Conference on X-Ray Lasers, edited by J. J. Rocca et al.
© 2002 American Institute of Physics 0-7354-0096-2/02/$19.00

scheme. Since their calculations are the most recent, we will compare our results with them.

CALCULATION DETAILS

The $(1s^2)$ $2s^2 2p^6$, $2s^2 2p^5 3\ell$, $2s^2 2p^5 4\ell$, $2s2p^6 3\ell$, and $2s2p^6 4\ell$ configurations of Fe XVII give rise to 89 fine-structure levels. To generate the wavefunctions, we have adopted the fully relativistic GRASP code of Dyall et al [4], and have included configuration interaction (CI) among the above basic 15 configurations. For the computations of Ω, we have employed the DARC program [6], which includes relativistic effects in a systematic way, in both the target description and the scattering model. The R- matrix boundary radius has been taken to be 4.0 au, and 30 continuum orbitals have been included for each channel angular momentum, for the expansion of the wavefunction. This allows us to compute Ω up to an energy of 300 Ry, more than sufficient for the determination of accurate excitation and de-excitation rate coefficients for electron temperatures up to 10^7 K. In order to obtain converged Ω at all energies and for all transitions, we have included the contribution of all partial waves with angular momentum $J \leq 40.5$. However, for some allowed transitions, especially towards the higher end of the energy range, even this large range is not sufficient for convergence. Therefore, to take account of higher neglected partial waves, a top-up based on the sum rules has also been included.

RESULTS AND DISCUSSION

In Table 1 we list our energies for the lowest 37 fine-structure levels of Fe XVII, because most of the available calculations are confined to these levels only. The available experimental values of Shirai et al [7] and NIST, and the theoretical results of Bhatia and Doschek (SS73: [2]) are also included in this table. The experimental energy order of NIST agrees with our calculations, but the SS73 values have different level ordering. Since SS, DW and JAJOM (which converts Ω from LS to LSJ coupling scheme) programs give different level ordering, there is a possibility of rearranging their level ordering, and a proposed level ordering of their energies is also included in Table 1. This will be helpful in our further discussion of radiative rates and collision strengths.

In Table 2 we compare our radiative rates for a few transitions with those of Hibbert et al [5], Cornille et al [3], and Bhatia and Doschek [2]. Generally different calculations agree within 25%, but for some weaker transitions the differences are up to a factor of two. This is mainly because of the inclusion of diffferent amount of CI in different calculations. However, while comparing the results of Bhatia and Doschek we are assuming our proposed level order as already discussed. If their original level order is considered then the discrepancies are very large (up to 5 orders of magnitude, especially for the 14-25 i.e. 3p $^1D_2$ - 3d $^3D_2^o$ transition for which our f- value is 1.51×10^{-2} whereas that of SS73 is 1.46×10^{-7}).

Table 1. Target levels of Fe XVII and their threshold energies (in Ry).

Index	Config.	Level	Expt.[7]	NIST	GRASP	SS73[a]	SS73[b]
1	$2s^2 2p^6$	$^1S_0$	0.0000	0.0000	0.0000	0.0000	0.0000
2	$2s^2 2p^5 3s$	$^3P^o_2$	53.3045	53.2966	53.1622	53.3300	53.3300
3	....	$^1P^o_1$	53.4437	53.4367	53.3059	53.4762	53.4762
4	....	$^3P^o_0$	54.2284	54.2269	54.0917	54.2471	54.2471
5	....	$^3P^o_1$	54.3195	54.3140	54.1813	54.3413	54.3413
6	$2s^2 2p^5 3p$	$^3S_1$	55.5276	55.5218	55.3881	55.5433	55.5433
7	....	$^3D_2$	55.7850	55.7788	55.6551	55.8192	55.8192
8	....	$^3D_3$	55.9038	55.8974	55.7720	55.9315	55.9315
9	....	$^1P_1$	56.6720	55.9804	55.8596	56.6992	56.0209
10	....	$^3P_2$	56.9383	56.1137	55.9902	56.1507	56.1507
11	....	$^3P_0$	56.5191	56.5155	56.4021	56.5668	56.5668
12	....	$^3D_1$	55.9869	56.6672	56.5441	56.9301	56.6992
13	....	$^3P_1$	56.9106	56.9061	56.7801	56.0209	56.9301
14	....	$^1D_2$	56.1201	56.9337	56.8086	56.9604	56.9604
15	....	$^1S_0$	57.8965	57.8894	57.9529	58.0879	58.0879
16	$2s^2 2p^5 3d$	$^3P^o_0$	58.9041	58.8983	58.7676	58.9478	58.9478
17	....	$^3P^o_1$	58.9754	58.9818	58.8392	59.0199	59.0199
18	....	$^3P^o_2$	59.1084	59.0977	58.9758	59.1592	59.1592
19	....	$^3F^o_4$	59.1124	59.1042	58.9842	59.1777	59.1777
20	....	$^3F^o_3$	59.1689	59.1612	59.0447	59.2277	59.2277
21	....	$^1D^o_2$	60.0922	59.2876	59.1739	60.2034	59.3555
22	....	$^3D^o_3$	60.1906	59.3666	59.2541	59.4420	59.4420
23	....	$^3D^o_1$	....	59.7080	59.6062	59.7822	59.7822
24	....	$^3F^o_2$	59.2934	60.0877	59.9701	60.1399	60.1399
25	....	$^3D^o_2$	60.1523	60.1618	60.0286	59.3555	60.2034
26	....	$^1F^o_3$	59.3722	60.1974	60.0705	60.2491	60.2491
27	....	$^1P^o_1$	60.6904	60.6904	60.6356	60.8157	60.8157
28	$2s2p^6 3s$	$^3S_1$	....	....	63.2048	63.3238	63.3238
29	....	$^1S_0$	....	....	63.6956	63.8049	63.8049
30	$2s2p^6 3p$	$^3P^o_0$	....	....	65.6263	65.7388	65.7388
31	....	$^3P^o_1$	65.6012	65.6012	65.6594	65.7724	65.7724
32	....	$^3P^o_2$	....	....	65.8295	65.9394	65.9394
33	....	$^1P^o_1$	65.9238	65.9238	65.9717	66.0842	66.0842
34	$2s2p^6 3d$	$^3D_1$	....	....	68.9168	69.0641	69.0641
35	....	$^3D_2$	....	....	68.9268	69.0765	69.0765
36	....	$^3D_3$	....	....	68.9459	69.1002	69.1002
37	....	$^1D_2$	....	....	69.3192	69.4618	69.4618

a: Bhatia and Doschek [2]   b: Proposed levels order of Bhatia and Doschek [2].

Table 2. Comparison of present radiative rates (A- values in s^{-1}) for some transitions of Fe XVII. ($a \pm b \equiv a \times 10^{\pm b}$).

i	j	GRASP	CIV3 [5]	SS89 [3]	SS73 [2]
1	3	9.572+11	9.317+11	9.542+11	9.463+11
1	5	8.408+11	8.020+11	8.184+11	8.168+11
1	17	9.209+10	9.880+10	8.074+10	8.923+10
1	23	5.878+12	5.770+12	5.620+12	5.765+12
1	27	2.516+13	2.324+13	2.554+13	2.516+13
1	31	4.095+11	3.733+11	3.784+11	3.763+11
1	33	3.324+12	3.137+12	3.317+12	3.327+12
2	6	3.380+09	3.299+09	3.249+09	3.288+09
2	7	2.599+09	3.528+09	2.588+09	2.603+09
2	9	3.731+08	3.737+08	3.717+08	3.940+08
2	10	4.522+09	4.418+09	4.447+09	4.433+09
2	14	1.173+08	1.193+08	1.199+08	1.175+08
2	28	9.588+10	6.337+10	9.526+10	9.526+10
3	11	7.828+09	7.870+09	7.764+09	7.737+09
3	15	1.397+10	1.202+10	1.373+10	1.368+10
3	29	6.689+10	4.098+10	6.548+10	6.532+10
6	18	5.532+09	5.406+09	5.732+09	5.744+09
8	32	8.121+10	4.704+10	8.057+10	8.044+10
9	16	3.012+08	3.106+08	2.891+08	3.162+08
10	33	3.919+10	2.358+10	3.892+10	3.882+10
12	27	5.846+09	5.630+09	5.876+09	5.906+09
14	24	1.373+09	1.425+09	1.420+09	1.435+09
14	25	1.255+09	1.123+09	1.219+09	1.245+09

In Table 3 we compare our results of Ω with those of Bhatia and Doschek [2] for resonance transitions, at 4 common energies of 75, 125, 175 and 250 Ry. According to their level identification, the differences between the DARC and DW values of Ω are significantly and unreasonably large for many transitions, and over the entire energy range. For $\sim 15\%$ transitions, the two sets of Ω differ by more than an order of magnitude. Since they too have included CI and relativistic effects while computing the values of Ω, such large differences between two different approaches of DW and R- matrix are neither expected nor can be explained. Therefore, as discussed earlier these discrepancies are mainly because of a mismatch between the level orders of the two calculations. On rearranging their level order, as given in Table 1, we find that only 16 transitions differ by more than an order of magnitude. All such transitions have their lower levels between 34 and 37, and half of these are allowed, and the differences between the two calculations are in accordance with

Table 3. Comparison of collision strengths for resonance transitions in Fe XVII. $a{\pm}b \equiv a{\times}10^{\pm b}$.

i	j	Present DARC Results				Earlier DW Results [2]			
E (Ry)		75.0	125.0	175.0	250.0	70.0	127.5	170.0	255.0
1	2	1.604-3	8.532-4	5.237-4	2.946-4	1.549-3	7.528-4	5.036-4	2.684-4
1	3	2.869-3	5.061-3	7.034-3	9.492-3	2.383-3	4.873-3	6.531-3	9.245-3
1	4	3.177-4	1.685-4	1.034-4	5.814-5	3.140-4	1.517-4	1.013-4	5.392-5
1	5	2.441-3	4.138-3	5.718-3	7.708-3	2.244-3	4.204-3	5.548-3	7.777-3
1	6	3.733-3	2.060-3	1.312-3	7.699-4	4.060-3	1.992-3	1.339-3	7.292-4
1	7	3.586-3	3.291-3	3.324-3	3.488-3	3.726-3	3.199-3	3.204-3	3.359-3
1	8	3.878-3	1.980-3	1.176-3	6.378-4	4.431-3	1.907-3	1.208-3	6.066-4
1	9	1.499-3	7.672-4	4.581-4	2.520-4	1.547-3	6.615-4	4.119-4	1.998-4
1	10	2.965-3	2.978-3	3.155-3	3.432-3	3.230-3	3.069-3	3.198-3	3.475-3
1	11	2.818-3	2.946-3	3.027-3	3.101-3	3.276-3	3.063-3	3.049-3	3.049-3
1	12	1.588-3	8.129-4	4.837-4	2.646-4	1.666-3	7.155-4	4.481-4	2.204-4
1	13	1.665-3	8.627-4	5.155-4	2.827-4	1.761-3	7.708-4	4.870-4	2.415-4
1	14	3.522-3	3.463-3	3.626-3	3.910-3	3.906-3	3.607-3	3.710-3	3.973-3
1	15	4.241-2	4.638-2	4.859-2	5.032-2	4.462-2	4.816-2	4.940-2	5.066-2
1	16	1.752-3	8.136-4	4.538-4	2.356-4	1.888-3	7.643-4	4.666-4	2.219-4
1	17	5.348-3	2.940-3	2.110-3	1.708-3	5.680-3	2.747-3	2.046-3	1.568-3
1	18	6.801-3	3.100-3	1.710-3	8.780-4	7.449-3	2.948-3	1.784-3	8.385-4
1	19	6.037-3	2.611-3	1.378-3	6.747-4	6.506-3	2.399-3	1.399-3	6.288-4
1	20	4.462-3	3.093-3	2.743-3	2.649-3	4.660-3	2.984-3	2.705-3	2.589-3
1	21	2.545-3	1.027-3	5.244-4	2.505-4	2.678-3	8.879-4	4.916-4	2.052-4
1	22	3.171-3	2.670-3	2.723-3	2.907-3	3.286-3	2.647-3	2.686-3	2.859-3
1	23	2.754-2	4.012-2	5.020-2	6.224-2	2.326-2	3.754-2	4.577-2	5.856-2
1	24	2.930-3	1.217-3	6.339-4	3.079-4	3.107-3	1.085-3	6.170-4	2.679-4
1	25	3.775-3	1.590-3	8.410-4	4.157-4	4.020-3	1.456-3	8.450-4	3.761-4
1	26	3.856-3	3.076-3	3.017-3	3.129-3	4.031-3	3.072-3	3.019-3	3.113-3
1	27	1.016-1	1.530-1	1.934-1	2.412-1	9.077-2	1.522-1	1.870-1	2.408-1
1	28	2.190-3	1.289-3	8.791-4	5.808-4	1.149-3	4.684-4	2.922-4	1.483-4
1	29	1.607-2	1.810-2	1.908-2	1.945-2	1.546-2	1.765-2	1.846-2	1.932-2
1	30	2.820-4	1.472-4	8.850-5	4.950-5	3.036-4	1.334-4	8.708-5	4.512-5
1	31	1.089-3	1.166-3	1.415-3	1.826-3	1.062-3	1.048-3	1.240-3	1.658-3
1	32	1.416-3	7.309-4	4.372-4	2.438-4	1.548-3	6.809-4	4.456-4	2.316-4
1	33	2.648-3	5.936-3	9.066-3	1.307-2	2.051-3	5.763-3	8.490-3	1.311-2
1	34	1.885-3	8.732-4	4.951-4	2.615-4	2.050-3	8.208-4	5.023-4	2.420-4
1	35	3.157-3	1.492-3	8.738-4	4.955-4	3.446-3	1.436-3	9.216-4	5.082-4
1	36	4.392-3	2.033-3	1.150-3	6.062-4	4.786-3	1.911-3	1.169-3	5.630-4
1	37	1.421-2	2.283-2	2.844-2	3.380-2	1.226-2	2.218-2	2.675-2	3.243-2

the wavefunctions i.e. if the f- value is larger than so is the value of Ω and vice versa. However, about 30% of the transitions still differ by more than 50%, but a majority of these have collision strengths $\leq 10^{-3}$. Furthermore, about half of such transitions are again allowed and the differences between the two calculations are in accordance with their corresponding f- values. Therefore, in conclusion we can say that there is good agreement between the R- matrix and DW calculations, and the differences are only due to the differences in the wavefunctions. However, accurate results for excitation rate coefficients, which are required in plasma diagnostics, cannot be accurately determined from either set of available data, because resonances in the threshold regions have not yet been resolved. Such calculations are still in progress, but a complete set of results for energy levels, radiative rates and collision strengths for *all* transitions along with detailed comparisons can be found in our forthcoming publication [1].

Our energy levels are expected to be accurate to better than 1%, and radiative rates and collision strengths are assessed to be accurate to 20%, especially for those transitions for which the magnitude of the f- and Ω values are high (≥ 0.01). For weaker transitions any accuracy assessments are insecure. The agreement between our present (GRASP and DARC) and earlier available SS and DW [2] results is satisfactory, provided their level ordering is slightly rearranged.

ACKNOWLEDGMENT

This work has been financed by EPSRC and PPARC of UK and DoE of USA, and we wish to thank Dr. Patrick Norrington for providing his code available to us prior to publication. KMA would like to thank Dr. Anand Bhatia for useful correspondence and discussion.

REFERENCES

1. Aggarwal, K. M., Keenan, F. P., and Msezane, A. Z., *Astrophys. J. Supl.* - in press.

2. Bhatia, A. K., and Saba, J. L. R., *Astrophys. J.* **563**, 434 (2001).

3. Cornille, M., Dubau, J., Faucher, P., Bely-Dubau, F., and Blancard, C., *Astron, Astrophys. Suppl.* **105**, 77 (1994).

4. Dyall, K. G., Grant, I. P., Johnson, C. T., Parpia, F. A., and Plummer, E. P., *Comput. Phys. Commun.* **55**, 424 (1989).

5. Hibbert, A., Le Dourneuf. M., and Mohan, M., *At. Data Nucl. Data Tables* **53**, 23 (1993).

6. Norrington, P. H. and Grant, I. P., *Comput. Phys. Commun.* - in preparation

7. Shirai, T., Funatake, Y., Mori, K., Sugar, J., Wiese, W. L., and Nakai, Y., *J. Phys. Chem. Ref. Data* **19**, 127 (1990).

Modeling of the Ni-like silver x-ray laser. Treatment of saturation and refraction

D. Benredjem[†], J. Kuba[*], R. Smith[*], G. J. Pert[¶], A. Klisnick[†], J. Dubau[†] and C. Möller[†]

[†] LIXAM, Université Paris-Sud, Centre d'Orsay, Bât. 350
91405 Orsay Cedex, France
[*] Lawrence Livermore National Laboratory, Livermore, CA 94550, USA
[¶] Department of Physics, University of York, York YO1 5DD, UK

Abstract. We present calculations for the 13.9 nm x-ray laser in Ni-like silver, in the transient collisional pumping scheme. A ray-trace algorithm is used to model 2D refraction. The electron density profile used in the ray-trace calculations is given by the hydro-code EHYBRID. Knowing the beam paths for various starting conditions (time, position), intensity of the XRL beam and population kinetics are then calculated in the saturation regime, by using a paraxial Maxwell-Bloch approach. For large intensities, interaction between the x-ray beam and the amplifying medium must be taken into account with a reliable modeling of the saturation regime. Our calculations for intensity combine a 3D radiative transfer problem with population kinetics. This approach is consistent, in the sense that beam amplification and population kinetics are treated simultaneously.

INTRODUCTION

The aim of this work is to study, in a self-consistent manner, the large amplification of the Ni-like silver 13.9 nm line. We take into account the interaction of the x-ray laser (XRL) field with the amplifying medium in population kinetics calculations. This is done through a paraxial Maxwell-Bloch approach combining the Maxwell wave equation and the Bloch relations, which provides a set of equations for the evolution of the Zeeman sublevels populations. A number of works have already used the Maxwell-Bloch formalism in specific investigations, the buildup of radiation [1], for example.

The required hydrodynamic data are obtained by using the time dependent Lagrangian hydrocode EHYBRID [2]. A post-processor has been used to account for refraction.

Our calculations model the experiment done at the Rutherford Laboratory [3], in the transient collisional excitation scheme. In this experiment, a slab target was irradiated by a driving laser consisting of a pulse of 300 ps duration (FWHM), followed by a short pulse of 1 ps duration. The long pulse is of low intensity (6×10^{11} W/cm^2) and the second pulse, arriving 200 ps later, is of high intensity (5.6×10^{14} W/cm^2).

CP641, X-Ray Lasers 2002: 8th International Conference on X-Ray Lasers, edited by J. J. Rocca et al.
© 2002 American Institute of Physics 0-7354-0096-2/02/$19.00

RAY-TRACING

In the modeling with EHYBRID, the atomic data base consists of 107 levels. Ionization is described by Griem's model [4]. A flux limitor of 0.1 was used, and the reflectivity at critical density was set to 0.7. The excitation rate coefficients were obtained by the code HULLAC [5] for all transitions between the $n = 4$ levels. For transitions involving higher levels, empirical formulae are used.

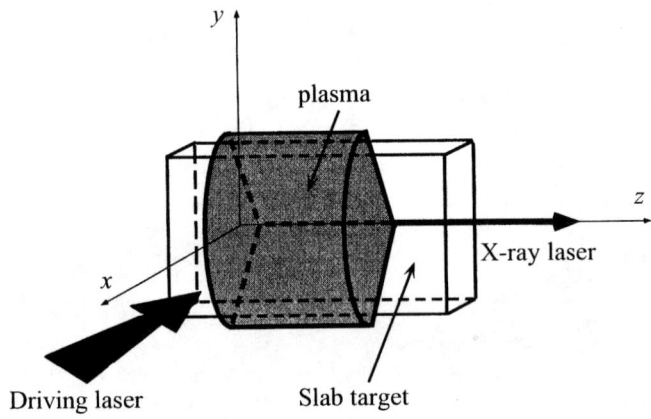

Figure 1. Geometry of the target irradiation

The trajectory of a beam in a medium with an electron density gradient $\nabla n_e(\mathbf{r},t)$, is fully described by the beam equation. Our description is based on the ray equation:

$$\frac{d}{ds}\left(n\frac{d}{ds}\mathbf{r}\right) = \nabla n(\mathbf{r}), \tag{1}$$

where n is the refraction index and $ds = \sqrt{dx^2 + dy^2 + dz^2}$. We consider the coordinates system represented in Fig. 1. In the beam approximation, one can derive the following relation from Maxwell's equations:

$$n = \sqrt{1 - \frac{n_e}{n_c}}, \tag{2}$$

where n_c is the critical (electron) density, $n_c = (4\pi^2 c^2 \varepsilon_0 m_e)/(e^2\lambda^2)$ $\approx 1.1\times10^{21}/(\lambda(\mu m))^2$. z being the coordinate with the largest variation, the paraxial

240

approximation allows us to write $\dfrac{d}{ds} \approx \dfrac{d}{dz}$, and the ray equation reduces to the following differential equations:

$$\frac{d\alpha}{dz} = \frac{d^2 x}{dz^2} = -\frac{1}{2n_c}\frac{\partial}{\partial x}n_e(x,y)$$

$$\frac{d\beta}{dz} = \frac{d^2 y}{dz^2} = -\frac{1}{2n_c}\frac{\partial}{\partial y}n_e(x,y),$$

(3)

where α and β are the angles made by the ray with the x and y axes, respectively. If we assume an exponential density profile with scale length L, i.e., $n_e(x) = n_{e0}\exp(-x/L)$, we will have $\dfrac{d\alpha}{dz} = \dfrac{n_{e0}}{2n_c L}\exp(-x/L)$.

The electron density, as it is simulated by the hydro-code EHYBRID, is a function of the coordinates x and t, i.e., $n_{e0}(x,t)$. For the variation along the y direction, we assume a Gaussian distribution. We then have:

$$n_e(x,y,t) = n_{e0}(x,t)\exp\left[-4\ln(2)\frac{y^2}{\delta^2}\right],$$

(4)

where δ is the focal line half-width. Similarly, the electron and ion temperatures as well as the ion density (even though we do not use them in the ray-trace calculations) would be expressed in the same way in this approximation. The electron density gradient in the y direction is then

$$\frac{\partial n_e}{\partial y} = -\frac{8\ln(2)n_{e0}}{\delta^2}y\exp\left[-4\ln(2)\frac{y^2}{\delta^2}\right].$$

(5)

The numerical ray trace code was used with the electron density profile given by EHYBRID. For each z value of the propagation, the ray trace code gives the electron density $n_e(x,y)$ for varying initial angles α_0, β_0 and initial coordinates x_0, y_0. To take the temporal evolution of the plasma into account, the data are updated every 0.1 ps, as given by EHYBRID.

We have calculated the x and y values at the exit from a 3 mm long plasma. The rays are launched from different x_0, at the time of peak gain predicted by EHYBRID and for a fixed y_0 value, namely 1 μm. The initial α value is either 0 mrad (Fig. 2a) or −5 mrad (Fig. 2b). The initial β value is taken to be 0 mrad in both cases. Figure 2a

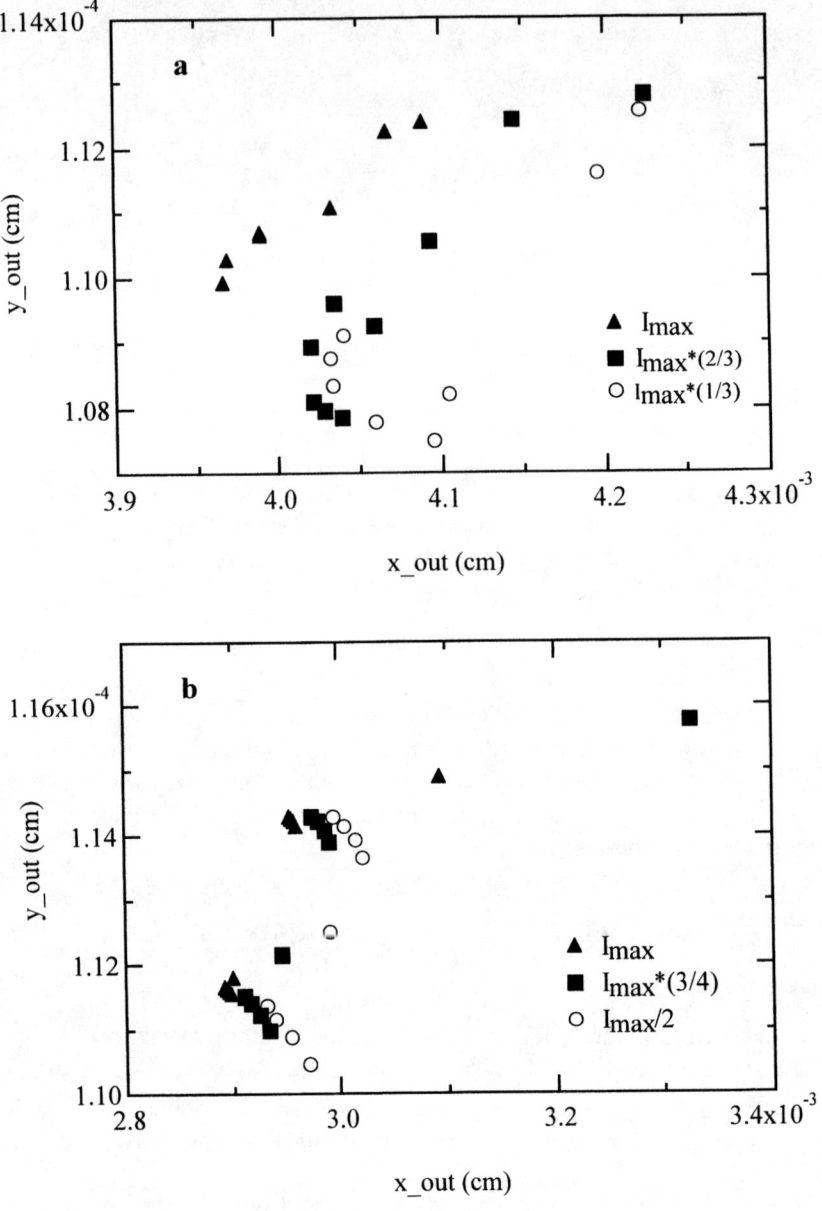

Figure 2. Coordinates of the beam at the exit, of a 3 mm long plasma for various values of x_0. The rays are launched at the time of peak gain predicted by EHYBRID. We have set $y_0=1$ μm and $\beta_0=0°$. I_{max} designates the largest intensity obtained by the Maxwell-Bloch approach. (a): $\alpha_0=0°$, (b): $\alpha_0=-5$ mrad.

shows that the most intense rays exit approximately 40 μm away from the target surface. For $\alpha = -5$ mrad, the rays exit at approximately 30 μm. We can see that the refraction in the lateral direction (y) is negligible. The 3-mm interval for the z variation is dictated by experiments which show saturation at this length.

In the above calculations, the intensity is given by the following relation which describes the small-signal regime as well as the saturation regime:

$$I(s) = \frac{\sigma}{G}\left\{\exp[G\Delta s - 1]\right\} + I(s - \Delta s)\exp[G\Delta s], \tag{6}$$

where the emissivity σ and the gain G are given by

$$\sigma = N_u h\nu_0 \frac{r^2}{4\pi L^2} A_{ul} \quad \text{and} \quad G(s) = \frac{G_{ss}}{1 + \dfrac{I(s)}{I_{sat}}}. \tag{7}$$

G_{ss} is the small-signal gain and I_{sat} the saturation intensity. Both quantities are given by EHYBRID. The emissivity depends on the radius r of the output region, the length L of the plasma column, and the Einstein coefficient for spontaneous emission between the two lasing levels u and l, N_u being the population density of the upper level.

When the intensity of the XRL beam is large, absorption and induced emission have an important effect on the populations of the lasing levels. The collisional-radiative equations should then account for these two processes. We have developed a satisfactory approach which is described below.

SATURATION

Let us consider an isolated transition between an upper level u and a lower level l. The monochromatic intensity of the XRL beam is solution of the radiative transfer equation:

$$\frac{\mathbf{k}}{k} \cdot \nabla I_{ul}(\nu, \mathbf{r}, t) = \sigma(\nu, \mathbf{r}, t) + G_{ul}(\nu, \mathbf{r}, t)I_{ul}(\nu, \mathbf{r}, t). \tag{8}$$

The gain, which involves absorption and stimulated emission only, i.e., first-order processes, is not sensitive to photon scattering, and can be written as

$$G_{ul}(\nu, \mathbf{r}, t) = h\nu \frac{B_{ul}}{c}\left[N_u(\mathbf{r}, t) - \frac{g_u}{g_l}N_l(\mathbf{r}, t) \right]\Phi_{ul}(\nu), \tag{9}$$

where B_{ul} is the Einstein coefficient for induced emission. The g's and the N's are the statistical weights and population densities of the levels, respectively. Φ designates the absorption profile. For the ranges of electron/ion densities and temperatures predicted by EHYBRID for the RAL experiment, the most important broadening mechanism is due to the Doppler frequency detuning. The emissivity in a small solid angle $\delta\Omega$ centred on the direction of the x-ray beam propagation can be expressed as :

$$\sigma(\nu,\mathbf{r},t) = \frac{\delta\Omega}{4\pi} N_u(\mathbf{r},t) A_{ul} h\nu \Phi_{ul}(\nu)$$
$$+ (1-\varepsilon)\delta\Omega N_l(\mathbf{r},t) \frac{B_{lu}}{c} h\nu \int d\nu' I_{ul}(\nu',\mathbf{r},t) R(\nu',\nu), \tag{10}$$

where B_{lu} is the Einstein coefficient for absorption. Calculations generally deal only with the first contribution in the rhs (first order). The second contribution is due to a second-order process consisting in a photon absorption at frequency ν' followed by an emission at frequency ν. It involves the number of photons of frequency ν', given by $N_l B_{lu} I_{ul}(\nu',z)$, times $R(\nu',\nu)$ –the frequency redistribution function- which represents the joint probability of absorbing at ν' and emitting at ν, summed over all ν'. The value of the multiplying factor $(1-\varepsilon)$ is determined by the relative rates of collisional deexcitation C_{ul} and spontaneous emission between levels u and l. In fact, we have, $\varepsilon = C_{ul}/[A_{ul}+C_{ul}]$. When collisional deexcitation overcomes spontaneous emission, we have $\varepsilon \approx 1$, and the contribution of frequency redistribution to σ is then negligible.

The radiative transfer equation is coupled to the population rate equations:

$$\frac{\partial}{\partial t} N_i(\mathbf{r},t) = -N_i(\mathbf{r},t)\Lambda_i(\mathbf{r},t) + \sum_{j\neq i} \lambda_{ji}(\mathbf{r},t)N_j(\mathbf{r},t), \tag{11}$$

where i is a label for lasing ion levels. Λ and λ are respectively, the total decay rate of the i level, and a rate of population increase. It is worth stressing that in the saturation regime, the population equations have to account for the interaction between the amplified radiation and the plasma.

Equations (8)-(11) are solved consistently. Figure 3 (below) shows the gain as a function of z for various initial values of α. Initial x and t values correspond to the cell and time of the peak gain predicted by EHYBRID. We see that saturation occurs for a z value of 2 mm. The small-signal gain is much smaller than the gain predicted by EHYBRID (140 instead of 800 cm^{-1}). Figure 4 represents the integrated intensity. The largest intensity is obtained with the smallest initial α value (-5 mrad) which correspond to rays initially directed towards the target and traveling a longer distance in the plasma than rays associated with positive α values.

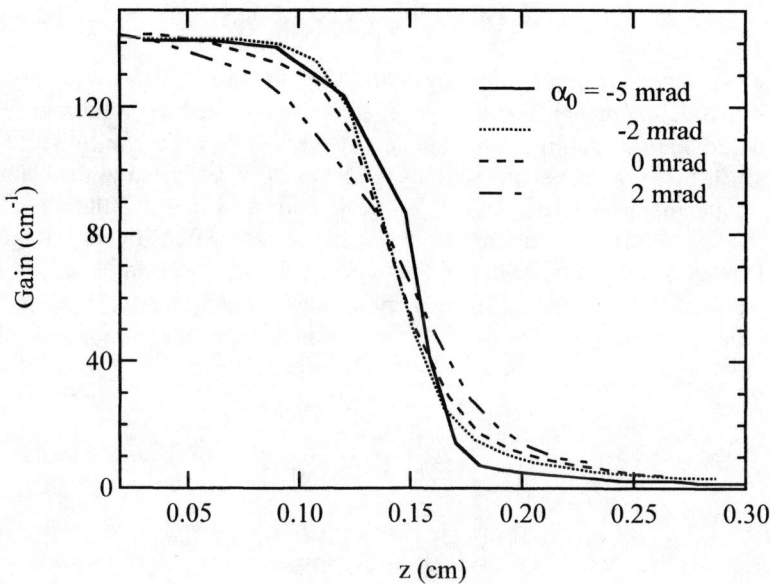

Figure 3. Gain calculated in the Maxwell-Bloch theory, as a function of z. The rays are launched at the time and cell (x) of the peak gain that is predicted by EHYBRID. We have set $y_0 = 1$ μm and $\beta_0 = 0°$.

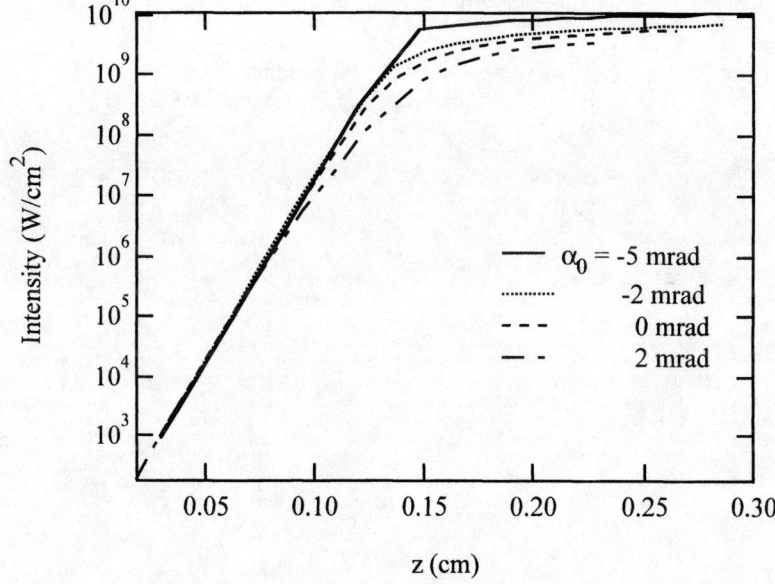

Figure 4. Intensity calculated in the Maxwell-Bloch theory, as a function of z. Same as in Fig. 3.

CONCLUDING REMARKS

We have calculated the output intensity of the 13.9 nm Ni-like silver x-ray laser line by a Maxwell-Bloch approach suitable for the treatment of the saturation regime. The required hydrodynamic quantities are obtained by a ray-trace code working as a post-processor to EHYBRID. To begin with, this work has been restricted to rays launched at the time and distance (from target) of peak gain as given by the EHYBRID modeling. In this particular case, the calculated gains and intensities are large, and saturation is reached for a smaller length than in experiments ($\approx$ 2 mm instead of 3 mm in most experiments). Our values for gain and intensity must be interpreted as upper limits. We plan a new set of calculations for other times of ray launching, according to the plasma's evolution.

ACKNOWLEGMENT

One of us (D.B.) is indebted to J.-C. Gauthier (LULI, École Polytechnique) for running the code HULLAC and providing the electron-ion collision cross sections.

REFERENCES

1. Strauss, M., *Phys. Fluids B* **1**, 907 (1989).
2. Pert, G.J., J. Comput. Phys. **39**, 251 (1980).
3. Klisnick, A, et al, Phys. Rev. A **65**, 033810 (2002).
4. Griem, H.R., *Plasma spectroscopy*, McGraw-Hill, New York, 1964.
5. Bar-Shalom, A, et al, *J. of Quantit. Spectrosc. and Radiat. Transf.* **71**, 169 (2001).

Capillary Discharge X-ray laser on Pd-like Xenon

E.P. Ivanova

Institute of Spectroscopy of R AS, 142190, Troitsk, Moscow region

Abstract. We have performed gain calculations for the possible laser transitions in Pd-like xenon. The theory is verified by comparison with the experimental gain observed in xenon plasma produced by the femtosecond laser. The optimum parameters are found to achieve x-ray lasing at 41.8 nm with gL>45 in xenon capillary discharge plasma. The necessary conditions can be reached in xenon plasma using some contemporary capillary discharges with rise time ~40 ns.

INTRODUCTION

In [1] the observation of gain of approximately exp (11) at 41.8 nm on the $4d^9 5d\ ^1S_0$ – $4d^9 5p\ ^1P_1$ transition in 8 times ionized xenon was reported. Xenon gas was pumped longitudinally by 10-Hz, 40 fs, circularly polarized laser pulse with energy 70 mJ, focal intensity 3×10^{16}W/cm^2. It was collisionally excited, with plasma 3.5-8.5mm length produced instantly by field induced tunneling. The EUV laser pulse duration was not measured in [1]. Differentially pumped cell contained 5 to 12 torr of xenon gas. Due to the low gas density, pump beam dispersion and depletion over the gain length was negligible. The optimum pressure was found to be 12 torr. At optimum pressure gain was reported as g=13 cm^{-1}. However the gain was predicted in the previous theoretical work [2] as g = 107-163 cm^{-1}.

At the same time in related work, Rocca et al. [3] using a capillary discharge, have demonstrated x-ray laser at 46.9 nm on the $2p^5 3p\ ^1S_0$ – $2p^5 3s\ ^1P_1$ in Ar IX. The product gL=25 was achieved in the following work of this group [4].

X-ray laser in Pd-like xenon on the transition 5d-5p, 0-1 is similar to that in Ne-like argon on the transition 3p-3s, 0-1; the rate of radiative depletion of the low active level is of the same order in both ions. But the electron collision strength of excitation of the upper $^1S_0$ level is few times larger in Pd-like xenon. This implies extremely economical scheme. The ratio of x-ray laser energy radiation to the excitation energy is appreciably larger in Pd-like xenon. On the other hand, at optimum plasma conditions Xe IX is overionized much faster than Ar IX. Rate of Xe IX ionization depends significantly on electron density and temperature.

EUV emission from xenon capillary discharge plasmas has been investigated as a function of current, pressure and tube radius in the experiments [5-8]; the capillary parameters were optimized aiming at the brightest radiation around 13.5 nm. This source of emission in this range may be suitable for imaging applications. In the experiments [5-8] the spectra around 41.8 nm were not mentioned, nevertheless the detailed investigations of [5-8] give the idea about plasma dynamics in different xenon filled capillary discharges. It will be discussed below. In this paper we perform gain calculations for the possible laser transitions in Xe IX. Our atomic-kinetic modeling is compared with the results of the experiment [1] and of the theory [2]. We suggest the

CP641, *X-Ray Lasers 2002: 8th International Conference on X-Ray Lasers,* edited by J. J. Rocca et al.
© 2002 American Institute of Physics 0-7354-0096-2/02/$19.00

optimum pressure and temperature to obtain the largest gL, and calculate gain history in a xenon capillary discharge with assumption that electron density (n_e), electron temperature (T_e), plasma kernel diameter (d) are the functions of time.

CALCULATIONS OF GAIN IN XENON PLASMA PRODUCED BY OFI

There are at least two reasons why femtosecond pulse-driven X-ray laser is a favorable approach to test theory: i) plasma density and geometry are known; ii) ionization balance is ~100% of Xe IX. However the electron energy distribution is non-Maxwellian. If plasma is ionized up to Xe IX state then the energy of 9% electrons is $T_e^{hot} \sim 500$ eV, that of other 91% is in the limits $T_e=10-100$ eV. The form of the

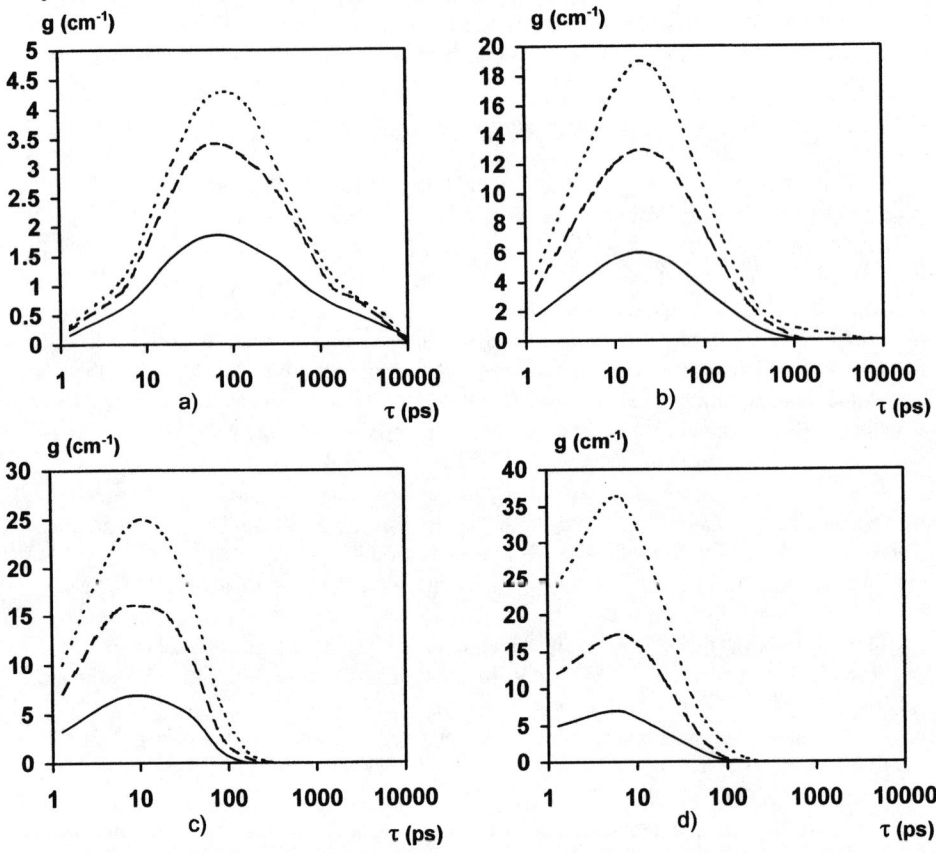

FIGURE. 1. Time evolution of gain calculated with assumption that plasma is created instantly. Plasma diameter d=1mm. a) $n_e=10^{17}$ cm^{-3}; b) $n_e=5\cdot10^{17}$ cm^{-3}; c) $n_e=10^{18}$ cm^{-3}; d) $n_e=2\cdot10^{18}$ cm^{-3}. For each n_e gain is shown for three T_e values: 50eV – solid line, 75eV – broken line, 100 eV – dotted line.

electron energy distribution is not important in the calculations of electron induced rate coefficients. We use two-hump Maxwellian form: 9% of electrons are of $T_e^{hot} = 500$ eV; by adjustment of our theoretical gain and that of [1] we will find T_e of 91% of electrons. Our method of rate coefficients and level populations calculations can be found elsewhere ([9], see references herein). The kinetic equations are calculated with the assumption that at the initial moment only the ground level in Xe IX is populated. Time evolution of gain is shown in Fig 1a,b,c,d for the set of n_e, T_e parameters. The largest gain at maximum corresponds to $n_e=2\cdot10^{18}$ cm^{-3} (~7 torr, Fig. 1d). Its time duration is ~60-80 ps. Appreciably longer laser action is possible at $n_e=10^{17}$-$5\cdot10^{17}$ cm^{-3} (0.4-2 Torr). It can be seen in Fig. 1a,b where FWHM of gain is 2000-300 ps. Gain decay is conditioned by collision mixing of level populations and also by overionization of Xe IX. Our gain calculations are in a satisfactory agreement with the results of the OFI experiment [1] at 12 Torr if T_e=70-75 eV of the first Maxwellian hump. Gain at maximum is 16 cm^{-1}, it turns to zero in ~40 ps.

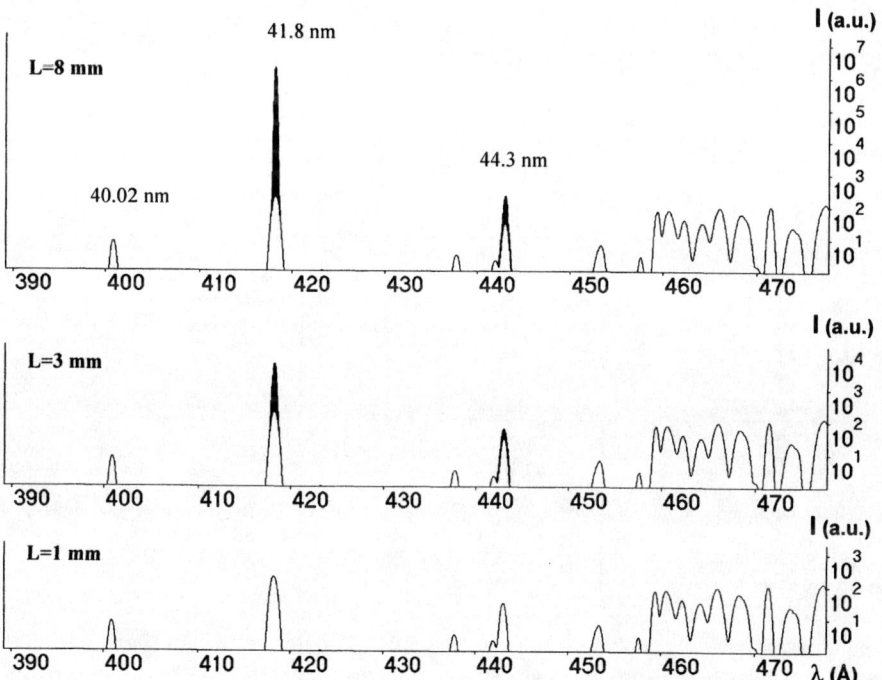

FIGURE 2. Theoretical spectra of Xe IX with accounting for the enhancement calculated for three values of plasma length: L=1, 3, 8mm. It is assumed that plasma is created instantly with Te=70 eV, n_e=3.4$\cdot10^{18}$ cm^{-3}, d=1 mm. Time duration of gain is ~40 ps.

The theoretical spectra of Xe IX with accounting for the enhancement are shown in Fig. 2 for three values of plasma length. The gain was calculated at $T_e=70$ eV, $n_e=3.4\cdot10^{18}$ cm^{-3}, d=1 mm. Note that gain dependence on d is unessential if d=0.3-1 mm. The line at 40.02 nm corresponds to the non-laser transition $5f\,^1P_1 - 5d\,^1P_1$ in Xe IX. A weak enhancement is possible on the optically self-pumped transition $5f\,^1P_1 - 5d\,^1D_1$ at 42.9 nm; its intensity in Xe IX is small. The spectra in Fig. 2 are calculated with gain averaged over time: g=12 cm^{-1}. This result is in a large disagreement with the prediction [2] made with similar plasma parameters. The gain for the transition from the upper level (u) to the low level (l) is calculated by the common expression: $g = A_{ul}\cdot(P_u - P_l(g_u/g_l))/(8\pi\nu_{ul})$; where A_{ul} is the radiative transition probability, P_u, P_l are the level populations of the upper and low levels; g_u, g_l are degeneracy's, ν_{ul} is the total line width which is conditioned by Doppler broadening and collision, radiative, Stark broadening for both levels. The reason of disagreement can be understood from the Table 1 where our atomic constants and the way of gain calculations are compared with those of [2].

TABLE 1. Comparison of the atomic constants and the gains.

	Lemoff et. Al. [2]	Present calculations
Radiative decay probability of transition: $4d^95p\,^1P_1 - 4d^{10}\,^1S_0$	$9.3\cdot10^{10}$ s^{-1}	$11\cdot10^{10}$ s^{-1}
Radiative decay probability of transition: $4d^95d\,^1S_0 - 4d^95p\,^1P_1$	$0.98\cdot10^{10}$ s^{-1}	$1.3\cdot10^{10}$ s^{-1}
Electron energy distribution	Non-Maxw.: 91% of electrons Of 10 to 100 eV, 9% -of 500 eV.	Two-hump Maxw.: 91% of electrons of $T_e=70$ eV, 9% - of $T_e^{hot}=500$ eV
Electron induced transitions: Rates per unit density. $4d^{10}\,^1S_0 - 4d^95p\,^1P_1$ $4d^{10}\,^1S_0 - 4d^95d\,^1S_0$ $4d^95d\,^1S_0 - 4d^95p\,^1P_1$ $4d^95d\,^1S_0 - 4d^94f\,^1P_1$ $4d^95d\,^1S_0 - 4d^95f\,^1P_1$	$1.3\cdot10^{-8}$ cm^3/s $1.5\cdot10^{-8}$ cm^3/s $1.7\cdot10^{-7}$ cm^3/s $3.2\cdot10^{-7}$ cm^3/s $1.7\cdot10^{-7}$ cm^3/s	$0.1\cdot10^{-8}$ cm^{-3}/s $0.4\cdot10^{-8}$ cm^3/s $1.9\cdot10^{-7}$ cm^3/s $1.5\cdot10^{-7}$ cm^3/s $1.9\cdot10^{-7}$ cm^3/s
Electron collision linewidth At $n_i=10^{17}$ cm^{-3}, $n_e=8\cdot10^{17}$ cm^{-2}	$1.36\cdot10^{11}$ s^{-1}	$0.9\cdot10^{11}$ s^{-1}
Doppler width	$0.56\cdot10^{11}$ s^{-1}	10^{11} s^{-1}
Inversion calculations	$N^*=N_i^2(R_{up}-1/3R_l)\cdot$ $(1/\tau_{up}+N_iR_{out})^{-1}$	Kinetic equation calculation For 55 levels of Xe IX
Gain	107 cm^{-1}	16 - 6 cm^{-1} at 4 - 25 ps

The radiative transition probabilities from the upper to the low active level and from the low active level to the ground level are in agreement within 20-30%. The rates of electron induced transitions and linewidths are in a satisfactory agreement, with the exception for the rate of collision excitation of the low and upper active levels $4d^95p$ $^1P_1$, $4d^95d\,^1S_0$ from the ground level. We conclude that the large disagreements of the inversion and gain are conditioned by the difference of the rates of electron induced transitions from the ground level. Also the way of the level population calculations is of importance.

CALCULATIONS OF GAIN IN XENON PLASMA PRODUCED BY CAPILLARY DISCHARGE

In a standard approaches the MHD and atomic-kinetic calculations are performed for the capillary discharge device with the certain parameters. Here we try to solve the reverse problem: to optimize the time of plasma compression so that the optimum ionization balance would correspond to the moment of the maximal compression. Moreover n_e, T_e at this moment must be of the optimum values at which gL-product is maximal. Note that at optimum conditions the moment of the maximal compression is close to the moment of the maximum current. From Fig. 1a-d we conclude that gL is of the greatest value at $n_e^{opt} = (2\text{-}3)\cdot10^{17}$ cm^{-3}, $T_e^{opt} = 60\text{-}70$ eV is chosen taking into considerations that rate of Xe IX ionization grows rapidly with T_e. We assume that xenon mass will be 20-25 times compressed during the pinching. We do not consider the period of the preionization ($\sim$2-3µs) and of the initial stage of pinching ($\sim$3 times of mass compression). Thus we assume that at the moment before the rise time the plasma conditions are reached: n_e=2.5$\cdot10^{16}$ cm^{-3}, T_e=10 eV, d=2 mm, and xenon is ionized up to Xe VIII stage. During rise time n_e, T_e grow: $n_e \rightarrow n_e^{opt}$=2$\cdot10^{17}$ cm^{-3}, $T_e \rightarrow T_e^{opt}$=65 eV; whilst d $\rightarrow$0.7 mm. Then T_e, n_e, d are constants during the period of the maximal compression. With these assumptions we estimate the functions of time for the portions of Xe VIII, Xe IX, and Xe X. The portions calculated with the optimum rise time are shown in Fig. 3a. It can be seen that the optimum ionization balance for Xe IX can be reached with rise time $\geq$38 ns. The corresponding g(t) at 41.8nm is shown in Fig 3b.

Some contemporary capillary discharge devices meet these conditions. In [8] xenon plasma emission was investigated by high voltage, high current discharge with 3 mm inner radius. In this case two peaks of intensity were observed during the first half period of current (Fig. 3b in [8]). The second peak was interpreted in [8] by a second shock wave. The first peak arose at $\sim$70 ns after the beginning of the main pulse, it was the moment of the maximum abundance of Xe XI, Xe XII. The maximum abundance of the Xe IX appeared at the initial stage of plasma compression, at $\sim$40-50 ns after the

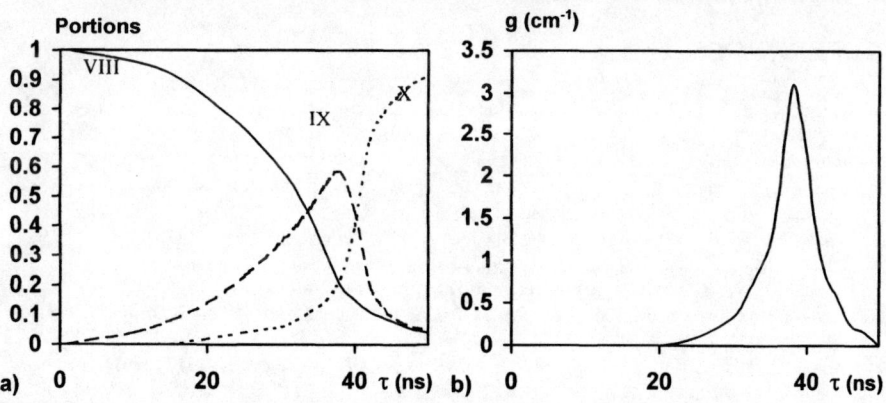

Figure 3. a) The portions of [Xe VIII], [Xe IX], [Xe X] during the compression, the ionization of Xe X is not accounted for: [Xe VIII]+[Xe IX]+[Xe X]=1; b) Gain dependence on time.

beginning of the main pulse. If the filling pressure was 10-20 Pa, then at 40-50ns the pressure in the column center can be estimated as 50-100 Pa (0.4-0.8 Torr, or $n_e\sim(1-2)\cdot10^{17}$ cm^{-3}). The maximum temperature in the pinch was estimated in [8] as 37.5eV by Corona/Saha model. This result is a contradiction to the estimations in the experiments [5,6,10]. It was shown that energies of only a few joules are sufficient to create current pulses in the range of several kiloamperes, which lead to a compression and a heating of the plasma to electron densities of more than 10^{17} cm^{-3} and temperatures of several tens of electron volts [10]. In [11] the x-ray laser in the Ni-like xenon at 13-14nm was theoretically modeled. Besides the enhancement was predicted at 11.3nm on the optically self-pumped transition 3d^{9}4f [J=1] – 3d^{9}4d [J=1]. We suggest that the second peak appeared in [8] when xenon was ionized up to XeXXVII (Ni-like stage). This suggestion is confirmed by the comparison of the totally emitted intensity obtained by integrating over the radial distribution with the intensity obtained behind the 100 μm aperture on-axis. This comparison was made in [8] for the filling pressure 18 Pa and the maximum current of 70 kA. The comparison indicated that for a sufficiently high current the maximum of the total intensity is already emitted before the pinch is reached, while the maximum intensity along the axis is emitted when the radius arrives at its minimum. The laser effect at the range 13.4-13.8nm was not obtained in [8]. A nonlinear intensity growth per length at 13.4-13.8nm in Xe XXVII could be noticeable at $n_e>10^{19}$, $T_e>200$ eV [11]. In the experiment [6] with a small capillary and 200ns rise time a bright emission on-axis was observed also at 11.3nm. This line was not identified unambiguously in any xenon ion. It could be attributed to the self-pumped inverted 4f-4d, 1-1 transition in Ni-like xenon predicted in [11].

CONCLUSION

X-ray laser at 41.8nm is quite achievable using some contemporary capillary discharges with rise time ~40 ns. Its investigation will boost a spring to shorter wavelengths x-ray lasers produced by xenon capillary discharge. On the other hand, it helps to comprehend the x-ray laser physics in the ions of the Pd-like sequence.

REFERENCES

1. Lemoff, B.E, Yin, G.Y, Gordon III, C.L, Barty, C.P.J., Harris, S.E., *Phys. Rev. Let.* **74** 1574-1577 (1995).
2. Lemoff, B.E., Barty, C.P.J., Harris, S.E., *Optics Letters* **19** 569-571 (1994).
3. Rocca, J.J., Tomasel, F.G., Marcony, M.C., Shlyaptsev, V.N., Chilla, J.L., Shapiro, B.T., Giudice, G., *Phys Plasmas* **2** 2547-2550 (1995).
4. Rocca, J.J., Clark, D.P., Chilla, J.L.A., Shlyaptsev, V.N., *Phys. Rev. Lett.*, **77** 1476-1479 (1996).
5. Klosner, M.A., Silfast, W.T., *Optics Letters*, **23** 1609-1611 (1998).
6. Klosner, M. A., Silvast W.T., *Applied Optics*, **40** 4849-4851 (2001).
7. Jushkin, L., Chuvatin, A., Zakharov, S.V., Ellwi, S., and Kunze, H.-J., *J. Phys. D: Appl. Phys.* **35**, 219-227 (2002).
8. Boboc, T., Bischoff, R., and Langhoff, H., *J. Phys. D: Appl. Phys.* **34** 2512-2517 (2001).
9. Ivanova, E.P., Zinoviev, N.A., *Quantum Electronics* **27** 207-214 (1999).
10. Bergmann, K., Schriever, G., Rosier, O., Muller, M., Neff, W., and Lebert, R., *Applied Optics* **38** 5413-5417 (1999).
11. Ivanova, E.P., Zinoviev, N.A., Knight, L.V., *Quantum Electronics* **31** 683-688 (2001).

Theoretical Investigation of Collisional X-ray Lasers in Pd-like Ions

E.P. Ivanova

Institute of Spectroscopy of Russian Academy of Sciences, 142190, Troitsk, Moscow region

Abstract. The spectroscopic constants of Pd-like ions along sequence (Z=50-63) are calculated. Wavelengths of the possible laser transitions are given. Gain calculations are performed in Eu XVIII at 20.6 nm, 22.9 nm, 14.3 nm with assumption that plasma is created instantly. Dramatic increase of gain in optically wide plasma is exhibited at 22.9 nm (optically self-pumped transition).

INTRODUCTION

The Pd-like x-ray laser scheme is most economical in compare with the Ne-like either Ni-like scheme. This is conditioned by: i) smaller energy necessary to obtain the Pd-like ionization stage; ii) stronger electron collision strengths of level excitations in Pd-like ion for the same wavelength x-ray laser as in Ne- or Ni-like ions. Gain coefficient calculation in the ions along the Pd-like isoelectronic sequence is of significant interest. For this purpose accurate spectroscopic constants are necessary: energy levels, probability of radiative and electron induced transitions. However the spectroscopic features of the Pd-like sequence are less known than those of Ne- or Ni-like sequence. In [1] the lines of three resonance transitions $4d^{10} - 4d^9 5p$ $^3P_1$, $^1P_1$, $^3D_1$ were identified through Dy XXI. The lines of two transitions $4d^{10} - 4d^9 4f$ $^3D_1$, $^1P_1$ were identified through Bi XXXVIII [1,2]. The energies of two transitions $4d^{10} - 5f\,^3D_1$, $^1P_1$ in Sb VI through Ce XIII were measured in [3-5]. In recent works [4-5] the complete energy structures of levels of the configurations $4d^9$ $5s$, $5p$, $5d$, $5f$, $6s$, $6p$ in Xe IX – Ce XIII were identified. The urgent problem is the investigation of heavier ions where amplification might be possible at λ=10.0-20.0 nm. The purpose of this paper is the calculation of spectroscopic constants along the isoelectronic sequence of palladium for Z=50-63 (Sn V-Eu XVIII). The laser effect is exhibited in Eu XVIII at 20.6 nm (5d-5p, 0-1), 22.9 nm (5f-5d, 1-1), and 14.3 nm (5d-4f, 0-1).

SPECTROSCOPIC CONSTANTS OF PD- LIKE IONS

We use relativistic perturbation theory with model potential of zero approximation (RPTMP) for calculation of energy levels, radiative transition probability (RTP) and electron collision strengths. The zero order wave functions are determined using the reference energies *5l (l=0-3)* of one electron above the core of 46 electrons

CP641, *X-Ray Lasers 2002: 8th International Conference on X-Ray Lasers*, edited by J. J. Rocca et al.
© 2002 American Institute of Physics 0-7354-0096-2/02/$19.00

$1s^2 2s^2 2p^6 3s^2 3p^6 3d^{10} 4s^2 4p^6 4d^{10}$ and those of one vacancy $4d_{5/2}$, $4d_{3/2}$ in the same core. The zero order energy of Pd-like ion is $E^{(0)} = E^{(0)}_{el} + E^{(0)}_{vac}$, where $E^{(0)}_{el}$ is the state energy of Ag-like ion, $E^{(0)}_{vac}$ is that of Rh-like ion. Energies are accounted from the ground level of Pd-like ion. Note that the accuracy of the final results depends significantly on the accuracy of one-electron and one-vacancy energies. Energy levels and ionization potentials of Ag-like ions are known from the experiments [6-7]. The energies of $4f_{5/2}$, $4f_{7/2}$ orbital change their positions relative to the other orbital energies along the Ag-like sequence, so that $4f_{5/2}$ energy level becomes the lowest at $Z>59$. The collapse of $4f_{5/2,7/2}$ orbital along the sequence made difficulties in the experimental identifications of their energies. Apparently, the energies of $4f_{5/2,7/2}$ levels were given in [7] with a large uncertainty. The splitting of the vacancy energies $E(4d_{3/2})-E(4d_{5/2})$ in the Rh-like ions for $Z=50$-58 was measured in the experiments [8-12]. We calculate this splitting at $Z>58$ by extrapolation procedure using model potential parameter. Ionization potentials of Pd-like ions are calculated by fitting procedure described in [13].

Energy levels of Pd-like ions are calculated by approach [14]. These results will be published separately where the reason of some deviation from the experimental data will be discussed. In general they are in good agreement with known experimental data [1-5]. The RTP calculations are performed using the approach [14,15]. The probabilities of the most strong resonance transitions to the ground state $4d^{10}\ ^1S_0$ are shown in Fig. 1. Sharp maxima and minima along the sequence are due to a strong configuration interaction.

Table 1 shows the cross sections of excitation (Ω) of $4d^9 5d\ ^1S_0$ state from the ground state by electron impact. For each ion the electron impact energy is chosen to

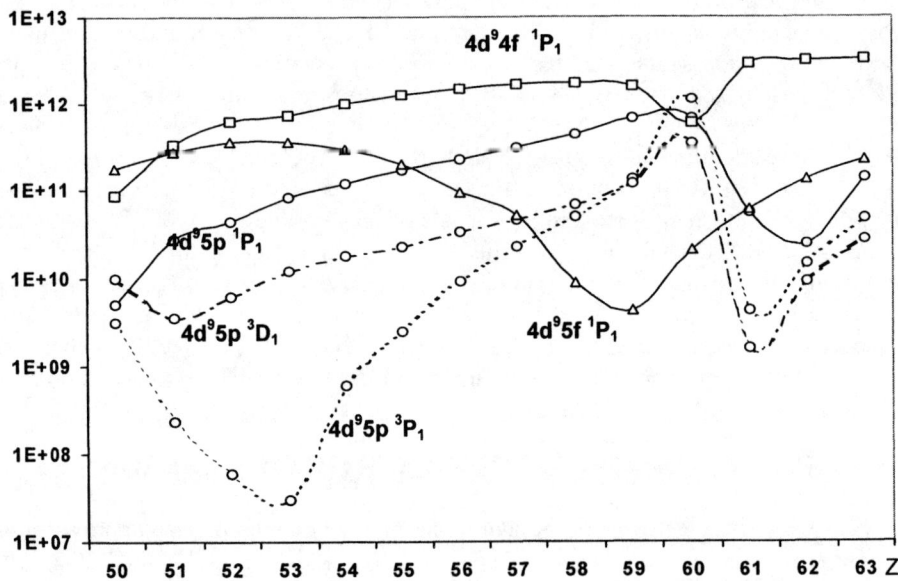

FIGURE 1. The probabilities (in s^{-1}) for the strongest resonance transitions ($\Delta J=1$) to the ground state along the Pd-like sequence.

be close to the ionization potential. One can see a smooth Ω diminution along the sequence. In Table 2 there are listed the wavelengths of the possible laser transitions and the RTP from the upper to the low active level.

Table 1. Cross sections of excitations by electron impact of $4d^9 5d$ $^1S_0$ level in Pd-like ions. Units: Eimp (eV), Ω (cm^{-2}).

Ion					Ion				
Sn V	Eimp	80	110	150	La XII	Eimp	200	240	290
	Ω	3.69-17	2.19-17	1.40-17		Ω	5.99-18	5.96-18	6.42-18
Sb VI	Eimp	90	120	160	Ce XIII	Eimp	220	260	300
	Ω	1.98-17	1.32-17	8.88-18		Ω	4.94-18	4.90-18	5.19-18
Te VII	Eimp	100	130	170	Pr XIV	Eimp	230	270	330
	Ω	1.21-17	8.21-18	6.15-18		Ω	4.20-18	4.14-18	4.31-18
I VIII	Eimp	110	140	180	Nd XV	Eimp	260	300	350
	Ω	2.13-17	1.56-17	1.03-17		Ω	3.52-18	3.56-18	3.66-18
Xe IX	Eimp	140	190	250	Pm XVI	Eimp	280	320	400
	Ω	1.13-17	1.16-17	1.17-17		Ω	3.09-18	3.15-18	3.16-18
Cs X	Eimp	160	200	250	Sm XVII	Eimp	310	360	450
	Ω	8.89-18	8.37-18	9.29-18		Ω	2.70-18	2.77-18	2.67-18
Ba XI	Eimp	190	220	280	Eu XVIII	E_{imp}	350	450	600
	Ω	7.20-18	6.98-18	7.83-18		Ω	2.12-18	2.02-18	1.95-18

Table 2. Wavelengths (λ, Å) and RTP (W, s^{-1}) from the upper to the low active level in Pd-like ions.

Transition $\rightarrow$	$^1S_0 - {}^3P_1$		$^1S_0 - {}^1P_1$		$^1S_0 - {}^3D_1$		$^1P_1 - {}^1D_1$		$^1S_0 - {}^1P_1$	
Ион $\downarrow$	λ	W	λ	W	λ	W	λ	W	λ	W
Sn V	771	1.44+08	827	3.07+09	859	1.17+08	843	4.8+07		
Sb VI	603	1.02+08	638	5.48+09	665	5.56+08	650	7.58+07		
Te VII	515	4.64+07	543	7.88+09	569	8.87+08	541	1.79+09		
I VIII	451	1.19+08	472	1.08+10	496	1.29+09	475	3.61+09		
Xe IX	401	3.87+08	418	1.37+10	443	1.64+09	429	6.16+09		
Cs X	361	8.98+08	375	1.75+10	400	1.76+09	393	9.12+09		
Ba XI	327	2.20+09	340	2.10+10	364	2.46+09	359	1.30+10		
La XII	300	3.92+09	312	2.51+10	335	2.88+09	331	1.70+10		
Ce XIII	276	6.97+09	287	2.96+10	310	3.24+09	311	2.16+10		
Pr XIV	255	1.44+10	266	3.39+10	288	3.34+09	289	2.66+10		
Nd XV	238	5.78+10	250	7.84+09	272	7.54+08	273	3.17+10		
Pm XVI	223	3.51+09	234	7.43+09	257	4.07+09	258	3.67+10		
Sm XVII	209	9.37+09	220	1.94+10	242	4.79+09	243	4.15+10	167	6.27+10
Eu XIII	195	1.47+10	206	2.64+10	228	5.33+09	229	4.65+10	143	6.83+10

GAIN CALCULATIONS IN EU XVIII

Gain calculations are performed with assumption that europium plasma is created instantly with definite parameters. The ion Eu XVIII is chosen for detailed consideration because of favorable set of spectroscopic constants for x-ray laser at 14.3 nm, at 5d-4f,

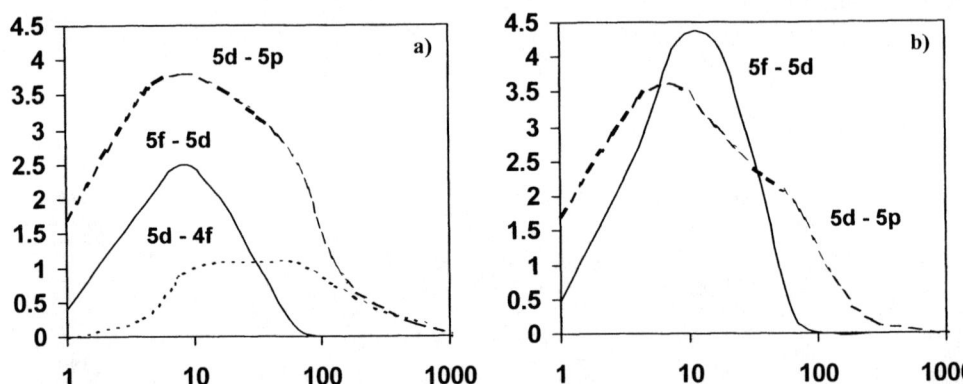

FIGURE 2. Time history of gain (cm^{-1}) in Eu XVIII at 14.3 nm (5d-4f, 0-1), 20.6 nm (5d-5p, 0-1), 22.9 nm (5f-5d, 1-1). It is assumed that plasma is created instantly with parameters: n_e=2.5·10^{18} cm^{-3}, T_e=250 eV. a) d=30 μm; b) d=1.5 mm.

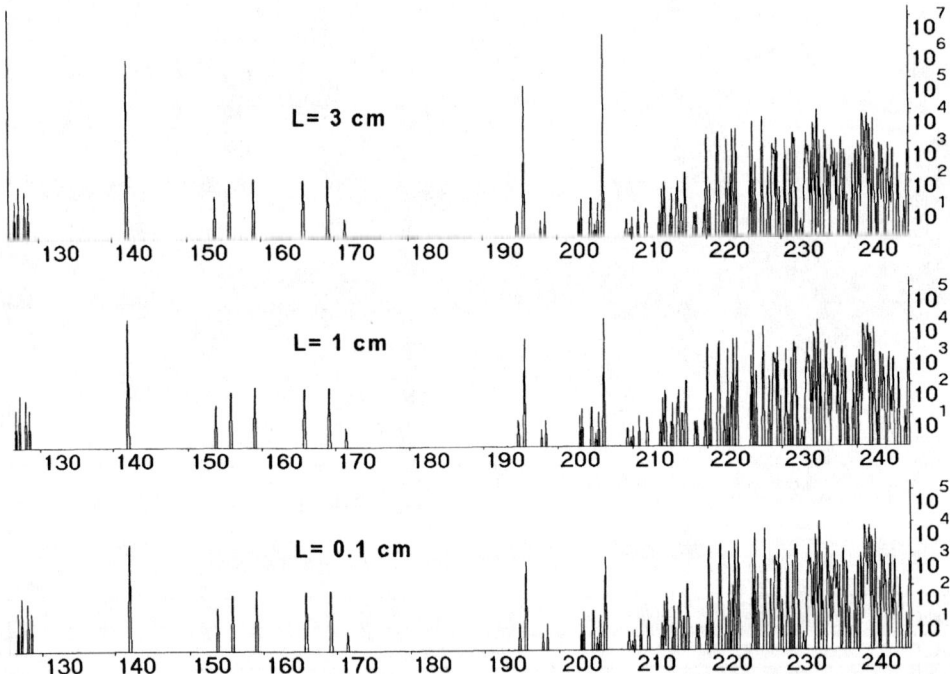

FIGURE 3. Model spectra of Eu XVIII with accounting for enhancement at three values of plasma length L: 0.1 cm, 1 cm, 3 cm. Plasma parameters: n_e=2.5·10^{18} cm^{-3}, T_e=250 eV, d=30 μm

256

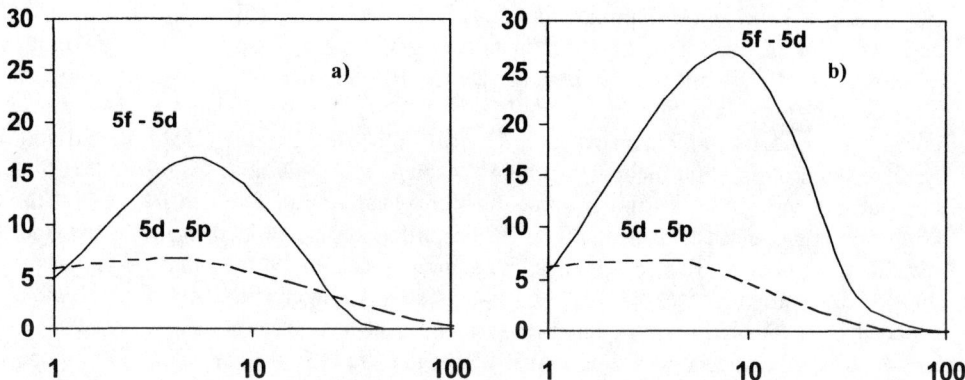

FIGURE 4. Time history of gain (cm^{-1}) in Eu XVIII at 20.6 nm (5d-5p, 0-1) and 22.9 nm (5f-5d, 1-1). It is assumed that plasma is created instantly with parameters: $n_e = 7.5 \cdot 10^{18}$ cm^{-3}, $T_e = 250$ eV.
a) d=100 μm; b) d=3.0 mm - a drastic increase of gain with d on the optically self-pumped transition.

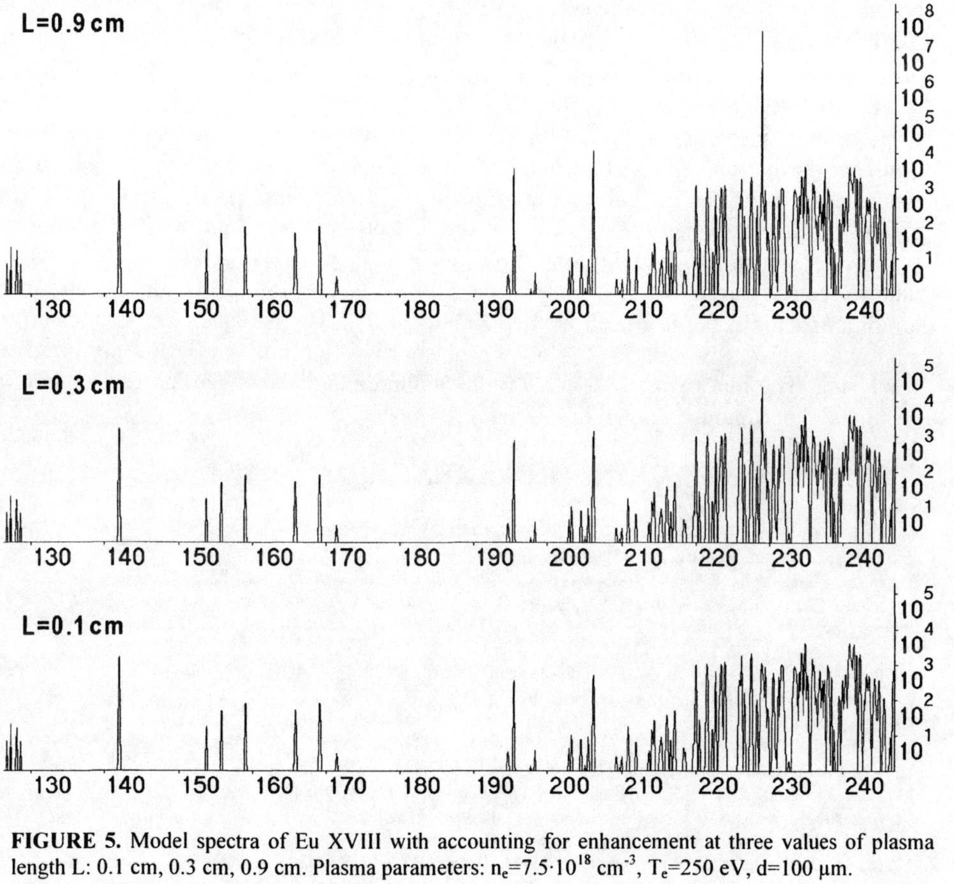

L=0.9 cm

L=0.3 cm

L=0.1 cm

FIGURE 5. Model spectra of Eu XVIII with accounting for enhancement at three values of plasma length L: 0.1 cm, 0.3 cm, 0.9 cm. Plasma parameters: $n_e = 7.5 \cdot 10^{18}$ cm^{-3}, $T_e = 250$ eV, d=100 μm.

0-1 transition. The same approach is used as in the work [16]; here 75 levels are accounted for in the set of kinetic equations for level populations. In Eu XVIII the amplification at 20.6 nm is noticeable at electron temperature $T_e=100eV$; the optimum electron density n_e is within the range $2.5 \cdot 10^{18}$-10^{19} cm^{-3}. Amplification at 14.3 nm is possible in optically thin plasma with $n_e \sim (2-3) \cdot 10^{18}$cm^{-3}, d<50 μm, T_e>150 eV, g~1cm^{-1}. Its time duration is ~300 ps and gain being a constant on this time scale (see Fig. 2a). Gain at 20.6 nm and 22.9 nm in optically thin plasma is also shown in Fig. 2a, the corresponding spectra with accounting for amplification are plotted in Fig.3. In wider plasma the gain at 14.3 nm disappears, the gain at 20.6 nm is of the same value, but the gain at 22.9 nm increases, and its duration becomes longer (Fig. 2b). This effect is especially pronounced in more dense plasma; it can be seen from comparison of Fig. 4a and 4b. The spectra corresponding to plasma parameters of Fig. 4a are plotted in Fig. 5.

CONCLUSION

In the Pd-like sequence with $Z \leq 63$ the most favorable set of spectroscopic constants for x-ray laser at the common 5d-5p, 0-1 transition is found in the ions from Xe IX through Pr XIV. It can be seen from Fig.1 and Tables 1, 2; in the region $60 \leq Z \leq 63$ both the radiative depletion of the low active level and RTP from the upper to the low active levels become smaller.

The important result is the exhibition of a dramatic increase of gain with plasma diameter on the optically self-pumped 5f-5d, 1-1 transition. It is shown in Fig. 4a, 4b where also increase of gain duration can be seen. This result brings to some suggestion for the experiments aimed at the fast production of plasmas with uniform parameters on a sufficiently long time scale. The idea is to use a clustered or nanostructured target of 2-3 mm wide and ~1.5 cm long. Plasma at optimum conditions should be produced by high intensity, short pulse laser. An expansion of such wide plasma will be negligible and plasma will be uniform during period of x-ray lasing. The number of output x-ray laser photons will be ~4 orders larger than in common experiments using a thin target.

REFERENCES

1. Sugar, J., Kaufman, V., *Physica Scripta* **26** 419-421 (1982).
2. Sugar, J, Kaufman, V., Rowan, W.L., *J. Opt. Soc. Am.* **10** 799-801 (1993).
3. Churilov, S.S., Joshi, Y.N. and Ryabtsev, A.N., *J. Phys. B: At., Mol,.Opt. Phys.* **27** 5485-95 (1994).
4. Churilov, S.S., Azarov, V.I., Ryabtsev, A.N., Tchang-Brillet, W.-U.L. and Wyart, J.-F., *Physica Scripta* 61 420-430 (2000).
5. Ryabtsev, A.N., Antsiferov, P.S., Nazarenko, A.V., Churilov, S.S., Tchang-Brillet W.-U.L., and Wyart, J.-F., *J. De Physique* 11 Pr2-317-319 (2001).
6. Moore, S., *Atomic Energy levels*, Institute of Basic Standards NBS, Washington D.C. (1978).
7. Kaufman, V. and Sugar, J., *Physica Scripta* **24** 738-741, 742-746 (1981).
8. Joshi, Y.N. and van Kleef, Th. A.M., *J. Opt. Soc. Am.* **70** 1344-1349 (1980).
9. Kaufman, V., Sugar, J., and Tech, J.L., *J. Opt. Soc. Am.* **73** 691-693, 1077-1079 (1983).
10. Van Kleef, Th.A.M., and Joshi, Y.N., *J. Opt. Soc. Am.* **71** 55-59 (1981).
11. Tech, J.L., Kaufman, V., J., and Sugar, J., *J. Opt. Soc. Am.* **1** 41-44 (1984).
12. Gayasov, R., and Joshi, Y.N., *Physica Scripta* **60** 225-227 (1999).
13. Ivanova, E.P., Glushkov, A.V., *Optika i Spektrosk. (Russian J.)* **58** 961-963 (1985).
14. Ivanova, E.P., Gulov, A.V., *At. Data Nucl. Data Tables* **49** 1-64 (1991).
15. Ivanov, L.N., Ivanova, E.P., Knight, L.V., *Phys. Letters* **A 206** 89-95 (1995).
16. Ivanova, E.P., *Capillary Discharge X-ray Laser on Pd-like xenon*, This Proceedings.

Computational Modelling of the Nickel-like Samarium X-Ray Laser

R. E. King* and G. J. Pert*

*University of York, Heslington, York, YO10 5DD, United Kingdom

Abstract. The Nickel-like Samarium X-Ray Laser (XRL) lasing at 73Å has exhibited saturation experimentally with a travelling-wave (TW) pump configuration. In this paper, the detailed computational modelling of this XRL is presented, and where possible the predictions of the code are compared with experiment. Simulation clearly shows how the development of the electron density and ionisation over the lifetime of the background pulse lead a optimum arrival time for the picosecond pumping pulse. Additionally, it is clear that although high gain is produced, it is often inaccessible to the XRL beam over the full length of the plasma.

INTRODUCTION

The progressive shortening of the pumping pulse duration in collisional excitation XRLs from the nanosecond to the picosecond regime [1] has resulted in predicted gain lifetimes of the order of the XRL pulse transit time over the target length [2][3]. To avoid gain depletion, travelling wave pumping, in which the pumping pulse is synchronised with the XRL pulse, has been adopted. The transient scheme was first demonstrated for Ne-like ions at the Max Born Institute, Berlin [4][5], and later extended to Ni-like ions with a travelling wave pump at the Rutherford Appleton Laboratory (RAL) [6], at Lawrence Livermore National Laboratory (LLNL) [7] and at Laboratoire pour l'Utilisation des Lasers Intense (LULI) [8]. The shortest saturated signal demonstrated using a picosecond pumping pulse was demonstrated at RAL, using Ni-like Samarium ions lasing at 73Å[9].

Pumping with picosecond lasers essentially de-couples the two stages of collisionally pumped XRLs. A series of long (~nanosecond) pulses preform much of the ionisation so that an abundance of the required ion stage is present in the plasma. Additionally, the transverse density gradients are allowed to relax to create a region of plasma suitable for the propagation of the XRL beam along the target length. The picosecond pumping pulse rapidly heats the plasma, giving rise to temperatures in excess of 1keV and strong monopole collisional excitation. This creates a population inversion and thus high gain, with simulations indicating that gain coefficients in excess of $100cm^{-1}$ are achievable on a picosecond timescale [3].

The theory of lasing with Ni-like ions has long been known. The upper $3d^9 4d$ lasing level is metastable against decay to the $3d^{10}$ closed shell ground state, creating a population inversion between the upper lasing level and a $3d^9 4p$ lower level. The basis of this paper will be the presentation and discussion of results from the computational simulation of the Ni-like Samarium XRL lasing at 73Å with direct comparison to experiments

CP641, *X-Ray Lasers 2002: 8th International Conference on X-Ray Lasers*, edited by J. J. Rocca et al.
© 2002 American Institute of Physics 0-7354-0096-2/02/$19.00

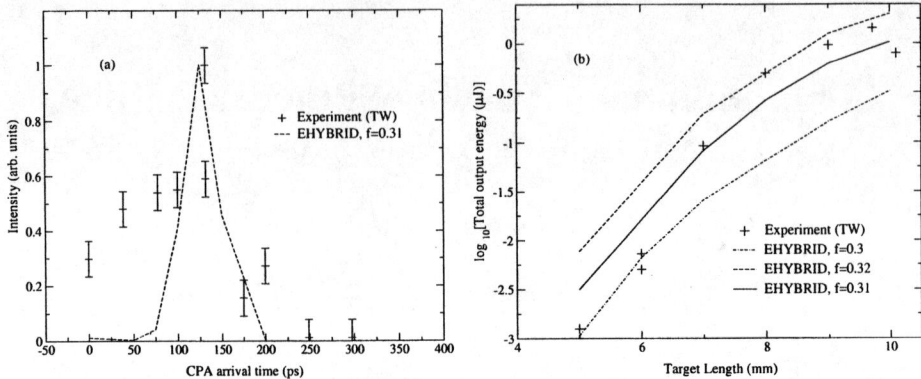

FIGURE 1. (a) Output XRL energy as a function of the arrival time of the CPA pulse with respect to the peak of the background pulse, and (b) output XRL energy as a function of target length for the optimum timing, as measured experimentally and as predicted by simulation.

performed by MacPhee et al [9] where possible.The experimental details are presented elsewhere [9][10], but in essence the pumping pulses used were a 280ps (FWHM) background pulse of irradiance 1×10^{13}Wcm^{-2}, followed by a 1ps CPA pulse of irradiance 3×10^{15}Wcm^{-2}. The optimum arrival time of the CPA pulse was found to be 130ps after the peak of the background pulse (Fig. 1(a)). For this configuration, a small signal gain (SSG) coefficient of 19cm^{-1} and a saturation intensity of $\sim 3 \times 10^{10}$Wcm^{-2} were measured, with the signal saturated for the longest target lengths ($\sim$ 1cm) used (Fig. 1(b)).

COMPUTATIONAL SIMULATION

Computational simulation is undertaken through a combination of the 1.5D hydro-atomic code EHYBRID [11][12], and a fully three dimensional RAYTRACE [13] code. The code EHYBRID self-consistently solves the equations describing the hydro-dynamic expansion and the atomic kinetics of the plasma, thus providing the space-time profiles of the physical characteristics of the plasma, for example the gain coefficient and density. The atomic dataset used in these simulations is adapted from the dataset used in previous work for the Ni-like Gadolinium XRL [14], with additional values taken from GRASP [15] and published data [16]. The output data from EHYBRID is post-processed by the RAYTRACE code, which includes saturation effects, providing the output characteristics of the output XRL beam, including the output energy.

RESULTS

Independently modelling the experimental conditions for the optimum pulse separation, simulation shows good agreement with experiment, with $f = 0.31$ (Fig. 1(b)). The fac-

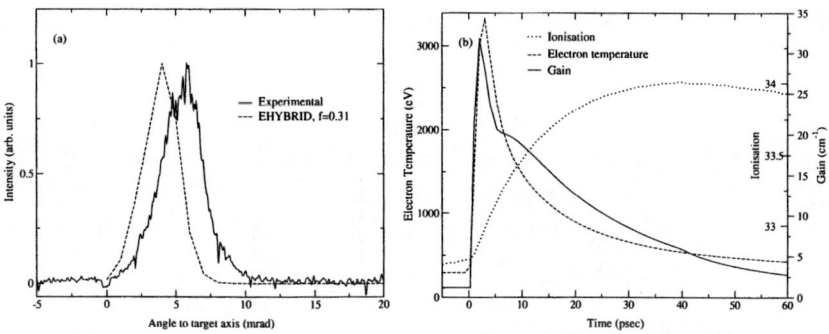

FIGURE 2. (a) XRL beam angular divergence and spread, taken from simulation and experiment. (b) Electron temperature, small signal gain, and ionisation as a function of time taken from EHYBRID .

tor f is a empirical factor pre-multiplying factor for the pump energy to account for a number of processes not modelled in the code, for example plasma reflectivity and beam focus defects. The value of f is consistent with modelling for the Ne-like Germanium XRL in travelling wave mode [10]. RAYTRACE predicts a SSG of 22cm^{-1} and a saturation intensity of $3 \times 10^{10}\text{Wcm}^{-2}$, in good agreement, within both the experimental error and the error associated with the simulation, with the values measured experimentally. The beam divergence is well reproduced by simulation, with the beam centroid is slightly underestimated (Fig. 2(a)).

The predicted electron temperature, SSG and ionisation as a function of time, taken from the cell with peak gain in a region of plasma with favourable density gradients, is shown in Fig. 2(b). Prior to the onset of the CPA pulse, the plasma is slightly under-ionised, with an abundance of predominately Cu-like ions. As the CPA pulse is switched on the plasma is rapidly heated, and a local gain coefficient of $\sim 30\text{cm}^{-1}$ is achieved. The gain is short lived as the plasma rapidly cools, however, at later times the plasma has near optimum ionisation, leading to a slow decay in the gain coefficient. Thus, the gain temporal profile has a spike of high gain superimposed onto a slow decay. This profile leads to extremely short, $\sim$ picosecond, output pulses. Recent experimental measurements for the Ni-like Silver XRL estimate a XRL output pulse duration in the picosecond regime [17].

The predicted optimum arrival time for the CPA pulse from simulation agrees well with experiment (Fig. 1 (a)). The series of figures in Fig. 3 shows the time evolution of the radial, i.e. parallel to the driving laser pulse, profile of the electron temperature and density, ionisation and SSG from simulation for the optimum timing. Just prior to the onset of the CPA pulse (Fig. 3(a)), there is a region of plasma about 20 microns wide and 40 microns from the target surface which has been ionised to close to the Ni-like ion stage. As the CPA pulse is switched on and the plasma is rapidly heated, this region of plasma shows high gain (Fig. 3(b)) at the time of peak XRL output. The density gradients in this region are favourable for the propagation of the XRL beam, leading to a high gain-length product and a lightly refracted beam. This gain is relatively short lived (Fig. 3(c)). The shorter relaxation times at high densities lead to rapid ionisation and high gain in these regions at later times (Fig. 3(d)), however the beam is quickly

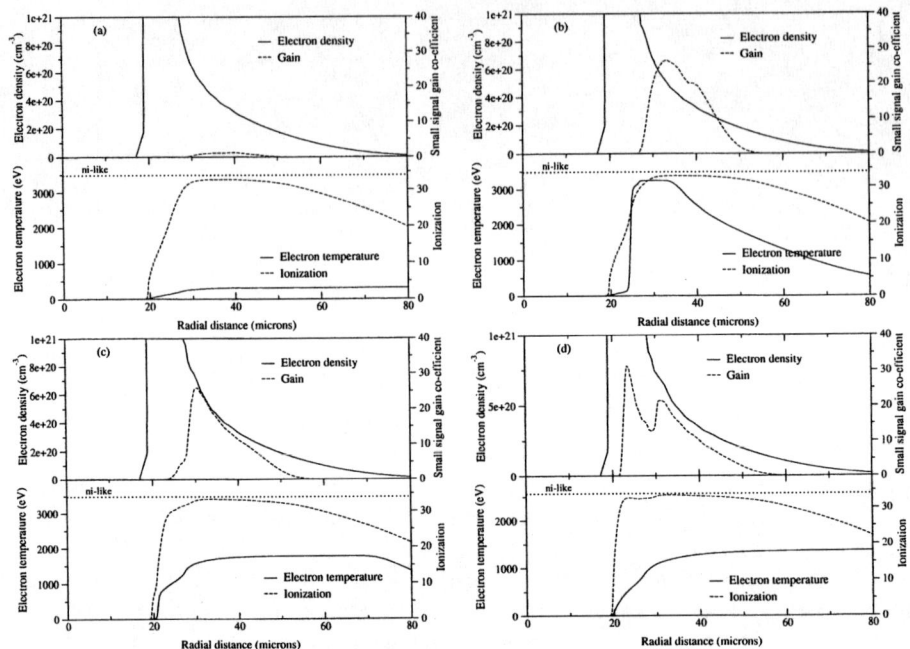

FIGURE 3. Radial spatial profile of electron temperature, ionisation, electron density, and small signal gain, taken from EHYBRID, for the optimum timing, (a) just prior to the CPA pulse, and (b) 3ps (c) 8ps and (d) 13ps after the onset of the CPA pulse.

refracted out of the gain region and the XRL output is low. Thus, although high gain is achieved at high densities, the density gradients are not favourable to the propagation of the XRL beam, and the region is, in effect, inaccessible in terms of producing high output.

The optimum timing is well reproduced by simulation (Fig. 1(a)), however the output energy is underestimated compared to experiment with the CPA pulse prior to and later than this timing. With the CPA pulse arriving early, the plasma is initially under-ionised, and the density gradients have not had sufficient time to develop (Fig. 4(a)). As before, at later times, 8ps after the CPA pulse is applied (Fig. 4(b)), high density regions begin to approach the Ni-like stage. Again, the steep density gradients in this region quickly refract the beam out of the gain region and, consequently, the gain-length product and hence the output XRL energy is low. However, it is clear that the XRL output from simulation will be highly dependent on the modelling of the ionisation rate through the lower ionisation stages and the accuracy of the density gradient in the gain region.

If the CPA pulse arrives later than the optimum timing, the region of plasma with suitable ionisation is located in low density regions. These region show gain as the CPA pulse is switched on, however, the gain is low in comparison to the case with optimum timing due to the lower ion density ($\sim n_e/28$) in the gain region (Fig. 5(a)). Again, the modest gain produced at higher densities (Fig. 5(b)) does not contribute significantly

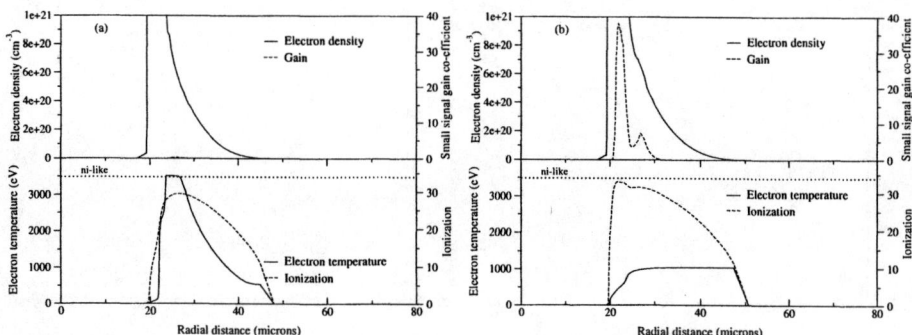

FIGURE 4. Radial spatial profile of electron temperature, ionisation, electron density, and small signal gain, taken from EHYBRID, with the CPA pulse timed to arrive at the peak of the background pulse, (a) just after, and (b) 8 picoseconds after the onset of the CPA pulse.

to the total output energy. The reason for the under-estimate of the output XRL energy from EHYBRID compared to experiment is not clear, however it should be noted the the experimental signals are small, and the XRL is operating in an unsaturated regime.

CONCLUSIONS

The predictions from computational simulation of the Ni-like Samarium XRL showed good agreement with experiment, with the output energy as a function of target length and of the arrival time of the CPA pulse accurately reproduced. The predicted beam divergence and angular spread also agreed well with experimental measurements. Examination of the detailed output from the code revealed the necessary conditions for high gain and output. It is clear that high density regions can exhibit high gain, however, the steep density gradients quickly refract the XRL out of the gain region, severely limiting the gain-length product and XRL output. The simulations highlight that the optimum arrival time for the CPA pulse is a compromise between allowing sufficient time for the density gradients to relax and a region of suitably ionised plasma to develop, and not allowing the plasma to expand to such an extent that the ion density is too low to support high gain.

ACKNOWLEDGEMENTS

This work is supported by the EPSRC, the EU and AWE as part of the UK X-Ray laser programme.

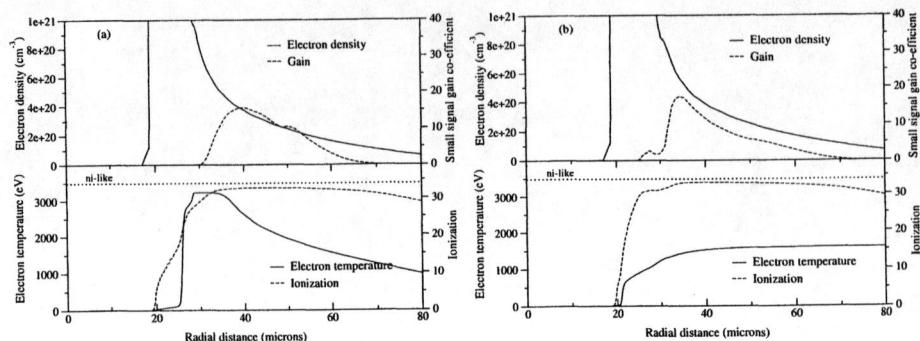

FIGURE 5. Radial spatial profile of electron temperature, ionisation, electron density, and small signal gain, taken from EHYBRID, with the CPA pulse timed to arrive 200ps after peak of the background pulse, (a) 3ps and (b) 10ps after the onset of the CPA pulse.

REFERENCES

1. Warwick, P., Lewis, C., McCabe, S., MacPhee, A., Bejhat, A., Kurkcuoglu, M., Tallents, G., Neely, D., Wolfrum, E., Healy, S., and Pert, G., *Optics Communications*, **144**, 192–197 (1997).
2. Nilsen, J., *Phys. Rev. A*, **55**, 3271–3274 (1997).
3. Healy, S., Janulewicz, K., Plowes, J., and Pert, G., *Optics Communications*, **132**, 442–448 (1996).
4. Nickles, P., Shlyaptsev, V., Kalachnikov, M., Scnürer, M., Will, I., and Sandner, W., *Phys. Rev. Lett.*, **78**, 2748–2751 (1997).
5. Kalachnikov, M., Nickles, P., Schnürer, M., Sandner, W., Shlyaptsev, N., Danson, C., Neely, D., Wofrum, E., Zhand, J., Behjat, A., Demir, A., Tallents, G., Warwick, P., and Lewis, C., *Phys. Rev. A*, **57**, 4778–4783 (1998).
6. Tallents, G., Lin, J., Zhang, J., Behjat, A., Demir, A., Güzelgöz, M., Lewis, C., MacPhee, A., Neely, D., Pert, G., Smith, R., Wark, J., Warwick, P., and Wolfrum, E., "The Optimisation of Soft X-ray Laser Output," in *OSA Tech. Digest Series*, 1997, vol. 7, p. 47.
7. Dunn, J., Osterheld, A., Shepherd, R., White, W., Shalyaptsev, V., and Stewart, R., *Phys. Rev. Lett.*, **80**, 2825–2828 (1998).
8. Klisnick, A., et al., "Generation of Intense Ni-like X-ray Lasers at LULI: From 130ps to 350fs Pumping Pulses," in *X-Ray Lasers 1998*, IOP Conf. Series, 1999, vol. 159, p. 107.
9. MacPhee, A., Lewis, C., Keenan, R., O'Rourke, R., Tallents, G., Pestehe, S., Strati, F., Pert, G., McCabe, S., Simms, P., Neely, D., and Allot, R., "Saturated Lasers at 196Å and 73Å Pumped with CPA Travelling Wave Excitation," in *Central Laser Facility Ann. Report*, 1998, pp. 49–51, RAL-TR-1999-062.
10. King, R., Pert, G., McCabe, S., MacPhee, A., Lewis, C., Keenan, R., O'Rourke, R., Tallents, G., Pestehe, S., Strati, F., Neely, D., and Allot, R., *Phys. Rev. A*, **64**, 053810 1–12 (2001).
11. Pert, G. J., *J. Fluid Mech.*, **131**, 401–426 (1983).
12. Holden, P., Healy, S., Lightbody, M., Pert, G., Plowes, J., Kingston, A., Robertson, E., Lewis, C., and Neely, D., *J. Phys. B*, **27**, 341–367 (1994).
13. Plowes, J., Pert, G., Holden, P., and Toft, D., *Optical and Quantum Electronics*, **28**, 219–228 (1996).
14. McCabe, S., and Pert, G., *Phys. Rev. A*, **61**, 033804 1–9 (2000).
15. Dyall, K., Grant, I., Johnson, C., Parpia, F., and Plummer, E., *C. Phys. Comm.*, **55**, 425–456 (1989).
16. Zhang, H., Sampson, D., and Fontes, C., *Atomic and Nuclear Data Tables*, **48**, 91–163 (1991).
17. Tallents, G., Abou Ali, Y., Edwards, M., King, R., Pert, G., Pestehe, S., Strati, F., Keenan, R., Lewis, C., Topping, S., Guilbaud, O., Klisnick, A., Ros, D., Clarke, R., Neely, D., and Notley, M., *These proceedings* (2002).

The Symplectic Algorithm for Use in A Model of Laser Field[*]

Xiaoyan Liu[1,2], Xueshen Liu[1], Peizhu Ding[1] and Zhongyuan Zhou[1]

1.Institute of Atomic and Molecular Physics, Jilin University, Changchun 130023, China

2.Mathematics department, Northeast Normal University, Changchun 130024, China

Abstract Using the asymptotic boundary condition the time-dependent Schrödinger equations with initial conditions in the infinite space can be transformed into the problem with initial and boundary conditions, and it can further be discrected into the inhomogeneous canonic equations. The symplectic algorithms to solve the inhomogeneous canonic equations have been developed and adopted to compute the high-order harmonics of one–dimensional Hydrogen in the laser field. We noticed that there is saturation intensity for generating high-order harmonics, which are agree with previous results, and there is a relationship between harmonics and bound state probabilities.

Key words: inhomogeneous equation, symplectic algorithm, high order harmonics and bound probability

1. INTRODUCTION

In recent 20 years with quick development of the laser technique especially short-pulse intense laser technique, the study of the laser-atom interaction is becoming one of the new active projects in physics. In this paper using the asymptotic boundary condition the time-dependent Schrödinger equations with initial conditions in the infinite space can be transformed into the problem with initial and boundary conditions, and it can further be discrected into the inhomogeneous canonic equations. The symplectic algorithms to solve the inhomogeneous canonic equations have been

---

[*] Supported by the National Natural Science Foundation of China and the special funds for major state basic research projects (G1999032804).

CP641, *X-Ray Lasers 2002: 8th International Conference on X-Ray Lasers,* edited by J. J. Rocca et al.
© 2002 American Institute of Physics 0-7354-0096-2/02/$19.00

developed and adopted to compute the high-order harmonics of one-dimensional Hydrogen in the laser field. The results show that there is saturation intensity. Before reaching this value the cutoff law works well, but exceeding this value the cutoff law is not suitable, the plateau has a tendency. Moreover there is a relationship between harmonics and bound state probabilities.

2.THE SYMPLECTIC ALGORITHM FOR LASER-ATOM

INTERACTION

The behavior of the atom H in an intense laser field can be described by the following time-dependent Schrödinger equation with initial conditions in the infinite space

$$
\begin{cases}
i\dfrac{\partial}{\partial t}\psi(x,t)=\left[-\dfrac{1}{2}\dfrac{\partial^2}{\partial x^2}+V(x)-\varepsilon(t)x\right]\psi(x,t) & (-\infty<x<\infty,\ 0\le t\le T) \\
\psi(0,x)=\varphi(x) & (-\infty<x<\infty)
\end{cases}
\tag{1.1}
$$

$$(1.2)$$

where the atomic potential is softcore potential[1] $V(x)=-\dfrac{1}{\sqrt{x^2+2}}$,

$\varepsilon(t)x=f(t)E_0x\sin(\omega_0 t)$, E_0 is the peak intensity, the pulse shape

$f(t)=\sin^2\omega t$; Initial state $\varphi(x)=Q(1+\sqrt{x^2+2})\exp(-\sqrt{x^2+2})$ is the ground state wave function.

As $|x|$ is sufficiently large and $|V(x)|<<|-\varepsilon(t)x|$, the solution $^W(x,t)$ of (1.1) (1.2) can be obtained using Fourier transformation when $V(x)$ is neglected. So

$^W(x,t)$ can be regards as the solution $\psi(x,t)$ of the equation (1.1) (1.2) when $|x|$ is

sufficiently large. Moreover when $X>>0$ $^W(\pm X,t)=\exp(\mp iAX-i\dfrac{q}{2})\varphi(\pm X-\alpha)$

can be obtained using phase integral, where $\alpha(t)=-\displaystyle\int_0^t A(t')dt'$,

$q(t)=\displaystyle\int_0^t A^2(t')dt'$ $\dfrac{\partial A(t)}{\partial t}=-\varepsilon(t)$ $A(0)=0$. Then the problem with initial conditions in the infinite space (1.1) (1.2) can be transformed the problem with initial conditions in the finite space as follows

$$i\frac{\partial}{\partial t}\psi(x,t)=\left[-\frac{1}{2}\frac{\partial^2}{\partial x^2}+V(x)-\varepsilon(t)x\right]\psi(x,t) \qquad (-X<x<X,t>0) \qquad (1.3)$$

$$\psi(x,t)|_{x=\pm X}=\phi(\pm X,t)=\exp(-iA(\pm X)-i\frac{q}{2})\varphi(\pm X-\alpha) \qquad (t>0) \qquad (1.4)$$

$$\psi(x,0)=\varphi(x) \qquad (-X<x<X) \qquad (1.5)$$

Suppose $\psi(x,t)=a(x,t)+ib(x,t)$, $\phi(x,t)=c(x,t)+id(x,t)$. Let time-step $\tau>0$, sufficiently large positive integral N, space-step $h=\frac{X}{N}$. Denote $t_k=k\tau$, $k=0,1,2,$ ...; $x_j=jh$, $j=\bullet\ N,\ \bullet\ N+1,\ ...,\ \bullet\ 1,0,1,\ ...\ N\ \bullet\ 1,N$; $g_j=g(x_j)$, $f_j(t)=f(x_j,t)$; $f_j^k=f(x_j,t_k)$; $2N\ \bullet\ 1$ dimension vector $A(t)=(a_{-N+1}(t),\ ...,a_{N-1}(t))^T$, $B(t)=$ $(b_{-N+1}(t),\ ...,b_{N-1}(t))^T$, $C(t)=(c_{-N}(t),0,...,0,c_N(t))^T$,$D(t)=(d_{-N}(t),0,...,0,d_{N-1}(t))^T$; $4N\ \bullet\ 2$ dimension vector $Z(t)=(A(t)^T\ B(t)^T)^T$,$Y(t)=(C(t)^T\ D(t)^T)^T$. The time-dependent Schrödinger equation (1.3) (1.4) can be discredited into the inhomogeneous canonic equations (1.6) by substituting symmetric difference quotient $\frac{\psi_{j+1}-2\psi_j+\psi_{j-1}}{h^2}$ for partial derivative $\frac{\partial^2}{\partial x^2}\psi(x_j,t)$,

$$\frac{d}{dt}Z=GZ-\frac{1}{2h^2}JY\ , \qquad (1.6)$$

where

$$G=\begin{bmatrix}0 & S\\ -S & 0\end{bmatrix}=JH, \qquad J=\begin{bmatrix}0 & I\\ -I & 0\end{bmatrix}, \qquad H=\begin{bmatrix}S & 0\\ 0 & S\end{bmatrix},$$

I and 0 are $2N$-1 order unit matrix and zero matrix, respectively. $2N$-1 order matrix $S=U+V$,

$$U=-\frac{1}{2h^2}\begin{bmatrix}-2 & 1 & 0 & 0 & \cdots & 0\\ 1 & -2 & 1 & 0 & \cdots & 0\\ 0 & & \ddots & & & \vdots\\ \vdots & & & \ddots & & 0\\ 0 & \cdots & 0 & 1 & -2 & 1\\ 0 & \cdots & 0 & 0 & 1 & -2\end{bmatrix},\ V=\begin{bmatrix}V_{-N+1}+E_{-N+1}(t)x_{-N+1} & & 0\\ & \ddots & \\ 0 & & V_{N-1}+E_{N-1}(t)x_{N-1}\end{bmatrix},$$

G is a infinitesimal symmetric matrix, the particular solution of "the inhomogeneous canonic equations" and the general solution of their corresponding homogeneous canonic equations can both generated by symplectic transformation. The symplectic algorithm for solving (1.6) can be obtained by using 2-order norm-conserved explicit symplectic scheme and trapezoid integral [3]. Therefore the wave function at every point (x_j, t_k) can be computed.

3. THE HIGH ORDER HARMONICS AND BOUND STATE PROBABILITY OF THE ONE–DIMENSIONAL HYDROGEN IN LASER FIELD

The intensity of high order harmonics spectrum is proportional to

$$|d(\omega)|^2 = \left| \frac{1}{T_2 - T_1} \int_{T_1}^{T_2} d(t) e^{-i\omega t} dt \right|^2 ,$$

where $d(\omega)$ is the Fourier translation of the dipole transition matrix element $d(t)$, operator $\vec{d}$ has three kind forms. In this paper $d(t)$ take the form

$$d(t) = < \psi(x,t) \left| \frac{\partial}{\partial x} (V(x) - E_0 x f(t) \sin \omega t) \right| \psi(x,t) > .$$

The normalized bound state $\phi_1, \phi_2, \cdots$ in free field can be obtained using the symplectic shooting method[4]. Then the ionization probabilities [5] $P_b(t) = \sum |\alpha_i(t)|^2$ of H atom in laser field can be computed. Where

$$\alpha_i(t) = \int_{-X}^{X} \psi(x,t) \phi_i(x) dx \quad i = 1, 2, \cdots .$$

Combined the computed wave function using symplectic algorithm with $d(t)$ and $\alpha_i(t)$, the harmonic spectrum $|d(\omega)|^2$ and bound state probability $P_b(t)$ can be obtained.

4. RESULTS AND CONCLUSION

Set time-step $\tau = 0.001$ bound point $X = 300$ space step $h = 0.05$. The Fig. 1 and Fig. 2 are the harmonic spectrum when peak intensity are 0.08 and 0.15, respectively pulse length is $5\,fs$ and laser frequency $\omega_0 = 0.055$. From Fig.1 and Fig.2 we can see that when $E_0 = 0.08$, the cutoff is around 40, in agreement with the predicted cutoff 39.6. But when $E_0 = 0.15$, the cutoff is around 86, not around the predicted cutoff 116.2. The plateau has a obvious decline (i.e. there are different strength in the harmonics). By changing the peak intensity the results can be obtained that a saturation intensity exists. Before reaching this value the cutoff law works well. But exceeding this value the cutoff law is not suitable and the plateau has a tendency.

Fig.3 and Fig.4 show the bound probability of the atom under the same case as the Fig.1 and Fig.2. it can be seen that there are relation between high order harmonics and bound probability. When the peak intensity is over some value, the bound probability deduce rapidly and even to zero. The phenomenon would happen that the cutoff law is not suitable and the plateau has a tendency.

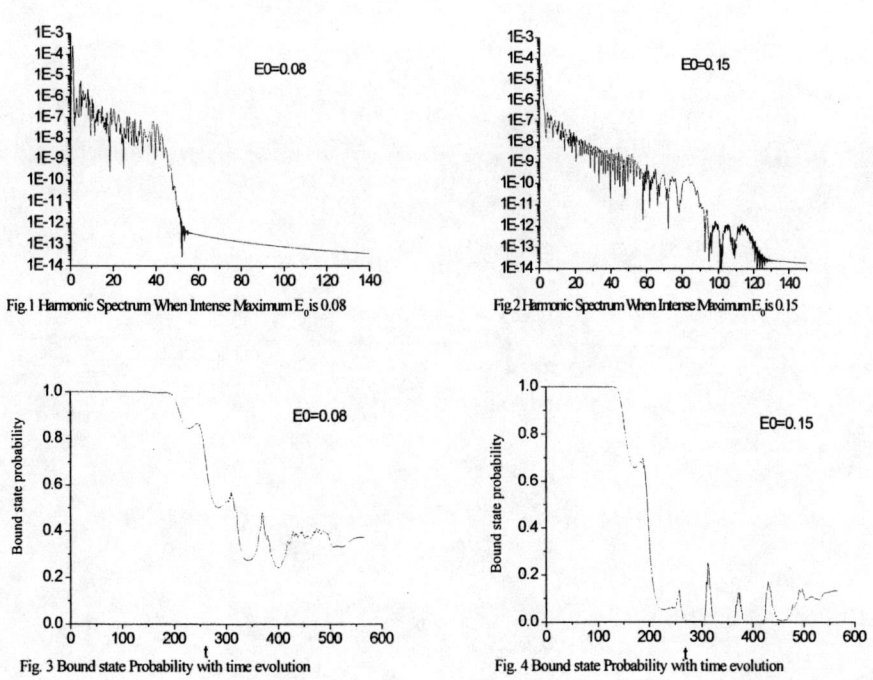

Fig.1 Harmonic Spectrum When Intense Maximum E_0 is 0.08

Fig.2 Harmonic Spectrum When Intense Maximum E_0 is 0.15

Fig. 3 Bound state Probability with time evolution

Fig. 4 Bound state Probability with time evolution

REFERENCE

[1] W. C. Liu and C.W. Clark. Closed-from solution of the Schrödinger Equation for a model one-dimensional Hydrogen atom. *J. Phys. B*, 1992, 25: L517.

[2] Xiaoyan Liu, Xueshen Liu, Zhongyuan Zhou and Peizhu Ding, The symplectic algorithm for solving the inhomogeneous canonic equation in one-dimensional laser field model. *Journal of Chinese Computational Physics*, 2002 ,19:1, 62.

[3] Zhongyuan Zhou and Peizhu.Ding Conservation Quantities of the Explicit Symplectic Scheme for Time-evolution of Quantum System(J), *Chinese Journal of Atomic and Molecular Physics* 1997 14 2 175.

[4] Xueshen Liu, Xiaoyan Liu, Zhongyuan Zhou and Peizhu Ding. ,Numerical Solution of One-dimensional Time-independent Schrödinger Equation by Using Symplectic Schemes(J), *Intern. J. Quantum Chem.*, 2000, 79:6, 343.

[5] Zhongyuan Zhou,Qiren Zhu and Peizhu Ding, Muti-phton Ionization Rate of Hydrogen in the Laser Fields (J) *Chinese Journal of High Power Laser And particle Beams* 2000,,12(2):169.

[6] Liu Xiaoyan, The Structure-preserving Algorithm for Use in the Intense Laser-atom Interaction, Doctor's thesis, Changchun, 2001, 7

Modeling and imaging of the Ni-like Pd X-ray laser and other new schemes

Joseph Nilsen, James Dunn, Raymond F. Smith, and Troy W. Barbee, Jr.

Lawrence Livermore National Laboratory, Livermore, CA 94551-9900

Abstract. The technique of using a nsec prepulse to create and ionize the plasma followed by a psec pulse to heat the plasma has enabled us to achieve saturated laser output for low-Z neon-like and nickel-like ions driven by small lasers with less than ten joules of energy. In this work we model recent experiments done using the COMET laser at Lawrence Livermore National Laboratory to illuminate slab targets of Pd up to 1.25 cm long with a 2 joule, 600 ps prepulse followed 700 psec later by a 6 joule, 6 psec drive pulse. The experiments measure the two-dimensional near-field and far-field laser patterns for the 14.7 nm Ni-like Pd x-ray laser line. This line has already demonstrated saturated output. The experiments are modeled using the LASNEX code to calculate the hydrodynamic evolution of the plasma and provide the temperatures and densities to the CRETIN code, which then does the kinetics calculations to determine the gain. Using a ray tracing code to calculate the near and far-field patterns, the simulations are then compared with experiments.

We also present several new schemes that we are modeling. The first scheme is Pd-like Nd that has a promising 5d - 5p laser line near 24.3 nm and 5p - 5s lines near 40 nm. Another potential scheme is Pt-like U with 6d – 6p laser lines near 22.5 and 25.8 nm. A Nd-like U laser scheme is also considered.

INTRODUCTION

Most researchers today use some variant of the prepulse technique [1] to achieve lasing in Ne-like or Ni-like ions. As a result the Ne-like 3p $^1S_0 \rightarrow$ 3s $^1P_1$ and Ni-like 4d $^1S_0 \rightarrow$ 4p $^1P_1$ laser lines now dominate the laser output. This technique illuminates solid targets with several pulses. The prepulse creates a large-scale-length plasma that is in the correct density range for gain and has sufficiently small density gradients for laser propagation. The main pulse then heats the plasma to lasing conditions. In this paper we model the Ni-like Pd experiments performed with the COMET laser at LLNL. The technique at COMET of using a nsec prepulse to create and ionize the plasma followed by a psec pulse to heat the plasma has enabled low-Z nickel-like ions to achieve saturated output when driven by small lasers with less than 10 joules of energy. For the Ni-like Pd experiments we model the 4d $^1S_0 \rightarrow$ 4p $^1P_1$ line at 14.7 nm using the LASNEX [2] and CRETIN [3] codes. These simulations are compared with the experimental measurements of the two-dimensional (2-D) near-field and far-field patterns.

CP641, *X-Ray Lasers 2002: 8th International Conference on X-Ray Lasers,* edited by J. J. Rocca et al.

In the second part of this paper we consider new schemes that may be possible using collisional excitation in isoelectronic sequences other than Ne-like and Ni-like.

PLASMA MODELING OF NI-LIKE PD

LASNEX one dimensional (1D) computer simulations of a Pd slab illuminated by a 2 J, 600 psec gaussian pulse followed 700 psec later by a 6 J, 6 psec gaussian pulse from a 1.05 µm Nd laser is used to model the Pd experiments done at COMET. The calculations assume a 120 µm wide by 1.6 cm long line focus with a 700 psec delay between the long and short pulse. The LASNEX calculations include an expansion angle of 15 degrees in the dimension perpendicular to the primary expansion so as to simulate 2D effects. Since the long pulse is defocused by a factor of 2 in the experiments the long pulse energy in the calculation is reduced to 1J.

CRETIN calculates the gains of the laser lines including radiation trapping effects for the six 4f and $4p \rightarrow 3d$ resonance lines in Ni-like Pd using the LASNEX calculated densities and temperatures as input. The atomic model of Pd used by the CRETIN code includes all 107 detailed levels for levels up to $n = 4$ in Ni-like Pd.

Two dominant laser lines are predicted; the $4d\ ^1S_0 \rightarrow 4p\ ^1P_1$ line at 14.7 nm and the $4f\ ^1P_1 \rightarrow 4d\ ^1P_1$ line at 17.0 nm. The Ni-like $4d\ ^1S_0 \rightarrow 4p\ ^1P_1$ line lases by monopole collisional excitation from the ground state populating the upper laser level. The $4f\ ^1P_1 \rightarrow 4d\ ^1P_1$ line lases because radiation trapping allows a large radiation field to build up on the $3d\ ^1S_0 \rightarrow 4f\ ^1P_1$ resonance line and populate the 4f upper laser state by the self-photopumping process [4]. The 17 nm line is less intense and is observed only weakly in the COMET experiments for Pd but it has been observed more strongly for lower Z materials such as Mo [4] under the COMET conditions and in Ag [5] using the P102 laser facility at Limeil. Only the 14.7 nm line is discussed in this

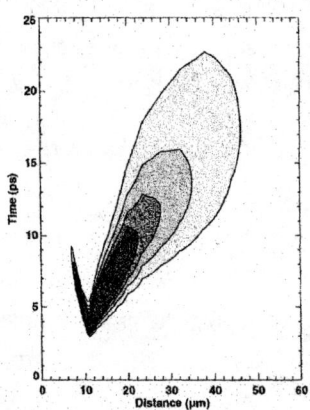

FIGURE 1. Gain contours versus space and time for the Ni-like Pd 14.7 nm line.

paper.

Figure 1 shows the calculated contours of the gain versus space and time for the 14.7 nm line under the nominal drive conditions described above. The short pulse laser peaks at 6 psec on this time scale so the evolution during the long pulse is not shown. The horizontal axis gives the distance from the target surface in the direction of the plasma expansion. The gray scale varies from 0 to 200 cm^{-1} with the gain contours representing gains greater than 200, 150, 100, and 50 cm^{-1}. The darkest contour represents gains greater than 200 cm^{-1}. The calculations predict a very strong gain region 10 to 30 µm from the target surface that is similar to what is observed in the experiments.

Given the importance of refraction a ray tracing code was used together with the time and space resolved gain data for the 14.7 nm line to predict the 1-D near-field and far-field spatial distribution for the 1.25 cm long Pd laser in the plasma expansion direction. To better match the observed gain we multiplied the gain by 0.25. The gain multiplier does not affect where the near-field pattern peaks spatially.

Doing a series of ray tracing calculations the simulations predict the peak of the near-field pattern moving from 18 µm from the surface for the 0.25 cm length to 36 µm for the 1.25 cm length. The time and space integrated small signal gain is 33 cm^{-1} for the short lengths and drops to 19 cm^{-1} for the longest length due to the refraction effect. Saturation is not included in our calculations. Figure 2(b) shows the normalized near-field pattern that is predicted for the 1.25 cm long Pd target. The pattern peaks 36 µm from the target surface with a full-width half-maximum width (FWHM) of 8.5 µm. Note the target surface is at 1µm.

Figure 2(a) shows the predicted 1-D far-field pattern. The peak deflection is only 3.5 mrad with a FWHM width of 1.5 mrad in the plasma expansion direction. The beam deflection is very similar to the 3 mrad measured in previous experiments [6]. The divergence is about half the 3 mrad measured in the 2-D far-field images shown in Fig. 3(a).

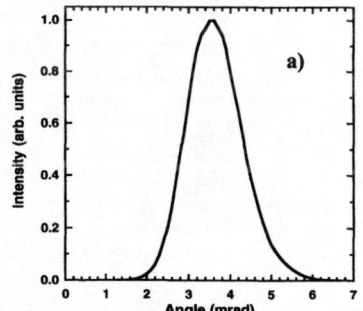 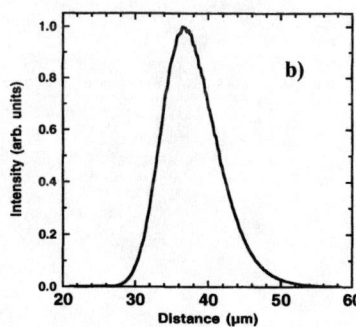

FIGURE 2. Calculated 1D far-field (a) and near-field (b) patterns for the Ni-like Pd 14.7nm laser.

IMAGING EXPERIMENTS IN NI-LIKE PD

The experiments were performed on the COMET laser system at LLNL. The laser operates at a wavelength of 1054 nm and utilizes the CPA technique to produce two beams of 500 fs and 600 ps (FWHM) pulse duration. For these experiments the short pulse was lengthened to 6 ps with an energy of 6 to 7 J while the long pulse energy was typically 1 to 2 J delivered in the line focus at the target chamber. The laser operates with a repetition rate of 1 shot every 4 minutes. The experiments had the short pulse arrive 700 ps after the peak of the long pulse. This pulse separation was found to be optimum for these experiments. The short pulse beam was focused to 120 µm by 1.6-cm line while the long pulse was defocused by a factor of two to produce a more uniform plasma. A traveling wave irradiation scheme was implemented using seven-segmented stepped mirror to mitigate against the transit time effects. We used a 1.25-cm long polished slab of Pd so that the line focus would overfill the target while maintaining uniform illumination along the entire length.

Having already demonstrated saturated output on the Ni-like Pd laser line at 14.7 nm in previous work [6], this paper presents the near-field and far-field images of the Ni-like Pd laser line at 14.7 nm. The far-field measurements used 2 flat Mo:Si multi-layer mirrors (a 0° and a 45° flat mirror) to direct the X-ray laser into the CCD camera which was 130.4 cm from the end of the target. The measured reflectivity of these two Mo/Si multilayer optics was 62% for the 0° mirror and 30% for the 45° mirror at the laser wavelength with a bandpass of 0.6 nm and 1.4 nm, respectively, assuming unpolarized X-rays. Thin Zr filters of thicknesses 190 nm and 300 nm were placed before the first multilayer mirror and the CCD camera to block the optical light and to protect the multi-layer mirror from damage. The X-ray back-thinned CCD camera has

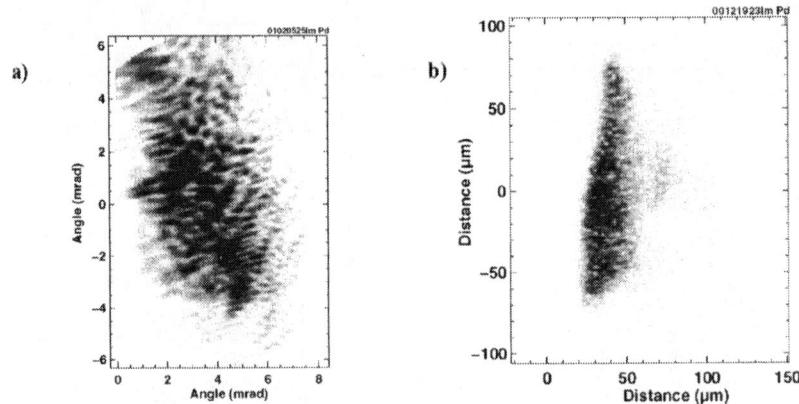

FIGURE 3. Far-field (a) and near-field (b) images of Ni-like Pd 14.7 nm laser line.

a 1024 by 1024 pixel array with a 24-µm pixel size.

Figure 3(a) shows the far-field pattern for the Ni-like Pd laser using 1.88 J in the long pulse and 6.67 J in the short pulse. The horizontal divergence is about 3 mrad in the plasma expansion direction while the vertical divergence is larger, about 6 mrad in the line focus direction. The absolute deflection angle is not measured but previous experiments using a spectrometer measured typical deflection angles of 3 mrad from the target surface [6].

The near-field X-ray laser images used a normal incidence spherical Mo:Si multi-layer mirror with a focal length of 11.75 cm that was placed 12.33 cm from the end of the Pd X-ray laser target to image the X-ray laser onto the CCD camera with a magnification of 20.1. A filter with 75 nm thick Al coated on 200 nm thick Lexan substrate was placed before the imaging mirror while a 295 nm thick Zr filter was placed before the CCD camera to block the optical light. The traveling wave mirror was set to 0.67 c for these images which slightly reduces the output of the x-ray laser but the output is still saturated.

Figure 3(b) shows the near-field pattern for the Ni-like Pd laser line at 14.7 nm using 1.27 J in the long pulse and 6.58 J in the short pulse. The horizontal direction is the plasma expansion direction. The laser emission is quite narrow in this direction. The target surface is estimated to be at 0 µm on this scale by observing the plasma emission but a better fiducial is needed. The laser emission in the vertical direction is determined by the width of the line focus and is about five times larger than in the expansion direction. Looking at the middle of the line focus the near-field pattern peaks about 35 µm from the target surface with a FWHM spatial extent of 24 µm. This compares very well with the calculated peak 36 µm from the target surface. However the calculations have a narrower spatial distribution with a FWHM spatial extent of 8.5 µm.

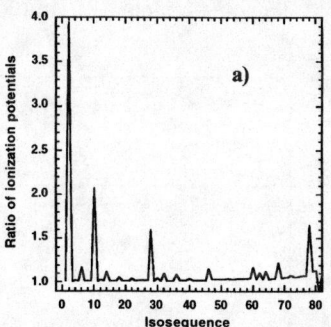 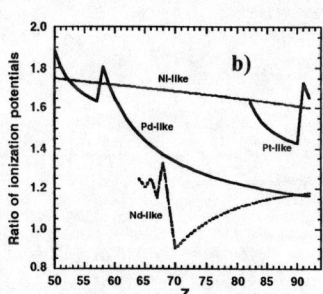

FIGURE 4. Ratio of ionization potentials (a) for adjacent iso-electronic sequences vs uranium isosequences. The ratio of ionization potentials (b) for adjacent iso-electronic sequences vs Z for the Ni-like, Pd-like, Nd-like, and Pt-like sequences.

NEW SCHEMES

Given the success of the Ne-like and Ni-like collisional excitation schemes we decided to look at how we could identify other candidate laser schemes. One of the features that make the Ne-like and Ni-like schemes so robust is that there is a large increase in the ionization potential comparing Ne-like to Na-like and Ni-like to Cu-like. This creates the stable ionization balance needed to get population into the ground state of the lasing ion which can then be excited by monopole collisional excitation to the upper laser state. Since Ne and Ni-like have closed n=2 and 3 shells the next obvious candidate for a lasing sequence is Nd-like, since that is the fully closed n=4 system. However, as Fig. 4(b) will show, there is very little ionization jump with Nd-like because the 5s electron is very close to the energy of the 4f electron. Taking U (Z=92) as an example, Fig. 4(a) shows the ratio of ionization potentials for adjacent sequences vs sequence. For example, for isosequence=2 (He-like) Fig. 4(a) plots the value of the ratio for He-like vs Li-like. Looking at this figure one sees the expected peaks at values of 2 (He-like), 10 (Ne-like), and 28 (Ni-like). By looking at this ratio for most sequences and ions we identified 46 (Pd-like) and 78 (Pt-like) as promising new schemes. In fact Pd-like Xe has already been demonstrated as a laser at 41.8 nm [7]. Fig. 4(b) shows the ratio of ionization potentials for adjacent sequences for Ni-like, Pd-like, Nd-like, and Pt-like versus Z. Keep in mind that the ratio for Ne-like to Na-like is typically 2 to 3 across all ions. Ni-like has a ratio of 1.6 to 1.8 for the ions shown while Nd-like is near 1, which means no ionization jump. For Pd-like and Pt-like there is a limited range that looks promising. Pd-like has a local maximum near Z=58 so we choose to look at Pd-like Nd (Z=60) because of its shorter wavelength. For Pt-like there is a maximum ratio at Z=91 but we decided to model Pt-like U (Z=92) because it is useable material and shorter wavelength.

To estimate what lines might lase and at what densities we created atomic models

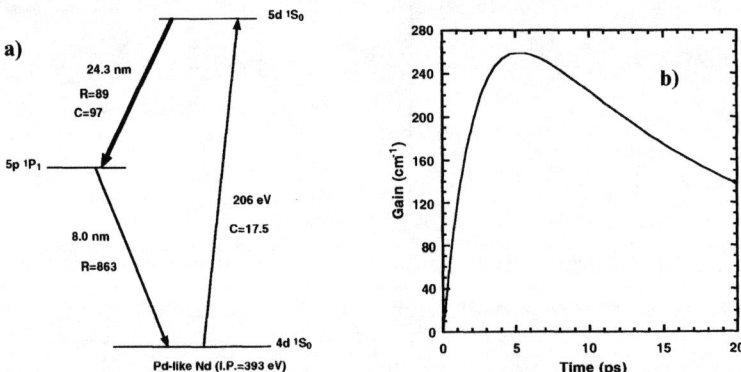

FIGURE 5. Energy level diagram (a) showing the lasing line and kinetic rates for Pd-like Nd laser and the plot (b) of gain vs time for the 24.3 nm laser line.

for Pd-like Nd and Pt-like U that consisted of only one sequence. An electron temperature near the excitation energy for the monopole excitation was chosen. Putting all the population in the ground state CRETIN was run with different densities to estimate what gain would be present on what lines.

Fig. 5(a) has an energy level diagram for Pd-like Nd showing the 5d $^1S_0 \rightarrow$ 5p $^1P_1$ transition that has predicted gain near 24.3 nm. The kinetic rates (R=radiative, C=collisional in nsec^{-1}) for an ion density of 10^{18} cm^{-3}, an electron temperature of 250 eV, and an ion temperature of 60 eV are also shown. Since this is the same transition that lases in Pd-like Xe at 41.8 nm we did a small correction to the calculated wavelength based on the actual wavelength for Xe. One should note that this transition is analogous to the Ni-like except it is for n=5 instead of n=4. CRETIN estimates a peak gain of 260 cm^{-1} that persists for 10 psec as shown in Fig. 5(b). CRETIN also predicts other lasing lines but this is the dominant lasing line in this wavelength range. CRETIN does calculate other strong gain lines but they are near 40 nm.

Creating a similar atomic model for Pt-like U, Fig. 6(a) shows the potential lasing lines and kinetic rates for an ion density of 10^{18} cm^{-3}, an electron temperature of 150 eV, and an ion temperature of 60 eV. There are two promising 6d – 6p laser lines at 25.8 and 22.5 nm. The wavelengths are calculated and then an estimated correction has been applied so the actual wavelengths could be off by as much as 1 nm. Using CRETIN to model the kinetics Fig. 6(b) shows plots of the gain for the 2 lines at the above conditions. The gain is largest on the 22.5 nm line and peaks near 200 cm^{-1} while the second line has about half the gain. Some preliminary experiments were tried with U on COMET but no sign of lasing was observed yet.

For Nd-like U we also estimated that the 5f $^1S_0 \rightarrow$ 5d $^1P_1$ line could have strong gain near 6.7 nm if one could achieve the correct ionization balance.

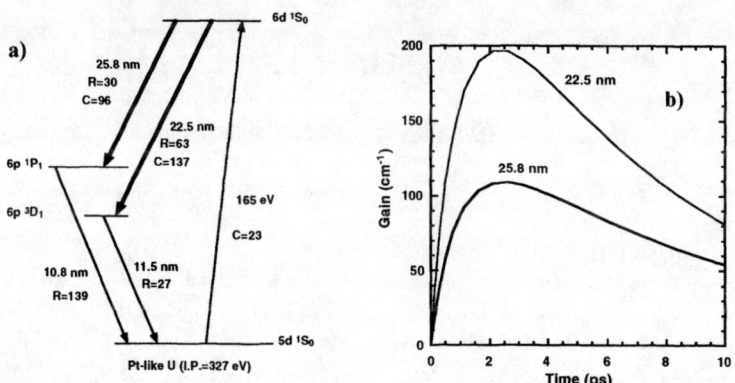

FIGURE 6. Energy level diagram (a) showing the lasing lines and kinetic rates for Pt-like U laser and the plot (b) of gain vs time for the 22.5 and 25.8 nm laser lines.

CONCLUSIONS

This work modeled recent experiments done using the COMET laser at LLNL. We used a 2 J, 600 ps prepulse followed 700 psec later by a 6 J, 6 psec drive pulse to illuminate a 1.25 cm long Pd slab target and produce a saturated 14.7 nm Ni-like Pd x-ray laser. The LASNEX code was used to calculate the hydrodynamic evolution of the plasma and provide the temperatures and densities to the CRETIN code, which then determined the gain from the atomic kinetics calculations including radiation transport effects. Ray-tracing calculations were done to estimate the near-field and far-field images that would be observed in this plasma. The measured two-dimensional near-field and far-field images were presented for the Ni-like Pd laser and these were compared with the simulations. The 1-D comparison was quite good in the plasma expansion direction.

We also presented modeling of several new laser schemes. The Pd-like Nd scheme has a promising 5d - 5p laser line near 24.3 nm and 5p - 5s lines near 40 nm. Modeling predicts potential 6d – 6p laser lines near 22.5 and 25.8 nm for Pt-like U. A Nd-like U laser is also possible near 6.7 nm.

ACKNOWLEDGMENTS

The authors would like to thank Richard A. Ward, Charles P. Verdon, and Albert L. Osterheld for their support. Work performed under the auspices of the US Department of Energy by the University of California Lawrence Livermore National Laboratory under Contract No. W-7405-Eng-48.

REFERENCES

1. J. Nilsen, B. J. MacGowan, L. B. Da Silva, and J. C. Moreno, Phys. Rev. A **48**, 4682 - 4685 (1993).
2. G. B. Zimmerman and W. L. Kruer, Comments Plasma Physics for Controlled Thermonuclear Fusion **2**, 51 – 61 (1975).
3. H.A. Scott, JQSRT **71**, 689-701 (2001).
4. J. Nilsen, J. Dunn, A. L. Osterheld, and Y. L. Li, Phys. Rev. A **60**, R2677 – 2680 (1999).
5. J. Kuba, A. Klisnick, D. Ros, P. Fourcade, G. Jamelot, J. L. Miquel, N. Blanchot, J. F. Wyart, Phys. Rev. A. **62**, 043808 (2000)
6. J. Dunn, Y. Li, A. L. Osterheld, J. Nilsen, J. R. Hunter, and V. N. Shlyaptsev, Phys. Rev. Lett. **84**, 4834 – 4837 (2000).
7. B. E. Lemoff, G. Y. Yin, C. L. Gordon III, C. P. J. Barty, and S. E. Harris, Phys. Rev. Lett. **74**, 1574 – 1577 (1995).

Saturation and Refraction in Amplified Spontaneous Emission X-Ray Lasers

G J Pert

Department of Physics, University of York, Heslington, York YO10 5DD, U.K.

Abstract. Recent developments in X-ray design have been based on plasmas pumped by very short (psec) pulses. In these systems the gain is both short lived and strong. Efficient operation requires that the pump pulse be applied in a travelling wave mode whereby the gain and output pulses are in synchronism. Under these conditions large overall gain lengths are readily achieved, and the output is rapidly saturated.

Saturation by reducing the local emission rate leads to non-linear coupling if more than beam is active. If the gain pulse has a short lifetime, the backward travelling pulse is weak and has a negligible effect. However as the gain lifetime increases, the backward wave becomes stronger, with a larger gain-length and may eventually saturate. In this case the backward saturation will reduce the gain seen by the forward wave, and the forward wave output will be reduced. In principal this may lead to oscillations in the output signal if the gain duration is sufficiently long. In the limiting case when the gain duration is very long, the output relaxes to the familiar steady state solution representing symmetric bi-directional output.

Refraction is endemic in collisionally pumped X-ray lasers, whereby the X-ray beam is deflected out of the gain region. Experimental systems use a pre-pulse generated plasma to provide a low density gradient plasma in which the laser may develop without suffering severe refraction. The refraction within the gain zone therefore defines a characteristic length, the refraction length over which amplification can occur. If the system is saturated over the refraction length, the resultant reduction in spontaneous emission at the head of the first refraction zone will reduce the input into the second and diminish its output. As a result a plot of output as a function of length may show characteristic overshoot/undershoot behaviour.

INTRODUCTION

X-ray lasers using collision pumping of Ne- and Ni-like ions in plasma generated from solid targets have proved remarkably effective in generating high energy outputs in reasonably clean beams. In recent years saturation has been routinely obtained with compartively short plasma lengths due to the relatively high gain achievable. This situation has been further eased by the introduction of so called 'transient' pumping schemes, whereby the plasma is rapidly heated by a short (CPA) pulse in a time much less than its natural hydrodynamic cooling time. The result is a high electron temperature, which strongly pumps the desired inversion levels resulting in high gains. It should be noted that this laser system has two attractive features, namely the gain is not too high ($\sim 100 cm^{-1}$) and the saturation irradiance large ($\sim 10^{10} W/cm^{-2}$). As a result, since the lasing medium takes the form of a large aspect ratio cylinder, the output can be directed into a collimated beam at high irradiance. Since the plasma creation and pumping processes can be separated, these devices may be constructed requiring relatively little input

CP641, *X-Ray Lasers 2002: 8th International Conference on X-Ray Lasers,* edited by J. J. Rocca et al.
© 2002 American Institute of Physics 0-7354-0096-2/02/$19.00

energy.

Since the X-ray medium is short lived, and X-ray mirrors relatively inefficient, these devices all operate in the 'amplified spontaneous emission mode' (ASE), whereby spontaneous emission from one end of the cylinder is amplified by stimulated emission as it travels along the laser. If the laser takes the form of a well defined cylinder of high aspect ratio, only those rays lying within a well-defined cone are strongly amplified; the remainder are lost before achieving significant strength. Since the Fresnel number of experimental X-ray lasers is generally large, the output cone is essentially defined by the geometrical acceptance solid angle of the lasing cylinder. This allows a simple picture of the idealised ASE system to be developed [1, 2], taking into account the effects of saturation and line narrowing. The model predicts differing behaviour for homogeneously and inhomogeneously broadened lines. Surprisingly experiment shows that although the output is dominantly Doppler broadened, the lines are not re-broadened in saturation. This effect is believed to be due to collisional relaxation of the ions, which induces behaviour characteristic of homogeneous broadening ([3]).

In 'transient' lasers, the gain has a duration comparable to or less than the X-ray transit time. The gain may therefore be varying during the generation of the output pulse. To overcome this problem,the travelling wave pumped configuration is used, in which the X-ray and gain pulses travel in synchronism along the axis of the laser. For homogeneous broadening the model is easily adapted to take account of time variations along the laser axis.

A major limitation of collisionally pumped X-ray lasers using a transversely irradiated plasma target is refraction due to the large transverse density gradients. To mitigate this effect, the plasma is created using a long pulse, low intensity beam with a long density gradient, preferentially in the required ionisation stage. A number of arrangements of the various possible pulse configurations have been successfully used. In each case the object is to allow the X-ray beam to traverse the full plasma length before exiting the gain region.

We consider the ASE laser to be represented by a uniform cylinder of cross section A and length ℓ, which emits laser radiation by spontaneous emission at a rate Σ per unit volume with a homogeneously broadened emission line profile, $f(\nu)$. As discussed earlier, we consider only radiation within a limiting solid angle Ω to be amplified. The spontaneous emission rate into the output beam is therefore $E = \Sigma \, \Omega/4\pi$ per unit volume. Radiation is amplified by the medium with gain $G(\nu)$ per unit length with the same line profile, $f(\nu)$, as spontaneous emission.

Both spontaneous emission and gain are reduced by the depletion of the upper state due to stimulated emission, when the beam irradiance is large. Within the random phase approximation, which is valid in experimental situations, we may express these reduction in terms of the saturation irradiation, I_s, such that for homogeneous broadening the general term takes the form

$$A(\nu) = \frac{A_0(\nu)}{(1 + \bar{I}/I_s)} \tag{1}$$

where the frequency integrated weighted line irradiance $\bar{I} = \int f(\nu)I(\nu)\,\mathrm{d}\nu$. In all cases of practical interest [2] the saturation irradiance may be assumed to be the same for both spontaneous emission and gain.

Time dependence is introduced into the model by allowing the spontaneous emission rate and gain to vary with a characteristic time τ, and propagate as a wave along the axis, z, with speed v, and temporal profile $A(t) = A_0 \psi[(t - z/v)/\tau]$ [4, 5]. The X-ray beam travels along the axis with speed c, which may not be equal to v. If we make the reasonable assumption that the level population rates are fast compared to the time scale for gain, i.e. the recovery time is short compared to the gain lifetime, we may assume that the saturation irradiance is quasi-steady, and approximately constant. In addition the spontaneous emission rate and gain have a similar temporal dependence, and therefore that the spontaneous emission irradiance $I_0 = E_0/G_0 = E(\nu)/[G(\nu)f(0)]$ is constant. The line profile is also approximately constant over the line.

TIME DEPENDENT UNI-DIRECTIONAL THEORY

In the travelling wave mode when the gain wave has a duration, short compared to the beam transit time, the laser beam will only amplify strongly in the direction of travel. The amplification is therefore only in the forward direction, and the backward beam is weak. We may therefore neglect any saturation effect of the backward beam on the forward, i.e. consider uni-directional propagation. In this case the irradiance of the exiting beam from the laser is given by the amplification function appropriate to the line profile function, $\alpha(X)$ ([2], namely $I(x) = I_0 \alpha(X)$, where the line centre gain-length product, $X = \int_0^\ell G(0)dz$, is given in terms of the small-signal line-centre gain-length, $x = G_0(0)\ell$, by the implicit relation

$$(1 - I_0/I_s) X \quad + \quad I_0/I_s \, \alpha(X) \quad = \quad x \tag{2}$$

If the small signal gain has the temporal profile $\psi(s)$ ($s = T/tau$), then integrating the gain along the characteristic of the X-ray beam yields the gain-length of the ray exiting at a delay T as

$$x(T) = G_0 \tau/\gamma \left[\Psi(T/\tau) - \begin{cases} \Psi[(T - \gamma\ell)/\tau] & \text{if } \ell < T/\gamma \\ 0 & \text{otherwise} \end{cases} \right] \tag{3}$$

where $\Psi(s) = \int_0^s \psi(s) \, ds$ and $\gamma = |\, 1/c - 1/v \,|$. Equations 2 and 3 are readily solved for known profile $\psi(s)$, gain-length $x = G_0\ell$ and mismatch ratio $\gamma\ell/\tau$. We note two characteristic regimes: the matched case $\gamma = 0$, and the strongly mis-matched $\gamma\ell/\tau \gg 1$.

MATCHED CASE The behaviour in this case with an exponentially decaying profile was discussed in some detail by [4]. We will therefore only note here some characteristic behaviour, and record some additional results.

The output pulse profile shows a dramatic effect. Whilst the output is unsaturated, the pulse width decreases with increasing gain-length due to stronger amplification at pulse maximum. In this regime the scaling of the reduction in pulse width depends on the nature of the pulse. For a pulse with a cusp at the peak (as in [4]) the pulse length

281

decreases as $\sim 1/G_0\ell$, whereas if the pulse is smooth at the peak, the output pulse near the peak, s_{max}, is approximately

$$\phi(s) \approx \exp\left(-\frac{1}{2}G_0 \mid \ddot{\psi}(s_{max}) \mid s^2\right) \tag{4}$$

Hence the pulse length scales as $1/\sqrt{G_0\ell}$. For a general pulse, a good guide to the change in behaviour occurs when the output pulse width approximately matches the point of inflexion in pulse profile.

Once the signal saturates, the pulse re-broadens as saturation at pulse maximum reduces the gain without a similar effect on the wings. In the limit when $I_{out} \gg I_s$, the output profile broadens to the gain $\phi(s) \approx \psi(s)$.

In a similar fashion the line width of the output beam is also reduced. If $G_0\ell \gg 1$, the output line profile is Gaussian with 1/e-half-width

$$\Delta v_{\frac{1}{2}} \approx \Delta v_{\frac{1}{2}}^0 \left(\frac{1}{2} G_0\ell \frac{\ddot{f}(0)}{f(0)}\right)^{-\frac{1}{2}} \tag{5}$$

Thus the line width decreases as $1/\sqrt{G_0\ell}$.

As both the line width and pulse length decrease with increasing gain-length up to saturation, it is possible for the signal to become band-width limited if $\Delta v_{1/2} T_{1/2} \approx 1$. For example for a 100Å laser with $\Delta\lambda/\lambda \sim 10^{-4}$ and $\tau \sim 5\,\text{psec.}$, $\tau\Delta v_{1/2}^0 \sim 15$, and the laser is band-width limited near saturation when $G_0\ell \sim 15$. In this case the signal is averaged in both frequency and time, i.e. over the line and pulse widths, and no further reduction in either occurs.

MISMATCHED CASE In the mis-matched case when the pulse has a finite duration τ, which is short compared to the mis-match time, $\gamma\ell$ separate sections of length τ/γ essentially overlap to produce the final of duration $\gamma\ell$. The small signal gain length as a function of time in this case is

$$x(T) = \begin{cases} \Psi(T/\tau) & T < \tau \\ \Psi_0 = \int_0^1 \psi(s)\,ds & \tau < T < \gamma\ell \\ \Psi_0 - \Psi[1 - (T - \gamma\ell)/\tau] & T > \gamma\ell \end{cases} \tag{6}$$

As a result the total output fluence

$$E(\ell) = E(\tau/\gamma) + I(\tau/\gamma)(\gamma\ell - \tau) \tag{7}$$

where $E(L)$ and $I(L)$ are the fluence and irradiance from a length L of the laser. If the output is strongly saturated $I(\tau/\gamma) \gg I_s$ then

$$E(\ell) \approx \Psi_0 I_s G_0\ell \tag{8}$$

and is independent of the mis-match parameter $\gamma\ell/\tau$.

UNI-DIRECTIONAL/BI-DIRECTIONAL TRANSITION

In the foregoing analysis we have assumed that any backgoing wave is weak. However if the gain pulse duration is long compared to the transit time, the backgoing wave will build up to a strength comparable to the forward. In consequence the initial phases of spontaneous emission for the latter will be come saturated. If the gain pulse is very long the system will settle into the standard steady-state bi-directional configuration, in which the backward and forward beams are equal and each saturates the initial stages of the other [1, 2]. Of particular interest is the transition from an initially uni-directional to bi-directional behaviour. This will occur if a travelling wave pulse of long gain duration propagates down the laser. The laser will initially establish itself in the forward uni-directional state. However the backward beam will gradually build up, eventually 'eating' into the forward as it reaches saturation at its exit and 'switching off' the forward spontaneous emission.

We have examined this behaviour in some detail, and the results have been reported elsewhere [5]. It is found that the decay of the forward beam and the growth of the backward are exponential, with a degree of overshoot, which is rapidly damped. The exponential decay rate is found to depend on the overall value of the gain-length product, the decay time increasing with increasing $G_0 \ell$

REFRACTION LIMITED LASER

Collisionally pumped lasers operate in a strong transverse refractive index gradient. Although considerable effort is made to mitigate the effects of refraction, it still remains perhaps the most critical parameter determining the overall performance of the laser. In its simplest role, refraction limits the output of the laser by deflecting rays from the spatially limited zone of gain to regions of the plasma where they can be no longer amplified. This defines a mode of *refraction limited operation* where rays are progressively lost from the amplifying pencil as they travel along the gain line up to a limiting distance - the refraction length. In this case the output will be nearly independent of length. If the laser is saturated over the refraction length, the laser may exhibit non-linear behaviour.

Refraction plays more subtle roles in the control of the optimum timing between the pre- and main pulses. For efficient operation it is essential that propagation through the plasma be allowed for rays in the zone of maximum gain. If the delay is too short, the gain occurs in a region of large refractive index gradient, and if too long it is satisfactorily located, but of low value [6].

We consider the progressive loss of rays from the lasing channel. Thus if we consider all the rays starting from a given cross section, we may imagine that only a fraction T(x) will reach the cross section $x = G_0 z$ within the gain zone, i.e. T(x) is the refraction loss function. In a real situation T(x) will be a complicated function of x, we have therefore consider a number of representative forms for T(x), which all show the same generic behaviour. Clearly T(x) must obey the general condition that

$$T(x) = \begin{cases} 1 & 0 \le x \le d \\ 0 & x > d \end{cases} \tag{9}$$

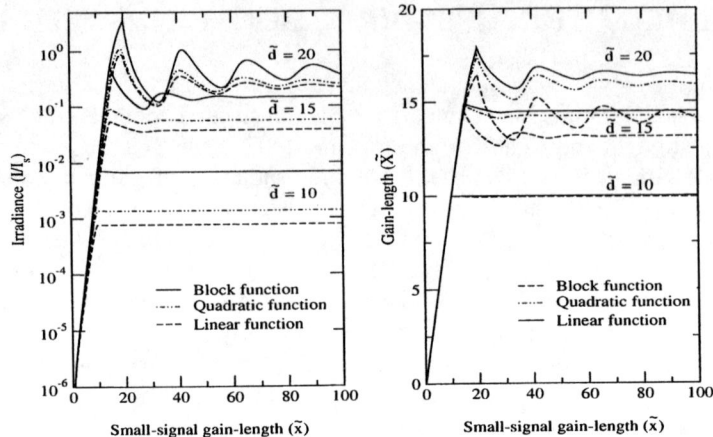

FIGURE 1. The variation of the output irradiance (I) from a refracting ASE laser of total small-signal gain-length ($\tilde{x}$) for various values of the refraction gain-length ($\tilde{x}_d$) encompassing unsaturated, saturation threshold and strongly saturated behaviour calculated using different forms of the saturation function. The output is plotted as a fraction of the saturation irradiance (I_s) for a Döppler broadened line with $I_0/I_s = 10^{-6}$.

where d is the refraction length (measured in gain-lengths), i.e. the distance over which all rays exit the gain-zone. The representative forms are

Block	$T(x) = 1$
Linear	$T(x) = 1 - x/d$
Quadratic	$T(x) = 1 - (x/d)^2$

If the laser is un-saturated over the refraction gain-length, $d < x_s \approx 15$, refraction limits the output to independent blocks corresponding to the length d. However if the laser is saturated over the length d saturation plays a important role, and the output is no longer constant along the laser. Over the first refraction length, d, the laser will saturate; the spontaneous emission providing the input into the second length d will therefore be reduced and the output of this length reduced. As a consequence the input to the third length is increased, and the average irradiance along the axis will oscillate, but with a decreasing amplitude, eventually reaching a quiescent state.

The model is readily adapted to consider this effect, the result being expressed as a complex integral equation, which can be readily evaluated numerically. Figure 1 demonstrates this behaviour in a steady-state ini-directional laser as the small-signal gain-length is increased for differing refraction gain-lengths. It can be clearly seen that if $d < x_s$, the irradiance is constant and saturation plays no role. However once the refraction length becomes comparable oscillation is observed, the amplitude increasing as the refraction gain-length increases.

Turning now to travelling wave systems, we expect that in the matched case the behaviour will be similar to the stady-state case described above, essentially averaging the steady state solutions for time varying gain. As a result the pulse length of the output will also oscillate with gain-length as re-broadening in the saturated regime is either

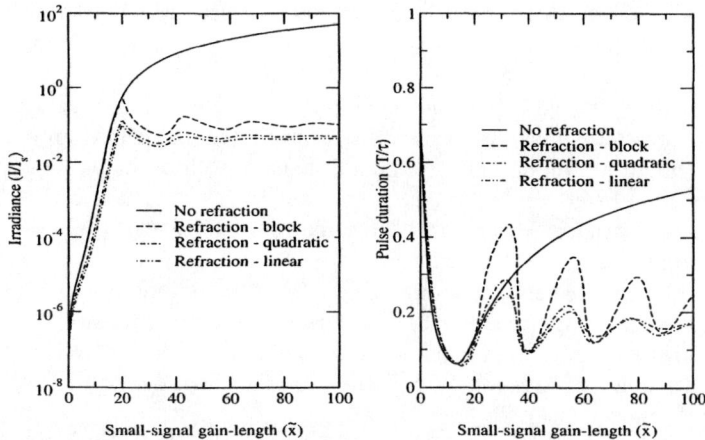

FIGURE 2. The variation of the output irradiance (I) and the pulse duration from a matched travelling wave ASE laser of total small-signal gain-length ($\bar{x}$) refraction gain-length ($\bar{x}_d$) = 20 calculated using different forms of the saturation function. The output is plotted as a fraction of the saturation irradiance (I_s) for a Döppler broadened line with $I_0/I_s = 10^{-6}$.

strong or weak depending on the level of saturation affecting a particular signal. Figure 2 confirms this is indeed the case. The results may also be compared with the output from the same system without refraction.

The un-matched case is more complex with the nature of the average over the gain profile depending on whether the gain wave leads or lags the X-ray wave, i.e. on the sign of the factor $\gamma = 1/v - 1/c$, in contrast to the non-refractive case. It is not proposed to discuss this condition further here.

CONCLUDING REMARKS

Simple models are useful for investigating the often complex behaviour of systems involving time variation, saturation and refraction. Saturation by acting as a non-linear coupling between different elements in a multiple beam configuration can induce complex and unexpected behaviour. This may involve overshoot and oscillation with subsequent decay to a quiescent state. The decay rates can be surprisingly slow.

Travelling wave systems are essentially uni-directional if the gain pulse duration is not too long. In this case saturation affects only the locally instanteous emission through the gain and (very improbably) spontaneous emission. As a result the main effect is an enhancement of those regions where the irradiance is still below the saturation threshold in comparison with those higher. Thus we find an effective increase in the extent of the beam not only temporally (pulse broadening) but also spatially in the cross section [4]. ASE lasers using inhomogeneously broadened lines, when the saturation is local within the line profile, similarly show re-broadened lines on saturation [2]. In contrast with homogeneously broadened lines, the reduction is gain is distributed over the full profile and the line width consequently continues to narrow into saturation. In this paper we have considered only homogeneously broadened systems in conformity with experimental observation.

The model can be extended to include the time development of bi-directional systems. These calculations show that the transiton from uni-directional behaviour is more complicated than might be initially expected, and involves both oscillation and overshoot. These effects arise from non-linear coupling of the backward beam with the forward. Similar behaviour resulting from the same effect is found when significant refractive loss occurs within the lasing channel.

Refraction plays more subtle roles in the control of the optimum timing between the pre- and main pulses. For efficient operation it is essential that propagation through the plasma be allowed for rays in the zone of maximum gain. If the delay is too short, the gain occurs in a region of large refractive index gradient, and if too long it is satisfactorily located, but of low value [6].

Comparison of the model against experiment and detailed simulation reveal the strengths and weaknesses of the model [7]. Its strength is its flexibilty. Thus it is possible very simply treat a wide range of differing conditions of gain, saturation, travelling wave parameters etc. and rapidly determine estimates for the expected behaviour. Its weakness is that it cannot take into account the full range of detailed effects involving spatial variation, refraction etc. and therefore the results are not an accurate representation of the working laser. None-the-less given a set of experimental data, the model can be used to derive approximate values of the characteristic parameters such as gain and saturation irradiance, which will characterise the general performance of the device.

ACKNOWLEDGEMENTS

This work was supported by EPSRC,AWE and the EU as part of the UK X-ray laser consortium.

REFERENCES

1. Casperson,L.W., *J Appl Phys* **48**, 256-262 (1977).
2. Pert,G.J., *J Opt.Soc.Am. B* **11**, 1425-1435 (1994).
3. Koch,J.A., MacGowan,B.J., DaSilva,L.B., Matthews,D.L., Underwood,J.H., Batson,P.J., Lee,R.W., London,R.A., and Mrowka,S., *Phys Rev A* **50**, 1877-1897 (1994).
4. King,R.E. and Pert,G.J., *C.R. Acad. Sci. Paris* **Serie IV 1**, 1093-1104 (2000).
5. Pert,G.J., *Optics Comm.* **191**, 113-123 (2001).
6. King R.E. 'Computational Modelling of the Nickel-like Samarium X-ray Laser' in *X-ray lasers 2002* in press
7. King,R.E.,Pert,G.J.,McCabe,S.P.,Simms,P.A.,MacPhee,A.G.,Lewis,C.L.S.,Keenen,R., O'Rourke,R.M.N.,Tallents,G.J.,Pestehe,S.J.,Strati,F.,Neely,D.,and Allott,R., *Phys Rev A* **64** 053810-1 - 053810-12 (2001).

Radiation-transfer modelling of double-pass saturated ASE systems

B. Rus and T. Mocek

Gas Lasers Department / PALS Centre, Institute of Physics, Academy of Sciences of the Czech Republic, Na Slovance 2, 18221 Prague 8, Czech Republic

Abstract. We numerically model radiation transfer in saturated ASE systems working in a double-pass regime. The goal is to obtain intensity and temporal characteristics of the output radiation emitted by a deeply saturated X-ray laser using a half-cavity. The model treats time-dependent, bi-directional amplification of axial emission in a refraction-free active medium, and accounts in full for radiation transfer of the spectral line profile. The atomic kinetics is modelled using a simplified three-level system with rates relevant to neon-like zinc, which is used as a testbed.. The counter-propagating emissions produce axially dependent depletion of the population inversion, generating cross-talk between the intensity in one direction and the emissivity in the opposite direction. As a consequence, the output signal exhibits a presence of strongly damped oscillations, the amplitude of which depends on a number of parameters.

PRINCIPLE OF THE NUMERICAL MODEL

Experimental realization of X-ray lasers characteristic by a large gain-length product requires an appropriate modelling in which saturated ASE is generated in both directions along the longitudinal axis [1]. In the context of a deeply saturated Ne-like Zn XRL [2], using ~450-ps pumping and working in the double-pass regime, we have developed a simple numerical code making it possible to model bi-directional amplification in both single- and double-pass systems. The model treats radiation transfer of a homogeneously-broadened lasing line in a refraction-free medium characterized by a small-signal emissivity j_0, gain g_0, and saturation intensity I_{sat}. The two intensities propagating in the counter axial directions follow a tandem of equations (cf. Fig.1):

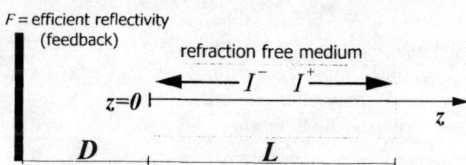

FIGURE 1. Geometry of the numerical model.

$$\frac{\partial I^\pm(z,t,\nu)}{c\partial t} \pm \frac{\partial I^\pm(z,t,\nu)}{\partial z} = \frac{j_0(t,\nu)}{1+\dfrac{I_{tot}(z,t)}{I_{sat}(t)}} + \frac{g_0(t,\nu)I^\pm(z,t,\nu)}{1+\dfrac{I_{tot}(z,t)}{I_{sat}(t)}} \qquad (1),$$

CP641, *X-Ray Lasers 2002: 8th International Conference on X-Ray Lasers*, edited by J. J. Rocca et al.
© 2002 American Institute of Physics 0-7354-0096-2/02/$19.00

where $I_{tot}(z,t) = \int\limits_{-\infty}^{+\infty} \left[I^{+}(z,t,v') + I^{-}(z,t,v') \right] dv'$.

The feedback F, introduced by the half-cavity mirror, enters to Eq.(1) as a boundary condition $I^{+}(z{=}0, t, v) = F\, I^{-}(z{=}0, t{-}2D/c, v)$.

SYSTEMS WITH STATIC POPULATION INVERSION

The differences of behavior between a unidirectional ASE system (described in below saturation regime by the Linford formula) and a bi-directional system may be illustrated for time-independent j_0, g_0, and I_{sat}. In relevance to the experimental conditions of the investigated Ne-like Zn [2], the simulations were carried out with $j_0{=}1.5{\times}10^4\,\mathrm{Wcm^{-3}}$ and $I_{sat}{=}1.5{\times}10^{10}\,\mathrm{Wcm^{-2}}$. The intrinsic spectral profile was gaussian with $\Delta\,v_{\mathrm{FWHM}}{=}50\,\text{Å}$.

Single-pass ASE amplifier

The simulations were run for two values of the small-signal gain: $g_0{=}7\,\mathrm{cm^{-1}}$ (the gain found experimentally), and $g_0{=}20\,\mathrm{cm^{-1}}$ (to examine behavior of an analogous ASE system which exhibits a very large small-signal gain-length product). The results are displayed in Figures 2 and 3.

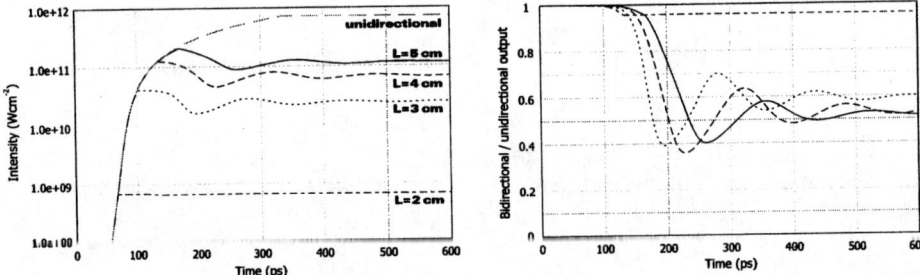

FIGURE 2. Comparison of the bi-directional and the unidirectional output of ASE systems for $g_0{=}7\,\mathrm{cm^{-1}}$. The individual curves correspond to bi-directional amplification for L=2, 3, 4, and 5 cm, respectively, and to a unidirectional ASE. The right graph shows the evolution of the ratio of the corresponding bi-directional output with respect to the single-pass ASE output.

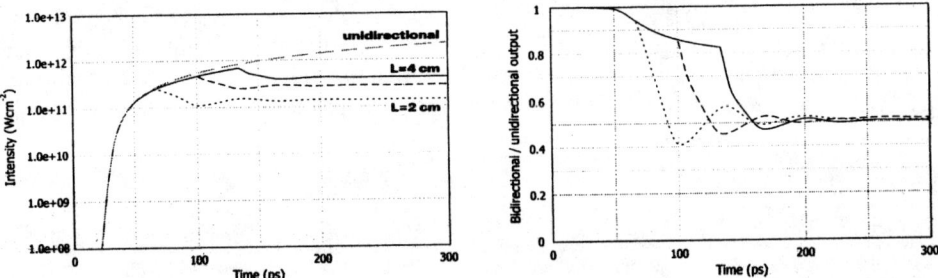

FIGURE 3. Comparison of the bi-directional and the unidirectional output of ASE systems for $g_0{=}20\,\mathrm{cm^{-1}}$, displayed in a manner analogous to Fig.2. The curves correspond to L=2, 3, and 4 cm.

Double-pass amplifier

Figures 4 and 5 show the output of the same systems as those modelled in Figs. 2 and 3, but working in the double-pass regime with a 100%-reflecting half-cavity mirror. The increasing mirror distance D introduces an increasing complexity into the damped oscillations of the output and increases the duration of the initial overshoot pulse. As expected, the half-cavity output for a given length L converges to a value equal to the output of a single-pass (bi-directional) system with a length $2L$ (cf. Figs 2 and 3).

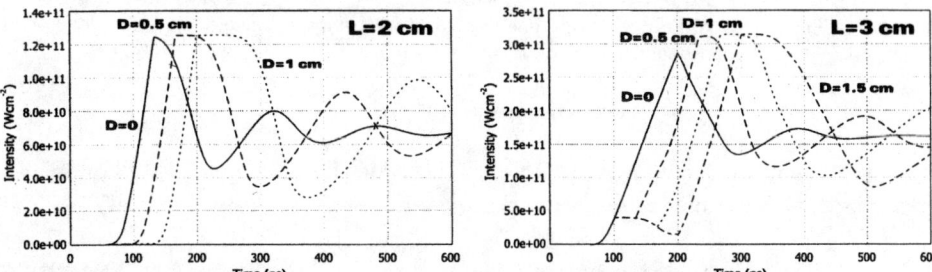

FIGURE 4. Output of a double-pass amplifier with $g_0=7$ cm^{-1} and an ideal mirror providing a feedback $F=1$. The left graph corresponds to the plasma length L=2 cm, with the distance plasma-mirror D=0, 0.5, and 1 cm; the right graph is for L=3 cm and D=0, 0.5, 1, and 1.5 cm.

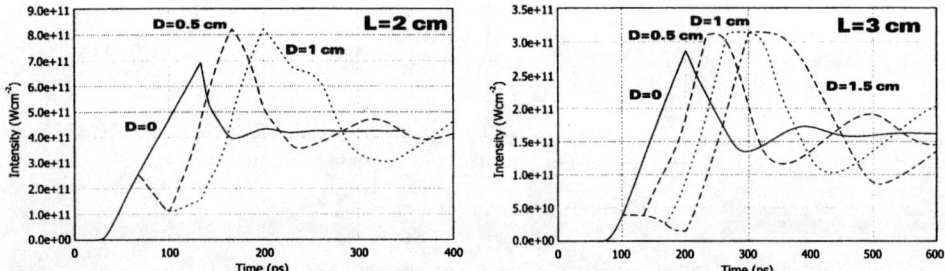

FIGURE 5. Output of a double-pass amplifier with $g_0=20$ cm^{-1} and an ideal mirror providing a feedback $F=1$. In analogy to Fig.4, the left graph corresponds to L=2 cm and D=0, 0.5, and 1 cm; the right graph corresponds to L=3 cm and D=0, 0.5, 1, and 1.5 cm.

SYSTEMS WITH DYNAMIC POPULATION INVERSION

Regarding the assumed time evolution of j_0 and g_0 in the zinc XRL of interest [3], the systems were modelled with a 50-ps flat-top j_0 linearly rising up to 1.5×10^4 Wcm^{-3} in 50 ps, and decreasing exponentially over 200 ps; g_0 included a rise edge of 50 ps to a peak value lasting 150 ps and followed by a linear decrease down to zero over 150 ps.

Single-pass ASE amplifier

Figure 6 shows the output of ASE systems with the peak g_0 value of 7 cm^{-1} and 20 cm^{-1}. For simplicity, I_{sat} is assumed constant in time as above, $I_{sat}=1.5 \times 10^{10}$ Wcm^{-2}.

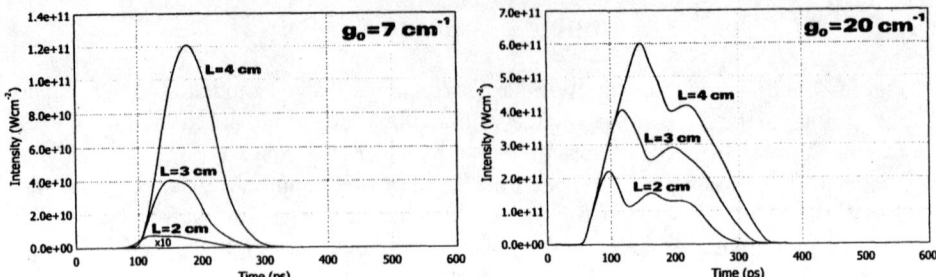

FIGURE 6. Single-pass ASE system with time-dependent emissivity and gain, for L=2, 3, and 4 cm; the left and right graphs correspond to peak values of $g_0=7\,\text{cm}^{-1}$ and $g_0=20\,\text{cm}^{-1}$, respectively.

Double-pass amplifier

Figure 7 shows the output of a half-cavity system with $g_0=7\,\text{cm}^{-1}$, for plasmas of lengths L=2 and 3 cm; the mirror, placed at different distances D, has a 10% reflectivity.

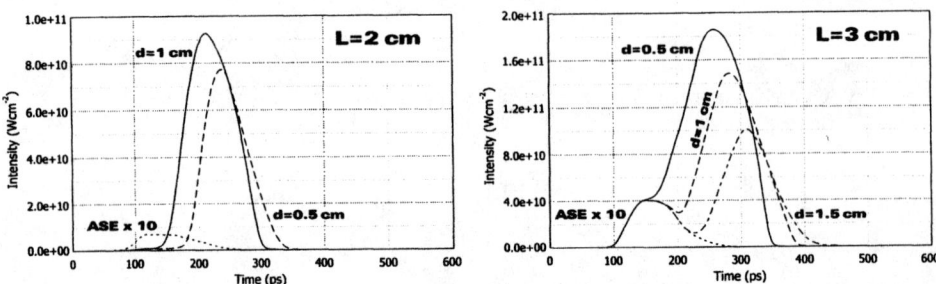

FIGURE 7. Double-pass system with time-dependent emissivity and gain, $g_0=7\,\text{cm}^{-1}$. The left and right graphs corresponds to L=2 and 3 cm; respectively. The feedback due to the mirror is $F=0.1$.

SUMMARY

The results of the modelling for bi-directional ASE single- and double-pass systems demonstrate both the validity of simple analytic estimations, and show as well non-trivial features of amplification in such systems. The model was applied to systems pertinent to multi-100-ps pumping, specifically to Ne-like zinc laser. As expected, the output of a bi-directional saturated ASE system equals to 1/2 of that produced by a unidirectional one (cf. Figs. 2 and 3). The bi-directional amplification introduces to the output pulse complex time features, and has a tendency to lengthen the output pulse.

REFERENCES

1. Pert, G.J., *Opt.Commun.*, **191**, 113-123, 2001.
2. Rus, B., *et al.*, these Proceedings.
3. Rus, B., Carillon, A., Gauthé, B., Goedtkindt, P., Jaeglé, P., Jamelot, G., Klisnick, A., Sureau, A., and Zeitoun, P. *J.Opt.Soc.Am.B* **11**, 564-573, 1994.

SATURATED AND SHORT PULSE DURATION X-RAY LASERS

G J Tallents[1], Y Abou-Ali[1], M Edwards[1], R E King[1], G J Pert[1], S J Pestehe[1], F Strati[1], R Keenan[2], C L S Lewis[2], S Topping[2], O Guilbaud[3], A Klisnick[3], D Ros[3], R Clarke[4], D Neely[4] and M Notley[4]

[1]*Department of Physics, University of York, York YO10 5DD, United Kingdom.*
[2]*School of Mathematics and Physics, Queen's University Belfast, Belfast BT7 1NN, United Kingdom.*
[3]*LIXAM, Université Paris-Sud, 91405 Orsay, France*
[4]*Central Laser Facility, Rutherford Appleton Laboratory, Chilton OX11 0QX, United Kingdom*

Abstract. The basis of a model of the relationship between gain and output laser intensity is reviewed and the measurement of the duration of X-ray lasing with a streak camera with 700 fs temporal resolution is described. Combined with a temporal smearing due to the spectrometer employed, we have measured X-ray laser pulse durations for Ni-like silver at 13.9 nm and Ne-like nickel at 23.1 nm with a total time resolution of 1.1 ps. An extension of the model is shown to consistently relate the measured X-ray laser pulse duration to estimates of the gain duration obtained by temporally resolving resonance line emission from states near in energy to the upper lasing level.

INTRODUCTION

X-ray lasing was first produced using single nanosecond duration pumping pulses. Streak camera measurements on these X-ray lasers showed that the lasing occurred during the leading edge of the pumping pulse with a duration of only a 100 ps or so [1, 2]. Subsequent experiments using two or more pumping pulses of duration ≈ 100 ps produced saturated output with X-ray laser output durations ≈ 30 - 40 ps [3, 4]. Here a pre-pulse was used to produce a pre-formed plasma so that the main pumping pulse interacts with a long scalelength plasma. This reduces refraction of the X-ray laser beam, increases the volume of the gain region and enhances absorption of the main pumping pulse energy [5, 6] so that the X-ray laser output and efficiency of X-ray laser production increases.

A more recent approach has been to use a background pulse of duration 100 – 300 ps on which is superimposed a pumping pulse of duration ≈ 1 ps [7, 8]. The background pulse pre-forms the plasma so that density gradients are shallow and also ensures that the ionisation of the plasma is such that most ions are in the ground state of the desired ion stage when the short pumping pulse arrives. The main pulse rapidly collisionally excites a population inversion with lasing occurring before excitation to the next higher ionisation stage can occur.

CP641, *X-Ray Lasers 2002: 8th International Conference on X-Ray Lasers*, edited by J. J. Rocca et al.

Gain durations with picosecond pumping pulses are short, typically 10 – 40 ps. This can be a problem with long lengths of plasma unless the pumping laser pulse is timed to arrive at the target coincident with the propagating X-ray laser pulse. The X-ray laser pulse usually travels close to the speed of light c in vacuum (33 ps mm^{-1}) unless the gain coefficient is very high and the group velocity is reduced [9]. Consequently, with short duration pumping pulses, travelling wave pumping where the laser pulsefront moves along the target length at c is needed.

Saturation of a laser occurs when the rate of stimulated emission from the upper lasing level is equal to or greater than the rate of de-population by other mechanisms. Saturation usually provides clear evidence that a laser scheme is working. The output and efficiency of laser production is enhanced and shot-to-shot variation of the laser output reduces with saturation. With short pulse pumping, the spectral and temporal bandwidth of the laser output are a minimum at saturation.

Saturated X-ray laser output was first observed with single nanosecond pumping pulses in 1992 [10]. Double pumping pulses of 75 ps duration have produced the record shortest wavelength saturated laser at 5.9 nm from Ni-like dysprosium ions [11]. The shortest wavelength laser producing saturated output with ≈ 1 ps pumping pulses was achieved with Ni-like samarium at 7.3 nm [12]. Pumping laser energies of a few joules have also been shown to be able to pump lasers to saturation. For example, Dunn et al [13] showed that 7 joules of laser energy could produce saturated output in Ni-like silver at 13.9 nm.

In this paper, the theoretical relationship between gain and the X-ray laser output is first reviewed. A measurement of the duration of X-ray lasing with a streak camera with 700 fs temporal resolution is then described. We have measured X-ray laser pulse durations for Ni-like silver and Ne-like nickel with a total time resolution of 1.1 ps. An estimate of the duration of the X-ray laser gain has been obtained by temporally resolving the spectrally integrated resonance line emission from states near in energy to the upper lasing level.

AMPLIFIED SPONTANEOUS EMISSION

X-ray lasers usually have a geometry consisting of a long length L of plasma with a population inversion over a small cross-section of area A. If we assume that the small signal gain coefficient is constant over the area A and along the length L and that refraction and diffraction can be neglected, the X-ray laser propagation then needs only to be treated in the one direction z along the plasma length. Spontaneous emission into a small solid angle Ω only is assumed to be amplified by stimulated emission. For gain durations much longer than the time for photons to travel the length L, there will be amplification of X-rays in the two directions along z

The X-ray irradiance in the forward, positive z direction $I_f(v)$ and in the backward, negative z direction $I_b(v)$ over a small frequency range v to $v+dv$ is given by the two appropriate equations of radiative transfer:

$$\frac{dI_f(\upsilon)}{dz} = G(\upsilon)I_f(\upsilon) + E(\upsilon)$$

$$\frac{dI_b(\upsilon)}{dz} = G(\upsilon)I_b(\upsilon) + E(\upsilon) \tag{1}$$

where $G(\nu)$ is the gain coefficient and $E(\nu)$ is the spontaneous emission rate into the solid angle Ω around the direction of propagation. Writing for the total irradiance $I(\nu) = I_f(\nu) + I_b(\nu)$ at any position z in the plasma, equations (1) can be added together to give

$$\frac{dI(\upsilon)}{dz} = G(\upsilon)I(\upsilon) + 2E(\upsilon). \tag{2}$$

For an homogeneously broadened X-ray laser transition of line profile $f(\nu)$, the gain coefficient $G(\nu)$ can be written in terms of the small signal gain coefficient at line centre g_0 :

$$G(\upsilon) = \frac{g_0}{\left(1 + \dfrac{I_{av}}{I_s}\right)} \frac{f(\upsilon)}{f(0)} \tag{3}$$

where I_s is the saturation irradiance for the X-ray laser line and I_{av} is the X-ray laser irradiance $I(\nu)$ averaged over the line profile $f(\nu)$ such that

$$I_{av} = \frac{\int I(\upsilon)d\upsilon}{\int \dfrac{f(\upsilon)}{f(0)}d\upsilon} = f(0)\int I(\upsilon)d\upsilon. \tag{4}$$

Here the line profile $f(\nu)$ is normalised so that its integral over all frequency is equal to one and frequency ν is measured from line centre. Solving equation (2) for $I(\nu)$, it follows that

$$I_{av} = I_0 f(0)\int \left[\exp(G(\upsilon)L) - 1\right]d\upsilon \tag{5}$$

where $I_0 = E(\nu)/G(\nu)$. Assuming spontaneous emission and gain have the same saturation behaviour and that they have the same line profile shape, the parameter I_0 is constant with varying average X-ray irradiance I_{av} and varying frequency ν. Spontaneous emission and gain will vary identically with local laser intensity and frequency if the population inversion is strongly inverted such that the upper quantum state population is much greater than the lower level population. Re-arranging equation (2), we can write

$$\frac{1 + 2\dfrac{I(\upsilon)}{I_s}}{I_0 + I(\upsilon)} dI(\upsilon) = g_0 dz.\tag{6}$$

Integrating each side from 0 to $I(v)$ and from 0 to L respectively, we have after integrating over all frequency and multiplying throughout by $f(0)$ the expression

$$G(0)L + 2f(0)\iint_0 \frac{I(\upsilon)}{I_s(I_0 + I(\upsilon))} dI(\upsilon)d\upsilon = g_0 L.\tag{7}$$

The second term containing the double integral over intensity and frequency is to a good approximation equal to I_{av}/I_s. We can write

$$G(0)L + 2\frac{I_{av}}{I_s} = g_0 L.\tag{8}$$

Equation (8) relates the actual gain length product $G(0)L$ at line centre to the small signal gain length product g_0L at line centre. It is particularly useful as I_{av} is a function only of the actual gain length product and the parameter I_0 (see equation 5). For Gaussian line profiles $f(v)$, the Linford approximation can be used for evaluating equation 5. With the Linford approximation

$$I_{av} = I_0 \frac{(\exp(G(0)L) - 1)^{3/2}}{(G(0)L\exp(G(0)L))^{1/2}}.\tag{9}$$

Equations (8) and (9) can be used to calculate the variation of X-ray laser output I_{av} as a function of length L for a particular small signal gain coefficient g_0 and ratio $R = I_0/I_s$. A simple spreadsheet such as Microsoft Excel can be used for this calculation. An easy procedure is to calculate values of the small signal gain length product g_0L in a spreadsheet column by varying the actual gain length product $G(0)L$ over some range such as 0 to 20. Assuming a small signal gain coefficient g_0 then gives L as a function of the variation of average X-ray irradiance (calculated using equation 9).

The parameter R is given by

$$R = \frac{I_o}{I_s} = \frac{\Omega}{4\pi} \frac{A_{ul} t_R}{2\pi\left(1 - \dfrac{g_u N_l}{g_l N_u}\right)}\tag{10}$$

where A_{ul} is the transition probability from the upper to lower laser level, t_R is the relaxation time for the upper laser level and N_u, N_l, g_u, g_l are the upper and lower laser level population densities and degeneracies respectively.

The relaxation time for the upper laser level is usually dominated by radiative decay to the lower laser level giving $A_{ul}t_R \approx 1$. For a large population inversion $N_l \ll N_u$, so

$$R \approx \frac{1}{2\pi}\frac{\Omega}{4\pi} \approx \frac{\Omega}{80}. \tag{11}$$

A typical solid angle Ω into which spontaneous emission can be amplified is determined approximately by the X-ray laser output area (say 45 μm × 45 μm) and the target length (say 5 mm). A value $\Omega \sim 8 \times 10^{-5}$ steradian giving $R \sim 10^{-6}$ from equation (11) can be expected.

Adjusting the small signal gain coefficient g_0 and possibly the ratio $R \approx \Omega/80$ enables the above simple model of ASE to be fitted to experimental data of X-ray laser output as a function of length, not just in the small signal regime as is often used with the Linford formula, but into the regime where saturation of the laser output strongly affects the X-ray laser output (see figure 1).

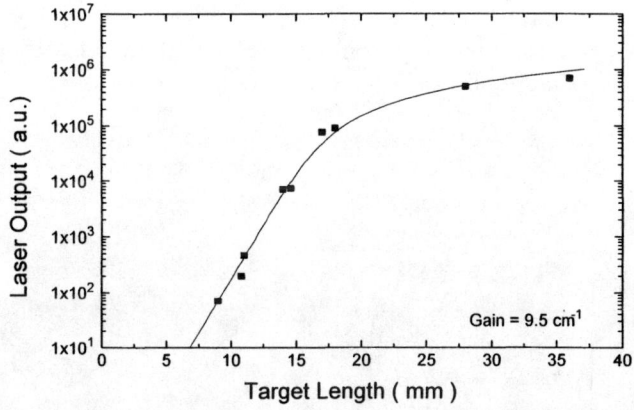

Figure 1. The measured variation with length of the output of Ni-like Sm at 7.3 nm pumped with double 75 ps pulses (data points). The solid curve is a fit of equation (8) (and using equation (9)).

The above model was initially developed by Casperson and Yariv [14] and Pert [15]. It has been extended by Strati and Tallents [9] to include time dependent effects associated with short durations of gain and mismatches between a travelling wave pump and the group velocity of the propagating X-ray laser pulse. Some of the results of the extended model are presented here under RESULTS.

EXPERIMENT

X-ray lasing was produced in Ni-like silver at 13.9 nm ($3p^6 3d^9 4d\ ^1S_0 \rightarrow 3p^6 3d^9 4p\ ^1P_1$) and Ne-like nickel at 23.1 nm ($3s^2 2p^5 3p\ ^1S_0 \rightarrow 3s^2 2p^5 3s\ ^1P_1$) by irradiating solid silver or nickel slabs with two beams of wavelength 1.06 μm from the VULCAN glass laser at the Rutherford Appleton Laboratory. For Ne-like nickel, a pre-plasma was formed with a background pulse of duration 280 ps and peak irradiance 2×10^{13} Wcm^{-2} in a line focus of length 16 mm and width 100 μm. The main pulse irradiance is enhanced by chirp pulse amplification (CPA) to give a duration of 1.2 ps and peak irradiance 7×10^{15} Wcm^{-2} in a line focus of length 12 mm and width 100 μm. So as not to over-ionise the plasmas, the laser energy for Ni-like silver experiments was reduced to give irradiances of approximately 5×10^{12} Wcm^{-2} and 2×10^{15} Wcm^{-2} for the background and main pulse respectively. The background pulse was produced from a component of the uncompressed main pulse and hence arrived on target without random jitter relative to the main pulse. The background pulse and main pulse were focussed onto the targets with separate focussing systems comprising a refracting lens (the background pulse) or a parabolic shaped mirror (the main pulse) to produce a spot focus which was imaged in each case by an off-axis spherical mirror to produce the line focus on the target [16].

An Axis-Photonique streak camera [17] positioned at the detection plane of a flat field spectrometer [18] was used to temporally resolve the X-ray laser output with a temporal resolution of 1.1 ps. A streaked slit diagnostic described in detail in these proceedings [19] was used to obtain an estimate of the gain duration by measuring emission of resonance lines emanating from upper quantum states close to the upper lasing level. The temporal resolution of the streaked slit camera was 6 ps.

RESULTS

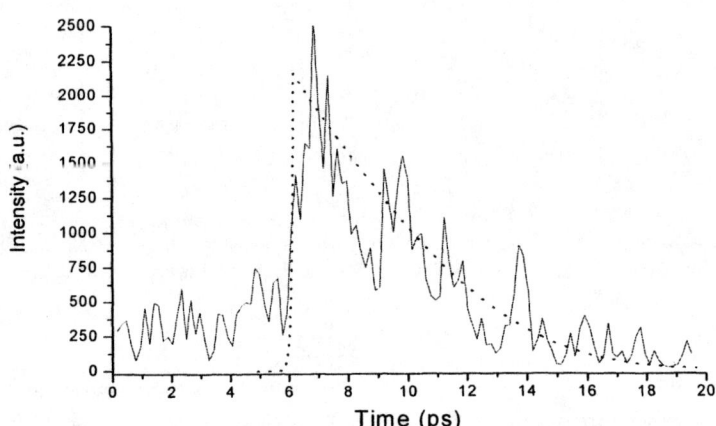

Figure 2. The temporal variation of X-ray lasing at 13.9nm and for Ni-like silver. A 10mm long silver target was irradiated with peak irradiance of 3.1×10^{12} Wcm^{-2} in the 280 ps background pulse and 1.2×10^{15} Wcm^{-2} in the 1.2 ps main pulse with peak-to-peak separation of 200 ps. The output was filtered with 1.2 μm plastic (CH). The FWHM X-ray laser duration was measured to be 3.7 ps. A model fit for $R = 10^{-6}$, peak small signal gain of 23 cm^{-1} and a gain duration of 22 ps is shown as a broken line.

The temporal variation of the Ni-like silver output at 13.9 nm as recorded with the streaked spectrometer using an image intensifier is shown in figures 2. The measured X-ray laser pulse duration (full-width at half-maximum FWHM) for silver is 3.7 ($\pm$ 0.5) ps. A similar measurement for Ne-like nickel at 23.1 nm gives an output duration of 10.7 ps. The temporal variation of X-ray laser output predicted by the model of Strati and Tallents [9] has been fitted to the experimentally measured X-ray laser output (figure 2).

Figure 3 shows the calculated X-ray laser output pulse duration as a function of peak small signal gain coefficient g_o for R parameters over the range $10^{-5} - 10^{-7}$ for our experimental conditions for Ni-like silver calculated using the model of Strati and Tallents [9]. For Ni-like silver, a peak gain coefficient $g_o = 18 - 27$ cm^{-1} will result in lasing of 3.7 ps duration as measured assuming the gain coefficient duration is 22 ps (as measured by the streaked slit camera) depending on the precise value of R. The temporal x-ray laser output predicted by the model [9] agrees well with the measured laser output assuming $R = 10^{-6}$, peak small signal gain of 23 cm^{-1} and a gain duration of 22 ps (see figure 2). Gain coefficients of $18 - 27$ cm^{-1} for Ni-like silver are consistent with the gain coefficients for Ni-like samarium measured by King et al [12] (≈ 19 cm^{-1}) and to the gain coefficient for Ni-like silver measured by Kuba et al [20] (≈ 19 cm^{-1}). Some more detail on the measurement of the above X-ray laser output durations is given in an accompanying paper [21].

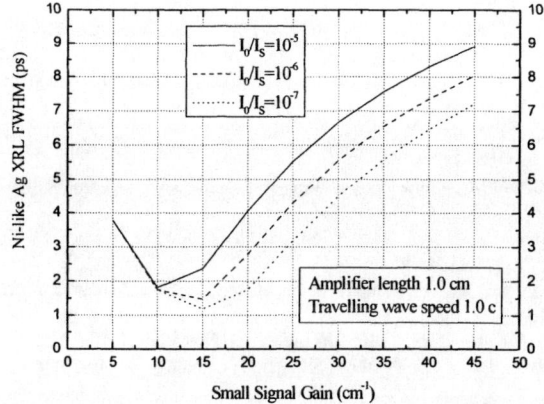

Figure 3. The X-ray laser pulse duration predicted by the model of Strati and Tallents [9] as a function of the peak small signal gain coefficient for different values of R (see text). A target length of 10 mm and gain duration of 22 ps as for Ni-like silver lasing in the experiment is assumed.

CONCLUSIONS

A review of a simple theory of amplified spontaneous emission useful for modelling X-ray laser output as a function of the amplifier length has been presented. An extension of the model has been applied to experimental time-resolved measurements of the duration of Ni-like silver output at 13.9 nm. The model consistently relates the measured X-ray laser pulse duration to estimates of the gain duration obtained by temporally resolving resonance line emission from states near in energy to the upper lasing level.

ACKNOWLEDGEMENTS

This work is financially supported by the U.K. Engineering and Physical Sciences Research Council, by the E.U. Training and Mobility of Researchers programme and by the Laboratoire Européen Associé for RAL/LULI collaboration. The assistance of the laser, target manufacture, engineering and other staff at the Central Laser Facility is gratefully acknowledged.

REFERENCES

1. O'Neill, DM, Lewis, CLS, D Neely, D, Uhomoibhi, J, Key, MH, MacPhee, A, Tallents, GJ, Ramsden, SA, Rogoyski, A, McLean, EA, and Pert, GJ, Optics Commun. **75**, 406 (1990).
2. Kodama, R, Neely, D, Dwivedi, L, Key MH, Krishnan, J, Lewis, CLS, O'Neill, D, Norreys, P, Pert, GJ, Ramsden, SA, Tallents, GJ, Uhomoibhi, J, and Zhang, J, Optics Commun. **90**, 95 (1992).
3. Zhang, J, MacPhee, AG, Lin, JY, Wolfrum, E, Smith, R, Danson, C, Key, MH, Lewis, CLS, Neely, D, Nilsen, J, Pert, GJ, Tallents, GJ, Wark, JS, Science **276**, 1097 (1997).
4. Zhang, J, MacPhee, AG, Nilsen, J, Lin, JY, Barbee, TW, Danson, C, Key, MH, Lewis, CLS, Neely, D, O'Rourke, RMN, Pert, GJ, Smith, R, Tallents, GJ, Wark, JS, and Wolfrum, E, Phys. Rev. Lett. **78**, 3856 (1997).
5. Lin, JY, Tallents, GJ, Demir, A, Healy, SB, and Pert, GJ, J. Appl. Phys. **83**, 1863 (1998).
6. Lin, JY, Tallents, GJ, Smith, R, MacPhee, AG, Wolfrum, W, Zhang, J, Eker, G, Keenan, R, Lewis,CLS, Neely, D, O'Rourke, RMN, Pert, GJ, Pestehe, SJ, and Wark, JS, J. Appl. Phys. **85**, 672 (1999)
7. Dunn, J, A L Osterheld, AL, Shepherd, R, White, WE, Schlaptsev, VN and Steward, RE Phys. Rev. Lett. **80**, 2825 (1998)
8. Lin, JY, Tallents, GJ, MacPhee, AG, Demir,A, Lewis, CLS, O'Rourke, RMN, Pert,GJ, Ros,D, and Zeitoun, P, 1999 Optics Commun. **166**, 211.
9. Strati, F and Tallents, G J, Phys. Rev. A, **64**, 013807 (2002).
10. Carillon, A, et al Phys. Rev. Lett. **68**, 2917 (1992).
11. Smith, R, Tallents, GJ, Zhang, J, Eker, G, McCabe, S, Pert, GJ, and Wolfrum, E, Phys. Rev. A**59**, R47 (1999).
12. King, RE, Pert, GJ, McCabe, SP, Simms, PA, MacPhee, AG, Lewis, CLS, Keenan, R, O'Rourke, RMN, Tallents, GJ, Pestehe, SJ, Strati, F, Neely, D, and Allott, R, Phys. Rev. A**64**, 053810 (2001).
13. Dunn, J, Li, Y, Osterheld, AL, Nilsen, J, Hunter, JR, and Shlyaptsev, VN, Phys. Rev. Lett. **84**, 4834 (2000).
14. Casperson, L, and Yariv, A, Phys. Rev. Lett. **26**, 293 (1971).
15. Pert, GJ, J. Opt. Soc. Am. B**11**, 1425 (1994).
16. Ross, IN, Appl. Optics **36**, 9348 (1987).
17. Axis-Photonique Inc., Varennes, Canada.
18. Kita, T, Harada, T, Nakano, N, and Kuroda, H, Appl. Optics **22**, 512 (1983).
19. Abou Ali, Y, et al *these proceedings.*
20. Kuba, J, Klisnick, A, Ros, D, Fourcade, P, Jamelot, G, Miquel, JL, Blanchot N, and Wyart JF, Phys. Rev. A**62**, 043808 (2000).
21. Edwards, M, et al *these proceedings.*

INVESTIGATION OF X-RAY LASER MEDIA

A Lasing Atom in Hot Plasma

Nir Bar-Gill*, Amiram Ron* and Mordechai Botton*

*Dept. of Physics, Technion, Israel

Abstract. In this paper we present a Hamiltonian-based model, which is used to derive transition and dephasing rates of a lasing atom in hot plasma. These rates are necessary in describing the lasing line characteristics using the semi-classical laser theory. Our derivation is corroborated by previously obtained expressions for the transition rate. Furthermore, we find expressions for the dephasing rates due to the interaction of the atom with an electron reservoir. We show that this dephasing is significant in plasma conditions relevant for X-Ray lasers.

INTRODUCTION

Interaction between atoms and hot plasma is used in certain cases as the core of a laser system. The evolution rates of the gain medium, the lasing ions, dictate the lasing line characteristics, such as line broadening and shift. These evolution rates include transition rates between atomic levels, and dephasing rates of the two lasing levels. The rates are usually introduced phenomenologically, although they have been calculated using *different* models for transition rates and for dephasing rates (see [1], [2], [3]). In [4], one model is used to derive both rates, giving results in the first contributing term. In this paper we propose a unified Hamlitonian-based model, from which we derive the atomic evolution rates, to all multipole orders of the interaction potential.

PROBLEM STATEMENT

As an example of a lasing atom in hot plasma, consider a capillary discharge X-Ray laser (see [5]). The model for this problem consists of two different reservoirs - a radiation reservoir and an electron reservoir. The lasing atoms are assumed to be a small system interacting with these reservoirs. It should be noted that we ignore the interaction of the lasing atoms with the plasma ions or among themselves. This is justified since these slow particles (compared to the electrons) interact at a much slower rate, and therefore have negligble effect.

For brevity, our atomic system will consist of levels $(1,0,0)$ and $(2,1,0)$ of a Hydrogen-like atom. General results can be obtained by introducing atomic oscillator strengths, which contain the dependance of the results on the specifics of the atomic system (see [1], [2]).

We use the Density Matrix technique to derive the atomic evolution equations in our model. Such an approach has been used in solving similiar problems ([4], [6]). Following Cohen-Tannoudji et. al. ([7]), we can write the Master Equation for the evolution of the

CP641, *X-Ray Lasers 2002: 8th International Conference on X-Ray Lasers,* edited by J. J. Rocca et al.
© 2002 American Institute of Physics 0-7354-0096-2/02/$19.00

atomic density matrix σ, interacting with a reservoir:

$$\frac{d\sigma_{ab}}{dt} = -i\omega_{ab}\sigma_{ab} + \sum_{c,d}\Gamma_{abcd}\sigma_{cd}. \tag{1}$$

Here ω_{ab} is the energy difference between states a and b, and Γ_{abcd} stands for the coupling between the atomic density matrix element σ_{ab} and the matrix element σ_{cd}. In this work we derive the elements of the rate matrix Γ for specific reservoirs - i.e. the radiation reservoir and the electron reservoir. The elements of this matrix include two separate quantities - transition rates and dephasing rates. Transition rates couple two diagonal density matrix elements (populations), while dephasing rates couple two off-diagonal density matrix elements (coherences). Rates coupling diagonal and off-diagonal terms are neglected here, based on the secular approximation that is used in the derivation of eq. (1), see [7]. The expression for the transition rate is:

$$\Gamma_{n\to m} = \Gamma_{nnmm} = \frac{2\pi}{\hbar}\sum_N W_N \sum_M |\langle M,m|H_{AB}|N,n\rangle|^2 \delta(E_N + \varepsilon_n - E_M - \varepsilon_m), \tag{2}$$

where N and M describe states of the reservoir, m and n are the states of the atomic system, W_N is the distribution function of the reservoir, H_{AB} is the interaction Hamiltonian, and E_N is the energy of state N. Note that the expression for the transition rate is actually Fermi's Golden Rule.

The dephasing rate is made up of two separate quantities - the adiabatic dephasing rate and the non-adiabatic dephasing rate. As the name suggests, the non-adiabatic dephasing rate is due to energy-changing transitions between the dephased levels and any other level of the atomic system:

$$\Gamma_{mn_{non-ad.dep.}} = \frac{1}{2}\left(\sum_{a\neq m}\Gamma_{ma} + \sum_{a\neq n}\Gamma_{na}\right). \tag{3}$$

The adiabatic dephasing rate is due to fluctuations of the dephased levels, caused by the interaction potential, without inducing level transitions:

$$\Gamma_{mn_{ad.dep.}} = \frac{2\pi}{\hbar}\sum_N W_N \sum_M |\langle M,m|H_{AB}|N,m\rangle - \langle M,n|H_{AB}|N,n\rangle|^2 \delta(E_N - E_M). \tag{4}$$

The adiabatic dephasing contributes only if the interaction potential has diagonal matrix elements, and if it is dependant on the internal atomic degrees of freedom.

DERIVATION OF TRANSITION AND DEPHASING RATES

We now use the results presented in the previous section to derive the evolution rates for an atom, interacting with a radiation reservoir and an electron reservoir.

The case of a radiation reservoir has been considered in the literature (e.g. [7]), and therefore will be presented briefly for completeness. The interaction potential between the radiation and the atom is well known, and includes a dipole term and a quadratic term. The atom is assumed to be stationary, as the motion of the atom induces Doppler shift and broadening of the laser radiation, but has a negligble effect on the evolution rates. It is easy to see that the quadratic term in the expression for the interaction potential does not contribute to either transition or dephasing rates, while the dipole term induces transition and non-adiabatic dephasing. The transition rate is given by $\Gamma_{mn} = \frac{1}{3\pi} \frac{\omega_0^3}{\hbar c^3} d^2 \langle n(\omega_0) \rangle$, where ω_0 is the energy difference between the atomic levels, d is the atomic dipole moment, and $\langle n(\omega_0) \rangle$ is the number of photons of energy ω_0. This result is the well-known Einstein coefficient for induced emission. The same calculation, assuming a vacuum radiation reservoir, leads to the Einstein coefficient for spontaneous emission. The dephasing rate is obtained by substituting the transition rate into eq. (3).

We now turn to the case of the electron reservoir. The interaction potential is assumed to be the Coulomb potential:

$$H_{AB} = \sum_i e^2 \left(\frac{1}{\mathbf{r_i} - \mathbf{R_s}} - \frac{Z}{\mathbf{r_i}} \right), \tag{5}$$

where e is the charge of the electron, $\mathbf{r_i}$ is the position of the ith electron of the reservoir, $\mathbf{R_s}$ is the position of the atomic electron, and Z is the charge of the atomic nucleus. Since this potential has terms of all multipole order (in terms of the atomic multipole), it will contribute to the transitions rates, and to both the adiabatic and non-adiabatic dephasing rates.

Putting again to work the density matrix approach, we calculate the transition and dephasing rates to any order in the multipole expansion of the interaction potential. It is instructive to present and analyze the expressions for the rates to the first contributing term. The transition rate in the dipole approximation is:

$$\Gamma_{m \to n} = \frac{2^3 \sqrt{2}}{3} \sqrt{\pi} n_0 \sqrt{m} \sqrt{\beta} \frac{e^4}{\hbar^2} \left(\frac{2^{7.5} a_0}{3^5 Z} \right)^2 e^{\frac{\beta \hbar \omega}{2}} K_0 \left(\frac{\beta \hbar \omega}{2} \right), \tag{6}$$

where n_0 is the electron density, m is the mass of the electron, $\beta = \frac{1}{K_B T}$ - K_B is Boltzmann's constant and T is the electron temperature, a_0 is Bohr's radius and K_0 is the modified Bessel function of order 0. The first contributing term to the adiabatic dephasing rate is the second order multipole in the interaction potential expansion, and it yields:

$$\Gamma_{mn_{ad.dep.}} = 46.9 \cdot 2^6 \sqrt{2} \pi^3 \sqrt{\pi} n_0 \sqrt{m} \sqrt{\beta} \left(\frac{e^4 a_0^2}{\hbar^2 Z^2} \right) \left(\frac{m a_0^2}{\beta \hbar^2 Z^2} \right) \tag{7}$$

It was verified numerically, that these first-order results are correct to within a few percent (compared to the exact solutions).

DISCUSSION

The results for the transition and adiabatic dephasing rates reflect their functional dependance on the parameters of the problem, such as the electron temperature and density.

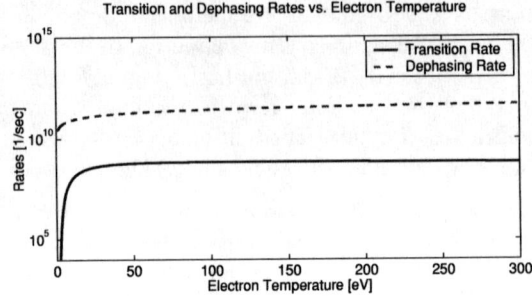

FIGURE 1. Transition and Dephasing Rates vs. Electron Temperature

For example, for the conditions of hot plasma in a capillary XRL, namely electron temperature of 150 [eV] and density of $10^{18}[cm^{-3}]$, we get $\Gamma_{dephasing} \simeq 100\Gamma_{m \to n}$. We see that both rates are linear in the electron density, and their dependance on the electron temperature is shown in Figure 1. We see that the rates are nearly independant of the electron temperature, and that $\Gamma_{mn_{dephasing}} >> \Gamma_{m \to n}$ for a wide range of temperatures.

In conclusion, we have presented a Hamiltonian-based method for calculating transition and dephasing rates of a lasing atom in hot plasma. The plasma is modeled as a radiation reservoir and an electron reservoir. We have found that while both reservoirs contribute to transition and non-adiabatic dephasing rates, the electron reservoir also contributes to adiabatic dephasing due to high-order multipole terms of its interaction potential. We have shown that adiabatic dephasing is dominant, while in some cases non-adiabatic dephasing could be of comparable magnitude (i.e. Neon-like XRL). Our approach is applicable to other problems as well, in which the evolution of a small (atomic) system is governed by its interaction with reservoirs

ACKNOWLEDGMENTS

We would like to thank Mr. Erez Berg, Mr. Roni Nemirovsky and Mr. Moshe Shuker for fruitful discussions.

REFERENCES

1. Sobel'man, I. I., Vainshtein, L. A., and Yukov, E. A., *Excitation of Atoms and Broadening of Spectral Lines*, Springer-Verlag, 1995.
2. Van Regemorter, H., *Astrophysics Journal*, **136**, 906–915 (1962).
3. Berman, P. R., and Lamb Jr., W. E., *Physical Review A*, **2**, 2435–2454 (1970).
4. Bommier, V., and Sahal-Bréchot, S., *Ann. Phys. Fr.*, **16**, 555–598 (1991).
5. Rocca, J. J., Shlyaptsev, V., Tomasev, F. G., Cortazar, O. D., Hartshorn, D., and Chilla, J. L. A., *Phys. Rev. Lett.*, **73**, 2192 (1994).
6. Lewis, E. L., *Physics Reports*, **58**, 1–71 (1979).
7. Cohen-Tannoudji, C., Dupont-Roc, J., and Grynberg, G., *Atom-Photon Interactions: Basic Processes and Applications*, John Wiley & Sons, 1998, ISBN 0-471-29336-9.

High-intensity regime of X-ray generation for innershell photopumping

F. Brandl[1,2], G. Pretzler[1,2], F. Rosmej[1], D. Habs[2] and E. Fill[1]

[1] *Max-Planck-Institut für Quantenoptik - D-85748 Garching, Germany*
[2] *Ludwig-Maximilians-Universität München - D-85748 Garching, Germany*

Abstract. - We investigate laser-produced plasmas as a source for innershell photopumping. Pulses from our ATLAS 10 titanium-sapphire laser (1 J/130 fs) are used to irradiate copper foils and the X-rays are detected with spatial resolution. The results demonstrate a dramatic reduction in the X-ray emitting spot size at intensities around 10^{19} W/cm^2 and a corresponding increase in the X-ray flux density. The distribution of the hot electrons is determined by recording Cerenkov radiation emitted in the visible region of the spectrum. The results demonstrate a change in the electron generation mechanism at relativistic intensities.

INTRODUCTION

Despite steady advances, the region of photon energies above 1 keV is still *terra incognita* to X-ray lasers. In order to make progress in this direction we investigate possible pump sources for an innershell X-ray laser. Such a source must have extremely high brightness ($>10^{31}$ photons/cm^{2}s) and very short pulse duration (<50 fs). In addition, only a small number of electrons are allowed to reach the irradiated medium in order to prevent pumping of the lower laser level.

Previous investigations have demonstrated that high pump intensity with low electron flux can be achieved by close-range irradiation if the sample is separated from the source by a narrow (≈ 100 μm) vacuum gap [1,2]. Self-generated fields of the electrons prevent them from reaching the medium irradiated. However, at intensities of $\approx 10^{18}$ W/cm^2, as used in these experiments, the spot from which the X-rays originated was considerably larger than the laser spot, this being attributed to electrons making large orbits in front of the target [3].

At higher intensities, at which the velocity of the electrons quivering in the laser field approaches the velocity of light, relativistic effects come into play. It is customary to normalize the electric field strength E_L according to

$$a = e\, E_L/m\omega_0\, c \qquad (1)$$

where m is the electron rest mass, ω_0 is the laser frequency and c is the velocity of light. The expression on the right side of eq. (1) is the ratio of the non-relativistic quiver velocity amplitude and the light velocity. The intensity is obtained as

CP641, *X-Ray Lasers 2002: 8th International Conference on X-Ray Lasers*, edited by J. J. Rocca et al.

$$I\lambda^2 = a^2 \times 1.37 \times 10^{18} \text{ W } \mu\text{m}^2 / \text{cm}^2. \tag{2}$$

The relativistic mass factor can be approximated by $\gamma = (1 + a^2)^{1/2}$ and thus significant relativistic effects can be expected for $a > 1$.

A first one results from the relativistic mass increase of the electrons. The plasma frequency decreases with increasing mass of the electrons which leads to an intensity dependent refractive index and to self-focusing and guiding of the laser beam [4].

A second effect is due to the Lorentz force on the electrons, which is proportional to $u_q \times B_L$, where u_q is the quiver velocity and B_L is the magnetic field of the electromagnetic wave. This force is directed in the forward direction and becomes appreciable when u_q approaches the velocity of light. It results in an acceleration of the electrons in the direction of the laser beam.

From these considerations it can be expected that titanium sapphire laser pulses ($\lambda = 790$ nm) should exhibit relativistic effects in the interaction with solid and gaseous targets if intensities above a few 10^{18} W/cm^2 are applied. In the experiments reported here the laser intensity was increased above 10^{19} W/cm^2 by adding a new amplifier to the ATLAS titanium-sapphire laser at our institute. Two types of diagnostics were used: a) a calibrated X-ray CCD with a knife edge for spatial resolution, b) a dielectric medium in immediate contact with the rear-side of the foil target, which generates Cerenkov radiation from relativistic electrons propagating through the foil. NaF, quartz, and BK7 glass were used as Cerenkov media.

EXPERIMENT

Pulses of 1 J/130 fs at 10 Hz were focused on target with an off-axis parabola. With wavefront aberrations corrected by means of two deformable mirrors [5], a spot 7 μm in diameter with an intensity of 2×10^{19} W/cm^2 was achieved.

The experimental arrangement used for diagnosing the X-ray emission and the Cerenkov radiation is shown in fig. 1. Copper targets were irradiated at an angle of incidence of 40^0 with p-polarized radiation. The intensity of the emission was measured by means of a CCD calibrated by a radioactive Mn^{55} source. The spatial extent of the X-ray emitting spot was determined by inserting a steel wedge in between the target and the CCD at a magnification of 10. The wedge was vertically oriented and, thus, the horizontal extent of the X-ray spot was detected.

X-ray diagnostics

By varying the pulse energy it was found that at intensities above 5×10^{18} W/cm^2 a new narrow X-ray emission region appeared in the middle of the usual X-ray spot, the size of which was of the order of the laser spot. A lineout of the half-shadow image, averaged over a few hundred pixel rows is shown in fig. 2a. A good fit to the data was obtained by using two error functions with significantly different slopes.

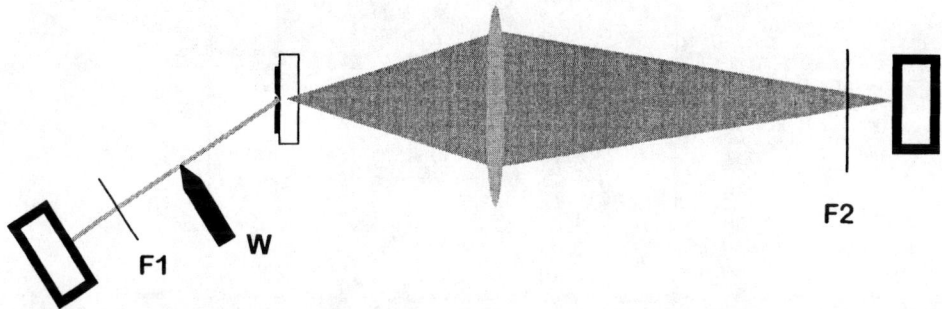

laser pulse from
ATLAS
Ti-Sapphire laser

F2

W

F1

FIGURE 1. Experimental arrangement for detecting X-rays and Cerenkov radiation emitted from a laser-produced plasma. Laser pulses are focused p-polarized at 40^0 onto target. The target consists of a 50 - 200 μm thick copper foil in immediate contact with a dielectric acting as a Cerenkov medium. NaF, fused silica and BK7 glass are used for this purpose. X-rays are detected by an X-ray CCD filtered by 5 μm of Ti (F1). Spatial resolution is provided by stainless-steel wedge W inserted half way into the emission. The magnification factor is 10. Cerenkov radiation is collected by an f/2 objective and then focused onto an intensified CCD. Filter F2 blocks the laser radiation and its second harmonic.

Differentiation yields a Gaussian intensity distribution as shown in fig. 2b. It is seen that the intensity is dominated by the narrow central spot and that the broad surrounding emission is only discernible by using a logarithmic display.

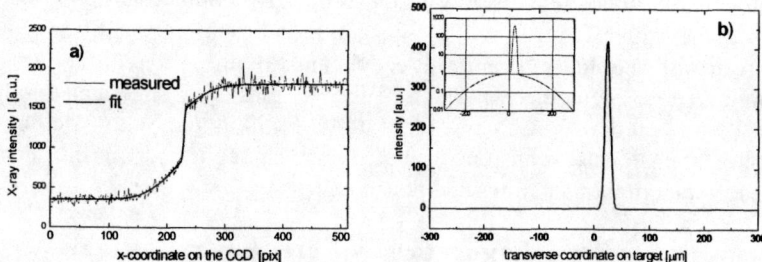

FIGURE 2.a) Lineout of half-shadow image and error functions used for fitting; **b)** Resulting X-ray spot in linear and logarithmic presentation.

Varying the intensity from below 10^{18} W/cm^2 to the maximum of 2 x 10^{19} W/cm^2 showed that the relative energy contained in the narrow emission increased with increasing laser intensity until at 2 x 10^{19} W/cm^2 the number of photons exceeded that from the broad region, resulting in an increase in intensity by more than an order of magnitude. The data determined in the new measurements connect smoothly to the ones obtained with the smaller version of the ATLAS laser [3], as shown in fig. 3. At non-relativistic intensities the data shows a maximum at around 10^{16} W/cm^2, which is attributed to an optimum electron temperature for generating Cu K$_\alpha$ radiation.

The steep increase in X-ray intensity as the laser intensity reaches relativistic values is evident.

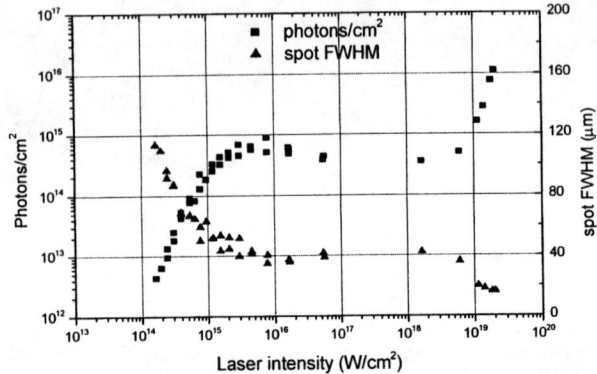

FIGURE 3. Areal photon density and spot size of X-rays emitted from the front side of copper foil target as a function of the laser intensity. Data at $>3 \times 10^{18}$ W/cm^2 apply to the narrow central spot.

The spectral composition of the X-rays generated was determined by means of a crystal spectrometer using a quartz crystal in second order. Kodak X-ray film was used as the detector. Screening with 5 cm lead bricks placed around the film box reduced the background to a low level. Spectra obtained at best focus of the laser and by defocusing the laser ±300 μm exhibit interesting features: At best focus the radiation consists almost exclusively of the copper K_α doublet at 1.5444 and 1.5406 Å. Upon defocusing the laser the K_α-lines are broadened and satellites appear at the short wavelength shoulders. This result is explained by the different modes of plasma expansion: At best focus the expansion of the pre-plasma is spherical, resulting in a low density and predominant interaction with the solid target. With a defocused laser pulse, one-dimensional expansion results in a much higher pre-plasma density and thus in the generation of satellites and plasma lines.

Electron diagnostics by Cerenkov radiation

Cerenkov radiation is generated if the velocity of a particle exceeds the phase velocity of light in a medium. The radiation is emitted at an angle given by

$$\cos(\Theta) = 1/\beta n, \tag{3}$$

where β is the particle velocity divided by c and n is the refractive index of the medium. The spectrum constitutes a continuum in the visible spectral region with a steady increase towards the UV. The short-wavelength limit is determined by absorption. The generation of Cerenkov radiation is quite an efficient effect, resulting in some hundred visible photons per particle [6]. Other mechanisms generating visible radiation from relativistic electrons include bremsstrahlung, transition radiation, luminescence and synchrotron radiation. However, their contribution to the visible emission observed turns out to be negligible (see discussion in [7]).

Equation (3) shows that for a beam with a wide range of energies the Cerenkov diagnostics probes electrons with a minimum β given by $1/n$ (emission on axis). In the experiment an f/2 objective was used for collecting the radiation. An intensified CCD with a time gate of 5 ns was employed as a detector. NaF, fused silica and BK7 glass were used as the Cerenkov media, which (at $\lambda = 500$ nm) have refractive indices of 1.33, 1.45, and 1.5, respectively, resulting in minimum β values of 0.75, 0.69 and 0.67. At pulse energies above a few 100 mJ corresponding to intensities above 2 x 10^{18} W/cm^2 a bright, round spot appeared on the back of the copper foil, the diameter of which was about 200 μm. A single shot was sufficient to obtain a well illuminated image. By increasing the intensity it was found that the spot became more and more asymmetric. In addition a dip in the center was seen to develop. Examples of the spatially resolved Cerenkov radiation demonstrating this effect are shown in fig. 4.

These features of the Cerenkov images suggest that they are generated by two different electron populations: The first propagates perpendicularly to the target surface, whereas the second is directed almost parallel to the laser beam. To evaluate the data quantitatively the GEANT code developed at CERN for high-energy physics [8] was used. GEANT is a Monte Carlo code containing a full set of particle-matter interaction effects and includes Cerenkov radiation.

In performing the simulations it was found that the features of the observed patterns could indeed be reproduced by assuming two electron populations differing in direction and temperature.

1 mm

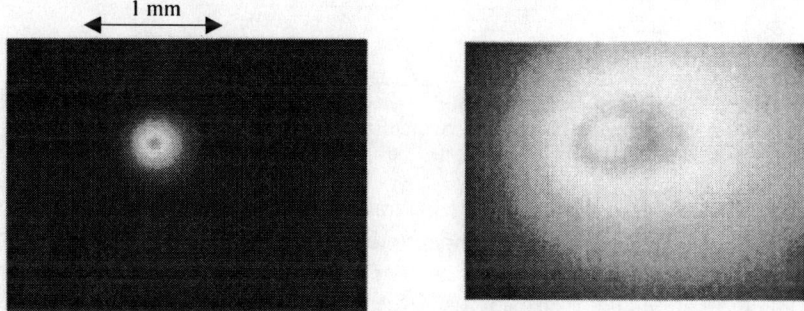

FIGURE 4. Spatially resolved Cerenkov radiation generated in BK7 glass. Left image: 0.17 J of laser energy; right image: 0.85 J of laser energy. The central dip and the asymmetry are clearly visible.

The dip in the center of the Cerenkov pattern required the assumption of a point-like electron source in front of the target. The angle of emission is 30^0 and is centered perpendicularly to the target surface. Reproducing the asymmetrically shifted peak required a second electron population with a higher temperature and propagating at an angle of 30^0 to the target normal. The number of electrons in the perpendicularly propagating population was approximately constant with the intensity, at a value of $(4.5 \pm 1.2) \times 10^{11}$. The number of oblique electrons increased approximately linearly with the laser intensity, reaching a value of 4.2×10^{11} at 1.5×10^{19} W/cm^2.

Assuming a one-dimensional Maxwellian energy distribution for the different electron populations yields the temperatures shown in fig. 5. In the figure the electron temperatures are compared with those obtained by applying semi-empirical scaling laws. The first of these, as proposed by Beg et al. [9], reads

$$T \text{ [MeV]} = 0.1 \, (I_{17} \, \lambda^2)^{1/3}. \tag{4}$$

This law applies to electrons generated by resonance absorption.

A different scaling law is obtained by invoking the relativistic ponderomotive potential U_p of the laser pulse, which may be approximated as [10,11]

$$U_p \text{ [MeV]} = 0.511 \, [(1+I_{17} \, \lambda^2/13.7)^{1/2} - 1). \tag{5}$$

In these equations I_{17} is the laser intensity in units of 10^{17} W/cm^2 and λ is in μm.

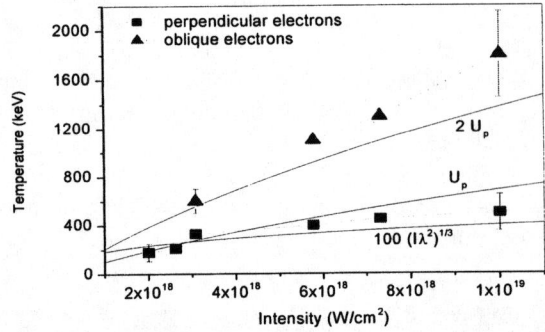

FIGURE 5. Temperature of the two different electron populations as obtained by matching GEANT simulations to two-dimensional Cerenkov radiation images. Data are compared to temperatures obtained by scaling laws invoking different electron heating mechanisms.

Comparison with the electron temperatures derived with the GEANT simulations shows good agreement of the temperatures of the perpendicular electrons with those derived from eq. (4). The temperature of the oblique electrons correlates with twice the value calculated from eq. (5). Recall that electrons generated by collisional ionization may be released at a time when the electric field strength is very low. These electrons gain an energy of twice the ponderomotive potential after leaving the focal region.

CONCLUSIONS AND OUTLOOK

It is shown in this work that at relativistic intensities (i.e. at a few 10^{18} W/cm^2 of Ti-sapphire laser radiation) the interaction of fs laser pulses with solids exhibits new features. Firstly, the spatial pattern of the X-ray emission shows a new, narrow spot close to the center of the original, broad source. At 10^{19} W/cm^2 this central spot fully determines the intensity of the X-ray emission. The spectrum of the X-rays consists mainly of Cu K_α radiation. Quantitative evaluation of the data, assuming an X-ray pulse duration given by the laser pulse duration, yields 10^{29} ph/cm^{2}s.

Concomitantly, the spatially resolved detection of Cerenkov radiation emitted by the electrons reveals that the appearance of the new spot is accompanied by a change in the Cerenkov pattern from symmetric to double-peaked, the second peak being shifted in the direction of the incident laser radiation. These observations are explained by relativistic effects which come into play at the intensities used in the new experiments.

The Cerenkov radiation experiments are being continued at this laboratory with the aim of determining the time duration of the Cerenkov pulse. This will be achieved by cross-correlating the Cerenkov pulse with the laser pulse by means of the optical Kerr effect. Since Cerenkov radiation is instantaneously emitted, this experiment will determine the time duration of the electron pulses generated.

ACKNOWLEDGEMENTS

The authors thank Prof. K. Witte for support and encouraging interest in this work and Dr. P. Thirolf for help in running the GEANT code. The ATLAS team is thanked for providing a reliable laser with an excellent beam quality. This work was supported by the Commission of the EU within the framework of the Association Euratom-Max-Planck-Institut für Plasmaphysik.

REFERENCES

[1] Pretzler, G., Schlegel, Th. and Fill, E., Laser and Particle Beams 19, 91- (2001).

[2] Pretzler, G., Schlegel, Th., Fill, E., and Eder, D., Phys. Rev. E 62, 5618- (2000).

[3] Eder, D.C., Pretzler, G., Fill, E., Eidmann, K., and Saemann. A., Appl. Phys. B 70, 211-217 (2000).

[4] Umstadter, D., Chen, S.-Y., Maksimchuk, A., Mourou, G. and Wagner, R. Science 273, 472-475 (1996).

[5] Baumhacker, H., Pretzler, G., Witte, K.J., Hegelich, M., Kaluza, M. Karsch, S., Kudryashov, A., Samarkin, V., and Roukossouev, A., Optics Letters, in print.

[6] Jelley, J.V. Cerenkov Radiation and its Applications, Pergamon Press, London 1958, pg. 72.

[7] Brandl, F.R., "Charakterisierung Laser-beschleunigter Elektronen mittels Cerenkov- und Röntgenstrahlung", Diploma thesis, University of Munich, Munich 2002.

[8] GEANT User's Guide, edited by Brun, R., Hansroul, M. and Lassalle, J.C., CERN Report No. DD/EE/82 (1982).

[9] Beg, F.N., Bell, A.R., Dangor, A.E., Danson, C.N., Fews, A.P., Glinsky, M.E., Hammel, B.A., Lee. P., Norreys, P.A., and Tatarakis M. , Phys. Plasmas 4, 447-457 (1997).

[10] Perry, M.D., and Mourou, G. Science 264, 917-923 (1994).

[11] Malka, G., and Miquel, J.L., Phys. Rev. Lett. 77, 75-78 (1996).

Effects Of Multi-Pulses On Intensity Of Resonance Lines Emitted From X-Ray Laser Media

A.Demir[1], G.J. Tallents[2] and G.J. Pert[2]

[1]-University of Kocaeli, Department of Physics, Kocaeli, 41200, TURKEY
[2]-University of York, Department of Physics, York, YO1 5DD, UK

Abstract: A post-processor to the EHYBRID fluid and atomic physics code has been developed to calculate the intensity of resonance lines emitted from X-ray laser media. Opacity allowing for the effect of Doppler decoupling has been taken into account in the evaluation of the line emission escaping from plasmas created by multi-pulses. Ne-like Fe, Ne-like Y and Ni-like Gd resonance line intensities have been simulated for various multi-pulse configurations using the code.

INTRODUCTION

Collisionally pumped x-ray lasers have been developed using a wide range of elements with saturated output down to 5.9 nm now achieved [1-6]. Significant reductions in the wavelengths for saturated output have been produced recently because the necessary driving laser energy for collisional x-ray lasers is greatly reduced by using pre-pulse and multi-pulse irradiation [6,7]. In particular, hydrodynamic simulations and experiments [8, 9] have shown that using either double 70 - 100 ps irradiation or short ≈ ps driving laser pulses superimposed on long pre-pulses reduces the required driving energy and increases the lasing output of collisional x-ray lasers. The initial low intensity pulse produces Ne-like or Ni-like ions over a large plasma volume and the high intensity second pulse creates the population inversions. The large volume of the gain region with low electron density gradients created in this way ensures that the X-ray laser beam propagates and amplifies without refracting out of the gain volume [8]. One of the efficient ways of producing lasing at shorter wavelength uses nickel-like ions, in which the pump intensity can be reduced significantly because the quantum efficiency of the nickel-like systems is much higher than that of the neon-like systems.

Resonance lines in the x-ray laser media are often optically thick as the lines have strong oscillator strengths and multi-pulse plasmas have large volumes. It is necessary to take into account opacity for the resonance line intensity calculations. Resonance lines emitted from the Li-like Ti [10] and Ne-like Ge x-ray laser media have been simulated previously [11-13]. For this paper, we simulate the Ni-like and Co-like Gd resonance line spectrum and the Ne-like Fe and Y resonance line spectra, taking into

CP641, *X-Ray Lasers 2002: 8th International Conference on X-Ray Lasers,* edited by J. J. Rocca et al.
© 2002 American Institute of Physics 0-7354-0096-2/02/$19.00

account opacity effect on the resonance lines. It is suggested that results of the simulation can be used to interpret transverse direction diagnostic of x-ray laser media, such as cross-slit cameras, pinhole camera and crystal spectrometers.

METHODS OF SIMULATION

The 1.5 dimension hydrodynamic with atomic physics code EHYBRID simulates laser interaction with a solid target as used to to create x-ray laser media. The code calculates the electron temperature, electron density, the fractional population of each ionic stage from the neutral to bare nuclei and using a collisional radiative treatment the populations of the excited levels of Ne-, F or Ni-, Co-like ions at each time step for 98 spatial cells [14]. The input parameters for EHYBRID are the FWHM of the pulse duration, the power of the laser pulse, and the length, width and thickness of the target. Atomic physics data for Ne-, F- or Ni-, Co-like ions is calculated using the code developed by Cowan [15].

Due to bulk plasma motion, photons emitted in a particular space are less likely to be absorbed at a distance away from the emitted region because of the Doppler shift of the absorption line profile relative to the emission profile. This is known as Doppler decoupling. The decoupling length d was calculated using [16]. The effect of opacity on the emitted resonance line intensities is taken into account using an escape factor approximation modified to allow for Doppler decoupling. The Ni-like and Co-like Gd, Ne-like Fe and Ne-like Y resonance line intensities are calculated using a post processor described above coupled to the EHYBRID code. The energies and oscillater strengths used in EHYBRID to calculate population densities via the collisional radiative processes are used to deduce the wavelengths and intensities of the resonance lines.

Laser pulses of 1.064µm wavelength focused on 100 µm wide and 1cm long slab targets have been simulated. Gadolinium has been pumped assumed to be with normally a double pulse configuration in which a prepulse, with 10% of the total energy, was focussed on target 2.2ns before the main pulse. The intensity of the main pulse is 4×10^{13} W/cm^2. However, the effects of varying the time delay between the pre-pulse and main pulse and the effect of different laser irradiances has also been studied. In the simulation of the Ne-like Fe and Y resonance line spectra, a 3ps short pulse is superimposed on a 700 ps long background pulse with another 10% pre-pulse 2ns early. Peak intensities assumed are respectively 2×10^{11} W/cm^2, 2×10^{12} W/cm^2 and 2×10^{15} W/cm^2 for the two long and the short pulse respectively.

RESULTS

Figure 1 shows the time and space integrated Co-like and Ni-like resonance line spectrum between 6 and 11.6 Å wavelength range. To create the spectrum, 112 Ne-like and 33 Co-like resonance lines have been simulated. The spectrum is dominated by the Ni-like resonance lines. Figure 2 shows the Ni-like and Co-like Gd resonance line intensities as a function of time. There is no resonance line emission arising from the pre-pulse. The Ni-like resonance emission peaks ~70 ps after the peak of the main pumping pulse at time 2.2ns. The Co-like emission peaks ~100 ps later than the Ni-like emissions and drops rapidly. The duration of the Ni-like resonance emissions ~40 ps.

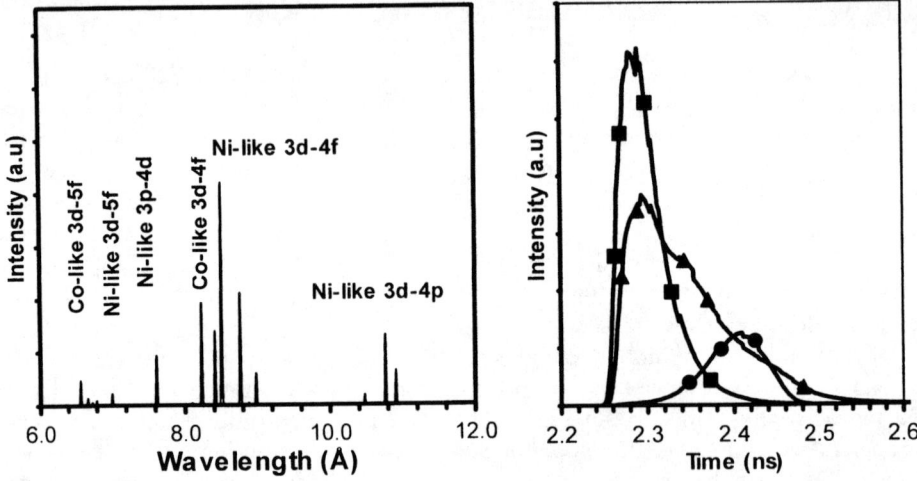

Figure 1. Time- integrated Ni-like and Co- like Gd resonance line spectrum.

Figure 2. The Ni-like and Co-like resonance line intensities as a function of time. ■ is the Ni-like 4f-3d at 8.507Å, ▲ Ni-like 4p-3d at 7.621 Å and ● Co-like 4f-3d at 8.229 Å

Figure 3 shows the ratio of the Co-like 4f-3d to Ni-like 4f-3d resonance lines for Gd as a function of the time delay between the pre-pulse and main pulse. The ratio reduces when the time delay between the pre-pulse and main pulse increases since the driving energy needed to ionise the larger plasma is constant. Figure 4 shows the ratio of the Co-like 4f-3d to Ni-like 4f-3d Gd resonance lines as a function of the main pulse intensity. For this simulation, the time delay between the pre-pulse and main pulse is 2.2 ns. The ratio increases with the main pulse driving laser energy.

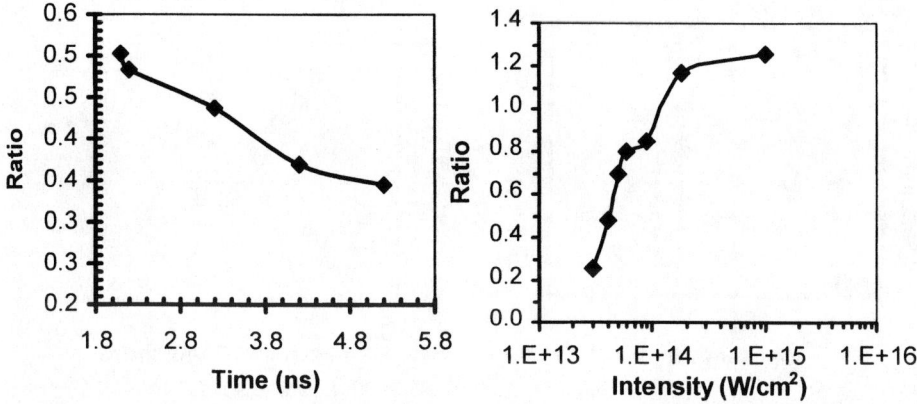

Figure 3. Ratio of Co-like 4f-3d to Ni-like 4f-3d Gd line intensities as function of time interval between prepulse and main pulse.

Figure 4. Ratio of the Co-like 4f-3d to Ni-like 4f-3d Gd line intensities as function of the main pulse intensity.

Figure 5 shows the time- and space- integrated Ne-like Y resonance line spectrum over the 4 and 6.8Å wavelength range. The F-like Y resonance lines are not simulated. Figure 6 shows the intensity of the Ne-like 3p-2s resonance line of Y as a function of the incidence time of the short 3ps pulse superimposed on the 700 ps long background pulse. The start time of the short pulse laser is measured from the incidence of the assumed Gaussian background pulse which has peak intensity at 350 ps. There is only a small resonance line emission when the short pulse interacts before 300ps.

Figure 7 shows the time-integrated Ne-like Fe resonance line spectrum between 11 and 17.2 Å wavelength range. The F-like Fe resonance lines are not simulated. Figure 8 shows the Ne-like resonance lines intensity as a function of time. There is no resonance line emission arising from the 700 ps long pulse before the short duration pulse is incident at 2.5ns . The resonance line emissions effectively starts after 2.5 ns. The Ne-like resonance emission peaks ~2.508 ns and drops rapidly. The duration of emission is ~10 ps.

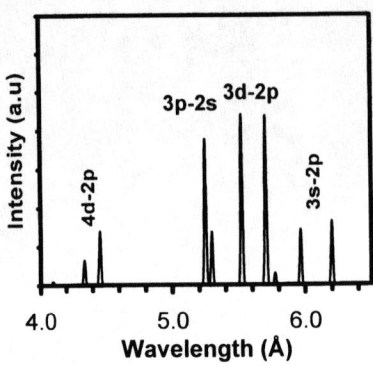

Figure 5. The time-integrated Ne-like Y resonance line spectrum.

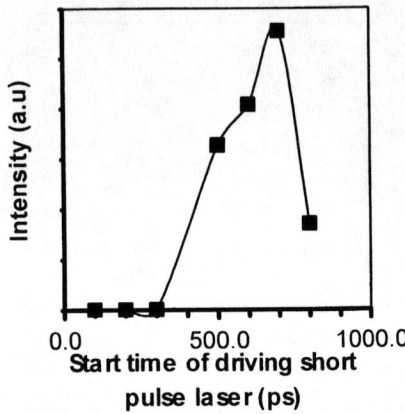

Figure 6. The intensity of the Ne-like 3p-2s resonance line as a function of the starting time of the 3 ps short pulse relative to the 700 ps long background pulse.

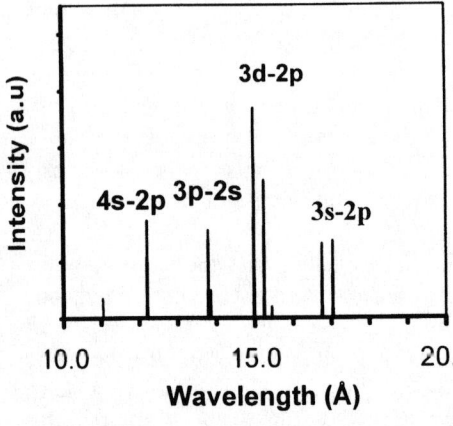

Figure 7. The time-integrated Ne-like Fe resonance line spectrum.

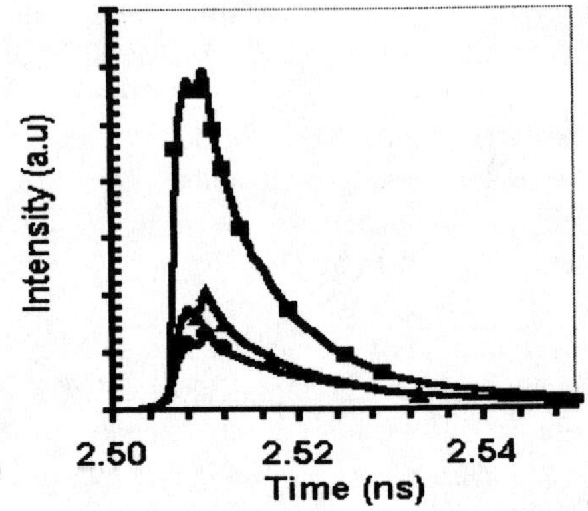

Figure 8. The intensity of the Ne-like Fe resonance lines as a function of the start time of the 3 ps short pulse relative to the 700 ps long background pulse. ■ Ne-like 3d-2p at 14.983Å , ▲ 3p-2s at 13.790Å and ●3s-2p at17.035 Å

316

CONCLUSION

X-ray spectra of Ni-like and Co like Gd resonance lines emitted from X-ray laser media created by irradiating solid Gd using two 75 ps duration pumping laser pulses separated by 2.2 ns have been simulated with a post-processor code coupled to the EHYBRID code. The duration of the Ni-like Gd resonance line emission ~40 ps. X-ray spectra of Ne-like Fe and Y resonance lines have been simulated with the post-processor code for a 3ps pulse pumping superimposed on a 700 ps background pulse. The duration of the Ne-like Fe resonance line emission is ~ 10 ps. The results of the post-processor code can be used to compare with transverse direction diagnostics of x-ray laser media such as crystal spectrometers, pinhole cameras and cross-slit cameras.

ACKNOWLEGMENTS

This work was partly supported by the Scientific and Technical Research Council of Turkey.

REFERENCES

[1] A Carillon *et al* Phys. Rev. Lett. **68**, 2917 (1992).
[2] J J Rocca *et al* Phys. Rev. Lett. **77**, 1476 (1996).
[3] J Zhang *et al* Phys. Rev. A **54**, R4653 (1996).
[4] J Zhang *et al* Science **276**, 1097 (1997).
[5] J Zhang *et al* Phys. Rev. Lett. **78**, 3856 (1997).
[6] J.Y.Lin et al J.App.Phys., **85**,672 (1999)
[7] R. Smith, *et al* Phys.Rev. A, 59,R47 (1999); H. Daido, *et al* Phys.Rev.Lett. 75,1074 (1995)
[8] A Behjat *et al* Optics Commun. **135**, 49 (1997); J Lin, *et al* J. Appl. Phys. **83**, 1863 (1998); S. McCabe and G.J.Pert, Phys.Rev. A **61**, 033804-1 (2000)
[9] P V Nickles, *et al* Phys. Rev. Lett. **78**, 2748 (1997).
[10] A Demir, *et al* Phys. Rev. E **55**, 1827 (1997).
[11] G.F. Cairns *et al* J.Phys. B:At. Mol. Opt. Phys. 29,4839 (1996)
[12] A. Demir *et al* Ins.Phys. Conf. Ser. **No159**, 467 (1999)
[13] S.J Pestehe and G.J. Tallents JQSRT **72**,853 (2002) P.J. Warwick, et al J. Opt. Soc. Am. B **15**, 1808 (1998)
[14] G J Pert J. Fluid Mech. **131**, 401 (1983); P B Holden, et al J Phys. B **27**, 341 (1994); S B Healy, *et al* Commun. **132**, 442 (1996).
[15] R.D. Cowan, J. Opt.Soc.Am., **58**,808 (1968)
[16] G B Rybicki and D G Hummer Astrophys J **219**, 654 (1978).

A Structured Target Geometry for Highly Efficient X-ray Lasers

Tianqing Jia, Ruxin Li, Chuanfu Cheng, Jingtao Zhang,

Hong Chen[*], Zhizhan Xu

Laboratory for High Intensity Optics, Shanghai Institute of Optics and Fine Mechanics, Chinese Academy of Sciences, P.O. Box 800-211, Shanghai,201800 China
[*]Pohl Institute of Solid State Physics, Tongji University, Shanghai 200092 China

Abstract. One-dimensional photonic crystal (1D PCs) with a defect layer is proposed as target for x-ray lasers. Its structure is shown as $AB...ABTBA...BA$, here T represents the defect layer made of the target material for x-ray lasing, on both sides of which are N bilayers of A (silica)and B (titania). The optical thicknesses of A, B and T are a quarter, a quarter and a half of the pumping laser wavelength, respectively. Here, the pumping laser is the localized mode of the 1D PCs. When irradiating perpendicularly on the novel target, the pumping laser intensity in the target layer will be enhanced by about 2 orders of magnitude. This is very important for the development of x-ray lasers (XRL) towards compactness, shorter wavelength and higher conversion efficiency.

INTRODUCTION

X-ray laser (XRL) was first demonstrated successfully in 1984 [1]. Since then, a lot of experimental and theoretical investigations have been devoted to develop XRL towards shorter wavelengths, compactness and higher conversion efficiency [6-10].

Great improvement in XRL has been achieved, however, there are still several problems. Conversion efficiency is still very low, only of the order of 10^6 [4, 6]. How to enhance it greatly is still an open task. Saturated x-ray lasers is successfully demonstrated for wavelengths >6 nm. However, it is far from gain saturation for wavelengths near or in the water window range. The pumping laser system normally occupies two standard optical tables. If the required pumping laser energy is further reduced, the XRL system will become more compact.

In the paper, we make a theoretical study on how to improve the XRL efficiency by controlling the spatial distribution of pumping laser intensity. The main idea is to realize the localization of pumping lasers, so that the intensity in target can be enhanced greatly. In order to achieve laser localization, we introduce the idea of photonic crystals (PCs)

CP641, X-Ray Lasers 2002: 8th International Conference on X-Ray Lasers, edited by J. J. Rocca et al.

into the target design.

Photonic crystal is a type of artificial structure, it possesses photonic conduction bands and band gap [11-12]. If a defect is made in the PCs, there will be a localized mode, the frequency of which lies in the photonic band gap. When light with this frequency irradiates on the PCs, it will be localized around the defect, which is similar to the case of electrons localized around defect states in crystals, hence the laser intensity is largely enhanced there [12].

1. PUMPING LASER DISTRIBUTION IN STRUCTURED TARGET

The novel target is designed as a 1D PCs with a defect layer (as shown inFig. 1), here T represents the defect layer made of target material, on both sides of which are6 bilayers of A and B. A and B are silica and titania, respectively. The optical thicknesses of A, B and T are a quarter, a quarter and half of the pumping laser wavelength (λ_0 = 1053 nm), respectively. Pumping laser is the localized mode of the 1D PCs, its frequency lies in the middle of photonic band gap. When the pumping laser beam irradiates perpendicularly on the structured target, it will considerably localized in T layer.

LiF has been used as target material [3]. The distribution of electric field amplitude of pumping laser is calculated with transfer matrix method, the results are shown in Fig. 2. The solid curve represents the distribution for glass as substrate, and the dashed curve, non-substrate. In the two cases, the peak amplitude of the electric field in the target layer increases by a factor of 17 and 13, respectively. In figure 2, the amplitude E_0 of the input laser electric field is normalized to a unit,D represents optical thickness.

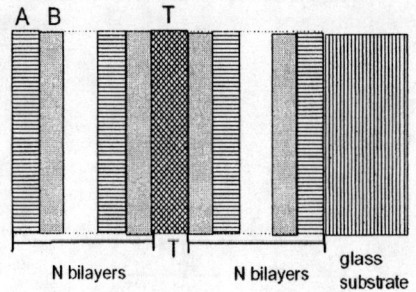

Fig. 1. Structured target.

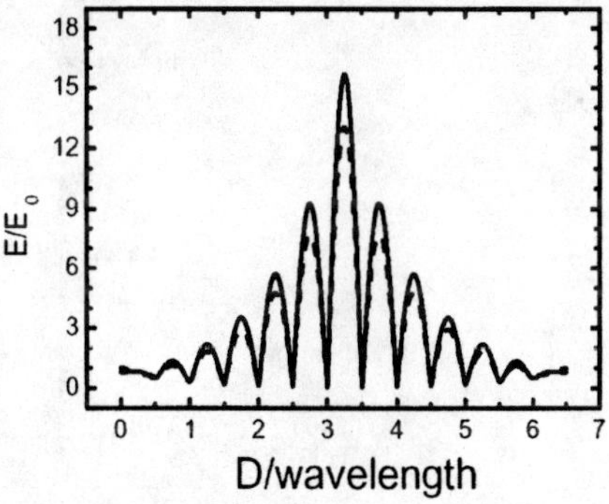

Fig. 2. The distribution of the pumping laser electric field in the structured target. The solid line, with silica glass as substrate, and the dashed curve, non-substrate.

319

For the selection of target materials, what we consider actually is the upper and the power energy levels and their life times, rather than the linear extinction coefficient (LEC) of the target materials. Therefore, we can substitute the semiconductors and metals by their compounds, and use them as target materials in XRL with electron collisional excitation scheme.

The solid curve in Fig. 3 shows the pumping laser electric distribution for AgCl as target material, and the dash curve shows the case of SiC as target. Compared with common slab target, the peak amplitude of pumping laser electric field in the target layer is enhanced by a factor of 11 and 8.5, respectively. The electric field in target layer embedded in 1D PCs is enhanced by an order of amplitude, therefore, the laser intensity is enhanced by about 2 orders, which will have great effect on the properties of XRL systems.

2. DISCUSSION AND CONCLUSION

Experimental and theoretical results indicate as pumping laser intensity increase, the gain coefficient increases, too [4-5]. The open squares in Fig. 4 show the gain coefficient of 23.6 nm x-ray lasers when slab Ge target is irradiated under Nd:glass lasers [4]. Theoretical simulation of x-ray lasers at 19.6 nm is shown with the solid diamonds [5]. The solid and dashed lines are approximations of the dependence of gain coefficient on pumping laser intensities with formula of $G \propto \ln(I_i / I_{th})$.

Saturated XRL has been obtained by applying the double-target method [2, 6]. However, it is very complicated to apply this method. When the structured target is applied, gain coefficient will be enhanced greatly, hence the target length for saturated

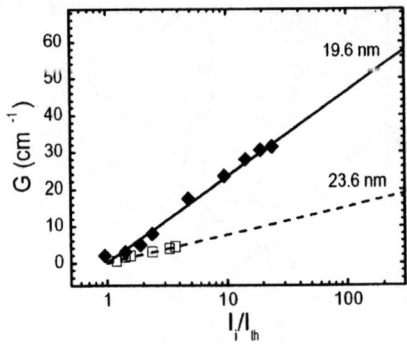

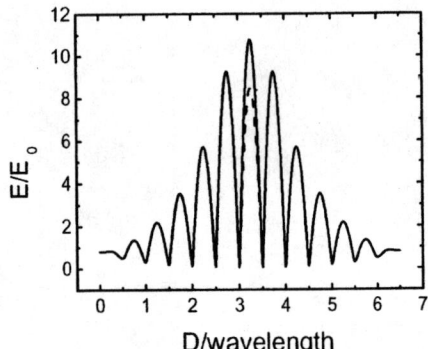

Fig. 3. The distribution of the pumping laser electric field in the structured target. The solid and the dashed curves represent for AgCl and SiC as target marterials, respectively.

Fig.4. Dependence of gain coefficient on the ratio of pumping laser intensity. The open and solid diamonds are experimental results and calculated values, respectively.

XRL will be decreased, too. Saturated XRL can be easily realized by using one target.

There is a power law dependence of x-ray laser intensity on pumping laser intensity $I_x \propto (I_i)^\gamma$ [9-10], here exponent fluctuates always around 2. The pumping laser intensity target layer is improved by about 2 orders, the XRL intensity will be enhanced by about 4 orders. Therefore, the XRL conversion efficiency will also be enhanced by about 4 orders.

Experimental results indicate the x-ray wavelengths from carbon target decrease by 0.66 nm and the photon energies increase by 60 eV when pumping laser intensity increases from 6×10^{14} to 10^{16} W/cm^2 [9-10]. The pumping laser intensity is greatly enhanced in the structured target, the wavelength of x-ray laser will be extended to shorter values.

In conclusion, by using the structured targets, pumping laser intensity will be increased by 2 orders. The XRL intensity will be enhanced by about 4 orders, and so dose the conversion efficiency. The wavelength of x-ray laser will be extended to shorter values.

ACKNOWLEDGEMENTS

This work is supported by the following foundations: Chinese National Natural Science Foundation under Grant No. 60108002; 69925513 and No.19974058; K.C. Wong Education Foundation, Hong Kong; Chinese Postdoctoral Science Foundation; Shanghai National Science Foundation under Grant No. 01ZA14070; Shanghai Foundation of Science & Technology under Grant No. 0159nm022; Chinese National Major Basic Research Project under Grant No. G1999075200.

REFERENCES

1. Matthews D.L., Hagelstein P.L., Rosen M.D., et al, *Phys. Rev. Lett.* **54**, 110-113 (1985).

2. Zhang J., MacPhee A.G., Lin J., et al., *SCIENCE,* **276**, 1097-1100 (1997).

3. Korobkin D.V., Nam C.H., Suckewer S., *Phys. Rev. Lett.* **77**, 5206-5209 (1996).

4. Neely D., Lewis C.L.S., O'Neill D.M., et al., *Opt. Commun.* **87**, 231-238 (1992).

5. Healy S.B., Janulewicz K.A., Plowes J.A., *Opt. Commun.* **132**, 442-450 (1996).

6. Dunn J., Li Y., Osterheld A.L., Nilsen J., *Phys. Rev. Lett.* **84**, 4834 (1997); Li Y., Duun J., Nilsen J., et al., *J. Opt. Soc. Am.* **B 17**, 1098-1102 (2000).

7. Zhang J., Key M.H., Norreys P.A., et al., *Phys. Rev. Lett.* **74**, 1335-1338 (1995).

8. Xu Z., Fan P., Ling L., et al., *Appl. Phys. Lett.* **63**, 1023-1025 (1993).

9. Teubner U., et.al., *J. Phys.* **B 29**, 4333-4340 (1996).

10. Andreev A.A., Teubner U., Kurnin I.V., and Forster E., *Appl. Phys.* **B 70**, 505-510 (2000); Altenbernd D., Teubner U., Gibbon P., et al., *J. Phys.* **B 30**, 3969-3973 (1997).

11. Moroz A., *Phys. Rev. Lett.* **83**, 5274-5277 (1999).

12. Hattori T., Tsurumachi N., and Nakatsuka H., *J. Opt. Soc. Am.* **B 14**, 348-352 (1997).

Calculations of Intense Laser Pulse Propagation and Soft X-Ray Lasing in Capillary Plasma Channels

Shosuke Karashima* and Takahisa Koike[†]

*Department of Electrical Engineering, Tokyo University of Science,
1-3 Kagurazaka, Shinjuku-ku, Tokyo 162-8601, Japan
[†]Department of Radiological Science, International University of Health and Welfare,
2600-1 Kiatakanemaru, Otawara-shi, Tochigi 324-8501, Japan

Abstract. When intense and short laser pulses pass through a preformed Z-pinch discharge plasmas, the wave guiding of the laser lights is created in performing the high density and high temperature plasma phase. The dynamics of the laser propagation is analyzed by the wave equation of the laser electric field coupled with the plasma hydrodynamic equations and the rate equations describing the plasma ion states. The optimum soft x-ray repetition lasing occurs in the $3s$-$3p$ transition in Ar IX, Ne-like plasmas, in the irradiated plasma channel. A gain coefficient 4.55 cm^{-1} without opacity was obtained by numerical computations.

INTRODUCTION

Optical guiding of intense laser pulses in plasma channels has play a key role in x-ray lasers, laser accelerators and inertial confinement fusion. We propose a theoretical model of the propagation of intense laser pulses irradiated in the discharge plasma channel. The intense and short laser pulses pass through preformed Z-pinch discharge plasmas. Guiding of these laser lights is created in performing the high density and high temperature plasma phase. The initial state of its preformed plasma takes the stagnation state after Z-pinch discharge. In the propagation process of the laser pulses the states of population inversion for repetition x-ray laser amplification through the plasma atomic processes may be generated in the plasma channel.

The propagation dynamics of laser pulses in the plasma channel is described by the wave equation of the laser electric field. The dynamical plasma phase is treated by the rate equations of the plasma ion states and the plasma hydrodynamic equations. These three relations are coupled each other. The effects of ponderomotive force on the momentum equation and tunneling ionization on the rate equations are taken into account in the model.

A Gaussian radial profile is assumed for the polarized laser field. The laser pulse guiding is analyzed through envelope of the laser spot size, phase shift, curvature and the amplitude of the laser field [1-3] along with the linear and nonlinear refractive index.

The present theoretical model applies to Ne-like Ar soft x-ray laser. The optimum soft x-ray lasing occurs in the $3s$-$3p$ transition in Ar IX plasmas [4,5]. The straight capillary

CP641, *X-Ray Lasers 2002: 8th International Conference on X-Ray Lasers*, edited by J. J. Rocca et al.
© 2002 American Institute of Physics 0-7354-0096-2/02/$19.00

containing the plasma channel is given by a 0.2 cm radius and several cm long. The envelope of laser field and the gain of soft x-ray lasing are evaluated.

MODELING AND FORMULATION

The repetition population inversion for lasing can be produced in the propagation processes through intense laser pulses in the plasma phase. The preformed plasma is created by means of capillary discharge method [4]. The propagation of short laser pulses in the plasma channel is analyzed based on the wave equation. The optimum conditions for soft x-ray lasing can be obtained from the plasma hydrodynamic equations and the coupled rate equations for the plasma ion charge states. These equations connect with each other through the plasma electron density and temperature.

The laser propagation in the plasma phase depends on free and bound electrons. The wave equation includes plasma current consisting of free electrons and a polarization current arising from the bound electrons. The polarization field P associates with the laser electric field E, and the plasma linear and nonlinear refractive index η_l and η_{nl}. The laser irradiating the capillary plasma is of the form:$E = E_0 / 2 \exp [i (kz - \omega t)] + c.c.$. The wave equation for this laser field, setting $\eta_l = 1$, becomes

$$\left(\nabla_\perp^2 + \frac{\partial^2}{\partial z^2} - \frac{1}{c^2}\frac{\partial^2}{\partial t^2} - \frac{\omega_p^2(r)}{c^2} + \beta \langle E \cdot E \rangle \right) E = 0, \tag{1}$$

where the nonlinear term β results in critical powers for focusing: $\beta = \beta_f + \beta_b$; $\beta_f = 1/2(e\omega_{p0} / mc^2\omega)^2$ and $\beta_b = (c/2\pi)(\omega^2/c^2) \eta_{nl}$, and ω_{p0} is the plasma frequency on axis ($r = 0$) and ω the characteristic laser frequency.

The preformed plasma channel provides an r-dependence for guiding of the laser pulse. The radial dependence of the plasma frequency is given by $\omega_p(r) = \omega_{p0} + [1 + (\Delta n/n_{p0})(r^2/r_c^2)]^{1/2}$; $n_{p0} + \Delta n$ is the density at the edge of the plasma channel $r = r_c$.

Here we change variables (z, t) in (z, ξ); $\xi = z - \eta_p ct$ with wave number $k = \eta_p \omega t$ and $\eta_p = (1 - \omega_{p0}^2/\omega^2)^{1/2}$. In terms of these independent variables, the wave equation for the complex amplitude $E_0(r, t)$ is given as follows:

$$\left(\nabla_\perp^2 + 2i\eta_p \frac{\omega}{c}\frac{\partial}{\partial z} + 2\frac{\partial^2}{\partial z \partial \xi} + (1 - \eta_p^2)\frac{\partial^2}{\partial \xi^2} + \frac{\partial^2}{\partial z^2} - k_c^2\frac{r^2}{r_c^2} + \beta E_0 \cdot E_0^* \right) E_0 = 0. \tag{2}$$

Finite pulse length effects are introduced for the second order operators. These terms can be simplified using the paraxial approximation. Here we assume the fundamental Gaussian amplitude for $E_0(r, t)$: $E_0 = b \exp[i\phi - (1 + i\theta)r^2/r_s^2]e$, where $b(z, \xi)$, $\phi(z, \xi)$, $\theta(z, \xi)$ and $r_s(z, \xi)$ stand for the amplitude, phase shift, curvature and spot size, respectively. Then we take the pulse amplitude to be $b_0(\xi) = b_0 \exp(-4\xi^2/L^2)$ and b_0 is the peak amplitude.

The population density of excited levels in ion species is calculated by the various atomic processes occurring in the plasma. We adopted the collisional-radiative model to obtain the population. Furthermore the effects of tunneling ionization rate due to the above threshold ionization (ATI) were taken into account. Then we can solve the coupled rate equations expressing the population density for each charge states in excited states.

We next consider the cylindrical plasma hydrodynamic equations, which consist of the continuity equation, the momentum conservation equation and the energy conservation for electrons and ions. The momentum equation includes the term of ponderomotive force: $-(n_e e^2/2m\omega)\nabla E^2$. The energy conservation equation for electrons includes the energy losses due to the resonance line radiation, collisional ionization, recombination processes and due to Bremsstrahlung radiation.

RESUTS AND DISCUSSION

The numerical calculations have been made using an wavelength $\lambda = 1.0$ μm, pulse length $L=20\mu$m, initial spot size $r_0=15\mu$m with pulse intensity 9.0×10^{16}W/cm^2, in the preformed Ar plasma electron density $n_e= 5.0 \times 10^{16}$ cm^{-3}. The normalized spot size r_s/r_0 is shown as a function of ξ and propagation distance z with finite pulse length effects in Fig.1. The pulse modulation is clearly apparent in the figure, as this fact was described [1,2] in the cases of with finite pulse length effects and in the absence of that.

Figure 2 shows the normalized laser beam intensity E_0^2/b_0^2 as a function of ξ/L and r/r_0. The direction of propagation is towards the right. The temporal and spatial distributions of electron densities are displayed in Fig. 3. The calculations show that the distributions were most enhanced at about 46 ns. The electron temperatures also show the same behavior as that. This behavior corresponds to the peak value of the calculated gain coefficients. The density profiles at radius above 50μm have a tendency to show larger plasma density gradient in the plasma column.

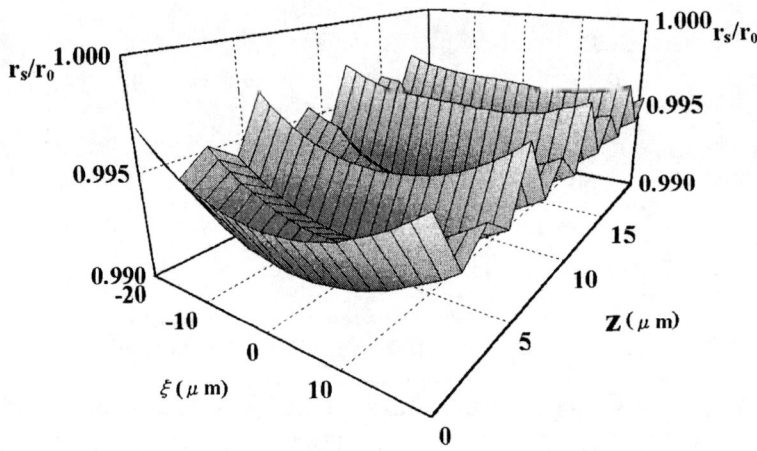

FIGURE 1. Surface plots of laser spot size r_s / r_0, as a function of ξ and propagation distance z.

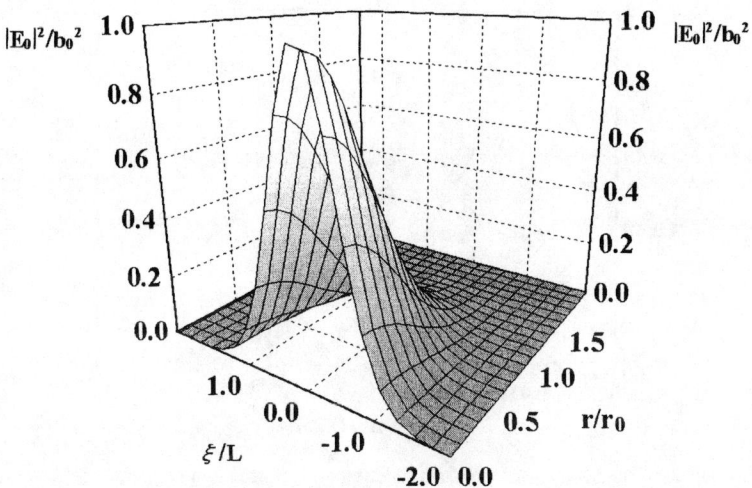

FIGURE 2. Surface plots of laser pulse normalized intensity as a function of ξ/L radial coordinate r/r_0.

The charge state fraction of Ne-like Ar plasma, Ar^{8+} is enhanced as shown in Fig.4. Accordingly the soft x-ray amplification occurs for $3s$-$3p$ transition in Ne-like Ar plasma. Here we have used Doppler broadening of the laser line and the Gaussian - shaped line profile for the gain calculations. The gain coefficient for this transition has been calculated in Fig.5 without opacity, solving for their excited level populations. The peak gain value is $g=4.55$ cm^{-1} at about 46 ns.

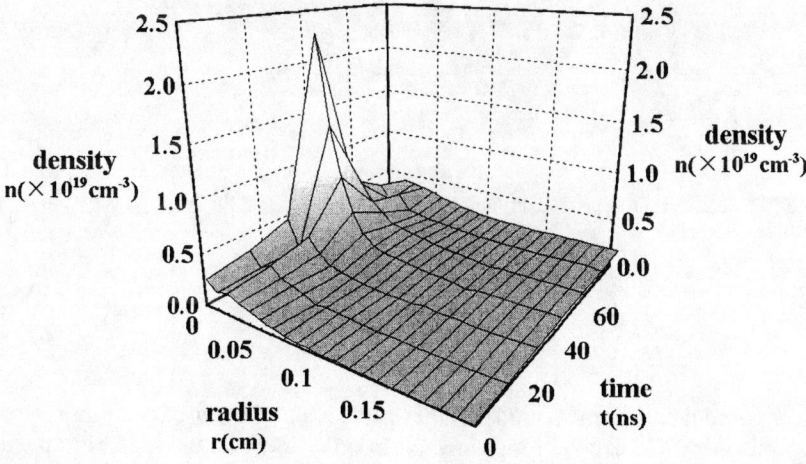

FIGURE 3. The temporal and spatial distributions of electron density ($n \times 10^{19}$cm^{-3}).

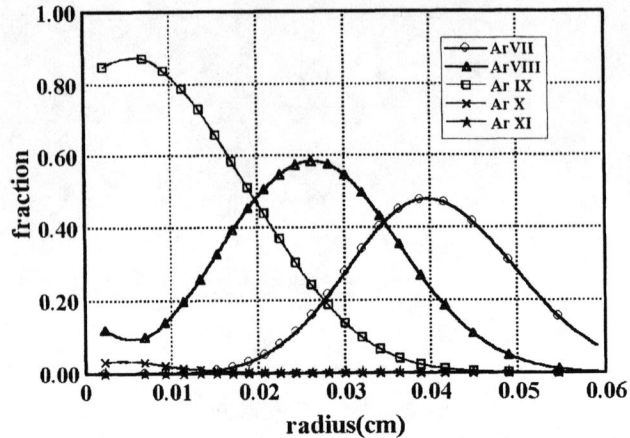

FIGURE 4. Charge state fraction as a function of distance from capillary central axis.

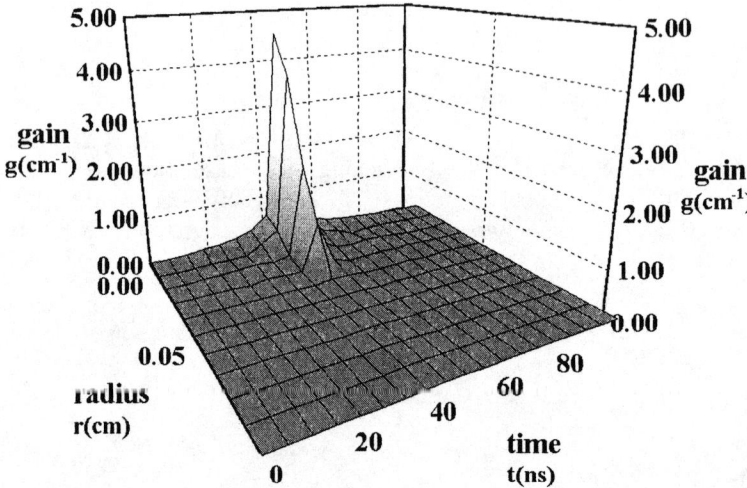

FIGURE 5. Temporal and spatial distributions of the gain coefficient for $3s$-$3p$ transition in Ne-like Ar plasma.

CONCLUSION

We proposed the theoretical model of the propagation of intense laser pluses irradiated in the preformed discharge plasma channel. In order to make the repetition population inversion for soft x-ray lasing, we considered a hybrid plasma channel model. From the simulation results of the stable propagation of incident laser pulses in the plasma channel, the model will be able to make repetition amplification of x-ray laser. In the future, the stably repetition rate x-ray laser will be more important for many kinds of applications.

The authors would like to thank Y. Makiuchi and K. Kobayasi for support numerical calculations.

REFERENCES

1. Sprangle, P., Hafizi, B., and Serafim, P., Phys. Rev. E **59**, 3614-3623 (1999).
2. Sprangle, P., Hafizi, B., and Serafim, P., Phys. Rev. Letters **82**, 1173-1176 (1999).
3. Esarey, E., and Leemans, W. P., Phys. Rev. E **59**, 1082-1095 (1999).
4. Moreno, C. H., Marconi, M. C., Shlyaptsev, V. N., Benware, B.R., Marcchietto, C. D., Chilla, J. L. A., Rocca, J. J., and Osterheld, A. L., Phys. Rev. A **58**, 1509-1514 (1998).
5. Karashima, S., Koike, T., and Makiuchi, Y., J. Phys. Soc. Jpn. **70**, 996-1001 (2001).

Sub-picosecond resolved investigation of dynamics of laser driven plasma by broad band chirped-pulse spectral interferometry

[a,b]Jiansheng Liu, [a]Ruxin Li, [a]Zhinan Zeng, [a]Zhizhan Xu and [b]Jingru Liu

[a]*Laboratory for High Intensity Optics, Shanghai Institute of Optics and Fine Mechanics, Chinese Academy of Sciences, P. O. Box 800-211, Shanghai 201800, P. R. China*
[b]*Northwest Institute of Nuclear Technology, Xian 710024, PR China*

Abstract: Owing to the intrinsic broad spectrum of ultrashort femtosecond laser, the linearly chirped laser can be used as a probe to real-time diagnose the dynamics of laser-driven plasma. In this paper, the probing methods and the probing resolution have been discussed. Using a spectral interferometry, the chirped-pulse spectral interferogram in laser-gas plasma has been obtained and the time-resolved electron density can be calculated.

INTRODUCTION

The interaction of ultrashort intense laser with matter would produce ultrafast dynamic processes such as the transient evolution of electron transportation of laser-driven plasma. The experimental knowledge of the dynamics of laser-driven plasma is important for us to understand the mechanics of laser-matter interaction and the scheme of x-ray laser generation in the laser-driven plasma. Recently[1,2], owing to the intrinsic broad spectrum of ultrashort femtosecond laser, the chirped laser can be used as a probe to real-time diagnose the dynamics of laser-driven plasma. In this paper, the probing methods and the probing resolution have been discussed. Using a time-resolved chirped-pulse spectral interferometry, the time profile of the electron density created by an intense short laser pulse has been measured.

PRINCIPLE OF MEASUREMENT

The principle of measuring the ultrafast dynamic processes using chirped pulse laser can be understood according to Fig. 1. After transmitting through a time-dependent process $T(t)$ which we call detect signal, the probe laser is modulated and thereafter recorded by a spectrometer which is a combination of a grating lens and a detector array such as 2-dimensional CCD. The ultrashort pulse laser may be assumed as an unchirped Gaussian

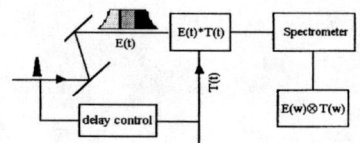

Fig. 1 The schematic diagram of single shot time-resolved measurement

pulse with a central frequency ω_0: $E(t) = \exp(-a_0 t^2 + j\omega_0 t)$. After being stretched by grating pairs, the electric field of the linearly chirped laser can be written in the form

$$E_1(t) = \exp(-at^2 + j(\omega_0 t + bt^2)) \tag{1}$$

When the chirped probe laser copropagates through the time-dependent $T(t)$, the transmitted probe laser can be written as $E_2(t) = E_1(t) \times T(t)$, and the measured spectral distribution corresponds to its Fourier transformation

$$\tilde{E}_2(\omega) = Fourier(E_1(t) \times T(t)) = \int E_1(t) \times T(t) \exp(-i\omega t) dt = \tilde{E}_1(\omega) \otimes \tilde{T}(\omega) \tag{2}$$

Where $\otimes$ is the convolution sign, and

$$\tilde{E}_1(\omega) = \exp(-\frac{1}{4}\frac{a}{a^2+b^2}(\omega-\omega_0)^2 - j\frac{1}{4}\frac{a}{a^2+b^2}(\omega-\omega_0)^2) \tag{3}$$

When equation (3) is introduced into the equation (2), we have

$$\tilde{E}_2(\omega) = \int \exp(-\frac{1}{4}\frac{a}{a^2+b^2}(\omega'-(\omega-\omega_0))^2 - j\frac{1}{4}\frac{b}{a^2+b^2}(\omega'-(\omega-\omega_0))^2)\tilde{T}(\omega')d\omega' \tag{4}$$

We define here $\omega_1 = \omega - \omega_0$. Since the detect signal $T(t)$ is a much slowly-varied process related to the unchirped pulse laser, its spectral bandwidth is much narrower than that of the laser $(<< \Delta\omega_0)$ ☐the above equation can be simply written as

$$\tilde{E}_2(\omega) \approx \exp(-\frac{1}{4}\frac{a}{a^2+b^2}\omega_1^2 - j\frac{1}{4}\frac{b}{a^2+b^2}\omega_1^2)\int \exp(-j\frac{1}{2}\frac{b\omega_1\omega'}{a^2+b^2})\tilde{T}(\omega')d\omega' \tag{5}$$

$$= \tilde{E}_1(\omega) \times T(\frac{1}{2}\frac{b\omega_1}{a^2+b^2}) = \tilde{E}_1(\omega) \times T(t)$$

Where $t = \frac{1}{2}\frac{b\omega_1}{a^2+b^2}$ ☐that means that there exists a linear relation between time and frequency domain. When the spectrometer has such a high resolution that the spectral response function can be assumed as a δ function, and γ is the detect efficiency of the spectrometer, the measured spectral distribution can be written as

$$N_2(\omega) = \gamma \int g(\omega' - \omega) \times \left| \tilde{E}_2(\omega') \right|^2 d\omega' = \gamma \times \left| \tilde{E}_2(\omega) \right|^2 \qquad (6)$$

The measured signal where the detect signal $T(t)$ switches off can also be written as : $N_1(\omega) = \gamma \times \left| \tilde{E}_1(\omega) \right|^2$. Therefore the detect signal $T(t)$ can be extracted by

$$T(t) = T(\frac{1}{2} \frac{b\omega_1}{a^2 + b^2}) = \sqrt{\frac{N_2(\omega_1)}{N_1(\omega_1)}} = \sqrt{\frac{N_2(\omega - \omega_0)}{N_1(\omega - \omega_0)}} \approx \sqrt{\frac{N_2(\omega_0 + 2bt)}{N_1(\omega_0 + 2bt)}} \qquad (7)$$

A more detailed analysis of temporal resolution can be done by assuming a Gaussian form of the detected signal, and the temporal resolution of this method can be given as $\tau_r \approx \sqrt{2N}\tau_0 = 8\ln 2\sqrt{N}/\Delta\omega_0$ if the spectral resolution of spectrometer is so high. It means that the temporal resolution is proportional to the square root of stretching fold N and inversely proportional to the spectral bandwidth of the probe laser.

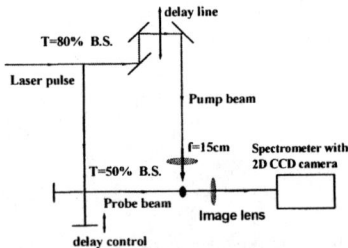

Fig. 2 The experimental setup

When an ultrafast process only change the phase of probe laser, e.g. when a probe laser transmits through laser-driven plasma or reflects from the critical surface etc., the measurement can be done using the technique of spectral interferometry as shown in Fig. 2. Similar to the analysis given above, the spectrum of the two laser beam is given by

$$\tilde{E}_2(\omega) = Fourier((E_1(t-\tau) + E_1(t)) \times T(t)) = \int (E_1(t-\tau) + E_1(t)) \times T(t)\exp(-i\omega t)dt$$
$$\approx \tilde{E}_1(\omega)\exp(j\omega\tau) \times T(\frac{b}{2}\frac{\omega - \omega_0}{a^2 + b^2} - \tau) + \tilde{E}_1(\omega) \times T(\frac{b}{2}\frac{\omega - \omega_0}{a^2 + b^2}) \qquad (8)$$

Assuming the phase function has a general form $T(t) = \exp(-j\phi(t))$, then the recorded spectral interference pattern is represented by the following form

$$N_2(\omega) = \gamma \left| \tilde{E}_2(\omega) \right|^2 = \gamma \left| \tilde{E}_1(\omega) \right|^2 (2 + 2\cos(\omega\tau - \phi(t))) = N_1(\omega)(2 + 2\cos(\omega\tau - (\phi(t-\tau) - \phi(t)))) \qquad (9)$$

EXPERIMENTAL RESULTS

The experiments were performed with a 10-Hz chirped-pulse amplification Ti:sapphire laser system, which could deliver laser pulses with a duration of 45 fs and an output energy as great as 450 mJ. However, in our experiments only 100-mJ laser pulses were used. The experimental arrangement is shown in Fig. 2. A linearly chirped pulse laser was deliberately generated from the laser system by introduction of a slight offset in the distance between the two gratings of the compressor. The pulse was then divided into two parts by a beam splitter with reflectivity of 20% and transmission of 80%. The higher-energy laser was used as pump laser and focused into air to produce plasma which would be investigated by the technique of spectral interferometry. The lower-energy laser was used as probe laser which was divided into two separate laser pulses by a Michelson interferometer setup, and the two laser pulses were separated by 1.5 ps and transmit through the plasma. The time delay between the pump and probe laser could be adjusted by the translation of the delay

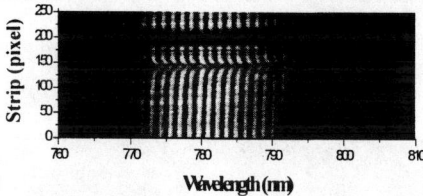

Fig. 3 Spectral interferogram in laser-gas plasma obtained by use of 200-fs chirped probe pulses separated by 1.5 ps.

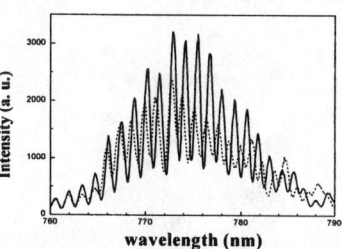

Fig. 4 The lineouts taken from Fig. 4. solid curve,

line and the synchronization between them was guaranteed. The two laser pulses which has transmitted through the plasma were imaged onto the slit of the spectrometer, and the spectral interferogram could be recorded on a single-shot basis. Fig. 3 shows a typical chirped-pulse spectral interferogram. Lineouts used to measure the fringe-shift change from Fig. 3 are shown in Fig. 4, and the calculated electron density as a function of time is shown in Fig. 5.

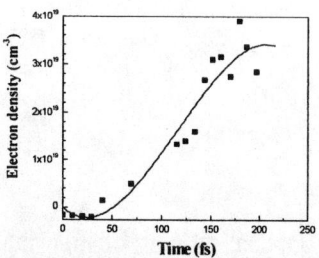

Fig. 5 Calculated electron density as a function of time (filled squares). The solid curve is polynomial fit curve.

ACKNOWLEDGMENTS

This work is supported by the major basic research project of Chinese Academy of Sciences, the Chinese High-Tech Program, the Chinese National Major Basic Research Development Program, the Chinese National Natural Science Foundation (Contracts No.19774058 and No.69925513).

REFERENCES

1. Chien, C. Y., Fontaine, B. La., Desparois, A., Jiang, Z., Johnston, T. W., Kieffer, J. C., Pepin, H., Vidal, F., and Mercure, H. P.,Opt. Lett. **25**, 578-580 (2000).
2. Jiang Zhiping, and Zhang Xi-Cheng, IEEE Journal of Quantum Electronics 36,1214-1222 (2000).

Demonstration of a Plasma Waveguide in a Fast Argon Capillary Discharge

B.M. Luther[*], Y. Wang[*], M.C.Marconi [*†], and J.J. Rocca[*]

[*]Department of Electrical and Computer Engineering, Colorado State University, Fort Collins, CO 80523, USA

[†]Department of Physics, University of Buenos Aires, Buenos Aires, Argentina

Abstract. We report results of the study of a plasma waveguide created by a fast argon capillary discharge. The rapid plasma compression that occurs following a fast current pulse generates an electron density profile with a minimum on axis which constitutes a suitable channel for the optical guiding of intense laser pulses. Interferometric measurements that directly map the evolution of the electron density profile of the waveguide are presented. Guided beam propagation of a 40 mJ, 2.5 ps, laser pulse by an 11 cm long capillary discharge is reported, along with measurements of the near field and far field distributions of the guided beam.

INTRODUCTION

Optical guiding of high intensity laser pulses over many Rayleigh lengths is of significant interest for the efficient longitudinal excitation of soft x-ray lasers with intense laser pulses. Plasma waveguides have the ability to provide long interaction lengths between the pump laser pulse and the laser media, overcoming the limitations imposed by the diffraction and ionization–induced refraction of the laser beam. Several methods for the generation of plasma waveguides have been described in the literature, including hydrodynamic expansion of a plasma following a laser-produced spark [1]; current induced capillary wall ablation [2-5], implosion of the plasma in a Z-pinch discharge [6], and slow discharge through a hydrogen filled micro-capillary [7]. All of these methods depend on the creation of an index of refraction that decreases radially from the axis of propagation of the beam, which in turn requires an electron density N_e that increases with radius.

In the field of x-ray lasers, ablative capillary discharges have been used to create plasma waveguides for collisional excitation [5] and recombination lasers [8]. An ablative capillary discharge plasma waveguide developed at Colorado State University was used at the Max Born Institute to demonstrate amplification (gxl=6.8) at 60.8nm by collisional excitation of Ne-like S in laser-heated sulfur capillary plasmas up to 3 cm in length [5]. In this discharge the current pulse ablates the

CP641, X-Ray Lasers 2002: 8th International Conference on X-Ray Lasers, edited by J. J. Rocca et al.
© 2002 American Institute of Physics 0-7354-0096-2/02/$19.00

capillary walls, forming high-density low temperature plasma near the walls, and increased current density and Ohmic heating on axis [4]. This results in an electron temperature distribution that decreases as a function of radius towards the walls and, due to pressure balance, a corresponding opposite density profile that forms a waveguide [9]. Herein we report the demonstration of a plasma waveguide created by a fast Argon capillary discharge of the type used to develop the collisional discharge-pumped soft x-ray lasers [10]. These discharges have been demonstrated to generate axially uniform plasma columns up to 36 cm in length that can be fired at repetition rates of several Hz [11]. In these discharges a shock wave originates in the vicinity of the capillary wall as a result of the strong current-induced JxB force and heating in the skin layer [12]. The shockwave propagates towards the axis forming a waveguide of continuously decreasing diameter until it collapses on axis. In the next section we present the results of interferometry measurements that show the formation and evolution of the waveguide, and also report the guiding of a 40 mJ, 2.5 ps laser pulse in an 11 cm long plasma column

EXPERIMENTS AND RESULTS

The common part of the apparatus used in all the experiments consist of a fast Ar-filled capillary discharge and a Titanium Sapphire short pulse laser. The discharge utilizes a 3.2 mm diameter, 11 cm long, ceramic capillary, filled with 250 to 1200 mtorr of Ar. Shorter capillaries were used for some interferometry experiments. The discharge current pulses typically had a peak amplitude of 15 to 20 kA, and a half-period of ~ 120 ns. The short laser pulses were generated with a KM Labs oscillator and amplified in a two stage Ti:Sapphire amplifier. The ~ 50 fs mode locked output of the oscillator is stretched to 170 ps and amplified to ~ 3 mJ/pulse in eight passes through a 1st stage amplifier. The output of the 1st stage amplifier is sent into a five pass amplifier and amplified to ~ 200 mJ. This beam is then compressed in a vacuum compressor giving ~ 150 mJ pulses. Pulse widths of ~ 2.5 ps were selected to conduct these experiments.

The formation of the plasma waveguide was measured by interferometry using the third harmonic (267 nm) of the compressed first stage amplifier. The third harmonic beam was sent through a Mach-Zehnder interferometer containing the capillary discharge. An imaging lens was placed after the combining beam splitter to image the capillary onto a CCD camera. The unused portion of the 267 nm beam from the beam splitter was sent to a photodiode to monitor the time of arrival of the probe pulse with respect to the discharge current pulse which was monitored with a Rogowski coil. Figure 1 shows a series of interferograms depicting the waveguide formation and early part of its evolution. The images show the propagation of the shockwave towards the axis. The annular plasma shell with density minimum on the center forms a waveguide of continuously decreasing diameter and increasing density, until it collapses on axis. Figure 2 shows the electron density profiles obtained from the interferograms for three different Ar pressures. They correspond to the early part of the plasma evolution, when the diameter of the column is large and

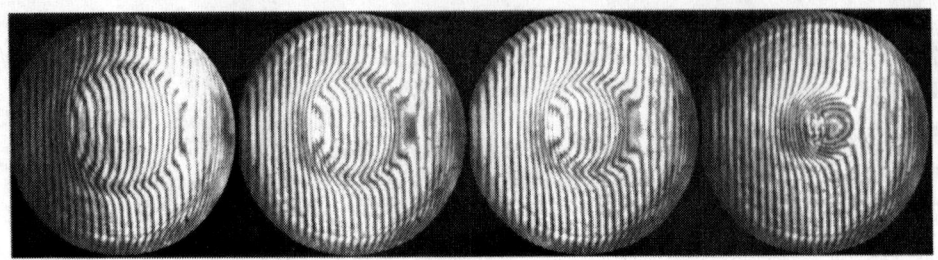

FIGURE 1. Time progression of interferograms showing the plasma column compression and waveguide development for a 250 mTorr Ar discharge. Times are 35.6, 37.1, 38.3, and 43 ns from the initiation of the current pulse respectively. Capillary diameter is 3.2 mm.

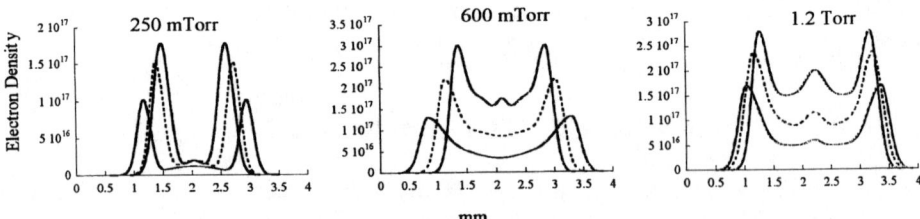

FIGURE 2. Time progression of electron density profiles (cm^{-3}) as determined from interferograms, at three Ar pressures. Electron density is increasing with time. Results correspond to the early stage of the plasma column evolution and waveguide formation.

the plasma density is in the range of $1\text{-}3\ 10^{17}\ cm^{-3}$. Latter in time the waveguides are narrower, with higher density on axis. A trend is observed towards an earlier increase of the electron density on axis with increasing Ar pressure.

The guiding ability of these plasma columns was tested using 40 mJ, 2.5 ps, 800 nm Ti:Sapphire laser pulses. Three experiments were performed: transmission, near field imaging, and far field imaging. For the transmission experiment a 1.2 mm pinhole was placed at the exit of the capillary. A 4 cm diameter Ti:Sp beam was focused at the entrance of the capillary using an f = 3.2 m effective lens, giving a calculated ω_0 = 40 μm, a Rayleigh length Z_r = 0.65 cm, and a guided length of approximately 16 Z_r. The beam was monitored before and after the capillary using beam splitters and photodiodes. Figure 3 shows the transmittance of the 40 mJ pulse, as a function of delay from the current pulse initiation, for a capillary filled with 250 mtorr Ar. The transmission can be seen to undergo an increase by a factor of five approximately 30 ns after the initiation of the current pulse. The transmission then falls to zero due to refraction and absorption losses as the plasma collapses on axis. This is consistent with the interferograms. It should be noticed that the plot shows only relative transmission and that a transmission of 1 in the plot does not imply 100% throughput. Nevertheless, the early waveguides that correspond to low plasma density were observed to transmit practically the entire beam energy.

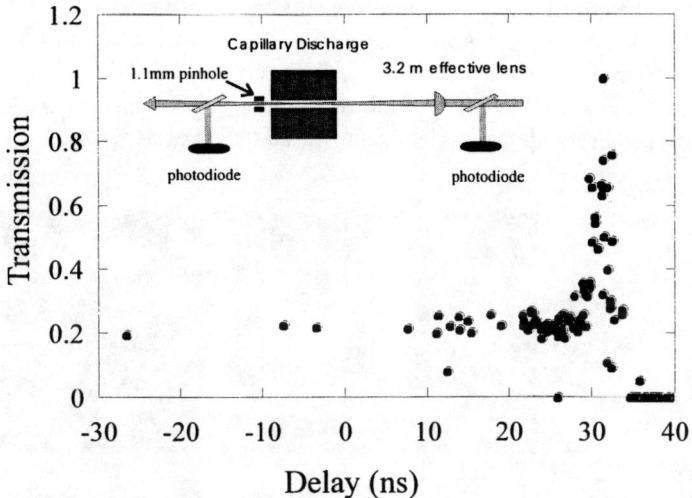

FIGURE 3. Transmission of 40mJ, 2.5 ps pulses through a 1.2 mm pinhole at the end of an 11 cm long plasma column as a function of delay with respect to the beginning of the current pulse. The inset shows the experimental setup. The Ar discharge pressure was 250 mtorr.

Strong beam guiding was also evident in the far and near field profiles of the transmitted beam. Figure 4a shows the far field burns of the 40 mJ pulses taken at 1 meter from the capillary exit. The two large burns on the lower left were taken without plasma present and measure ~ 16mm in diameter. Assuming a beam with $M^2=1$ the beam divergence associated with these burns are computed to correspond to a beam waist of $\omega_0 \sim 32$ μm. The small (numbered) burns, which correspond to transmission of the plasma guided beam, are measured to have a diameter of ~ 4 mm. This results in a calculated $\omega_0 \sim 127$ μm. It should be emphasized that these measurements are only approximate since the $1/e^2$ value cannot be accurately determined from the burns, and the M^2 value of the laser beam is not included. The multiple burn patterns in Fig. 4 also illustrate the reproducibility of the guiding over several shots. A more accurate measurement of the guided beam profile was obtained imaging the exit of the capillary. The near field beam profiles were obtained using a 0.5 m lens to image the capillary exit onto a CCD camera. Figure 4b shows the near field beam profile corresponding to a 600 mtorr Ar discharge with no plasma present and Fig. 4c shows the near field profile of the guided beam with a HWHM ~ 150 μm . The near field profiles are consistent with the interferometrically determined electron density profiles. For a parabolic electron density profile it can be shown that the guided beam radius r_{beam} is related to the electron density profile by [4]:

$$r_{beam} = (r_w^2/\pi r_e \Delta N)^{1/4} \qquad (1)$$

where ΔN is the difference in the electron density between the base of the parabola and its value at an arbitrary position r_w, and r_e is the classical electron radius. The 600 mtorr cuts from Fig.2 give values of: r_w between 0.12, and 0.072 cm, and ΔN between 9.2×10^{16} and 1.3×10^{17} cm^{-3}, which result in r_{beam} between 200 and 145 μm, in reasonable agreement with the guided beam with a radius of ~150um.

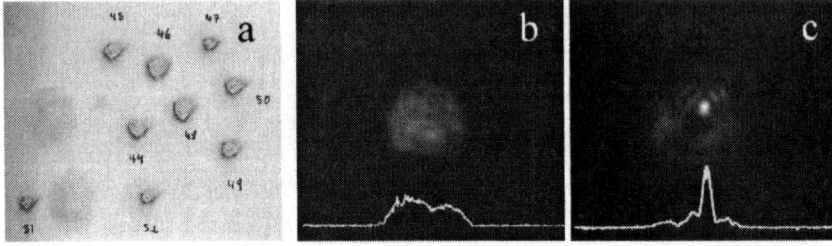

FIGURE 4. a). Far field burns for 40 mJ, 2.5 ps pulses taken 1m from the capillary exit. The large burns to the left were obtained with the discharge off. The smaller (numbered) burns correspond to guided pulses. Near field images and beam profiles of unguided (no plasma) (b) and guided (c) beams for 600 mtorr discharge taken near the time of maximum beam transmission through the capillary.

Significantly smaller guided beams were observed in 250 mTorr discharges. The stronger shockwaves and slower increase of the axial plasma density at this pressure result in the formation of narrower and denser waveguides close to the pinch time. Figure 5a shows the evolution of the interferometrically measured electron density profile in a 250 mTorr discharge for the time interval that shortly precedes the collision of the shock wave on axis. Figure 5b compares the near field images of an unguided pulse (top) to that of a pulse guided at a time near the end of the compression. The intensity distribution (Fig. 5c) of the guided pulse corresponds to the lowest mode of the waveguide and has a measured e^{-2} radius of ~ 40 μm (~ 50 μm

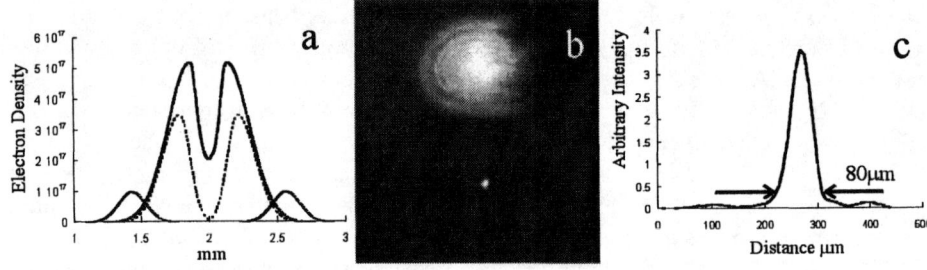

Figure 5. Measurements for a 250 mtorr discharge at times of a few ns before the maximum plasma column compression. a) measured evolution of the electron density (cm^{-3}) profiles. b) Near field image of unguided (top) and guided (bottom) beam ~ 1ns before the time of maximum compression of the plasma column. c) Intensity distribution of guided 1 mJ beam after propagating ~ 48 Rayleigh lengths through a 5.5 cm long capillary corresponding to the near field image shown in b).

FWHM diam.). This measured spot size corresponds well with the beam radius of 42 μm computed from (1) using the electrons density profile of the latest time (top trace) shown in Fig. 5a. Near field images at earlier times show different TEM waveguide modes whose n,m values rapidly decrease as the plasma column compresses.

CONCLUSIONS

We have measured electron density profiles with a transient minimum on axis in a fast Ar capillary discharge that are suitable for guiding intense laser pulses over long distances. The compressing plasma column forms a waveguide that sustains guided beams of progressively smaller spot diameter. Guided propagation of beams with 50 μm FWHM diameter is observed shortly before the shockwave collides on axis. Guided propagation of 40 mJ, 2.5 ps pulses through an 11 cm long plasma column was demonstrated. Far field burns and near field images are consistent with the beam sizes calculated from interferometrically measured electron density profiles.

ACKNOWLEDGEMENTS

We would like to acknowledge the contributions of Alex Klix, Luca Furfaro, Jorge Filevich, and V.N. Shlyaptsev, and thank Sterling Backus, Henry Kapteyn, and Margaret Murnane for their help with the Titanium Sapphire system. This work was funded by the National Science Foundation. We also gratefully acknowledge a grant from the W. M. Keck Foundation.

REFERENCES

1. C. G. Durfee and H.M. Milchberg, Phys. Rev. Lett. 71, 2409, (1993); C.G. Durfee III, J. Lynch, and H.M. Milchberg, *Phys. Rev. E*, **51**, 2368-2389 (1995)
2. Y. Ehrhlich, C. Cohen, Z. Zigler, J. Krall, P. Sprangle and E. Esaray. Phys. Rev. Lett. 77, 4186, (1996).
3. J.J. Rocca, F.Tomaserl, M. C. Marconi, J.L.A. Chilla, C.H. Moreno, B.R. Benware, V.N. Shlyaptsev, J. Gonzalez, and C. Macchietto. P 164. SPIE vol 3156, (1997).
4. P. Sprangle, E.Esarey, J. Krall, and G. Joyce, *Phys. Rev. Lett.*, **69**, 2200-2203 (1992)
5. K. A. Janulewicz, J. J. Rocca, F. Bortolotto, M. P. Kalachnikov, V. N. Shlyaptsev, W. Sandner, and P. V.Nickles, Physical Review A 63, 033803 (2001).
6. T. Hosokai, M. Kando, H. Dewa, H. Kotaki, S. Kondo, N. Hasegawa, K. Nakajima, K.Horioka *Opt. Lett.*, **25** (2000)
7. D.J. Spence and S.M. Hooker, Phys. Rev. E, 63, 015401(R) (2000)
8. Y. Avitzour, I. Gelzner, Y. Ping and S. Suckewer, in these proceedings.
9. M.C. Marconi, C.H. Moreno, J.J. Rocca, V.N. Shlyaptsev and A.L. Osterheld, , Physical Review E **62,** 7209, (2000).
10. B.R. Benware, C.D. Macchietto, C.H. Moreno and J.J. Rocca, Phys. Rev. Lett. **81**, 5804, (1998).
11. C.D. Macchietto, B.R. Benware and J.J. Rocca, Optics Lett. **24**, 1115, (1999).
12. J.J. Gonzalez, M. Frati, J.J. Rocca, V.N. Shlyaptsev and A.L. Osterheld. Phys. Rev. E, 65, 026404, (2002).

Investigation of X-ray Amplification in Neon Clusters

T. Mocek[1], J.J. Park[2], C.M. Kim[2], H.T. Kim[2], D.G. Lee[2], K.H. Hong[2], and C.H. Nam[2]

[1]*Department of Gas Lasers/PALS Research Center, Institute of Physics, Na Slovance 2, Prague 8, Czech Republic*

[2]*Department of Physics and Coherent X-Ray Research Center, KAIST, Taejon 305-701, Korea*

Abstract. We have experimentally investigated the feasibility of X-ray lasing in Li-like Ne ions produced by the interaction of intense, 28-fs laser pulses with gaseous clusters of neon. The idea of the proposed recombination ($n=3\text{-}4$) X-ray laser is based on adiabatic cooling of fast expanding small-sized Ne clusters. The driving laser beam was focused into a cryogenically cooled gas jet of Ne, and time-integrated spectra were simultaneously measured in transverse and longitudinal direction with respect to the laser axis. The transverse spectrum from Ne at the temperature of about -137 °C showed dominant emission from highly placed levels ($n=4$) in Ne^{7+} ions whilst the resonance lines ($n=2\text{-}3$) at 8.8, 9.8, and 10.3 nm remained very weak suggesting enhanced recombination from Ne^{8+}. In the longitudinal spectrum, the lasing line candidate corresponding to the 3d-4f transition of Ne^{7+} at 29.24 nm has been observed as a distinct, but not dominant.

INTRODUCTION

Tremendous advances in ultrashort laser technology has made it possible to produce femtosecond pulses at the multiterrawatt power level and thus has opened up the opportunity to scale down X-ray laser schemes. In particular, inversion schemes for recombination and collisionally excited soft X-ray lasers, based on optical field ionization (OFI), have been proposed and demonstrated [1,2]. While the repetition rate of these sources already reached 10 Hz, the number of generated photons is still limited by the low density (< 20 Torr) of the amplifying medium. A new challenging topic is how to increase the density of lasant material and thus to boost up the X-ray laser output. The use of clusters as a novel gain medium for X-ray lasers has been proposed [3,4] as an efficient method to produce highly charged ions at modest laser intensities. The main characteristic of a gaseous cluster is the near-solid-density inside the cluster while the average density is low which leads to an enhanced absorption of laser energy via collisional and resonance absorption. In particular, a small-sized cluster could be a promising medium for a recombination X-ray laser. Our previous data [5] suggested that a rapid expansion of neon cluster plasma accompanied by adiabatic cooling may lead to strong recombination of Ne^{8+} ions into highly-placed excited levels of Ne^{7+}. Population inversion can be expected on several $n=3\text{-}4$

CP641, *X-Ray Lasers 2002: 8th International Conference on X-Ray Lasers,* edited by J. J. Rocca et al.
© 2002 American Institute of Physics 0-7354-0096-2/02/$19.00

transitions in Ne^{7+} with wavelengths ranging from 26 nm up to 30 nm. To date, there has been no experimental study on X-ray laser using cluster target.

EXPERIMENTAL SETUP

The experiment has been performed at KAIST with a 3-TW Ti:sapphire laser system delivering 28-fs, 50-mJ (on the target) pulses at a fundamental wavelength of 825 nm. The laser is based on chirped pulse amplification technique and has been described in detail elsewhere [6]. The prepulse to main pulse ratio better than 10^{-5} has been achieved [7] to avoid presence of a prepulse which could result in the fragmentation of clusters and creation of an underdense plasma before arrival of the main heating pulse. The linearly polarized laser beam was focused into a cryogenically cooled gas jet of neon with a f=45 cm spherical mirror, yielding the maximum peak intensity of about 7 x 10^{16} W/cm^2. The gas jet was produced by a solenoid-driven pulsed valve fitted with a 0.3 mm diameter circular nozzle. The neon gas was cooled before expansion by passing through a reservoir filled with cold nitrogen gas which was forced under pressure from a separate liquid nitrogen bath. Using this arrangement gas could be gradually cooled from the room temperature down to about -137 °C. In contrast to the case of nitrogen we have not observed any significant drop of backing pressure in the valve that would indicate formation of liquid droplets instead of clusters under these conditions. The valve was operated at a backing pressure of 30 bar and a repetition rate of 10 Hz. The laser illuminated the gas jet perpendicularly with respect to the flow of the gas and the laser focus could be placed at a variable height ranging from 0.1 mm up to ~ 2 mm (0.2 mm being the typical value) above the nozzle tip. The time-integrated soft X-ray emission was measured transversely and longitudinally with respect to the laser axis, by means of two space-resolving, flat-field XUV spectrometers [8]. The spectral resolution was approximately 0.02 nm near 20 nm, and the resolving power was about 1000. The transverse spectrograph was operated without any filter and its wavelength range was from 4 nm up to ~ 40 nm. In the case of axial spectrograph, a 0.4-µm-thick aluminum filter was placed in front of CCD to block the stray laser light from the pump laser and the detectable spectral range was from 18 nm to 38 nm. The measured spectra were corrected for the diffraction efficiency of the grating, reflectivity of the toroidal mirror, quantum efficiency of the CCD, and filter transmission.

RESULTS

Figure 1(a) shows transverse spectra from neon obtained at (a) room temperature and (b) at −137 °C. The gas jet backing pressure was 30 bar and the signal was integrated over 10 shots. The spectrum in Fig. 1(a) is dominated by emission from Ne^{3+}, Ne^{4+}, and Ne $^{5+}$ charge states. This finding is consistent with the peak value of laser intensity (~7 x 10^{16} W/cm^2) which is sufficient to strip neutral atoms up to Ne^{6+} ions exclusively by OFI. It is then followed by recombination to the lower ionization states and emission of spectral lines up to Ne^{5+}. In sharp contrast, the spectrum obtained at –137 °C (Fig. 1(b)) shows very strong lines from Ne^{7+} (Li-like) as a result of collisional heating and ionization in near-solid-density neon clusters. A remarkable feature observed in Fig. 1(b) is the dominant emission on lines from n=4 and 5 levels of Ne^{7+}, when compared to the resonance lines of Ne^{7+}: 2s-3p at 8.81 nm, 2p-3d at 9.82 nm, and 2p-3s at 10.3 nm. It suggests that the adiabatic cooling of fast-expanding small-sized neon clusters resulted in an enhanced three-body recombination from Ne^{8+} and a population inversion between the level s n=4 and 3 might be established.

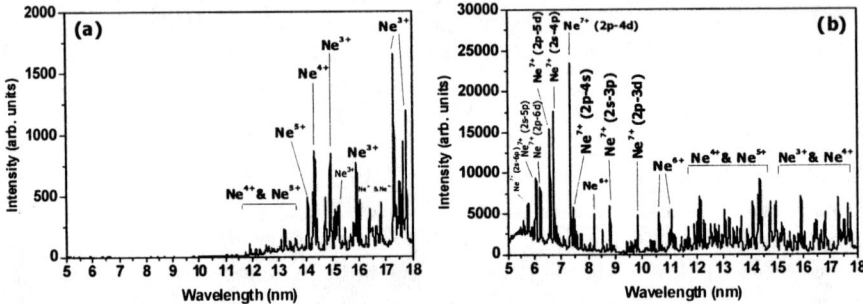

FIGURE 1. Transverse spectrum of neon (p=30 bar) irradiated by 28-fs laser pulses (10 shots) at (a) room temperature of +23 °C compared with the spectrum at (b) -137 °C.

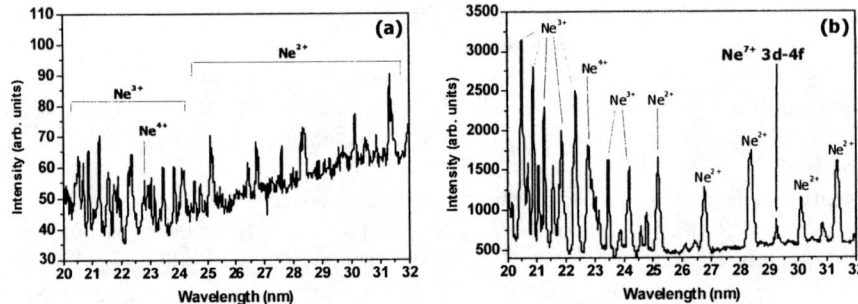

FIGURE 2. Axial spectra from neon (p=30 bar, 10 shots) obtained at (a) T = +23 °C and (b) T = -137 °C, respectively.

To explore the possibility of X-ray lasing in neon cluster plasma, we have measured axial spectra under various experimental conditions. In particular, we were looking for stimulated emission on the following transitions in Ne^{7+}: 3s-4p at 26.03 nm, 3p-4d at 28.23 nm and 28.26 nm, 3d-4f at 29.24 nm, and 3p-4s at 29.87 nm and 29.91 nm, respectively. The alignment of axial spectrograph was checked by inserting a small iris into the path of the driving laser beam to make favorable conditions for generation of high order harmonics. For neon at the room temperature and backing pressure of 20 bar we have observed bright, blue-shifted high harmonics up to the 47th order. When the iris was fully opened and pressure increased to 30 bar, only weak lines from low-charged neon ions (Ne^{2+} and Ne^{3+}) were identified in the axial spectrum (Fig. 2(a)). Figure 2(b) shows axial spectrum obtained at the temperature of –137 °C. The main differences between the spectrum in Fig. 2(b) and that in 2(a) are: 1) much higher signal level due to high density, hot, spatially larger plasma, and 2) presence of the Ne^{7+} 3d-4f candidate line at 29.24 nm. Note that this line was absent in transverse spectrum obtained under same conditions. While changing experimental parameters such as pressure, height of the focus, laser energy, etc. this line became either weaker or completely disappeared. None of the other candidate lines was found in the spectrum. Our data suggest that population inversion could bee achieved over a limited plasma length, however the conditions for amplification and propagation of X-ray beam were not optimized in this preliminary experiment. Of particular importance is the deflection and absorption due to very high plasma density. Relativistic self-focusing or channeling might be a solution of this problem. We will address these issues in future investigations using an improved alignment method and focusing geometry.

SUMMARY

In summary, a preliminary experimental study aimed at demonstration of soft X-ray lasing in neon cluster plasma has been performed. By focusing a 28-fs laser pulse at an intensity of ~7 x 10^{16} W/cm^2 into a cryogenically cooled gas jet of neon a plasma column was produced. The transverse spectrum from laser irradiated neon clusters exhibited dominant emission from highly placed levels of Ne^{7+} ions while the resonance lines remained very weak suggesting enhanced recombination from Ne^{8+}. In the axial spectrum, one of the lasing line candidates, namely the Ne^{7+} 3d-4f at 29.24 nm has been observed as a distinct, albeit not dominant. Further experiments with an upgraded setup are underway to optimize plasma conditions and to demonstrate amplification.

ACKNOWLEDGMENTS

This research was supported by the Ministry of Science and Technology of Korea through the Creative Research Initiative Program. The visit of T. Mocek to KAIST was supported by the Korea Science and Engineering Foundation and by the Czech Ministry of Education, Youth and Sports through the Laser Plasma Research Center (project LN00A100).

REFERENCES

1. Nagata, Y., Midorikawa, K., Kubodera, S., Obara, M., Tashiro, H., and Toyoda, K., *Phys. Rev. Lett.* **71**, 3774-3777 (1993).
2. Lemoff, B.E., Yin, G.Y., Gordon III, C.L., Barty, C.P.J., and Harris, S.E., *Phys. Rev. Lett.* **74**, 1574-1577 (1995).
3. Chichkov, B.N., Egbert, A., Meyer, S., Wellegehausen, B., Aschke, L., Kunze, H.J., and Kato, Y., *Jpn. J. Appl. Phys.* **38**, 1975-1978 (1999).
4. Sagisaka, A., Honda, H., Kondo, K., Suzuki, H., Nagashima, K., Kawachi, T., Nagashima, A., Kato, Y., and Takuma, H., *Appl. Phys. B* **70**, 549-554 (2000).
5. Mocek, T., Kim, C.M., Shin, H.J., Lee, D.G., Cha, Y.H., Hong, K.H., and Nam, C.H., *Phys. Rev. E* **62**, 4461-4464 (2000).
6. Cha, Y.H., Kang, Y.I., and Nam, C.H., *J. Opt. Soc. Am. B* **16**, 1220-1223 (1999).
7. Cha, Y.H., Hong, K.H., and Nam, C.H., *Opt. Commun.* **185**, 413-418 (2000).
8. Choi, I.W., Lee, J.U., and Nam, C.H., *Appl. Opt.* **36**, 1457-1466 (1997).

Advances toward inner-shell photo-ionization x-ray lasing at 45 Å

Stephen J. Moon, Franz A. Weber, Peter M. Celliers, and David C. Eder

Lawrence Livermore National Laboratory, Livermore, CA 94551

Abstract. The inner-shell photo-ionization (ISPI) scheme requires photon energies at least high enough to photo-ionize the K-shell, ~286 eV, in the case of carbon. As a consequence of the higher cross-section, the inner-shell are "selectively" knocked out, leaving a hole state $1s2s^22p^2$ in the singly charged carbon ion. This generates a population inversion to the radiatively connected state $1s^22s^22p$ in C+, leading to gain on the 1s-2p transition at 45 Å. The resonant character of the lasing transition in the single ionization state intrinsically allows much higher quantum efficiency compared to other schemes. Competing processes that deplete the population inversion include auto-ionization, Auger decay, and in particular collisional ionization of the outer-shell electrons by electrons generated during photo-ionization. These competing processes rapidly quench the gain. Consequently, the pump method must be capable of populating the inversion at a rate faster than the competing processes. This can be achieved by an ultra-fast, high intensity laser that is able to generate an ultra-fast, bright x-ray source. With current advances in the development of high-power, ultra-short pulse lasers it is possible to realize fast x-ray sources based that can deliver powerful pulses of light in the multiple hundred terawatt regime and beyond. We will discuss in greater detail concept, target design and a series of x-ray spectroscopy investigations we have conducted in order to optimize the absorber/x-ray converter – filter package.

INTRODUCTION

Since the first demonstration of x-ray lasing at Lawrence Livermore National Laboratory in 1985 [1], and elsewhere [2], based on the utilization of high-power pump lasers designed and built for fusion research, new and more efficient pumping techniques have been vigorously sought after. In the last decade, many important experiments have been conducted and good progress has been made towards table-top x-ray lasers. The advent of a new generation of chirped pulse amplification techniques has made possible the design and fabrication of powerful and ultra-short pulse duration pump lasers that seem to have the capability to shrink source dimensions from occupying whole rooms to table-top size. Many applications, however, do indeed require x-ray laser wavelengths shorter than currently attainable with table-top sized schemes since scaling to wavelengths shorter than 100 Å proves rather difficult [3] – [7].

Inner-shell photo-ionization x-ray lasers represent a class of x-ray lasers previously unrealized and first proposed by Duguay and Rentzepis [8]. Several authors have since adopted that inner-shell photo-ionization scheme [9, 10], and conclude that operating on a Kα transition of a low Z element constitutes an attractive way to efficiently pump an x-ray laser in the sub 50 Å regime. All these new schemes still need sub 50 femtosecond and greater than 1 J in an optical pulse to drive the transition. Advances in powerful ultrashort pulse laser (USPL) technology prompted us to investigate that group of schemes with the application of carbon as the lasant material in greater detail. The choice of this material is by virtue of its electronic properties with particular respect to the Auger recombination rate and lasing wavelength at a wavelength of 45 Å.

Here, in section 2 we discuss physics of the target and present our modeling results. We describe our method for a traveling wave setup suitable to pump inner-shell photo-ionization schemes in section 3. The report on our experimental findings on target output and laser performance is given in section 4. We summarize in section 5.

TARGET PHYSICS

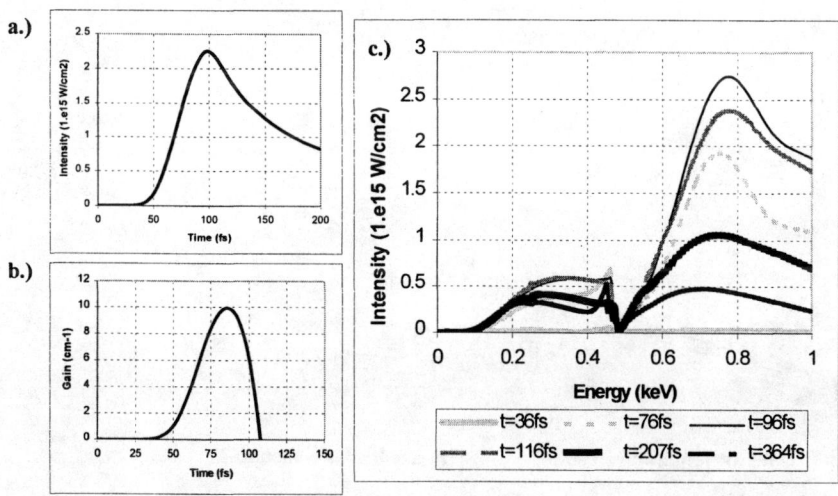

Figure 1. (a.) Shows the simulated temporal history of the source, spectrally integrated. (b.) Is the calculated gain in carbon lasant (10^{18} cm^{-3} foam) at 45 Å. (c.) Shows the simulated spectral emission for various times.

Results of our modeling effort are shown in figures 1 a through c. An USP high intensity optical laser of 1×10^{17} W/cm^2 and 40 fs pulse duration is incident on a foil consisting of 200 Å layered on 1000 Å of Ti. The laser is at normal incidence on the gold absorber, the front side of the target, and the incoherent x-ray source, which pumps the carbon lasant, is due to back-side emission. The modeling was performed using the hydrodynamics/atomic code LASNEX [11]. The code includes many physics models necessary to simulate a multidimensional radiative hydrodynamics problem, including laser-matter interaction, radiation transfer, electron thermal diffusion by conduction, and simple description of non-local thermodynamic equilibrium atomic kinetics. Reference 11 includes a description of the physics models used in LASNEX.

The energy from the optical laser is deposited in a self-consistent manner by solving the wave equation for the laser electro-magnetic field and the atomic kinetics and x-ray emission are calculated with an average-atom atomic model, which includes spin orbit coupling. Careful optimization of the converter/filter assembly parameters yielded the values indicated above. Furthermore, the ionization velocity in the absorber material ensures that the thin gold layer is completely ionized at the end of a ~ 40 fs optical pump pulse, and at the same time the absorption occurs at near solid density. Figure 4 a shows the calculated backside emission from the foil structure and figure 1 b shows the corresponding gain of 9.9 cm^{-1} for low-density carbon foam at 10^{18} cm^{-3}. Figure 1 c depicts spectral emission from the target structure for various times.

TRAVELING WAVE PUMPING OF AN ULTRASHORT PULSE X-RAY LASER

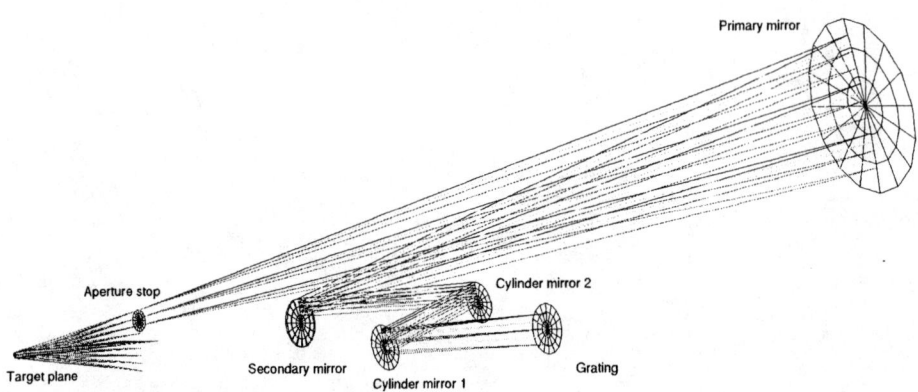

Figure 2. USPL traveling wave pumping system in line focus geometry.

The use of traveling optical waves as a method of delivering optical energy is not new to laser science [12] – [16]. Even before chirped pulse amplification (CPA) came on the scene, Bor [17] employed a traveling wave technique to produce optical pulses with a 1 picosecond duration. Indeed, the field of collisionally pumped XRLs has also seen the implementation of traveling wave pumping (TWP) schemes, although at much longer pulse durations, in order to enhance XRL performance. Previous analysis [18] for these schemes shows that TWP may be advantageously applied whenever the pump time divided by the target length turns out to be less than 33 ps/cm.

The physical characteristics of the required traveling wave optical pumping system are largely determined by the physics of the x-ray laser transition as described in section 2. The very rapid rise time of the broad-bandwidth x-ray source, faster than the upper state lifetime in carbon, dictates, because of the competing processes and subsequent depopulation of the lower laser state, zero gain for times greater than approximately 80 fs (see figure 1 b). This time constraint is true for every incremental length of the lasing volume. Thus, we find that the stringent pump requirements of the ISPI XRL scheme as a consequence also impose very stringent requirements on the quality and fidelity of the focusing system. A traveling wave focusing scheme is therefore essential to pump an XRL amplifier target.

The pump light (optical energy of USPL) must be delivered to the target at an intensity of 1.0×10^{17} W/cm^2 in a spot size of approximately 30 μm that illuminates the target along a linear track of ~ 10 mm length and traveling at the velocity of light c (index of refraction ~ 1.000) along the target surface. The arrival time deviation from a linear variation at any point along the trajectory must be less than 50 fs corresponding to a distortion level around 10^{-4} or better. The pulse duration of the focused beam must be a faithful representation of the incident pulse from the drive laser. Also attendant is the vertical spatial distortion that, accordingly, must be limited to better than +/- 15 μm over 10 mm or $\sim 10^{-3}$ over the laser length. These are not unreasonable numbers for generic optical systems; however, the utilization of a high power USPL with its 50 to 100 nm bandwidth mandates all-reflective optical surfaces in order to eliminate not only the chromatic distortions that lead to pulse temporal broadening [19], but also distortions due to the intensity induced nonlinear index of refraction variations (B-Integral). As a result, only a vacuum diffraction grating succeeds as a dispersive element out of a variety of traditional methods (prisms, Fabry-Perot interferometers, and gratings [20], [21]) to tilt wave fronts of optical pulses by angular dispersion.

The schematic of the TWP setup is depicted in figure 2. The novel feature of the design is a holographically-generated grating that serves both, to produce the pulse front tilt needed for the traveling wave, and also to correct residual aberrations in the focusing system. We designed the system to project an image of an incident 50 mm beam onto the 10 mm target surface at 5x demagnification. Most of the focusing power is accomplished using a reflective inverted off-axis Schwarzschild microscope configuration (primary mirror: 1500 mm concave radius, secondary mirror: 309 mm convex radius). These two mirrors are placed in a classic Schwarzschild confocal configuration – and the design is optimized for 5:1 demagnification. The grating is centered on, and perpendicular to, the chief axis of the Schwarzschild system; the grating image is then projected onto the target plane, which is also centered on and perpendicular to the chief axis. In our design we use an off-axis subset of the cylindrically symmetric Schwarzschild configuration, hence the rays traveling through our focusing system arrive obliquely on all the optics. In order to produce a pulse front traveling at velocity c across the target surface using an imaging system operating with magnification M, it can be shown that the grating must be illuminated at an angle $q = \sin^{-1}(M)$ from the grating normal (while the diffracted beam leaves the grating approximately at the normal). In our case $M = 0.2$ and $q = 11.54°$. The pulse front intersects the grating surface along a vertical line; this line must be focused onto the target at a point, which requires the focusing system to be astigmatic. We accomplish this by using two additional cylindrical mirrors that are configured to produce the required line focus on the target. Using the combination of the two spherical and two cylindrical elements a ray trace analysis can optimize this system to

345

produce a line focus that is a few times diffraction-limited at the target plane. The trajectory of the pulse front produced by imaging a flat grating onto a flat target will not have a constant velocity, but will have a small quadratic deviation from this. (A planar pulse front originating from the grating reaches the target plane as a cylindrical pulse front because it must pass through a focus before reaching the target.) We can adjust the shape of the pulse front to correct for this by using a slightly cylindrical surface at the grating plane. The required grating surface curvature depends on the magnification and focal length of the Schwarzschild system.

The remaining geometrical aberrations in the system can be corrected by generating the grating holographically. To accomplish this the system must be back-illuminate with a diffraction limited line focus at the target plane. The grating can then be produced by interfering the back propagating beam with the $11.54°$ incident beam at the grating surface. The back propagating beam incorporates the residual aberrations in the focusing system, thus allowing the production of a grating pattern that will correct for these aberrations. During operation slight adjustments in the phase velocity of the pump at the target can be accomplished by slight tilts of the grating about an axis perpendicular to the plane of incidence.

EXPERIMENTAL RESULTS

In order to effectively and accurately investigate the emission from the x-ray converter it is necessary to field a spectrometer, which covers wavelength regions on both the long and short wavelength side of the carbon K-edge. That requirement puts severe limitations on the choice of materials to be employed in the dispersing element of the detection system.

The experimental investigation was conducted at the LLNL Falcon USPL facility. The laser is based on a Ti:Sapphire oscillator and uses CPA to produce nominally 500 mJ pulses of 35 fs individual duration at a repetition rate of 1 Hz. We used the laser's regenerative amplifier pumped at saturation level. After initial determination of output levels all shots were set up in a p-polarized configuration and a 45 degree angle of incidence on target was maintained. The center wavelength of the laser l = 820 nm was focused on target by means of an off-axis parabola at f/2. In our first set of experiments we have used maximum energy through the four pass amplifier and air compressor and have been able to achieve around 100 mJ on target. On every shot we monitored the unconverted beam energy as well as the level of amplified spontaneous emission (ASE) by means of a calorimeter and a 170 ps fast-rise time photodiode, respectively. A typical ASE profile rose linearly in time until the arrival of the main pulse.

The CCD as well as a soft x-ray pinhole camera (SXR PHC) were filtered in order to block stray light and unwanted optical reflections. We used 12.5 μm of beryllium for the SRX PHC whereas we installed a total of 2kÅ of aluminum on top of 2 kÅ of Lexan in the spectrometer. The spectral composition of backside x-ray emission for various focal distances is shown in figure 3. For verification of the focal spot size on the target front side we employed an x-ray pinhole camera comprising a multiple pinhole array coupled to a soft x-ray sensitive CCD as described above. The diagnostic was operated at a magnification of 15. Data analysis on both, the pinhole images and the static spectroscopy data was performed by subtracting dark images in order to correct for detector temperature-dependent noise in addition to the subtraction of high energy photon and electron generated background. We then integrated the signal over the regions of interest and adjusted for analyzer reflectivity, filter transmission functions, and detector quantum efficiency.

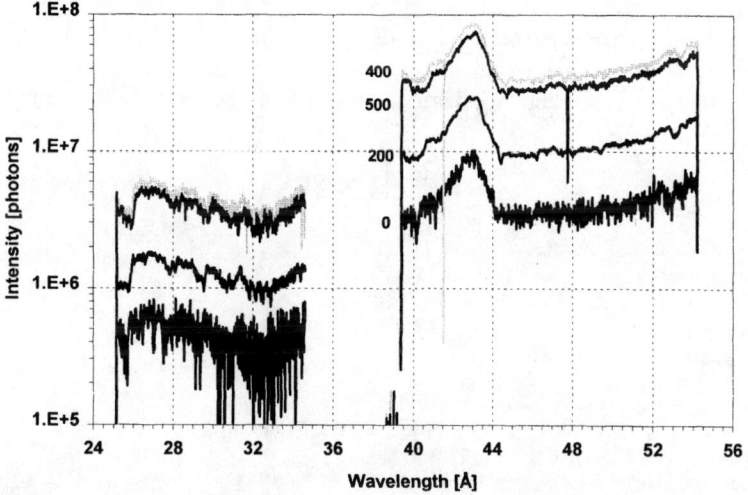

Figure 3: Time integrated backside emission increases away from best focus.

We found the spot size at best focus to be around 23 μm at FWHM, which corresponds to a laser intensity on target of 6.9×10^{17} W/cm^2. The best focus is located at a position measuring 100 μm away from the initial target surface due to a distance calibration offset.

CONCLUSIONS

We have set out to pursue a longstanding goal of x-ray laser research, the realization of the inner-shell photo-ionization (ISPI) pumping scheme in order to demonstrate lasing at wavelengths not yet possible by table-top energy sources. We have shown an efficient x-ray converter/filter package and studying its backside x-ray emission. For that purpose we have devised and fabricated a versatile soft x-ray spectrometer universally suitable for analysis of various x-ray sources based on laser produced plasmas. A soft x-ray pinhole camera has been used as an additional diagnostic to verify focal spot properties and target intensities. We found a maximum conversion efficiency of 1.2×10^{-4} at a position of 300 μm out of best focus. We did not see any evidence for increased backside x-ray emission as a result of intentional prepulses delivered to the target, independent of pulse delay. In addition, we have shown how to create a line focus using an ultrashort optical laser pulse in traveling wave geometry employing a holographically produced grating for compensation of optical aberrations in the relay imaging system, which are introduced by the use of standard spherical and cylindrical components.

ACKNOWLEDGEMENTS

This work was performed under the auspices of the U.S. Dept. of Energy by the University of California Lawrence Livermore National Laboratory, through the Institute for Laser Science and Applications, under Contract No. W-7405-Eng-48.

REFERENCES

1. D. L. Matthews et al., Phys. Rev. Lett. **54**, 110 (1985).
2. S. Suckewer et al., Phys. Rev. Lett. **55**, 1753 (1985).
3. J.J. Rocca et al., Phys. Rev. Lett. **77**, 1476 (1996).
4. B.E. Lemoff et al., Phys. Rev. Lett. **74**, 1574 (1995).
5. Y. Nagata et al., Phys. Rev. Lett. **71**, 3774 (1993).
6. P. V. Nickles et al., Phys. Rev. Lett. **78**, 2748 (1997).
7. J. Dunn et al., Phys. Rev. Lett., **80**, 2825 (1998).
8. M. A. Duguay and P. M. Retzepis, Appl. Phys. Lett. **10**, 350 (1967)
9. H. C. Kapteyn, Appl. Opt. **31**, 4931 (1992).
10. S. J. Moon and D. C. Eder, Physical Review A **57**(2), Feb. 1998.
11. G.B Zimmerman and W.L. Kruer, Comments Plasma Phys. Controlled Fusion **2**, 51-61 (1975).
12. M. E. Mack et al., Appl. Phys. Lett. **15**, 166 (1969).
13. R. Wyatt and E. E. Marinero, Appl. Phys. **25**, 297 (1981).
14. H. J. Polland et al., Appl. Phys. B **32**, 53 (1983).
15. M. H. Sher et al., Opt. Lett. **12**, 891 (1987).
16. C. P. J. Barty et al., Phys. Rev. Lett., **61**, 2201 (1988).
17. Zs. Bor, S. Szatmari, and A. Muller, Appl. Phys. B **32**, 101 (1983).
18. J. C. Moreno, J. Nilsen, and L. B. Da Silva, AIP Conf. Proc. **332**, 21 (1994).
19. Z. L. Horvath et al., Opt. Eng. **32**, 2491 (1993).
20. J. R. Crespo Lopez-Urrutia et al.: Proc. SPIE **2012**, 258 (1993).
21. Z. S. Bor et al., Opt. Eng. **32**, 2501 (1993).

Gain Calculations for Inner-Shell Lasing by Electron Collisional Ionisation

L. M. Upcraft

Laboratoire d'Optique Appliquée, ENSTA,Chemin de la Huniere, 91761 Palaiseau, FRANCE

Abstract. Current high power femtosecond lasers have been shown to produce electron pulses which may be appropriate for the pumping of X-Ray lasers through collisional ionisation. Non-radiative Coster-Kronig type decay processes may be fast enough to form an inverted state and allow X-ray lasing within the biologically interesting "water window". Calculations of the atomic processes in metallic Ti, Mn and Cu are presented that potentially useable gains on the $M_1 - L_3$ transition.

INTRODUCTION

The most successful and reliable X-Ray lasers are currently those which use electron collisional excitation of a valence shell electron in either Ne-like or Ni-like ions. The shortest wavelength possible with these schemes is 2.8 nm in Ni-like U and thus moving to shorter wavelengths requires an examination of other possible transitions. Lasing on transitions to atomic states which have an electron 'hole' in an inner shell have been widely investigated within the context of photopumped schemes but have not yet been demonstrated at X-Ray wavelengths. A conceptually simpler approach would be to use electrons to collisionally ionise the lasing atom but this has two significant problems. Firstly, the lifetime of the upper level at X-ray energies is very short, typically less than 1fs for K_α transitions and a few 10's of fs for the L series transitions. For amplification to occur, ionisation of the target must occur over similar timescales which places considerable demands on the pumping source. However, current high power, high repetition rate Ti:Saph lasers have been shown [1-3] to generate high energy electron pulses on ~100 fs timescales and may therefore serve as appropriate sources. Secondly, the collisional ionisation cross sections for bound inner shell states are smaller than those for higher lying states and thus creating an inversion (i.e. more holes in the lower level state than the upper) is not directly possible with electrons. However, non-radiative Auger type transitions may act to form a transient inversion from an initially non-inverted state. In particular, Coster-Kronig (where the ejected electron is from the same main quantum level as the decay electron) can be several orders of magnitude faster than radiative decay rates.

With the advances in short electron pulses demonstrated by Ti:Saph lasers, it is therefore of interest to examine systems which may allow lasing on short wavelength transitions driven by electron collisional excitation. Chukhovskii *et al* [4] has previously developed a semi-analytical/empirical model which predicts gain on K_α

CP641, X-Ray Lasers 2002: 8th International Conference on X-Ray Lasers, edited by J. J. Rocca et al.

transitions of light elements but did not consider the atomic processes in detail. Kim et al [5] solved the time-dependent atomic rate equations for a number of elements but considered that the ionising electrons were monochromatic in energy which was matched to the peak of the ionisation cross section for the initial hole state. While this is desirable to create the maximum population of the initial state, it is unrealistic in terms of the electron energy distributions known from experiments [1-3]. In this work we have modelled the temporal evolution of atomic states in Ti, Mn and Cu subjected to an ionising electron beam of arbitrary form by solving the coupled rate equations of the form;

$$\frac{dn_i}{dt} = n_j A_{j \to i} - n_i Q_{i \to k} - n_i C_{i \to l} \tag{1}$$

where A, Q and C are the radiative, non-radiative and collisional ionisation rates respectively between all possible atomic/ionic states which include states i, j, k and l. Radiative rates are calculated from the atomic code of Cowan [6] while the non-radiative rates are taken from Chen et al [7]. The collisional ionisation rates are integrated over the required energy distributions for each of K, $L_{1,3}$ and $M_{1,5}$ sub shells and use cross sections calculated after the method of Deutsch et al [8].

RESULTS

For each atom (Ti, Mn and Cu), calculations are performed assuming a solid density which is initially un-ionised. In all cases, the only transitions which show gains are the $M_1 \to L_{2,3}$ lines which have wavelengths of ~3.1 nm, ~2.2 nm and ~1.5 nm for Ti, Mn and Cu respectively. In this case is it instructive to look at the collisional ionisation cross sections of the lasing shells. These are shown for Cu in figure 1.

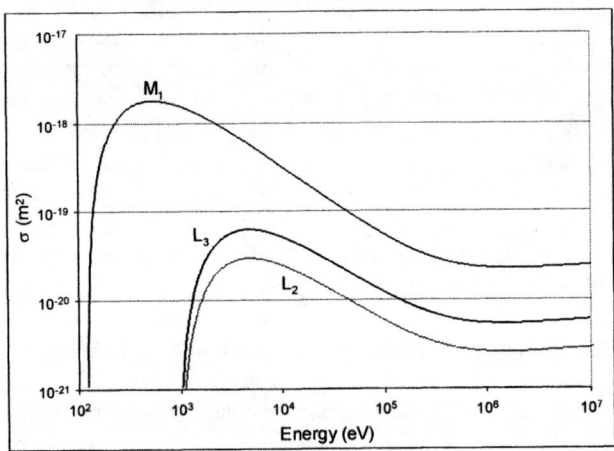

FIGURE 1. Inner shell electron collisional ionisation cross sections, σ, for the M_1, L_2 and L_3 sub shells of Cu as a function of electron energy.

Examining the cross section curves shown in figure 1 then it is clear that in order to minimise the number of holes produced in the upper lasing state, M_1, then the electron energy should not be near the peak of the cross section at around 600 eV. However, the assumption of Kim *et al* [5] that the optimum conditions occur at the peak of the cross section for the lower lasing state (i.e. $L_{2,3}$) is found to be incorrect. The peak of the L_2 cross section occurs at an energy of 4500 eV and at this point the ratio of the M_1 to L_2 cross section is 20.6. At higher energies, this ratio drops and at 1 MeV is around 8.8 thus implying that using much higher energies will ease the constraints upon forming an inversion.

Calculations have been carried over a parameter space in which the electron pulse duration is assumed to have a temporal half width of $2.5 \le t_{1/2} \le 75$ fs, a maxwellian temperature distribution of $2.5 \le T_e \le 100$ keV and a flux density of $1.0 \times 10^{17} \le \rho_e \le 1.0 \times 10^{19}$ cm^{-3} s^{-1}. Over this entire range, Ti failed to show any gains above 10^{-3} cm^{-1} for any transition. This is attributable to the fact that it has only 2 3d electrons with which Coster-Kronig transitions can act rapidly enough to form inversion. Since both of these are very weakly bound (~5.0 eV in metallic Ti) they are easily ionised before they can be used to establish an inversion. The case is somewhat less extreme in Mn with 5 3d electrons and gains of around 10 cm^{-1} are calculated for much of the phase space examined. The largest gains predicted, and hence the most favourable case for experimental observation, are those in Cu and given the limited space available, discussion is henceforth restricted to this element.

Figure 2 shows the calculated gains at electron temperatures of 5 keV and 100 keV as functions of the electron flux and pulse half width. These graphs confirm the expected behaviour as discussed in terms of the ionisation cross sections in that higher temperatures result in larger gains, although it must be remembered that gains calculated here are the result of integration over a maxwellian temperature distribution.

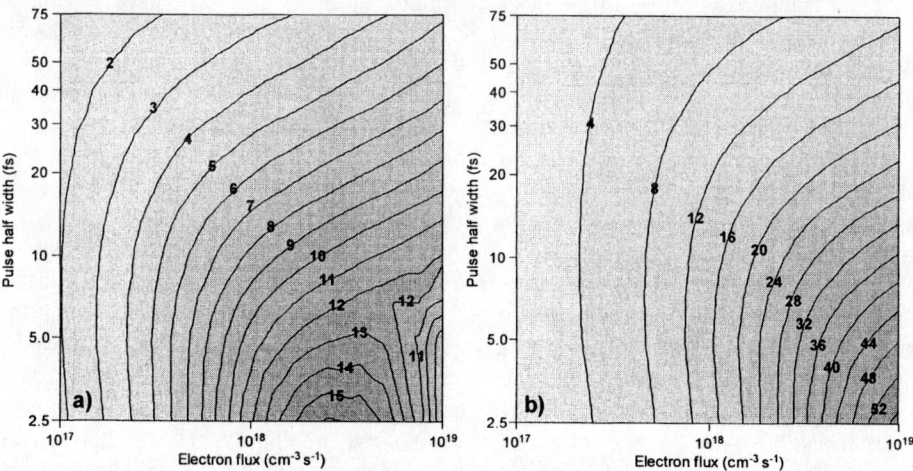

FIGURE 2. Contour plots of the gain coefficients (cm^{-1}) for the $M_1 \rightarrow L_3$ transition in Cu as functions of the electron flux and pulse half width for two different electron temperatures of a) 5 keV and b) 100 keV.

The trend shown in figures 2a and b for a fixed temperature is clear in that the largest gains are predicted for shorter pulse durations and larger electron fluxes. Although the largest gains are predicted for the shortest examined pulse half widths of 2.5 fs, significant gains are shown for currently more reasonable pulse half widths of 10 fs as shown in figure 3.

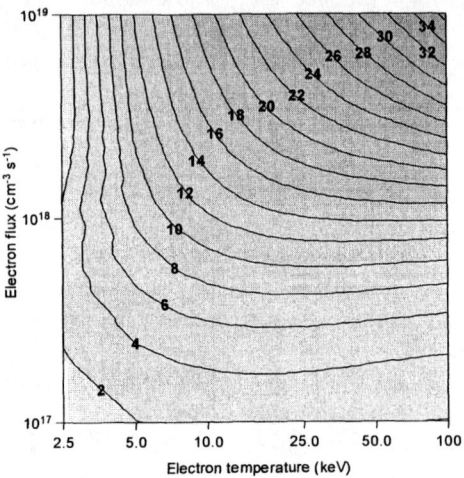

FIGURE 3. Contour plots of the gain coefficients (cm^{-1}) for the $M_1 \rightarrow L_3$ transition in Cu as functions of the electron temperature and electron flux a pulse half width of 10 fs.

A common feature of both the Mn and Cu calculations, is that although gains are also predicted on the $M_1 \rightarrow L_2$ transition, they are typically an order of magnitude or more smaller than for the $M_1 \rightarrow L_3$ transition. Given that the ionisation cross sections for the formation of L_2 and L_3 holes is similar (figure 1) this is perhaps suprising. However, an examination of the Coster-Kronig rates shows that many of the $L_2 \rightarrow M_xM_y$ rates are up to an order of magnitude larger than the $L_3 \rightarrow M_xM_y$ rates. This means that the L_2 hole states are filled much faster than the L_3 resulting in much smaller gains.

The calculations presented so far have been performed using an assumed electron energy distribution which is maxwellian. However, recent high power femtosecond laser experiments [2, 3] have shown that the distribution is more accurately described by two approximately maxwellian components. The bulk of the distribution is typically a few 10's of keV, while up to around 5% of the electrons are of a much higher temperature of around 100 keV. Thus calculations have made using the estimated distribution of Pretzler *et al* [3]. This photopumping experiment involved a 130 fs Ti:Saph pulse incident upon copper in which the electron distribution was found to consist of 97% electrons at a temperature of 20 keV with the remaining fraction at 200 keV. The electron flux is unknown in this experiment but can be crudely estimated at around 10^{18} cm^{-3} s^{-1} from figures given. Gain calculations using this distribution are shown in figure 4 but indicate that significant shortening of the pulse duration would still be necessary to achieve modest gains of ~20 cm^{-1}.

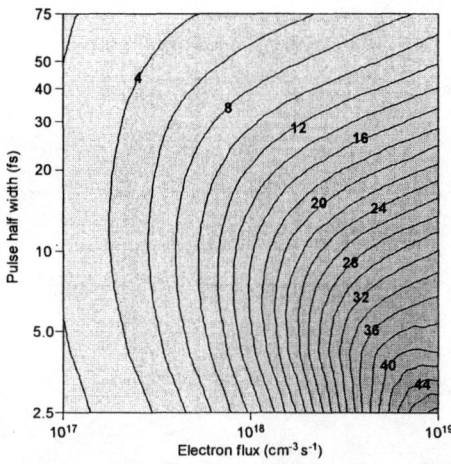

FIGURE 4. Contour plots of the gain coefficients (cm^{-1}) for the $M_1 \rightarrow L_3$ transition in Cu as functions of the electron pulse half width and flux. The temperature distribution is taken as 97% 20 keV and 3% 200 keV.

CONCLUSIONS

Atomic calculations on the inner shell processes of atoms subjected to electron collisional ionisation predict that gain coefficients of around 20 cm^{-1} on the $M_1 \rightarrow L_3$ transition at a wavelength of 1.53 nm may be achievable in copper using electrons generated from currently available high power femtosecond lasers if the electron pulse duration can be reduced to around 20fs. Ti and Mn are less suitable due to having fewer M shell electrons which are unable to create the transient inversion on these timescales before being collisionally ionised.

ACKNOWLEDGMENTS

This work was supported by the European X-Ray Laser Network programme.

REFERENCES

1. Rousse, A., Audebert, P., Geindre, J. P., Falliès F., Gauthier, J. C., Mysyrowicz, A., Grillon, G. and Antonetti, A., *Phys. Rev. E.,* **50** 2200-2207 (1994)
2. Li, Y. T., Zhang, J., Chen, L. M., Mu, Y. F., Liang, T. J., Wei, Z. Y., Dong, Q. L., Chen, Z. L., Teng, H., Chun-Yu, S. T., Jiang, W. M., Zheng, Z. J. and Tang, X. W., *Phys.Rev; E.,* **64**, 046407 (2001).
3. Pretzler, D., Schlegel, Th. and Fill, E., *Phys.Rev; E.,* **62**, 5618-5623 (2000).
4. Chukhovskii, F. N., Gibbon, P., Uschmann, I. and Förster, E., *Opt. Comm.,* **154** 313-318 (1998).
5. Kim, D., Toth, C. and Barty, C. P. J., *Phys. Rev. A,* **59**, R4129-4132 (1999).
6. Cowan, R. D., *Theory of Atomic Structure and Spectra,* University of California Press, 1981
7. Chen, M. H. and Crasemann, B., *At. Data. and Nuc. Data Tables,* **24**, 13-37 (1979).
8. Deutsch, H., Margreiter, D., and Mark, T. D., *Z. Phys. D.,* **29**, 31-37 (1994).

FREE ELECTRON LASERS AND OTHER
ACCELERATION-DRIVEN X-RAY SOURCES

The Low-Energy Undulator Test Line: A SASE FEL Operating from 660 to 130 nm

S. G. Biedron for the LEUTL † and FEL Exotica†† Teams

†The LEUTL Team: S.G. Biedron[*], M. Borland, P. Den Hartog,
R. Dejus, M. Erdmann, Z. Huang, K.-J. Kim, J. Lewellen, Y. Li,
A. Lumpkin, O. Makarov, S.V. Milton, E. Moog, V. Sajaev,
and G. Wiemerslage

Advanced Photon Source, Argonne National Laboratory, Argonne, IL 60439, USA
[*]*and MAX-Laboratory, University of Lund, Sweden S-221 00*

††Recent work in exotic schemes has been performed by the following
individuals; D. Attwood[1], S.G. Biedron[2,3], R. Bartolini[4], F. Ciocci[4],
G. Dattoli[4], W.M. Fawley[1], G. Felici[4], H.P. Freund[5], J. Lewellen[2], Y. Li[2],
S.V. Milton[2], H.-D. Nuhn[2], P.L. Ottaviani[6], A. Renieri[4], and J. Rocca[7]

[1]*Lawrence Berkeley National Laboratory, Berkeley, CA 94720, USA*
[2]*Advanced Photon Source, Argonne National Laboratory, Argonne, IL 60439, USA*
[3]*and MAX-Laboratory, University of Lund, Sweden S-221 00*
[4]*ENEA, Divisione Fisica Applicata, Centro Ricerche Frascati, C.P. 65, 00044 Frascati, Rome, Italy*
[5]*Science Applications International Corporation, McLean, VA 22102, USA*
[6]*ENEA, Divisione Fisica Applicata, Centro Ricerche E. Clementel, Via Don Fiammelli 2, Bologna, Italy*
[7]*Colorado State University, Fort Collins, CO 80523 USA*

PACS Codes: 41.60, 42.65.Ky, 52.59.-f
Keywords: Free-Electron Lasers, Harmonic Generation, Frequency Conversion, Intense Particle Beams and Radiation Sources

Abstract. There is a strong desire for short wavelength (down to 1 Å), short pulsewidth (<100 fs), high-brightness, transverse, and longitudinally coherent light pulses for use by the synchrotron radiation community. Much effort is ongoing worldwide to advance this desire both experimentally, in theory and design, and politically. One of the ongoing experimental efforts is the low-energy undulator test line (LEUTL) at the Advanced Photon Source (APS) at Argonne National Laboratory (ANL). This experiment is based upon the self-amplified spontaneous emission (SASE) process, a method to attain a next-generation light source. This presentation gives an overview concerning the history and results of next-generation light sources, the results of the LEUTL SASE FEL, and the description of the upcoming first user experiment on LEUTL. We will also briefly review exotic schemes for future, next-generation light sources based upon FELs including biharmonic undulators and the possibility of interfacing of traditional x-ray lasers with FELs.

CP641, *X-Ray Lasers 2002: 8th International Conference on X-Ray Lasers,* edited by J. J. Rocca et al.
© 2002 American Institute of Physics 0-7354-0096-2/02/$19.00

WHAT ARE THE DESIRED CHARACTERISTICS OF A NEXT-GENERATION LIGHT SOURCE?

The synchrotron radiation sources of the past and present may be defined as follows. First-generation machines were electron synchrotrons and storage rings that were built for other purposes (e.g., high-energy and nuclear physics), but whose bending magnet radiation was parasitically used by synchrotron radiation "users." This radiation could cover many wavelength regimes due to the nature of the bending magnet emission. However, it had rather large photon source sizes as the electron beam emittance in these machines was large, and neither intended nor ideal for synchrotron radiation applications. The second-generation machines were dedicated machines for synchrotron radiation users and employed bending magnets as the primary source of synchrotron radiation. The beam emittances were designed by the machine builders to be relatively small in order to provide users with a smaller source size and higher brilliance. Third-generation machines are dedicated for synchrotron radiation users and were designed ab initio to accommodate many so-called insertion device magnets such as undulator and wiggler magnets. Undulator magnets have the feature that they generate relatively narrow spectral lines, and this enhances the overall photon brilliance.

Currently, the synchrotron radiation community employs many third-generation light sources, such as the Advanced Photon Source (APS) at Argonne National Laboratory (ANL), USA, the European Synchrotron Radiation Facility (ESRF) in Grenoble, France, and SPring-8 in Harima Science Garden City, Japan. Such sources are capable of producing photons with energies of ~1-100 keV at peak spectral brightnesses of 1×10^{23} photons·s^{-1}·(0.1% bandwidth)$^{-1}$·mm^{-2}·mrad^{-2}. Although these short-wavelength, high-brightness machines have proven successful in discovering previously unobtainable structural information in a variety of scientific areas, the ability to obtain dynamical (temporal) information, particularly in relation to the biological sciences, is limited to time scales longer than ~10 ps. To obtain dynamical information at shorter time scales, one must produce and use x-rays in the 1-Å-wavelength regime with pulse lengths on the sub-ps timescale. It is also preferable that these pulses be fully coherent longitudinally in order to insure a narrow spectral bandwidth. Also, a diffraction-limited source with full transverse coherence is desirable. Finally, there is interest of peak radiation intensities and brightnesses many orders of magnitude higher than are available from contemporary machines. This will enhance spatial resolution and may even lead to the possibility of single molecule structure determination. Longer wavelength next-generation light sources are also important and desired by the scientific community; one example of which will be discussed below.

WAYS TO OBTAIN A NEXT-GENERATION LIGHT SOURCE

Numerous mechanisms to produce such high, peak-brightness photon pulses have been discussed [1], many belonging to the category of single-pass, high-gain free-electron lasers (FELs). However, for a number of reasons, oscillator (multiple-pass) and amplifier (referring to a one-undulator amplifier) FELs [2] alone are not well suited for these short wavelengths. The oscillator (multiple-pass) method is unsuitable for the x-ray (1-Å) regime, because mirrors of sufficient reflectivity are unavailable. In addition, such mirrors would be required to tolerate extreme power densities based on high peak- and average- powers. It is also the consensus that the storage-ring approach to the next-generation light source operating in the x-ray (1-Å) regime is unrealistic, as the necessary high electron bunch brightness and short pulse duration are presently unachievable in a storage ring [3].

The mechanisms of self-amplified spontaneous emission (SASE), amplifier, two-undulator harmonic-generation scheme (TUHGS), and high-gain harmonic generation (HGHG) FELs, as well as the nonlinear harmonics generated in each, belong to a genré of linear-accelerator-driven sources [1]. The SASE mechanism can provide the photon power and pulse duration required by the next-generation light source users. However, SASE is a random process that starts up from noise; therefore, its output is neither fully longitudinally coherent nor transform limited, and $\Delta\omega\Delta t > 1$, where $\Delta\omega$ is the frequency spread of the photon pulse and Δt is the pulse duration. An amplifier FEL requires a coherent (electromagnetic) input seed signal with a wavelength equal to that of the desired output radiation; however, this is presently unavailable in the 1-Å regime. The TUHGS and HGHG processes use a fully coherent seed at a subharmonic of the desired radiation output and can therefore generate a less noisy output photon pulse with full longitudinal coherence. The method of nonlinear harmonic generation provides the possibility of generating substantial photon powers at quite short wavelengths, while utilizing an electron bunch of lower energy and using one of the above methods. Finally, modular approaches to the next-generation light source provide the possibility of a coherent source in the x-ray regime based upon "stacked" or progressive modules of the above-described systems.

THE LOW-ENERGY UNDULATOR TEST LINE (LEUTL)

The low-energy undulator test line (LEUTL) of the APS is a test bed for the next generation of synchrotron light sources and is being configured to be an extreme ultraviolet user facility [4]. Its first experiment is an FEL based upon the SASE phenomenon and is currently tunable in wavelength from 660 to 130 nm. The repetition rate of the drive laser for the photocathode-rf gun is 6 Hz, which defines the repetition rate of the FEL. The FEL symbiotically (or epiphytically) shares with normal APS operations the injector (linear accelerator) necessary to produce the high-brightness beam required for high FEL gain. The APS linear accelerator has three electron guns: the photocathode-rf gun, used for LEUTL operation or storage-ring operation, and two thermionic-rf guns, used for storage-ring operation and for diagnostic check-out through the LEUTL line. The 2856-MHz APS linear accelerator is capable of accelerating electrons from the photocathode-rf gun to nearly 600 MeV [5,6], and incorporates a magnetic bunch compressor can compress the electron pulses down to a few hundred fs [7]. The chosen-energy electron beam is then transported through a transfer line from the linear accelerator to the LEUTL tunnel proper. There, a total of eight fixed-gap undulator devices, as well as various photon and electron beam diagnostics before the first and after every undulator thereafter, are employed by the FEL in the 50-m-long LEUTL hall. Each linearly-polarized undulator is 2.4 meters in length and uses ~160 permanent-magnet pole pieces to generate a undulator period of 3.3 cm. Beyond the LEUTL hall are additional diagnostics in an end-station area outside of the radiation enclosure. After first achieving "saturation" in September 2000 at both 530- and 385-nm wavelengths, we have been examining details of the FEL's performance such as z-resolved radiation spectra, nonlinear harmonic emission [8], and electron beam microbunching as evidenced by coherent transition radiation [9]. We have also recently pushed the FEL's operating wavelength down to 130 nm by increasing the electron beam energy. A typical peak power for FEL operation, for the case of 130 nm, is ~500 MW. The first LEUTL FEL user experiment will be installed in the next few months and will be discussed briefly in a later section. Results of this experiment are expected before December 2002. The LEUTL system is shown schematically in Figure 1.

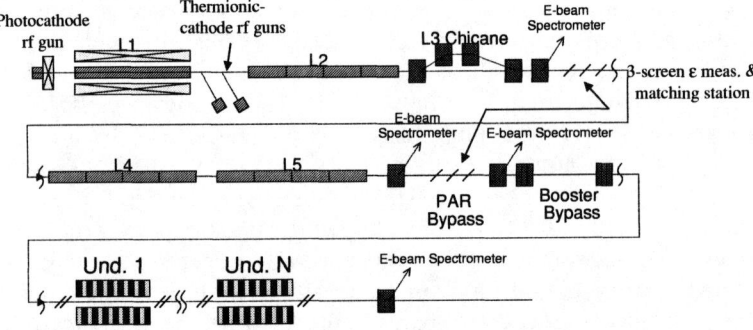

FIGURE 1. Schematic layout of the APS linac and undulator line of the LEUTL system.

LEUTL RESULTS

As described above, SASE devices have a random output component. It is therefore important to characterize the statistical fluctuations of the experimental data collected at the optical diagnostic located after each undulator [10]. Figure 2 shows one of the most basic LEUTL SASE measurements for an operational wavelength of 530 nm; optical energy as a function of distance along the LEUTL undulator line. The three lines represent the statistics of this system. The center line and square points indicate the 50th percentile of the optical energy, while the outlying lines represent the 25th and 75th percentiles of the optical intensity. These numbers are drawn from a sample size of 100 – 200 images per station. The nature of the intensity fluctuations clearly changes as the optical power saturates, as expected. In particular, the fluctuation becomes less and the 50th and 75th percentile lines become closer, as one would expect. After saturation, little additional power can be extracted, but the occasional "poor shot" can still result in a much lower than average intensity. This is also apparent in the statistics of the measured optical energy at each station. The crosses represent the normalized standard deviation of the output energy at each station. After saturation, the deviation dramatically decreases. The effects of electron beam jitter as well as SASE fluctuations affect this experimental data.

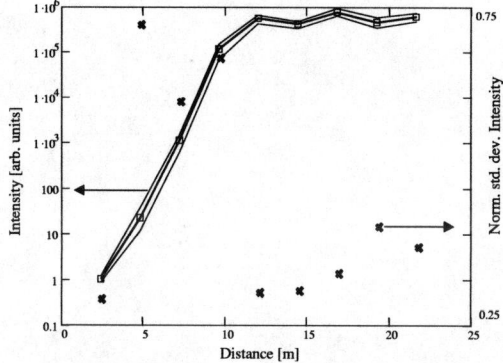

FIGURE 2. Measured optical intensity as a function of distance along the undulator line, normalized to the intensity at the first diagnostic station. The black crosses are the normalized standard deviations of the intensities measured at each diagnostic station. Electron beam properties are: peak current, 266 A; bunch length, 0.30 ps (rms); normalized emittance, 8.4 μm (rms); energy spread, 0.10% (rms). Theoretical gain length (power e-folding length) is 0.59 m, measured gain length is 0.57 m.

THE FIRST APS FEL USER EXPERIMENT

The first LEUTL user experiment is being led by Michael Pellin of the Chemistry Division at ANL [11]. This experiment will involve single photon ionization and resonant ionization to threshold (SPIRIT) and will use the high VUV pulse energy and tunable wavelength from LEUTL to uniquely study a few specific materials and systems. As an example, trace quantities of light elements (H, C, N, O) in semiconductors with 100 times lower detection limit than the current state-of-the-art can be investigated. In addition, LEUTL can facilitate the examination of organic molecules with minimal fragmentation to permit cell mapping by mass, polymer surface understanding, modified (carcinogenic) DNA mapping, and determination of photoionization thresholds. SPIRIT and LEUTL will also characterize the excited states of molecules, such as the cold wall desorption in accelerators and the sputtering of clusters. First results from SPIRIT are expected by December 2002. The setup of the SPIRIT experiment is shown in Figure 3.

EXOTIC NEXT-GENERATION LIGHT SOURCE CONCEPTS

A group of researchers has been meeting frequently since early 1998 to compare FEL simulation codes and discuss and develop exotic schemes to enhance the capabilities of next-generation light sources. Most recently we have discussed the possibility of enhancing the power of two wavelengths simultaneously using a bi-harmonic undulator [12] and the possibility of seeding with a soft x-ray FEL [13].

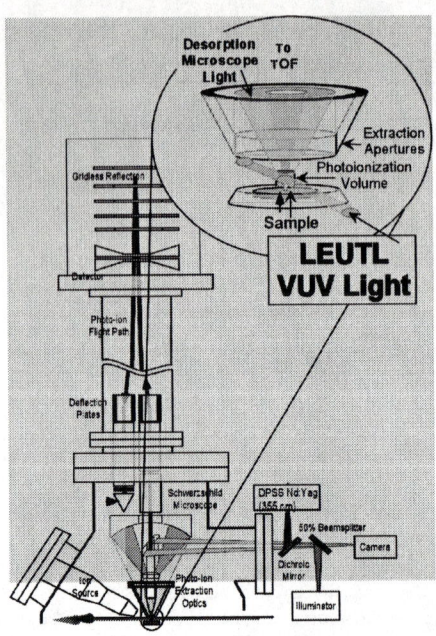

FIGURE 3. The SPIRIT experimental setup.

We performed a comparison of the high-gain FEL output from a biharmonic undulator, monoharmonic undulator, and two monoharmonic undulators in the TUHGS configuration. A biharmonic undulator has fields that are tuned such that the FEL emission is driven for a certain wavelength and a higher harmonic simultaneously. We examined three cases (monoharmonic, biharmonics, and TUHGS with monoharmonic undulators) and also a variation of the third case (TUHGS) by elongating the second undulator. The electron beam and undulator parameters of the cases can be examined in Table 1. In this presentation only the results of the code MEDUSA are shown (Table 2), although other cases were treated with other codes.

CONCLUSION

In summary, a next-generation light source would contain many of the following qualities, depending upon the needs of the users: tunable to ultrashort wavelengths, including the x-ray regime to 1.0 Å; full longitudinal coherence, full transverse coherence; ultra-short pulse durations (<100 fs); high peak powers; and as well as being as small and as cost effective as reasonably possible.

Table1: Electron beam and undulator parameters for the three cases.

Undulator type	Monoharmonic	Biharmonic	TUHGS
Electron beam energy (MeV)	935	1078	935
Normalized electron beam emittance (π mm mrad) [x,y]	1, 1	1, 1	1, 1
Electron beam peak current (A)	850	850	850
Energy spread (%)	0.05	0.05	0.05
Undulator(s) resonant wavelength	At fundamental only	At fundamental and third harmonic	First at fundamental only and second at third harmonic
Third harmonic undulator period (cm)	-na-	2.0	2.0
Peak magnetic field (kG)	1.767	1.767	First 1.767
Undulator period (cm)	6.0	6.0	6.0
Peak magnetic field (kG) of third harmonic undulator	-na-	5.301	Second 5.301
Fundamental wavelength (nm)	13.36	13.36	13.36
Third harmonic wavelength (nm)	4.45	4.45	4.45
Ninth harmonic wavelength (nm)	1.48	1.48	1.48

Table 2: Comparison of types of high-gain FELs

Type	Fundamental (14.45 nm)	Third (4.45 nm)	Ninth (1.48 nm)	Saturation z (m)	Energy (MeV)
Monoharmonic	398 MW	0.64 MW	50 kW	~60	935
Biharmonic (Perpendicular)	1270 MW	52 MW in y-pol.	5 kW	~60	1078
TUHGS (Undulator 1 ends at 63 m)	398 MW (z=63m)	0.424 MW (z=63m) 300 MW (z=84m)	0.373 MW (z=84m)	~80	935
TUHGS (Undulator 1 ends at 51 m)	11.8 MW (z=51m)	59.6 W (z=51m) 460 MW (z=74m)	0.15 MW (z=74m)	~74	935

Recently, there has been impressive progress made in the area of high-gain SASE FEL devices operating at wavelengths ranging from the near infrared to the far ultraviolet. The first experiment LEUTL [660-130 nm], the Tesla Test Facility (TTF) at DESY Hamburg [400-100 nm], and the Visible to Infrared SASE FEL (VISA) at Brookhaven National Laboratory (BNL) [840 nm] have demonstrated the single-pass, high-gain (SP HG) FEL using the SASE process to saturation [13-15]. In addition, a multistage method, the high-gain harmonic generation (HGHG) SP HG FEL was demonstrated [16] at BNL [5.3 microns] to saturation.

The LEUTL project was the first to demonstrate the SASE process to saturation in the tunable range from the visible to the VUV (660-130 nm). Currently, the first user experiment is being installed with the first results are expected by December 2002. Numerous beam and user proposals have been received and are currently under review.

Based upon theory and these recent experiments, a number of high-brightness, laser-like photon sources for users are being proposed. These light sources will operate from the ultraviolet to hard x-ray wavelength regimes and are based on the many methods of SP HG FELs. Some present proposals include LCLS-SLAC (0.15 nm, CDR prepared, first light 2008), TTF-2 (VUV user facility, 2004), TESLA/FEL DESY (0.1 nm, TDR complete), SCSS-SPring8 (> 2 nm, approved and funded), SDL-BNL (> 100 nm, reached saturation), SPARX-Italy (> 1 nm), and FERMI –Trieste.

ACKNOWLEDGMENTS

The work of S.G. Biedron, M. Borland, R. Dejus, M. Erdmann, Z. Huang, K-.J. Kim, J. Lewellen, Y. Li, A. Lumpkin, S.V. Milton, E. Moog, V. Sajaev, and G. Wiemerslage is supported at Argonne National Laboratory by the U.S. Department of Energy, Office of Basic Energy Sciences under Contract No. W-31-109-ENG-38. The activity and computational work for H. P. Freund is supported by Science Applications International Corporation's Advanced Technology Group under IR&D subproject 01-0060-73-0890-000. The activity of W.M. Fawley is supported at Lawrence Berkeley National Laboratory by the U.S. Department of Energy under Contract No. DE-AC03-76SF00098. The work of H.-D. Nuhn is supported by the U.S. Department of Energy, Office of Basic Energy Sciences, Division of Material Sciences, under contract No. DE-AC03-76SF00515. The work of R. Bartolini, F. Ciocci, G. Dattoli, G. Felici, P.L. Ottaviani, and A. Renieri is supported by ENEA the Italian Agency for New Technologies, Energy, and the Environment.

REFERENCES

1. See for example *Introduction to the Physics of FELs*. Laser Handbook, vol.6. Ed. W.B. Colson, C. Pellegrini, and A. Renieri. (Amsterdam:North Holland, 1990) and references 4-45 in S. G. Biedron, *Toward Creating a Coherent Next-Generation Light Source* (Alnarp, Sweden:Reproenheten, 2001).
2. See for example references 46-53 in in S. G. Biedron, *Toward Creating a Coherent Next-Generation Light Source* (Alnarp, Sweden:Reproenheten, 2001).
3. 10th ICFA, Workshop on 4th Generation Light Sources, ESRF, Grenoble, January 22-25, 1996.
4. S. V. Milton et al., originally published in Science Express as 10.1126/science.1059955 on May 17, 2001 and Science 292 (2001) 2037-2041.
5. S.V. Milton et al., Proc. SPIE **3614** (1999) 86.
6. S.G. Biedron, et al., "The Operation of the BNL Gun-IV Photocathode Gun at the Advanced Photon Source," Proc. of the 1999 Particle Accelerator Conference (1999) pp. 2024-2026.
7. M. Borland, J. W. Lewellen, "Initial Characterization of Coherent Synchrotron Radiation Effects in the APS Bunch Compressor," Proceedings of the 2001 Particle Accelerator Conference, P. Lucas, S. Webber, 4, IEEE, (2001) pp. 2839-2841.
8. S.G. Biedron et al., "Measurements of Nonlinear Harmonic Generation at the Advanced Photon Source's SASE FEL," Nuclear Instruments and Methods in Physics Research A 483 (2002) 94.
9. A.H. Lumpkin et al., Phys. Rev. Lett. 88 (2002) 234801.
10. J.W. Lewellen et al., Nucl. Instrum. Methods A483 (2002) 40.
11. M. Pellin, private communication.
12. S.G. Biedron et al., Proc. SPIE 4632 (2002) 50-65.
13. S.G. Biedron et al., "Simulation of the fundamental and nonlinear harmonic output from an FEL amplifier with a soft x-ray seed laser," Proceedings of 7th European Accelerator Conference (EPAC 2000), 26-30 June, Vienna (2000) 729.
14. V. Ayvazyan et al., Phys. Rev. Lett. 88 (2002) 104802.
15. A. Tremaine et al., Nucl. Instrum. Methods A483 (2002) 24.
16. L.-H. Yu et al., Science 289 (2000) 932.

The Linac Coherent Light Source[†]

John N. Galayda

Stanford Linear Accelerator Center, Menlo Park, California 94025

Abstract. A collaboration of scientists from SLAC, UCLA, Los Alamos National Laboratory, Brookhaven National Laboratory, and Argonne National Laboratory have proposed to build the Linac Coherent Light Source (LCLS) facility, a free-electron laser (FEL) on the SLAC site, spanning photon energies 0.8-8 keV. The laser output will be 8-10 GW with pulse lengths 230 fsec or less. The LCLS will offer unprecedented experimental opportunities in the areas of atomic physics, chemical dynamics, plasma physics, nanoscale dynamics, and biomolecular imaging. SLAC has proposed to begin engineering design of the laser in 2003, leading to project completion in 2008. The laser produces x-rays by the self-amplified spontaneous emission (SASE) process: an intense, highly collimated pulse of 14.5 GeV electrons, traveling through a 122 m-long undulator magnet system, is induced by its own synchrotron radiation to form sub-nanometer-scale bunches. The bunching process enhances the coherence and hence the intensity of the emitted synchrotron radiation. The process is analogous to the instability of a high-gain amplifier; the "noise" signal that seeds the instability is the shot noise in the electron beam.

SELF-AMPLIFIED SPONTANEOUS EMISSION

Synchrotron radiation is produced by an energetic charged particle as it is deflected by an externally applied magnetic field. The properties of synchrotron radiation produced in storage rings and free-electron lasers may be derived from the Lienard-Wiechert potentials of a relativistic accelerating charge[1]. Quantum fluctuations in the emission of synchrotron radiation contribute a random component to the momentum of radiating electrons; however, the properties of the synchrotron radiation and the mechanism for its amplification in a free-electron laser may be rigorously understood in a classical treatment[2]. A description of the self-amplified spontaneous emission (SASE) process, along with a rather complete list of references to key publications in free-electron laser research, may be found in Chapter 4 of the Linac Coherent Light Source Conceptual Design Report[3].

A useful radiation source, and one of particular relevance to free-electron lasers (FELs), is produced by passing an electron beam through an *undulator* magnet. The field of such a magnet varies sinusoidally in z with peak field B_o and period λ_u:

$$\overline{B}(x, y = 0, z) = (0, B_0 \cos(k_u z), 0); \qquad k_u = 2\pi / \lambda_u \qquad (1)$$

[†] Work supported by DOE contract DE-AC03-76SF00515

CP641, *X-Ray Lasers 2002: 8th International Conference on X-Ray Lasers,* edited by J. J. Rocca et al.
© 2002 American Institute of Physics 0-7354-0096-2/02/$19.00

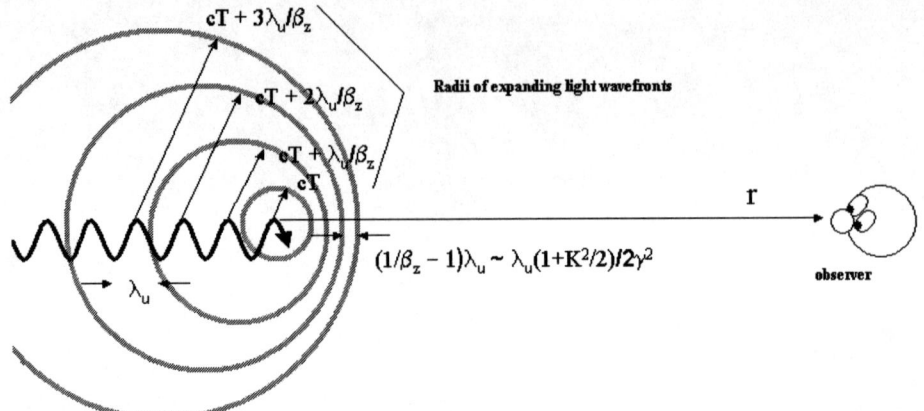

FIGURE 1. An electron moving in an undulator with period λ_u may be envisioned as emitting one wavefront of synchrotron radiation for each undulator period traversed. The electron travels a distance $\beta_z \lambda_u$ as the wavefronts travel a distance λ_u, resulting in radiation with wavelength $(1/\beta_z - 1)\lambda_u$, as seen by an observer on-axis. The factor K appears as a result of replacing β_z^2 with $\beta^2 - <\beta_x^2>$, and applying equation 2.

An electron entering the undulator magnet with high momentum in the z direction will be deflected so as to oscillate in the x direction. Here we assume hat the maximum deflection of the electron is sufficiently small that its motion is, to a good approximation, sinusoidal. Following Hofmann's exposition[4], the transverse deflection $\beta_x(t')$ of the electron is customarily expressed in terms of the undulator strength parameter K, defined below:

$$\dot{\beta}_x(t') \cong -\left(\frac{ec\beta_z B_0}{mc\gamma}\right)\cos(k_u \beta_z ct') \tag{2}$$

$$\beta_x(t') \cong -\left(\frac{ec\beta_z B_0}{mc\gamma}\right)(k_u \beta_z c)^{-1} \sin(k_u \beta_z ct') \equiv -\frac{K}{\gamma}\sin(k_u \beta_z ct')$$

The electric field of undulator radiation produced at retarded time t' by a single electron travels a distance r to an observer, arriving at time t as illustrated in figure 1. In the far-field approximation, the on-axis field is

$$E_x(t) \cong \left(\frac{4e^2 B_0 c\gamma^3}{4\pi\varepsilon_o (1+K^2/2)^2 mc^2}\right)\frac{1}{r}\cos\left(\frac{2\pi}{\lambda_u}\frac{2\gamma^2}{(1+K^2/2)}\beta_z ct + \text{const.}\right) \tag{3}$$

One can see that the radiation wavelength is equal to the undulator period λ_u, shortened by the Doppler shift factor $(1+K^2/2)/2\gamma^2$ for β near 1.

If synchrotron radiation from N electrons reaches the observer at time t, one may add their contributions to the observed electric field, each including its appropriate phase factor in the cosine function. It is convenient to replace the cosine in equation 3 by $e^{-i\omega t}$ so that the total electric field produced by N electrons may be expressed as

$$E_{tot}(t) = E_x(t) \sum_{j=1}^{N} \exp(i\phi_j) \tag{4}$$

In a non-laser synchrotron light source such as the SPEAR ring at SLAC, the electrons are randomly distributed, on a length scale of the order of the synchrotron radiation wavelength. In this case <E(t)> = 0 and the average power is given by <E²>, and is proportional to N. However, if the electrons were distributed in short bunches with separation equal to the wavelength of the undulator radiation, the summation over phase factors can approach its maximum value of N. Since the radiated power is proportional to N², and N can be of the order of 10^{10}, an enormous increase in radiated power can be realized, as compared to a conventional synchrotron source.

A free-electron (FEL) laser exploits the dynamics of electron beams to bring about the bunching necessary to achieve dramatic power enhancement. The bunching results from the interaction of the electrons with their own synchrotron radiation in addition to the undulator magnet field. Random fluctuations in the electron distribution produce corresponding fluctuations in the spontaneous synchrotron radiation. The strongest of these fluctuations begin to bunch the electrons, further intensifying the radiation field. In essence, the process amplifies a particular Fourier component of the shot noise in the electron beam current, determined by the wavelength of the undulator radiation. For this reason the process is called *self-amplified spontaneous emission* (SASE). Under ideal conditions the process can fully bunch the electrons, at which point the output power reaches its saturation level. The gain of the SASE process is characterized by the distance the electron beam travels as the synchrotron radiation power is amplified a factor of e. This distance, the *gain length*, varies from 2.4 meters for 0.8 keV photons to about 5.8 meters for 8 keV x-rays in the Linac Coherent Light Source (LCLS) design. Saturation of the SASE process should be complete in less than 20 gain lengths.

SASE is a single-pass phenomenon. For a properly conditioned electron beam, a resonant cavity for the synchrotron radiation is not necessary. However a SASE FEL can serve as an amplifier for a light pulse from an external source. Furthermore, the bunching of electrons by SASE can produce Fourier components of the beam current at harmonics of the fundamental FEL wavelength. This results in strongly enhanced emission of synchrotron radiation at these harmonics. Schemes to enhance harmonic generation by seeding with an external laser or with synchrotron radiation is presently the subject of active study.

It is quite challenging to produce an electron beam with properties necessary for a SASE FEL. For an 8 keV FEL such as the Linac Coherent Light Source, peak currents in the range 1,000-4,000 amperes are necessary. The electron beam must remain within the synchrotron radiation fan in order to be bunched. This implies that the electron beam emittance must be comparable to that of the photon beam, $\lambda/4\pi$. Finally the energy spread of the electron beam must be small, of the order of 10^{-5}. For lasing at 8 keV, these conditions can be satisfied by adding a high-quality electron source to the SLAC linac, and accelerating this beam to 14.35 GeV.

The temporal coherence of a SASE FEL is limited by its high gain; only L_g/λ_u wavefronts are produced in one gain length L_g. Hence the bunching process can proceed independently for wavetrains of this length, within a single 230 fsec electron bunch. Of course, if a SASE FEL is used to amplify a seed pulse, the temporal coherence of the seed is imprinted on the FEL output.

THE LINAC COHERENT LIGHT SOURCE PROJECT

A SASE FEL is, at this time, the most straightforward means of creating gigawatt-level coherent light pulses in the 0.1 nanometer wavelength range. However, the necessary electron accelerator and undulator system are far beyond the "tabletop" paradigm in size and cost. In 1992[5], C. Pellegrini proposed the use of the last kilometer of the SLAC linac to construct a 0.1 nm SASE free electron laser. This makes possible a cost savings of about $300M compared to construction of a dedicated linac. A study group was organized at SLAC to determine the feasibility of this proposal. This study group expanded, and now includes researchers from Lawrence Livermore National Laboratory, Los Alamos National Laboratory, Brookhaven National Lab, Argonne National Laboratory and the University of California-Los Angeles. The LCLS Collaboration completed a design study of an x-ray FEL, *LCLS Design Study Report*[6] in 1998. Since then, the Collaboration has conducted experimental investigations of high-performance electron guns and successfully completed experimental tests of the FEL process. Funds for R&D were provided by DOE in the amount $1.5M/year from 1999 to 2002. A more detailed design for the LCLS has been published in the LCLS Conceptual Design Report[7]. The LCLS Construction Project will begin in fiscal year 2003, when it enters the Project Engineering Design phase. It is expected that, in FY2005, it will be possible to order items with long delivery times, such as the undulator magnets and components for the photoinjector. Civil construction will begin in 2006. Commissioning tests of the accelerator systems will begin in 2006, and laser tests will begin in 2008. By the end of 2008, the laser and all experiment facilities will be complete. The Total Estimated Cost for construction of the LCLS is $221M. When it is complete, the LCLS will provide spatially coherent beams of x rays from 0.8 keV to 8 keV. Peak output powers of about 8 GW are expected. Initially, the LCLS will produce x-ray pulses with duration 230 fsec or less. Upgrades to pulses in the 10 fsec range are under development. **Table 1** lists some key parameters of the LCLS.

Though the extraordinary x-ray beam of the LCLS will itself be the subject of study at first, much effort has already been devoted to plans for use of the LCLS for research in atomic physics, chemistry, materials science, plasma physics and biology[8]. The LCLS will be used as a "pump" to create extraordinary states of matter, such as atoms with all inner shell electrons stripped away; or "warm, dense plasma," solids raised instantly to temperatures approaching 10 eV. The LCLS will also serve as a probe of chemical dynamics. By determining interatomic separations as they evolve on a femtosecond time scale, the LCLS may be used, in effect, for "freeze-frame" photography of atoms as they form and break molecular bonds. The intensity of the focused x-ray beam will be such that structure determination of samples as small as a single virus particle or perhaps a single protein molecule become feasible.

TABLE 1. Characteristics of the LCLS x-ray beam

Photon Beam Parameters	Lower Limit	Upper Limit	Units
Photon energy	0.8	8	KeV
Photon wavelength	1.5	0.15	Nanometers
Coherent photons/pulse	10.2	1.8	$X10^{12}$
Peak brightness	0.78×10^{32}	12.1×10^{32}	Photons/(sec mm^2 mrad^2 0.1% BW)
Average brightness(120 Hz)	0.39×10^{21}	2.8×10^{21}	Photons/(sec mm^2 mrad^2 0.1% BW)
Peak coherent power	10.6	8	GW
Transverse beam size, rms	37	27	Microns
Beam divergence, rms	3.2	0.39	Microradians
Electron Beam	**Lower Limt**	**Upper Limit**	**Units**
Electron beam energy	4.54	14.35	GeV
Electron pulse duration	230	230	Femtoseconds
Electron beam peak current	3,400	3,400	Amperes
SASE gain length	2.4	5.8	Meters

A physical description of the LCLS begins with the injector linac, a 150 MeV accelerator that will be located in a "spur" tunnel that joins the main linac tunnel at the 2 kilometer point("Linac 0" in figure 3.). A key component in the injector linac and, indeed, for the whole of the LCLS is the photocathode RF gun. The gun is a 2856 MHz, 1.6-cell resonant structure that was designed, constructed and tested by collaborators from SLAC, Brookhaven and UCLA[9]. This gun is capable of producing a 1-nanocoulomb pulse of 7 MeV electrons, 10 psec long, with emittance necessary to support the SASE process at 0.15 nanometers. Current from this gun is switched by illuminating the copper cathode with a 500 microjoule pulse of UV light. It is expected that a Ti:sapphire laser system will be used for this purpose in the LCLS. This laser must produce a pulse with very flat profile both temporally and transversely. A very fast risetime and falltime are important to produce a uniform charge density (and hence correctable space-charge forces) in the electron beam. Commercially available lasers routinely meet LCLS power specifications, with the exception of the required 120 Hz repetition rate.

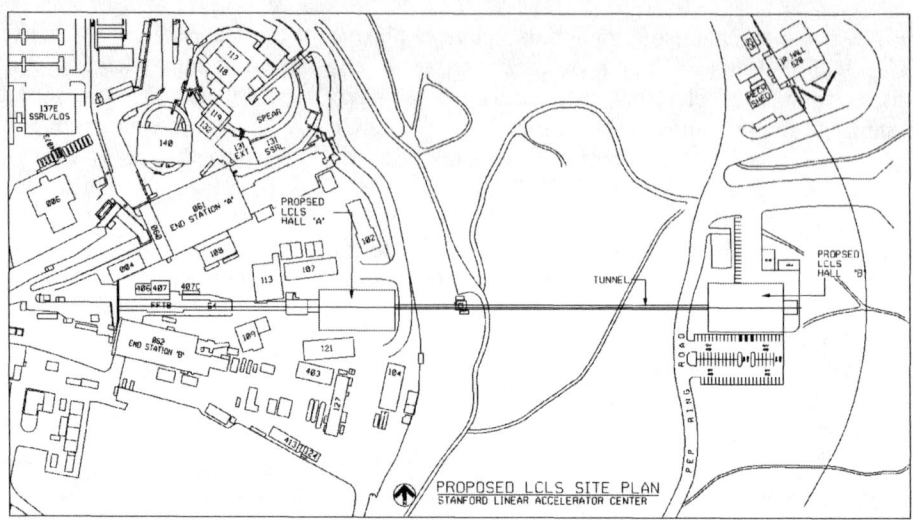

FIGURE 2. Site plan for the Linac Coherent Light Source. The linac provides electrons to End Station "A" as well as to the LCLS undulator, housed in the Final Focus Test Beam (FFTB) building. Experiment Hall A is attached to the FFTB. X rays from the LCLS may be directed through a tunnel to Experiment Hall B, 322 meters from the undulator.

The 150 MeV electron beam is injected to the SLAC linac to be accelerated to a maximum of 14.35 GeV. During the acceleration process, the 100-ampere current pulse must be compressed in several stages to reach 3,400 amperes. Pulse compression in the electron beam is achieved in close analogy with compression of a laser pulse: an energy "chirp" is applied to the electron beam, so that electrons at the rear of the bunch are more energetic than those in the front. When the electron beam passes through a chicane of four bending magnets, the more energetic electrons follow a more direct trajectory and nearly overtake the leading electrons. This conceptually simple process is made difficult by a phenomenon closely related to the SASE mechanism; in passing through the chicane bends, the electrons emit synchrotron radiation. For radiation wavelengths comparable to the length scale set by the bunch and any current fluctuations in the bunch, there can be a high degree of coherence in the synchrotron radiation and hence an enhanced energy loss and spreading in the electron beam. Coherent synchrotron radiation effects can amplify current fluctuations and energy errors in the bunch[10]. With careful design, compression can be accomplished without undue damage to the electron beam properties. The LCLS design incorporates two bunch compressors, one at the 250 MeV point and the other at the 4.5 GeV point in the linac(Bunch Compressor 1 and Bunch Compressor 2 in figure 3.). The compressive effects of other bends in the beam path must be controlled, however, including those at the 150 MeV point and the "dogleg" bends at the end of the linac. After acceleration, the electron beam is passed through a 121-meter undulator channel. The channel, designed at the Advanced Photon Source of Argonne National Laboratory, incorporates thirty-three 3.4-meter long hybrid permanent magnets, which must meet stringent field requirements to maintain the necessary collinearity between the electron beam and the photon beam. The channel incorporates focusing magnets and a variety of diagnostics for the electrons and x rays.

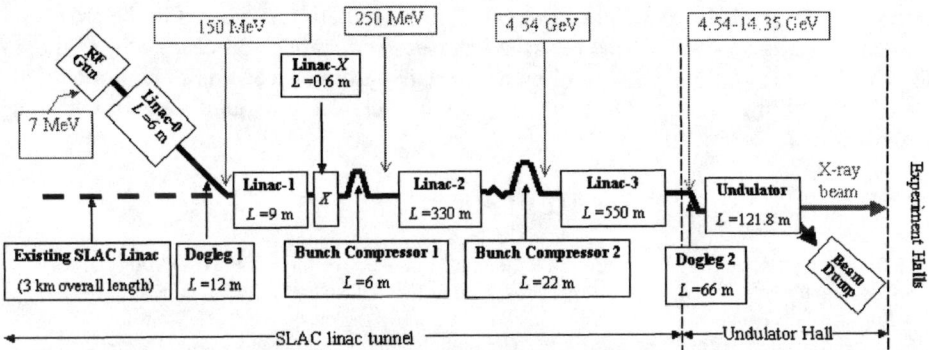

FIGURE 3. Schematic layout of the LCLS. Electron beam energies at various stages of the accelerator are indicated at the top of the figure. Bends in the electron beam path, i.e. the "doglegs" and bunch compressors, are designed to compress the electron bunch, raising its peak current from 100A to 3,400A.

An 8 GW pulse of coherent radiation emerges from the undulator channel, with an r.m.s. beam size of 30 microns and an opening angle of 0.5 microradian at 0.15 nanometers. The ability of optics to withstand this beam will itself be an area of research once the LCLS is operational; design of robust optics for 1.5 nanometer radiation is especially challenging. The x-ray transport, optics and diagnostics system for the LCLS will be constructed at Lawrence Livermore National Laboratory[11]. The optical device located closest to the source point will be a differentially-pumped gas cell for controlled attenuation of x-ray beam intensity. Prototypical optical elements to focus, monochromatize and split the x-ray beam will be included in the scope of the LCLS Project.

LCLS experimental facilities will be located in two experiment halls, as indicated in Figure 2. Hall A is located 40 meters from the end of the undulator and will house up to four x-ray enclosures. The x-ray beam may pass through Hall A to Hall B, located 322 meters from the undulator exit. The divergence of the x-ray beam is such that materials traditionally used for x-ray optics at synchrotron sources may be employed. Low-Z materials are favored for optics in the Near Hall.

FUTURE DEVELOPMENTS FOR THE LCLS

Ongoing efforts to shorten the LCLS x-ray pulse dureation and to improve the temporal coherence will continue into the post-commissioning phase. The LCLS Project goals, listed in Table 1, call for a pulse duration of 230 fsec for the electron beam. The x-ray pulse waveform need not replicate that of the electron beam. It is likely to contain spikes of power with varying amplitude and coherence length of the order of 100-200λ[12]. Indeed, the x-ray pulse may even be considerably shorter than the electron current pulse. Since the SASE process starts from shot noise in the beam, there can be considerable variation in light output from pulse to pulse. After SASE lasing has been achieved, several methods for control of the properties of the x-ray pulse will be explored by seeding the SASE process with a lower-power x-ray pulse. Seeding can be used to improve temporal coherence[13], to reduce the duration of the fully amplified x-ray pulse[14], or to generate harmonics of the seed pulse[15]. The seed pulse can itself be extracted from synchrotron radiation produced in the first few undulators of the LCLS. However, seed pulses from other sources, such as those described at this conference, would be an attractive alternative; it is to be hoped that the LCLS will provide a point of convergence for heretofore disjoint areas of research in x-ray lasers.

REFERENCES

1. Hofmann, A., "Properties of Synchrotron Radiation", in the proceedings of the *CERN Accelerator School – Synchrotron Radiation and Free Electron Lasers*, edited by S. Turner, CERN 98-04, 3 August 1998
2. Schroeder, C., Pellegrini, C., and Chen, P., "Quantum Fluctuations in Free-Electron Lasers", in *Quantum Aspects of Beam Physics, - 18th Advanced ICFA Beam Dynamics Workshop, Capri, Italy 15-20 October 2000*, edited by Pisin Chen, World Scientific Publishing Co., May 2002 (ISBN 981-02-4950-0)
3. Chapter 3, "FEL Physics", *Linac Coherent Light Source (LCLS) Conceptual Design Report*, SLAC-R-593, April 2002, http://www-ssrl.slac.stanford.edu/lcls/CDR/
4. A. Hofmann, op. cit, section 4
5. Pellegrini, C., "A 4 to 0.1 FEL based on the SLAC Linac", in the proceedings of the *Workshop on Fourth Generation Light Sources,, February 24-27 1992, SSRL-92-02*, pp. 364-375, 1992
6. Arthur, J. et al., *LCLS Design Study Report*, SLAC-R-0521,
7. *Linac Coherent Light Source (LCLS) Conceptual Design Report*, SLAC-R-593, April 2002, http://www-ssrl.slac.stanford.edu/lcls/CDR/
8. Shenoy, G. and Stohr, J., *LCLS – the First Experiments*, October 2000 http://www-ssrl.slac.stanford.edu/LCLS/papers/LCLS_Experiments_2.pdf,
9. Palmer, D. T. , et al., "Emittance Studies of the BNL/SLAC/UCLA 1.6-cell photocathode gun", in *Proceedings of the 1997 Particle Accelerator Conference, Vancouver, B.C., Canada, 12-16 May 1997*, p.2687
10. Heifets, S., Krinsky, S. and Stupakov, G., CSR Instability in a Bunch Compressor, SLAC-PUB-9165, March 2002
11. *LCLS Conceptual Design Report*, Chapter 9, op. cit.
12. *LCLS Conceptual Design Report*, Chapter 4, pg. 4-9, op. cit.
13. Feldhaus, Saldin, E. I., Schneider, J. R., Schneidmiller, E. A., "Possible Application of X-ray Optical elements for reducing the spectral bandwidth of an X-ray SASE FEL, *Opt. Commun.* **140**, pp. 341-352, 1997
14. Schroeder, C. B., Pellegrini, C., Reiche, S., Arthur, J. and Emma, P., "Chirped-Beam Two-Stage SASE-FEL for High Power Femtosecond X-ray Pulse Generation", , in *Proceedings of the 2001 Particle Accelerator Conference*, P. Lucas and S. Webber, editors, IEEE Operations Center, Piscataway, NJ, 2001 pp. 2757-2759.
15. Brefeld, W., Faatz, B., Feldhaus, J., Korfer, M., Kryzwinski, J., Moller, T., Pflueger, J., Rossbach, J., Saldin, E., Schneidmiller, E. A., Schreiber, S., and Yurkov, M. V., "Development of a Femtosecond X-Ray SASE FEL at DESY", *Nuc. Inst. Meth. A*, 483, pp 75-79 (2002).

Characteristics of Relativistic Nonlinear Thomson scattering of an intense laser field as ultrashort x-ray source

K. Lee*, Y. H. Cha*, J. M. Han*, Y. Rhee*, M. S. Sim¶, B. H. Kim¶, and D. Kim¶

*Laboratory for Quantum Optics, Korea Atomic Energy Research Institute,
P. O. Box 105, Deokjin-Dong, Yuseong-Ku, Daejeon, 305-600, Korea.
¶Physics Department, Pohang Univ. of Science and Technology,
San 31, Hyoja-Dong, Nam-Ku, Pohang, 790-784, Korea.

Abstract. Radiations from a single electron irradiated by an ultra-intense laser pulse were extensively investigated by a numerical simulation. The characteristics of the radiation such as spectral distribution, radiation intensity, and angular distribution show that the relativistic nonlinear Thomson scattering, or Larmor radiation could be a good ultrashort radiation source toward x-ray region. The effect of ions which generates Bremsstrahlung radiation degrading the characteristics of Larmor radiation is also discussed.

INTRODUCTION

Since the advent of the CPA(Chirped Pulse Amplification) laser [1], new regime of the high intense field - matter interaction has been expanded explosively in fundamental physics and applications. The rapidly developing ultrashort intense laser technology [2] also has opened up a new branch of relativistic plasma physics. Among them, the generation of subfemtosecond pulse is considered as one of the most important potential applications.

In a EUV (Extreme Ultra-Violet) regime, the train of 250 attosecond pulse has been experimentally demonstrated [3] and Hentschel et al. [4] observed a 650 attosecond Gaussian x-ray pulse in a high order harmonic generation. A femtosecond x-ray laser pulse [5] using fast Coster-Kronig decay process was investigated. Some schemes incorporating a scattering process of intense laser pulse by free electrons have been explored such as the relativistic Doppler shift which arises from backscattering of laser radiation from a counterstreaming relativistic electron beam [6,7] and the harmonic frequency upshift [6]. The first one is referred as Compton backscattering and the other as relativistic nonlinear Thomson scattering.

The harmonic spectrum of the relativistic nonlinear Thomson scattering for a single electron has been intensively investigated in analytical ways [8]. Recently such a prediction has been experimentally verified by observing angular patterns of harmonics with relatively low laser intensity of $\sim 10^{17}$ W/cm^2 [9]. Esarey et al. [6] has

CP641, X-Ray Lasers 2002: 8th International Conference on X-Ray Lasers, edited by J. J. Rocca et al.

developed a comprehensive theory for the nonlinear Thomson scattering considering plasma effect and presented parameters for producing 9.4 ps x-ray pulse of 310 eV photon energy with ultrahigh laser intensity of ~10^{20} W/cm^2. Ueshima et al. [10] has suggested a few methods to enhance the radiation power with a particle-in-cell simulation even with higher intensity. Very recently, Kaplan et al. [11] proposed a scheme for zeptosecond (10^{-21} sec) radiation named as lasertron.

In this thesis, we present angular distribution, spectral distribution, and radiation power in detail for a single electron irradiated by a 20 fs FWHM (Full Width Half Maximum) laser pulse. The results show that with 10^{20} W/cm^2 laser intensity, 2 attosecond pulse with photon energy from 100 eV up to 1000 eV can be generated. Considering the radiation power and the angular distribution, the radiation by a linearly polarized pumping laser has superior aspects.

FORMULATION

The motion of single electron in a high intensity laser field is described by the following relativistic equation of motion,

$$m_e \frac{d}{dt'}(\gamma \vec{v}) = -e(\vec{E} + \vec{\beta} \times \vec{B}), \tag{1}$$

where m_e is the electron mass, $\vec{v}$ the velocity, γ the relativistic factor, and $\vec{E}$ and $\vec{B}$ are the electric and magnetic field of an incident laser respectively. With the motion obtained in the above equation, the angular radiation power [12] emitted by an electron and detected far away from the electron toward the direction, $\hat{n}$ at a time, t is then calculated as

$$\frac{dP(t)}{d\Omega} = |A(t)|^2, \tag{2}$$

$$A(t) = \sqrt{\frac{e^2}{4\pi c}} \left[\frac{\hat{n} \times \{(\hat{n} - \vec{\beta}) \times \dot{\vec{\beta}}\}}{(1 - \vec{\beta} \cdot \hat{n})^3} \right]_{t'}, \tag{3}$$

where t' is the electron's time or retardation time and the relation with t is given by

$$t = t' + \frac{x - \hat{n} \cdot \vec{r}(t')}{c}. \tag{4}$$

Then the angular spectral intensity can be obtained from the Fourier transform of A(t),

$$\frac{d^2 I}{d\omega d\Omega} = 2|A(\omega)|^2, \tag{5}$$

$$A(\omega) = \frac{1}{\sqrt{2\pi}} \int_{-\infty}^{\infty} A(t) e^{-i\omega t} dt. \tag{6}$$

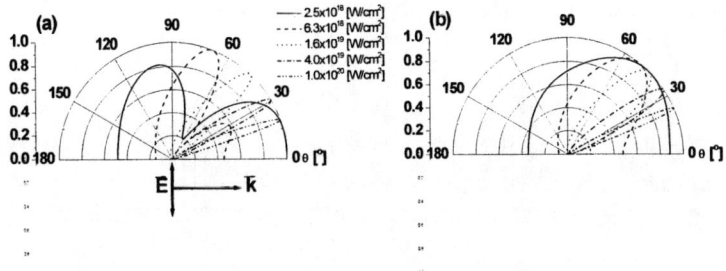

FIGURE 1. θ Distributions at φ=0° of the radiation power are plotted for (a) linearly and (b) circularly polarized pumping lasers.

RESULTS AND DISCUSSIONS

Radiations were evaluated for various conditions of laser peak intensity (I) and for two-different polarizations. For the pumping laser, a Gaussian envelope function was propagated toward z direction with a 20 fs FWHM(Full Width Half Maximum) and a 800 nm central wavelength. Since the quivering amplitude of an electron amounts to the central wavelength with 10^{20} W/cm², the electron can be exerted by the pondermotive force which is considered as an electron acceleration mechanism. But for the radiation, such a deviation may not much alter the radiation characteristics, thus the plane wave approximation was adopted for this calculation of radiation.

The time integrated θ distribution of the radiation at φ=0° are plotted in Fig. 1, in which to observe and compare the angular distributions, each radiation is normalized to its maximum. As the laser intensity increases, the radiation is directed more toward z-axis with narrower divergence. The increase of laser intensity enhances not only a total radiation power but also a radiation flux. These characteristics of the radiations can be considered beneficial as a radiation source. In φ distribution, the radiations are directed to the direction of the laser electric field.

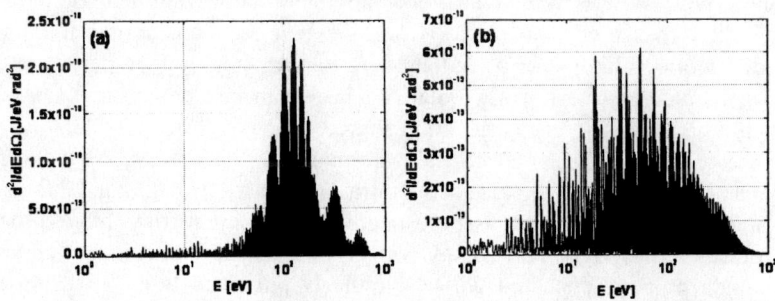

FIGURE 2. The angular spectral intensity for I=10^{20} W/cm² at the direction of maximum radiation (see Fig. 1) are plotted for (a) linearly and (b) circularly polarized pumping lasers.

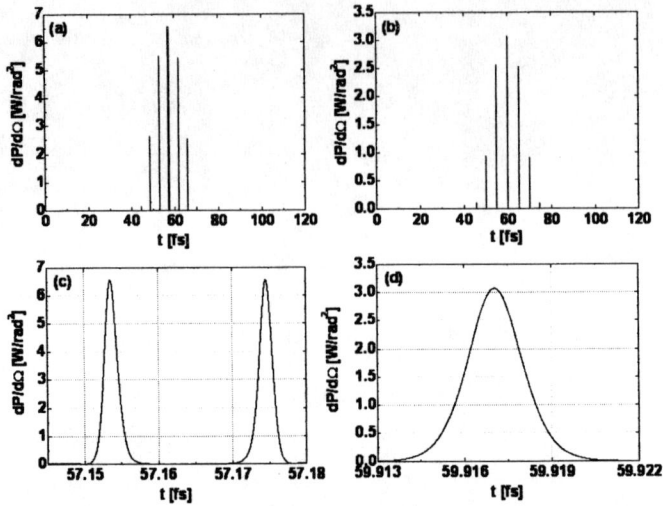

FIGURE 3. The angular radiation power for $I=10^{20}$ W/cm^2 at the direction of maximum radiation (See Fig. 1) are plotted for (a) linearly and (b) circularly polarized lasers. To investigate the pulse shape, the highest peaks are magnified for (c) linearly and (d) circularly polarized lasers.

For the laser intensity of $I=10^{20}$ W/cm^2, the angular spectral intensities at the direction of maximum power are plotted in Fig. 2. First of all, the relativistic motion of the electron generates high energy photons above 100 eV up to 1000 eV. From the results, it can be noticed that the linearly polarized pumping laser generates high energy photons more while with the circularly polarized pumping laser, the spectrum is broadly distributed from low energy. One interesting thing in the spectrum is the appearance of modulation, which is expected something to do with Doppler shift due to the electron's motion.

The angular powers at the maximum direction are plotted in Fig. 3. The spike-like radiation with interval of the laser period of 2.7 fs can be noticed from the figures. The highest peak power, 6.5 W/rad^2 for the linear polarization amounts to 0.65 mW/cm^2 when measured 1 m away from the source. Considering the narrower angular distribution, high energy radiation spectrum, and higher peak power, the radiation generated by the linearly polarized laser can be considered to have more potential as a high intensity x-ray radiation source.

In Fig. 3 (c) and (d), the highest peaks are magnified to observe the pulse shapes. The estimated FWHM's of the pulse show ultrashort pulses of 2.0 as and 2.2 as for the linearly and circularly polarized lasers respectively. One interesting thing is that there are two pulses with a very narrow interval of 21 as with a linearly polarized laser while a single pulse is appeared with a circularly polarized laser. The single pulse during single cycle of the pumping laser with circular polarization can be easily understood considering the circular motion of the electron, more accurately a helical motion. The double pulse structure with a linearly polarized laser is caused by the disappearance of $\dot{\vec{\beta}}$ in the mean time, where there is no radiation. The shortness of the

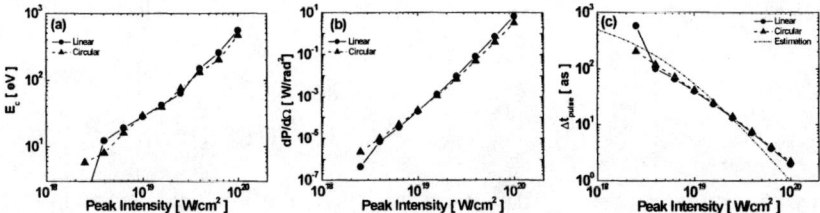

FIGURE 4. The dependence on the laser intensities for some characteristic parameters are plotted for (a) critical photon energy, E_c at which the spectral intensity reduces one tenth of its maximum, (b) angular power, and (c) FWHM of the pulse duration for highest peaks, wherein the "Estimation" is obtained from Eq. (8). All the parameters plotted are obtained at the direction of maximum radiation.

radiation pulse is caused by the relativistic motion of the electron, which causes a time contraction in a view of a detector as,

$$\frac{\Delta t}{\Delta t'} \approx 1 - \hat{n} \cdot \vec{\beta}, \tag{7}$$

which comes from Eq. (4). In above equation, $\Delta t'$ is the time interval that the electron emits radiations and Δt is the time interval that a detector receives the radiations. Considering the dynamics of an electron above equation gives the following estimation of the pulse width for a relativistic high intensity laser,

$$\frac{\Delta t}{\Delta t'} \approx \frac{4}{a^4}. \tag{8}$$

where $a = 8.5 \times 10^{-10} \lambda_{\mu m} I^{1/2}$ is the normalized field strength. By considering $\Delta t'$ as a quarter-period of the laser oscillation, Eq. (8) estimated the pulse width, 1.3 as for $I=10^{20}$ W/cm^2 (a=6.8).

The variations of the critical photon energy (E_c), the peak angular power, and the pulse width on the laser intensities are plotted in Fig. 5, from which the scaling to the laser intensity is obtained approximately as

$$E_c \propto I^{1.5}, \tag{9}$$

$$\frac{dP}{d\Omega} \propto I^{4.7}, \tag{10}$$

$$\Delta t_{FWHM} \propto I^{-1.4}. \tag{11}$$

In the case of the pulse width, the simulation shows weaker dependence on the laser intensity than the simple estimation of Eq. (8).

The effect of ions has been also estimated by comparing the laser field and the coulomb field exerted by ions in plasma, which causes Bremsstrahlung radiation. For the ion field, average ion field is multiplied by the number of ions which could be encountered during the path of an electron. This analysis leads to the estimation of a critical density, n_c below which the effect of the ion field can be neglected.

$$n_c[\text{cm}^{-3}] \approx 4.5 \times 10^7 \frac{\alpha}{z} \frac{\sqrt{I[\text{W}/\text{cm}^2]}}{l_{path}[\text{cm}]}. \tag{12}$$

For a hydrogen plasma ($z=1$), 10 % perturbation of the ion field to the laser field ($\alpha=0.1$) gives $n_c = 6 \times 10^{18} \text{cm}^{-3}$ for $I=10^{20}$ W/cm^2. In above equation, electron's path length, l_{path} was calculated by numerical integration.

CONCLUSION

The characteristics of the relativistic nonlinear Thomson scattering has been investigated by a numerical calculation. The highly relativistic motion of an electron in an intense laser field generates high energy photon up to 1 keV in a narrow angular divergence toward the direction of laser propagation. In addition, ultrashort pulse of 2 attosecond has been obtained, which shows a potential application to the ultrashort x-ray source. An estimation of the effect of the ion field shows that during the motion of an electron by a pumping laser, the ion field can be neglected up to a rather high density of 10^{18}cm^{-3}.

ACKNOWLEDGMENTS

This work has been supported in part by Korea Research Foundation Grant (KRF-2000-015-DP0175), by POSTECH research fund, by ADD and by electron Spin Science Center founded by Korea Science and Engineering Foundation.

REFERENCES

1. M. D. Perry and G. Mourou, *Science* **264**, 917 (1994).
2. G. A. Mourou, C. P. J. Barty, and M. D. Perry, *Phys. Today* **51**, 22 (1998).
3. P. M. Paul, E. S. Toma, P. Breger, G. Mullot, F. Auge, Ph. Balcon, H. G. Muller, and P. Agostini, *Science* **292**, 1689 (2001); E. Hertz, N. A. Papadogiannis, G. Nersisyan, C. Kalpouzos, T. Halfmann, D. Charalambidis, and G. D. Tsakiris, *Phys. Rev. A* **64**, 051801 (2001).
4. M. Hentschel *et al. Nature* **414**, 509 (2001).
5. C. Toth, D. Kim, and C. P. J. Barty, *Inst. Phys. Conf. Ser.* 309 (1998).
6. E. Esarey, S. K. Ride, and P. Sprangle, *Phys. Rev. E* **48**, 3003 (1993).
7. F. V. Hartemann, *Phys. Plasmas* **5**, 2037 (1998).
8. Vachaspati, *Phys. Rev.* **128**, 664 (1962); L. S. Brown and T. W. B. Kibble, *Phys. Rev.* **133**, A705 (1964); E. S. Sarachik and G. T. Schappert, *Phys. Rev. D* **1**, 2738 (1970).
9. S.-Y. Chen, A. Maksimchuk, and D. Umstadter, *Nature* **396**, 653 (1998); S.-Y. Chen, A. Maksimchuk, E. Esarey, and D. Umstadter, *Phys. Rev. Lett.* **84**, 5528 (2000).
10. Y. Ueshima, Y. Kishimoto, A. Sasaki, and T. Tajima, *Laser and Particle Beams* **17**, 45 (1999).
11. A. E. Kaplan and P. L. Shkolnilov, *Phys. Rev. Lett.* **88**, 074801 (2002).
12. J. D. Jackson, *Classical Electrodynamics*, 2nd ed., Wiley, New York, 1975, Chap. 14.

Gigawatt, femtosecond VUV pulses from a SASE FEL: Photon beam characterisation and first applications

K. Tiedtke* for the TTF FEL-Team[†]

* *HASYLAB at DESY, Hamburg, Germany*
[†] *Electronic address: http://tesla.desy.de*

Abstract. Parallel to the enormous progress in optical and conventional X-ray lasers there have also been tremendous advances in the field of Free Electron Lasers (FELs) based on the principle of Self-Amplified Spontaneous Emission (SASE). At the TESLA Test Facility (TTF FEL) at DESY, a linac-driven SASE FEL has produced short pulses with GW peak power in the wavelengths range of 80-120 nm. The radiation pulse length has been adjusted between 30 fs and 200 fs. Currently an energy upgrade of the TTF linear accelerator to 1 GeV is being prepared which will make radiation wavelengths down to 6 nm available for users.

INTRODUCTION

Before the experimental verification of synchrotron radiation [1] in 1947, only very limited X-ray sources were available. Synchrotron radiation, emitted by relativistic electrons or positrons undergoing a centripetal acceleration, spans the spectral range from the visible to hard X-rays. The development of storage rings for dedicated synchrotron radiation research has led to 3rd generation facilities with special magnetic devices, so-called wigglers and undulators, and an increase by more than ten orders of magnitude in flux over the last forty years. The radiation emitted by these devices is based on spontaneous radiation of many electrons uncorrelated in space and time. As a consequence, the radiation power scales linearly with the number of electrons, and the radiation exhibits only limited coherence in space and time. The idea of a SASE FEL is to improve the electron beam quality such that a large number of electrons in the bunch radiates in phase. In this way coherent radiation is generated which can be, in principle, proportional to the number of electrons squared.

The SASE process starts from shot noise in the first part of the undulator. An electron bunch with a high peak current, a low emittance, and a small energy spread ($< 0.1\%$) emits spontaneous radiation thus building up a radiation field. The radiation wavelength λ_n of the fundamental ($n = 1$) and its harmonics ($n \geq 2$) is related to the period length λ_u of a planar undulator by

$$\lambda_n = \frac{\lambda_u}{2n\gamma^2} \left(1 + \frac{K^2}{2}\right), \tag{1}$$

where $\gamma = E_e/mc^2$ is the relativistic factor of the electrons and $K = eB_u\lambda_u/2\pi m_e c$ the undulator parameter with peak magnetic field B_u. If the phase space of the electron bunch

CP641, *X-Ray Lasers 2002: 8th International Conference on X-Ray Lasers,* edited by J. J. Rocca et al.
© 2002 American Institute of Physics 0-7354-0096-2/02/$19.00

is well matched to that of the radiation field, the interaction of the radiation field with the electrons lead to a steadily growing density modulation of the bunch (micro-bunching) which enhances the power and the coherence of radiation. Starting from the shot noise of the electron beam, the radiation power $P(z)$ grows exponentially with the distance z along the undulator

$$P(z) = P_0 \exp(z/L_g) \qquad (2)$$

where L_g is the power gain length and P_0 the effective input power. At the end of this process, when the electron bunch is completely modulated, laser saturation sets in and no further gain is possible.

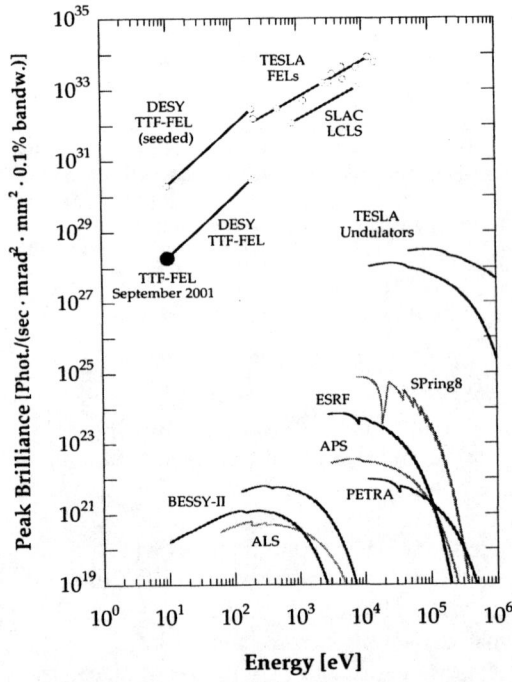

FIGURE 1. Expected spectral peak brilliance of proposed SASE FELs as well as the spontaneous emission from these devices in comparison with state-of-the-art synchrotron radiation sources. The large dot marks the experimental result from the TTF FEL.

Compared with state-of-the-art synchrotron radiation sources, one expects full transverse coherence, larger average brilliance and, in particular, up to eight orders of magnitude larger peak brilliance (see Fig.1) at pulse lengths of about 100 fs FWHM.

The present paper reports on the first demonstration of such a spectacular performance of a SASE FEL for wavelengths around 100 nm, i.e. in a wavelength range not directly accessible by conventional lasers. In the beginning, the tunability of the SASE FEL radiation has been demonstrated in proof-of-principle experiments between 80 and 180 nm [2]. Maximum output power in the gigawatt range has been obtained in September 2001 around 100 nm [3] and this saturation regime has been extended to a wavelength range of 80-120 nm in the following months [4]. Along with a detailed characterisation of the

FEL beam parameters, first exploratory experiments with the high power FEL photon beam have been performed on clusters [5] and solid state materials [6].

EXPERIMENTAL SETUP

The experimental results presented in this paper have been achieved at the TESLA Test Facility (TTF) Free-Electron Laser [7] at the Deutsches Elektronen-Synchrotron DESY. The TESLA (TeV-Energy Superconducting Linear Accelerator) collaboration consists of 46 institutes from 11 countries and aims at the construction of a 500 GeV (center-of-mass) e^+/e^- linear collider with an integrated X-ray laser facility [8]. Major hardware contributions to TTF have come from Germany, France, Italy, and the USA.

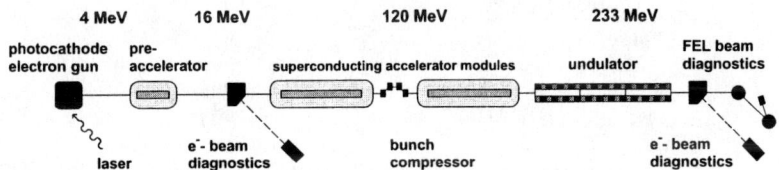

FIGURE 2. Schematic layout of Phase 1 of the SASE FEL at the TESLA Test Facility at DESY, Hamburg. The linac contains two 12.2 m long cryogenic modules each equipped with eight 9-cell supercinducting (SC) accelerating cavities. The total length is 100 m.

The layout of the TTF FEL in Phase 1 is sketched in Fig.2. A laser-driven radio-frequency (rf) photocathode electron gun provides electron bunches with a few nC of charge synchronized with the accelerator rf. After acceleration to energies around 240 MeV and a bunch compression to obtain the required high peak currents, the electrons pass a very precise undulator, where the FEL radiation is generated. The main parameters for FEL operation in Phase 1 are given in Table1.

EXPERIMENTAL RESULTS

Since the first observation of SASE in the VUV region around 109 nm [9] the performance of the FEL has been gradually improved. By varying the energy of the linear accelerator, full wavelength tunability in a wide range from 80 to 180 nm has been demonstrated. Recently, saturation has been achieved in the wavelength range from 80 to 120 nm. Furthermore, first exciting results on the interaction of the high power FEL photon beam with solids [6] and clusters [5] have been obtained.

Photon Beam Characterization

For the characterisation of the FEL radiation at saturation, measurements of the radiation energy, spectral characteristics, angular divergence, and statistical properties have

TABLE 1. Main parameters of the FEL operation in Phase 1 of the TESLA Test Facility (TTF 1)

Electrons:	
beam energy	220–270 MeV
bunch charge	2.7–3.3 nC
charge in lasing part of the bunch	0.1–0.3 nC
peak current	(1.3 ± 0.3) kA
rms energy spread	(150 ± 50) keV
rms normalized emittance	$(6 \pm 2)\,\pi$ mm·mrad
bunch spacing	0.44 / 1 μs
number of bunches in a train	up to 70
rf pulse repetition rate	1 Hz
Undulator:	
undulator period λ_u	27.3 mm
peak magnetic field	0.46 T
total magnetic length	13.5 m
Photons:	
wavelength range (with saturation power)	80–120 nm
pulse energy	30–100 μJ
pulse duration (FWHM)	50–200 fs
peak power level	1 GW
average power	up to 5 mW
spectral width (FWHM)	1%
spot size at undulator exit (FWHM)	250 μm
angular divergence (FWHM)	260 μrad

been performed. The photon beam diagnostic setup provides the necessary instrumentation such as a narrow-band monochromator, thermopile detector, and a microchannel-plate (MCP)-based detector. For a more detailed analysis of the TTF FEL operation mode, numerical simulations with the three dimensional, time-dependent FEL code FAST [10] have been performed.

Figure 3 depicts the measured average energy in the radiation pulse along the TTF FEL undulator. The interaction length of the electron beam and the electromagnetic radiation has been increased gradually using electromagnetic correctors installed inside the undulator. The measurement of the radiation pulse energy E has been performed by means of a MCP-based detector installed 12 m behind the undulator. By suppressing the FEL interaction along the whole undulator the detector shows the expected level of spontaneous emission of about 2.5 *nJ* collected from the full undulator length. With increasing interaction length the radiation energy grows exponentially and reaches a saturation value between 30 and 100 μJ, depending on the accelerator tuning. It should be noted that the gain curve is superimposed on that of the ever present spontaneous emission collected from the full undulator length. The analysis of the gain curve shown in Figure 3 yields a power gain length of $L_g = (67 \pm 5)$ cm. Some examples of the spectral distribution of the FEL radiation are presented in Figure 4. The single-shot spectra were recorded with a monochromator of 0.2 nm resolution equipped with an intensified CCD camera that images a fluorescent screen in the focal plane. The curve in the upper right corner represents the spectrum averaged over 100 consecutive pulses.

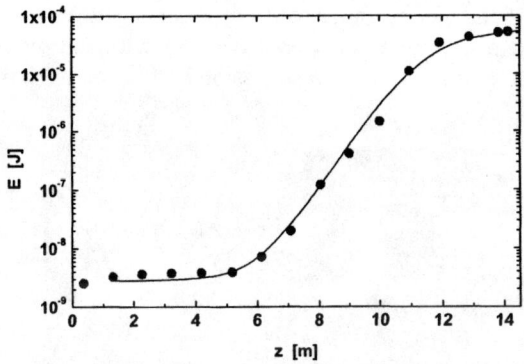

FIGURE 3. Average radiation pulse energy as it develops along the undulator. Circles: experimental results. Solid curve: numerical simulations with code FAST [10]

Each single-shot spectrum shows an ensemble of peaks, which reflects the number of longitudinal modes in the radiation pulse [11], as it is expected for SASE FEL starting from shot noise. The FWHM spectral width $\Delta\omega$ of each peak in the single shot spectra is related to the radiation pulse length by $\tau_{rad} \cong 2\pi/\Delta\omega$. For the case under study τ_{rad} is

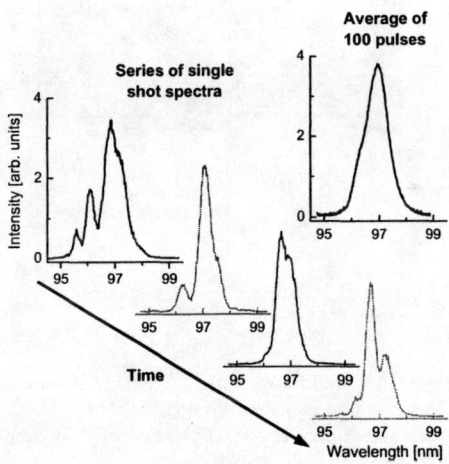

FIGURE 4. Measured spectral distribution of the FEL radiation. Each single shot spectrum exhibits an ensemble of a few peaks corresponding to the fluctuation of the number of longitudinal modes of FEL radiation. Since the center frequencies of these spikes are random (within the bandwidth of the FEL), they are smeared out in the averaged spectrum shown in the upper right corner.

about 50 fs. By varying the bunch compression, it was possible to tune the pulse length from 50 -250 fs. Figure 5 shows an example of two characteristic FEL spectra taken for different settings of the electron bunch compression and hence reflecting different FEL pulse lengths.

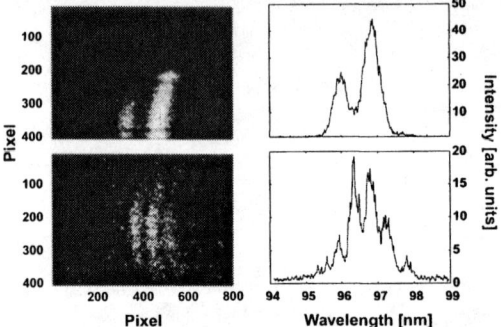

FIGURE 5. Spectra from short (top) and long (bottom) FEL pulses. On the left hand side, the CCD image of the dispersed FEL radiation in the exit plane of the monochromator is shown in a false color code. The dispersive direction is in the horizontal. On the right hand side, the spectra are evaluated quantitatively along the horizontal center line of the CCD image. It is seen that the number of optical modes are different: for the short pulse setting (top) there are, in average 2.6 modes, in the long pulse setting (bottom) there are 6 modes in average.

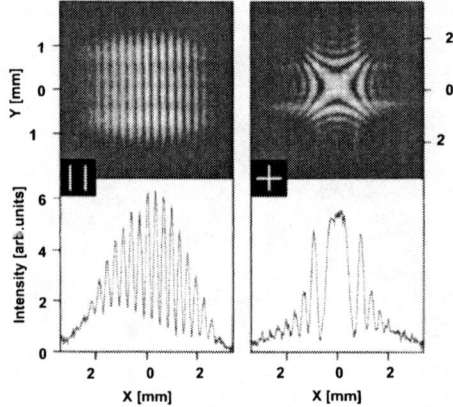

FIGURE 6. Diffraction patterns of two different slit arrangements illustrating the transverse coherence of the FEL radiation. **Left**: double slit, each slit 2 mm (vert) × 200μm (horiz.), horizontal slit separation 1 mm. **Right**: crossed slits, each slit 4 mm× 100μm. **Bottom:** horizontal cuts through the center of the diffraction patterns above.

The transverse coherence of the FEL radiation is reflected in Figure 6 which depicts the diffraction patterns originating from two different slit arrangements, a double slit and two crossed slits which were located 12 m behind the undulator. The images have been recorded with a CCD camera viewing a Ce:YAG fluorescen screen in about 3 m distance behind the slits. The high fringe visibility in the left figure is a clear indication of the almost full transverse coherence of the FEL radiation.

First applications

Intense radiation from femtosecond optical lasers has opened several new branches in physics and stimulated many fields of research e.g. ultra fast dynamics, non-linear optics, and plasma physics. On the contrary only little is known about the interaction of intense, ultra-short VUV pulses with matter.

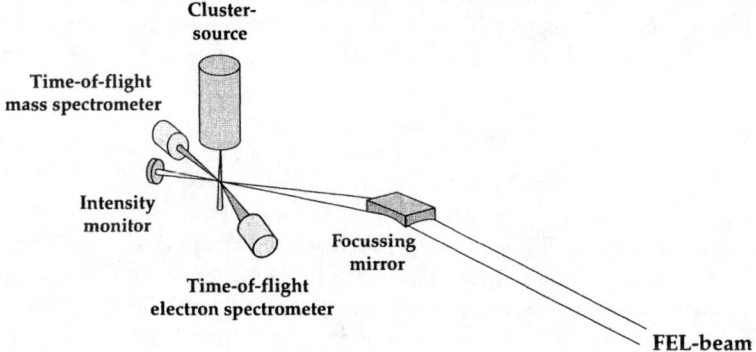

FIGURE 7. Schematic illustration of the cluster experiment.

Figure 7 illustrates the experimental set-up of the cluster experiment at the TTF FEL [5]. A pulsed beam of atoms, molecules or clusters was prepared by expanding gas at high pressure through a small nozzle. The size of the clusters could be controlled by variation of the pressure. The FEL radiation was focused with an elliptical mirror onto the cluster beam. The diameter of the focal spot was approximately 20 μm, resulting in a power density of up to $1 \cdot 10^{14}$ W/cm^2. Ions and electrons produced in the interaction zone were detected with time-of-flight (TOF) methods.

Figure 8 shows an averaged TOF-spectrum resulting from the ionisation of nitrogen

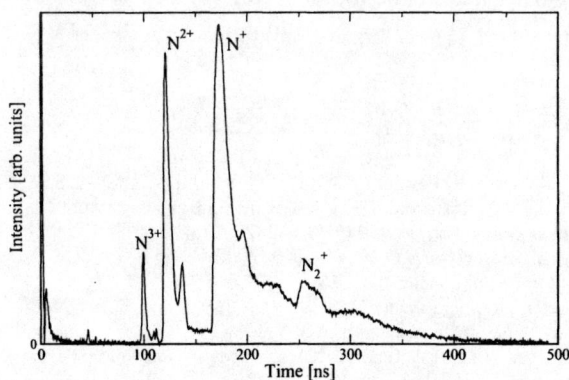

FIGURE 8. Averaged time-of-flight mass spectrum of ionisation products of nitrogen clusters [5]. The spectrum has been recorded after ionisation with FEL radiation at 98 nm wavelength with a power density up to $1 \cdot 10^{14}$ W/cm^2.

clusters comprising 2500 atoms. The energy of a single FEL photon at 98 nm wavelength is well below the first dissociation energy of 24.3 eV. However, at a power density of 10^{14} W/cm^2 the cluster completly disintegrates into atomic and molecular ions. For the atomic ions charge states up to N^{3+} are visible. The broad N^+ peak as well as the double N^{2+} peak are clear indications for high kinetic energies of the fragments giving evidence for a Coulomb explosion. For more details see Ref. [5].

In addition to the gas phase experiments the interaction of the focused VUV radiation with solids has been investigated [6]. The ablation of different bulk samples and thin films have been analyzed by a TOF spectrometer as a function of the radiation power density.

SUMMARY AND OUTLOOK

It has been shown that the FEL at the TESLA Test Facility at DESY is an outstanding, presently unrivaled radiation source. The FEL experiments in Phase 1 including the detailed characterisation of the VUV radiation ended in March 2002. In TTF Phase 2 the maximum energy of the TTF linac will be increased to at least 1.0 GeV by adding additional accelerator modules. This facility will then serve a dual purpose: Besides offering a final test of the TESLA components it is also the basis for a world wide unique user facility providing FEL radiation from VUV up to the soft X-ray region. The minimum wavelength will be about 6 nm in the fundamental and reach the water window in the higher harmonics. The user facility will be available for scientific users in 2004.

ACKNOWLEDGEMENTS

The authors gratefully acknowledge that all work presented from the TESLA Test Facility at DESY is the result of a joint effort of all members of the TESLA collaboration. A list of participating institutes can be found in Ref. [9, 12]. The author is indebted to the group of Th. Möller (HASYLAB) for providing figures 7 and 8 prior to publication.

REFERENCES

1.   F. Elder, A. Gurewitsch, R. Langmuir, and H. Pollock, Phys. Rev. **71**, 839 (1947)
2.   J. Rossbach et al., Nucl. Instr. and Meth. A **475**, 13 (2001)
3.   V. Ayvazyan et al., Phys. Rev. Lett. **88**, 104802 (2002)
4.   V. Ayvazyan et al., Eur. Phys. J. D **20**, 149-156 (2002)
5.   T. Möller et al., to be published
6.   J. Krzywinski et al., to be published
7.   W. Brefeld et al., Nucl. Instr. and Meth. A **393**, 119 (1997)
8.   F. Richard, et al. (eds.),TESLA Technical Design Report, DESY2001-011
9.   J. Andruszkow et al., Phys. Rev. Lett. **85**, 3825 (2000)
10.   E. L. Saldin, E. A. Schneidmiller, and M. V. Yurkov, Nucl. Instr. and Meth. A **429**, 233 (1999)
11.   E. L. Saldin, E. A. Schneidmiller, and M. V. Yurkov, Opt. Commun. **148**, 383 (1998)
12.   http://tesla.desy.de

HIGH-ORDER HARMONICS AND OTHER

HIGH-BRIGHTNESS X-RAY SOURCES

Subfemtosecond Pulse Generation and Measurement

Michael Hentschel*, Reinhard Kienberger*, Markus Drescher*†, Georg Reider*, Christian Spielmann* and Ferenc Krausz*

*Institut f. Photonik, Technische Universitaet Wien, Gusshausstr. 27/387, A-1040 Wien, Austria
†Fakultaet f. Physik, Universitaet Bielefeld, D-33615 Bielefeld, Germany

Abstract. We report the generation and measurement of isolated soft-x-ray pulses ($\lambda_x = 14$ nm) with a duration of $\tau_x = 650 \pm 150$ attoseconds by using few-cycle intense visible/near-infrared ($\lambda_0 = 750$ nm) laser pulses. For the temporal characterization of the x-ray pulses, a cross-correlation technique relying on laser field assisted x-ray photoemission from Krypton atoms was employed. The experimental results bear direct evidence of the x-ray pulse being synchronized to the field oscillations of the visible light pulse with attosecond precision and of bound-free electronic transitions from the 4p state of krypton responding to 90-eV excitation on an attosecond time scale. As a first demonstration of attosecond metrology, the synchronized single sub-fs x-ray pulses were used for tracing the electric field oscillations in a visible light wave with a resolution of better than 150 as.

INTRODUCTION

The interest in ultrafast microscopic processes is the driving force behind the development of sources and measurement techniques that allow time-resolved studies at ever shorter timescales. In the classical scheme of time resolved experiments, a short excitation pulse setting the process going and a short probe pulse for taking snapshots of subsequent stages of its evolution constitute the basis for gaining experimental access to fast-evolving microscopic processes. Recent advances in ultrafast optics made laser pulses as short as a few femtoseconds [1]-[6] available for these pump-probe measurement techniques.

Currently available laser pulse durations are practically at the limit set by the laser field oscillation cycle, which lasts somewhat longer than one femtosecond (1 fs = 10^{-15} s) in the visible spectral range. Coherent extreme ultraviolet (XUV) and x-ray sources [7]-[10] offer, owing to their short wave cycle, the potential for producing electromagnetic radiation of attosecond duration [11]-[14]. Using few-cycle visible laser pulses [15] for both, the generation of isolated soft-x-ray harmonic pulses and their temporal characterization in a novel cross-correlation technique [16], we have demonstrated the existence of sub-femtosecond electromagnetic pulses for the first time [17]. In this paper we review these recent developments, concentrating on the measurement of ultrashort XUV/x-ray pulses. For an extended review on the generation of such pulses see e.g. Ref.[18].

CP641, *X-Ray Lasers 2002: 8th International Conference on X-Ray Lasers,* edited by J. J. Rocca et al.
© 2002 American Institute of Physics 0-7354-0096-2/02/$19.00

XUV/X-RAY PULSES BY HIGH-ORDER HARMONIC GENERATION

In our experiments coherent high harmonic generation (HHG) is accomplished by focusing 0.5-mJ linearly polarized $\approx$ 750-nm laser pulses of 7 fs duration at a 1 kHz repetition rate into a volume of Ne gas (see photograph in Fig. 2) [19]. The pulses originate from a Ti:sapphire CPA system combined with a hollow fiber chirped mirror compressor [20, 21]. The neon atoms are exposed to peak electric fields of $\approx 8.8 \times 10^{14}$ W/cm^2 over an effective interaction length of $\approx$ 3 mm. They emit coherent soft-x-ray harmonic radiation up to photon energies exceeding 100 eV [19].

CROSS CORRELATION OF X-RAY PULSES WITH A FEW-CYCLE LASER FIELD

The temporal characterization of broadband (multi-eV-bandwidth) XUV/soft-x-ray pulses produced by HHG-sources turns out to be a challenging task, since conventional auto-correlation schemes based on second order nonlinearities require photon fluxes orders of magnitude higher than those available from existing sources. We have recently proposed and demonstrated a novel cross-correlation technique based on x-ray photoionization in the presence of an intense few-cycle laser field [16, 17]. It is based on controlling the final kinetic energy of x-ray-generated photoelectrons by the light field. The features (width and center of gravity) of the resulting photoelectron spectra as a function of the relative delay between x-ray and laser pulse are determined by a convolution of the laser light field and the x-ray pulse envelope. Careful deconvolution allows to determine the duration of the x-ray pulse and its timing jitter with respect to the light field on an attosecond time scale.

In our experiment, photoelectrons from atoms exposed to a strong light field and an x-ray pulse simultaneously are detected with a time-of-flight (TOF) spectrometer. The detection cone is aligned orthogonally to the electric field vector of the linearly polarized laser field. In the strong field limit, a quasi-classical model summarized in the following section gives proper account for the modifications of the photoelectron spectrum by the laser field.

In the framework of this model, the photoelectron is first ejected by a short x-ray pulse with a distribution of initial momenta known from conventional photoionization studies [22]. Subsequently, its kinetic energy is influenced by the light field. For x-ray pulse durations τ_x very short as compared to $T_0/2$, the model predicts that depending on the phase of the light field at the instant of 'birth' of the electron, a momentum component along the electric field vector is added to the initial momentum of the electron, resulting in a shift of the photoelectron angular distribution in momentum space (Fig. 1). The width of the x-ray photoelectron energy spectrum ΔW, which is equal to the bandwidth of the x-ray pulse spectrum in the absence of the light field, increases with increasing momentum shift.

As the analysis in the next section shows, scanning the instant of birth of the photoelectron through the light field oscillations by changing the relative delay t_d between the

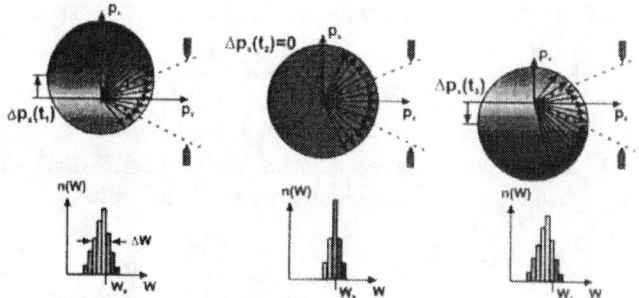

FIGURE 1. Principle of measuring light-field/x-ray-intensity cross correlation with attosecond resolution. The photoelectrons created initially with isotropic momentum distribution by the absorption of an x-ray photon pick up a momentum from the strong laser light field. Photoelectrons detected within a cone aligned orthogonally to the direction of the light field vector (x direction) display a kinetic energy spread at t_1 and t_3 while remaining unaffected by the light field at t_2.

light pulse and x-ray pulse results in a modulation of the center of gravity as well as the width $\Delta W(t_d)$ of the x-ray photoelectron spectrum with a period equal to half the light oscillation period T_0. For sub-femtosecond x-ray pulse measurement we shall exploit the field-induced spectral broadening [17] because this effect is enhanced by increasing the detection angle as revealed by Fig. 1, resulting in a significantly enhanced signal yield and modulation amplitude of $\Delta W(t_d)$. By contrast, a modulation in the kinetic energy shift gets increasingly washed out with increasing detection angle, deteriorating the temporal resolution.

Light-field-controlled x-ray photoemission

Our method of measuring a cross-correlation between an x-ray pulse and a visible laser pulse can be summarized as laser-light-assisted x-ray photoionization. The basic idea is to overlap the two pulses with variable delay with respect to each other in a gas medium and measuring the kinetic energy of the photoelectrons stripped off the atoms by the x-ray pulse. Taking advantage of the semi-classical approach outlined above, the motion of an x-ray-induced photoelectron 'born' at a delay time t_d with respect to the peak of the laser pulse is governed by the classical equation of motion. Along the direction of laser polarization (x) this equation reads as

$$m v_\parallel = qE = -eE_L(t)\cos\omega_L t. \tag{1}$$

In the adiabatic limit ($dE_L/dt \ll E_L\omega_L$) the femtosecond laser pulse changes the velocity of the electron by

$$\Delta v_\parallel = \frac{e}{m}\frac{E_L(t_d)\sin\omega_L t_d}{\omega_L} = \sqrt{\frac{4U_p(t_d)}{m}}\sin\omega_L t_d, \tag{2}$$

$$\Delta v_\perp = 0 \tag{3}$$

in the directions parallel and perpendicular to the laser polariaztion, respectively. Here

$$U_p = \frac{e^2 E^2}{4m\omega_L^2} \qquad (4)$$

is the time averaged electron wiggling energy. This change of the initial velocity of the electron results in a modification of the final kinetic energy W_f as given by

$$W_f \approx W_0 + 2U_p(t_d) \sin^2 \omega_L t_d \cos 2\theta$$
$$+ \sqrt{8 W_0 U_p(t_d)} \sin \omega_L t_d \cos \theta, \qquad (5)$$

where $W_0 = \hbar\omega_x - W_b$ is the initial kinetic energy of the photoelectron, ω_x denotes the x-ray photon energy and W_b stands for the atomic binding energy of the electron liberated ($\hbar\omega_x \gg W_b$) and θ is the angle between the final momentum of the electron and the laser electric field vector.

In order to comply with measurements, Eq. 5 needs to be generalized for a finite emission time of the photoelectrons (corresponding to the finite x-ray pulse duration) and for a distribution of initial kinetic energies over an energy range (reflecting the bandwidth of the x-ray pulse). In addition, the distribution of final kinetic energies as a function of θ has to be integrated over the detection cone to obtain a realistic prediction for the measured x-ray photoelectron spectrum modified by the laser field.

For an x-ray pulse short compared to half the laser period, i.e. $\tau_x \ll T_0/2$, and a fast (attosecond) response of the electronic transitions involved in the process, the predicted modulations in the final kinetic energy distribution versus t_d survive these integrations and provide a sub-femtosecond probe for the measurement of the x-ray pulse duration. A finite x-ray pulse duration or a timing jitter of the x-ray pulse (relative to the phase of the light field) of any origin results in a reduced depth of the resultant modulation of the width and center of gravity of the photoelectron energy spectrum. In fact, an x-ray pulse duration or a timing jitter exceeding $T_0/2$ (~ 1.25 fs in our case) smears out the modulation completely. The claim of a sub-femtosecond pulse duration relies on this line of argumentation: the modulation depth of $\Delta W(t_d)$ provides a reliable upper limit, on an attosecond time scale, for x-ray pulse duration and timing jitter.

Setup

Figure 2 displays the setup of our experiment. The x-ray pulses exiting the neon harmonic source co-propagate with the laser pulses down the beam delivery tube. After 150 cm they hit a 150 nm thick zirconium foil with an aperture of 2 mm mounted on a nitrocellulose membrane of 5 μm thickness. This filter, virtually dispersion-free, produces an annular laser beam with the x-ray beam in the center. The energy in the laser beam is adjusted by an iris and measured with a photodiode. Both the laser and the x-ray pulses impinge on a curved mirror with a focal length of 35 mm. It is coated with a Mo/Si multilayer stack designed to reflect photons around 90 eV energy within a 5 eV band. This bandwidth is broad enough to support x-ray pulses as short as 0.4

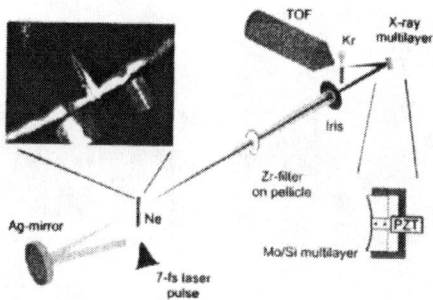

FIGURE 2. Schematic of the experiment. The focused 7-fs laser beam interacts with neon atoms to produce high-harmonic radiation. The laser and the highly collimated x-ray beam co-propagate through a 2-m beamline towards the measurement (see text).

fs. The mirror substrate consists of two concentric parts, matching the impinging laser and x-ray beam. The central piece is sitting on a piezo stage adjustable in transverse and longitudinal direction ensuring spatial and temporal overlap of the pulses. In the focal plane a nozzle is situated supplying the target atoms (krypton) which are photoionized by the x-ray pulses in presence of the laser field. All parts in the vicinity of the spectrometer are conductively coated and properly grounded to avoid electrostatic charging.

The entrance aperture of the time-of-flight photoelectron spectrometer with a diameter of 5 mm is placed in a distance of approximately 5 mm from the interaction region. The relatively large numerical aperture (NA ~ 1) of our TOF spectrometer is due to a carefully designed mesh-free collecting electrostatic lens system optimized for the energy region of 50-100 eV, which has resulted in an order-of-magnitude improvement of detection efficiency as compared to conventional TOF tubes.

Measurement

Figure 3 depicts a series of Kr-4p photoelectron spectra as a function of delay t_d. The plot comprises some 120 spectra of the 4p feature, each normalized to the same number of counts, resulting in a constant area under the spectral distribution functions. Owing to the relatively large solid angle of detection the overall measurement time could be limited to ~ 5 hours. The data clearly bring to light a quasi-periodic evolution of the photoelectron energy spectrum at a period of $\sim T_0/2$. The modulation period of the data is close to 1.25 fs in the wings of the correlation function $\Delta W(t_d)$, in agreement with a carrier wavelength of ~ 750 nm of the light pulse. However, it decreases to ~ 1 fs near zero delay, indicative of a pronounced blue shift of the light pulse at its peak, the origin of which we attribute to self phase modulation in the HHG target.

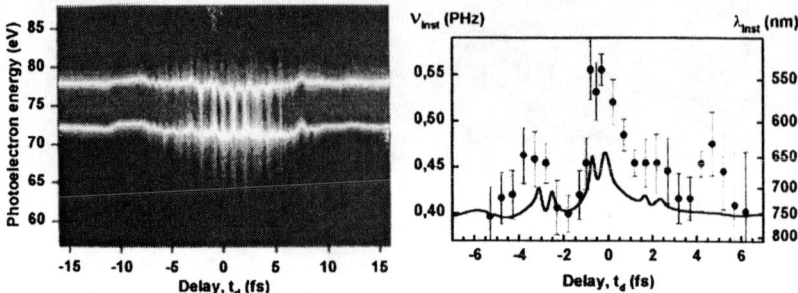

FIGURE 3. Left: Photoelectron spectra of Kr-4p electrons as a function of the delay t_d between the light wave and the x-ray pulse. The width of the spectral feature oscillates at twice the laser frequency. From the modulation depth one can set an upper limit to the x-ray pulse duration of 650 as. Right: Instantaneous frequency of the laser field derived from the oscillating spectral electron feature (dots) and calculated with a numerical code(line).

ISOLATED SUB-FEMTOSECOND X-RAY PULSES

For evaluating the x-ray pulse duration from our light-field/x-ray-intensity cross-correlation data we introduce a measure for the depth of modulation of $\Delta W(t_d)$, the fringe visibility

$$FV = (\Delta W_{max} - \Delta W_{min})/(\Delta W_{max} + \Delta W_{min} - 2\Delta W_\infty), \qquad (6)$$

where ΔW_{max} and ΔW_{max} stand for adjacent local extremal values of $\Delta W(t_d)$ as a function of the delay time and $\Delta W_\infty = \Delta W(t_d \to \infty)$, and the local fringe spacing FS. The central portion of the correlation function $\Delta W(t_d)$, with maximum FV and minimum FS, provides optimum resolution for determining the x-ray pulse duration. A sinusoidal fit to the two central modulation periods yields FS = 960 as ± 30 as and FV = 0.30 ± 0.03.

For comparison with the measured data (Fig. 3), we simulated the light-field-induced variation of the Kr-4p photoelectron spectrum versus t_d based on the quasi-classical model described above. The duration τ_x of our Gaussian model x-ray pulse was used as the only fit parameter. The visible light pulse was also modelled with a Gaussian intensity envelope peaking at 10^{14} W/cm^2, carrying a quadratic frequency sweep to account for the frequency upshift present at the center of the measured correlation function $\Delta W(t_d)$. Best agreement with the data was achieved for x-ray pulse durations in the range of 600 as $\leq \tau_x \leq$ 700 as, yielding a fringe visibility of FV = 0.30 ± 0.05 in the correlation function near $t_d = 0$ as. Assuming pulse durations of τ_x = 500 as and τ_x = 800 as, respectively, the FV is enhanced and suppressed with respect to the measured value well beyond the experimental error of 10%. Therefore, τ_x = 650 as ± 150 as represents a safe estimate for the duration of our 90-eV x-ray pulse. The sensitivity of the method is underlined by the computed FV decreasing by a factor as large as three (from 0.30 to 0.10) as the x-ray pulse duration is increased from 650 as to 900 as, with the modulation in $\Delta W(t_d)$ disappearing for $\tau_x \geq$ 1 fs.

An essential question is whether this sub-femtosecond pulse is accompanied by one or more satellite pulses spaced by $\sim T_0/2$ as observed in a recent experiment [12]. A concomitant of the emergence of equidistant satellite pulses is the appearance of spectral sidebands (at twice the photon energy of the driver laser) of rapidly increasing magnitude with increasing weight of the satellite pulse(s). The absence of any significant modulation in our x-ray harmonic spectrum reflected by the Mo/Si multilayer limits the possible satellite fluence to less than 10% of the total x-ray fluence. Further confirmation of the low weight of x-ray satellite pulses is supplied by results summarized in the following section.

PROBING ATOMIC RESPONSE AND LIGHT FIELD OSCILLATIONS

The agreement of the measured x-ray pulse duration and that obtained from our numerical calculations [23] within the experimental error suggests that the timing jitter of the x-ray pulse with respect to the phase of the visible light wave must be very small ($\ll 1$ fs) even on a sub-femtosecond time scale. As a consequence, our experiment bears evidence of the sub-femtosecond x-ray pulse being locked to the carrier wave of its generating few-cycle light pulse with attosecond precision, indicating a surprising robustness of the HHG process against random shifts of the absolute phase [18].

With its duration extracted from the central part of $\Delta W(t_d)$ in Fig. 3, the x-ray pulse may now also be used to probe the evolution of the electric field in the few-cycle light pulse. The sweep of instantaneous frequency ν_{inst} (or wavelength λ_{inst}) in the visible light pulse can be evaluated from the modulation in $\Delta W(t_d)$ by fitting a sinusoidal oscillation of adjustable period to our data. The dots in Fig. 3 show the carrier frequency sweep evaluated in this manner, revealing a dynamic blue shift from a carrier wavelength of ~ 750 nm to ~ 550 nm. To understand this finding we have to remember that the light pulse used here is derived from the one generating the high harmonics.

The line in Fig. 3 is obtained from propagating a bandwidth-limited 7-fs laser pulse through the volume of neon gas emitting the high harmonics in a numerical calculation [24] and exhibits a dynamic frequency shift at the pulse center originating from self-phase modulation due to ionization of the neon gas in the harmonic source. The measured $\sim 30\%$ dynamic frequency blue shift at the center of the pulse is larger than predicted by our numerical study but reflects qualitatively the predicted behaviour. The presence of this substantial frequency sweep in the light pulse employed for controlling x-ray photoemission provides a further strong argument for the isolated nature of our x-ray pulse: any satellite pulse of notable energy would significantly suppress fringe visibility near $t_d = 0$, where the modulation period of the cross-correlation trace $\Delta W(t_d)$ is substantially shorter than the predicted temporal spacing of possible satellites. The achieved direct probing of the field oscillations in a light wave as implemented here is a demonstration of attosecond metrology and may permit complete measurement of the electric field of absolute-phase-stabilized few-cycle light [25, 26].

ACKNOWLEDGMENTS

We thank M. Uiberacker (TUW) for assistance in the preparation of measurements and Y. Lim and U. Kleineberg for manufacturing the x-ray multilayer mirror as well as U. Heinzmann (UB) for the collaboration. Invaluable theoretical support of T. Brabec, G. Tempea (TUW) and P. Corkum (NRC Canada) is gratefully acknowledged. This work has been sponsored by the Austrian Science Fund (grants Y44-PHY and F016) and by the European ATTO network.

REFERENCES

1. G. Steinmeyer, D. H. Sutter, L. Gallmann, N. Matuschek & U. Keller, *Science* **286**, 1507-1512 (1999).
2. A. Baltuška *et al.*, *Appl. Phys. B* **65**, 175 (1997).
3. M. Nisoli *et al.*, *Appl. Phys. B* **65**, 189 (1997).
4. U. Morgner *et al.*, *Optics Letters* **24**, 411 (1999).
5. D. H. Sutter *et al.*, *Optics Letters* **24**, 631 (1999).
6. A. Shirakawa *et al.*, *Appl. Phys. Lett.* **74**, 2268 (1999).
7. A. L'Huillier & P. Balcou, *Phys. Rev. Lett.* **70**, 774-777 (1993).
8. C. J. Joachain, M. Dörr & N. J. Kylstra, *Adv. At., Mol., Opt. Phys* **42**, 225 (2000).
9. P. Salières, A. L'Huillier & M. Lewenstein, *Adv. At., Mol., Opt. Phys* **41**, 83 (1999).
10. J. J. Macklin, J. D. Kmetec & C. L. Gordon III, *Phys. Rev. Lett.* **70**, 766-769 (1993).
11. N. A. Papadogiannis et al., *Phys. Rev. Lett.* **83**, 4289 (1999).
12. P. M. Paul, E. S. Toma, P. Breger, G. Mullot, F. Augé, Ph. Balcou,H. G. Muller & P. Agostini, *Science* **292**, 1689-1692 (2001).
13. M. Ivanov,P. B. Corkum, T. Zuo & A. Bandrauk, *Phys. Rev. Lett.* **74**, 2933-2936 (1995).
14. I. P. Christov, M. M. Murnane & H. C. Kapteyn, *Phys. Rev. Lett.* **78**, 1251-1254 (1997).
15. Ch. Spielmann, N. H. Burnett, S. Sartania, R. Koppitsch, M. Schnürer, C. Kan, M. Lenzner, P. Wobrauschek & F. Krausz, *Science* **278**, 661-664 (1997).
16. M. Drescher, M. Hentschel, R. Kienberger, G. Tempea, Ch. Spielmann, G. A. Reider, P. B. Corkum & F.Krausz, *Science* **291**, 1923-1927 (2001). Published online February 15, 2001; 10.1126/science.1058561.
17. M. Hentschel, R. Kienberger, Ch. Spielmann, G. A. Reider, N. Milosevic, T. Brabec, P. B. Corkum, U. Heinzmann, M Drescher & F. Krausz, *Nature* **414**, 509-513 (2001).
18. T. Brabec, & F. Krausz, *Rev. Mod. Phys.* **72**, 545-591 (2000).
19. M. Schnürer, Z. Cheng, M. Hentschel, G. Tempea, P. Kálmán, T. Brabec, and F. Krausz, *Phys. Rev. Lett.* **83**, 722 (1999).
20. S. Sartania, Z. Cheng, M. Lenzner, G. Tempea, Ch. Spielmann, F. Krausz & K. Ferenc, *Opt. Lett.* **22**, 1562 (1997).
21. M. Hentschel, Z. Cheng, F. Krausz & Ch. Spielmann, *Appl. Phys. B* **70**, 161-164 (2000).
22. U. Becker & D. A. Shirley, *VUV and Soft X-Ray Photoionization* 152 (Plenum, New York, 1996).
23. J. Itatani, F. Quéré, G. L. Yudin, M. Yu. Ivanov, F. Krausz. & P. B. Corkum, Submitted for publication.
24. N. Milosevic, A. Scrinzi, & T. Brabec, *Phys. Rev. Lett* (submitted for publication).
25. M. Lewenstein, Ph. Balcou, M. Yu. Ivanov, A. L'Huillier, & P. B. Corkum, *Phys. Rev. A* **49**, 2117-2132 (1994).
26. A. Poppe, R. Holzwarth, A. Apolonski, G. Tempea, Ch. Spielmann, T. W. Hänsch & F. Krausz, *Appl. Phys. B* **72**, 373-376 (2001). Published online December 13, 2000; 10.1007/s003400000526.

Observation of Blueshift on Harmonic Generation from Subpicosecond Laser Produced Solid Surface Plasmas

A. Ishizawa[*], M. Baba, M. Turu, S. Sakakibara, T. Kanai, and H. Kuroda

Institute for Solid State Physics, University of Tokyo, 5-1-5 Kashiwanoha, Kashiwa, Chiba 277-8581, JAPAN

[*] *ishizawa@issp.u-tokyo.ac.jp*

R. A. Ganeev

NPO Akadempribor, Academy of Sciences of Uzbekistan, Tashkent 700143, Uzbekistan

T. Ozaki

NTT Corporation, NTT Basic Research Laboratories, 3-1, Morinosato Wakamiya, Atsugi-shi, Kanagawa pref., 243-0198 Japan

Abstract. Harmonic generation from solid surface plasmas is studied using a 475 fs Nd: Glass laser system. We observe that for a 45 degrees angle of incidence, the second harmonic is blue shifted by 1.6 nm, and the fifth harmonic is blue shifted by 5.1 nm for p-polarization at the intensity of 1×10^{17} Wcm^{-2}. Furthermore, we find out that the blueshift distance is increased with orders of harmonic. The blueshift is interpreted as the collisionless absorption due to the anomalous skin effect. We also observe that the ratio between the harmonics intensities produced by p-polarized and s-polarized pump were 8 and 2 for the second and fifth harmonics, respectively at pump intensities I below 10^{17} W cm^{-2}. The same ratio decreased to 0.8 and 0.5 for $I > 2 \times 10^{17}$ W cm^{-2}.

The availability of compact, high-intensity subpicosecond lasers with chirped pulse amplification (CPA) has opened the new field of study of laser-matter interactions with solid targets. Thus, using compact high-intensity laser system, it is now possible to study laser-matter interaction for the fast ignition scheme for inertial confinement ignition, harmonic generation from solid surface, and ultrashort x-ray generation, etc. The studies of laser-plasma interactions have been performed for several decades. At the high laser intensities above 10^{16} Wcm^{-2}, the electron temperature is more than one keV and the main laser absorption is related to collisionless absorption. Collisionless absorption mechanisms include vacuum heating, J x B heating, sheath inverse Bremstrahlung, and the anomalous skin effect. Recently, Hansen et al. reported for the first time that the second harmonic is blueshifted when the laser intensity exceeds 5×10^{16} Wcm^{-2} and the blueshift is due to the absorption by the anomalous skin effect [1]. However, no detailed study of the blue shifts in the other harmonic orders, or the dependence on the laser intensity at the incidence angle of 45 degrees has yet been reported. It was commonly observed that efficiency of

CP641, *X-Ray Lasers 2002: 8th International Conference on X-Ray Lasers,* edited by J. J. Rocca et al.

harmonic generation for the p-polarized laser pump was much larger than that for the s-polarized laser beam. Only a few experimental studies have been reported on the effects of polarization on the harmonic radiation. However, the cause of the disagreement between some recent theoretical and experimental results is not yet clear.

In this paper, we present blue shifts of up to about 2 nm in the second harmonic and for the first time about 5 nm in the fifth. We carry out more detailed model-calculations of the electron temperature in our experimental condition using a 1D hydrodynamic code HYADES [2]. We find that the blueshift in our experiment can be well explained from the absorption mechanism due to the anomalous skin effect. We also show that the harmonic dependence on the pump laser polarization decreases when laser intensity exceeds 10^{17} W cm^{-2}.

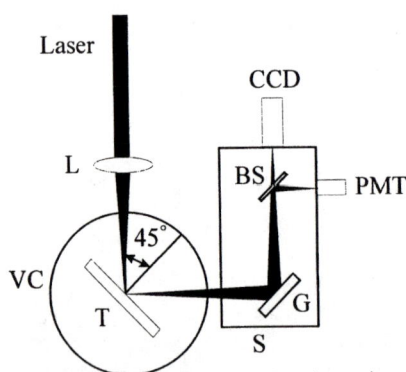

Fig.1 The experimental setup. L, achromatic lens; VC, vacuum chamber; T, aluminum deposited target; S, spectrograph; G, grating; BS, beam splitter (CaF$_2$)

Figure 1 shows the experimental setup. The experiments were carried out with a CPA Nd: Glass laser system operating at a wavelength around 1060 nm. The laser delivered 100 mJ energy in 475 fs pulse and produced a peak intensity on the target of 2 x 10^{17} Wcm^{-2} in the vacuum the surface of a solid target with a 10 cm focal length achromatic lens at an incidence angle of 45 degrees relative to target normal with either s- or p-polarization selected using a half wave plate. We can find the optimum position of the lens by measuring the second harmonic signal using the CCD camera.

Figure 2(a) shows that the second harmonic is blue shifted by 1.6 nm for p-polarization in our experiment. Figure 2(b) shows the dependence of the wavelength shift of the second harmonics on the incidence laser intensity in our experiment. The peak frequency of the second harmonic emission is blueshifted for >2 x 10^{16} Wcm^{-2}. It is found that the increase of this shift width with an increase in the laser intensity is small in comparison with the experimental results by Hansen et al. [1]. Unfortunately, we cannot detect the shift below 1 x 10^{16} that the fifth harmonic light is blueshifted by 5.1 nm for p-polarization at the intensity of 1 x 10^{17} Wcm^{-2}. This shift is much larger

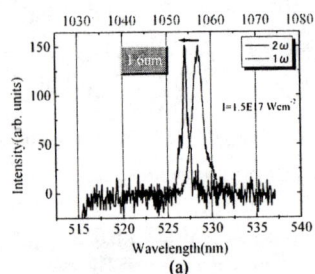 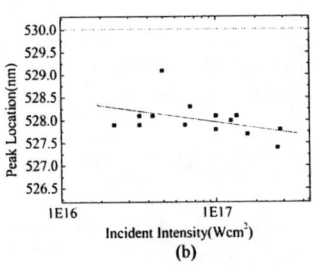

Fig.2 (a) The second harmonic spectrum (black line) and the incident laser spectrum (gray line) at the laser intensity of 1.5 x 10^{17} Wcm^{-2} in our experiment. (b)Location of the second harmonic peak as a function of incident intensity in our experiment.

in magnitude than the Doppler shift seen in the spectrum of the fundamental. Therefore, the hydrodynamic motion of plasma cannot explain these results. The calculation results by Rozmus et al. [3] show that at low temperature (Te<300 eV), the collision absorption mechanism is dominant, and at higher temperature (Te>800 eV), the collisionless absorption mechanism due to the anomalous skin effect becomes dominant.

The electronic temperature profile in our experiment condition was calculated using a 1D hydrodynamic code HYADES [2]. As these results, it is found that the electron temperature in the skin depth from the critical density surface is much higher than 800 eV. Therefore, there is possibility that from our experimental and calculation results the harmonic frequency shift is due to the collisionless mechanism. We assume the blue shift of harmonics as follows. We treat the skin depth in the collisionless absorption as the phase shifter. When the driving laser irradiates the solid target, the electron's phase becomes the same phase of the driving laser field. Each electron is reversed in the direction and returns to the plasma. It is considered that the electron pumps up to the upper level at virtual time in the multiphoton absorption process. Therefore, the electron's phase is accumulated and returns to the plasma and locked by this process. The phase locked electrons coupled in each resonance layer and then high order harmonics are generated. In an upcoming paper, we will report a more detailed analysis of the experimental data [4].

Figure 3 shows the second (a) and fifth (b) harmonic yield dependence on the pump laser polarization. We found that for both harmonics, the harmonic yield dependence on the pump laser polarization decreases when the laser intensity exceeds

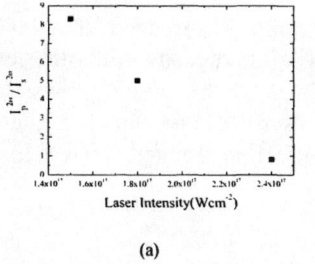
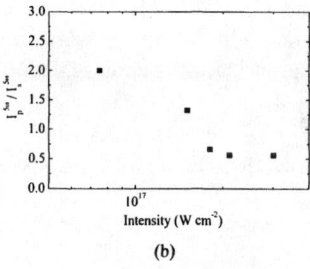

(a) (b)

Fig.3 The (a) second and (b) fifth harmonic intensity dependence on the pump laser polarization at the intensities of $\geq 1 \times 10^{17}$ W cm^{-2} in our experiments.

1×10^{17} W cm^{-2}. Here I_p/I_s is the ratio of p-polarization induced and s-polarization induced intensities of generated harmonics. Second and fifth harmonic radiation generated by s-polarized pump were 8 and 2 times smaller compared to that for the p-polarized laser radiation at the intensities below I = 10^{17} W cm^{-2}, while the same ratio decreases to 0.8 and 0.5 for I > 2×10^{17} W cm^{-2}. Some different processes are involved in decreasing of I_p/I_s ratio with the increase in laser intensity. One of them is the rippling caused by Rayleigh-Taylor-like instability at the critical density surface. In previous work, we performed investigation of the spatial distribution on harmonic generation from solid surface plasmas by using a 2.2 ps, 1054 nm laser pulse at the maximum intensities of 2×10^{17} W cm^{-2} [5,6]. It is found that the divergence of the harmonics gradually increases when the pump intensity was increased. It is considered that the factor, which causes the divergence of the harmonics to gradually increase, is thought to be Rayleigh-Taylor-like instability at the critical surface. At an intensity lower than 1×10^{15} W cm^{-2}, the harmonics are generated in the specular direction and the divergence is comparable to the pump laser because the plasma surface is still free from Rayleigh-Taylor-like instability. The plasma surface gradually leads to a rippling of the plasma surface when the pump intensity is increased. The rippling blurs the distinction between s- and p-polarization. Another possible explanation for the decrease in I_p/I_s decreasing is Faraday rotation effect due to the influence of the spontaneous magnetic fields generated within the plasma. There is a possibility that the latter mechanism of the magnetic field generation influence the Faraday rotation of the pump laser polarization. In previous work we investigated the Faraday rotation of a picosecond pump laser pulse [7]. The pump laser is focused on an Al deposited target at an incidence angle of 75 degrees. Each component of polarization of laser light reflected from the critical surface was measured. As the results, it was not found how many magnetic fields generated. There is a possibility that large magnetic fields generate at the plasma-vacuum interface to be shorter than the pump laser wavelength and last for the duration of the pulse. In an upcoming paper, we will report a more detailed analysis of the experimental data [8].

In conclusion, investigations have been performed on the harmonic generation from solid surface plasmas using the subpicosecond Nd: Glass laser and several interesting phenomena have been observed. The blue shift of the harmonics by collisionless process has been observed. These results are important demonstrations of the large harmonic blueshift that can be provide a tool to realize a tunable coherent subpicosecond light source. It was found that the harmonic yield dependence on the pump laser polarization decreases with the growth of laser intensity when the latter exceeds 1×10^{17} W cm^{-2}. This decrease was explained by the rippling caused by Rayleigh-Taylor-like instability at the critical density surface, and Faraday rotation effect due to the influence of the spontaneous magnetic fields generated within the plasma.

REFERENCES

1. C. T. Hansen, *et al.*, *Phys. Rev. Letters* **83**, 5019-5022 (1999).
2. A. M. Rubenchik, *et al.*, Appl. Surf. Sci. **129**, 193-198 (1998).
3. W. Rozmus, *et al.*, Phys. Plasmas **3**, 360-367 (1996).
4. A. Ishizawa, *et al.*, accepted in May 2002 in Phys. Rev. E.
5. A. Ishizawa, *et al.*, IEEE J. Quantum Electron. **35**, 60-65 (1999).
6. A. Ishizawa, *et al.*, IEEE J. Quantum Electron. **36**, 665-668 (2000).
7. A. Ishizawa, *et al.*, "Aspect of Harmonic Generation from Solid Surface Plasma by Using Picosecond Laser," in *Frontier of Laser Physics and Quantum Optics*, edited by Zhizhan Xu et al., Springer, pp. 125-137,
8. A. Ishizawa, *et al.*, to be published in IEEE J. Quantum Electron..

Small-scale Coherent EUV Light Sources from High-Harmonic Generation

R.A. Bartels, S. Backus, C. Lei, A. Paul, I.P. Christov[†], M.M. Murnane, and Henry C. Kapteyn

Department of Physics and JILA
University of Colorado and National Institute of Standards and Technology
Boulder, CO 80309-0440
Kapteyn@jila.colorado.edu

[†]*Department of Physics, Sofia University, Sofia, Bulgaria*

ABSTRACT

Coherent EUV light can be generated using the process of high-harmonic upconversion of pulses from a high-intensity femtosecond laser. Recent advances in ultrafast laser technology, combined with the development of efficient techniques for high-harmonic generation (HHG), now make routine small-scale, table-top experiments using coherent EUV light. New advances, such as the use of pulse shaping and quasi-phase matched geometries, give new flexibility in the design of coherent EUV sources.

High-harmonic generation is now a well-established technique for generating coherent light in the vacuum- and extreme-ultraviolet regions of the spectrum. In HHG, the nonlinear response of atoms in the process of ionizing is used to "combine" photons of a driving laser, making it possible to generate light with photon energies as high as 0.5 keV. The use of a hollow-waveguide geometry for the HHG[1, 2] process makes it is possible to obtain reasonable conversion efficiency of laser light to x-rays even using relatively low, sub-millijoule driving pulse energies. This allows for the use of very high-repetition rate 5-10 kHz lasers[3] to drive the process, resulting in significant ($\sim 10^{12}$ ph/sec @ 40 eV) flux levels.

In recent measurements, we have shown that HHG emission in a phase-matched geometry has full-spatial coherence,[4] resulting in a beam of EUV light with spatial characteristics comparable to a helium-neon laser. The waveguide configuration is also experimentally extremely convenient, dramatically reducing gas load on the vacuum system and making it possible to construct small-scale EUV "beamlines" on an optical tabletop. Figure 1 shows a series of interferograms resulting from illumination of double-pinholes of varying separation. A detailed analysis of these data demonstrate that the light possesses full spatial coherence, to the limits of our measurement ability,

CP641, *X-Ray Lasers 2002: 8th International Conference on X-Ray Lasers,* edited by J. J. Rocca et al.
© 2002 American Institute of Physics 0-7354-0096-2/02/$19.00

over the entire beam. It is important to note also that these images were take over extended exposure times corresponding to many thousands of shots, demonstrating the long-term wavefront stability of this source.

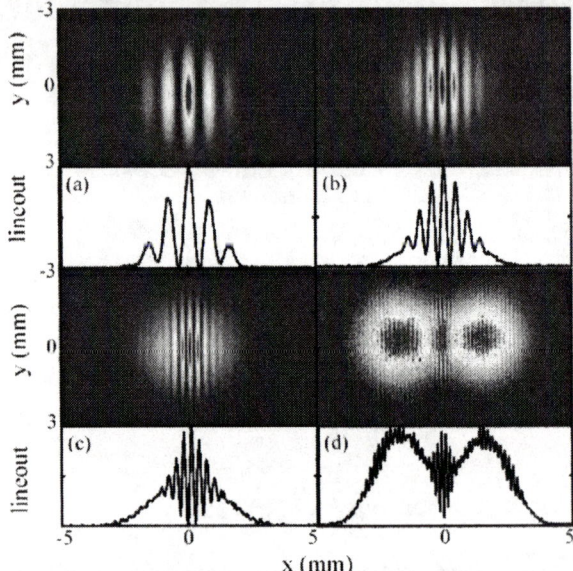

Figure 1: Interferogram images of the EUV beam diffracted by pinhole pairs of various separations, together with lineouts of the images. (a) = 142 μm, (b) = 242 μm, (c) = 384 μm, (d) = 779 μm separation.

The small scale and continuous operation of this coherent EUV source allows for rapid experimental progress. In recent work, we have demonstrated its utility for both time-resolved and non-time-resolved work. Using an extremely simple setup, we have generated Gabor holograms using EUV light.[4] Figure 2(a) shows an image of a tapered optical fiber (nonlinear scanning optical microscope tip) taken using a very simple geometry: the EUV light from a hollow fiber directly illuminates an object. After several meters of propagation, the interference pattern is recorded on an EUV-sensitive CCD. The only spectral filtering necessary is the use of an aluminum filter that reflects the light from the driving laser, while allowing the high-harmonic light, the wavelength range of 30-50 nm, to pass. Figure 2(b) shows a computer reconstruction of this image, demonstrating the reconstruction of the object shape. Future improvements promise to give sub-100 nm resolution in combination with sub-10 fs time resolution for stroboscopic imaging.

Other work shows that the spectral characteristics of the HHG emission can be manipulated using a variety of techniques.[5, 6] By altering the shape of the driving pulse in very subtle ways, we can selectively enhance emission in *one* harmonic order, as illustrated in figure 3. In this experiment, an evolutionary algorithm was used to find an optimum pulse shape, testing several thousand different pulse shapes before converging on an optimum. The result is an order of magnitude increase in the intensity of the selected harmonic order. This selective enhancement results from a pulse only marginally longer (21 fs) than the transform limit of 17 fs. The enhancement results from the fact that HHG is unique as a nonresonant purely-electronic nonlinearity *that has a finite nonlinear response time.* In the "recollision" model of HHG, the EUV emission is the result of an ionized electron recolliding with its parent ion after propagating as a free electron for <1/2 cycle of the optical field of the driving laser.[7] This attosecond time-scale dynamics makes it possible to manipulate the timing, and the quantum phase, of the electron recollision. Using a nonlinear "chirp" makes it possible to obtain a constructive interference of the EUV emission from a sequence of such recollision bursts. Detailed theoretical models of this process[6, 8] agree with experimental data, and provide a persuasive confirmation of our understanding of the attosecond electronic dynamics of high-harmonic generation.

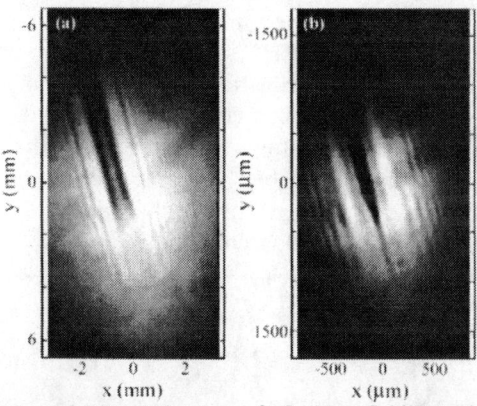

Figure 2:   Reconstruction of Gabor holograms of of a Near-field Scanning Optical Microscope (NSOM) tip, generated with our EUV beam. Fig. 2(a) corresponds to the hologram, while fig. 2(b) is the numerically-reconstructed image.

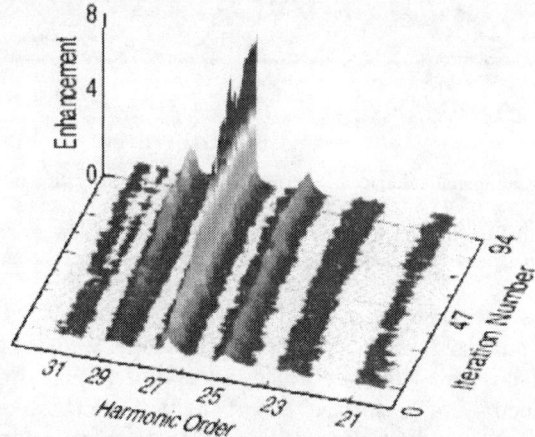

Figure 3: Optimization of a single (27th) harmonic in argon while suppressing adjacent harmonics, by using an evolutionary algorithm to search for an optimum pulse shape. The peak enhancement for the 27th harmonic is a factor of 8, while the energy enhancement is a factor of 4.6. The contrast ratio between the 27th harmonic and adjacent harmonic increases by a factor of 4.

In very recent work, we have demonstrated that it is possible to apply sophisticated quasi-phase matching concepts to the process of high-harmonic generation. One limitation of HHG arises from the fact that EUV emission is a result of an ionization process. As the gas ionizes, the increasing free-electron density changes the phase-matching conditions, limiting efficient phase-matched generation to relatively low levels of ionization, and lower-energy harmonic photons. Previously, we had proposed[9, 10] that this phase-mismatch could be ameliorated by periodically modulating the intensity of the driving laser, such that HHG at the highest harmonic orders can be effectively turned on-and-off. In this way, constructive interference can be maintained over a large propagation distance, even in the presence of high levels of ionization, and higher-order harmonics can be generated for otherwise-equivalent conditions. This periodic intensity modulation can be accomplished by periodically modulating the fiber diameter, allowing us to verify theoretical predicitons of QPM in recent experiments. Figure 4 shows HHG spectra for both a modulated, and a constant-diameter fiber, with all other conditions remaining the same. These data demonstrate that this periodic modulation results in a very substantial shift of the HHG spectrum to shorter wavelengths. Another interesting characteristics of the QPM HHG emission is that there are no longer distinguishable harmonic spectral peaks. This is consistent with theoretical models that predict that the quasi phase-matching process can result in the generation of a single, isolated attosecond-duration pulse.

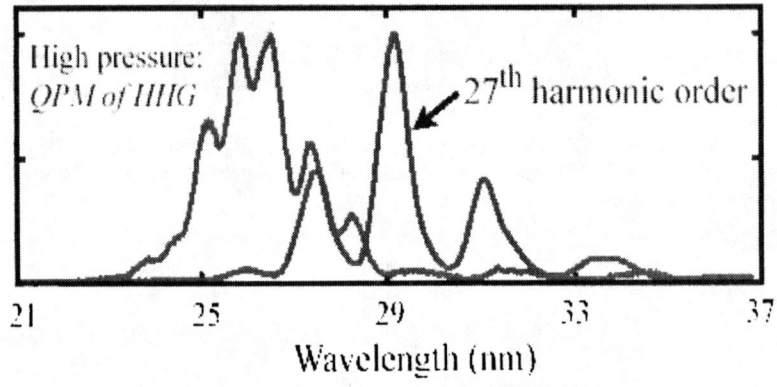

Figure 4: Experimental evidence of quasi-phase-matched generation of EUV harmonic light. In this experiment, a hollow fiber of 140 μm diameter was bulged in 10 places with a period of 1 mm. When emission from this fiber is compared with that of a straight fiber, the spectrum (left curve) is shifted to shorter wavelengths, and individual harmonic peaks merge into a near-continuum.

The very short pulse duration of the HHG source can be used for new types of time-resolved studies. In time-resolved spectroscopy, we have studied the dynamics of O_2 molecules on the Pt (111) surface, observing for the first time changes in the electronic structure of a surface adsorbate with femtosecond time-resolution, by observing changes in the valence band photoelectron spectrum.[11] In conclusion, coherent sources based on high-harmonic generation provide the first practical, fully coherent table-top-scale source of light in the EUV region of the spectrum.

Bibliography

[1]   A. Rundquist, C. G. Durfee III, S. Backus, C. Herne, Z. Chang, M. M. Murnane, and H. C. Kapteyn, "Phase-matched Generation of Coherent Soft X-rays," *Science*, vol. 280, pp. 1412-1415, 1998.

[2]   C. G. Durfee, A. R. Rundquist, S. Backus, C. Herne, M. M. Murnane, and H. C. Kapteyn, "Phase matching of high-order harmonics in hollow waveguides," *Physical Review Letters*, vol. 83, pp. 2187-2190, 1999.

[3]   S. Backus, R. Bartels, S. Thompson, R. Dollinger, H. C. Kapteyn, and M. M. Murnane, "High-efficiency, single-stage 7-kHz high-average-power ultrafast laser system," *Optics Letters*, vol. 26, pp. 465-467, 2001.

[4]   R. A. Bartels, A. Paul, H. Kapteyn, M. Murnane, S. Backus, I. P. Christov, Y. Liu, D. Attwood, and C. Jacobsen, "Generation of Spatially Coherent Light at Extreme Ultraviolet Wavelengths," *Science*, pp. in press, 2002.

[5]   R. Bartels, S. Backus, E. Zeek, L. Misoguti, G. Vdovin, I. P. Christov, M. M. Murnane, and H. C. Kapteyn, "Shaped-pulse optimization of coherent emission of high-harmonic soft X-rays," *Nature*, vol. 406, pp. 164-166, 2000.

[6]   I. P. Christov, R. Bartels, H. C. Kapteyn, and M. M. Murnane, "Attosecond time-scale intra-atomic phase matching of high harmonic generation," *Physical Review Letters*, vol. 86, pp. 5458-5461, 2001.

[7]   K. C. Kulander, K. J. Schafer, and J. L. Krause, "Dynamics of Short-Pulse Excitation, Ionization and Harmonic Conversion," in *Super-intense laser-atom physics*, vol. 316, *NATO Advanced Science Institutes Series*, B. Piraux, A. L'Huillier, and K. Rzazewski, Eds. New York: Plenum Press, 1993, pp. 95-110.

[8]   R. Bartels, S. Backus, I. Christov, H. Kapteyn, and M. Murnane, "Attosecond time-scale feedback control of coherent X-ray generation," *Chemical Physics*, vol. 267, pp. 277-289, 2001.

[9]   I. P. Christov, H. C. Kapteyn, and M. M. Murnane, "Quasi-phase matching of high-harmonics and attosecond pulses in modulated waveguides," *Optics Express*, vol. 7, pp. 362-367, 2000.

[10]   I. P. Christov, "Control of high harmonic and attosecond pulse generation in aperiodic modulated waveguides," *Journal of the Optical Society of America B-Optical Physics*, vol. 18, pp. 1877-1881, 2001.

[11]   M. Bauer, C. Lei, K. Read, R. Tobey, J. Gland, M. M. Murnane, and H. C. Kapteyn, "Direct observation of surface chemistry using ultrafast soft-x- ray pulses," *Physical Review Letters*, vol. 8702, pp. 5501-U59, 2001.

405

Phase and Polarization Control of the Harmonic Emission: Towards Attosecond Pulses

H. Merdji*, Y. Mairesse*, E. Priori*, M. Kovacev*, P. Agostini*, P. Breger*, P. Montchicourt, P. Salières*, B. Carré*, O. Tcherbakoff¶, E. Mével¶, and E. Constant¶.

*CEA-DSM-DRECAM, Centre d'Etudes de Saclay, 91191 Gif-sur-Yvette, France.
¶CELIA, Université Bordeaux 1, Cours de la Libération, 33405 Talence, France

Abstract. In the first section we demonstrate the generation of four phase-locked harmonic pulses separated in time using frequency-domain interferometry. The spectra present a high sensitivity to a change of the relative phase on an attosecond time scale. The spectral resolution and the control of this relative phase could be used to perform high resolution measurements. In the second section we demonstrate temporal gating technique to isolate attosecond pulses. Recent results of such a gating showing a spectral broadening are reported. We show that this enlargement can unambiguously be attributed to a temporal confinement by a factor 2 of the harmonic emission.

INTRODUCTION

Most exciting applications of high harmonics generation are linked to its temporal structures. The pulse of an electromagnetic wave cannot be shorter than its period (2.67 fs at 800nm). Thus high harmonic generation is a direct approach to reach the attosecond range.

First the control of the phase on these time scales offers a powerful tool to perform Fourier transform spectroscopy with an unprecedented resolution. Another exiting application could be to use multiple time delayed beams to perform Ramsey-like spectroscopy in the XUV range. It has been recently demonstrated that frequency-domain interferometry can be performed with high order harmonics by generating two temporally-separated phase-locked harmonic pulses [1]. Frequency-domain interferometry is a widely used technique offering a variety of possibilities as a tool for probing but also manipulating and controlling ultrafast phenomena, for instance in solid state and plasma physics. The extension of the technique considering two temporally-separated harmonic pulses to multiple phase-locked harmonic pulses will lead to highly contrasted spectral fringe patterns and an increased phase sensitivity. In the first section of this paper we will describe the extension to four phase-locked XUV pulses and will show the extreme sensitivity of this method to a change of the relative phase on an attosecond time scale.

CP641, X-Ray Lasers 2002: 8th International Conference on X-Ray Lasers, edited by J. J. Rocca et al.
© 2002 American Institute of Physics 0-7354-0096-2/02/$19.00

Secondly the generation and the control of attosecond pulses open up the field of the so called atto-science. Recently harmonics 11-19 of a 40 fs, 800nm pulse generated in argon were found to have a linear relationship between their phases. The corresponding time-domain profile is a periodic train of 250 attosecond pulses separated by half the optical cycle 1.35 fs [2]. Even if such a train could be useful for some applications it would be more useful to generate a single attosecond pulse. One straightforward, but demanding solution to this challenge is to start from a sub-10fs infrared pulse. Such pulses are indeed produced from Ti:S lasers in a few groups which master the most sophisticated techniques of pulse recompression [3]. The other approach is to retain the "long" 50 fs pulses which are the standard in many laboratories, generate the train, and subsequently select a single attosecond pulse, taking advantage of the high sensitivity of the harmonic generation process to polarization of the fundamental light. This paper summarizes recent results in polarization gating. The harmonic generation is extremely sensitive to the laser degree of ellipticity so that by modulating in time the laser degree of ellipticity it is possible to confine temporally the harmonic emission [4-7]. By combining two time delayed left and right circularly polarized pulses we were able to control the linearly polarized temporal gate duration. The second section of this paper reports preliminary results in polarization gating.

SECTION 1: PHASE CONTROL ON AN ATTOSECOND TIMESCALE

The four phase-locked harmonic pulses have been generated using four IR laser pulses separated in time as shown in Fig. 1. The first double laser pulse is created by using the group velocity difference on the two axes of a birefringent plate rotated at 45° from the laser polarization. A polarizer placed after the plate projects both components on the same axis. The calibrated thickness of the plate fixes the time delay τ between the two pulses at 120fs.

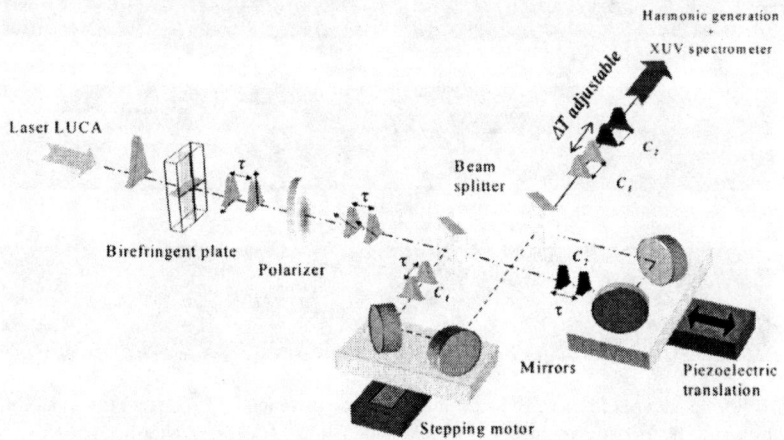

FIGURE 1. Diagram of the setup used for the generation of four laser pulses delayed in time.

This couple of laser pulses is sent into a Michelson interferometer where two replicas (C_1 and C_2) are produced with the same time delay τ between the pulses. The length of the first arm of the interferometer is controlled by a stepping motor, whereas the second one is moved by a piezoelectric translation with nanometric precision. It is then possible to control very precisely the optical paths of (and thus the time delay ΔT between) the two laser pulse couples C_1 and C_2. These four IR laser pulses are focused in a pulsed xenon jet where they generate four harmonic pulses that are sent into a spectrometer. The spectral fringe pattern obtained after dispersion by the grating is recorded on microchannel plates mounted in the focal plane of the spectrometer at grazing incidence (12°) in order to increase the resolution to 0.1Å.

When the two arms of the Michelson are slightly misaligned, the two couples are not spatially superposed and we observe two identical spectral fringe patterns corresponding to the interference between the two pulses of each couple, with a fringe period fixed by the same time separation τ (Fig. 2a). In this case the observed fringe patterns are independent of the time delay ΔT between C_1 and C_2. A realignment of the Michelson allows a spatial superposition of the harmonic couples. A dramatic change of the fringe pattern is observed, as shown in Fig. 2b), which gives evidence for the generation of four phase-locked harmonic pulses separated in time. The delay ΔT between C_1 and C_2 was chosen to be 2τ corresponding to an equal separation of the four pulses of τ. Due to this regular time separation, an effective fringe narrowing is obtained (factor 2 with respect to the 2x2 sources case). By slightly varying the delay to $\Delta T = 2\tau + T_q/2$ ($T_q/2$ is the harmonic half period ~ 120 as for the 11th harmonic, corresponding to a mirror displacement of 18nm), we observe a clear change of the fringe pattern, that becomes more regular with a fringe period twice as small as in Fig. 2a) (Fig. 2c).

(a)

(b)

(c)

FIGURE 2. Fringe patterns for the 11th harmonic generated in xenon by two laser pulse couples separated by τ=120 fs. (a) The Michelson is misaligned. (b) and (c) The Michelson is aligned and the delay ΔT between the two couples is set respectively to 2τ and $2\tau + T_q/2$.

In Fig. 3, we show the corresponding spectral profiles. In the case of a single couple, the modulation of the harmonic spectrum follows $\cos^2(\omega\tau/2)$. When the two couples interfere, an additional modulation comes into play in $\cos^2(\omega\Delta T/2)$. When $\Delta T \sim 2\tau$, the same periodicity as in the case of a single couple is observed, with a significantly narrowed central peak surrounded by satellite structures (Fig. 3a). When $\Delta T \sim 2\tau + T_q/2$, the periodicity seams to decrease to half the former value, with regular peaks of equal intensity (Fig. 3b). In both cases, there is a good agreement with the results of the fit using the theoretical formula of interference between four phase-locked pulses with the corresponding delays (dashed lines). This demonstrates the high sensitivity of our experimental setup to a change of the relative phase on an attosecond time scale: a phase shift between the two couples of a half harmonic period (~ 120 as) results in a strong modification of the whole fringe pattern, whereas for the simple two pulses case (variation of τ), there is only a general shift of the whole fringe pattern in the envelop, which is much more difficult to diagnose unambiguously due to the experimental fluctuations.

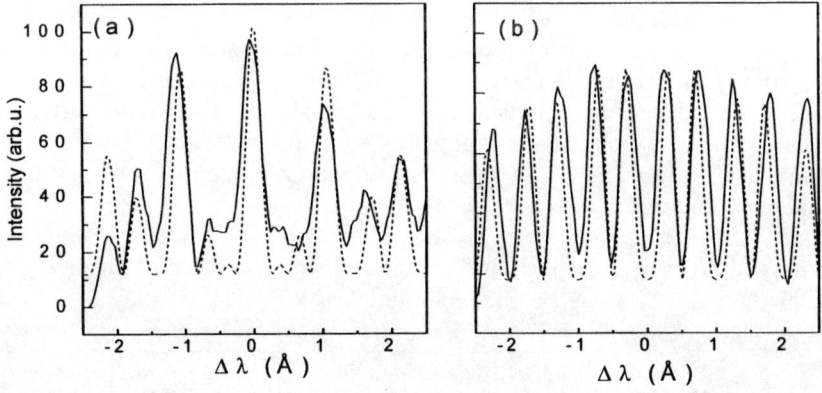

FIGURE 3. Spectral profiles (solid lines) obtained from the interference of two couples (H11) delayed of about 2τ (a) and $2\tau + T_q/2$ (b). In dashed lines, are indicated the fits from the theoretical formula. In (a) is indicated the spectral profile obtained in the case of a single couple (squares).

SECTION 2: TEMPORAL CONFINEMENT OF THE HARMONIC EMISSION THROUGH POLARIZATION GATING

The idea to take advantage of the rapid dependence of high harmonics generation on the ellipticity of the pump light was proposed by Corkum et al. in 1994 [4]. This dependence is intuitively clear since harmonic generation is forbidden in circular polarization by angular momentum conservation. The harmonic yield was measured to depend on the ellipticity ε as the exponential function $\exp(-\beta\varepsilon^2)$ with $\beta \approx 35$ for harmonic 21. Hence if the pump pulse polarization is changed as a function of time from circular to linear and back to circular the resulting harmonic pulse would last essentially the duration of the linear polarization. The first suggestion [4] was to combine two pulses with slightly different frequencies linearly polarized along perpendicular directions. A variant of this idea was actually implemented by the Lund group [6]. The spectra of the resulting harmonics were expected to show a broadening but, on the contrary, the spectral widths were found to be narrowed. Careful investigations demonstrated that the chirp of the harmonics is in fact responsible for this situation. Both the intensity-dependent dipole phase and the ionization chirp play a role although the first factor was shown to dominate.

The scheme described above can be substantially simplified, as proposed by several people [5,7], by using two properly delayed circularly polarized pulses. A first implementation was attempted at the CELIA laboratory [7] with 35 fs pulses. The changes observed in the harmonic spectra (spectral broadening in the cutoff and spectral squeezing in the plateau) were consistent with a 7 fs temporal confinement of the high harmonic generation. In the present experiment carried out on the LUCA laser at Saclay, the initial linearly polarized 55 fs pulse (intensity, FWHM) is fed into a Michelson interferometer. The polarization direction of the pulses in each arm is controlled independently by two quarter-wave plates, equivalent to two half-wave plates due to the double pass. The two pulses leave the interferometer and pass through a quarter-wave plate whose optical axis orientation governs the resulting polarization. If the latter is set at 45° from two mutually orthogonal linearly polarized pulses, a time-dependent ellipticity is determined: the light polarization is first circular, becomes exactly linear during a short time and evolves to circular again ("narrow gate"). The width of the gate depends on the temporal profiles of the pulses and the delay introduced between them by the interferometer. The radiation is then focused inside a 3mm long argon jet where harmonics are generated. Harmonics are spectrally resolved by a grating flat-field spectrometer and detected on a pair of micro channel plates or a windowless photomultiplier.

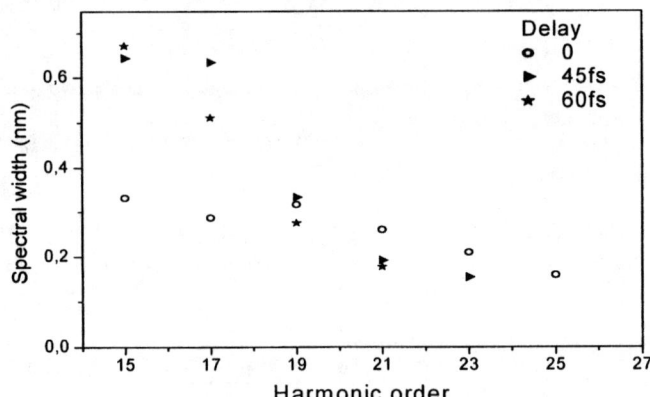

FIGURE 4. Spectral broadening as a function of the harmonic order for different time delay between the two circularly left and right polarized pulses.

The chirp due to self phase modulation during propagation of the pump pulse between the output of the compressor and the harmonic chamber entrance window increased the pulse duration to 75 fs. Fig. 4 shows measurements of the spectral width of harmonics for different time delay between the two left and right circularly polarized IR pulses. A clear broadening is seen when we generate the harmonics with a temporal gate. The maximum broadening recorded varied from about a factor 2 at harmonic 15 to essentially 0 at harmonic 23. Numerical simulations using a simple model for harmonic generation and the ellipticity dependence discussed above are in progress. First results show that compared to the initial harmonic duration the temporal emission is confined by a factor 2 for a delay of 60 fs between the two circular left and right polarized IR pulses. However more investigations are needed for a complete understanding of the present results especially an evaluation of the role of the intrinsic harmonic chirp and the laser chirp is currently in progress.

CONCLUSION

In the first section, we have performed the generation of four phase-locked harmonic pulses separated in time. The spectral profiles obtained from frequency-domain interferometry present a high sensitivity to a change of the relative phase on an attosecond time scale, indicating that such a control is achieved on the harmonic emission. Interesting perspectives arise from the application of this technique. The sensitivity of this technique could be useful for e.g. the precise diagnostic of dense plasmas. Moreover, pump probe experiments with two spatially separated harmonic couples (see Fig. 2a) would benefit from an absolute reference pattern or the possibility to probe large plasmas with relative density measurements.

In the second section we have presented preliminary results of the temporal confinement of the harmonics emission using the high sensitivity of the generation process to the laser ellipticity. This method can be used to generate attosecond pulses or to improve the temporal resolution in XUV pump probe experiments. The main difficulty of this method is linked to the fact that when we increase the delay, on one hand the temporal gate duration decreases together with its intensity and, on the other hand, an increasing amount of laser energy ionizes the medium without contributing to the harmonic generation. Thus the generation of an isolated attosecond pulse seems to be possible only when we start with a 15-20 fs laser pulse which is however easily achievable compared to the 5-6 fs required in ref [3].

REFERENCES

1. P. Salières *et al.*, Phys. Rev. Lett. **83**, 5483 (1999).
2. P. M. Paul et al. Science **292**, 1689 (2001).
3. M. Drescher et al. Science **291**, 1923 (2001); M. Hentschel et al. Nature **414**, 509 (2001).
4. P. B. Corkum et al. Opt. Lett. **19**, 1870 (1994).
5. V.T. Platonenko and V.V. Strelkov Quantum Electronics **28**, 749 (1998).
6. C. Altucci et al. Phys. Rev. A **58**, 3934 (1998).
7. E. Constant PhD thesis (1997). E. Constant, E. Mevel and O. Tcherbakov ATTO workshop Milan (2001).

Generation of High Intensity Coherent XUV Light Using High-Order Harmonics

Katsumi. Midorikawa*, Eiji. Takahashi, and Yasuo. Nabekawa

Laser Technology Laboratory, RIKEN
2-1 Hirosawa, Wako-shi,Saitama 351-0198, Japan
e-mail: kmidori@postman.riken.go.jp

Abstract. The peak power of 0.13 GW was generated at 62.3 nm using high order harmonics of 35fs Ti:sapphire laser pulses. These pulses were focused into 14-cm-long static gas cell containing 0.6 Torr Xe. The peak brightness achieved was 3 x 10^{28} photon /mm^2/$mrad^2$/s. Also, the average brightness of this coherent XUV light is comparable to that of synchrotron orbit radiation despite a low repetition rate of 10 Hz.

INTRODUCTION

High-order harmonic generation (HHG) using ultrashort high-peak-power lasers is very attractive as a coherent short-wavelength light source. High-order harmonic (HH) is now being applied to some fields such as solid-state and plasma physics and extreme-ultraviolet (XUV) spectroscopy for atoms and molecules.

For further development of applications, one of the most important issues is XUV pulse energy and beam quality. Recently, through the use of a capillary waveguide[1-3] or a self-guided beam[4-6] in a static gas cell under the phase-matched condition, conversion efficiencies were improved and good beam quality was obtained. However, these efficiencies were obtained at pump energies lower than a few mJ, and the obtained harmonic energy was of nanojoule order. Last year, we reported the energy scaling of HH under the phase-matched condition using a long interaction length and a loosely focused beam in Ar.[7, 8] In that experiment, the total output energy in the spectrum region of 34.8 to 25.8 nm (the corresponding order of 23rd to 31st) attained was 0 .7 μJ, and the maximal 27th harmonic (29.6 nm) energy was found to be 0 .33 μJ with 0.7 mrad (FWHM) beam divergence.

In this study, by using the previous energy scaling procedure in Ar, we demonstrated the generation of microjoule XUV light at wavelengths of 73.6 to 42.6 nm in Xe. This high-power ultrashort XUV pulse is expected to boost nonlinear optics in the XUV region.

As has been pointed out in a previous publication,[2] to increase the output HH energy, the coherence length L_{coh} should be much longer than the absorption length

CP641, *X-Ray Lasers 2002: 8th International Conference on X-Ray Lasers*, edited by J. J. Rocca et al.

L_{abs} for the given harmonic order q. The coherence length is given by $\pi/\Delta k$, where Δk $=k_q - qk_0$ indicates the amount of phase mismatch between the harmonic field and the atomic dipoles. If the condition is $L_{coh} \gg L_{abs}$, the maximum harmonic yield is attained with an interaction length L_{int} that is longer than three times the absorption length. To extend the interaction length between the laser pulse and the target medium, the Rayleigh length must be extended. Increasing the Rayleigh length leads to the increase in the spot size of the pump pulse. This allows us to increase the pump pulse energy, which is indispensable for high-energy output.

EXPERIMENT

Figure 1 shows the experimental setup. In the experiment, to increase both the interaction length and the pump pulse energy, a loosely focused pumping geometry with an f =5000 mm plano-convex lens was employed. The pump pulse was supplied by a 10 Hz Ti:sapphire laser system based on a chirped pulse amplification. In this system, the pulse width was 35 fs and the wavelength was centered at 810 nm. The pump pulse was loosely focused by a plano-convex lens, and delivered into the target chamber through a CaF$_2$ window. We set the focus around the entrance pinhole of the interaction cell. The interaction cell had two pinholes on each end surface of the bellow arms. The diameter of the pinholes was 1.2 mm. These pinholes isolated the vacuum and gas-filled regions. The interaction length was variable from 0 to 150 mm in the interaction cell. Xe gas was statically filled in the interaction cell. The generated harmonics illuminated a 150 μm(H) x 12 mm (V) slit of the spectrometer. Harmonic signals were observed with a flat-field normal-incident XUV spectrometer with a platinum-coated concave grating blazed at 60 nm (1200 lines/mm). This spectrometer equipped with a microchannel plate (MCP) can cover the spectral range from 30 to 80 nm. A CCD camera detected two-dimensional fluorescence from a phosphor screen placed behind the MCP. Therefore, we measured spectrally resolved far-field profiles of HHs. The absolute energy of HH was measured directly with an unbiased silicon XUV photodiode (XUV-100). The photodiode was inserted in front of the spectrometer, and the output signal was recorded directly on an oscilloscope.

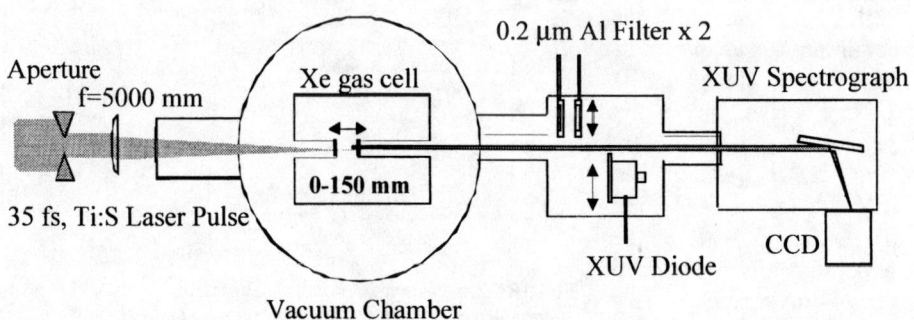

FIGURE 1. Experimtal setup

RESULTS

A typical XUV harmonic spectrum for L_{int} = 14 cm and 0.6 Torr Xe is depicted in Fig.2. The 11th and 13th harmonic signals reached maxima around 0.6 Torr. The pump pulse energy and the truncated diameter were set at 14 mJ and 14 mm, respectively. The spot size (1/e radius) of the pump pulse at the focus was measured to be 210 μm with an attenuated pump pulse having 14 mm beam diameter. Therefore, the geometrical phase shift was estimated to be 5.85 x 10^{-2}q cm^{-1} at the entrance pinhole. To balance between the Gouy phase and the dipole phase, we used the negative dispersion of the target gas in this experiment. In 0.6 Torr Xe, the dispersion of the target gas was estimated to be 1.19 cm^{-1} for the 11th harmonic, 1.15 cm^{-1} for the 13th harmonic, 1.14 cm^{-1} for the 15th harmonic, 1.15 cm^{-1} for the 17th harmonic, and 1.18 cm^{-1} for the 19th harmonic.[9] Considering only the geometrical phase and gas dispersion in neutral atom, phase matching was not always satisfied perfectly because the Gouy phase changes along the propagation distance. However, the 11th and 13th harmonic signals were expected to satisfy the phase-matched condition at around 0.6 Torr within the interaction length of 14 cm. The absorption length for 0.6 Torr Xe was estimated to be 1.35 cm for 11th, 1.58 cm for 13th, 1.89 cm for 15th, 3.65 cm for 17th, and 4.3 cm for 19th harmonic from Ref. 10. Therefore, the 11th to 15th order harmonics satisfied the absorption limited HHG condition. On the other hand, the 17th and 19th order harmonics did not satisfy the above condition yet, because of low absorption.

The inset in Fig.2 shows the far-field spatial profile of the 13th harmonic. The output beam divergence was measured to be 0.5 mrad (FWHM) with a Gaussian-like profile. Of course, other order harmonics showed similar beam quality. The beam divergence of HH is related to a noncollinear phase matching. This type of phase matching is called the Cerenkov phase matching.[3] In the off-axis direction, the phase mismatch is given by Δk = $k_q - qk_0 - k_{cere}$, Δk_{cere} = $q\pi\theta^2/\lambda_0$, where θ is the noncollinear angle of HH. Assuming $k_q - qk_0 = 0$, the beam divsrgence of the 13th harmonic in our condition was estimated to be 0.6 mrad, since the off-axis FWHM angle is given by $sinc^2(\Delta k_{cere}L_{med}/2)$. The calculated result is in good agreement with the experimentally obtained beam divergence. This agreement also supports the fact that the phase

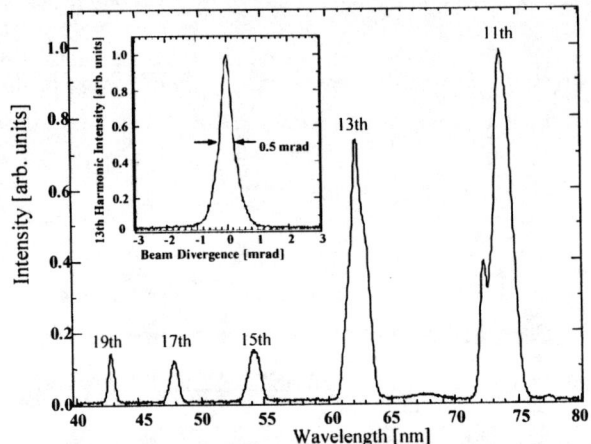

FIGURE 2. Measured harmonic spectrum distribution. This profile is normalized by the 11th harmonic signal. The inset shows a normalized far-field 1D spatial profile of the 13th harmonic in 0.6 Torr X gas.

matched condition is satisfied on the propagation axis of the pump pulse.

The absolute output energy of HH was measured using an XUV photodiode. The values of spectral sensitivity and quantum efficiency of the XUV photodiode were referred to Ref. 11, and calibrated with 266 nm Q-switched YAG laser pulses in this experiment. To block the pump laser, an Al filter of which thickness was 200 nm with mesh-back was inserted in front of the photodiode. For the measurement of XUV energy, two Al filters were employed to avoid

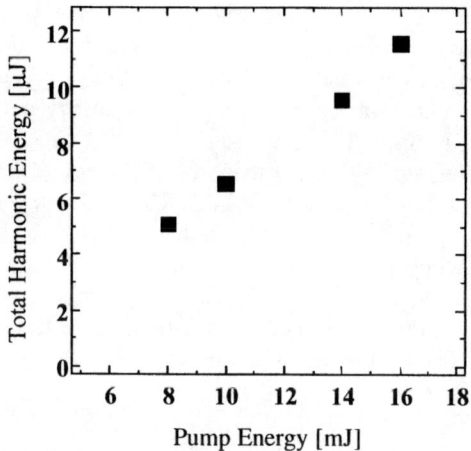

FIGURE 3. Total harmonic energy as a function of the pump energy.

the saturation of the HH signal. When we measure the HH energy, the filter transmission rate must be carefully treated. Since the transmission rat is very low in this wavelength region, a slight difference of the estimated transmission rate leads to a large difference in HH energy. Although the filter transmission rate can be estimated from data published in Ref. 12, the Al filter transmission rate gradually changes due to oxidation. Therefore, we carefully measured the filter transmission rate. The measure total output energy from the 11th to 19th as a function of the pump energy was shown in Fig. 3

Figure 4 shows the output energy distribution and conversion efficiency of the harmonics with a 16 mJ pump energy. The total average harmonic energy between the 11th to 19th harmonic was estimated to be 11.5 µJ from the filter transmission rate. From the relative harmonic strength distribution, the maximal XUV light energy was estimated to be 7 µJ for the 11th harmonic (72.7nm), 4.7 µJ

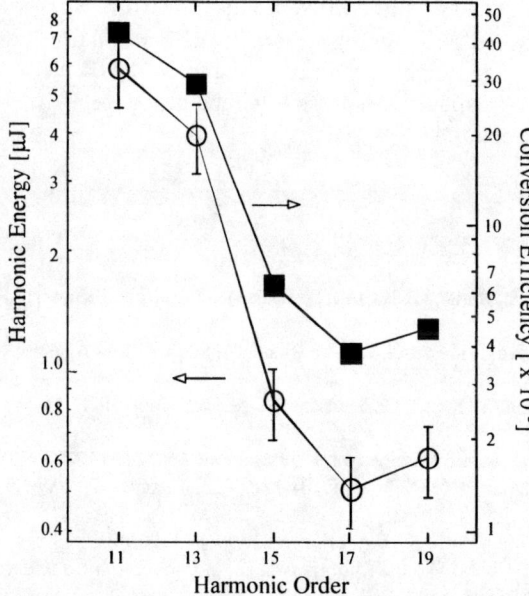

FIGURE 4. Experimentally obtained energy yield per shot and conversion efficiency of HH for a 14-cm-long medium in 0.6 Torr Xe gas. Solid squares and open circles correspond to the maximal conversion efficiency and the output energy, respectively.

for the 13th harmonic (62.3 nm), and 1 µJ for the 15th harmonic (54 nm). The conversion efficiencies attained were 4.3 x 10^{-4} and 3.0 x 10^{-4} at the 11th and 13th harmonics, respectively. However, at the 16 mJ condition, beam divergence increased to 0.7 mrad. Moreover, when a higher energy pump pulse was used, total output energy decreased.

Thus far, the best conversion efficiency of 4 x 10^{-5} at the 15th harmonic (52.5 nm) in Xe gas has been reported in Ref. 2. In our experiment, the conversion efficiency at the 15th harmonic was estimated to b 6.4 x 10^{-5} for 16 mJ pumping. The conversion efficiency at the 15th harmonic was improved by a factor of 1.6, compared with that in Ref. 2. Moreover, it should be noted that our pump energy is 10 times larger than that of Ref. 2. From the good beam quality (see the inset in Fig.1) and the high conversion efficiency achieved, it is found that our energy scaling procedur[7, 8] is also applicable for HHG in Xe.

SUMMARY

We generated 10-µJ coherent XUV light at wavelengths between 73.6 to 42.6 nm using HHs. The peak power of this coherent XUV light was estimated to be 0.13 GW at 62.3 nm, assuming the same pulse width as the pump of 35 fs. From the relative harmonic strength distribution, the maximal XUV light energy was estimated to be 7 µJ for the 11th harmonic (72.7 nm), 4.7 µJ for the 13th harmonic (62.3 nm), and 1 µJ for the 15th harmonic (54 nm). Th peak brightness of this low-emittance 62.3 nm light was estimated to be 3 x 10^{28} photon/mm^2/mrad^2/s, assuming a beam diameter of 60 µm at the exit of the gas cell. Also, the average brightness of this coherent XUV light is comparable to that of synchrotron orbit radiation despite a low repetition rate of 10 Hz. This XUV energy is the highest ever reported using HHs.

REFERENCES

1. Rundquist, A., Durfee III, C. G., Chabg, Z., Herne, C., Backus, S., Murnane, M. M., and Kapteyn, H. C., *Science* **280**, 1412 (1998).
2. Constant, E., Garzella, D., Breger, P., Mevel, M., Dorrer, Ch., Le Blanc, C., Salin, F., and Agostini, P., *Phys. Rev. Lett.* **82**, 1668 (1999).
3. Durfee III, C. G., Rundquist, A. R., Backus, S., Herne, C., Murnane, M. M., and Kapteyn, H. C., *Phys. Rev. Lett.* **83**, 2187 (1999).
4. Tamaki, Y., Itatani, J., Nagata, Y., Obara, M., and Midorikawa, K., *Phys. Rev. Lett.* **82**, 1422 (1999).
5. Schnurer, M., Cheng, Z., Hentschel, M., Tempa, G., Kalman, P., Brabec, T., and Krausz, F., *Phys. Rev. Lett.* **83**, 722 (1999).
6. Tamaki, Y., Itatani, J., Obara, M., and Midorikawa, K., *Phys. Rev. A* **62**, 063802 (2000).
7. Takahashi, E., Nabekawa, Y., Otsuka, T., Obara, M., and Midorikawa, K., in *Pacific Rim Conference on Lasers and Electro-Optics* Technical Digest Series (2001).
8. Midorikawa, K., Takahashi, E., Tamaki, Y., Nagata, Y., and Nabekawa, Y., *Soft X-Ray Lasers and Application IV, Proc. SPIE* **4505**, 189 (2001).
9. X-Ray Form Factor, Attenuation and Scattering tables in *The National Institute of Standards and Technology (NIST)*. http://physics.nist.gov/

10. Henk, B. L., Gullikson, E. M., and Davis, J. C., *Atomic Data and Nuclear Data Tables* **54**, 181 (1993).
11. Krumrey, M., Tegeler, E., Goebel, R., and Kohler, R., *Rev. Sci. Instrum.* **66**, 4736 (1995).
12. Powell, F. R., Vedder, P. W., Lindblom, J. F., and Powell, S. F., *Opt. Eng.* **29**, 614 (1990).

High-order harmonics as a continuously tunable coherent femtosecond x-ray source

Chang Hee NAM, Hyung Taek KIM, Kyung-Han HONG, Dong Gun LEE, and Jung-Hoon KIM

Dept. of Physics & Coherent X-ray Research Center, KAIST, Daejeon 305-701, Korea

Abstract. With the application of appropriately chirped laser pulses, harmonic chirp can be coherently controlled so that sharp harmonics be produced. Using the strong blueshift property and coherently controlling harmonic generation process, we demonstrated a continuously tunable high-order harmonic generation, without losing spectral sharpness.

1. INTRODUCTION

Atoms under intense laser field can emit very high-order harmonics of a driving laser frequency as a consequence of the periodic modulation of an electron motion. Due to the inversion symmetry of gaseous atoms, harmonics only in odd orders are allowed. This makes the harmonic x-rays an x-ray source at fixed frequencies with an interval of $2\omega_0$ (ω_0 = laser frequency). Though the harmonic x-ray source can be operated at numerous frequencies, it can be a limitation in actual applications. In order to overcome this problem, we have investigated the property of strong harmonic blueshift [1] induced when driven by intense laser pulses of 30 fs or shorter, which covered sufficiently the frequency interval between adjacent odd harmonics. With the application of appropriately chirped laser pulses [2], continuous tuning of harmonic wavelengths was achieved without losing spectral sharpness.

2. SPECTRAL STRUCTURE OF HIGH-ORDER HARMONICS

Spectral structure analysis shows that high-order harmonics possess chirped spectral structure. It is known that harmonics have dynamically induced negative chirp due to the intensity-dependent harmonic phase. This intensity-dependent harmonic phase changes rapidly in time when harmonics are generated in an intense femtosecond laser field, increasing the harmonic chirp with the decrease of laser pulse duration. In addition, harmonics generated in intense laser field can have positive chirp due to the self-phase modulation (SPM) of a laser pulse [2]. The laser pulse propagating in an ionizing medium can acquire positive chirp due to SPM, and this induced laser chirp is directly transferred to the temporal structure of harmonics and

CP641, X-Ray Lasers 2002: 8th International Conference on X-Ray Lasers, edited by J. J. Rocca et al.
© 2002 American Institute of Physics 0-7354-0096-2/02/$19.00

positively chirped harmonics are emitted. Thus, harmonics can have either negative or positive chirp, depending on experimental conditions.

The interaction between intense laser field and an atom can be described using a time-dependent Schrödinger equation (TDSE). As the intense laser field can provide an electric field comparable to or stronger than the Coulomb field in an atom, a perturbation approach cannot be used and TDSE needs to be solved numerically. The numerically obtained spectrum with femtosecond laser pulses shows a complicated structure as the harmonic order increases, making it nearly impossible to distinguish individual harmonic. For clear understanding of numerical result, we apply the coherent sum method to select only short trajectory components of harmonics and the Wigner distribution (WD) function in the analysis of high-order harmonics to reveal the harmonic structure in time-frequency domain [3]. The WD of high-order harmonic spectrum calculated for neon atom exposed to a laser pulse of duration 27 fs and intensity 1×10^{16} W/cm^2 is shown in Fig. 1. The temporal variation of harmonic frequency indicated by the dotted lines shows the negative chirp of harmonics in time. In spite of a large frequency spread greater than 2 ω_0, corresponding to the interval between adjacent harmonics, individual harmonics are readily distinguishable in this time-frequency analysis. The WD clearly shows that harmonics are blueshifted and negatively chirped. For example, in the case of the 85th harmonic, as time evolves it sweeps from 90 ω_0 to 87 ω_0, with a peak near 87.8 ω_0.

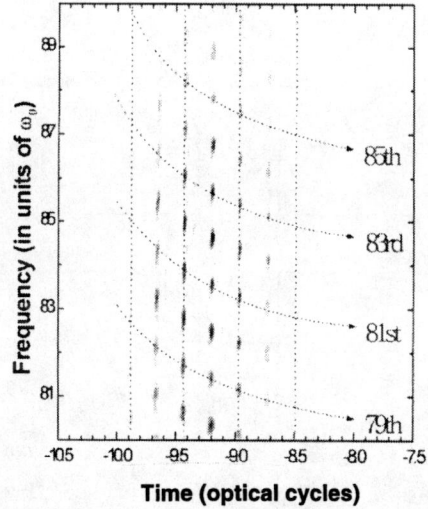

FIGURE 1. Wigner distribution of single-atom high-order harmonics spectrum from a neon atom irradiated with a 27-fs, 817 nm laser pulse of intensity I= 1×10^{16} W/cm^2. Positive values are colored black, and negative values are colored white (in arbitrary units).

When applied laser intensity is sufficiently strong, the laser pulse propagating in a rapidly ionizing medium experiences a strong SPM due to the rapid change of refractive index in time [4]. Figure 2 shows the Wigner distribution of laser pulses after propagating 0.7-mm long Ne gas layer. As the laser intensity increases above the saturation intensity, the harmonic generation occurs only in the leading edge of the pulse due to the depletion of atoms. And, moreover, only the positively chirped leading edge of the pulse is important in the high harmonic generation process due to the plasma-induced phase mismatch in the remaining part of pulse. It is seen from Fig. 2 that the amount of the positive chirp (the slope of the Wigner distribution) in the leading edge increases as the laser intensity increases. When this SPM-induced positive chirp becomes large enough, it can overcompensate the dynamically induced negative chirp, changing the overall sign of the harmonic chirp from negative to positive. Therefore, positively chirped harmonics can be generated due to strong SPM.

Generation of high-order harmonics from atoms in an intense laser field is basically a coherent interaction process between the laser field and atoms. Atoms respond to every optical cycle of the laser field, which is directly reflected in the characteristics of generated harmonics. Thus, active manipulation of temporal structure of the laser pulse can lead us to coherently control the high-order harmonic generation process. In other words, the spectral and temporal structure of harmonics can be controlled by manipulating the temporal structure of the driving laser pulse. This property can be applied to achieve the continuous tuning of harmonics without losing spectral sharpness.

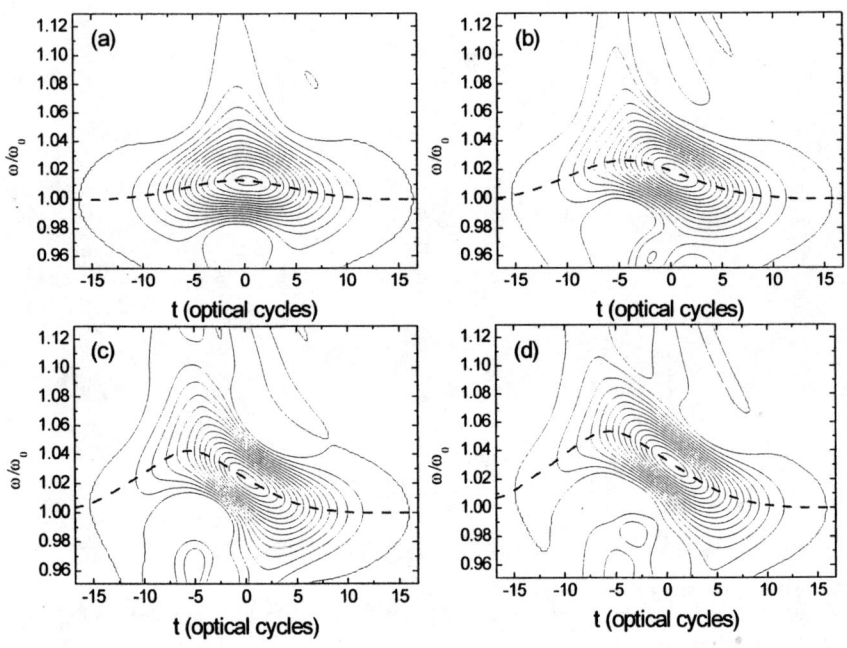

FIGURE 2. Wigner distribution of the 30-fs, 800-nm laser pulse propagating through a 0.7-mm long neon gas of density 1×10^{18} cm^{-3}. The laser intensity is (a) I= 1×10^{15} W/cm^2, (b) I= 3×10^{15} W/cm^2, (c) I= 6×10^{15} W/cm^2, and (d) I = 1×10^{16} W/cm^2.

3. CONTINUOUSLY TUNABLE HARMONICS

Experiments for the continuous tuning of high-order harmonics were performed with a femtosecond terawatt Ti:sapphire laser operating at 10 Hz [5]. The pulse duration and center wavelength were 26 fs and 820 nm, respectively. The pulse duration and amount of laser chirp were measured with a second-harmonic generation frequency resolved optical gating (FROG) method. The laser beam was focused into a gas jet with a nozzle of 0.5-mm diameter. The peak gas density was 3×10^{19} cm^{-3}, and the full width at half maximum of the gas density profile was about 0.6 mm. The generated harmonics were spectrally resolved by a flat-field extreme ultraviolet (XUV) spectrometer, and detected by a back-illumination x-ray CCD. The laser chirp condition was controlled by changing the grating separation in the pulse compressor, and this process does not change the laser spectrum and provides a predictable temporal structure of the final laser output.

A strong blueshift of the high harmonics was generated in the intense femtosecond laser field used in these experiments. The high harmonic spectrum from helium atoms driven by a chirp-free 26-fs laser pulse is shown in Fig. 3. The peak laser intensity was 2×10^{16} W/cm^2, well in excess of the saturation intensity for optical-field ionization of helium under a 26-fs laser pulse (3×10^{15} W/cm^2). Under this intense laser field, helium atoms were rapidly ionized in the leading edge of the laser pulse, allowing strong harmonic generation only in the leading edge. The harmonics emitted in this case were strongly blueshifted. The harmonic of 51st order was blueshifted more than two times the laser frequency, the harmonic of 81st order more than four times the laser frequency, and the harmonic of 101st order more than six times the laser frequency (see Fig. 3). As the harmonic order was increased, the harmonic blueshift became large enough for continuous tuning of harmonic wavelengths over the frequency interval between adjacent harmonics.

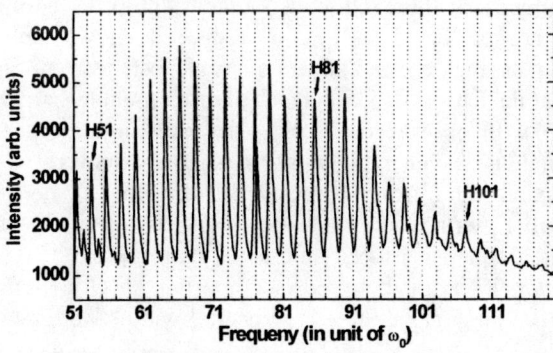

FIGURE 3. Strongly blueshifted harmonic spectrum from helium atoms of density 3×10^{19} cm^{-3} driven by a chirp-free 26-fs laser pulse with an intensity of 2×10^{16} W/cm^2.

Both the nonadiabatic effect [1] and SPM of the driving laser pulse contribute to the blueshift of harmonic wavelengths observed in Fig. 3. The nonadiabatic effect is basically a single atom effect, because it comes from the atomic response to the rapidly rising intense laser field. On the other hand, the blueshift due to SPM is a result of the propagation of the driving laser pulse through the harmonic generation medium. In an ionizing medium the refractive index decreases with time, which causes SPM of the propagating laser pulse. The SPM then causes a blueshift of the laser spectrum and, as a result, the high harmonics are also blueshifted. Because the SPM effect increases with medium density, the adoption of a high-density medium enhances the blueshift as well as increasing the harmonic signal level, facilitating the wavelength tuning of harmonics in a strong signal level.

When laser intensity above the saturation intensity for optical-field ionization is applied to atoms, harmonic emission occurs in the leading edge of the laser pulse before all neutral atoms are ionized, and the nonadiabatic response of the atoms to a rapidly rising laser field results in a wavelength blueshift of the high-order harmonics [1]. This nonadiabatic effect can induce a large blueshift of harmonic wavelengths. However, the control of the laser intensity for the harmonic blueshift was accompanied by a modification of the spectral shape and brightness of harmonics, not a promising feature for realistic applications of harmonic x-rays. To achieve continuous tuning, the harmonic blueshift should be much larger than twice the laser frequency. In addition, the harmonic chirp [2,6,7], dynamically and SPM-induced harmonic chirp, results in a broadening of the harmonic spectrum, which should be appropriately compensated for to achieve sharp harmonics. Consequently, we investigated the wavelength tunability of high harmonic x-rays on the basis of the harmonic blueshift in helium atoms using chirped femtosecond laser pulses.

The wavelength tunability of high-order harmonics was achieved by controlling the incident laser energy and laser chirp. Although adjustment of the grating distance for the control of laser chirp also affected the harmonic wavelengths, a sharp harmonic at a specific wavelength could be generated through iteration of these two steps, achieving simultaneous control of the spectral position and spectral width of the harmonics. Spectral images of the continuously tuned high-order harmonics obtained from helium atoms are presented in Fig. 4. Figure 4 shows that the harmonic wavelengths decrease as the laser parameters are adjusted from Spectrum (I) to Spectrum (V), and that the harmonic frequencies of order greater than or equal to 73 in Spectrum (I) overlap with or exceed those of the next higher-order harmonics in Spectrum (V). Continuous wavelength tuning of sharp harmonics was achieved by changing the laser parameters from a negatively chirped 110-fs, 11-mJ pulse in Spectrum (I) (Fig. 5) to a positively chirped 44-fs, 1.0-mJ pulse in Spectrum (V). Here we tuned the harmonic frequencies such that the 79th harmonics of the 5 spectra in Fig. 4 had equal frequency intervals between neighboring spectra, demonstrating the continuous wavelength tuning capability. The spectral widths were maintained within $0.7\omega_0$ by adjustment of the laser chirp condition. This represents a direct proof that it is possible to generate a given harmonic at a specific wavelength by controlling the incident laser energy and laser chirp, a critical step in the realization of a continuously tunable high harmonic x-ray source.

Wavelength (nm)

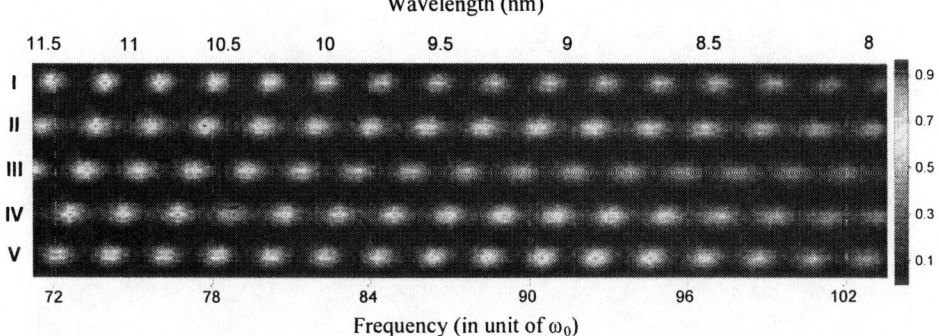

FIGURE 4. Continuously wavelength-tuned high-order harmonics generated from helium atoms driven by intense femtosecond laser pulses. The laser energy and chirped pulse duration were (I) 11 mJ and – 110 fs, (II) 5.9 mJ and -87 fs, (III) 4.1 mJ and -47 fs, (IV) 1.8 mJ and 26 fs, (V) 1.0 mJ and +44 fs, respectively.

4. CONCLUSION

We have shown that high harmonic generation processes could be coherently controlled using chirped femtosecond laser pulses to produce sharp and strong harmonic spectra. At the laser intensity below the saturation intensity of OFI harmonics are negative chirped due to the dynamically induced chirp. In contrast, when the laser intensity exceeds the saturation intensity for OFI, the harmonic chirp can change its sign from negative to positive due to the SPM-induced positive laser chirp. On the basis of this coherent control technique with chirped laser pulses for the generation of sharp and strong harmonic and the nonadiabatic effect, which induces a large blueshift of high-order harmonics, we demonstrated the wavelength tunability of high harmonic x-rays in helium atoms without losing spectral sharpness.

ACKNOWLEDGMENTS

This research was supported by the Ministry of Science and Technology of Korea through the Creative Research Initiative program.

REFERENCES

1. H. J. Shin, D. G. Lee, Y. H. Cha, K. –H. Hong, and C. H. Nam, *Phys. Rev. Lett.* **83**, 25449 (1999).
2 D. G. Lee, J. –H. Kim, K. H. Hong, and C. H. Nam, *Phys. Rev. Lett.* **87**, 243902 (2001).
3. L. Cohen, Proc. IEEE **77**, 941 (1989).
4. C. Kan, N. H. Burnett, C. E. Capjack, and R. Rankin, *Phys. Rev. Lett.* **79**, 2971 (1997).
5. Y. H. Cha, Y. I. Kang, and C. H. Nam, *J. Opt. Soc. Am. B* **16**, 2872 (1999).
6. Z. Chang et al., *Phys. Rev. A* **58**, R30 (1998).
7. J. –H. Kim, H. J. Shin, D. G. Lee, and C. H. Nam, *Phys. Rev. A* **62**, 055402 (2000).

Characterization of Surface Plasma in Surface Harmonic Experiments using Satellite Structures in the Reflected Pump Spectrum

T. Ozaki, K. Oguri, T. Nishikawa and H. Nakano

NTT Basic Research Laboratories, NTT Corporation
3-1 Morinosato Wakamiya, Atsugi, Kanagawa 243-0198, Japan

Abstract. We investigate the spectral structure of high-intensity ultrashort laser pulses reflected from silicon target. The fundamental frequency spectrum involves complicated redshifts and blueshifts due to the laser-plasma interaction. On the other hand, we find that simulations accurately reproduce the observed second harmonic spectrum. Simulation results also infer that the second harmonic spectrum is a good candidate to evaluate the scale length of the preplasma produced on the target surface.

INTRODUCTION

Solid surface harmonics, which are generated as a result of the interaction between high intensity ultra-short laser pulses with solid targets [1,2], offer several characteristics different from gas harmonics. One distinctive feature of this phenomenon is the simultaneous generation of odd and even harmonics. This harmonic generation method is also robust against ionization, since the nonlinear medium itself is plasma. However, works [3] have shown that solid surface harmonics is not fully free from ionization effects. From extensive investigations using simulation and experiment, it has been shown that preplasma produced on the surface of the target can considerably change the conversion efficiency and divergence of the generated harmonics. This preplasma typically has electron density scale length shorter than the wavelength λ of the pump laser, and also changes on the time scale of the pump laser, which in our case is 100 fs. In the present paper, we show that the spectral shape of the reflected pump spectrum can be used to infer the scale length of this preplasma, especially with the second harmonics. We also report the observation of satellitelike peak in the reflected fundamental spectrum, which cannot be explained by simulations using 50 fs pump lasers.

EXPERIMENTAL

55 fs duration Ti:sapphire chirped-pulse amplification laser system (manufactured by New Femto Lasers, Beijing) is used in this work. At a repetition rate of 10 Hz, the system is capable of delivering a maximum energy of 50 mJ per pulse. These pulses

CP641, *X-Ray Lasers 2002: 8th International Conference on X-Ray Lasers,* edited by J. J. Rocca et al.
© 2002 American Institute of Physics 0-7354-0096-2/02/$19.00

are focused onto silicon wafer targets using off-axis paraboloid multi-layer mirror. The focal length of the off-axis paraboloid is 150 mm, and the spot diameter of the focus is measured to be 25μm × 46 μm for 45° incidence angle. Experiments were also performed for incidence angle of 22.5°, but the spot diameter was adjusted so that the peak laser intensity at focus would be the same for the two angles for the same pump energy. We measure the spectrum of the pump laser reflected from the silicon target in the specular direction. The spectrum is monitored by a commercial spectrometer (Ocean Optics Model USB2000), allowing rapid, single-shot data acquisition. The spectral resolution of the spectrometer was measured to be about 2 nm at the fundamental wavelength.

RESULTS

Reflected Fundamental Pump Spectrum

In Fig.1, we show the spectrum of the pump reflected from the target surface, for pump energies between 1.4 mJ and 13.5 mJ. The corresponding intensities of the pump laser are 2.8×10^{15} W cm^{-2} and 2.7×10^{16} W cm^{-2}, respectively. The spectrum of the incident pump laser is also shown in the figure as dotted lines. The results show that for irradiation energy of 1.4 mJ, the reflected pump spectrum closely resembles that for the incident pump. As the pump laser intensity is increased, a satellitelike peak appears near the 800 nm wavelength. Both the intensity and spectral width of this peak increases with increase in the pump intensity. We also notice that the spectrum as a whole is redshifted at high intensities.

Experiments were also performed at a smaller incidence angle of 22.5°. In Fig.2, we compare the reflected pump spectrum observed from experiments using the two incidence angles. The focal spot size was finely adjusted to keep the peak pump intensity constant for the same pump energy. We find that, contrary to the results for 45° incidence angle, those for 22.5° angles does not show any spectral shifts relative to the incoming laser.

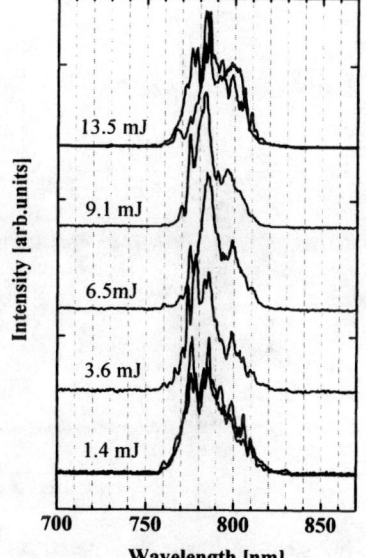

FIGURE 1. Spectrum of reflected pump from silicon target.

Particle-in-cell (PIC) simulations are performed to model the experimental results. One-dimensional relativistic electromagnetic PIC code LPIC++ [4] is used in the present work. The density scale length and temperature of the surface plasma is calculated using the HYADES code [5]. The results of the calculations are shown in

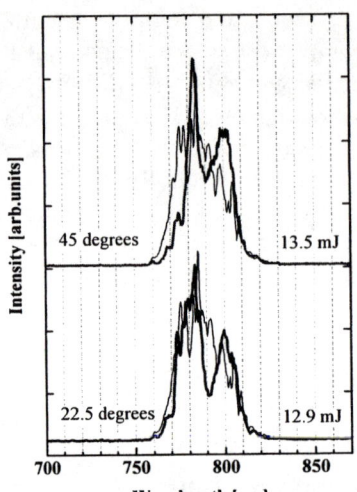

FIGURE 2. Reflected pump spectrum for 45° and 22.5° incidence angles.

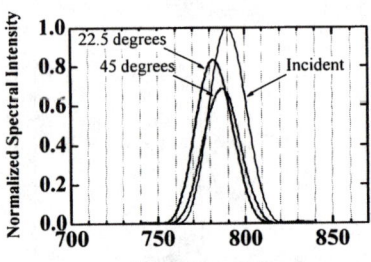

FIGURE 3. PIC simulation results performed with experimental conditions.

Fig.3. Simulation shows that the 22.5° reflected spectrum has a larger blueshift compared with the blueshift for 45°. However, there is a disagreement between experiment and simulation in the spectral positions of the reflected pump spectrum relative to those for the initial pump. The reflected pump spectrum obtained from simulation both shows blueshift compared with the incident pump, where as the experimental results show none or redshift.

The blueshift of the reflected spectrum observed in the simulation is due to the change in the oscillation frequency of the electron in the plasma sheath with oscillation amplitude [6]. For smaller incidence angle, the relative effect of resonance absorption is reduced, and collision absorption is also small for plasma with high electron temperature of several hundred eV. In such cases, free electrons can be driven into the high-density region of the plasma, resulting in anharmonic oscillation of the electrons, and thus the blueshift of the reflected laser spectrum. On the other hand, the redshift of the experimentally observed spectrum compared with simulations is most probably attributed to Doppler shift resulting from the motion of the turning point of the pump laser.

Although the overall position of the reflected pump spectrum can be well explained by the above mechanisms, the peak observed at 800 nm is not reproduced in our simulations. Therefore further work is necessary to identify the cause for these structures, including the possibility of interaction with ion plasma waves [7].

Second Harmonic Spectrum

The second harmonic spectrum generated in the specular direction was also observed using the spectrometer. For this purpose, color glass filters were used to cut the fundamental wavelength, and the second-order diffraction was used to observe the harmonics. The experimentally observed spectrum for an incidence angle of 22.5° is shown in the lower trace of Fig.4, for a peak pump intensity of 2.6×10^{16} W cm^{-2}. The half wavelength of the pump laser peak is at 393 nm, and so the actual second harmonic is blushifted by a wavelength of 3 nm. The spectrum also possesses sharp structures on the long-wavelength side. PIC simulation modeling of the experiment

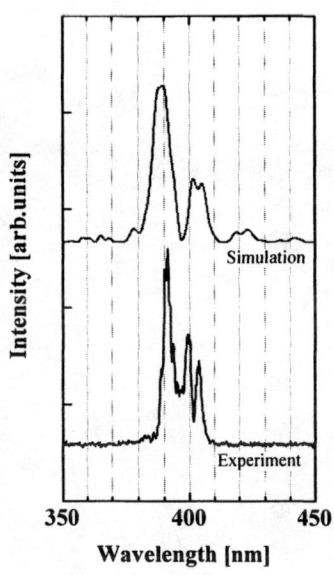

FIGURE 4. Second harmonic spectrum obtained from experiment and simulation.

was performed, assuming surface plasma with density profile with a scale length of λ/4. The calculated spectral profile for the second harmonic is shown in the upper trace of Fig.4. It is found that both blueshift as well as the long wavelength structures are well reproduced in this case, contrary to the results for the fundamental wavelength. Detailed simulations have also shown that the magnitude of the blueshift changes with preplasma scale lengths. These results infer that the second harmonics will be a better candidate for evaluating surface plasma scale lengths.

CONCLUSIONS

We have investigated the characteristics of the high-intensity pump spectrum reflected from silicon target. The fundamental frequency spectrum is blushifted due to anharmonic oscillation of free electrons within the plasma sheath, while simulations infer a redshift. For the second harmonics generated in the specular direction, simulations accurately reproduce the experimentally observed spectrum.

REFERENCES

1. Norreys, P. A., Zepf, M., Moustaizis, S., Fews, A. P., Zhang, J., Lee, P., Bakarezos, M., Danson, C. N., Dyson, A., Gibbon, P., Loukakos, P., Neely, D., Walsh, F. N., Wark, and Dangor, A. E., *Phys. Rev. Lett.*, **76**, 1832-1835 (1996).
2. Zhang, J., Zepf, M., Norreys, P. A., Danger, A. E,, Bakarezos, M., Danson, C. N., Dyson, A., Fews, A. P., Gibbon, P., Key, M. H., Lee, P., Loukakos, P., Moustaizis, S., Neely, D., Walsh, F. N., and Wark, J. S., *Phys. Rev. A*, **54**, 1597-1603 (1996).
3. Zepf, M., Tsakiris, G. D., Pretzler, G., Watts, I., Chambers, D. M., Norreys, P. A., Andiel, U., Dangor, A. E., Eidmann, K., Gahn, C., Machacek, A., Wark, J. S., and Witte, K., *Phys. Rev. E*, **58**, R5253-5256 (1998).
4. Lichters, R., Pfund, R. E., and Meyer-ter-Vehn, J., "LPIC++ : A parallel one-dimensional relativistic electromagnetic particle-in-cell code for simulating laser-plasma-interaction", Smith, C. D., and Jones, E. F., MPQ Reports **225**, Max-Plank Institut fur Quantenoptik, Garching, 1997.
5. Larsen, J. T., and Lane, S. M., *J. Quant. Spectrosc. Radiat. Transf.* **51**, 179-186 (1994).
6. Hansen, C. T., Wilks, S. C., and Young, P. E., *Phys. Rev. Lett.*, **83**, 5019-5022 (1999).
7. Drake, R. P., and Marjoribanks, R. S., *Phys. Plasmas*, **9**, 267-274 (2002).

The role of absolute phase of few-cycle laser field in the generation and measurement of attosecond high-order harmonic pulses

Zhinan Zeng, Ruxin Li and Zhizhan Xu

Laboratory for High Intensity Optics, Shanghai Institute of Optics and Fine Mechanics, Chinese Academy of Sciences, P.O. Box 800-211, Shanghai 201800, China

Abstract. The effect of the absolute phase of the few-cycle driving laser pulse on the generation and measurement of high-order harmonic attosecond pulses is investigated theoretically. We find that the position of the generated attosecond soft-x-ray pulse in cutoff region is locked to the oscillations of the driving laser field, but not to the envelope of the laser pulse. This property ensures the success [M. Hentschel *et al*, Nature vol. 414, 509 (2001)] of the width measurement of attosecond soft-x-ray pulse based on the cross-correlation between the attosecond pulse and its driving laser pulse. However, there is a timing jitter of the order of tens of attoseconds between the attosecond pulse and the driving laser field. This sets a limit for the measurement with the shorter attosecond pulses if we cannot control the absolute phase of the few-cycle laser field. Also, we propose a novel method to detect the absolute phase of the driving laser field by measuring the spatial distribution of the photoelectrons induced by the attosecond soft-x-ray pulse and its driving laser pulse.

INTRODUCTION

The phase-sensitivity of the high-order harmonics for very short driving laser pulse has been investigated theoretically in recent years [1,3]. Whether the harmonics in the cutoff are resolved or not and the different periodicity of spectrum have been discussed under the absolute phase [2] of 0 and $\pi/2$ [1]. In the investigation of the time profile of the harmonics, they find that there is the timing jitter for the attosecond pulses generated by the laser pulses with different absolute phases. In Ref. [3], the authors also believe that the timing jitter and the intensity fluctuations introduced by

CP641, *X-Ray Lasers 2002: 8th International Conference on X-Ray Lasers*, edited by J. J. Rocca et al.

the phase sensitivity will present a severe limitation in the attosecond measurement. Very recently a soft-x-ray pulse of the duration of 650±150as has been generated and measured [4]. With this attosecond pulse, the authors traced the electronic dynamics with a time resolution of ≤150as. In the experiment, they also observed an attosecond synchronism of the soft-x-ray pulse with the carrier wave of the few-cycle laser pulse. They speculated that the observed attosecond timing stability of subfemtosecond x-ray pulse to its few-cycle driver is due either to a robustness of the high-harmonic generation process against the modification of the absolute phase of the driving laser, or to a substantially reduced x-ray yield for values of the phase other than the optimum [4]. They believed that the latter is more likely and ascribed their experimental results to this possibility.

We reinvestigate the effect of the absolute phase of the few-cycle driving laser field on the generation and measurement of the high-order harmonic attosecond pulse. It is found that the position of the generated attosecond soft-x-ray pulse is locked to the oscillation of the driving laser field, but not to the envelope. This property ensures the success of the width measurement of the attosecond soft-x-ray pulse based on the cross-correlation method [4]. However, we find that this method becomes ineffective for the pulse duration down to about 100as due to the timing jitter. We also propose a novel method to detect the absolute phase of the driving laser field by measuring the spatial distribution of the photoelectrons induced by the attosecond soft-x-ray pulse and the driving laser field.

THE EFFECT OF THE ABSOLUTE PHASE ON THE HHG

The approach here is similar to that in the Ref. [1]. We calculate the dipole acceleration by solving numerically the time-dependent Schrödinger equation (TDSE). The laser pulse in our calculation is described by $E(t) = E_0(t)\cos(\omega_L t + \varphi_L)$, where ω_L is the frequency of the laser field and φ_L the absolute phase. Here $E_0(t) = E_0\exp(-t^2/\tau^2)$ denotes the envelope of the laser pulse, where E_0 is the peak amplitude of the laser pulse and $(2\ln2\tau)^{1/2}$ is its duration. The wavelength of the laser pulse is 800nm. The dipole acceleration will then be Fourier transformed to get the spectrum, and a time-frequency analysis (the Gabor analysis [6]) then provides the time profile of the harmonics.

In Fig.1, we show the time profiles of attosecond pulses for different absolute phases, obtained with a laser pulse of 7fs and $3.51×10^{14}$W/cm^2. The atom considered is helium, whose ionization potential is 0.8996a.u. (about 24.5eV). The energy of the soft-x-ray pulse is between 94eV and 100eV. In Fig. 1, the numbers (1-8) on the phase-axis denote the absolute phase of 0 to $7\pi/8$ at an interval of $\pi/8$. First, we can see the same results as shown in Fig. 2 of Ref. [1] that there is only one single emission for $\varphi_L = 0$ (phase = 1), whereas for $\varphi_L = \pi/2$ (phase = 5), the harmonics is emitted twice. Between these two cases, we can see the timing jitter and the harmonic intensity fluctuations as discussed in previous work [1,3].

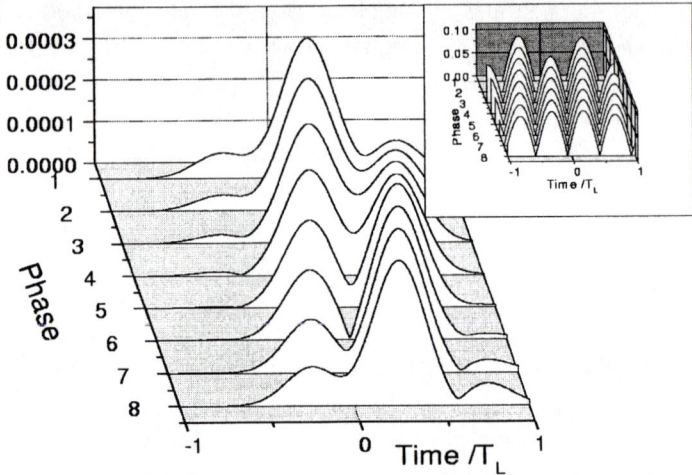

Figure 1. The time profile of the attosecond pulses generated with a laser pulse of 7fs, 800nm and 3.51×10^{14} W/cm^2 with different absolute phases. The inset: $|E(t)|$ of the driving laser.

Comparing Fig. 1 with its inset which shows $|E(t)|$ of the driving laser field, we can see that when the absolute phase of the laser pulse changes, the position of each half-cycle will be shifted. In Fig. 2, we show the absolute position of the attosecond soft-x-ray pulse and the relative position between the attosecond pulse and the laser oscillation as a function of the absolute phase. From Fig. 2(b) we can see that the variation of the relative position with different absolute phase is very small, which means that the attosecond pulse is always generated at almost the same position of each half-cycle of the laser pulse. The variation change of the relative position is less that $0.18T_L$ (T_L is the laser period), or about 48as. This result provides a good explanation for the experiment reported in Ref. [4]. Although the synchronism of the soft-x-ray pulse with the carrier wave of the few-cycle driving laser pulse permit the cross-correlation measurement, if we cannot control the absolute phase of the laser pulse, the randomness of the absolute phase will still limit the attosecond measurement when it is less than 100as.

From Fig.1 we can also see that the relative intensity of the two adjacent soft-x-ray pulses changes with different absolute phase. Fig. 3 shows the intensity ratio of two adjacent soft-x-ray pulses generated around the envelope peak of laser pulse. The logarithms of the intensity ratios are linearly for different absolute phases. This means that the effect of the absolute phase on the intensity ratio is exponential. When $\varphi_L = 0$, the intensity ratio of two adjacent attosecond pulses is very large (more than 10 as shown in the figure). So it appears that there is only one emission in the time profile. When $\varphi_L = \pi/2$, the intensity ratio of two adjacent attosecond pulses is near unity, and two emissions in the time profile appear.

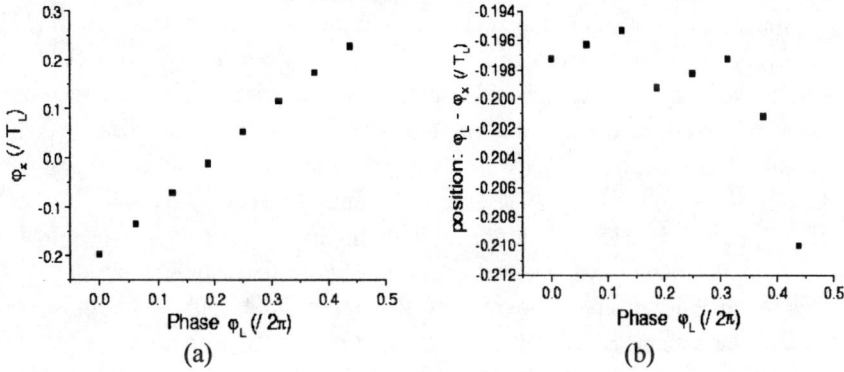

(a)   (b)

Figure 2. The time profile of the attosecond pulse is shown in Fig. 1. Here we show the position of the attosecond pulse. (a) The position of the attosecond pulse as a function of the absolute phase. (b) The relative position of the attosecond pulse and the laser oscillation as a function of the absolute phase.

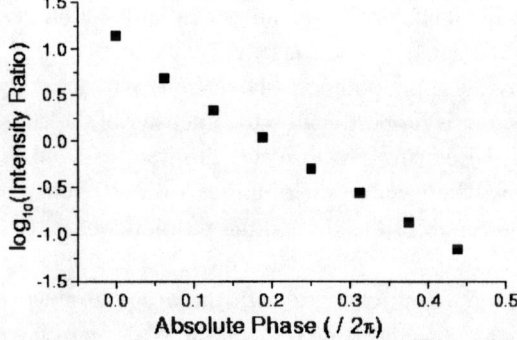

Figure 3. The intensity ratio of two adjacent attosecond pulses (shown in Fig. 1) around the envelope peak of the laser pulse. The logarithms of the intensity ratios are almost in a line.

A SCHEME TO DETECT THE ABSOLUTE PHASE

Since the relative intensity of two adjacent soft-x-ray pulses changes with the absolute phase, so if we produce free electrons using these two adjacent soft-x-ray pulses, the relative yield of the photoelectrons will be decided by the relative intensity of the attosecond pulses and also be decided by the absolute phase of the laser pulse. Based on this fact we propose a method to measure the absolute phase of the laser pulse. After the electron is set free by an x-ray photon, its motion is governed by the laser pulse. We can relate the final value $v_{//,f}$ and the initial value $v_{//,i}$ of the velocity component parallel to the laser polarization with the equation $v_{//,f} = (4U_p(t_d)/m)^{1/2}$ sin $(\omega_L t_d + \varphi_L) + v_{//,i}$ [5] (ω_L and φ_L are defined above. $U_p(t)$, t_d and m are the ponderomotive energy, the time at which the electron is set free and the electron mass as in Ref. [5].). Because the time interval of two adjacent soft-x-ray pulses is $T_L/2$, the electrons produced by these two adjacent attosecond pulses will acquire a velocity component in the opposite direction, e.g., if the final electron velocity in the first pulse is $V_1 = V + V_{1i}$, the final electron velocity in the nearly second one will be almost $V_2 = -V + V_{1i}$ because the time interval is $T_L/2$ (corresponding to π) and $dE_0(t)/dt \ll E_0(t)\omega_L$, even for a 7-fs pulse [5]. Here V_{1i} is the initial velocity of the photoelectron and V denotes the velocity component acquired from the driving laser field. They have the same energy distribution after moving in the laser field, but different spatial distribution. In the scheme shown in Fig. 4(a), we can see that these electrons will come out with different directions. With a microchannel plate (MCP), one can record the position and the density distribution of these electrons for obtaining the absolute phase.

In the simulation, we choose electrons with initial kinetic energy between 74eV and 80eV (The initial kinetic energy of the free electron is decided by $W_0 = mv_i^2/2 = \hbar\omega_x - W_b$ [5], where W_b is the ionization potential energy of the atom. For simplicity, $W_b = 20eV$ is chosen in the calculation.). The laser pulse parameters are as given above. The electron is set free by the x-ray photon instantaneously in an arbitrary direction. The yield ratio of the electron is proportional to the intensity of the attosecond pulse. Then the MCP can detect the electrons with different directions and Fig. 4(b) shows the result. The inset shows the electron distributions with different absolute phases. The position X = 500, the center of the slit and the position where the electron is set free by the x-ray photon were in a line. Fig. 4(b) shows the position of the peak density of the electron distribution which changes with the absolute phase of the laser pulse. When the absolute phase changes from 0 to 2π, the position of the peak density of the electron distribution oscillates around the position X=500 like a cosine function.

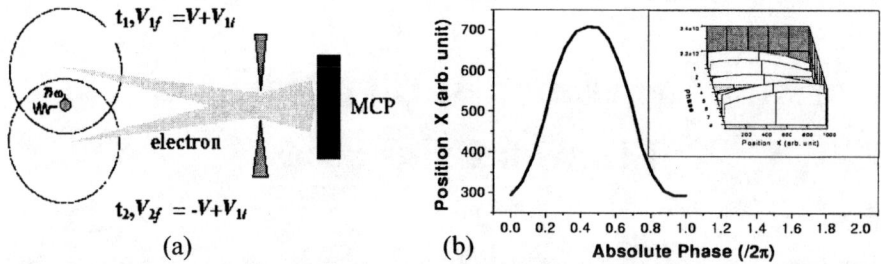

Figure 4. (a) A scheme to measure the absolute phase. (b) The position of the peak density of the electron distribution which changes with the absolute phase of the laser pulse. The inset: 1D Simulation for the electron distribution on the MCP.

CONCLUSION

In conclusion, we performed a detailed analysis for the phase-sensitivity of the high harmonic generation using the few-cycle pulse. We find that the position of the attosecond soft-x-ray pulse is locked to the laser oscillation with a timing jitter of less than 48as which gives a good explanation for the experiment reported in Ref. [4]. The randomness of the absolute phase will still limit the attosecond measurement although it is less than 100as if we cannot control the absolute phase accurately. Also we find that the intensity ratio of two adjacent attosecond pulses changes with the absolute phase as an exponential function. Based on the intensity ratio of two adjacent soft-x-ray pulses that can change with the absolute phase, we propose a method to measure the absolute phase of a driving laser field.

ACKNOWLEDGMENTS

This work is supported by the National Science Foundation of China under Grant Nos. 19974058 and 69925513 and the Chinese National Major Basic Research Project under Grant No. G1999075200.

REFERENCES

1. Bohan, A., *et al*, *Phys. Rev. Lett.* **81**, 1837 (1998).
2. Brebec, T., and Krausz, F., *Rev. Mod. Phys.* **72**, 545 (2000); Paulus, G. G., Grasbon, F., *et al*, *Nature* **414**, 182 (2001)
3. Tempea, G., *et al*, *J. Opt. Soc. Am. B* **16**, 669 (1999).
4. Hentschel, M., *et al*, *Nature* **414**, 509 (2001).
5. Drescher, M., *et al*, *Science* **291**, 1923 (2001).
6. Antoine, P., *et al*, *Phys. Rev. A* **51**, 1750 (1995).

Theory for Generation of High Harmonics in the Electron Exit Process

Jingtao Zhang[1], Wei Yu[1], Zhizhan Xu[1], Xiaofeng Li[2], Panming Fu[2],

DongSheng Guo[3], Richard R. Freeman[4]

[1] Laboratory for High Intensity Optics, Shanghai Institute of Optics & Fine Mechanics,
Chinese Academy of Sciences, Shanghai 201800, China
[2] Laboratory of Optical Physics, Institute of Physics,
Chinese Academy of Sciences, Beijing 100080, China
[3] Dept of Physics, Southern University and A&M College,
Baton Rouge, Louisiana 70813
[4] Dept of Applied Science, University of California-Davis, Davis, California 95616

Abstract. Experiments manifested that plasma interacting with a strong laser field can produce both even and odd high harmonics. Applying a nonperturbative quantum electrodynamics theory, we treat plasma states in a laser field as Volkov states. Our study shows when charged particles exit the single-mode laser field, spontaneous emissions as high harmonics with both even and odd orders occur due to the energy-momentum conservation. In multiphoton ionization (MPI) Volkov states play a role as intermediate states subject to a subsequent transition to plane waves according to the scattering theory of Guo, Aberg, and Crasemann (GAC). We identify the intermediate Volkov states in MPI as plasma states. By applying Gao et al.'s recent formula for MPI, derived from GAC theory with the inclusion of spontaneous emissions, we obtain spectra for high harmonics with both even and odd orders generated in the photoelectron exit process.

INTRODUCTION

When an atom is imposed in an intense laser field, the atomic bound electrons will be excited and finally get rid of the Coulomb potential of the parent ion thus be ionized. Moving in the laser field along with the electric field, the ionized electron finally leave the field thus be totally free, or recombines with its parent core before it free. Accordingly, there are two kinds of harmonics observed in experiments: One only exhibits odd order harmonics, the other shows both even and odd order harmonics. The former was shown in many ATI experiments and is successfully described by

CP641, *X-Ray Lasers 2002: 8th International Conference on X-Ray Lasers*, edited by J. J. Rocca et al.
© 2002 American Institute of Physics 0-7354-0096-2/02/$19.00

recombination model [1-5]. The latter was found in plasma irradiated by laser light, in which the harmonic generation is treated as scattering process [6-8].

In the non-perturbative Quantum Electrodynamics theory, the electron in the intense laser field is described by a quantized Volkov state [9,10]. Gao *et al.* [4,5] studied the HHG in the recombination mechanism and showed the plateau structure as well as the cut-off law. On the other hand, it has also found that [11]: When the ionized electron exits the laser field, the spontaneous emitted mode is required by the energy-momentum conversation. The spontaneous emitted mode has discrete spectrum thus belongs to the harmonics. The main purpose of this paper is to present the numerical results from the transition rate in Ref. [11] and to compare with existing experimental measurements.

TRANSITION FORMULA WITH SPONTANEOUS EMISSION

The transition rate formula for the angular distribution of a given ATI peak, with the inclusion of the spontaneous emission, is given by [11]

$$\frac{d^4W}{d^2\Omega_{P_f}d^2\Omega_{k'}} = \frac{e^2\omega^{9/2}}{(2m_e)^{1/2}(2\pi)^5}(j-u_p-E_b/\omega)^{1/2}(j-u_p)^2$$
$$\sum_q (u_p-j+q)\Phi(P_f-qk+k')X_q(P_f,k')|^2,$$

(1)

where $X_q(P_f,k')$ is defined by

$$X_q(P_f,k') = \frac{1}{\omega}X_{-j}(Z,\eta)\sum_{j'}\frac{1}{u_p-j'}X_{-j'}(Z_f,\eta)$$
$$[-(\vec{P}_f+(j-q-u_p)\vec{k})\cdot\vec{\varepsilon}^{*}x_{q-j+j'}(Z_{k'})$$

(2)

$$+e\Lambda\vec{\varepsilon}^{*}\cdot\vec{\varepsilon}^{*}x_{q-j+j'+1}(Z_{k'})+e\Lambda\vec{\varepsilon}^{*}\cdot\vec{\varepsilon}^{*}x_{q-j+j'-1}(Z_{k'})]$$

in which $X_{-j'}(Z_f,\eta)$ are the generalized phased Bessel functions, and the arguments are given in Ref. [11]. In Eq.(1), j is the transferred-photon number in the ionization process and signifies the order of ATI peaks. The wavefunction $\Phi(P_f-qk+k')$ is the Fourier transform of the initial wavefunction. Each term in the summation over q in Eq.(1) corresponds to the contribution of spontaneous emission with frequency $\omega'=(u_p-j+q)\omega$. The electron is ionized into the laser field by absorbing j photons and exits the field by emitting $j-q$ photons back to the field and converting these photons into a large photon with energy ω' as a spontaneous-emission mode. The spontaneous emission has a discrete spectrum as:

$\Delta\omega' = \omega\Delta(u_p - j + q) = \omega$, thus has the characteristic of harmonics.

NUMERICAL RESULTS AND DISCUSSION

The angular distribution of a given order harmonic is obtained by the integration over the solid angle of the photoelectrons for all of the ATI channels. The emission rate of a given harmonic is given by the integration over the solid angle of the harmonic mode. In our calculations, the wavelength of the incident laser is chosen as 1064-*nm* and the laser is linearly polarized.

Numerical study shows that the emitted photons propagate most along the laser incident direction. The calculated HHG spectra show plateau structure with sharp cut-off at the end of the plateau for high laser intensities. For stronger incident laser fields, the plateau becomes longer and the emission rate is greatly enhanced. Some results can be concluded. First, the HHG spectra exhibit plateau and cutoff structures, which is determined by the characters of the phased Bessel functions, and the cutoff order is the

relation $q_{max} = \varepsilon_b + 2u_p + Z_f$; Each ATI channel has its own emission spectrum, but

with different probabilities. Second, the harmonics propagate mostly along the laser incident direction. Third, the even order harmonics show comparable strength with that of odd harmonics. Figs. 1-2 show the calculated ATI spectra and the corresponding HHG spectra along the laser incident for several laser intensities. Why the even order harmonics are found only in the laser-plasma experiments but not in most ATI experiments? The reasons are multiple. First, the conversion efficiency of this kind harmonic is very low, which generally two or three orders lower than the first kind of harmonics at same laser intensity. Second, this kind of harmonics is generated in the process of the ionized electron exiting the laser field, thus the yield of

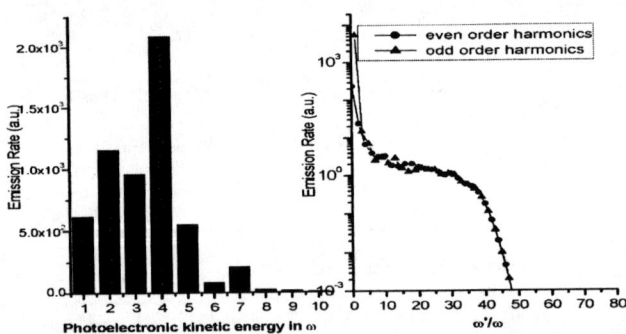

Figure 1. The calculated HHG spectrum along the laser propagation (rigth) and the corresponding photoelectron kinetic enregy distribution (left) for laser intensity 100TW/cm². Other

parameters: $E_b / \omega = 10.4, \xi = 0, \lambda = 1064nm$.

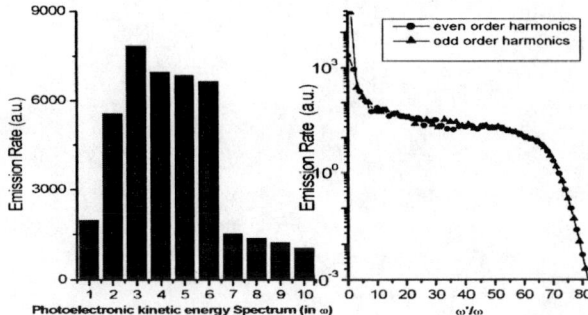

Figure 2. Same as those in Fig.1, but for laser intensity 200TW/cm^2.

harmonic is affected by the total ionization rate. While in most ATI experiments, the ionization is not high, especially for the ultra-short laser pulses. On the other hand, in plasma experiments, two ultra-intense laser beams are used: One is to ionize to the atomic bound electrons, the other is to irradiate the ionized electrons, and the laser intensity is far more than that used here, which means that the ionization is very high, then both odd order and even order harmonics are observed.In Summary: Based on the emission rate formula given by Ref. [11], we calculated the HHG spectra when the ionized electron exits form the laser field. The calculated HHG spectra show that the even order harmonics are of the comparable strength with that of odd order harmonics. This study provides a new mechanism of even-odd harmonics generated in laser-plasma interaction.

ACKNOWLEDGEMENTS

This work is supported by: K.C. Wong Education Foundation, Hong Kong; Chinese National Natural Science Foundation under Grant Nos. 60108002, Chinese National Key Basic Research Foundation under Grant No. G1999075200.

REFERENCES

1. Huillier, A. L., and Balcou P., *Phys. Rev, Lett.*70, 774-7 (1993); *Phys. Rev. A* 48, R3433-7(1993).

2. Becker W, Lohr A., Kleber M., and Lewenstein M., *Phys. Rev. A*, 56, 645-55(1997).

3. Lewenstein M., *et al.*, *Phys. Rev. A* 49, 2117-32 (1994).

4. Gao J., Shen F., and Eden G., *Phys. Rev. Lett.* 81, 1833-6 (1998); *Phys. Rev. A* 61, 043812(21) (2000).

5. Gao L.-H., Li X. F. *et al*, *Chin. Phys. Lett.*, 16, 502-4(1999); *Phys. Rev. A* 61, 063407(5) (2000).

6. Slamin Y. I., Faisal F. H. M., Phys. Rev. A 55, 3964-67(1997); *Phys. Rev. A* 54, 4383-95(1996).

7. Bamber C, *et. al*, *Phys. Rev. Lett.* 76, 3116-9(1996); Lappas G., *et. al*, *Phys. Rev. A* 47, 1327-35(1993).

8. Esarey E., Ride S. K. and Sprangle P., *Phys. Rev. E* 48, 3003-21(1993).

9. Guo D-S. and Aberg T., *J. Phys. A*, 21, 4577-91(1988,); *ibid*, 25, 3383-97(1992).

10. Guo D-S., Åberg T. and Crasemann B., *Phys. Rev. A*, 40, 4997-5005(1989).

11.Gao J., Guo D.-S. and Wu Y. S., *Phys. Rev. A*, 61, 043406(5)(2000).

INTENSE INCOHERENCE X-RAY
AND XUV SOURCES

Electron-based EUV and ultrashort hard-x-ray sources

A. Egbert, B. Mader, B. Tkachenko, and B.N. Chichkov

Laser Zentrum Hannover e.V., Hollerithallee 8, 30419 Hannover, Germany
ch@lzh.de

Abstract. A brief review of our progress in the realization of femtosecond laser-driven ultrashort hard-x-ray sources is given. New results on the development of electron-based compact EUV sources for "at-wavelength" metrology and next generation lithography are presented.
AIP Conference Proceedings

INTRODUCTION

In this paper, electron-induced characteristic emission from solid targets is used for the development of ultrashort hard-x-ray and compact EUV sources.

Ultrashort hard-x-ray source. There is a strong demand for the development of ultrashort hard-x-ray sources for applications in time-resolved diffraction studies, control of chemical reactions, medical imaging, etc. At present, picosecond and sub-picosecond x-ray pulses in the keV energy range are generated by synchrotrons [1, 2] or by focusing high-power (terawatt) femtosecond laser radiation onto solid targets [3-7] or liquid jets [8, 9]. These hard-x-ray sources are very expensive and complex for practical applications. In our work we address an alternative concept allowing to develop a compact, high-repetition rate hard-x-ray source. This concept is based on the combination of low-intensity femtosecond laser pulses with an x-ray diode. The femtosecond laser pulses induce photoemission of electrons from a metal cathode which then are accelerated towards a high-Z target material by applying an external high voltage electric field. Hard-x-rays are produced by Bremsstrahlung of accelerated electrons and characteristic line emission in the target. For the realization of this kind of x-ray source, the laser system requirements are significantly reduced as compared to the currently demonstrated laser-plasma x-ray sources. This work extends the results of Rentzepis and coworkers [10-12], Derenzo et al. [13] and, more recently, of Girardeau-Montaut et al. [14] who have demonstrated x-ray diodes driven with nano- and picosecond laser pulses. The combination of femtosecond pulses with the x-ray diode allows to considerably improve the previous results and to develop efficient, stable, high-repetition rate ultrashort x-ray sources.

EUV source. Extreme ultraviolet (EUV) lithography at 13.5 nm has been internationally accepted as the successor to optical lithography for large-scale semiconductor chip manufacturing. At present, laser-produced plasmas and discharge-produced plasmas are considered as the only practical EUV sources for the implementation of next generation

CP641, *X-Ray Lasers 2002: 8th International Conference on X-Ray Lasers,* edited by J. J. Rocca et al.

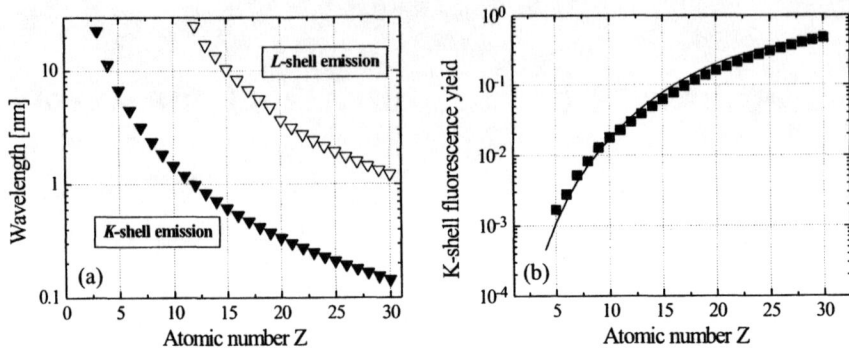

FIGURE 1. Dependencies of K- and L-shell emission wavelengths (a) and K-shell fluorescence yield (b) on the atomic number Z. The solid line in (b) represents a fit corresponding to $Z^4/(Z^4 + \text{const})$.

lithography. Usually, it is overlooked that one can produce EUV radiation directly from solid targets avoiding all plasma-related drawbacks. This can be done by using electron-induced characteristic emission from solids. The EUV source operating on this mechanism can provide an important alternative to the very complex and expensive synchrotron EUV sources and can be applied for small-scale lithographic manufacturing and EUV metrology. This source is developed in this paper and is characterized by a stable and debris-free operation; it is compact, "easy-to-calibrate", and low-cost.

The principle experimental setup for both sources is analogous to a conventional x-ray tube. Electrons are produced by photoemission from a metal cathode or are generated by a tungsten filament and accelerated in a high-voltage electric field towards a solid target (anode). In the hard-x-ray and EUV sources only the target materials are different. In Fig. 1a, dependencies of the wavelengths of the characteristic K- and L-shell emission on the atomic number Z are shown. As can be seen, the K_α line of beryllium ($Z = 4$) and the L-shell radiation of silicon ($Z = 14$) are located at 11.4 nm and 13.5 nm, respectively. It is well-known that for atoms with inner-shell vacancies, the fluorescence yield reduces for lighter elements due to the competition between the radiative and Auger decays. As an example, in Fig. 1b the K-shell fluorescence yield as a function of the atomic number Z is shown [15]. The solid line represents a fit corresponding to $A/(A + W) \simeq Z^4/(Z^4 + \text{const})$, where A and W are the radiative and Auger decay probabilities. In spite that this reduction of the fluorescence yield is disadvantageous for the generation of characteristic emission with Be and Si targets, investigations on this subject are very important for EUV-related applications.

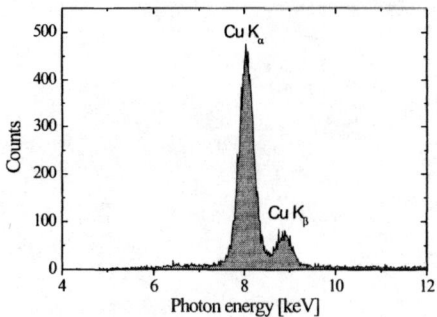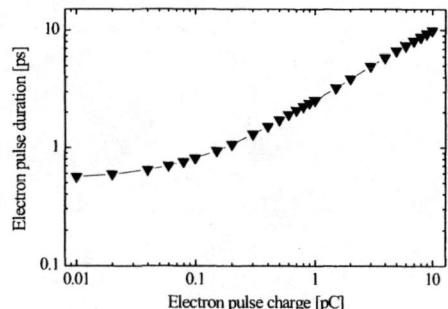

FIGURE 2. Hard-x-ray spectrum obtained with a copper target at an accelerating voltage of 50 kV (left). Calculated electron pulse duration after propagation of 1 cm distance in a 50 kV accelerating field as a function of the total electron charge. The calculations are performed for 100-fs laser pulses and a focal spot size of $A = 5 \times 10^{-4}$ cm^2.

ULTRASHORT HARD-X-RAY SOURCE

In our experiments commercial Ti:sapphire laser systems delivering 120 fs and 160 fs laser pulses at 780 nm with pulse repetition rates of 1 kHz and 250 kHz, respectively, are used to produce photoelectrons from metal cathodes. The laser pulses are focussed onto the surface of a metal cathode to a spot size of $A \simeq 5 \times 10^{-4}$ cm^2. The cathode consists of a quartz substrate coated on the backside with a 100-nm thick metal film. The copper anode (target) has positive potential which can be varied continuously up to 100 kV, whereas the cathode is set to ground potential. The distance between the metal cathode and the anode is 1 cm. The hard-x-ray spectra are recorded using an Si-photodiode-based energy dispersive x-ray detector (AMPTEK, XR-100 CR) operating in a single-photon counting regime. This allows to measure the absolute number of hard-x-ray photons. In Fig. 2(left), a hard-x-ray spectrum obtained with a copper target at an accelerating voltage of $U = 50$ kV is shown.

The evolution of the electron pulse during the propagation from the cathode to the target has been modelled using a space-charge tracking algorithm (ASTRA) [16], which has been developed at DESY (German electron synchrotron facility, Hamburg). The code takes into account pulse broadening due to the initial energy distribution and space-charge effects. In Fig. 2(right), the calculated temporal duration of the electron pulse, after propagation of 1 cm distance, is presented as a function of the total pulse charge. These calculations are performed for 100 fs laser pulses, a laser focal spot size of $A = 5 \times 10^{-4}$ cm^2 at the cathode surface, and an accelerating voltage of $U = 50$ kV. For the total electron pulse charge of $Q \simeq 0.1 - 1$ pC, which corresponds to our experimental conditions, the generated electron and hard-x-ray pulses should have a temporal duration below 5 ps.

A comparison of the x-ray flux corresponding to copper K-lines generated with our laser-driven x-ray diode and that obtained with laser-plasma x-ray sources [4-9] is

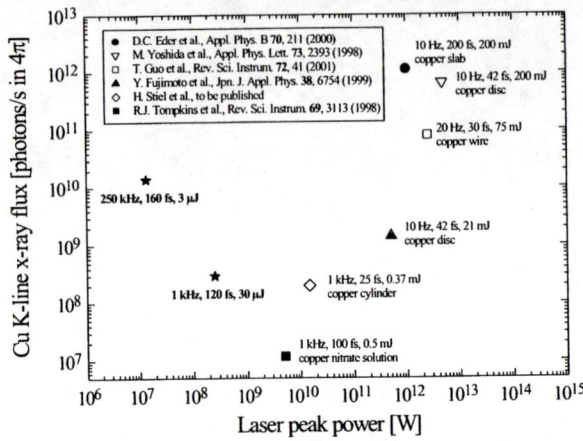

FIGURE 3. Comparison between the x-ray fluxes corresponding to copper K-lines generated in this work (shown by stars) and that obtained with available laser-plasma x-ray sources. Laser parameters, target materials, and references are given in the figure.

illustrated in Fig. 3 as a function of laser peak power. Our results are marked by stars. As can be seen, at a repetition rate of 250 kHz more than 10^{10} hard-x-ray photons/s (emitted in 4π) are generated with a laser pulse energy as small as 3 μJ. These results are comparable with that obtained with high-power laser systems. Note, that using more efficient photocathodes, a significant increase of the x-ray source efficiency and average power can be expected.

COMPACT EUV SOURCE

In the EUV source, the cathode is a tungsten filament surrounded by a Wehnelt cylinder. The electrons released from the filament are accelerated in the high voltage electric field (up to 20 kV) towards Be or Si target. This experimental setup allows to reach a maximum tube current of 3 mA.

For the characterization of the generated radiation, an EUV spectrograph developed and fabricated by JENOPTIK Mikrotechnik GmbH is used. This spectrograph was absolutely calibrated at the Physikalisch Technische Bundesanstalt (PTB) using the BESSY II synchrotron facility. The measured spectrograph sensitivity varies between 0.7 and 1.4 counts per photon in a photon energy range from 70 to 130 eV. In Fig. 4 typical beryllium (a) and silicon (b) spectra are shown.

For a reference, in Fig. 4 calculated reflectivity curves (for 80 layers) for molybdenum/beryllium (a) and molybdenum/silicon (b) multilayer mirrors are also presented.

In Fig. 5, our results on the measurements of the absolute conversion efficiencies

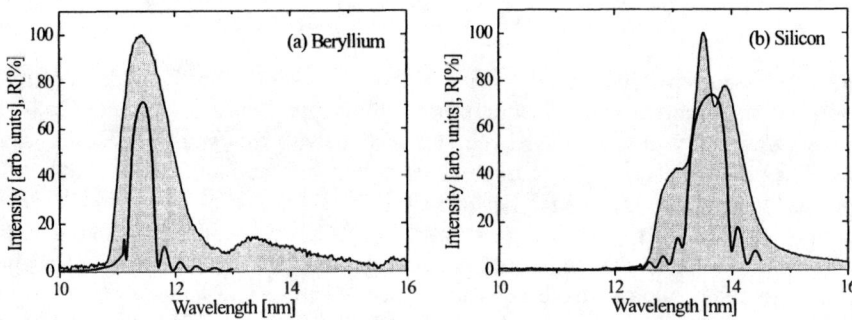

FIGURE 4. Typical EUV spectra obtained with beryllium (a) and silicon (b) targets. Molybdenum/beryllium (a) and molybdenum/silicon (b) multilayer reflectivity curves are shown by solid lines.

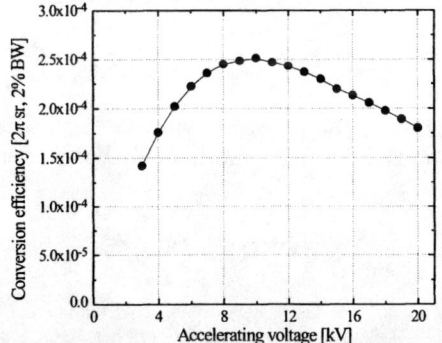

FIGURE 5. Absolute conversion efficiencies from electrons into 13.5 nm EUV photons (emitted in 2π sr and 2% bandwidth) for the silicon target.

for the silicon target are demonstrated. These efficiencies are defined as the ratio of the number of generated EUV photons emitted at 13.5 nm (in 2% bandwidth and a solid angle of 2π sr) to the number of electrons. At an accelerating voltage of 10 kV the conversion efficiency reaches a maximum value of 2.5×10^{-4} and reduces at higher accelerating voltages. At present, the measured average power generated at 13.5 nm (in 2% bandwidth and a solid angle of 2π sr) using 1.5 mA electron current and 10 kV accelerating voltage is 34 μW (or 2×10^{12} photons/s). This is already sufficient for some metrological applications since the radiation is produced without any debris and one can use very small distances between the EUV source and the object.

CONCLUSION

An electron-based femtosecond laser-driven hard-x-ray source has been demonstrated. This source has been realized by combining a high-repetition rate femtosecond laser system with an x-ray diode. It has been shown that with this source picosecond and sub-picosecond hard-x-ray pulses can be generated. A comparison with available laser-plasma hard-x-ray sources has been performed.

Electron-induced EUV radiation from solid targets has been studied using an experimental setup similar to that of an ordinary x-ray tube. Conversion efficiencies from electrons into EUV photons have been measured. A compact, stable, debris-free, cw EUV radiation source at 13.5 nm with a power of 34 μW or 2×10^{12} photons/s (in 2% bandwidth and a solid angle of 2π sr) has been demonstrated. The EUV radiation power can be absolutely calibrated and varied by controlling the electron current and the accelerating voltage. Since no debris are produced, this source (called "EUV tube") promises to become a very important tool for EUV optics characterization and "at-wavelength" metrology.

ACKNOWLEDGMENTS

We are grateful to K. Flöttmann, K. Gäbel, T. Mißalla, P. Nickles, G. Schriever, M.C. Schürmann, U. Stamm, and H. Sticl for support, joint measurements, and discussions.

REFERENCES

1. R. W. Schoenlein, W. P. Leemans, A. H. Chin, P. Volfbeyn, T. E. Glover, P. Balling, M. Zolotorev, K.-J. Kim, S. Chattopadhyay, and C. V. Shank, Science **274**, 236 (1996).
2. R. W. Schoenlein, S.Chattopadhyay, H. H. W. Chong, T. E. Glover, P. A. Heimann, and C. V. Shank, A. A. Zholents, M. S. Zolotorev, Science **287**, 2237 (2000).
3. A. Rousse, C. Rischel, and J.-C. Gauthier, Rev. Mod. Phys. **73**, 17 (2001).
4. T. Guo, Ch. Spielmann, B. C. Walker, and C. P. J. Barty, Rev. Sci. Instrum. **72**, 41 (2001).
5. M. Yoshida, Y. Fujimoto, Y. Hironaka, K. G. Nakamura, K. Kondo, M. Ohtani, and H. Tsunemi, Appl. Phys. Lett. **73**, 2293 (1998).
6. Y. Fujimoto, Y. Hironaka, K. G. Nakamura, K. Kondo, M. Yoshida, M. Ohtani, and H. Tsunemi, Jpn. J. Appl. Phys. **38**, 6754 (1999).
7. D. C. Eder, G. Pretzler, E. Fill, K. Eidmann, and A. Saemann, Appl. Phys. B **70**, 211 (2000).
8. R. J. Tompkins, I. P. Mercer, M. Fettweis, C. J. Barnett, D. R. Klug, Lord G. Porter, I. Clark, S. Jackson, P. Matousek, A. W. Parker, and M. Towrie, Rev. Sci. Instrum. **69**, 3113 (1998).
9. H. Stiel, G. Korn, and A. Thoss, to be published.
10. T. Anderson, I. V. Tomov, and P. M. Rentzepis, J. Appl. Phys. **71**, 5161 (1992).
11. T. Anderson, I. V. Tomov, and P. M. Rentzepis, J. Chem. Phys. **99**, 869-875 (1993).
12. I. V. Tomov, P. Chen, and P. M. Rentzepis, Rev. Sci. Instrum. **66**, 5214-5217 (1995).
13. S. E. Derenzo, W. W. Moses, S. C. Blankespoor, M. Ito, and K. Oba, IEEE Trans. Nucl. Sci. **41**, 629-631 (1994).
14. J.-P. Girardeau-Montaut, B. Kiraly, C. Girardeau-Montaut, and H. Leboutet, Nucl. Instrum. Meth. Phys. Res. A **452**, 361-370 (2000).
15. M. O. Krause, "Atomic Radiative and Radiationless Yields for K and L Shells", J. Phys. Chem. Ref. Data **8**, 307 (1979).
16. K. Flöttmann, http://www.desy.de/~mpyflo.

XUV Emission From Small Fibre Z-Pinch

Daniel Klir, Pavel Kubes, and Jozef Kravarik

Czech Technical University, Technicka 2, 16627 Prague 6, Czech Republic

Abstract. Z-pinch experiments were performed on a small device (3-5 kJ, 20-30 kV, 100 kA, 600 ns) with carbon fibres of 20 and 120 µm diameters and 8 mm in length. XUV radiation was detected in the radial and axial direction using the filtered PIN diodes, the XUV grazing incidence spectrograph, and the VUV pinhole camera. The XUV pulses with FWHM of 10-100 ns were emitted from the plasma corona at 150-300 ns after the current breakdown. These pulses originated from two or three parts of the fibre with approximately 300 µm diameter and 1-2 mm in length. Spectral lines of carbon and oxygen ions were observed with temporal resolution in the 2-25 nm region. The electron temperature 60-80 eV was determined from C V and C VI ions while the electron temperature estimated from O VI ions was 15 eV. The presence of a dense core of the fibre could have an important influence on plasma dynamics and hence on XUV emission.

INTRODUCTION

During the past two decades laser-produced plasma has become the most successful and advanced technique for XUV and soft X-ray lasing [1]. The need to develop more efficient and compact coherent X-ray sources has led plasma physicists to study electrical discharges. Capillary discharges are most commonly used for that purpose [2]. However, the inversion population of levels was observed also in various geometrical configurations of z-pinches employing various pumping mechanisms. The uniformity of lasing medium remains one of the most serious difficulties.

In this paper results from the experiment with a carbon fibre z-pinch are presented. Our experimental results with aluminium wires can be found in [3,4]. The aim of our work is the study of processes in discharge plasma and also the investigation of the population inversion in H- and Li- like ions. The emphasis is put on the stabilization effect and influence of a dense core on plasma evolution and subsequent recombination.

EXPERIMENTAL SET-UP

The experiments were carried out with carbon fibres of 20 and 120 µm diameters and 8 mm in length at a small z-pinch device. The z-pinch capacitor banks of 12 µF were charged up to the voltage of 20-30 kV. The total stored electrical energy was 3-5 kJ; the current peaking at 100 kA and the current rise time being 600 ns.

CP641, *X-Ray Lasers 2002: 8th International Conference on X-Ray Lasers,* edited by J. J. Rocca et al.
© 2002 American Institute of Physics 0-7354-0096-2/02/$19.00

XUV radiation was detected in the radial and axial direction using filtered (Al 0,8 μm) PIN diodes and the XUV grazing incidence spectrograph. Gratings with 600 and 1200 grooves per 1 mm provided a useful spectral range of 2-7 and 7-24 nm. Time integrated spectra were registered on the photographic film UV-4. Temporal resolution was enabled by a four-frame MCP camera with 3-6 ns gating duration and 10 ns frame separation. The MCP intensifier was also used for VUV pinhole imaging of the coronal plasma.

RESULTS

The XUV pulses with FWHM of 10-100 ns and with the total energy reaching 100 mJ (in the Al transmission window and in 4π solid angle) were emitted from the plasma corona randomly at 150-300 ns after the current breakdown. The long-lasting and low-power emission came after this relatively short pulse. With respect to the axis of the fibre, the XUV emission was similar in both longitudinal as well as transverse directions. There was a noticeable drop in the circuit dI/dt just prior to the XUV pulse (See Fig. 1). One of the main differences between the fibres of 20 and 120 μm diameters was the duration of the XUV pulse. In the case of the thinner fibre of 20 μm diameter, the XUV pulse was shorter (FWHM of 10-30 ns) and the total emitted energy was lower.

As shown in Fig. 2, the spectral lines of O V and O VI ions were dominant in the range of 11-24 nm. However, the spectral lines of H- and He-like carbon were also observed. The intensities of O VI corresponded to Boltzmann distribution. The appropriate electron temperature was 10-15 eV. The population of excited states of O V ions, on the other hand, did not fit Boltzmann distribution. It seemed to be due to significant recombination of O VI ions. It can be seen in Fig. 2 (b) that the intensive continuum-like pedestal was produced during the peak of the XUV pulse, while the intensities of the lines did not show any significant changes in comparison with the time before and after the pulse. The origin of the observed continuum has not yet been explained.

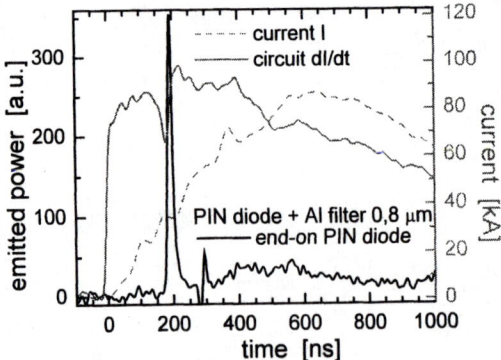

FIGURE 1. Oscilogram, shot no. 010509-3, carbon fibre 120 μm, U=27 kV

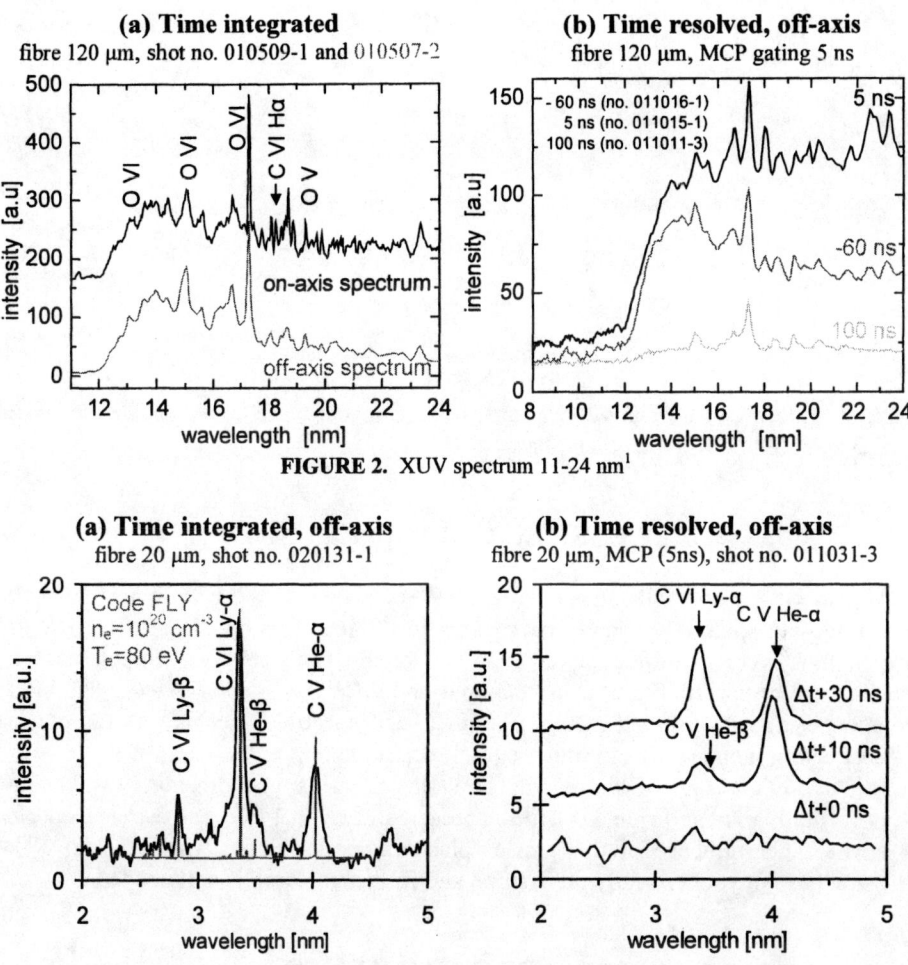

FIGURE 2. XUV spectrum 11-24 nm[1]

FIGURE 3. XUV spectrum 2-5 nm

The spectrum in the spectral region of 2-5 nm contained the lines of H- and He-like carbon ions as displayed in Fig. 3. The emission of the C VI Ly-α line was the most intensive usually few ns before the maximum of XUV emission and it lasted for 20 ns. K-shell spectra were analysed with the aid of non-LTE atomic physics code FLY. The maximum electron temperature and electron density of the "hottest" region were estimated 80 eV and 10^{19}-10^{20} cm^{-3} respectively. The emission in this spectral region was rather "shot-by-shot" dependent.

Figure 4 shows VUV pinhole images. Achieved spatial resolution was 200 μm. The XUV pulses originated from two or three parts of the fibre with ≈300 μm diameter and 1-2 mm in length. The average electron density 10^{21} cm^{-3} was estimated from the Saha

---

[1] The moment of the maximum PIN diode signal is the point of reference (time "zero") in Fig. 2 and 4.

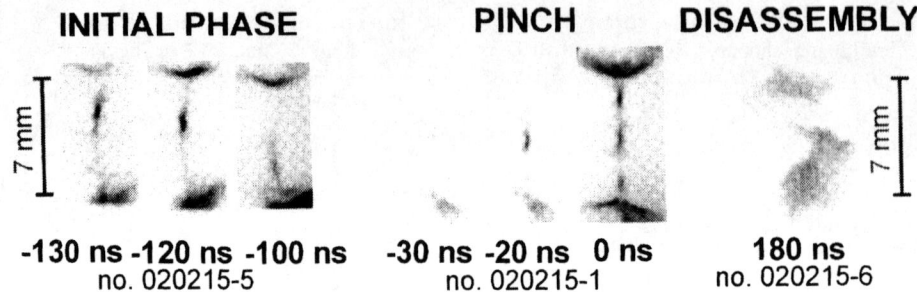

FIGURE 4. Side-on VUV pinhole images

equilibrium between O V and O VI ions. It is clear that there could not be the whole fibre evaporated during the pinch phase.

DISCUSSION AND CONCLUSION

We have experimentally investigated XUV emission of the carbon fibre z-pinch discharge with spatial, temporal and spectral resolution. Spectral lines of carbon and oxygen ions were identified in the 2-25 nm region. The electron temperature and density were estimated. Significant variations and differences between axial and radial radiation were not observed in the time-integrated spectra. It seems that the plasma channel is inhomogeneous and shows strong recombination properties.

The dense core lasted for a few hundred ns. This phenomenon has become experimentally evident [5] and it appears because of an initial electrical current explosion of a wire. The influence of this dense "plasma" on discharge dynamics and XUV emission has not yet been fully understood and further research is still needed.

ACKNOWLEDGMENTS

This research has been supported by the research program No. J04/98:212300017 "Research of Energy Consumption Effectiveness and Quality", the CTU in Prague, by the programs INGO LA055 "Research in Frame of the ICDMP", and LN00A100 "Research Center of Laser Plasma", the Ministry of Education, Czech Republic.

REFERENCES

1. *X-ray Lasers 2000*, edited by G. Jamelot, C. Moller, and A. Klisnick, *J. Phys. IV* **11**-Pr2 (2001).
2. Rocca, J.J. et al., *Phys. Rev. Lett.* **73**, 2192-2195 (1994).
3. Kubes, P., Kravarik, J., and Klir, D., *Czech. J. Phys.* **52**, D127-D132 (2002).
4. Kubes, P. et al., *Plasma Physics Report* **28** (4), 296-302 (2002).
5. Kalantar, D.H., and Hammer, D.A., *Phys. Rev. Lett.* **71**, 3806-3809 (1993).

Enhanced X-ray Emission from Nitrogen Clusters Ionized by Intense, Ultrashort Laser Pulses

T. Mocek[1], J.J. Park[2], C.M. Kim[2], H.T. Kim[2], D.G. Lee[2], K.H. Hong[2], and C.H. Nam[2]

[1]*Department of Gas Lasers/PALS Research Center, Institute of Physics, Na Slovance 2, Prague 8, Czech Republic*
[2]*Department of Physics and Coherent X-Ray Research Center, KAIST, Taejon 305-701, Korea*

Abstract. We report soft X-ray spectra from nitrogen clusters irradiated by 28-fs laser pulses at an intensity of about 7×10^{16} W/cm^2. While the spectrum obtained at room temperature showed entirely transitions from N^{4+} ions, new lines originating from N^{5+} and N^{6+} charge states appeared with cooling. By lowering the pre-expansion gas temperature we have observed strong, nonlinear increase of X-ray emission on lines of N^{4+}, N^{5+}, and N^{6+}. The generation of highly charged ions of N^{5+} and N^{6+} is explained by collisional processes in near-solid-density clusters. A preliminary experiment on the feasibility of a charge-exchange-pumped X-ray laser has been performed in longitudinal geometry.

INTRODUCTION

Recently, there has been much progress in understanding the interaction of intense, femtosecond laser pulses with gaseous clusters which represent an intermediate state of matter between molecules and bulk solids. Experimental studies have shown that clusters are very efficient in absorbing laser energy [1] due to the high local density ($\sim 10^{22}$ cm^{-3}) inside the cluster which results in rapid heating via collisional processes (inverse bremsstrahlung). This mechanism is absent in the case of low-density gas ($< 10^{19}$ cm^{-3}) because the electron-ion collisions are not frequent on the time scale of the laser pulse duration. Efficient generation of X-rays from such compact, high-repetition-rate, debris-free source is highly attractive for potential applications in XUV lithography, microscopy, etc. Cluster formation can be achieved by adiabatic expansion of gas from high-pressure valve into vacuum. The gas becomes supersaturated and solid density droplets bonded by van der Waals forces are formed. The size of clusters can be estimated with the help of an empirical Hagena parameter [2] which scales as $pT^{-2.3}$, where p is the backing pressure, and T is the pre-expansion gas temperature. While majority of published data have been obtained with high pressure valves at room temperature, there are just few experimental results using cooled targets.

The use of clusters as a novel gain medium for soft X-ray lasers has been recently proposed [3] as a method to produce highly charged ions at modest laser intensities. In

CP641, *X-Ray Lasers 2002: 8th International Conference on X-Ray Lasers*, edited by J. J. Rocca et al.
© 2002 American Institute of Physics 0-7354-0096-2/02/$19.00

one particular scheme, a mixture of clusters and atomic/molecular gas of nitrogen has been considered [4]. When the H-like nitrogen ion (produced by collisional ionization of nitrogen cluster) collides with the molecular nitrogen, a state-selective charge exchange process into the $n=5$ level of He-like nitrogen can take place, and a population inversion between the $n=5$ and some of the lower levels might be established.

EXPERIMENTAL SETUP

The experiments were performed at KAIST with a 3-TW Ti:sapphire laser system [5] delivering 28-fs pulses at a fundamental wavelength of 825 nm. The linearly polarized laser beam was focused into a cryogenically cooled gas jet of nitrogen with a f=45 cm spherical mirror, yielding the maximum peak intensity of about 7×10^{16} W/cm^2. The gas jet was produced by a solenoid-driven pulsed valve fitted with a 0.3 mm diameter circular nozzle. The laser illuminated the gas jet perpendicularly with respect to the flow of the gas and the laser focus was typically placed at 0.2 mm (adjustable up to 2 mm) above the nozzle tip. The nitrogen gas was cooled before expansion by passing through a reservoir filled with cold nitrogen gas. At temperatures around -100 $^\circ$C we have observed a drop of backing pressure in the valve indicating that liquid droplets rather than clusters have been formed under these conditions. The valve was operated at a backing pressure of 15 bar. The time-integrated soft X-ray emission was measured simultaneously in two directions using flat-field XUV spectrometers [6]. The transverse spectrometer was positioned to view the plasma perpendicularly with respect to the driving laser, while the longitudinal spectrometer was aligned along the laser axis and a 0.4-μm-thick aluminum filter was placed in front of CCD detector to block the stray laser light.

RESULTS

Figure 1(a) shows transverse spectrum of nitrogen obtained at the room temperature. The signal was integrated over 10 laser shots. In Fig. 1(a) the spectrum is dominated by Li-like (N^{4+}) lines. The applied laser intensity of ~7×10^{16} W/cm^2 is sufficient to strip neutral atoms up to He-like (N^{5+}) ions by the optical field ionization (OFI) since the threshold intensity to produce N^{5+} by OFI is ~3×10^{16} W/cm^2. As the polarization of the driving laser is linear, average electron temperature of about 26 eV is estimated. Such a cold OFI plasma is dominated by three-body recombination processes feeding electrons into the upper excited levels of the next lower ionization state N^{4+}. Therefore the X-ray spectrum in Fig. 1(a) shows mainly lines belonging to transitions in N^{4+}. By contrast, Fig. 1(b) obtained at -85 $^\circ$C shows a very different spectrum. The signal level of X-ray line emission significantly increased and the spectral distribution has changed in such a way that the lines belonging to N^{5+} and even N^{6+} (H-like) appeared in the spectrum. Note that the threshold intensity to generate N^{6+} ions by OFI is ~8×10^{18} W/cm^2, which is about two orders of magnitude higher than in our experiment. The generation of highly charged ions clearly indicates that high electron temperature was achieved and the plasma parameters were fairly different from the previous case (Fig. 1(a)) of cold recombining OFI plasma. The change in X-ray spectrum is a clear evidence of collisional processes taking place in near-solid-density nitrogen clusters.

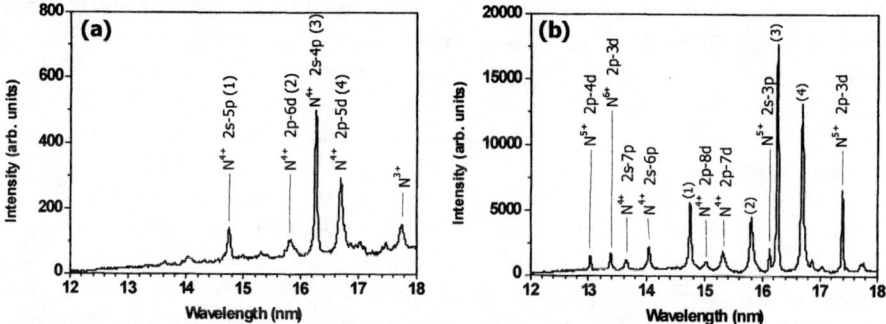

FIGURE 1. Transverse spectrum from nitrogen (p = 15 bar) irradiated by 28-fs laser pulses (10 shots) at (a) room temperature of +23 °C compared with the spectrum at (b) -85 °C.

The role of cluster size in the interaction has been investigated through a temperature scan by measuring the dependency of X-ray line emission on the pre-expansion gas temperature. Figure 2 shows the relative peak intensity values for three selected radiative transitions belonging to different charge states of nitrogen: 2p-5d of N^{4+} at 16.69 nm, 2p-3d of N^{5+} at 17.39, and 2s-3p/2p-3d of N^{6+} at 13.38 nm, as a function of temperature. In all cases the X-ray yield rapidly increases with cooling, and finally saturates at a temperature which exhibits dependency on the ionization state. From the experimental data we inferred that the emission on the N^{4+} line saturates around -85 °C being ~50 times stronger than that at the room temperature. A remarkable feature in Fig. 2 is the stepwise appearance of subsequent charge states. At the room temperature the highest charge state in the spectrum is N^{4+}, however, as the temperature is lowered down to –25 °C, lines belonging to the next higher stage N^{5+} first appear and start to grow. Finally, when the temperature reached about –50 °C, the N^{6+} line appeared.

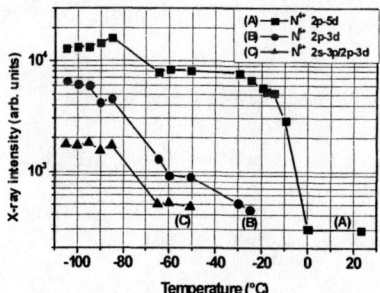

FIGURE 2. Measured peak intensity of the N^{4+} 2p-5d line at 16.69 nm (A), N^{5+} 2p-3d line at 17.39 nm (B), and N^{6+} 2s-3p/2p-3d line at 13.38 nm (C), respectively, as a function of pre-expansion gas temperature.

Spatially resolved spectra obtained at low temperatures also revealed that the size of plasma increased. The experimental results are qualitatively similar to our previous results on Ar clusters [7] and can be well explained using the hydrodynamic model

[8]. While at the room temperature the target consists of pure molecular gas, by decreasing the pre-expansion temperature nitrogen clusters are formed in the jet (lower temperature = larger clusters). The high local density within the cluster makes collisional ionization dominant process. Enhanced collisional heating results in generation of highly ionized hot plasma which undergoes expansion followed by a long-lasting X-ray emission from an underdense plasma.

To check the feasibility of a charge-exchange-pumped soft X-ray laser, we have measured axial spectra under various experimental conditions. The signal observed at room temperature was on the noise level only. The first lines belonging to N^{4+} appeared around $-35\ ^{\circ}C$, and finally the spectrum at $-115\ ^{\circ}C$ showed strong signal on the N^{5+} 2p-3d line due to clusters. As suggested in [4] we were looking for stimulated emission on the n=5-3 transition in N^{5+}, however none of the candidate lines (35.55 nm and 35.59 nm) has been found in the spectrum. An important issue is the high density of the medium which could be responsible for large refraction. Therefore it is planned to improve the alignment system together with the detection part to make it possible to scan precisely the angle under which is the driving laser beam focused on the target.

CONCLUSION

In conclusion, we have shown that soft X-ray emission from a cryogenically cooled nitrogen jet irradiated by intense, femtosecond laser pulse exhibits features characteristic for highly ionized, dense cluster plasma dominated by collisional processes. We have observed a nonlinear increase of soft X-ray emission on lines from N^{4+}, N^{5+}, and N^{6+} charge states and attributed these phenomena to efficient collisional heating of growing nitrogen clusters.

ACKNOWLEDGMENTS

This research was supported by the Ministry of Science and Technology of Korea through the Creative Research Initiative Program. The visit of T. Mocek to KAIST was supported by the Korea Science and Engineering Foundation and by the Czech Ministry of Education, Youth and Sports through the Laser Plasma Research Center (project LN00A100).

REFERENCES

1. Ditmire, T., Smith, R.A., Tisch, J.W.G., and Hutchinson, M.H.R., *Phys. Rev. Lett.* **78**, 3121-3124 (1997).
2. Wörmer, J., Guzielski, V., Stapelfeldt, J., and Möller, T., *Chem. Phys. Lett.* **159**, 321-326 (1989).
3. Chichkov, B.N., Egbert, A., Meyer, S., Wellegehausen, B., Aschke, L., Kunze, H.J., and Kato, Y., *Jpn. J. Appl. Phys.* **38**, 1975-1978 (1999).
4. Sagisaka, A., Honda, H., Kondo, K., Suzuki, H., Nagashima, K., Kawachi, T., Nagashima, A., Kato, Y., and Takuma, H., *Appl. Phys. B* **70**, 549-554 (2000).
5. Cha, Y.H., Hong, K.H., and Nam, C.H., *Opt. Commun.* **185**, 413-418 (2000).
6. Choi, I.W., Lee, J.U., and Nam, C.H., *Appl. Opt.* **36**, 1457-1466 (1997).
7. Mocek, T., Kim, C.M., Shin, H.J., Lee, D.G., Cha, Y.H., Hong, K.H., and Nam, C.H., *Appl. Phys. Lett.* **76**, 1819-1821 (2000).
8. Ditmire, T., Donnelly, T., Rubenchik, A.M., Falcone, R.W., and Perry, M.D., *Phys. Rev. A* **53**, 3379-3402 (1996).

X-Ray Generation Enhancement Mechanism from Femtosecond-Laser-Produced Plasma with Nanostructure-Array Targets

Tadashi Nishikawa[*][1], Hidetoshi Nakano[*], Naoshi Uesugi[†],
Masashi Nakao[¶], Kazuyuki Nishio[§], and Hideki Masuda[§]

[*]NTT Basic Research Laboratories and [¶]NTT Photonics Laboratories, NTT Corporation,
3-1, Morinosato Wakamiya, Atsugi-shi, Kanagawa 243-0198, Japan
[†]Tohoku Institute of Tech., 35-1 Yagiyama Kasumicho Taihaku-ku, Sendai-shi, Miyagi 982-8577, Japan
[§]Tokyo Metropolitan University, 1-1 Minamiosawa, Hachioji-shi, Tokyo 192-0397, Japan

Abstract. Nanostructure-array targets are very attractive for enhancing x-ray intensity generated from femtosecond-laser-produced-plasma. A 40-fold soft x-ray fluence enhancement and a 9-fold pulse peak intensity enhancement is obtained from the 500-nm interval nanohole-array target when it's nanohole diameter is 450 nm. The relatively short x-ray pulse duration of 19 ps can be kept due to the target structure having high local density and nanometer-sized spaces.

INTRODUCTION

Nanostructured targets are attractive for enhancing x-ray intensity generated from laser-produced-plasma. Efficient hot electron, fast ion, and thermonuclear neutron production with moderate laser intensity has also been reported. Several types of nanostructured target have been reported [1]. These include grating, colloidal, porous, and velvet targets. In hard x-ray energy regions, large x-ray intensity enhancement of one or two orders of magnitude has been achieved while keeping the x-ray pulse duration short. But in the soft x-ray energy regions, the attained x-ray conversion efficiency enhancement has been small or x-ray pulse duration expands too much.

In order to overcome this problem, we adopted a nanohole-array structured target [2]. The structure was made by utilizing the self-organization process of the anodic oxidation of an aluminum plate [3]. The new features of this target are that the structure is aligned to the laser incident direction, a large nanostructure depth, and a high local density. Large x-ray fluence enhancement of around 30-fold was achieved by the nanohole-array alumina target over the entire soft x-ray wavelength region from 5 to 20 nm. The x-ray pulse duration was 17 ps. Another feature of this target is the controllability of the nanohole diameter and the nanowall thickness. Therefore, by measuring the targets with various nanohole-array size, the origin of the x-ray generation enhancement can be investigated and guidelines for designing nanostructured targets can be established.

---

[1] E-mail: nisikawa@will.brl.ntt.co.jp

CP641, *X-Ray Lasers 2002: 8th International Conference on X-Ray Lasers*, edited by J. J. Rocca et al.
© 2002 American Institute of Physics 0-7354-0096-2/02/$19.00

EXPERIMENT

The nanohole interval is determined by the applied voltage used for anodization. However, in order to achieve good self-ordering, a specific solution must be selected. For 25, 40, and 200 applied voltages, sulfuric acid, oxalic acid, and phosphoric acid solution were selected respectively, and 63, 100, and 500 nm nanohole interval targets could be made [3]. The nanohole layer depth was around 40 μm. The nanohole diameters after the anodization are 25, 35, and 100 nm, respectively. These diameters can be widened by dipping the target in phosphoric acid solution after anodization. Therefore, the nanohole-array target with different nanohole diameter and different nanowall thickness can be made. The target was mounted on an *xyz* translation stage in a vacuum chamber and was rastered to expose a fresh surface at each laser shot. Laser pulses from a titanium-sapphire-laser-based amplifier system were focused on the target by using a lens with a focal length of 200 mm at normal incidence. The incident laser pulse duration was 100 fs and the wavelength was 790 nm. The laser beam spot diameter on the target was 30 μm and the peak laser intensity was 1.4×10^{16} W/cm^2. The soft x-ray pulse shape was measured at each single laser shot with an x-ray streak camera mounted at a 45 degree angle to the target normal. The measured x-ray energy region after the gold toroidal mirror was from 0.06 to 0.2 keV and the time resolution was 3 ps.

RESULTS

Figure 1 shows the pore-widening etching time dependence of target's surface structure and soft x-ray pulse shape for the 100 nm interval nanohole-array target. The solid line shows the soft x-ray pulse shape generated from the anodic nanohole alumina target and the dotted line shows that from a conventional plane aluminum plate. Without the pore widening process and at the nanohole diameter of 35 nm, only less than 2-fold soft x-ray fluence enhancement was obtained. However, by dipping the target in phosphoric acid solution for 82 minutes and widening the nanohole diameter to 90 nm, a 10-fold soft x-ray fluence enhancement and a 4-fold pulse peak intensity enhancement could obtained. Between 0 and 82 minutes, soft x-ray pulse duration is also slightly expanded from 6 to 11 ps. When the etching time increases further, the nanohole-array structure destructs and spaces larger than one-micrometer are formed. In this case, the generated x-ray pulse becomes a double-pulse shape structure. The time interval between the double x-ray pulse is proportional to the size of the space. This indicates that the secondary x-ray pulse component is generated from the colliding plasma at spaces surrounded by the nanowalls. We believe that this plasma collision process also contributes to enhance x-ray generation when the nanohole-array structure is preserved.

Figure 2 shows the etching time and nanohole diameter dependence of soft x-ray pulse shape for the 500 nm interval nanohole-array target. Just after the anodization and when the nanohole diameter was 100 nm, less than 2-fold soft x-ray fluence enhancement was obtained. However, by widening the nanohole diameter from 100 nm to 450 nm, the soft x-ray pulse peak intensity and the soft x-ray fluence increases. The soft x-ray pulse duration is also expanded as the nanohole diameter increases.

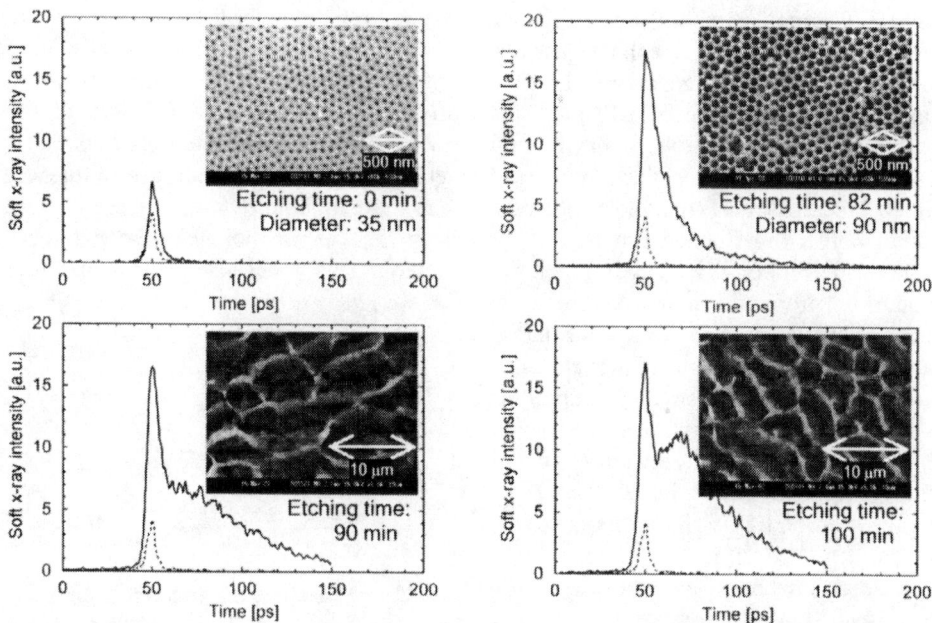

FIGURE 1. Etching time dependence of target's surface structure (scanning electron microscope image) and soft x-ray pulse shape for the 100 nm interval nanohole-array target.

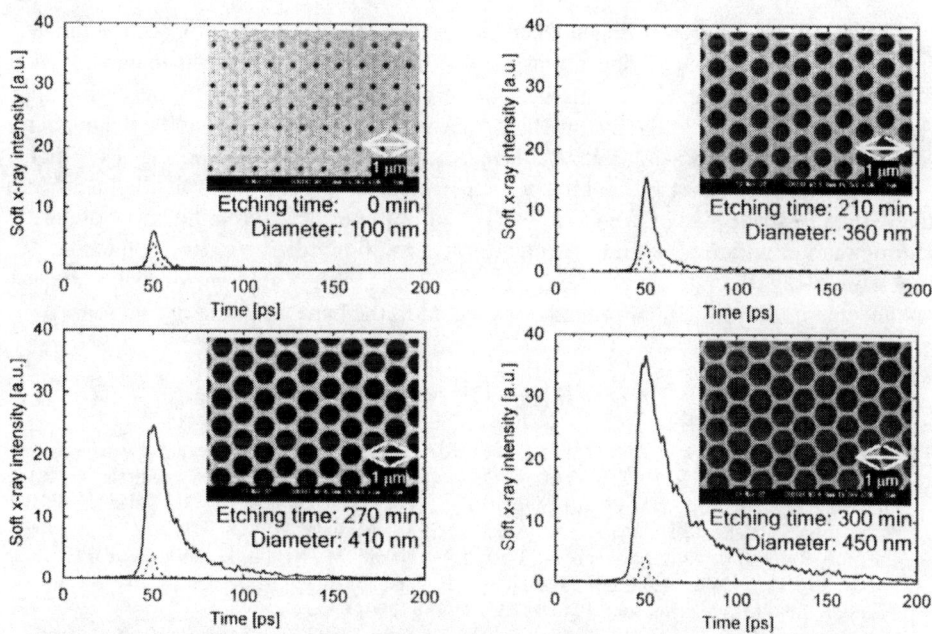

FIGURE 2. Etching time dependence of target's surface structure (scanning electron microscope image) and soft x-ray pulse shape for the 500 nm interval nanohole-array target.

After 300 minutes etching time and when the nanohole diameter becomes 450 nm, a 40-fold soft x-ray fluence enhancement and a 9-fold pulse peak intensity enhancement can be obtained compared with the plane surface aluminum plate. The soft x-ray pulse duration in this case is 19 ps. The relatively short x-ray pulse duration can be kept due to the target structure having high local density and nanometer-sized spaces.

Figure 3 summarizes the nanohole diameter dependence of soft x-ray fluence enhancement for each nanohole-interval. When the nanohole diameter is optimized, higher soft x-ray fluence can be obtained from the lager nanohole interval target, which can accommodate larger nanohole diameters. This is because plasma collision process is promoted at the nanometer-sized spaces. However, even when the nanohole diameter of the 500-nm interval target is larger than that of the smaller interval target, the soft x-ray intensity is smaller, provided that the nanohole wall thickness (nanohole interval - diameter) is larger than the skin depth of around 100 nm. This is because a nanostructure wall that is thicker than the skin depth prevents full interaction with the femtosecond laser pulse.

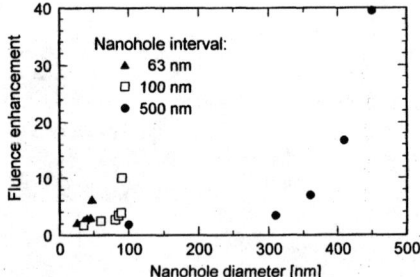

FIGURE 3. Nanohole diameter dependence of soft x-ray fluence enhancement.

CONCLUSION

The nanohole-array size dependence of soft x-ray pulse generation characteristics from femtosecond-laser-produced plasma was investigated. A nanostructure wall that is thinner than the skin depth enlarges the region of interaction with the laser pulse. And plasma collisions at the nanometer spaces cause the x-ray emission enhancement. We found that a large x-ray intensity can be obtained while keeping the x-ray pulse duration relatively short by making a nanostructured target with wall thickness less than 100 nm, space size of around a few 100 nm. In our case, with the nanohole-array alumina target with a 500 nm hole interval and a 450 nm diameter, we obtained a 40-fold soft x-ray fluence enhancement and a 9-fold soft x-ray pulse peak intensity enhancement. Further enhancement is expected in the hard x-ray energy region.

REFERENCES

1. Murnane, M.M. et al., *Appl. Phys. Lett.* **62**, 1068-1070 (1993); Nishikawa, T. et al., *Appl. Phys. Lett.* **70**, 1653-1655 (1997); Volkov, R.V. et al., *Quantum Electron.* **28**, 1-2 (1998); Kulcsár, G. et al., *Phys. Rev. Lett.* **84**, 5149-5152 (2000); Volkov, R.V. et al., *Quantum Electron.* 31, 241-246 (2001).
2. Nishikawa, T., Nakano, H., Uesugi, N., Nakao, M., and Masuda, H., *Appl. Phys. Lett.* **75**, 4079-4081 (1999); Nishikawa, T., Nakano, H., Uesugi, N., Nakao, M., Nishio, K., and Masuda, H., *J. de Physique IV* **11**, Pr2 425-428 (2001); Nishikawa, T., Nakano, H., Oguri, K., Uesugi, N., Nakao, M., Nishio, K., and Masuda, H., *Appl. Phys. B* **73**, 185-188 (2001).
3. Masuda, H., and Satoh, M., *Jpn. J. Appl. Phys.* 35, L126-L129 (1996); Masuda, H., Hasegawa, F., and Ono, S., *J. Electrochem. Soc.* **144**, L127-L130 (1997); Masuda, H., Yamada, H., Satoh, M., Asoh, H., Nakao, M., and Tamamura, T., *Appl. Phys. Lett.* 71, 2770-2772 (1997); Masuda, H., Yada, K., and Osaka, A., *Jpn. J. Appl. Phys.* 37, L1340-L1342 (1998).

APPLICATIONS OF X-RAY LASERS
AND OTHER BRIGHT X-RAY SOURCES

Soft X-Ray Microscopy and EUV Lithography: An Update on Imaging at 20–40 nm Spatial Resolution

D. Attwood, E. Anderson, G. Denbeaux, K. Goldberg,
P. Naulleau, and G. Schneider

Center for X-Ray Optics, Lawrence Berkeley National Laboratory, Berkeley, CA 94720 USA

Abstract. Major advances in both soft x-ray microscopy, at wavelengths from 0.6 to 4 nm, and EUV lithography, at wavelengths between 13 and 14 nm, are reviewed. In the XRL-2000 proceedings[1] we reported soft x-ray microscopy resolved to 25 nm, in static two-dimensional imaging, with applications to biology, magnetic materials, and various "wet" environmental samples. In this 2002 update we report significant extensions to three-dimensional tomographic imaging,[2] dynamical studies of magnetic[3] and electronic devices,[4] and static two-dimensional microscopy poised for extension to below 20 nm spatial resolution.[5] In the XRL-2000 proceedings we reported EUV lithographic imaging of 50 nm lines/100 nm spaces in static microfield (~100 μm) exposures. In this 2002 update we report scanned full-field (25 mm by 32 mm) images at better than 100 nm lines/100 nm spaces,[6] static microfield exposures down to 50 nm lines/50 nm spaces, and isolated lines to 39 nm wide at 0.1 NA.[7] With soon to be available 0.3 NA optics,[8] we expect to print isolated lines, in static micro exposures, at 16–20 nm width in 2003. These results will demonstrate EUV lithography's ability to meet not only the ITRS Roadmap[9] 45 nm node (26 nm isolated lines in resist) in 2007, but also the 32 nm node (18 nm isolated lines in resist) in 2009, both of which the semiconductor industry is now preparing for.[10]

1. SOFT X-RAY MICROSCOPY

High resolution soft x-ray microscopy is based on the use of Fresnel Zone plate lenses with narrow, accurately placed outer zones,[11] and a properly aligned and illuminated microscope. In the XRL-2000 proceedings we reported on state-of-the-art zone plates of 25 nm outer zone width (Δr), of 25 nm thick nickel zones (1:1 aspect ratio),[5] achieving a record 23 nm spatial resolution, albeit with modest diffraction efficiency. Using new nanofabrication techniques for both zone plate lenses and test patterns, soon to be reported results[12] will show clearly resolved images to 20 nm lines and spaces, with expectations to move below 20 nm in the coming months.

While pushing ahead with improved spatial resolution in two-dimensional soft x-ray imaging, great progress has also been made with three-dimensional (3D) micro-tomography of cryofixed biological materials[2,13] and, separately, two-dimensional (2D) dynamical studies for the physical sciences.[3,4] The soft x-ray microtomography

CP641, *X-Ray Lasers 2002: 8th International Conference on X-Ray Lasers,* edited by J. J. Rocca et al.
2002 American Institute of Physics 0-7354-0096-2

technique was developed by the Göttingen group[14] in Germany, and has recently been extended to further applications in Berkeley.[2,13,15] Figure 1 illustrates the use of a rotating capillary tube, containing the sample under study, to affect micro-tomographical imaging in the soft x-ray microscope. To accommodate the need for greater working distance, increased depth of focus, and improved efficiency special zone plate lenses with somewhat larger onto zone width ($\Delta t = 40$ nm) and thickness are used, with a concomitant but modest loss of resolution. Figure 2 shows 2D slices from a full 3D tomographic image of a single cryofixed drosophila melanogaster cell. At the conference this was shown as a 3D rotating image of the cell.*

Dynamical studies using soft x-ray microscopy have also been introduced in the past year, to date for applications in the physical sciences. These studies include observations of magnetic materials switching, by in-plane applied magnetic fields at a

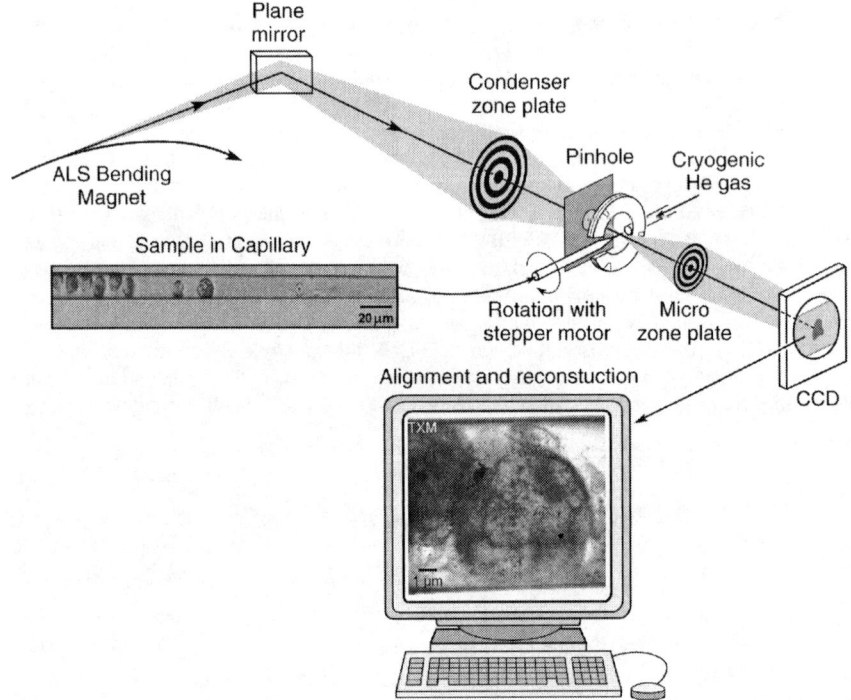

FIGURE 1. Soft x-ray tomography at high spatial resolution obtained with a rotating sample holder in the full field microscope at LBNL's Advanced Light Source. Rotating 3D images of the cell were shown at the conference. Courtesy of Dr. Gerd Schneider.[2]

---

* Contact Dr. Gerd Schneider, Center for X-Ray Optics, Lawrence Berkeley National Laboratory (GRSchneider@lbl.gov) regarding the CD movie clip displaying the full tomographic data.

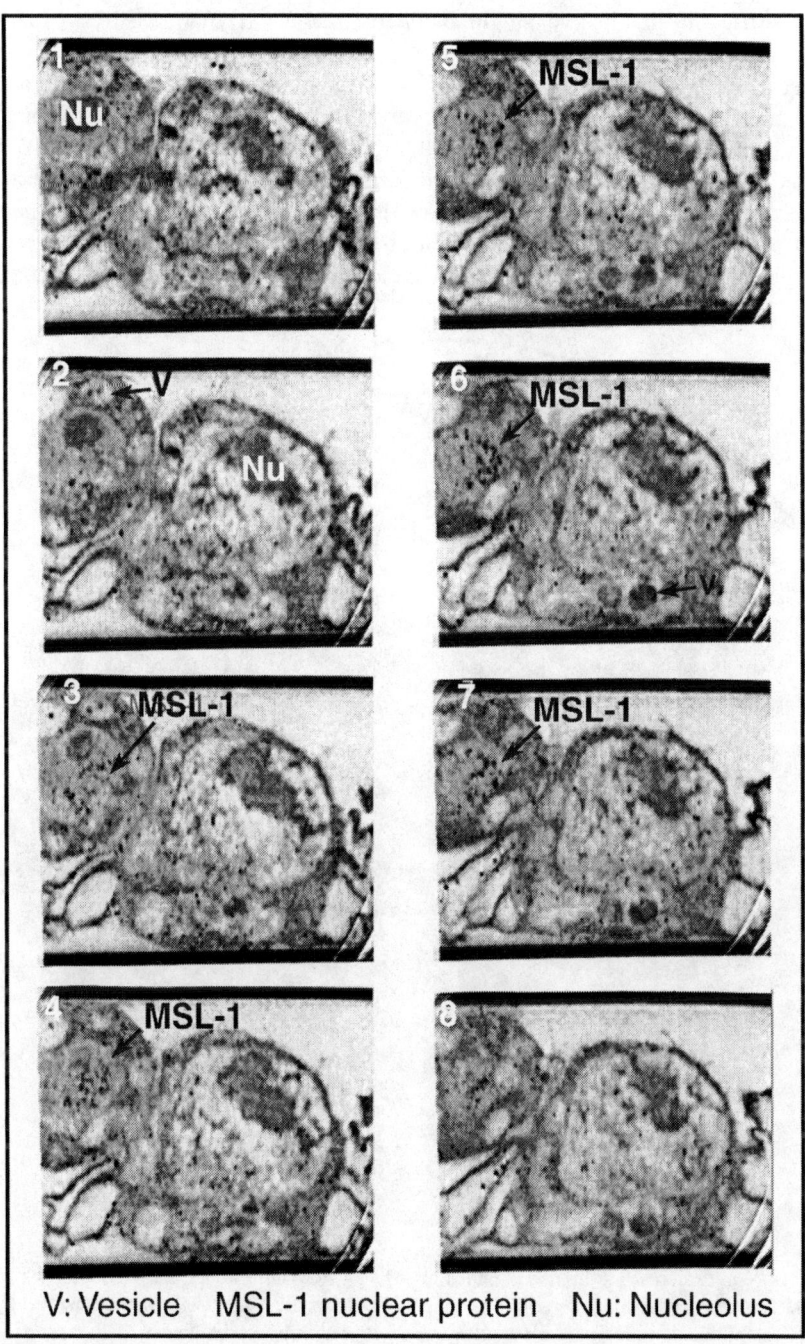

FIGURE 2. Eight transverse 2D slices through a 3D tomographic image of a single biological cell, obtained with cryofixation of the sample, and use of the microscope shown in Figure 1. Courtesy of Dr. Gerd Schneider.[2]

resolution sufficient to observe single magnetic grains.[3] Dynamic studies of void formation in high current density nanochip "vias" (layer to layer electrical interconnects) have also been conducted. These voids correspond to atomic transport driven by intense and continued electron bombardment within the narrow conduction pathway. Figure 3 shows a soft x-ray micrograph (1.8 keV, 0.6 nm wavelength) of a side thinned nanoelectronic shop where void formation and movement within a copper interconnect ("via") are studied. Movies of void formation and movement, obtained by sequential soft x-ray imaging,[4] were shown at the conference. Correlation with high electron energy TEM studies indicates that void formation tends to occur near grain

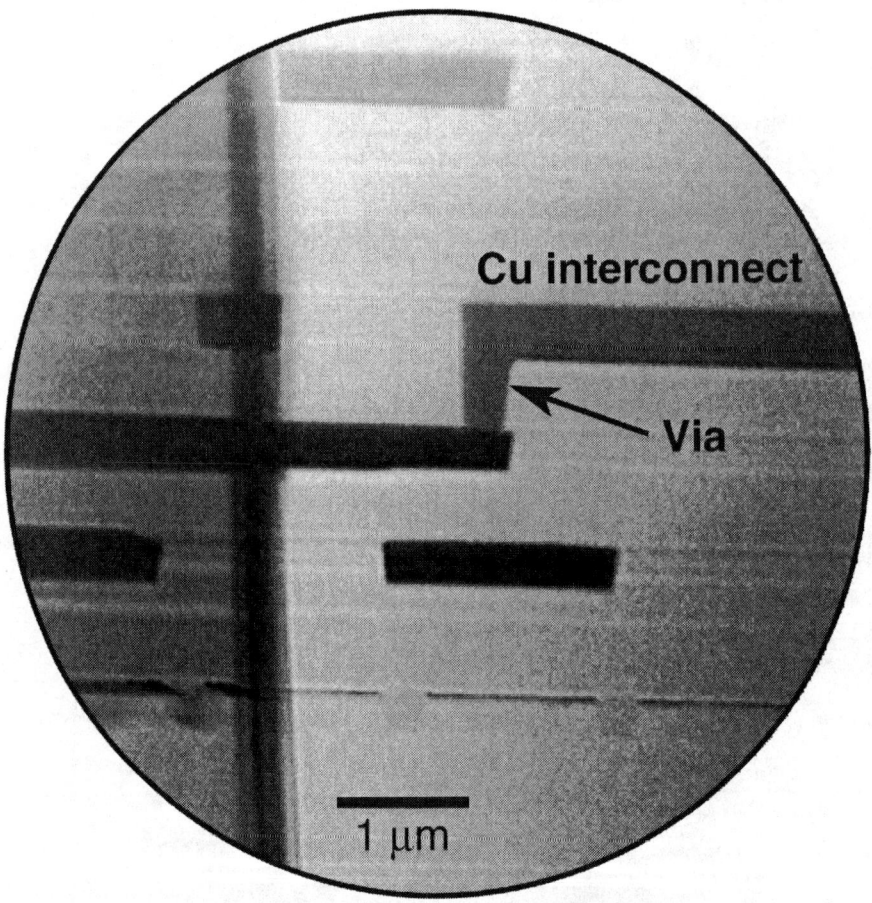

FIGURE 3. X-ray microscope image, obtained at 1.8 keV, looking through the vertical stack of interconnected layers of a computer chip. The layers of electronic patterns are interconnected by copper "vias." Density voids in the Cu appear due to electron bombardment at high current density. The growth and movement of voids within the Cu via, leading to catastrophic failure in some cases, is studied dynamically using sequential images. Movies of the void formation and movement where shown at the conference. Courtesy of Dr. Gerd Schneider.[4]

boundaries within the metallic material, with void migration along these boundaries toward the surface or materials interface, eventually leading to complete structural failure at sufficiently high current density and time duration.[4] For a summary of the most recent advances in high resolution soft x-ray microscopy, with broad applications to the physical and life sciences, see the proceedings of International X-Ray Microscopy Conference, XRM-2002.

2. EXTREME ULTRAVIOLET (EUV) LITHOGRAPHY

Lithography is the process of copying patterns, used currently with deep ultraviolet (DUV) lasers (KrF at 248 nm; Arf at 193 nm) and sophisticated multi-element (20 lenses) optical de-magnification cameras ("steppers") to print the most advanced computer chip patterns. A typical electronic device today (Pentium 4 by Intel, Athlon by AMD) consists of about twenty such overlaid and interconnected patterns, each up to 25 mm × 32 mm in size, containing a total of about fifty million transistors. For the recent (2001) "130 nm node" (half the period of the finest optical image of equal lines and spaces patterns), isolated lines of 90 nm width were printed in resist (narrowed by near-threshold resist processing) and further narrowed by materials processing to 65 nm during transfer of the pattern to a conducting material (the "metalization" step). Figure 4 shows a recently updated[9] table, the International Technology Roadmap for Semiconductors (ITRS, December 2001), giving the various nodes, half-pitch optical

First year of volume production	2001	2003* 2004	2005* 2007	2007* 2010	2009* 2013	2011* 2016
Technology Generation (Dense lines, printed in resist)	130 nm	90 nm	65 nm	45 nm	32 nm	22 nm
Isolated Lines (in resist) [Physical gate, post-etch]	90 nm [65 nm]	53 nm [37 nm]	35 nm [25 nm]	25 nm [18 nm]	18 nm [13 nm]	13 nm [9 nm]
Chip Frequency	1.7 GHz	4.0 GHz	6.8 GHz	12 GHz	19 GHz	29 GHz
Transistors per chip (HV) (3 × for HP ; 5 × for ASICs)	100 M	190 M	390 M	780 M	1.5 B	3.1 B
DRAM Memory (bits)	510 M	1.1 G	4.3 G	8.6 G	34 G	69 G
Gate CD Control (3σ, post-etch)	5 nm	3 nm	2 nm	1.5 nm	1.1 nm	0.7 nm
Field Size (mm × mm)	25 × 32	25 × 32	22 × 26	22 × 26	22 × 26	22 × 26
Chip Size (mm) (2.2 × for HP ; to 4 × for ASIC)	140	140	140	140	140	140
Water Size (diameter)	300 mm	300 mm	300 mm	450 mm	450 mm	450 mm

*Semiconductor Industry Association (SIA), December 2001.   *Possible 2-year cycle.

FIGURE 4. The international Technology Roadmap for Semiconductors[9] (ITRS) describes present and anticipated features of future nanoelectronic "chips." EUV lithography is expected to be used for the 45 nm and 32 nm "nodes." Shown above are representative expectation for both the published three-year cycle of dates, and the two-year cycle of dates projected by some leading edge manufacturers.

patterns anticipated, line widths in resist, isolated metal lines, anticipated number of transistors, clock speed, maximum field size, etc. The table is shown for both the released three-year cycle and also for a two-year cycle preferred by some leading edge manufacturers. It is anticipated that EUV lithography will be used in high volume manufacturing for the 45 nm and 32 nm nodes, in the years 2007 and 2009, respectively, following a two year cycle. The stepper is expected to use a six mirror, 0.25 NA stepper, operating at a 13–14 nm wavelength. The leading candidate EUV source, at this time, is a laser produced plasma expected to employ a high average power (~10 kW) diode pumped Nd laser and a liquid Xe jet, the latter selected largely for its low level of debris production. Alternate EUV source candidates within this spectral window are actively being pursued. These include a variety of electrical discharge plasmas, which typically offer good efficiency and power, but as yet unacceptably high levels of debris. The final source selected for production steppers will have to produce 60–80 watts of clean (debris free), collectable, in-band (within the multilayer bandpass of a nominally ten mirror stepper) EUV power with tight specifications for long life-time at high repetition rate (6–10 kHz). For an updated review of EUV source requirements and progress consult the International Sematech website concerning the most recent EUV Source Workshop[16] (November 2001, Matsue and October 2002, Dallas).

The most recent (Feb. 2002) summaries of EUV lithography were presented at the SPIE Microlithography Conference, in Santa Clara, CA. The published Proceedings[17] include reports on print experiments, optics, coatings, masks, resists, sources, environmental chemistry, etc. Scanned patterns[6] were presented with 100 nm lines and spaces printed across a full field of 26 mm by 32 mm, using a laser produced Xe jet plasma. Smaller "microfield" exposures, typically 100 µm across, were printed statically to 50 nm lines and spaces, with special illumination,[7] and isolated lines were printed to 39 nm, as shown in Figure 5. The 39 nm and 50 nm lines were printed using synchrotron undulator radiation, which is useful for metrology and special exposure studies, but is not a candidate source for production printing. The 39 nm isolated lines were printed in resist with 0.1 NA, four bounce optics. This bodes well for future production steppers which will employ 0.25 NA, six bounce optics, and thus should achieve 16 nm line widths in resist. Based on these expectations, anticipated progress with the source, resist sensitivity and line edge roughness (LER), and advances in defect free masks, it is expected that EUV lithography will be available for high volume production beginning in 2007. For those working in the EUV source area there are many opportunities to contribute, not only to the stepper (print) source, but also to development of a variety of specialized sources for metrologies of the optics, coatings, mask and mask blanks, etc.

CONCLUSION

Substantial progress has been presented in both soft x-ray microscopy and extreme ultraviolet lithography, in each case showing high quality images with clearly defined features in the 20–40 nm regime. The soft x-ray microscopy has now been extended to

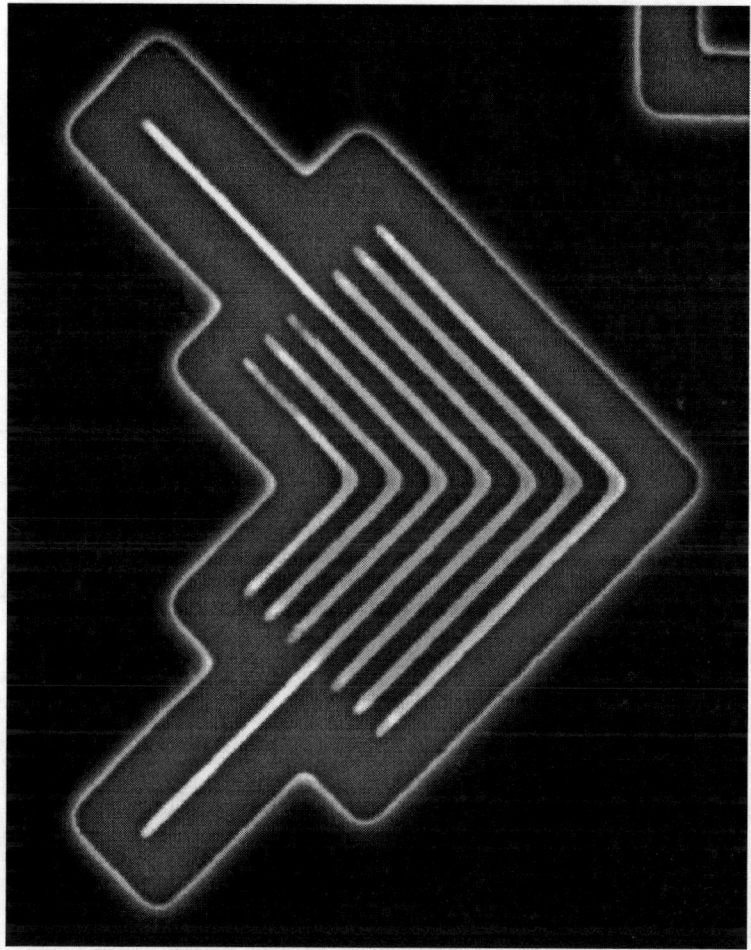

FIGURE 5. A static microfield EUV image of isolated and semi-isolated 39 nm wide lines, printed using the 0.1 NA ETS set 2 optics at 13.4 nm wavelength. Courtesy of Dr. Patrick Naulleau,[7] Lawrence Berkeley National Laboratory.

3D tomographical imaging and is poised to address significant issues in the physical and life sciences, while EUV lithography is rapidly maturing to the point that it is the favored technology for high volume manufacturing of 20 GHz microprocessors in the 2007 to 2009 time frame.

Acknowledgements. We are pleased to acknowledge support from the DOE Office of Basic Energy Sciences, the Extreme Ultraviolet LLC, as well as colleagues at LBNL's Center for X-Ray Optics and Advanced Light Source, the EUV Virtual National Laboratory, and the Institute of Röntgen Microscopy, Universitat Göttingen.

REFERENCES

1. D. Attwood, "Applications of Short Wavelength Radiation: Soft X-Ray Microscopy and EUV Lithography," J. Phys. IV (France) 11, part 2, 443 (2001); also as X-Ray Lasers 2000 (EDP Sciences, Les Ulis, France, 2001).

2. G. Schneider, E. Anderson, S. Vogt, C. Knöchel, D. Weiss, M. Legros, and C. Larabell, "Computed Tomography of Cryogenic Cells," to be published.

3. P. Fisher, T. Eimüller, S. Glöck, G. Schütz, s. Tsunashima, M. Kumazawa, N. Takagi, G. Denbeaux and D. Attwood, "High Resolution Imaging of Magnetic Domains with Magnetic Soft X-Ray Microscopy," J. Magn. Soc. of Japan 25, 186 (2001); P. Fischer et al., "Magnetic Domains in Nanostructural Media Studied with M-TXM," J. Synch. Rad. 8, 325 (2001).

4. G. Schneider, M. Meyer, G. Denbeaux, E. Anderson, B. Bates, A. Pearson, D. Hambach, E. Stach, and E. Zschech, "Dynamical X-Ray Microscopy Investigation of Electromigration in Passivated Inlaid Cu Interconnect Structures," Appl. Phys. Lett. (2002), accepted for publication.

5. G. Denbeaux, E. Anderson, W. Chao, T. Eimuller, L. Johnson, M. Kohler, C. Larabell, M. Legros, P. Discher, A. Pearson, G. Schutz, D. Yager, and D. Attwood, "Soft X-Ray Microscopy to 25 nm with Applications to Biology and Magnetic Materials," Nucl. Inst. Meth. A. 467, 841 (2001); G. Denbeaux, E. Anderson, W. Chao, X-Ray Micrsocopy 2002 (Grenoble, France).

6. D. Tichenor, W.C. Replogle, S.H. Lee, W.P. Ballard, A.H. Leong, G.D. Kubiak, L.E. Klebanoff, S. Graham, J.E.M. Goldsmith, K.L. Jefferson, J.B. Wronosky, T.G. Smith, T.A. Johnson, H.Shields, L.C. Hale, H.N. Chapman, J.S. Taylor, D.W. Sweeney, J.A. Folta, G.E. Sommargren, K.A. Goldberg, P. Naulleau, D.T. Attwood, and E.M. Gullikson, "Performance Upgrades in the EUV Engineering Test Stand," SPIE 4688, 72 (2002).

7. P. Naulleau, K.A. Goldberg, E.H. Anderson, D. Attwood, P. Batson, J. Bokor, P. Denham, E. Gullikson, B. Harteneck, B. Hoef, K. Jackson, D. Olynick, S. Rekawa, F. Salmassi, K. Blaedel, H. chapman, L. Hale, R. Soufli, E. Spiller, D. Sweeney, J. Taylor, C. Walton, G. Cardinale, A. Ray-Chaudhuri, A. Fisher, G. Kubiak, D. O'Connell, R. Stulen, D. Tichenor, C.W. Gwyn, P-Y Yan, and G. Zhang, "Static Microfield Printing at the Advanced Light Source with the ETS Set-2 Optic," SPIE 4688, 64 (2002).

8. J. Taylor, G. Sommargren, D. Phillion, S. Baker, D. Sweeney, E. Gullikson, U. Dinger, G. Sentz, F. Eisert, P. Kürz,.S. Burkart, M. Weiser, S. Schulte, S. Stacklies, R. Hudyma, and P. Gabella, "Fabriation and Metrology of High-NA Images Optics for the EUV Micro-Exposure Tool (MET)," SPIE 4688 (2002).

9. International Technology Roadmap for Semiconductors-2001, (Semiconductor Industry Association, Austin, TX 78741, December 2001); http://public.itrs.net.

10. ASM Lithography, the Dutch stepper company, has signed agreements for EUV beta tool (stepper) deliveries in 2005. Production tools are expected to follow in time for high volume microprocessor production in 2007.

11. D. Attwood, Soft X-Rays and Extreme Ultraviolet Radiation: Principles and Applications (Cambridge University Press, Cambridge UK, 1999).

12. E. Anderson, W. Chao, A. Liddle (CXRO/LBNL), private communication.

13. C. Larabell, et al., X-Ray Microscopy-2002 (Grenoble, France, August 2002); to be published.

14. D. Weiss, G. Schneider, B. Niemann, P. Guttman, D. Rudolph and G. Schmahl, Ultramicroscopy 84, 185 (2000).

15. J. Thieme, X-Ray Microscopy-2002 (Grenoble, France, August 2002); to be published.

16. EUVL Source Workshop (International Sematech, March 2002, Santa Clara, CA); www.sematech.org

17. R.L. Engelstad, editor, Emerging Lithographic Technologies VI (SPIE, Bellingham, WA 98227); www.spie.org.

X-ray Holographic Microscopy in China

Jianwen Chen, Hongyi Gao, Honglan Xie, Peiping Zhu and Zhizhan Xu

Shanghai Institute of Optics and Fine Mechanics , The Chinese Academy of Sciences, P.O. Box 800-211, Shanghai 201800, China

Abstract. Development state for X-ray holographic microscopy and holographic tomography is presented in China.

Holography was proposed by Gabor[1] for his aim to eliminate the spherical aberration of electron optics. After four years Baze[2] first suggested Gabor holography for X ray imaging. However until the early 1970s Aoki and Kikuta and their collaborators[3] first recorded X-ray holograms. In 1986, Howells et al. [4] recorded X ray in-line holograms using a storage ring X-ray source and a toroidal grating monochromator and reconstructed them by using a He:Cd laser with 1~2μm resolution at Brookhaven National Laboratory. With the development of undulator, Jacobsen *et al.* [5] recorded Gabor X-ray holograms using the undulators and obtained the reconstructed image with resolution of 56nm using numerical reconstruction method. Simultaneously, a digital twin imaging elimination method in soft X ray in-line holography was developed by Joyeux et al. [6] at Srsay Cedex, France. In 1992, McNulty et al.[7] recorded lensless Fourier transform X-ray holograms in digital form with a CCD camera, and the specimen image was obtained by numerical reconstruction with   resolution of 60nm.

It is in middle 70's that China began to study X-ray microscopy using Laser plasma as X-ray sources. However, the research on X-ray holography and holographic tomography had not developed in China until 1995 when the National Synchrotron Radiation Laboratory (NSRL) in Hefei, China was set up.

In 1997, the first X-ray hologram[8] of biological specimen was performed on the bend magnet in NSRL by a group from Shanghai Institute of Optics and Fine Mechanics (SIOM), the Chinese Academy of Sciences under the cooperation with colleagues of NSRL. The X-ray source is come from an 800MeV electron storage ring with characteristic wavelength of 2.4nm. The continuous X-ray spectrum is monochromatized using a zone plate and filtered by a pinhole with 30μm diameter in order to obtain partial coherent X rays, shown as Fig1. The zone plate can be shifted in axis in the range of ±110mm. The distance of the zone plate-to-source is 1.06m and of the zone plate-to-pinhole is 600mm.   For this condition the pinhole provides a X-ray beam of wavelength λ=2.3nm, and the intensity 10^6~10^7 photons/sec.

The experimental setup is typical Gabor in-line holography. The main advantage is simplicity itself. The specimen is placed in line with X-ray source, which provides both the reference waves and the specimen illumination. The distance R from

CP641, X-Ray Lasers 2002: 8th International Conference on X-Ray Lasers, edited by J. J. Rocca et al.

quasi-monochromatic pinhole to the specimen is about 700mm. The specimen is some granules which adhere to a Si_3N_4 film. The recording medium is methyl methacrylate acid photoresist (PMMA) which is placed at recording distance of about 400μm from the specimen. The exposure process is performed in an vacuum chamber with the pressure $<2\times10^{-3}$ Pa. When the electron beam intensity is 100mA, the exposure time is 30 minuets.

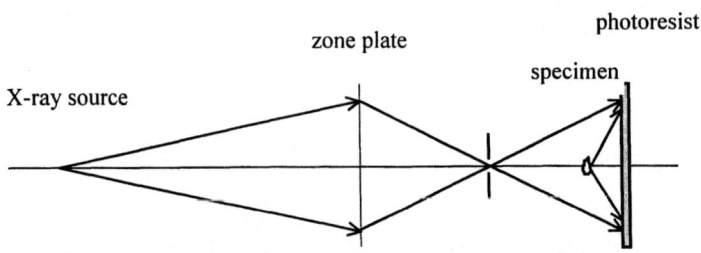

FIGURE1. The diagram of experimental setup.

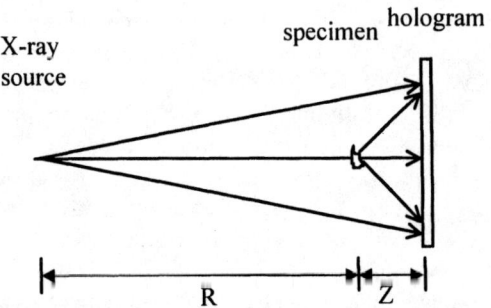

FIGURE2. The scheme of Gabor in-line holography

In 1999[9], an Atomic Force Microscope (AFM) is used to read the X-ray hologram of minute granules recorded on photoresist. Digitized hologram is reconstructed by numerical method and the hologram can be processed easily to eliminate or reduce some nonlinear response, aberration and distortion in the recording process. The detailed process of numerical reconstruction has been analyzed by Jacobsen et al..

The optical amplitude $\varphi(x,y)$ at an image plane distance z away can be calculated by Fresnel-Kirchhoff diffraction expression

$$\varphi(x,y) = \iint \tau(\xi,\eta)\frac{\exp(ikr)}{i\lambda r}d\xi d\eta \qquad (1)$$

where $\tau(\xi,\eta)$ is the amplitude transmittance at the hologram plane, and

$$r^2 = z^2 + (\xi - x)^2 + (\eta - y)^2 \qquad (2)$$

In the Fresnel approximation, Eq.(1) can be written as

$$\varphi(x, y) = \frac{1}{i\lambda z} \exp(i2\pi \frac{z}{\lambda}) \exp(i\pi \frac{x^2 + y^2}{\lambda z})$$

$$\cdot \iint \tau(\xi, \eta) \exp(i\pi \frac{\xi^2 + \eta^2}{\lambda z}) \exp(-i2\pi \frac{\xi x + \eta y}{\lambda z}) d\xi d\eta \qquad (3)$$

With an N×N square pixel discretized hologram, we denote the hologram pixel size as $\Delta\xi$ and the reconstructed image pixel size as Δx. According to[5]

$$\frac{\Delta x \cdot \Delta\xi \cdot N}{\lambda z} = 1 \qquad (4)$$

the Fresnel integral can be calculated by Fourier transform, we have

$$\varphi(\Delta x \cdot P, \Delta x \cdot Q) = Const \cdot DFT\left\{ \tau(\Delta\xi \cdot K, \Delta\xi \cdot L) \exp\left[i\pi \frac{\Delta\xi^2}{\lambda z}(K^2 + L^2) \right] \right\} \qquad (5)$$

where $DFT\{\ \}$ denotes discrete Fourier transform. In addition, if the sampling step is satisfied with $\Delta x = \Delta\xi = \Delta$, the size of reconstructed image is the same as the hologram. On this condition, the discretized step is

$$\Delta = \sqrt{\frac{\lambda z}{N}} \qquad (6)$$

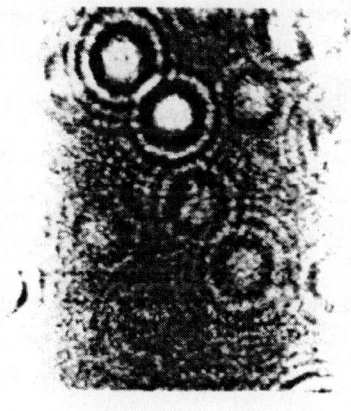
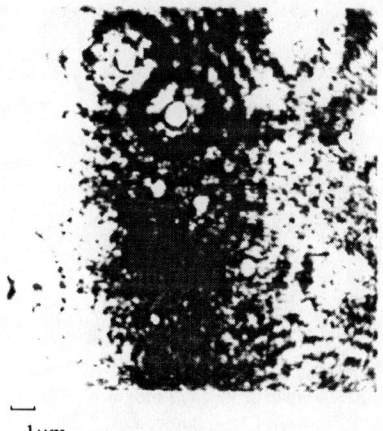

1μm

Fig.3a: X-ray holography of dust granules, readout by AFM

Fig.3h: Reconstruction of the hologram by numerical method

As described above, the holograms are magnified using an Atomic Force Microscope (AFM, Nanoscope IIIa, Digital Instrument Inc.). The film negative is then digitized using a scanning microdensitometer. The hologram of the granules and its reconstructed images are shown in Fig3. From the reconstructed image we think that $0.3 \mu m$ transverse resolution is achieved in this experiment. The longitudinal resolution is worse than the transverse resolution. The holographic fringes are visible clearly.

Almost at the same time, another group of NSRL in Hefei combining X-ray holography with CT (computer tomography) technology, made three-dimension reconstruction with high resolution for the biological specimen[10]. The experimental condition is similar to above mentioned.

According to two imaging methods of holography, holographic tomography (HT)also has two methods: the Gabor in-line holographic tomography and the Fourier transform holographic tomography. The former is selected by the group of NSRL in Hefei to study HT. The tens Gabor in-line holograms were recorded with 5^0 step size within angles between -40^0 and $+50^0$. These holograms are taken as projection data at each angle after they are digital reconstructed with the computer . Then by tomographic reconstructing from these projection data , three dimensional density distribution of the specimen can be attained.

ACKNOWLEDGEMENTS

This project was partly supported by Chinese High-Tech Program, Chinese National Major Basic Research Development Program (No. G1999075200).

REFERENCES

[1] D. Gabor, A new microscopic principle, Nature, 1948, 151: 777-778.
[2] V.A. Baze, A study in diffraction micorscopy with special reference to X-ray, J. O. S. A., 1952, 42(10): 756-762.
[3] S. Aoki and S. Kikuta, X-ray holographic microscopy, Jpn. J. Appl. Phys., 1974, 13(9): 1385-1392.
[4] M. Howells, C. Jacobsen, J. Kirz et al., X-ray holography at improved resolution: a study of zlymogen gramoles, Science, 1987, 238: 514-517.
[5] C. Jacobsen, M. Howells, J. Kirz et al., X-ray holographic microscopy using photoresists, J. Opt. Soc. Am., 1990, A7(10): 1847-1861.

[6] G. Koren,   F.Polack,D.Joyeux, Iterative algorithm for twin-image elimination in in-line holography using finite-support constraints, J. Opt. Soc. Am, 1993, A10(3): 423-433.

[7] I. McNulty, J. Kirz, c. Jacobsen, E.H. Anderson, M.Howells, D. Kern, High-resolution imaging by Fourier transform X-ray holography, *Science*, 1992, 256: 1009-1012.
[8] Zhang Yuxuan, Chen Jianwen et al., In-line soft X-ray holographic image and digital reconstruction at "water window" wavelength region. Acta Optica Sinica, 1997, 17(11):1599~1600.
[9] Jiang Shiping, Chen Jianwen et al., Soft X-ray holographic microscopy with sub-micrometer resolution. Acta Optica Sinica, 1999, 19(12):1709~1710.
[10] Zhang Yuxuan and Zhang Xinyi, Three-dimensional X-ray microscopic imaging using Synchrotron Radiation, Progress in Physics, 2001, 21: 12 (in Chinese).

Coherence properties and applications of a transient collisional excitation type silver soft x-ray laser at Advanced Photon Research Center

Hiroyuki Daido*, Keisuke Nagashima*, Maki Kishimoto*, Masataka Kado*, Tetsuya Kawachi*, Noboru Hasegawa*, Momoko Tanaka*, Kouta Sukegawa*, Renzhong Tai*, Peixiang Lu*, Huajing Tang*, Mamiko Nishiuchi*, Akira Sasaki*, Takashi Arisawa*, Yoshiaki Kato*, Kazumichi Namikawa[+], and Bedrich Rus**

*Advanced Photon Research Center, Kansai Research Establishment, Japan Atomic Energy Research Institute (JAERI), 8-1, Umemidai, Kizu, Kyoto, 616-0215 Japan
[+] Department of Physics, Tokyo Gakugei University, 4-1-1 Nukui-Kita, Koganei, Tokyo, 184-8501 Japan
** PALS Research Centre, Institute of Physics, Na Slovance 2, 18221 Prague 8, Czech Republic

Abstract. A ~10J, pico-second laser pumped transient collisional excitation type silver soft x-ray laser at 13.9 nm routinely delivers collimated saturated amplification soft x-ray laser pulses with an energy of 20 μJ at the Advanced Photon Research Center. The soft x-ray laser pulse contains reasonable spatial coherence length of 0.1 mm at the place 1 m away from the x-ray laser source. The pulse width and energy are a few ps and a few tens of μJ, which is sufficient for imaging with pico-second time resolution. We describe the three kinds of subjects including the spatial coherence measurement of the silver x-ray laser using a multi-slit array, the wave front division type soft x-ray interferometer using a Lloyd's mirror and the measurement of domain size distributions in a single BaTiO₃ ferroelectrics crystal at room temperature. As for the interferometer experiment, we have obtained a 50 μm diameter wire image with fringe shifts. As for the speckle measurement, the spatial correlation and the spatial power spectrum, which are related to the domain size distribution, have been characterized.

CP641, *X-Ray Lasers 2002: 8th International Conference on X-Ray Lasers,* edited by J. J. Rocca et al.
© 2002 American Institute of Physics 0-7354-0096-2/02/$19.00

INTRODUCTION

Recently, the transient collisional excitation soft x-ray lasers have proven to be the most powerful soft x-ray source for various applications because these lasers, which did not need big pumping lasers such as the laser fusion systems, provided high quality short pulse x-ray with sufficient pulse energy for imaging [1]. In this paper, we would like to describe x-ray laser applications, which utilize coherence property of the x-ray laser beam. The outline of this paper is as follows; firstly, we will introduce briefly about Ni-like Ag soft x-ray laser beam at 13.9 nm wavelength which is quite useful for applications because we can use very high quality soft x-ray mirrors with reflectivity of more than 60 % based on the Molybdenum-Silicon multi-layers. Secondary, we will describe the coherence measurement for the Ag soft x-ray laser using a multi-slit array as a diffraction element. Thirdly, we will describe specific application experiments, which include soft x-ray laser interferometer using a Lloyd's mirror configuration as well as soft x-ray speckle pattern measurement for instantaneous observation of Ferro-electric domain structures.

SOFT X-RAY LASER BEAM CHARACTERIZATION

At Advanced Photon Research Center, the saturated amplification of the Nickel-like Silver soft x-ray laser has been developed which could deliver 7ps pulse with 25 µJ energy at the wavelength of 13.9 nm [2]. Figure 1 shows a typical near field pattern of the silver x-ray laser gain region which was measured using a configuration also shown in Fig. 1. In this spectral range we usually use pumping laser pulses like this such as 2 ps pulses separated by 1.2 ns which overlap with the very weak pedestal pulses with ns duration. The traveling wave pumping using a 6-step mirror makes a much larger gain coefficient than that of non-traveling wave pumping such as 35 cm^{-1} for the traveling wave, while 24 cm^{-1} for the non-traveling wave pumping. The traveling wave pumping has given us the saturated amplification laser which is quite important for the applications. The output energy was around 25 µJ. Note that sometimes the near field pattern was composed of 2 bright spots distributed vertically where vertical corresponds to the direction parallel to the x-ray laser target surface [3].

Figure 2 shows the experimental setup of a far-field pattern and a spatial coherence measurement [4]. The soft x-ray laser pulse propagates and diffracted by the multi-slit array. If we measure a far-field pattern, we place the x-ray turning mirror instead of the slit array adjacent to the slab target. The beam can be detected by the CCD detector, which coupled with a Zirconium filter. We put

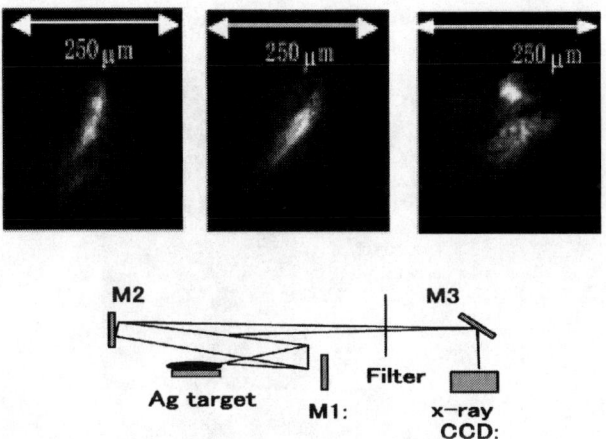

Fig.1 Schematic diagram of the near field pattern measurement and the observed pattern. Also shown are the near field images of the Ag x-ray laser at the end of the gain region. (a) The near-filed image for the pumping pulses composed of a 600ps pulse as a pre-pulse and a 1.5 ps main pulse without the traveling-wave pumping technique. (b) The near filed image for the pumping pulses composed of a 1.5 ps pulse as a pre-pulse and a 1 .5 ps main pulse without the traveling wave pumping technique. (c) The near filed images for the pumping pulses composed of a 4 ps pulse as a pre-pulse and a 4 ps main pulse with traveling wave pumping technique. In the figures (a)-(c), the pumping laser incidents from the left.

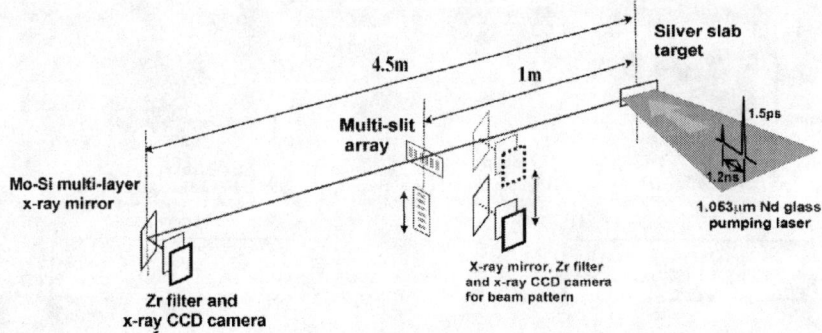

Fig.2 Experimental setup for the beam pattern and spatial coherence measurement by the single-wire-diffraction of the 13.9 nm Ni-like Ag x-ray laser.

the molybdenum-Silicon multi-layer turning mirror to cut the undesired x-ray especially for the hard x-ray, which is not easy to remove by the filters. We can measure fringe patterns at the CCD. The measurement principle is that if we measure a diffraction pattern and we also obtain the corresponding

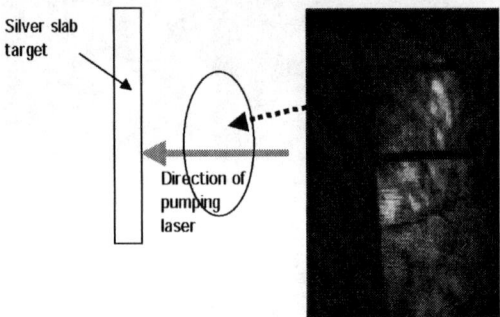

Fig.3 The far-field pattern of the 13.9 nm Ni-like Ag x-ray laser under the traveling wave pumping scheme. We use the pumping pulses of a pre-pulse and a main pulse, both of which have the same pulse duration of 1.5ps.

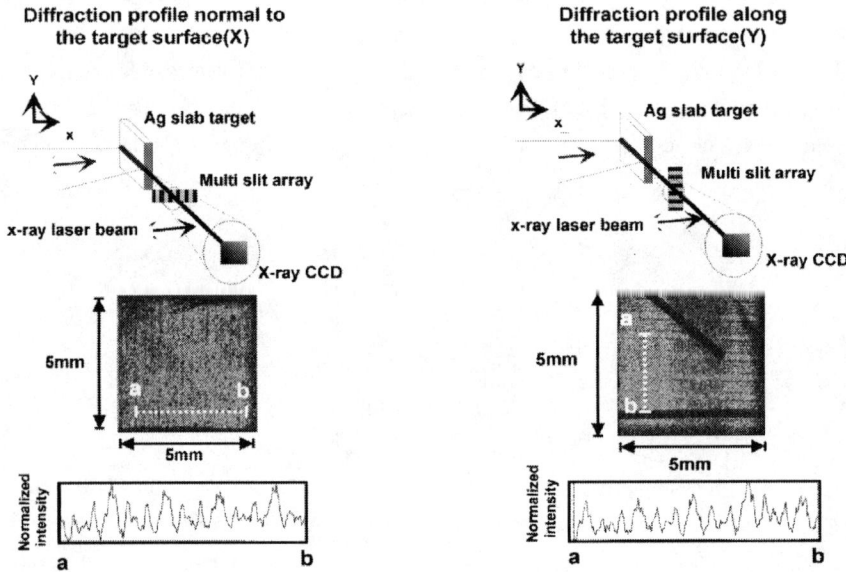

Fig.4 Diffraction patterns of the multi-slit array irradiated by the x-ray laser with the direction of the slit-array vertical (a) and parallel (b) to the target surface.

calculated pure coherent pattern, we can obtain the complex coherence factor, which gives us the degree of spatial coherence. Figure 3 shows the far field pattern of the nickel-like silver x-ray laser. Due to the split double sources in the near-filed pattern obtained by the traveling wave pumping as shown in Fig.1, fringes in the pattern are clearly visible. The beam divergence angles for the horizontal and the vertical directions of the target surface are around 7 and 20 mrad, respectively. According to the divergence angles, the intrinsic coherence length is estimated to be 1 μm. Figure 4 shows the typical diffraction patterns of the silver x-ray laser caused by the slit array detected on the back-illumination type soft x-ray CCD for both directions such as vertical and horizontal in terms of the target surface. Figure 5 shows the modulus of the complex coherence factor of the 13.9 nm x-ray laser, which are compared with those calculated with the assumption of quasi-monochromatic incoherent disc sources with different diameters. The curve, which fits best to the experimental result is that of double incoherent disc sources. The calculated result of the splitting two-disk source model is in disagreement with the experimental result. The two disk sources should have certain amount of mutual coherence, which is also supported by the fringe patterns in the far field patterns as shown in Fig. 3. Anyway, the spatial coherence length is ~100 μm at 1 m away from the x-ray laser source.

X-RAY LASER APPLICATION EXPERIMENTS

We will describe the application experiments which we have performed at the Advanced Photon Research Center. Firstly, we introduce a ps x-ray laser interferometer using a Lloyd's mirror, which is the wave-front division type interferometer [5]. The experiment was performed in collaboration with Dr. Rus from Institute of Physics at Prague who stayed at the Advanced Photon Research Center for 3 months, recently. Figure 6 shows the schematic diagram of the experimental setup for a Lloyd's mirror interferometer using a 13.9 nm soft x-ray laser. We put the spherical mirror to make diverging x-ray beam for the interferometer. The Lloyd's mirror is composed of a super polished BK7 glass plate whose size is 100 x 25 x 15 mm with a surface roughness of less than 1 nm. We set two types of configurations in which the distances between the Lloyd's mirror and the CCD are 230 cm and 510 cm, respectively, depending upon the required spatial resolution. The free space fringes detected by the back-illumination type CCD at 230 cm from the Lloyd's mirror show straight fringes which means that the x-ray laser wave-front keeps high quality. The only the problem is speckles in the pattern. If we place the wire as a sample for holographic arrangement, we can observe diffraction patterns which are made by the interference between the free space reference wave and the scattered wave from the object. Figure 7 shows the photo of the diffraction pattern detected by the CCD detector. You can see the fringe shift in the shadow region of the wire whose diameter is 50 μm. We have also tested the diffraction patters caused by fine powders made of tin oxide on the Lloyd's mirror. The shift by 1/2 fringe corresponds to around 0.5 μm.

The result suggests that the resolution leads

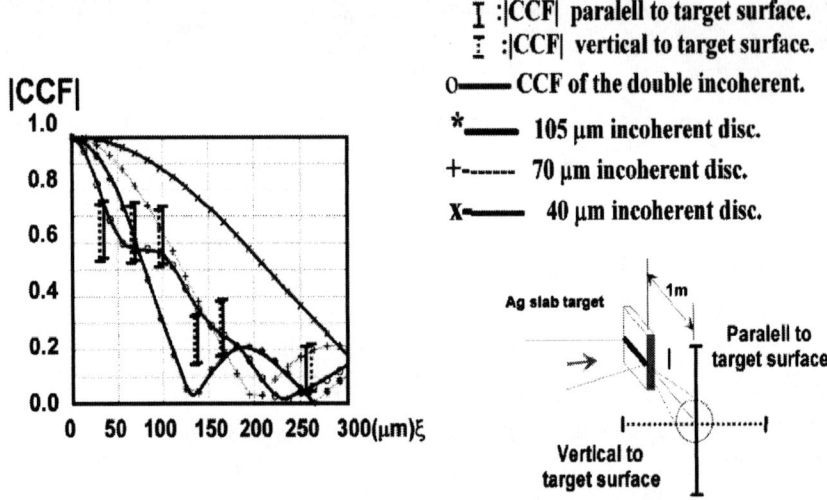

I :|CCF| paralell to target surface.
I :|CCF| vertical to target surface.
O——— CCF of the double incoherent.
*———— 105 μm incoherent disc.
+------- 70 μm incoherent disc.
X——— 40 μm incoherent disc.

Fig.5 The profile of the Modulus of the Complex coherence factor measured from the diffraction profile of the x-ray laser and comparison with the various quasi-monochromatic incoherence disc model.

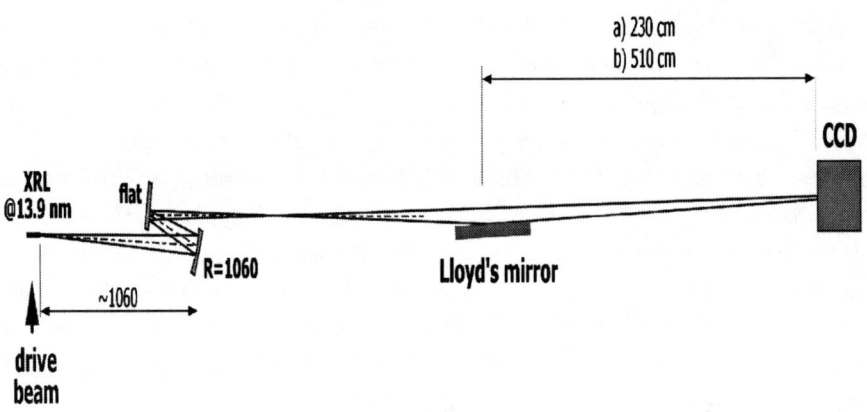

Fig.6 Wave-front division type x-ray laser interferometer using Lloyd's mirror, which is made of super-polished BK7 glass. The size of the mirror is $100 \times 25 \times 15$ mm and the roughness of the surface is less than 1 nm.

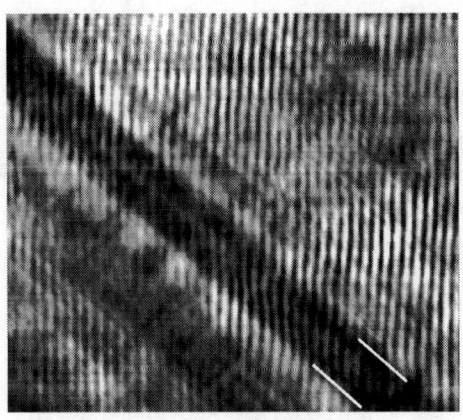

Fig. 7 An interferogram obtained by the Lloyd's mirror interferometer illuminated by 13.9 nm soft x-ray laser. The distance between the two white central bars corresponds to the edges of the shadow of the 50 μm diameter wire. In side the bar, fringe deformation is visible due to the interference between the straight x-ray and scattered x-ray.

to an order of 100 nm. We have demonstrated the pico-second x-ray laser interferometry with holographic capability. The surface roughness of an order of 100 nm can be potentially measured using this technique. Details can be found in these proceedings [6].

The next subject is the soft x-ray speckle patterns for instantaneous measurement on Ferro-electric domain structures. The silver x-ray laser pulse is turned by the multi-layer mirror, which acts as a reflector as well as a polarizer whose Brewster angle is around 45 degrees. Then the almost linearly polarized pulse passes through the filter to cut the visible light and the pulse also passes through the crossed slits whose area is 400 μm x 100 μm. After passing through the 100 μm-width-slit, the x-ray pulse contains almost full spatial coherence in the horizontal direction. Then the pulse hits the Ferro-electric sample. The phase front and amplitude of the pulse is modified by the characteristic surface response of each domain distributed irregularly along the one specific direction. Then the speckle patterns are produced on the CCD according to the Fresnel-Kirchihoff integral. On the basis of the ideas, we have performed the experiments resulting in the fruitful results. If the temperature increases more than the Curie temperature, the domains disappear and mirror like reflection occurs. If the x-ray laser pulse reflects at the group of C-domains only, the reflection surface does not distort the phase of the coherent beam. On the other hand, if the x-ray laser pulse hits the a+c domains, the detected beam pattern is asymmetry caused by the different optical properties of each domain. The measured speckle patterns at various temperatures show very interesting features. Especially at around the Curie temperature of 122 degrees Centigrade, the size of each domain seems to reduce and almost disappear. The surface responds to like a mirror surface, but we can still find slight difference with a mirror. The fact implies that very small domains still survive. The details will be found in Ref. [7].

SUMMARY

A saturated amplification Ni-like Silver soft x-ray laser at 13.9 nm has been developed. We have measured the energy of ~25 μJ in 7ps which is sufficiently high and short the for various applications. Coherence measurement of the Silver laser has been performed using a multi-slit array, resulting in transverse coherence length of ~100 μm at 1m from the source. Two types of application experiments have been performed including the soft x-ray interferometer using a Lloyd's mirror including interferograms of fine structures. We have shown the prospect for surface holography with spatial resolution of less than 100 nm. We have also shown soft x-ray speckle measurement technique for observation of Ferro-electric domain structures.

REFERENCES

1. Daido H, Reports on Progress in Physics (to be published).
2. Kawach T. et al, Phys Rev. A (submitted for publication) and also in these proceedings.
3. Tanaka M. (submitted for publication).
4. Tang H. et al, Jpn. J. Appl. Phys. (submitted for publication).
5. Rocca J J et al. Opt. Letters **24**, 420-422 (1999).
6. Rus B. et al, in these proceedings
7. Tai R. Z. et al, submitted for publication

A picosecond 14.7 nm x-ray laser for probing matter undergoing rapid changes

J. Dunn [a], R.F. Smith [a], J. Nilsen [a], A.J. Nelson [a], T.W. Van Buuren [a],
S.J. Moon [a], J.R. Hunter [a], J. Filevich [b], J.J. Rocca [b], M.C. Marconi [b, c],
V.N. Shlyaptsev [d]

[a] Lawrence Livermore National Laboratory, Livermore, CA 94551-9900
[b] Dept. of Electrical and Computer Engineering, Colorado State University, Fort Collins, CO 80523
[c] Dept. of Physics, University of Buenos Aires, Argentina.
[d] University of California Davis-Livermore, Livermore, California 94551

Abstract. With laser-driven tabletop x-ray lasers now operating in the efficient saturation regime, the source characteristics of high photon flux, high monochromaticity, picosecond pulse duration, and coherence are well-matched to many applications involving the probing of matter undergoing rapid changes. We give an overview of recent experiments at the Lawrence Livermore National Laboratory (LLNL) Compact Multipulse Terawatt (COMET) laser using the picosecond 14.7 nm x-ray laser as a compact, ultrafast probe for surface analysis and for interferometry of laser-produced plasmas. The plasma density measurements for known laser conditions allow us to reliably and precisely benchmark hydrodynamics codes. In the former case, the x-ray laser ejects photo-electrons, from the valence band or shallow core-levels of the material, and are measured in a time-of-flight analyzer. Therefore, the electronic structure can be studied directly to determine the physical properties of materials undergoing rapid phase changes.

INTRODUCTION

A major goal in the pursuit of x-ray laser research has been to achieve higher efficiency, reduced size, low cost and a high repetition rate for the pumping scheme. In particular, operation in the efficient saturation regime for a tabletop-sized source is essential for applications and the future development of the field. In the last few years, the most noted examples of saturated, tabletop schemes have been based on collisional excitation pumping although the methods have been varied. The fast capillary discharge scheme produced intense Ne-like Ar lasing at 46.9 nm on the 3p $^1S_0 \rightarrow$ 3s $^1P_1$ transition at high repetition rate, high average power operation [1]. The picosecond laser-driven transient schemes have been demonstrated to give in excess of 10 µJ output/pulse on the Ni-like Pd 4d $^1S_0 \rightarrow$ 4p $^1P_1$ line at 14. 7 nm [2]. More recently, an x-ray laser operating at 10 Hz for the Pd-like Xe ion 5d $^1S_0 \rightarrow$ 5p $^1P_1$ transition at 41.8 nm was driven into saturation using 35 fs irradiation of a xenon gas cell [3]. The experimental demonstration of the second category of laser scheme, the picosecond-driven transient collisional schemes using chirped pulse amplification (CPA) laser systems, was shown first for Ne-like Ti 3p $\rightarrow$ 3s transition at 32.6 nm using 10 J of laser energy [4].

CP641, *X-Ray Lasers 2002: 8th International Conference on X-Ray Lasers,* edited by J. J. Rocca et al.
© 2002 American Institute of Physics 0-7354-0096-2/02/$19.00

At LLNL over the last 5 years, much theoretical and experimental effort on the COMET facility has developed this picosecond, tabletop x-ray laser mainly because of the high efficiency, relatively high repetition rate and unique combination of source characteristics. Various applications, including picosecond x-ray laser interferometry of dense plasmas and probing of the electronic structure of materials, are discussed.

APPLICATIONS OF A PICOSECOND X-RAY LASER

Typical COMET x-ray laser characteristics are listed, with the beam brightness information specifically for the Ni-like Pd line at 14.7 nm, in Table 1.

TABLE 1. COMET x-ray laser source parameters

Source Parameters	COMET X-ray Laser	Importance [b]
Laser Pump Energy (J)	5 - 10	
X-ray Laser Energy (μJ)	25	
Photons/Shot	2×10^{12}	x
Shot Rate (Hz)	0.004	
Wavelength (nm)	12 – 47	
$\Delta\lambda/\lambda$	$< 10^{-4}$	x
Source Size (μm^2)	25×100	
Divergence ($mrad^2$)	2.5×10	
XRL Pulse Duration (ps)	2 – 25	x
Peak Brightness, B [a]	1.6×10^{25}	x
Average Brightness, B [a]	1.3×10^{11}	

[a] [Units of ph. mm^{-2} $mrad^{-2}$ s^{-1} (0.1% BW) $^{-1}$].

[b] "x" denotes considered important x-ray laser source parameter.

High photon flux/shot, high monochromaticity, and short pulse duration when combined with small source area and beam divergence properties of the 14.7 nm line [5] give ultra-high peak brightness ~ 10^{25} ph. mm^{-2} $mrad^{-2}$ s^{-1} (0.1% BW) $^{-1}$. This assumes a 2 ps pulse duration. The range of x-ray laser pulse durations is expected to be dictated by the gain lifetime [6] and by the pulse length of the picosecond driving laser beam. A recent experiment reported the x-ray laser pulse to be 2 ps (FWHM) for a 1.3 ps-driven Ni-like Ag 13.9 nm x-ray laser [7]. Overall, the 14.7 nm peak brightness is 5 – 6 orders of magnitude higher than 3rd generation synchrotron undulator sources. The 46.9 nm fast capillary discharge x-ray laser operates at similar high peak brightness and has the further advantage, with 10 Hz repetition rate, of high average brightness, 5×10^{14} ph. mm^{-2} $mrad^{-2}$ s^{-1} (0.1% BW) $^{-1}$. From Table 1, this is more than 10^3 times higher than the Pd line brightness. Third generation synchrotron

TABLE 2. Summary of x-ray laser applications with desired x-ray laser characteristics [a]

Application	Average Power	Brightness	Δλ/λ	Short Pulse [b]	Coherence	Tunability	Polarization	Short Wavelength
Biology:								
Microscopy	-	Yes	-	Yes	-	-	-	2.4 – 4.4 nm
Holography	-	Yes	-	Yes	Yes	-	-	Yes
Diffraction	-	Yes	-	Yes	-	-	-	0.1 – 1 nm
Plasmas:								
Interferometry [c]	-	Yes	Yes	Yes	Yes	-	Maybe	-
Non-linear Optics	Maybe	Yes	-	Yes	Yes	-	Yes	-
PlasmaGeneration	-	Yes	-	Yes/ -	Yes	-	-	-
Warm Dense Matter	Yes	Yes	Yes	Yes	-	Yes	-	0.1 nm
Atomic and Molecular Physics:								
Photoionization	Sometimes	Yes	Yes	Yes	-	Desirable	Sometimes	Yes/ -
Multiphoton Processes	-	Yes	-	Yes	-	Yes	-	Yes
Material Science and Chemistry:								
Semi-conductors	Yes	Yes	Yes	Yes	Maybe	Yes	-	1 – 30 nm
Photoemission	-	Yes	Yes	Yes		Yes/ No	-	1 – 60 nm
Spectroscopy [c]								
Industrial:								
EUVL Printing	Yes	-	-	-	-	-	-	13.5 nm
EUV Metrology	Yes	-	-	-	Yes	-	-	13 – 14 nm

Note: The symbol "-" in row for a given application refers to characteristic that is not required or applicable.

[a] After Table in ref. [8].
[b] Short pulse defined as femtosecond to picosecond regime.
[c] Work described in this paper.

483

undulator sources still have higher average brightness of $0.5 - 6 \times 10^{18}$ ph. mm^{-2} mrad^{-2} s^{-1} (0.1% BW)$^{-1}$ at $50 - 10$ nm, respectively. The collisional x-ray lasers have discrete wavelengths while synchrotrons are tuneable.

Table 2 summarizes some of the possible applications using x-ray lasers with the corresponding x-ray characteristics. This is based on a table published 10 years ago [8] and includes some updates and minor changes reflecting recent developments. The table is not meant to be exhaustive covering all possible applications, but lists relevant areas of interest in biology, high temperature plasmas, atomic and molecular physics, material science, chemistry and industrial semiconductors. For example, x-ray laser microscopy of biological samples first demonstrated by Da Silva et al. [9] requires the characteristics of peak brightness, short pulse duration and wavelength in the "water window". The Extreme Ultraviolet Lithography (EUVL) project for printing semiconductor masks specifies high average power, $30 - 100$ W at 13.5 nm, with high uniform spatial and constant temporal source requirements [10]. Recently, x -ray lasers have been focused at high intensity to form laser-produced plasmas [11]

Laser-driven x-ray lasers using COMET type facilities are unique sources on account of their high photon flux/pulse, high monochromaticity, and ps pulse duration in the $12 - 47$ nm regime. Presently, these x-ray lasers have excellent peak brightness capability compared to other sources but the average power or brightness parameter, while still high, requires further development. Therefore, applications using high peak brightness are a good match. In addition, the short pulse duration is essential for probing transient events in high temperature plasmas, material phase changes or chemical processes. For example as shown in Table 2, x-ray laser interferometry of high temperature laser-produced plasmas, when the additional requirements of transverse and longitudinal coherence are satisfied, is an excellent application for this source. The ultra-high peak brightness is necessary to overcome x-ray self-emission from the plasma being probed. Likewise, the same x-ray source properties, high photon flux/pulse, high monochromaticity and ps pulse, are important for a surface analysis probe to study the valence band and shallow core electronic structure.

These applications are described in the next sections with both sets of experiments performed on the COMET facility. Optimization of the laser conditions for generating the saturated, 14.7 nm, Ni-like Pd x-ray laser have been detailed previously [2, 5]. For these experiments a 600 ps long pulse followed by a 6 ps short pulse beam, separated by 700 ps peak-to-peak was chosen to given optimum x-ray laser output [12].

PICOSECOND X-RAY LASER INTERFEROMETRY

The first soft x-ray laser interferometry experiments were performed on the NOVA laser by Da Silva and co-workers with the 15.5 nm Ne-like Y laser using an amplitude-division Mach-Zehnder interferometer equipped with multi-layer coated thin foils as beamsplitters [13]. Large 1-mm-scale plasmas were probed to a maximum electron

10x mag.　　　　　　　　　　　　　　　　　**22x mag.**

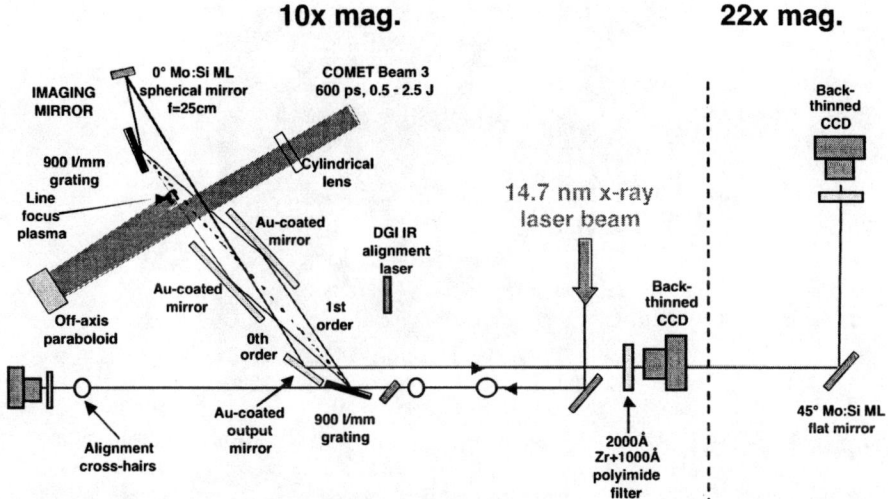

Fig. 1　Layout for interferometry experiments. Setup for 10× magnification imaging and for higher 22× imaging with additional 45° multilayer optic (to right of vertical dashed line).

density of 2×10^{21} cm^{-3} and to within 25 μm of the initial target surface. More recently, Rocca and co-workers developed a Mach-Zehnder type instrument at 46.9 nm using diffraction gratings as beam splitters [14]. Diffraction gratings are robust, use well-established technology and can be utilized for various wavelengths, as short as 2 nm. They also have a high throughput which makes them a good match to the present generation of tabletop picosecond laser-driven x-ray lasers. A version of the diffraction grating interferometer (DGI) was developed and optimized by Colorado State University for the ps Ni-like Pd 14.7 nm x-ray laser. Figure 1 shows the setup for the laser-produced plasma interferometry experiment with more details found in ref. [5]. Experiments to date have shown that even with multiple modes present in the x-ray laser beam, excellent interference fringe visibility of 0.72 ± 0.12 can be achieved. Al plasmas with lengths 0.1 – 0.5 cm were irradiated by a line focus at 10^{11} - 10^{12} W cm^{-2} with a 600 ps, 1054 nm laser pulse were probed. Two dimensional (2-D) plasma density profiles were measured and compared to 1-D, 1.5-D and 2-D hydrocode simulations [15]. Two campaigns to probe plasmas have been conducted. The first used 10× magnification imaging giving pixel limited spatial resolution of 2.5 μm at the plasma. Lateral plasma expansion and features within 10 μm of the target surface were observed demonstrating the effectiveness of picosecond soft x-ray probes for large millimeter-scale plasmas.

Figure 2 is an example of an interferogram for a 0.5 cm long plasma heated by a 600 ps, 1054 nm laser pulse. A peak density of 3×10^{20} cm^{-3} was measured and was limited by the detector to resolve fringe structure close to the target where the scalelengths are short. In principle, the x-ray laser interferometry technique should be able to probe

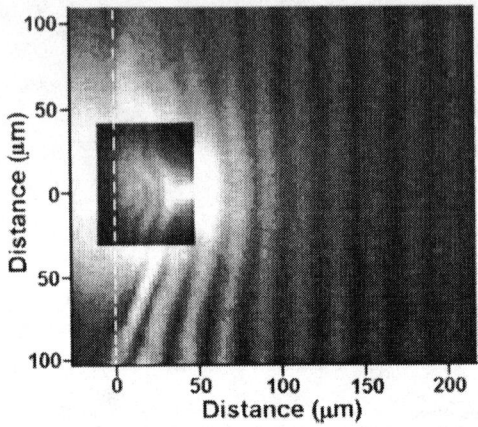

Fig. 2   A picosecond 14.7 nm x-ray laser interferogram of a 0.5 cm Al plasma recorded at 0.7ns
after the peak of 0.4 J, 600 ps pulse. Laser line focus is a 0.6 cm × 40-μm and target surface
is at 0 μm. Two different look up tables are used for clarity.

several orders of magnitude higher density with substantially less refraction compared
to other UV or optical interferometric techniques. To address this issue, a second more
recent experiment with higher 22× magnification gave substantially improved spatial
resolution of 0.6 μm. This data is under analysis and will be reported in the near future.

PICOSECOND X-RAY LASER PROBING OF MATERIALS

Various compact, pulsed extreme ultraviolet (EUV) and soft x-ray sources, for
example higher order harmonic (HOH) [16, 17] sources for valence band and laser
produced plasma (LPP) line and continuum emission [18] for core levels, have been
used for the direct measurement of the electronic structure of material surfaces. Static
chemical shifts in the Si $2p$ core level electronic structure for SiO_2 and Si_3N_4 samples
were demonstrated using 2.5 ns LPP, 255 eV x-ray pulses after integrating for ~100
shots [18]. The HOH sources have the advantage of shorter duration since the process
is driven by ultrafast pulses, e.g. less than 60 fs [16], at a high repetition rate which
makes them very attractive for dynamic pump-probe experiments. However, the
presence of multiple harmonics requires further wavelength selection with diffraction
gratings [16] or multilayers [17]. The lower photon fluence/shot requires many shots
to achieve good signal statistics. The 84.5 eV Ni-like Pd x-ray laser, Table 1, is highly
monochromatic with a combination of high photon fluence and ps duration to perform
dynamic optical pump-x-ray probe electronic structure measurements on a single-shot.
Some preliminary results are presented to demonstrate this capability.

Figure 3(a) shows the experimental setup. No optical pump was used to heat the
sample. The x-ray laser probe is collimated by a normal incidence Mo:Si multilayer
spherical mirror and relayed along the beamline by a 45° Mo:Si multilayer flat mirror.
The narrow reflectivity window of the mirrors selects the x-ray laser wavelength and

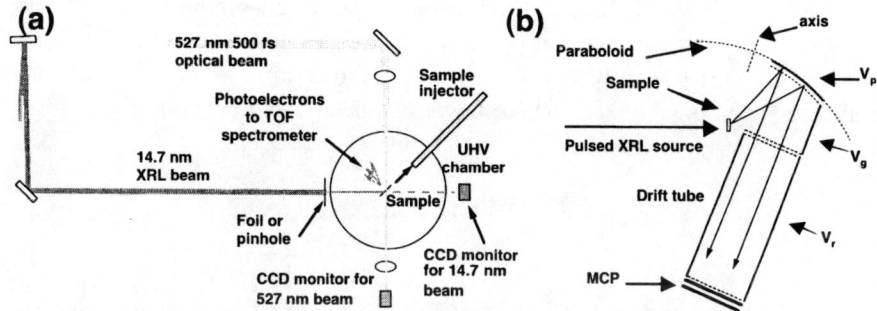

Fig. 3 (a) Layout for optical pump – x-ray laser probe of material surfaces. (b) Schematic of paraboloid mirror analyzer time-of-flight photoelectron spectrometer.

minimizes other plasma x-rays reaching the sample. A pinhole or thin filter isolates the beamline from the experimental ultra-high vacuum (UHV) chamber. The UHV chamber, operated at 5×10^{-7} mbar pressure, contains the sample to be probed and the photoelectron spectrometer, shown in Fig. 3(b). Photoelectrons, with a total binding energy and sample work function less than 84.5 eV photon energy ejected from the sample with a certain kinetic energy (KE), are collected by the paraboloid mirror and collimated before entering the drift tube [19]. A bias applied to the paraboloid mesh V_p and the retardation grid V_r can be used in tandem as a bandpass filter to control the energy and time of arrival of the electrons on the microchannel plate (MCP) detector. The signal is digitized using a fast 3 GHz oscilloscope. Figure 4 (a) and (b) show the time-of-flight spectra recorded for an *in-situ* sputter-cleaned Ta reference sample illuminated with $10^9 - 10^{10}$ x-ray laser photons. The valence band and two shallow core levels ($4f_{7/2}$ and $4f_{5/2}$ with binding energy of 22 and 24 eV, respectively) are expected to arrive between 100 and 150 ns. Peaks corresponding to these features are observed in the spectrum sitting on a background slope of secondary photo-emission (earlier than 100 ns). The secondary emission is believed to be produced by scattered photons and higher KE electrons and will be reduced further through the use of filters

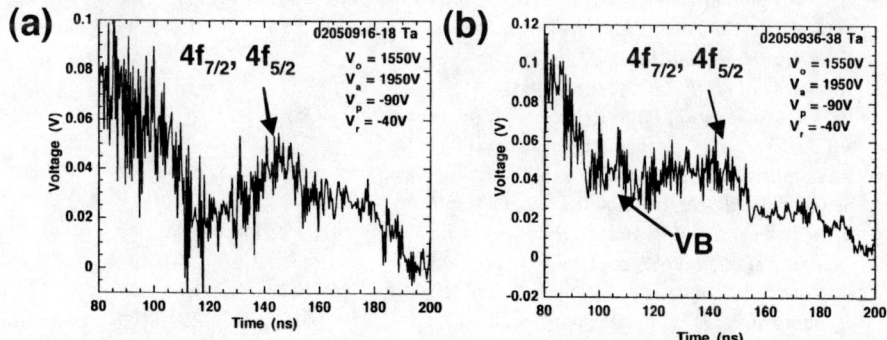

Fig. 4 (a), (b) Time-of-flight photoelectron spectra from a sputter-cleaned Ta sample for two sets of x-ray laser shots. Valence band and $4f_{7/2}$ and $4f_{5/2}$ shallow core levels are labeled.

and baffles in the setup. In addition, electrical noise, apparent on the spectra of Fig. 4 (a) and (b), is registered on the detector and is associated with the plasma generating the x-ray laser. This presently limits the signal to noise ratio and requires averaging several shots, even though there are sufficient photoelectrons for a single shot record. This is under investigation and new results will be reported soon.

ACKNOWLEDGMENTS

The authors would like to thank Andy Hazi and Al Osterheld for their continued support. JD would like to thank Dennis Matthews for permission to reproduce and modify the table from ref. 8. This work was performed under the auspices of the US Department of Energy by the University of California Lawrence Livermore National Laboratory under Contract No. W-7405-Eng-48 and by US Department of Energy Grant No. DE-FG03-98DP00208.

REFERENCES

1. J.J. Rocca *et al.*, *Phys. Rev. Lett.* **77**(8), pp. 1476-1479 (1996); B.R. Benware, C.H. Macchietto, C.H. Moreno, and J.J. Rocca, *ibid.* **81**(26), pp. 5804-5807 (1998).
2. J. Dunn *et al.*, *ibid* **84**, pp. 4834 – 4837 (2000).
3. S. Sebban *et al.*, *ibid.* **86**, pp 3004-3007 (2001).
4. P.V. Nickles *et al.*, *ibid.* **78**(14), pp. 2748-2751 (1997).
5. J. Dunn, R.F. Smith, J. Nilsen, J.R. Hunter, T.W. Barbee, Jr., V.N. Shlyaptsev, J. Filevich, J.J. Rocca, M.C. Marconi, H. Fiedorowicz, and A. Bartnik, SPIE Proc. **4505**, 62 (2001).
6. F. Strati and G. J. Tallents, *Phys. Rev. A* **64**, 013807 (2001); J. Dunn, A.L. Osterheld, Y. Li, J. Nilsen, and V.N. Shlyaptsev, in "Short Wavelength Lasers and Applications", ed. J.G. Eden and J.J. Rocca, *IEEE Journal of Selected Topics in Quantum Electronics* **5**(6), 1441 – 1446 (1999).
7. A. Klisnick *et al.*, *Phys. Rev. A.* **65**, pp.033810-1 – 4 (2002).
8. Proceedings of Workshop on "Applications of X-ray Lasers", Jan 12 - 14, 1992, San Francisco, CA. Ed. by R. London, D. Matthews, S. Suckewer, Table on page xii (1992).
9. L.B. Da Silva *et al.*, *Science* **258**, 269 (1992).
10. D.T. Attwood, in "X-ray Lasers 2000", 7th International conference on X-ray Lasers, St. Malo, France June 19 – 23, ed. G. Jamelot, C. Möller, A. Klisnick, *J.Phys. IV* **11**, Pr2-443 (2001).
11. J.J. Rocca *et al.*, *ibid* **11**, Pr2-459 (2001).
12. J. Dunn *et al.*, *ibid.* **11**, Pr2-19 (2001).
13. L. B. Da Silva *et al.*, *Phys. Rev. Lett.* **74**, pp.3991-3994 (1995).
14. J. Filevich *et al.*, *Opt. Lett.* **25**, pp. 356 (2000).
15. R.F. Smith *et al.*, "Picosecond X-ray Laser Interferometry of Dense Plasmas", *Phys. Rev. Lett.* **89**, pp.065004-1(2002); R.F. Smith *et al.*, these proceedings.
16. A. Rettenberger and R. Haight, *Phys. Rev. Lett.* **76**, pp. 1912 – 1915 (1996).
17. M. Bauer *et al.*, *ibid.* **87**, pp. 025501-1 (2001).
18. H. Kondo, T. Tomie and H. Shimizu, *Appl. Phys. Lett.* **69**, 182 (1996); H. Kondo, T. Tomie and H. Shimizu, *ibid.* **72**, pp. 2688 (1998).
19. G. Liu, J.J. Barton, C.C. Bahr and D.A. Shirley, *Nucl. Instr. Meth. A***246**, pp. 504-506 (1986).

Density depression in laser-created plasma unveiled with table-top soft x-ray laser interferometry

J. Filevich[1], J.J. Rocca[1], E. Jankowska[1a], E.C. Hammarsten[1],
M.C. Marconi[1b], S.J. Moon[2], V.N. Shlyaptsev[3].

[1]Dept. of Electrical and Computer Engineering, Colorado State University, Fort Collins, CO 80523
[2]Lawrence Livermore National Laboratory, Livermore, CA 94551
[3]Dept. of Applied Science, University of California Davis-Livermore, Livermore, CA 94551

Abstract. Soft x-ray laser interferograms of laser-created plasmas generated at moderate irradiation intensities (1×10^{11}- 7×10^{12} W cm^{-2}) with $\lambda = 1.06$ μm light pulses of ~13 ns FWHM duration and narrow focus (~30 μm) reveal the unexpected formation of an inverted density profile with a density minimum on axis and distinct plasma sidelobes. Model simulations show that this strong 2-dimensional hydrodynamic behavior is essentially a universal phenomenon that is the result of plasma radiation induced mass ablation and cooling in the areas surrounding the focal spot. These measurements, which mapped plasma densities up to 0.9×10^{21} cm^{-3}, demonstrate the use of a portable soft x-ray laser interferometer as a high resolution tool for the study of high density plasma phenomena and the validation of hydrodynamic codes.

INTRODUCTION

The understanding of the dynamics of laser-created plasmas is of fundamental and practical interest. The hydrodynamic motion of plasmas created by laser irradiation of solid targets is conventionally known to result in electron density distributions with maximum density along the axis of the irradiation beam. However, in plasmas generated with high irradiation intensities the ponderomotive force has been observed to cause a density depression or cavity in the electron density profile. Early interferometry experiments of laser-created plasmas at irradiation intensities of 3×10^{14} W cm^{-2} showed a flattening of the interfering fringes in the sub-critical region that, for an axis-symmetric plasma, is indicative of a density depression [1]. Density depressions induced by the ponderomotive force have also been observed in numerous other high-intensity laser experiments, in agreement with simulations. Some of the most recent studies include the formation of plasma channels and laser-hole boring in underdense and overdense plasmas motivated by the fast ignitor concept in inertial confinement fusion [2,3]. These experiments involved laser intensities of 1.7×10^{15} W cm^{-2} and 2×10^{17} W cm^{-2} respectively. In addition, for several cases involving short laser pulses the saturation of the heat flux, refraction, or channeling of the laser radiation due to relativistic self-focusing at ultrahigh fluxes are found to be

CP641, *X-Ray Lasers 2002: 8th International Conference on X-Ray Lasers*, edited by J. J. Rocca et al.
© 2002 American Institute of Physics 0-7354-0096-2/02/$19.00

responsible for density suppression [4,5]. In this paper we report the observation of a pronounced density minimum on axis in both line-focus and spot-focus plasmas generated at intensities as low as 10^{11} W cm^{-2}, which cannot be explained by the ponderomotive force, the effects of laser radiation refraction, electron heat saturation, nor the influence of plasma instabilities. Our case is different, yet universal enough to exist in a wide parameter space. In fact, in retrospective, evidence of these effects may be inferred from published visible laser plasma interferograms mapping much smaller electron densities of the order of ~10^{18} cm^{-3} [6,7]. However, this kind of two-dimensional (2D) plasma behavior was not clearly revealed from the data nor understood. In our measurements the 2D effect is distinctively uncovered by the use of soft x-ray laser interferometry, which for the case of the spot-focus plasma experiment allowed us to map the density profile up to ~10^{21} cm^{-3}, close to the critical density. Simulations show that plasma radiation and heat conduction enlarge the ablation region creating a plasma density profile with large distinctive side-lobes and a minimum on axis.

Soft x-ray laser interferometry is a powerful new tool for the diagnostics of dense plasmas, that allows probing of plasma densities and scale-lengths larger than those that can be probed with optical lasers. Interferometry measurements of dense large scale plasmas conducted using a laboratory soft x-ray laser pumped by the Nova laser [8,9] significantly exceeded the density values probed with optical lasers, but were limited to several shots per day by the low repetition frequency of the pump laser. The results reported herein were obtained with a new soft x-ray interferometry tool based on a portable soft x-ray laser and a Mach-Zehnder interferometer configuration in which diffraction gratings are used as beam splitters [10]. The high repetition rate of the probe laser allowed us to complete several series of interferograms of laser-created plasmas that map the entire temporal plasma evolution.

EXPERIMENTAL SETUP

The experimental set up used in the measurements comprises a λ=1.06 μm Q-switched Nd-YAG laser (Quanta Ray GCR-190) used to irradiate the target, the soft x-ray laser probe, and a soft x-ray interferometer. The Ne-like Ar 46.9 nm capillary discharge laser was described in previous publications [11-13]. The interferometer is based on a Mach-Zehnder configuration in which gold-coated diffraction gratings positioned at an incidence angle of 79 degrees are used as beam splitters [10]. It has the advantages of high throughput (~ 6% per arm excluding imaging mirrors) and increased robustness against plasma debris. The diffraction gratings were designed to split the incoming soft x-ray laser beam evenly into two orders. Elongated gold-coated grazing incidence mirrors are used to direct the beams towards the second grating, where they are recombined to produce the interference pattern. Spherical mirrors coated with Si/Sc multilayers that provide a normal incidence reflectivity of ~ 40 percent [14] were used to image the plasma onto a CCD/MCP detector with a magnification of either 51.2× or 24.7×. The plasmas were generated by focusing the beam of the Nd-YAG laser onto a polished electrolytic copper target placed in the zero order arm of the interferometer. The interferometer was pre-aligned using a λ= 824 nm semiconductor laser having a temporal coherence length similar to that of the soft x-ray laser (~ 200 μm). To ensure that the soft x-ray and the semiconductor laser beams follow the same path the diffraction gratings were ruled with two vertically separated regions with line densities of 300 lines/mm and 17.06 lines/mm for the soft x-ray beam and IR laser beam respectively. The detector consisted of a micro-channel plate-phosphorous screen-CCD combination. The micro-channel plate was gated to further reduce the background radiation produced by the plasma self-emission. The fringe visibility of the interferograms obtained without the plasma present exceeds 50 percent across the entire 400 × 500 μm^2 field of view.

MEASUREMENTS AND SIMULATIONS

Figure 1 shows a series of interferograms that depict the evolution of a line focus plasma created at an irradiation intensity of 1×10^{11} W cm^{-2} by focusing 0.63 J - 13 ns FWHM duration laser pulses into a ~ 30 μm wide line 1.8 mm in length. These interferograms were obtained with a magnification of 24.7×. Fig. 2 shows the

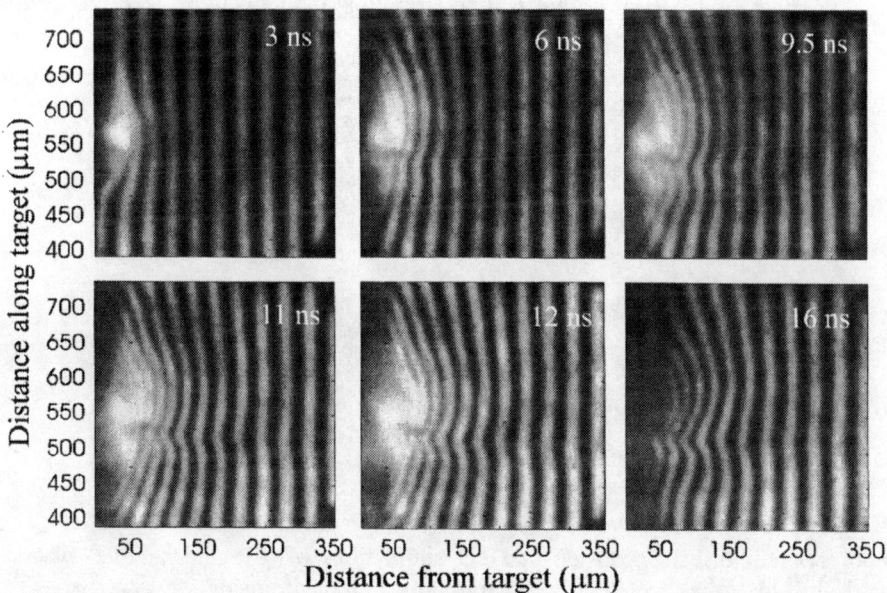

Figure 1. Sequence of soft x-ray interferograms describing the evolution of 1.8 mm long plasma generated focusing a 13 ns FWHM Nd-YAG laser pulses with 0.6 J energy onto a ~ 30 μm wide line on a copper target. The times delays are measured respect to the beginning of the laser pulse. The heating laser is incident from the right.

corresponding density maps obtained from the fringe shifts assuming the plasma is uniform along the probe beam axis. The earlier interferograms (see the 3-ns frame) show an expanding plasma with a density distribution that presents a maximum on axis. However, as time evolves and the pump laser intensity increases, the plasma density profile acquires a concave shape with a density minimum on axis, which is already clearly developed at 6 ns after initiation of the current pulse, and that becomes more pronounced in the subsequent several-ns. The asymmetry observed in some of the interferograms is due to a slight misalignment of the target. At times after termination of the Nd-YAG laser pulse, the decreasing degree of ionization of the plasma causes significant absorption of the probe beam by photo-ionization. Measurements were conducted to verify that the observed density profile is not a consequence of the intensity profile of the pump laser. It should be noticed that due to the large scale-length of the plasma and large density gradients an ultraviolet laser probe (eg. 4th harmonic of Nd:YAG at 266 nm) would be very strongly refracted.

To understand this unusual electron density profile we conducted simulations with the hydrodynamic code LASNEX [15]. The severe distortion of the plasma motion inferred from our soft x-ray laser interferograms can be well described only by two-dimensional (2D) simulations. Both single-dimension (1D) and 2D modeling rule out the possibility that the ponderomotive force might be the primary cause of the observed plasma behavior, as it is known to take place for larger fluxes [1] or with long wavelength laser irradiation [16]. The computed evolution of the plasma density profile is shown in Fig. 3. The resulting physical picture consists of a relatively high temperature (~36 eV) central region surrounded by a lower temperature (~10 eV) plasma on either side. The on-axis density distribution and velocity fit well the results

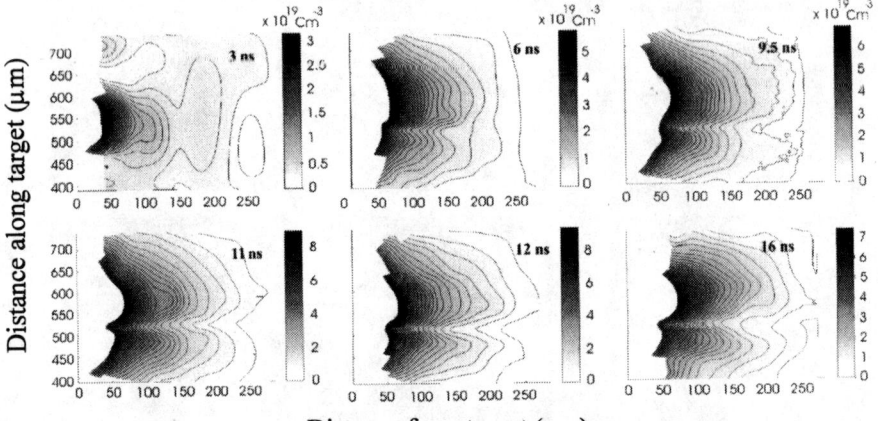

Figure 2. Electron density profiles corresponding to the interferograms of Fig. 1.

of both 1D cylindrical expansion and 2D simulations of a laser irradiated plasma. Instead, the side-lobes are entirely 2D formations created outside the laser-irradiated target region by the build-up of new cold material resulting mainly from XUV plasma radiation induced evaporation. In addition 2D simulations show that radiation cooling contributes in substantially lowering the temperature of the sidelobes. The lobes have slower axial expansion ($<1\times10^6$ cm s^{-1}) as compared to the hotter central plasma region ($5_{-}6\times10^6$ cm s^{-1}). After several ns the pressure balances in the lateral direction within a distance ~100 μm from the axis, as dictated by the sounds speed (~10^6 cm s^{-1}) and plasma lifetime (~10^{-8} s), hence forming a density depression on axis and lobes in the colder areas. To illustrate that the sidelobes are mainly caused by outside spot ablation, we conducted simulations in which we constrained the target ablation outside ablation, we conducted simulations in which we constrained the target ablation outside the area illuminated by the laser while maintaining otherwise identical physics. Simulations show that in this case the expansion is almost classical and presents significantly smaller density inhomogeneities. Furthermore an additional simulation shows that if all radiation effects are excluded the expansion becomes classical and the sidelobes disappear, in which case the temperature in the conical expansion does not drop as dramatically along the surface.

Our modeling shows that this 2D density behavior is essentially a universal effect. It includes the case of larger focal spots, but in this case it is not as pronounced. Some indications to this effect can be found in the experiment of Bol'shov et al. [6]. Due to the fact that dense plasmas are very efficient sources of XUV radiation even at low plasma temperatures, this kind of plasma behavior should be seen in many experiments. However, because it takes a relatively long time ~4-10 ns for the off-spot material to evaporate and the colder plasma to expand, this effect was not previously clearly identified in shorter pulse experiments. The longer pulse duration in the present experimental situation allows the establishment of the pressure balance, such that the hotter central plasma corresponds to a lower density.

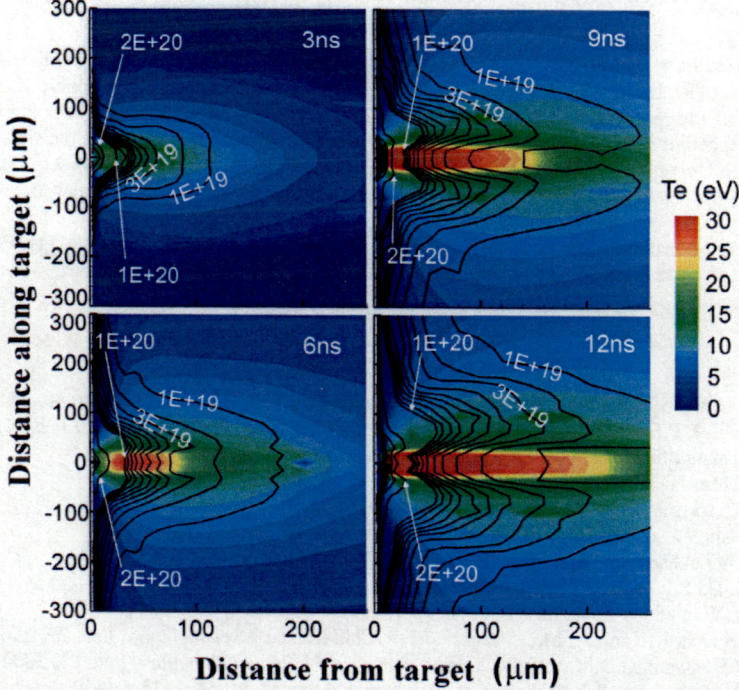

Figure 3. Simulated electron density (line contours) and temperature (filled contour) profiles of the line focus plasma of figures 1-2 computed using LASNEX. The heating laser is incident from the right.

A similar but even more pronounced density depression was observed in plasmas generated at increased irradiation intensity by focusing the Nd:YAG laser onto a spot <30 μm in diameter. The formation of pronounced density peaks off-axis is again seen. The electron density measured, at the time of maximum laser intensity, reached a value of 9×10^{20} cm^{-3} (which amounts to about 90% of the critical density for the $\lambda = 1.06$ μm pump laser) at 27 μm from the target near. For more details on the point focus experiments that include series of interferograms and a discussion see the paper by Rocca et al. in these proceedings [17].

CONCLUSIONS

In conclusion, we have observed 2D hydrodynamic effects of large magnitude in laser-created plasmas generated at relatively low irradiation intensities leading to the formation of concave electron density profiles. Simulations show that the build up of cold mass outside the focal spot, combined with pressure balance results in the formation of the observed density depression. The results constitute the first demonstration of the use of table-top soft x-ray laser interferometry in the observation of new high density plasma phenomena. Soft x-ray laser interferometry with compact table-top lasers is positioned to become an important high-resolution diagnostic, not only to measure density profiles in high density plasmas, but also for the development and validation of hydrodynamic codes and plasma theory.

The authors would like to thank G.B.Zimmerman, D.C.Eder (LLNL), N.G.Kovalsky and A.E.Stepanov (TRINITI), A. Ya. Faenov (VNIIFTRI), Yu.A.Zakharenkov for valuable discussions. This work was supported by U.S. Department of Energy grant No. DE-FG03-98DP00208 and by the National Science Foundation. Part of this work was performed under the auspices of the U.S. Dept. of Energy by the University of California, Lawrence Livermore National Laboratory under contract No. W-7405-Eng-48. We also gratefully acknowledge the support of the W.M. Keck Foundation.

[a] E. Jankowska permanent address is : Dept. of Physics, Wroclaw University of Technology, Poland.
[b] M.C. Marconi permanent address is: Dept. of Physics, University of Buenos Aires, Argentina.

REFERENCES

1. D.T.Attwood, D.W.Sweeney, J.M. Auerbach and P.H.Y. Lee, Phys. Rev. Lett. **40**, 184, (1978).
2. S. Wilks, P.E.Young, J.Hammer, M.Tabak and W.L. Kruer. Phys. Rev. Lett. **73**, 2994, (1994).
3. K. Takahashi et.al . Phys. Rev. Lett. **84**, 2405, (2000).
4. C.E.Max, C.F.McKee, and W.C.Mead, Phys.Rev.Lett., **45**, 28 (1980).
5. G.S.Sarkisov, Phys. Rev. E, **59**, 7042 (1999).
6. Bol'shov et al, Sov.Phys. JETP **65** 1160 (1987).
7. Yu.A.Zakharenkov et al., Sov. Phys. JETP, **70**, 547 (1976).
8. L.B. Da Silva, et. al, Phys. Rev. Lett. **74**, 20, 3991 (1995).
9. A.S. Wan et. al, Phys. Rev. E, **55**, 6293, (1997).
10. J.Filevich, K.Kanizay, M.C.Marconi, J.L.A.Chilla, and J.J.Rocca. Optics Lett. **25**, 356, (2000).
11. B.R.Benware, C.D.Macchietto, C.H.Moreno, and J.J.Rocca, Phys. Rev. Lett. **81**, 5804 (1998).
12. C.D. Macchietto, B.R. Benware, and J.J. Rocca. Optics Lett. **24**, 1115, (1999).
13. Y.Liu, M.Seminario, F.Tomasel, J.J.Rocca and D.T.Attwood, Phys. Rev. A **63**, 033802 (2001).
14. Yu. A. Uspenskii et. al Optics Lett. **23**, 771-773, (1998).
15. G.D. Zimmerman and W.L. Kruer. Comments Plasma Phys. Controlled Fusion **2**, 51, (1975).
16. V.Yu.Baranov at al, Laser and Particle Beams **14** 347 (1996).
17. J.J. Rocca et al. Proceedings on 8th Int. Conf. on X-Ray Lasers, May 2002, Aspen, Colorado, USA.

X-ray Double-Exposure Holography

Hongyi Gao, Jianwen Chen, Honglan Xie and Zhizhan Xu

Shanghai Institute of Optics and Fine Mechanics, The Chinese Academy of Sciences, P.O. Box 800-211, Shanghai 201800, China

Abstract. Double-exposure holography for measuring phase information in X-ray fields is suggested in this paper. The double-exposure method cannot only obtain directly the phase information, but also gives qualitative and quantitative results. This means also that one can obtain simultaneous information about both the amplitude and phase in an experiment.

In 1952, Baze [1] first suggested Gabor holography for X-ray imaging. In 1986, Howells et al. [2], recorded X-ray in-line holograms using a storage ring X-ray source and a grating monochromator and reconstructed them using visible laser at Brookhaven National Laboratory. With the development of undulator, Jacobsen et al. [3], recorded Gabor X-ray holograms and obtained the reconstructed image with resolution of 56 nm using a numerical reconstruction method. Simultaneously, a digital twin imaging elimination method in soft X-ray in-line holography was developed by Joyeux et al. [4] at Srsay Cedex, France.

In 1992, I. McNulty et al. [5] recorded the first X-ray Lensless Fourier Transform hologram in Brookhaven National Laboratory, and the reconstructed it by digital method with the resolution up to 60 nm. This is the first X-ray off-axis hologram. On the base of their working, we suggest a method called X-ray double-exposure holography for measuring phase information.

As we know that holography is a two-step imaging process. In the first step, the information of an object wave is stored. In the second step, the hologram is reconstructed using coherent visible light or digital method. For weak phase object, in general, reconstruction of the phase difference amplification is concerned using a Mach-Zehnder interferometer as show in Fig.1. In a sense, it will depend on any artificial disturbance introduced in the reconstruction step [6]. Double-exposure holographic interferometer interference takes place between the wave fronts reconstructed.

The principle is to record two holograms on one film, where the first recording is the interference fringe without specimen, and the second with specimen. The first hologram "froze" the phase information of system while the second recorded the phase information from both the system and the object. The difference between two recordings gives out pure phase information of the object.

Double-exposure holography has the merits of eliminating the aberration and the noise of system [7], which provides us with an accurate way to obtain the information

CP641, *X-Ray Lasers 2002: 8th International Conference on X-Ray Lasers*, edited by J. J. Rocca et al.
© 2002 American Institute of Physics 0-7354-0096-2/02/$19.00

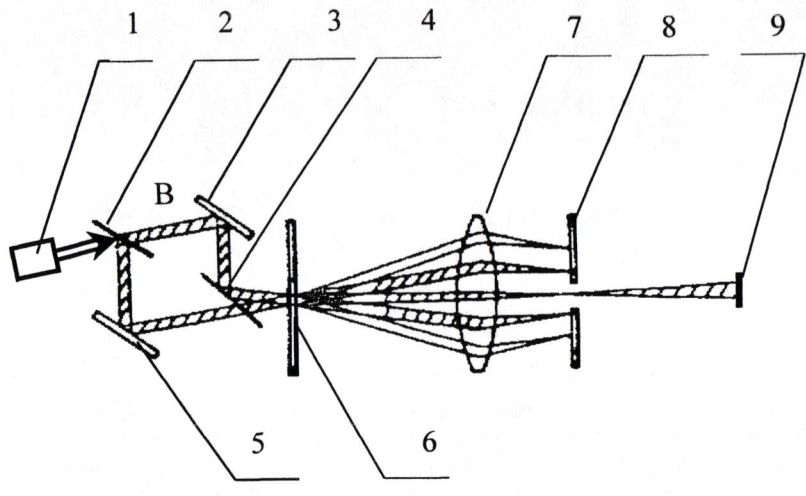

FIGURE 1. Mach-Zehnder interferometer for phase difference amplification
1—He-Ne Laser; 2, 4—50% splitter; 3, 5—total reflector; 6—hologram; 7—lens; 8—diaphragm; 9—CCD system

of phase variation between two different states or to make some measurements of the dynamic distortion. Combining with CCD as a recording device and computer as a reconstruction facility, the process becomes more convenient and smart. The sensitivity of double-exposure holography is determined by the minimum detectable displacement of the interference fringes. This is always a few fractions of fringes width and is not high enough for precise measurements. A method for phase difference amplification [8] of double-exposure holograms by means of carrier fringes with high frequency is given. This method keeps the main advantages of the double-exposure method, and the sensitivity can be enhanced by orders of magnitude.

ACKNOWLEDGEMENTS

This project was partly supported by Chinese High-Tech Program, Chinese National Major Basic Research Development Program (No. G1999075200).

REFERENCES

1. V.A. Baze, A study in diffraction micorscopy with special reference to X-ray, J. O. S. A., 1952, 42(10): 756-762.
2. M. Howells, C. Jacobsen, J. Kirz et al., X-ray holography at improved resolution: a study of zlymogen gramoles, Science, 1987, 238: 514-517.
3. C. Jacobsen, M. Howells, J. Kirz et al., X-ray holographic microscopy using photoresists, J. Opt. Soc. Am., 1990, A7(10): 1847-1861.
4. G. Koren, F.Polack, D.Joyeux, Iterative algorithm for twin-image elimination in in-line holography using finite-support constraints, J. Opt. Soc. Am, 1993, A10(3): 423-433.
5. I. McNulty, J. Kirz, C. Jacobsen, E. H. Anderson, M. Howells, D. Kern, High-resolution Imaging by Fourier Transform X-ray Holography, Science, 1992, 256, 1009-1012.
6. J. W. Chen, Some problems of reconstruction in electron holography, Optical Engineering, 1993, 32(10):2593.
7. J. W. Chen, G. Matteucci, A. Migliori, G. F. Missiroli et al., Mapping of Microelectrostatic fields by means of electron holography: Theoretical and experiments, Physical Review A, 1989, 40, 3136-3145.
8. Shufen Fu and J. W. Chen, Phase Difference Amplification of Double-exposure Holograms, Optics Communication, 1988, 67, 417-420.

Soft X-ray Laser Interferometry/Shadowgraphy of Exploding Wire Plasmas

E. Jankowska[a], E.C. Hammarsten, B. Szapiro[b], J. Filevich,
M.C. Marconi[c], J.J. Rocca

Dept. of Electrical and Computer Engineering,
Colorado State University, Fort Collins, CO 80523

Abstract. We present the first results from soft x-ray laser interferometry measurements of current-driven thin wire explosions obtained using a capillary discharge pumped 46.9 nm laser and an amplitude division interferometer based on diffraction gratings. We have obtained series of high-resolution soft x-ray interferograms/shadowgrams that depict the initial stage of the evolution of exploding Al wires 15 μm and 25 μm in diameter. The images show a dense vapor core that completely absorbs the probe beam during the initial part of the explosion, and a surrounding plasma shell where both a shift of the interference fringes and partial absorption of the soft x-ray laser probe beam are observed. The excitation of the 25 μm diameter wires at a current rate of 30 A/ns is observed to result in the uniform expansion. However, an increase of the rate of energy deposited per unit mass is observed to give rise to significant instabilities. The expansion velocity of the wire core was determined from the variation of the measured absorption width of the soft x-ray laser beam. The determination of the electron density and vapor density profile requires the combined analysis of the soft x-ray absorption and fringe shift data.

INTRODUCTION

The study of the physics of thin exploding metal wires is of significant current interest. Cylindrical arrays of thin exploding metal wires have been used at Sandia National Laboratories to form plasmas in fast Z-pinch implosions in which up to 200 TW of x-ray radiation were generated [1]. Progress has been made in the diagnostics of the initial phase of these plasmas [2-4]. Exploding wires driven by kA current pulses are characterized by a dense central core of neutral vapor surrounded by a plasma shell [2]. The electron density distribution in the coronal region of single exploding wires and wire arrays have been measured using optical laser interferometry [3,4]. Those experiments mapped the coronal plasmas in regions with plasma density values of the order of 10^{17} cm^{-3} - 10^{18} cm^{-3}. The metal vapor density has been measured with x-ray radiography using an x-pinch radiation source [2]. The probing of the exploding wire plasmas with soft x-ray lasers can potentially provide additional information on the early part of the evolution, where the density is high. In this paper we report the first results of the study of thin Al wire explosions using a table-top 46.9 nm capillary discharge soft x-ray laser probe.

CP641, *X-Ray Lasers 2002: 8th International Conference on X-Ray Lasers,* edited by J. J. Rocca et al.
© 2002 American Institute of Physics 0-7354-0096-2/02/$19.00

EXPERIMENTAL SETUP

The experimental setup used to probe the plasma is shown schematically in Fig.1. The system consists of a soft x-ray laser probe, the exploding wire assembly, and an amplitude division soft x-ray interferometer based on diffraction gratings (DGI). The interferometer consists of a Mach-Zehnder configuration where the laser beam is split and recombined by diffraction gratings. The zero and first order beams diffracted on the first grating form the two arms of the interferometer. The elongated mirrors redirect the beams toward the second grating where they are recombined to form the interference pattern. The interferometer was pre–aligned with a $\lambda = 824$ nm semiconductor laser. The soft x-ray DGI is described in more details in a previous publication [5,6] and in other papers in these proceedings [7,8].

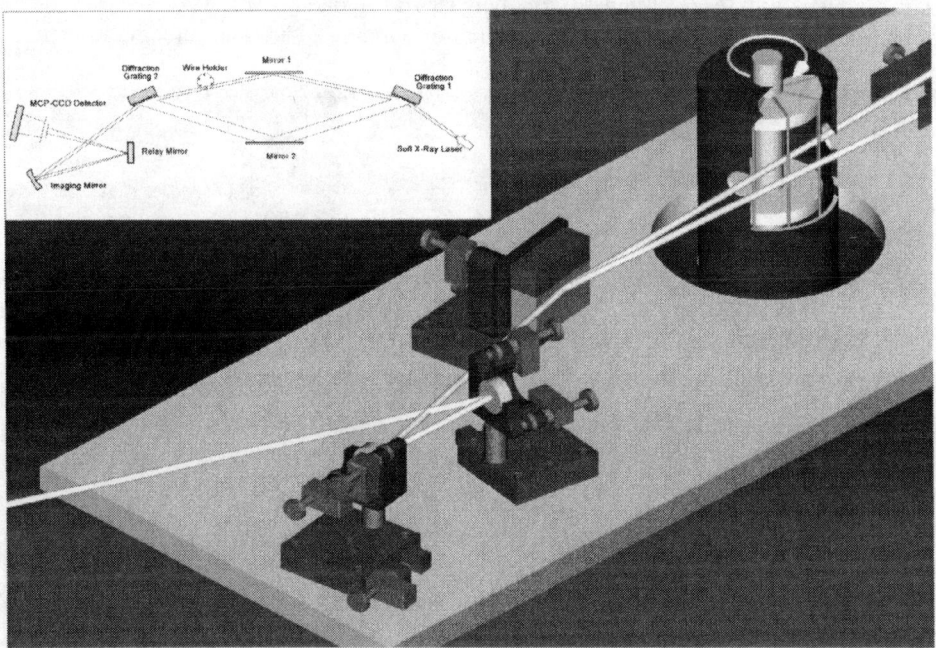

FIGURE 1. Soft X-ray diffraction grating interferometer (Insert) used to study current-driven wire explosions. The lower figure is a partial schematic of the interferometer showing the wire assembly second diffraction grating, and imaging and relay mirrors

A neon-like argon capillary discharge pumped laser emitting a single laser line at a wavelength of 46.9 nm was used as the probe beam [6]. For this experiment, laser pulse energies of ~0.1 mJ and pulse duration of ~1.2 ns FWHM were used. The laser beam has a divergence ~4.5 mrad. The short pulse, high brightness and good spatial coherence [9] of the x-ray laser allowed us to obtain good quality interferograms with fringe visibility that, when the plasma is not present, exceeds 50 percent across the entire field of view. The exploding wire plasmas were generated by heating Al wires 15 μm or 25 μm in diameter with current pulses of either 4.7 kA or 6.0 kA peak amplitude, with an average current increase rate of 30 A/ns and 42 A/ns respectively. A motorized wire holder (top right on Fig.1) was developed to be able to realize several single wire explosions without having to interrupt vacuum. This wire holder was placed on the path of the zero order beam between the elongated mirror and the second grating. The flat relay mirror and a spherical imaging mirror of 30 cm focal length, both coated with Si/Sc multilayers [10], were used to image the exploding wire with a magnification of 25X onto a gated CCD/MCP detector. All the measurements were conducted under vacuum at a pressure of $< 2 \times 10^{-5}$ Torr.

MEASUREMENTS

A series of interferograms illustrating the early stages of the explosion of a 25 μm Al wire excited by a current pulse that increases at a rate of 30 A/ns (4.7 kA peak amplitude at 155 ns) is shown in Fig. 2. The times indicated are measured relative to

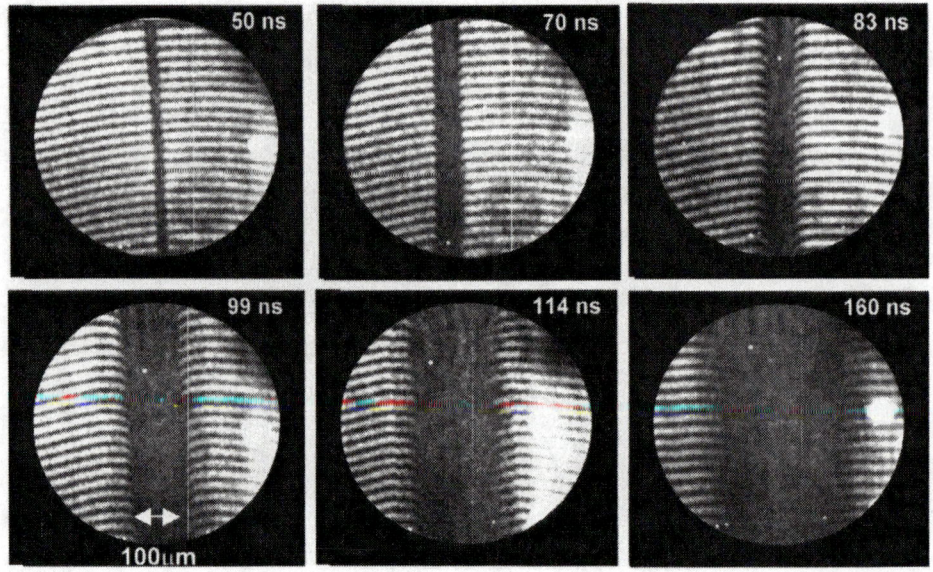

FIGURE 2. Series of the interferograms illustrating the explosion of a 25 μm Al wire excited with a current pulse of 4.7 kA peak current with 155 ns rise time.

the initiation of the current pulse. The relatively good shot to shot reproducibility of the wire explosion allows to map the evolution of the plasma by obtaining one interferogram per each wire explosion. The first frame shows the wire shadow at a very early stage of the explosion and the subsequent frames illustrate the rapid expansion. The central part of the vaporized wire, which contains a large density of neutral Al atoms that can be photo-ionized by the probe laser beam, completely absorbs the probe beam. It should be noticed that the 26.5 eV photons of the probe beam are also capable of photo-ionizing singly ionized Al and therefore this species also contribute to the beam absorption. Such absorption effectively creates a sequence of shadowgrams from which the velocity of expansion of the wire core can be measured. The explosion of the 25 μm wire driven by the 30 A/ns current pulse is observed to be relatively uniform along the length of the wire, indicating the absence of significant instabilities. Figure 3 shows the measured variation of the diameter of the absorbing core region as a function of time. The radial expansion is observed to reach an asymptotic velocity of 3.4 μm/ns.

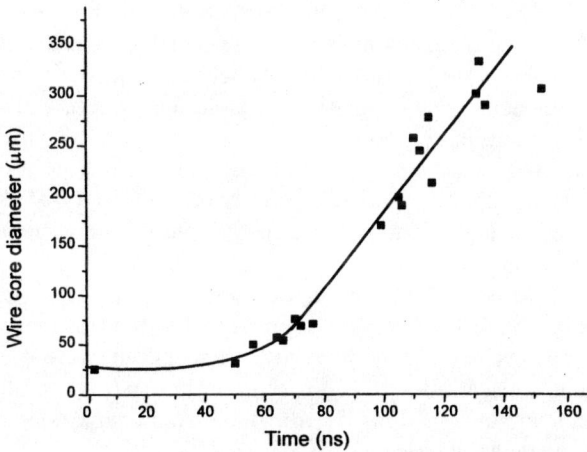

FIGURE 3. Expansion of the core region of an exploding Al wire 25 μm diameter excited by 30 A/ns current pulse.

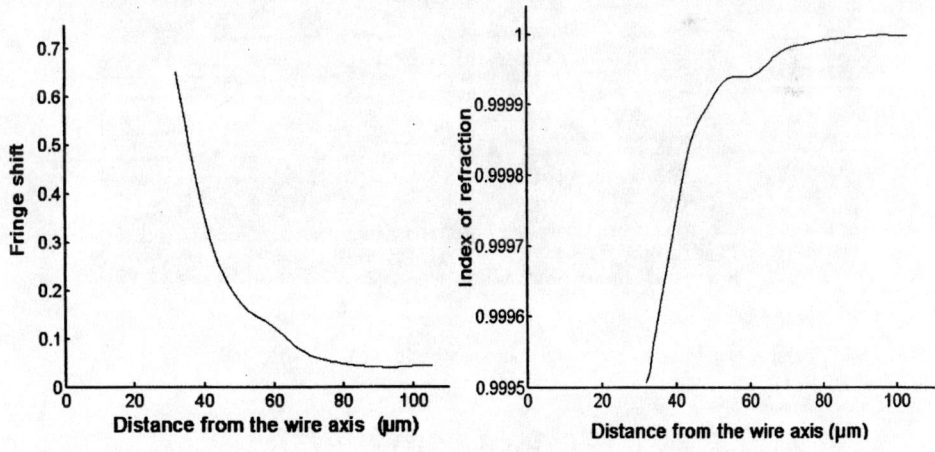

FIGURE 4. Fringe shift (a) and refractive index distribution (b) for the interferogram shown in Fig.2 corresponding to a delay of 83 ns.

501

At the periphery of the exploding wire core the absorption decreases and a deflection of the interference fringes is observed. As an example, figure 4 shows the measured fringe shift as a function of the distance from the wire axis for the case of the interferogram corresponding to an 83 ns delay in Fig 2. The variation of the index of refraction as a function of the radius obtained by Abel inversion of the fringe shift data is shown in the same figure. Due to the presence of neutral atoms the analysis of this data is significantly more complicated than that of interferograms corresponding to highly ionized plasmas where the fringe shifts are completely dominated by the free electrons contribution. This is particularly true for the wavelength of the probe beam used in this experiment, 46.9 nm, because the contribution of free electrons and neutral atoms to the refractive index have the same sign. Nevertheless it is in principle possible to separate the contributions due to neutral atoms from those of the free electrons by simultaneously analyzing Abel inverted absorption, and the fringe shift data obtained from the interferograms.

Experiments were also conducted for different excitation currents and wire diameters. In contrast with the relatively good axial uniformity of the 25 μm wire explosions illustrated in Fig. 2, an increase in the rate of energy deposited per unit mass was observed to give rise to significant instabilities. Both the images corresponding to an increase of the excitation current rate to 42 A/ns for the 25 μm diameter wire (pulse with 6.0 kA peak amplitude at 144 ns) and those corresponding to 15 μm diameter wires excited by 30 A/ns current pulses were often observed to display significant instabilities. This observation is exemplified by the interferograms/shadowgrams of Fig. 5, in which significant non-uniformities are observed to develop along the wire axis.

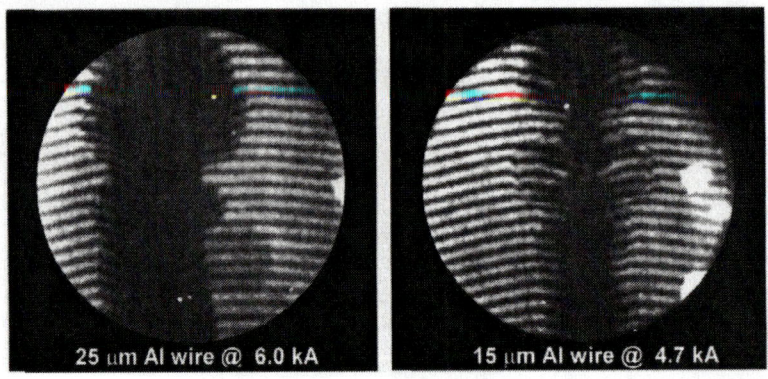

FIGURE 5. Plasma instabilities observed for 25 μm wire (6.0 kA, 42 A/ns) and 15 μm wire (4.7 kA, 30 A/ns)

CONCLUSIONS

We have used a capillary discharge-pumped table-top soft x-ray laser to probe the evolution of current-driven thin Al wire explosions utilizing interferometry/

shadowgraphy techniques. The central core of the explosion contains a large density of neutral metal vapor that absorbs the laser beam during the early part of the evolution, limiting the region of visible fringe shifts to the periphery of the central core. Al wires with 25 μm diameter excited with current pulses of 4.7 kA (30 A/ns) were observed to present a stable expansion, with relatively uniform behavior along the wire axis. In contrast, an increase in the energy deposition rate (obtained by either increasing the current or decreasing the wire diameter) resulted in significant instabilities. The velocity of expansion of the core region was measured to be 3.4 μm/ns for 25 μm diameter wires excited by 4.7 kA current pulses. The determination of the electron density and vapor density profile requires the combined analysis of the soft x-ray absorption and fringe shift data.

ACKNOWLEDGMENTS

This work was supported by U.S. Department of Energy grant No. DE-FG03-98DP00208 and by the National Science Foundation. We also gratefully acknowledge the support of the W.M. Keck Foundation.

[a] E. Jankowska permanent address is: Dept. of Physics, Wroclaw University of Technology, Poland.
[b] B. Szapiro permanent address is: Dept. of Physics, The University of the South, Sewanee.
[c] M.C. Marconi permanent address is: Dept. of Physics, University of Buenos Aires, Argentina.

REFERENCES

1. C.Deeney, M.R.Douglas, R.B.Spielman, T.J.Nash, D.L.Peterson, P.L.Eplattenier, G.A.Chandler, J.F.Seamen, K.W.Struve, Phys. Rev. Lett. **81**, 22, 4883 (1998).
2. S.A.Pikuz, V.A.Romanova, N.Y.Baryshnikov, Min Hu, B.R.Kusse, D.B.Sinars, T.A.Shelkovenko, D.A.Hammer, Rev. Sci. Instrum. **72**, 1098 (2001).
3. D.B.Sinars, T.A.Shelkovenko, S.A.Pikuz, Min Hu, V.M.Romanova, K.M.Chandler, J.B.Greenly, D.A.Hammer, B.R.Kusse, Phys. of Plasmas 7, 2, 429 (2000).
4. S.V.Lebedev, R.Aliaga-Rossel, S.N.Bland, J.P.Chiittenden, A.E.Dangor, M.G.Haines, I.H.Mitchell, Phys. of Plasmas **6**, 5, 2016 (1999).
5. J.Filevich, K.Kanizay, M.C.Marconi, J.L.A.Chilla, and J.J.Rocca. Opt. Lett. **25**, 356, (2000).
6. B.R.Benware, C.D.Macchietto, C H.Moreno, and J.J.Rocca, Phys. Rev. Lett. **81**, 5804 (1998).
7. J.Filevich, J.J.Rocca, E.Jankowska, E.C.Hammarsten, M.C.Marconi, S.J.Moon, V.N.Shlyaptsev, Proceedings on 8th International Conference on X-ray Lasers, May 2002, Aspen, Colorado, USA.
8. J.J.Rocca, B.Luther, M.C Marconi, T.Whiteaker, D.A.Braley, J.Filevich, E.C.Hammarsten, A.Rahman, B.T.Szapiro, Y.Wang, E.Jankowska, M.Grisham, T.Richter, V.N.Shlyaptsev, S.J.Moon, Proceedings on 8th InternationalConference on X-ray Lasers, May 2002, Aspen, Colorado, USA.
9. Y.Liu, M.Seminario, F.Tomasel, J.J.Rocca and D.T.Attwood., Phys. Rev. A **63**, 033802 (2001).
10. Y.Yu, A.Uspenskii, V.E.Lebashov, A.V.Vinogradov, A.I.Fedorenko, V.V.Kondratenko, Yu, P.Pershing, E.N.Zubalcv, and V.Yu Fcdotov, Optics Lett. **23**, 771 (1998).

Ablation of Organic Polymers and Elemental Solids Induced by Intense XUV Radiation

Libor Juha[1,3], Ansgar R. Präg[1], Josef Krása[1], Andrea Cejnarová[1], Božena Králiková[1], Jiří Skála[1], Dagmar Chvostová[1], Vladimír Vorlíček[1], Jacek Krzywinski[2,3], Andrzej Andrejczuk[2], Marek Jurek[2,3], Dorota Klinger[2], Ryszard Sobierajski[3,4], Henryk Fiedorowicz[5], Andrzej Bartnik[5], Ladislav Pína[6], Jozef Kravárik[7], Pavel Kubeš[7], Yuri L. Bakshaev[8], Andrei S. Chernenko[8], Valeri D. Korolev[8], Mikhail I. Ivanov[9], Marek Scholz[10], Leszek Ryc[10], Krzysztof Tomaszewski[10], Richard Viskup[11], Frederick P. Boody[12]

[1]Institute of Physics, Czech Academy of Sciences, Na Slovance 2, 182 21 Prague 8, Czech Republic
[2]Institute of Physics, Polish Academy of Sciences, Al. Lotników 32/46, PL-02-668 Warsaw, Poland
[3]Deutsches Elektronen-Synchrotron DESY, Notkestrasse 85, 22603 Hamburg, Germany.
[4]Warsaw University of Technology, Pl. Politechniki 1, PL-00-661 Warsaw, Poland
[5]Military University of Technology, Ul. Kaliskiego 2, PL-00-908 Warsaw, Poland
[6]Czech Technical University, Břehová 7, 115 19 Prague 1, Czech Republic
[7]Czech Technical University, Technická 2, 166 27 Prague 6, Czech Republic
[8]Russian Research Center "Kurchatov Institute", 123182 Moscow, Russia
[9]Research Institute of Pulsed Systems, 115304 Moscow, Russia
[10]Institute of Plasma Physics and Laser Microfusion, Hery 23, PL-00-908 Warsaw, Poland
[11]University of Bratislava, Mlynská Dolina, 84215 Bratislava, Slovakia
[12]Ion Light Technologies GmbH, Lessingstrasse 2c, 93077 Bad Abbach, Germany

Abstract. The ablation efficiency of organic polymers (polymethylmethacrylate – PMMA, poly-tetrafluoroethylene – PTFE, polyethyleneterephtalate – PET, and polyimide – PI) and elemental solids (aluminum and silicon) by single pulses of extreme ultraviolet (XUV) radiation emitted from Z-pinch, plasma-focus, and laser-produced plasmas was investigated. The ablation characteristics measured for these plasma-based sources will be compared with those obtained for irradiation of samples with XUV radiation generated by a free-electron laser (FEL). The Z-pinch was driven by the S-300 pulsed-power machine (Kurchatov Institute, Moscow) and the plasma focus was realized in the PF-1000 machine (Institute of Plasma Physics and Laser Microfusion, Warsaw). Higher temperature plasma than with the discharge plasmas was obtained by focusing the near-infrared (fundamental frequency) beam from the PALS high-power iodine laser system (Czech Academy of Sciences, Prague) on the surface of a metallic slab target or into single- and double-gas-puff target placed in a vacuum interaction chamber. Much softer ($\lambda = 86$ nm), but monochromatic and more intense, radiation is provided in many (repetition rate = 1 Hz) short pulses by the SASE-FEL at the Tesla Test Facility (DESY-HASYLAB, Hamburg). The role of nonthermal processes in XUV ablation will be evaluated. Mechanisms based on the nonthermal processes will be proposed.

CP641, *X-Ray Lasers 2002: 8th International Conference on X-Ray Lasers*, edited by J. J. Rocca et al.
© 2002 American Institute of Physics 0-7354-0096-2/02/$19.00

1. INTRODUCTION

Earlier experiments using XUV radiation showed that interaction of short-wavelength radiation with solid materials in vacuum induces transfer of macroscopic amounts of material into vacuum (reviewed in [1]). Short wavelength radiation sources can have either low (synchrotrons) or high (FEL and hot dense plasma) peak power. Low-peak-power sources remove material components from the irradiated surface by photon-induced desorption (often called direct dry etching). Each XUV photon can break any chemical bond and its energy is usually higher than any crystal's cohesive energy. XUV photons excite electrons from inner atomic shells, followed by an Auger cascade of other electrons. This leads to formation of electron-depleted regions in the material, which rapidly decomposes by Coulomb explosion. Thus, photons absorbed in the near-surface region may split the sample material into tiny pieces, which are ejected into vacuum. Low peak intensity irradiation removes material only from the surface and a very thin near-surface layer.

The situation is different with high-peak-power, high-energy pulsed sources. The sample receives a high local radiation dose (given by pulse energy and radiation absorption length in the material) and dose rate (due to the short pulse duration). Thus, a large number of events which cause radiation-induced structural decomposition occur almost simultaneously, in a relatively thick layer of irradiated material. Since a significant part of the radiation energy absorbed in the material is thermalized, the layer is suddenly overheated as well as simultaneously chemically altered by the radiation. The overheated and fragmented sample region represents a new phase, which tends to blow off into the vacuum.

We present ablation of organic polymers (PMMA, PTFE, PET, PI) and elemental solids (Al, Si) induced by intense XUV radiation emitted from discharge and laser-produced plasmas. PMMA and PTFE samples were irradiated at high intensities by coherent 86-nm FEL radiation. Samples were irradiated over a wide range of photon energies, pulse energies, and pulse durations.

2. EXPERIMENTAL

Details of the organic polymer samples have been presented elsewhere [1]. Aluminum was irradiated as a 40 nm thick layer deposited on 12 μm and 1.5 μm thick mylar foils and as a 400 nm thick foil (Goodfellow). 5x5-mm^2 silicon (001) chips were made from 420 μm thick wafers cut from a single crystal (Institute of Physics, Prague). PMMA and PTFE pieces, 13x32-mm^2, were cut from sheet [1], mounted in a 27x32-mm^2 stainless steel frame, and irradiated without masks by XUV FEL radiation (TTF-FEL [2-4]). Samples irradiated with other sources were placed behind a contact mask. Discharge plasmas used included a Z-pinch plasma formed from various loads (a single 120-μm dia. copper wire for most experiments) and driven by the S-300 pulsed-power machine [5], and a deuterium plasma focus, driven by the PF-1000 facility [6].

Higher temperature plasma was obtained by focusing a near-IR laser (PALS [7]) on planar metal targets or into single and double gas puff targets. Single stream gas-puff targets were formed by pulsed Xe injection through a 2-mm dia. circular nozzle. The double-stream gas puff targets utilized two coaxial nozzles [8]. An annular outer nozzle produced a hollow cylinder of He, suppressing sideways expansion of the Xe. Double-stream gas puffs have been shown [8] to improve IR to XUV conversion efficiency, compared to conventional single-stream gas puffs. Charged particle emission from plasma generated on a solid-state target surface was stopped short of the sample surface by a gas fill in the interaction chamber and by thin Si_3N_4 windows, both of which stop ions while transmitting a significant portion of the photons. For large-scale gas-puff targets, ion emission is suppressed by the outer gas envelope surrounding the plasma. Absence of ion current was checked using ion collectors. For discharge plasmas, ion bombardment of the sample surface was reduced by suitable choice of irradiation geometry.

Depth profiles of the structures formed by ablation and expansion of irradiated material were measured with a surface profiler (Alpha-Step 500, Tencor; USA). Raman spectra were made with an Ar^+ laser (514 nm) microbeam in the usual backscattering geometry (Renishaw Ramascope Model 1000), which enables probing chosen places on the sample surface.

3. RESULTS & DISCUSSION

Figs. 1a,b show that radiation that can pass through 1.1 cm of expanding high pressure Xe (longer λ UV-Vis, harder x-rays) and the reactive components in recombining Xe plasma cannot (Fig. 1a) produce deep, well developed structures in PET like those formed by XUV radiation emitted through the relatively thin sidewalls of a gas puff (Fig. 1b). Figs. 1b,c demonstrate significantly higher conversion efficiency of laser IR radiation to short-wavelength radiation, which efficiently ablates polymers in double stream gas puff targets (Fig. 1c; hor 3.4cm) than in conventional, single gas puffs (Fig. 1b; hor 3.4cm).

The XUV ablation behavior of PET near threshold, irradiated by longer XUV pulses emitted from discharge plasma, is shown in

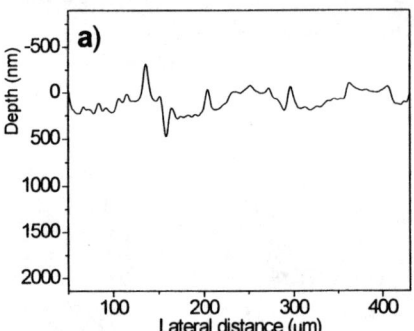

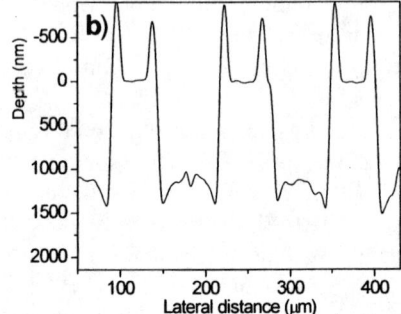

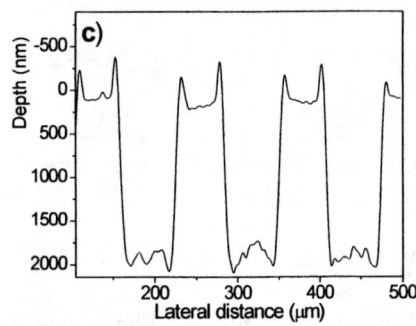

FIGURE 1. Ablation of PET samples irradiated through a steel-mesh mask by single XUV pulses from a Xe plasma, formed by focusing a laser into, a) and b), Xe single gas puff targets and c) Xe-He double gas puff targets.

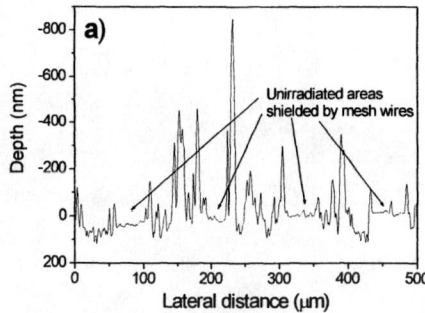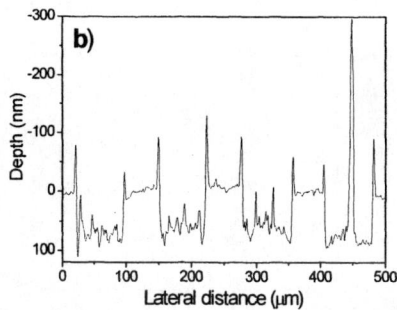

FIGURE 2. Ablation of PET by single XUV pulse from copper plasma driven by S-300 pulsed power machine. Samples were placed behind mesh contact masks at a) 54 cm and b) 49 cm from the source.

Fig. 2. The qualitative difference between the structures in Figs. 2a and 2b is dramatic, although the difference in XUV fluence at the sample surface was <20%. Above threshold, clean deep structures were ablated in PMMA and PTFE as well as PET but not PI.

Almost all our experiments show that PMMA is ablated less efficiently than PTFE. This is shown in Fig. 3 using a plasma focus as the XUV source. Nonthermal effects of the radiation on polymer structure play a key role in the ablation mechanism, i.e. ablation can be explained by the formation of radiation-chemical scissions of the polymer chain followed by blowoff of low-molecular fragment fluid into the vacuum. It is well known [9] that PTFE is less resistant to ionizing radiation than PMMA, PET, and PI.

Structures with hydrodynamic character are seen in irradiated areas (Fig. 2b, right) or near mask–sample contacts (Figs. 1,2). We believe they are relics of hydrodynamic transport, from the irradiated squares into vacuum, of a "fluid" formed from polymer chains undergoing multiple radiation scissions and suddenly heated by intense XUV.

Massive "mound" structures can be seen in the irradiated squares between the mesh wires (Fig. 2a), especially near threshold and for harder x-rays [1]. They are formed by expansion of irradiated and heated polymer regions (similar to relics above). Expansion of heated irradiated PMMA was first described many years ago [10] and is attributed to formation of numerous small bubbles. Highest bubble concentrations were

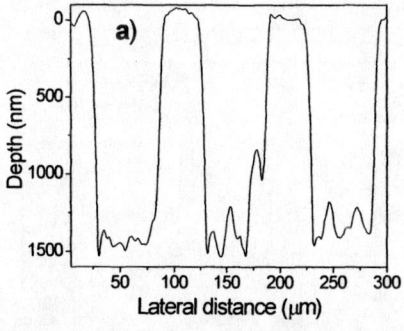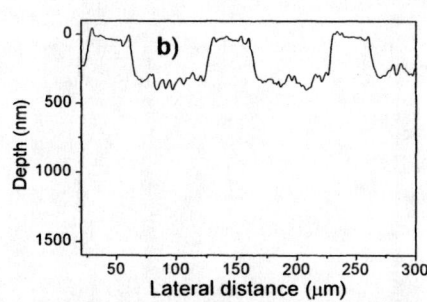

FIGURE 3. Ablation of a) PTFE and b) PMMA by seven accumulated XUV pulses from the plasma focus. The 1st pulse had Al-wire surrounded by 2.5 torr D_2, the 2nd to 5th pulses had a 2.5-torr D_2 gas fill without the wire, and, for the 6th and 7th pulses, the deuterium pressure was 3.0 torr.

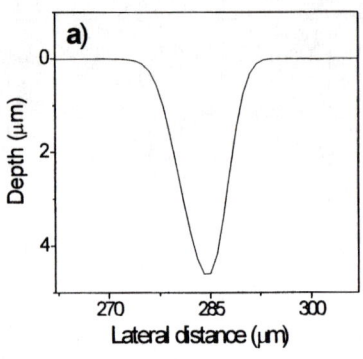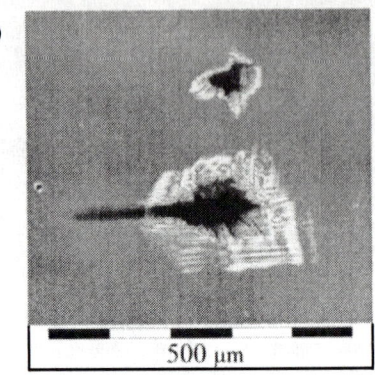

FIGURE 4. Craters formed in PMMA by 5-µJ pulses of 86-nm FEL radiation. Profile (a) and Nomarski micrograph (b, top) for 380-pulse crater. Micrograph (b, bottom) of deep hole drilled by <1850 pulses.

observed near the ablation threshold. At the highest fluences, the carpet of bubbles disappeared and only individual bubbles were seen. Similar expansion to what we observed in PET and PMMA samples has recently been reported for PTFE irradiated by XUV radiation from an undulator [11]. Remarkably, we observed expansion in PMMA and PET but not in PTFE. For future microstructuring applications of short-wavelength FELs, it seems to be very important that we did not observe such bubbles and mounds at PMMA surfaces irradiated with coherent FEL radiation (Fig. 4).

Relatively deep holes were drilled in PMMA by FEL pulses (Fig. 4). Since the XUV beam's incidence angle to the sample surface was 55°, the profile shown in Fig. 4a is only a lower estimation of crater depth. Using optical microscopy, we found that 380 pulses ablated a 25-µm layer of PMMA (~70 nm per pulse). This is roughly one order of magnitude more than the attenuation length of 86-nm radiation in PMMA [12].

Aluminum layers were completely ablated under almost all irradiation conditions, consistent with the very high XUV ablation efficiency reported for Al [13-16]. In contrast, Si undergoes XUV ablation only if exposed to short pulses at high fluences. Although we did not observe Si ablation with discharge plasma-based XUV sources, the high-power laser driven sources allowed us to realize ablation (Fig. 5).

In Raman spectra of the irradiated Si surfaces, a band at 607 cm^{-1} appeared. The

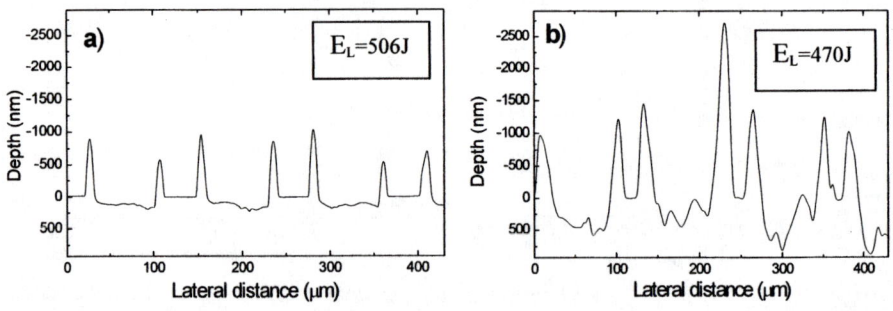

FIGURE 5. Ablation of Si irradiated through steel-mesh mask by single XUV pulse from Xe plasma, formed by focusing laser) into a) single (Xe,) and b) double (Xe-He) gas puff targets.

spectra showed Si amorphization at the unirradiated surface between the sample areas exposed to double-stream gas puff XUV emission, i.e. higher XUV fluence (Fig. 5b). The XUV ablation efficiency of Si is not as high as that of organic polymers, which is caused by its structure, so that thermal processes begin to play an important role.

4. CONCLUSIONS

It has been demonstrated that all four sources used can provide enough XUV radiation, emitted in sufficiently short pulses, to ablate aluminum and the organic polymers PTFE, PMMA, and PET. Polymer layers with a thickness of several microns can be ablated by a single shot under optimum irradiation conditions. Silicon, an inorganic, covalently bound crystalline material, is much more resistant to XUV ablation than organic polymers. PMMA was ablated less effectively than PTFE. This is very likely a result of the higher radiation stability of PMMA compared to PTFE. PI is less prone to XUV ablation than PMMA, for the same reasons. These findings and the hydrodynamic like structures observed in irradiated areas of the polymer samples support a XUV ablation model based on the assumption of radiation-induced scissions of the polymer chains, resulting in formation of a fluid-like phase. Removal of macroscopic amounts of the material is then realized by expansion, ejection, and vaporization of this phase (overheated by the dissipated fraction of absorbed radiation) into vacuum.

ACKNOWLEDGMENTS

This work was funded by the Czech Ministry of Education within the framework of programs INGO (Grant LA055) and National Research Centers (Grant LN00A100).

REFERENCES

1. Juha, L., et al., *Surf. Rev. Lett.* **9**, in press (2002).
2. Ayvazyan, V., et al., *Phys. Rev. Lett.* **88**, 104802 (2002).
3. Treusch, R., et al., *Nucl. Instrum. Meth. Phys. Res.* **A467-8**, 30-33 (2001).
4. Andrejczuk, A., Hahn, U., Jurek, M., Krzywinski, J., Pelka, J., Reniewicz, H., Schneidmiller, E. A, Sobala, W., Sobierajski, R., Yurkov, M., and TTF FEL team, *HASYLAB Annu. Rep.* 117-120 (2001).
5. Bakshaev, Y.L., et al., *Rev. Sci. Instrum.* **72**, 1210-1213 (2001).
6. Scholz, M., Miklaszewski, R., Gribkov, V.A., and Mezzetti, F., *Nukleonika* **45**, 155-158 (2000).
7. Jungwirth, K., et al., *Phys. Plasmas* **8**, 2495-2501 (2001).
8. Fiedorowicz, H., et al., *Opt. Commun.* **184**, 161-167 (2000).
9. Swallow, A. J., *Radiation Chemistry*, Longman, London, 1973, p. 224.
10. Ross, M., and Charlesby, A., *Atomics* **4**, 189-194 (1953).
11. Maida, O., et al., *Jpn. J. Appl. Phys. - Part 1* **40A**, 2435-2439 (2001).
12. Ferincz, I. E., Toth, C., and Young, J. F., *J. Vac. Sci. Technol.* **B14**, 828-832 (1997).
13. Anderson, A.T., et al., *Fusion Technol.* **30**, 757-763 (1996).
14. Filippov, N. V., et al., *Phys. Lett.* **A211**, 168-171 (1996).
15. Anderson, A. T, and Peterson, P. F., *Exp. Heat Transfer* **10**, 51-65 (1997).
16. Bakshaev, Yu. L., *Plasma Dev. Oper.* **7**, 157-161 (1999).

Application Prospects for Soft X-ray Lasers

CLS Lewis*, R Keenan*, SJ Topping*, S Hubert*, HW Van derHart*, D Riley*, GJ Tallents[#], RE King[#], GJ Pert[#]

* *School of Mathematics and Physics, Queen's University Belfast, Belfast BT7 1NN, UK*
[#] *Department of Physics, University of York, York YO10 5DD, UK*

Abstract. Collision-pumped soft x-ray laser devices operating in quasi-steady-state mode in the 5-25 nm wavelength region now routinely produce saturated outputs, corresponding to about $10^{13} - 10^{14}$ photons in pulses of ~50 psec duration. These partially coherent beams are extremely bright, single-shot sources of radiation with a range of demonstrated and potential applications. Here, we review some recent developments in production and characterisation of Ne-like and Ni-like soft x-ray lasers pumped by the Vulcan laser at the Central Laser Facility, UK. Stemming from this, we examine some application options including radiography, interferometry, Thomson scattering and non-linear optics.

X-RAY LASER BEAM CHARACTERISATION

Alignment and target handling procedures have been developed and improved to make the 5-beam pumped XRL setup for the CLF Vulcan laser a robust and routine facility for generating saturated output XRL beams. We have concentrated on Ge and Ni targets which provide typically 1-5 mJ of energy in ~30-50 psec duration pulses at wavelengths 19.6 nm and 23.1 nm respectively. These beams are the basic starting point for our programmes to develop and demonstrate a range of application experiments using XRLs. As beam spatial profiles are often relevant to applications we have also developed a high magnification imaging system based on two x-ray multilayer mirrors (XRMs), where a first mirror images to an "interaction plane" at ~2X magnification and a second mirror image relays this plane to a back-thinned CCD at up to ~20X magnification (ie overall magnification of up to ~40X, with an intrinsic spatial resolution in the plasma amplifier exit plane region of a few microns). We have systematically varied target irradiation conditions to look for correlations with "good beam structure" in the interaction plane. We have also systematically evaluated shot-to-shot reproducibility under various conditions and find that even though beam structure can be poor in some cases it is nevertheless reasonably reproducible.

Saturated Outputs

To produce saturated output beams for application purposes stripe or slab targets are irradiated by 5 beams of the Vulcan Nd:glass laser operating at 1.05 µm. Line foci ~100 µm wide and ~20 mm long are generated using a standard spherical lens and off-axis spherical mirror combination. Each beam, delivering ~40J in ~80psec pulses, is

CP641, X-Ray Lasers 2002: 8th International Conference on X-Ray Lasers, edited by J. J. Rocca et al.

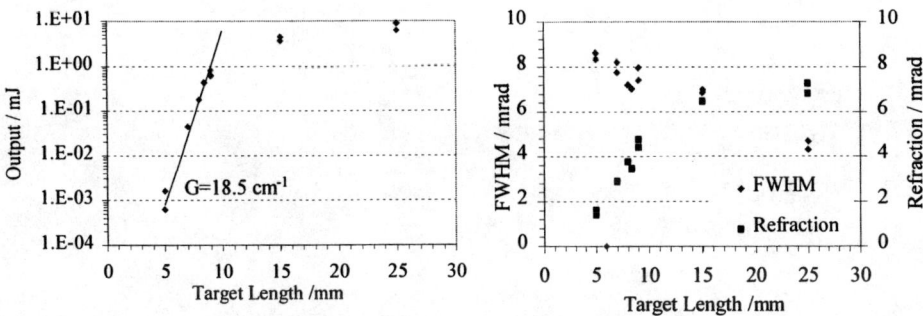

Figure 1. Characteristics of XRL beam from Ge target showing (left) that ~5mJ output from ~20mm targets is reproducibly available, being strongly saturated and (right) that the horizontal divergence of the beam is ~5mrad from ~20mm targets. The beam centre exits at ~5mrad from the target surface.

focussed into a 25 mm x 100 µm line focus. The beams converge on the target centred at angles 0°, ±30° and ±60° wrt the horizontal target normal and are axially and temporally displaced to give an average and uniform intensity on target of up to 5×10^{13} W/cm^2, and in a quasi travelling wave pump mode. The central beam at 0° is used as a 20%/2nsec prepulse beam and is generated in the laser area by splitting the oscillator pulse into two before the preamplifiers of Vulcan.

Output energies are estimated from best information available on mirror reflectances, filter transmissions, grating efficiencies and CCD detector quantum efficiencies. Based on our own, in situ, measurements of Al filter absorption coefficients at our XRL wavelengths [1] we conclude that energies up to ~5mJ are readily produced at 19.6 and 23.1 nm from Ge and Ni targets respectively. Beam pointing and divergence are monitored using flat-field spectrometers, and divergence can also be estimated from near-field imaging using the image relaying system discussed next. Characteristics for Ge targets operating strongly into saturation at 19.6 nm are illustrated in Figure 1. Ni targets exhibit a similar performance at 23.2 nm.

Beam Profiles

Beam profiles are monitored using an image relaying arrangement based on XRMs [2] whereby the plasma amplifier exit plane is imaged at ~2X magnification to an "interaction" plane where the XRL beam would normally be used in an application and the distribution in this plane is examined by imaging it at higher magnification (10-20X typically) to a back-thinned CCD detector with 13µm or 25µm pixels. The setup is illustrated in Figure 2. We find that best XRL beam quality is obtained when the target is moved ~100µm away from the focussing optics after best optical pump beam obscuration has been observed on 100µm stripes during target alignment [3]. This is likely due to 3-D effects in coupling 4 off-normal beams into an expanding plasma plume and coalescing the 4 gain channel zones potentially available. Although beam quality is generally poorer than we have previously observed with double-passed

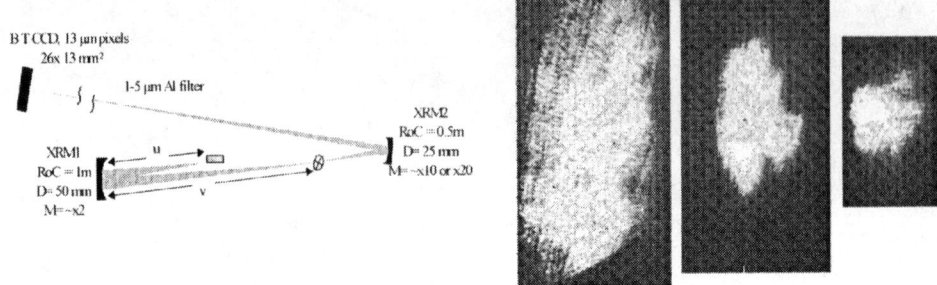

FIGURE 2. The basic imaging geometry (left) uses two XRMs with u~75cm and v~150 cm. The cross-hair position denotes a typical "interaction plane" at Z=0. The three images (right) correspond to beam profiles at planes defined by Z=0, Z=10 and Z=20 cm recorded by moving the second imaging mirror appropriate amounts. The smallest image (Z=0) corresponds to ~150μm scalelength at the target.

and coupled amplifiers which benefit from beam injection, the advantage here is ease of alignment and robustness of the arrangement for applications. We also note that the system is fairly reproducible, in the sense that poor beam quality is not randomly generated. This is illustrated in Figure 3 where several pairs of images taken on different shots and in a variety of planes relative to the "interaction plane" are shown.

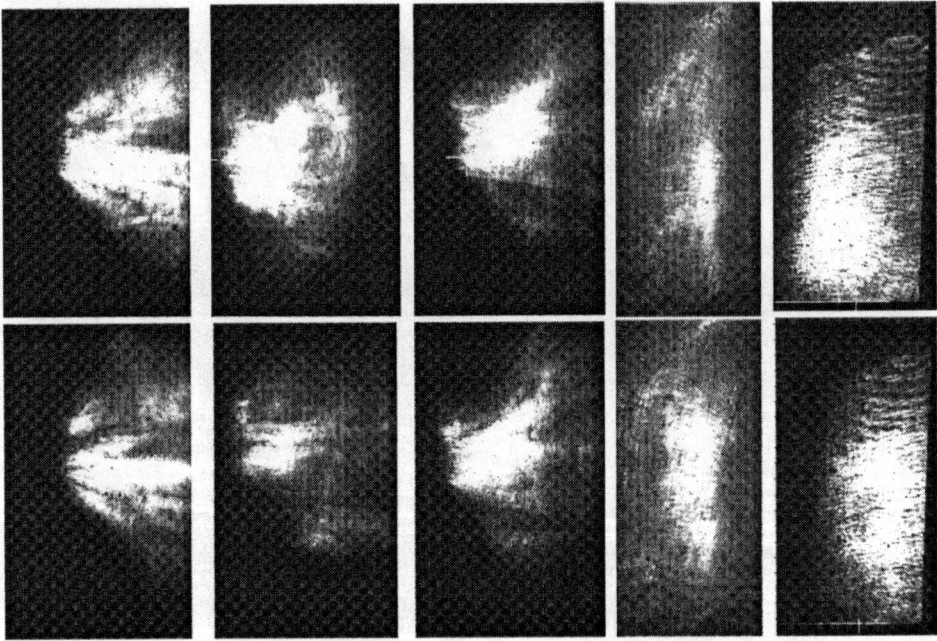

FIGURE 3. Top/bottom images are pairs taken under similar pump conditions and correspond to a range of target displacements relative to the focussed pump beams and to various distances from the target exit aperture (ie near-field patterns). Characteristic dimension of the centre images is ~150 μm.

512

XRL APPLICATIONS : DEMONSTRATED AND POTENTIAL

We have opted to confine our considerations to collision-pumped XRLs operating in the quasi-steady-state (QSS) mode; ie preplasmas pumped by ≤100 psec optical pulses. This is simply due to the relatively lesser problems anticipated in signal detection and signal noise isues arising from these devices in comparison to transient excitation devices (TCE) which will typically have ~1000X fewer photons available per pulse. A QSS XRL with photon energy ~60 eV (ie λ~20 nm) provides ~10^{14} photons/mJ in a single pulse of 20-50 psec duration. All applications foreseen at present with these XRLs (QSS and TCE types) are intrinsically single-shot events with little prospect for signal averaging until true hi-rep, table-top pump lasers emerge.

Radiography and Interferometry

The generic techniques associated with radiography include simple projection shadowgraphy and backlighting a sample which is imaged to the detector plane. In either case the scope of XRL radiography is illustrated by considering ~10^{14} photons in a primary XRL pulse (ie~1 mJ in ≤50 psec) which probes a sample of opacity τ and is eventually recorded as an image covering ~10^6 pixels in a CCD with 13μm pixels (ie magnified image size of 13mm x 13mm). The counts per pixel can be expressed as $C = C_0 e^{-\tau}$ where C_0 also includes any filter/mirror attenuation needed. It is straightforward to show that counts of ~10^4 per pixel are expected for attenuations equivalent to 3-4 μm of Al above the L-absorption edge and hence that the CCD image will be sensitive to variations in average opacity of such samples at the Δτ ~ 0.01 level. Since opacity depends on the sample mass absorption coefficient (μ), the sample density (ρ) and the sample thickness (t) through $\tau = \mu\rho t$, then radiography of such samples using ~60 eV XRLs can detect changes in μ, ρ and t at the ~0.1% level. Experiments have already demonstrated feasability in a variety of scenarios including laser-accelerated thin foils undergoing Rayleigh-Taylor mass redistribution, shock propagation and break-out in thin foils and simple cold, inert foils to measure cold mass absorption coefficients [1].

An example of the high sensitivity of probing under these conditions can be seen in the pair of images on the right-hand-side of Figure 3. The nearly horizontal "fringes" apparent near the tops of the images arise not from intrinsic XRL beam structures but from ripples in the 3 μm Al foil filter used to control CCD exposure. For a rippled (sinusoidal) cold Al foil with ripple amplitude A, thickness t and spatial wavelength λ, the transmission pattern will have a recorded visibility V~1/3 when the condition $A/\lambda \sim \{\ln2/(2\pi^2[\mu\rho t])\}^{1/2}$ is met. For Figure 3, this implies the 3μm foil (with average μρt~7) had ripples with an amplitude of ≤ 5 μm and a period of ~50 μm.

When radiography is carried out in a low absorption scenario then the added power of interferometry can be tapped by providing a reference beam to generate fringe patterns. Several techniques in the soft X-ray regime have been considered and demonstrated by others. We have begun development of a new interferometer design based on transmission gratings to provide two XRL beamlets from diffracted orders. The basic concept is shown in Figure 4 along with some preliminary measurements.

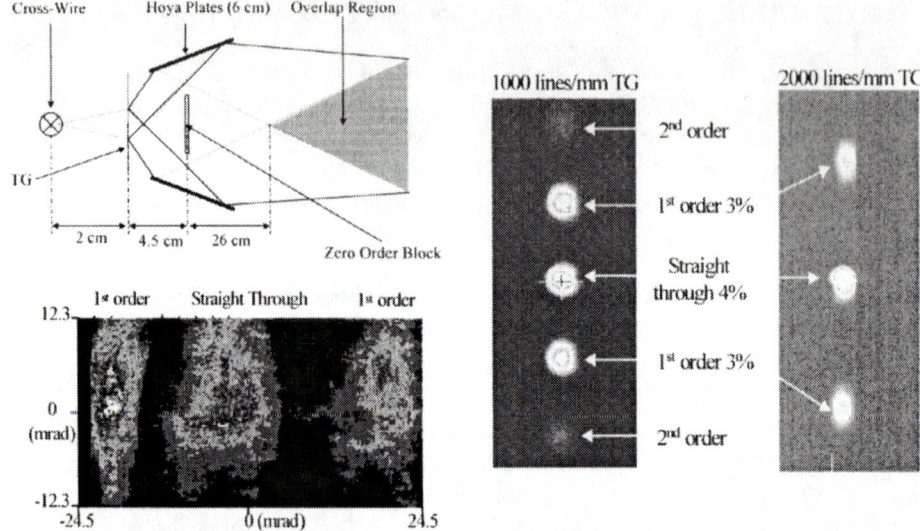

FIGURE 4. Basic geometry (top left) for generating interference fringes with ± diffracted orders from a transmission grating (TG). The crosswire position denotes the position of an aperturing pinhole needed to ensure high spatial coherence in the incident beam. Efficiency measurements (right) for several gratings using a 10 Hz YAG laser X-source and an XRM operating at 23nm show about 30% of the transmitted radiation falls in each of 0 and ±1 orders. Preliminary measurements with the Ne-like Ni XRL incident on a TG, without grazing angle reflecting plates have been made (low left).

Thomson Scattering

We have evaluated the prospects of using an XRL with $\sim 10^{14}$ photons in a ~ 50 psec pulse as a probe beam for Thomson scattering from a range of plasmas which can be referred to as "warm, dense plasmas". This appears to offer exciting possibilities for simultaneously measuring density and temperature in radiatively heated or shock-heated plasma material. A preliminary experiment, depicted in Figure 5, was not successful in observing a Thomson signal but provided information which enables more detailed modelling and planning for a future experiment.

The information to be gained from Thomson scatter can in principle include not only the electron and ion temperatures and their densities but also the electron distribution function itself. Use of an XRL is potentially exciting as it allows probing of warm, dense plasma matter conditions not possible at longer probe wavelengths in the UV. The advantage of an X-ray laser over a thermal laser-plasma source is that it can have a low divergence and hence high brightness as well as having a narrow relative bandwidth of $\sim 10^{-4}$ compared to $\sim 10^{-2}$ for, say, a He-like ion resonance transition.

The most important parameter governing Thomson scatter is the scatter parameter, α, given by $\alpha = [2k_0\lambda_D\sin(\theta/2)]^{-1}$ where k_0 is the initial photon wave-number, θ is the scatter angle and λ_D the Debye length. For the particular case where $\alpha > 1$ occurs, the

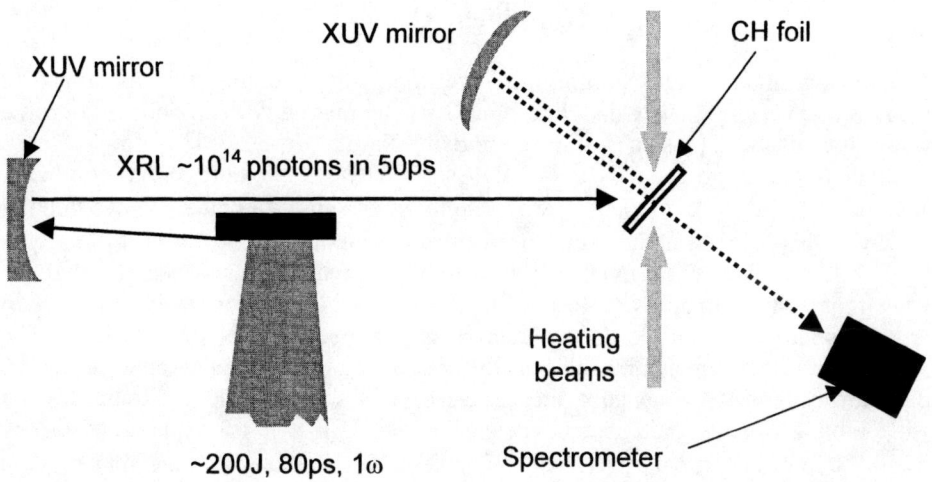

FIGURE 5. Basic geometry for Thomson scattering at θ~165°. The sample target could be directly or indirectly heated. The spectrometer could be time-resolving or gated to enhance S/N ratios.

scatter momentum is small and the scatter is sensitive to density fluctuations on a spatial scale greater than the Debye length. The scatter spectrum is then determined by the collective plasmon oscillations of the plasma and contains resonances shifted by the plasma frequency either side of the incident photon frequency. For the conditions illustrated in Figures 5 and 6, α.~1.8 and the peak is shifted ~5Å which is resolvable. Calculations suggest the optical depth to the XRL is ~0.1 and that self emission is sufficiently low to enable detection of the Thomson scatter signal [4]

FIGURE 6. Simulation output from HYADES code indicating conditions present in a 3μm CH foil

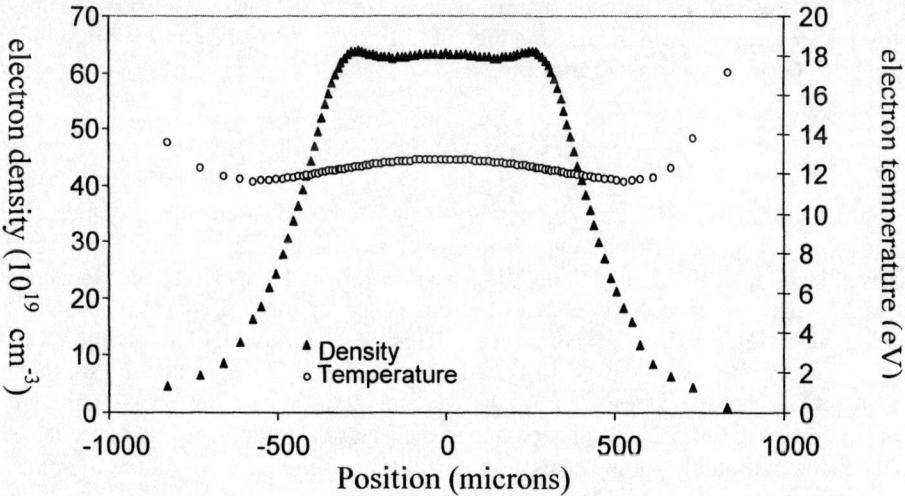

about 1 nsec after irradiation by 0.53 μm/80 psec laser pulses at ~5x10^13 Wcm^{-2}.

Non-linear Optics

We have made some progress in preparing for a definitive experiment to demonstrate near-frequency doubling of an XRL by mixing with an optical laser in a resonantly enhanced plasma medium comprising Na-like Ar ions [5].

Modelling has progressed to evaluate sensitivity of the third order non-linear susceptibility, $\chi^{(3)}(\omega)$, to optical wavelength in a sum-difference 4-wave mixing scheme. This resulted in the choice of a Vulcan beam at 2ω ($\lambda=0.53\mu m$) to mix with two XRL beams at 23.1 nm generated from a Lloyds mirror which devided the wavefront of the primary XRL beam. Beams were angled appropriately to mix in an argon gas-jet target, ionised to be predominantly composed of Na-like Ar ions. The angles were set by calculations designed to optimise phase-matching conditions. To date no positive experimental evidence has been observed to demonstrate efficient wave mixing which should generate detectable radiation at ~11.5 nm at an efficiency of ~0.1% from the primary XRL, which itself is $\sim 10^{-5}$ efficient from the optical pump laser energy [6].

The programme is now supplemented by detailed R-matrix Floquet calculations which will provide more insight into the expected behaviour, including multiphoton ionisation rates, of the complex non-linear interactions in these types of experiment [7]. The R-matrix Floquet approach is well suited to highly charged ions in strong fieldsA single colour code has been used to date to calculate two and three photon ionisation rates for Ar VIII with the XRL beam at an intensity of 10^{15} Wcm^{-2} and typical predictions are illustrated in Figure 7. It is planned to develop the code to run reliably at more realistic practical intensities (ie $\sim 10^{12}$ Wcm^{-2}) and to run two colour simulations (ie optical and XRL wavelengths interacting simultaneously). Both aims are non-trivial.

Crucial to eventual success in NLO experiments is a knowledge of the spatial coherence of the XRL beam and hence an ability to measure it. The interferometer design discussed earlier and based on transmission gratings on 50 nm thick silicon nitride substrates will be used for such measurements. The system is also potentially useful for generating two mutually coherent diffracted XRL beams by amplitude splitting for future wave-mixing experiments.

CONCLUSIONS

Bright, saturated XRLs can be readily generated and used for a variety of applications. Progress in some applications will require higher rep rate drivers and improved beam quality. Although not discussed earlier, we have clearly seen evidence that our 5-beam pumped XRL is probably composed of multiple quasi-independent beamlets with high individual coherence. This is based on the observation that aperturing down the 2X imaging XRM produces "best focus" images that are speckled in appearance and more resricted in the vertical than in the horizontal (stronger refraction plane) direction. This may be due to the lack of injection-seeding or feedback and the fact that multiple pump beams are more likely to generate gain-guiding channels through refraction effects in inhomogeneous plasmas.

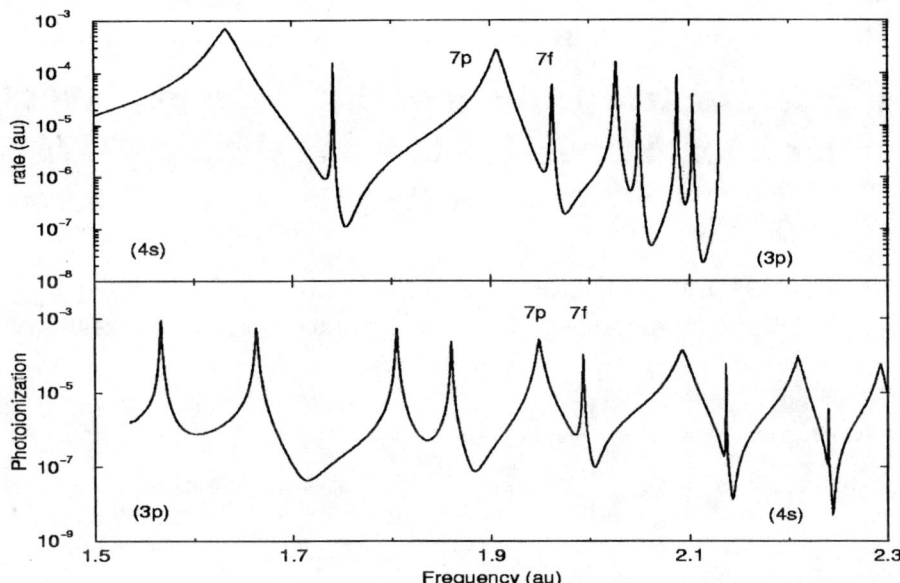

FIGURE 7. Two photon ionisation rates (10^{-9} au $\equiv 10^8$ sec^{-1}) from the 4s state in Ar^{7+} (top) and three photon ionisation rates from the 3p ($m_L = 0$ sublevel) state in Ar^{7+} (bottom) calculated for an X-ray laser intensity of 10^{15} Wcm^{-2} ; from reference [7]. At around 52eV (1.92 au) the 7p state is reached by 2 photons and the 4s state is nearly reached by one photon, providing the basis of the scheme to use the Ne-like Ni XRL. Due to the high intensity used dressed energy effects lead to the 3p and 4s states exchanging character near 53 eV.

ACKNOWLEDGEMENTS

The expert assistance from laser and target area staff at the Central Laser Facility is, as always, essential to progress in these complex experiments and we acknowledge all their help. The work has been supported by EPSRC under contract GR/N36806 and the European Union under TMR Network Contract EU/TMR-EEC CT98-0185.

REFERENCES

1. Wolfrum E et al. *J. Phys B : At. Mol. Opt Phys*. **34**, L565-570 (2001) ; Keenan R, Lewis CLS et al. *J. Phys B : At. Mol. Opt Phys*. (in press 2002).
2. Supplied by X-ray Optics Group, PN Lebedev Institute, Moscow.
3. Keenan R et al. in *These Proceedings* AIP 8th ICXRL edited by JJ Rocca (2002).
4. Riley D et al. *IEEE Trans Plas Sc*. [ICOPS Banff, Canada May 2002]. ed CE Capjack (2003); Khattak et al. in Central Laser Facility RAL Annual Report 2002-2002
5. Muendel MH and Hagelstein PL *Phys. Rev. A* **44**, 1953 (1991) ; Lewis CLS et al *SPIE* **3776**, 282 (1999)
6. Topping SJ et al. in *These Proceedings* AIP 8th ICXRL edited by JJ Rocca (2002).
7. Van der Hart HW and Feng L . *J. Phys B : At. Mol. Opt Phys*. **35**, 1185-1195 (2002) ; Burke PG et al. *J. Phys B : At. Mol. Opt Phys*. **24**, 761 (1991)

Interferometric probing of thin Niobium layers under high electrical field using the zinc X-ray laser at PALS

T. Mocek[1], D. Ros[2], B. Rus[1], D. Joyeux[3], A.R. Präg[1], M. Kozlová[1],
A. Carillon[2], D. Phalippou[3], F. Ballester[4], E. Jacques[4], M. Boussoukaya[4],
and G. Jamelot[2]

[1]Department of Gas Lasers/PALS Research Center, Institute of Physics, Na Slovance 2, Prague 8,
Czech Republic
[2]Laboratoire de Spectroscopie Atomique et Ionique (LSAI), Bâtiment 350, Université Paris-Sud,
91405 Orsay, France
[3]Institut d'Optique Théorique et Appliquée, Laboratoire Charles Fabry, Bâtiment 503,
Université Paris-Sud, 91403 Orsay, France
[4]Commisariat a l'Energie Atomique (CEA), Saclay, 91191 Gif-sur-Yvette, France

Abstract. We report on interferometric investigation of perturbed surfaces under high electric field by using the 21.2 nm zinc soft X-ray laser operating in a half cavity configuration. DC electric fields of up to 100 MV/m were applied to the surface of a niobium sample. Both positive and negative polarizations of the electric field were examined. Single shot, high-quality X-ray interferograms were recorded with a double-mirror Fresnel interferometer and surface maps were obtained from the experimental data by holographic reconstruction.

INTRODUCTION

The optimized laser-plasma based soft X-ray lasers (XRL) operating in the *quasi steady state* (QSS) regime, i.e. pumped by multi-100-ps laser pulses, are recently capable to provide high-quality X-ray beam, multi-millijoule output energy, peak spectral brightness of up to 10^{27} photons s^{-1} mm^{-2} $mrad^{-2}$, and low divergence [1]. These particular advantages make the QSS XRLs an ideal tool for X-ray interferometric investigations of solid state surfaces with an intrinsic accuracy of a few nanometers. X-ray interferometric technique can provide a nanoscale mapping of polished metallic surfaces submitted to a strong electric field and is able to reveal phenomena inaccessible by any other current diagnostic. In contrast to conventional techniques such as scanning electron microscopy or atomic force microscopy, the X-ray interferometric records can be made *in situ*, i.e. during the very action of the electric field. The first successful utilization of XRL for a metal-surface mapping by means of X-ray interferometry has been reported by F. Albert *et al.* [2]. The main motivation of the experiment described here was to get better understanding of phenomena such as field electron emission and breakdown that limit the performance of current superconductive cavities of particle accelerators.

CP641, *X-Ray Lasers 2002: 8th International Conference on X-Ray Lasers,* edited by J. J. Rocca et al.
© 2002 American Institute of Physics 0-7354-0096-2/02/$19.00

EXPERIMENTAL SETUP

The experiment was performed at PALS (Prague Asterix Laser System) [3,4] using the Ne-like zinc soft X-ray laser at 21.2 nm operating in deep saturation through double-pass amplification [1]. The active medium of XRL is a 3-cm long plasma column produced from a slab target by a separately delivered prepulse and a main pulse of an iodine laser (1.315 μm). The temporal delay between the two pulses was 10 ns and the pulse duration was 450 ps. To generate plasma with reduced lateral density gradients, the ~130-μm wide focus of the main pulse (pump irradiance ~2.8 × 10^{13} Wcm^{-2}) is placed on top of a much broader (~700 μm) prepulse focus. The X-ray laser emission emerges as a narrowly collimated beam possessing a high spatial quality and respective horizontal and vertical divergences of 3.8 (±0.5) mrad and 5.8 (±0.5) mrad. With a small-signal gain of 7.0 (±0.5) cm^{-1}, the double-pass XRL beam is deeply saturated and provides ~4 mJ of energy per pulse which corresponds to a peak power in excess of 40 MW.

The interferometric measurements were performed with a Fresnel double-mirror wavefront-splitting interferometer [5], located at 3 m from the plasma (Fig. 1). The interference pattern is produced by reflecting the X-ray beam at grazing incidence angle of 6 degrees off a tandem of adjacent plane mirrors inclined to each other at a small angle (5.2 mrad). Upon crossing over, the reflected beam halves produce an interference pattern. The resulting fringes were detected by a phosphor-coated CCD camera located in the overlapping region of the beam halves, ~70 cm downstream from the Fresnel double mirror center. To increase the detected fringe spacing, the CCD camera was inclined to view the fringes under a near-grazing incidence angle.

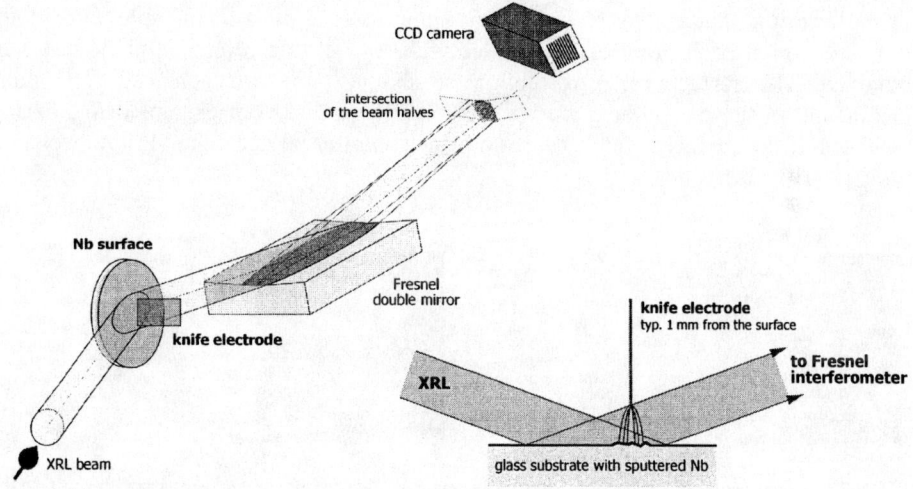

FIGURE 1. Schematic of the Fresnel double mirror X-ray interferometer and the detail of Niobium coated electrode showing the principle of surface probe technique.

RESULTS

Figure 2 shows a reference interferogram (no electric field applied) produced by the XRL beam and the Fresnel double-mirror interferometer. The apparent fringe uniformity and high contrast demonstrate the high spatial coherence of this X-ray laser. The visibility of fringes in the central part of the beam amounts up to ~0.5. It is also seen that fringes extend over the whole beam intensity profile while appearing equally on its periphery, which indicates that the whole beam wavefront possesses an appreciably high degree of coherence.

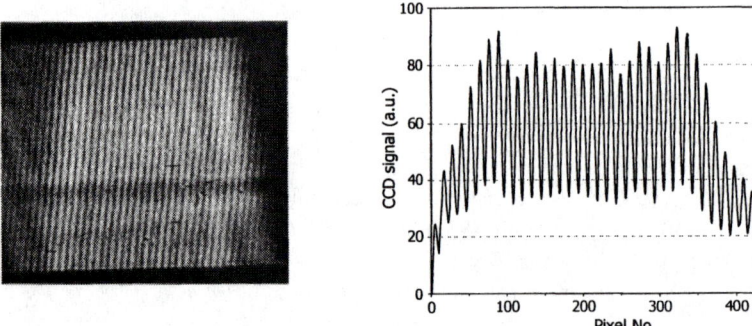

FIGURE 2. Typical example of an X-ray interferogram (no E-field) and its lineout produced by the 21.2 nm zinc XRL.

Figure 3 shows a sequence of X-ray interferograms of the Nb surface modified by a progressively increasing electric field along with the reconstructed surface maps. From the fringe pattern we have evaluated the phase shift at the sample surface and the elevation at each point of the surface with respect to a reference plane has been obtained. The surface relief resolution of about 2 nm was achieved. The noise originating in the phosphor grain was removed by a fast Fourier transform filtering. Changes of the surface relief, occurring prior and after an accidental microdischarge, were clearly observed.

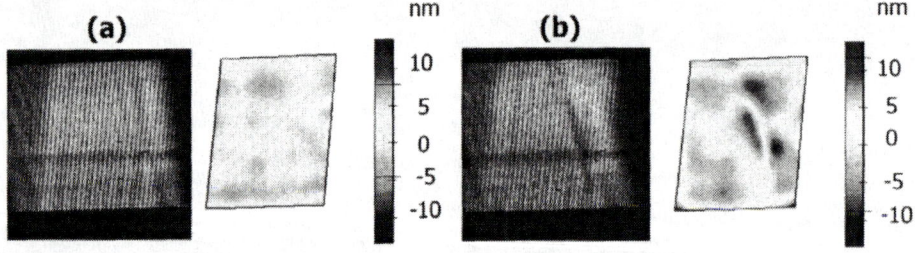

FIGURE 3. Examples of recorded interferograms and reconstructed surface maps for increasing electric field: (a) no voltage, (b) 37.5 MV/m (neg. pol.), (c) 50 MV/m (neg. pol.), and (d) 78 MV/m (pos. pol.).

During the experimental campaign we have collected more than 40 high-quality interferograms for various conditions. The analysis of the experimental data is currently in progress. In particular, only small deformations of the surface were observed when positive voltage was applied while in the case of negative polarity large modifications (accompanied by breakdown) were measured. The corresponding images show local reflectivity alterations across the cathode surface suggesting a local increase of its roughness.

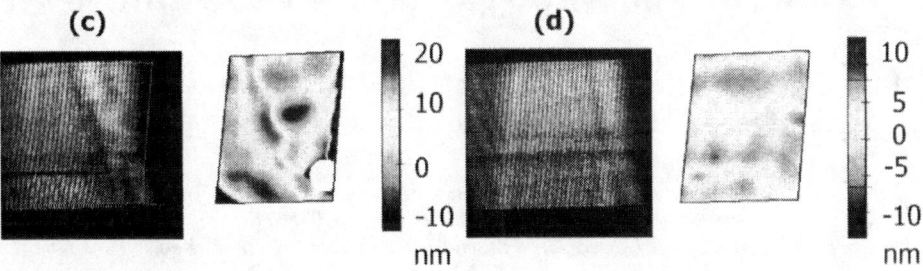

FIGURE 3. Continued.

SUMMARY

In summary, we have successfully used the 21.2 nm soft X-ry laser of PALS as a tool in a routinely running application experiment on in-situ interferometric probing of thin Nb surfaces subjected to large electric fields of up to 100 MV/m. The obtained interferograms demonstrate very good spatial coherence and high quality of the XRL beam wave front. Detailed analysis as well as physical interpretation of the experimental data is underway. Our next planned application experiments will include e.g. interferometric probing of laser produced plasmas and solid surfaces subjected to strong radiation fields.

ACKNOWLEDGMENTS

This work was financially supported by the EU Transnational Access to Research Infrastructures grant HPRI-00108, by the National Research Centers project LN00A100, and by the Czech Academy of Sciences grant A1010014.

REFERENCES

1. Rus, B., *et al.*, "Multi-millijoule highly coherent X-ray laser at 21 nm as a routine tool for applications," in this Proceedings.
2. Albert, F., Zeitoun, P., Jaeglé, P., Joyeux, D., Boussoukaya, M., Carillon, A., Hubert, S., Jamelot, G., Klisnick, A., Phalippou, D., Ros, D., Zeitoun-Fakiris, A., Phys. Rev. B **60**, 11089-11094 (1999).
3. Rus, B., Rohlena, K., Skála, J., Králiková, B., Jungwirth, K., Ullschmied, J., Witte, K.J., and Baumhacker, H., Laser Part. Beams **17**, 179-194 (1999).
4. Jungwirth, K., Cejnarová, A., Juha, L., Králiková, B., Krása, J., Krouský, E., Krupičková, P., Láska, L., Mašek, K., Mocek, T., Pfeifer, M., Präg, A., Renner, O., Rohlena, K., Rus, B., Skála, J., Straka, P., and Ullschmied, J., Phys. Plasmas **8**, 2495-2501 (2001).
5. Svatos, J., Joyeux, D., Phalippou, D., and Polack, F., Opt. Lett. **18**, 1367-1369 (1993).

Picosecond Fourier holography using a Lloyd's mirror and an X-ray laser at 13.9 nm

B. Rus[1], H. Daido[2], H. Tang[2], M. Nishiuchi[2], M. Kishimoto[2], M. Tanaka[2], T. Kawachi[2], N. Hasegawa[2], K. Nagashima[2], T. Arisawa[2], Y. Kato[2]

[1] Gas Lasers Department / PALS Centre, Institute of Physics, Academy of Sciences of the Czech Republic, Na Slovance 2, 18221 Prague 8, Czech Republic
[2] Advanced Photon Research Center, Japan Atomic Energy Research Institute, Kansai Research Establishment, 8-1 Umemidai, Souraku-gun, Kyoto, 619-0215 Japan

Abstract. Holographic capability of a Lloyd's mirror wavefront-splitting interferometer was experimentally investigated. The radiation source was a TCE Ni-like Ag X-ray laser at 13.9 nm, routinely operated at JAERI. The X-ray beam was focused down to a point located a few centimeters in front of the close edge of the Lloyd's mirror, in order to produce a near-spherical diverging wave a portion of which, upon reflection by the mirror, recombines with the portion of the wave propagating in free space. The holographic capability of the setup was examined in two configurations. In the first configuration, pertinent to a transmission holography, a test object consisting of a 50-μm wire was used to produce a diffracted wave encoded as holographic information in the generated interferogram. In the second configuration, SnO_2 powder with grains of ≈100 nm, deposited on the Lloyd's mirror surface, was used to demonstrate a reflection hologram. The data obtained in both cases show that the Lloyd's mirror is a convenient setup for holographic probing with high spatial resolution.

INTRODUCTION

One of application techniques exploiting X-ray lasers (XRL), which has been recently implemented in a few laboratories, is wavefront-splitting XUV interferometry [1-4]. It was applied to interferometric probing of laser plasmas [3], and of solid surfaces deformed by intense electrical fields [5,6].

While imposing rather severe requirements on the spatial coherence and the wavefront quality of the incident beam, the advantage of wavefront-splitting interferometry is its inherent holographic capability. This is given by the fact that in this arrangement a reference wave is made to recombine, under a given nonzero angle, with a wave scattered by the investigated object and carrying both amplitude and phase information about this object. In the XUV region, an arrangement making it possible to perform conceptually simple holographic experiments is Lloyd's mirror, schematically represented in Figure 1. The XRL beam is focused down to a point acting as a source of spherical wavefront. A portion of the spherical wave is reflected from a glass mirror

CP641, *X-Ray Lasers 2002: 8th International Conference on X-Ray Lasers*, edited by J. J. Rocca et al.
© 2002 American Institute of Physics 0-7354-0096-2/02/$19.00

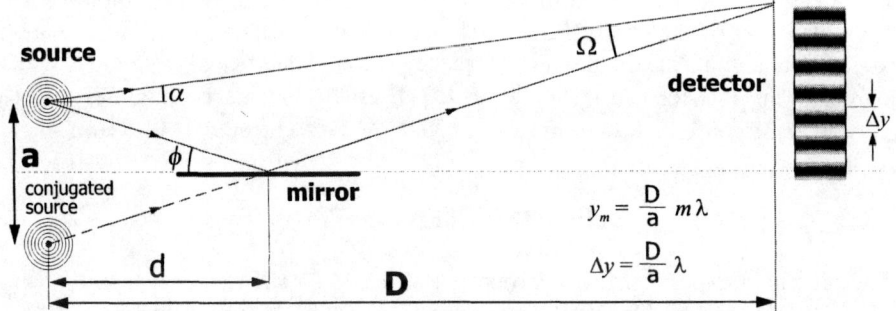

FIGURE 1. Basic geometry of the Lloyd's mirror interferometer. The spatial period of the diffraction fringes is identical to the equivalent Young experiment, with a minimum on the axis due to a π phase shift introduced by the reflection on the mirror surface.

and interferes with a portion of the wavefront proceeding directly to the detector in free space; the quality (straightness) of the fringe system depends on the quality of the spherical wave. The fringe separation can be written in terms of the angles involved as

$$\Delta_{fr} = \frac{\lambda}{2\sin(\phi-\alpha)\,2} \cong \frac{\lambda}{\phi-\alpha} \tag{1}.$$

HOLOGRAPHIC CAPABILITY OF THE LLOYD'S MIRROR

The holographic information about the object in the XUV domain can be recorded using the Lloyd's mirror in two ways, both presenting an analogy to the Fourier transform setup in optical holography [7] (the reference is supplied as a spherical wave, and each object point is encoded in a spatial frequency unique to this point). In the first case, Fig.2(a), the examined information is the amplitude and phase transmittance of a thin object placed in the path of the directly propagating wave. In the second case, Fig.2(b), information about a three dimensional object sitting on the mirror is recorded.

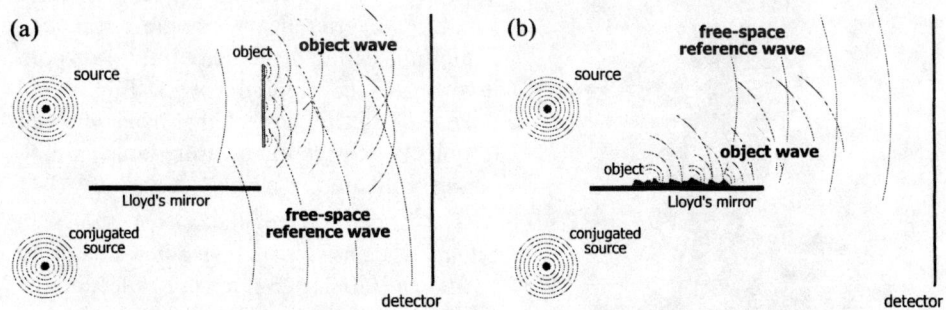

FIGURE 2. X-ray holography using the Lloyd's mirror sctup: (a) transmission hologram of a semi-transparent "2-D object" with the reference wave supplied by the conjugated source; (b) reflection hologram of a surface object with the reference wave coming from the real source.

In both configurations, the information about the object is encoded as amplitude and phase modulation of the carrier fringes. The complete "transverse" information about the object containing a maximal spatial frequency B may be retrieved from the recorded hologram if the interference angle $\Omega \geq B \lambda$ [7], cf. Fig.1. This condition is directly applicable to the configuration from Fig.2(a), yielding here Ω (mrad) $\geq 14\, B\, (\mu m^{-1})$.

EXPERIMENTAL SETUP

The presented experiment was carried out at the JAERI X-ray laser facility, using a transiently pumped Ni-like XRL emitting at 13.9 nm [8]. It delivers typically $\approx 20\,\mu J$ in pulses with a duration of a few picoseconds, in a beam with typical divergence of 10×20 mrad (h×v). The employed experimental arrangement is schematically shown in Figure 3. A near-spherical wavefront is produced by a single concave spherical mirror working a few degrees off-axis. Since interferograms were recorded by a CCD having 25-μm pixels, the mean angle between the reference and reflected rays was deliberately

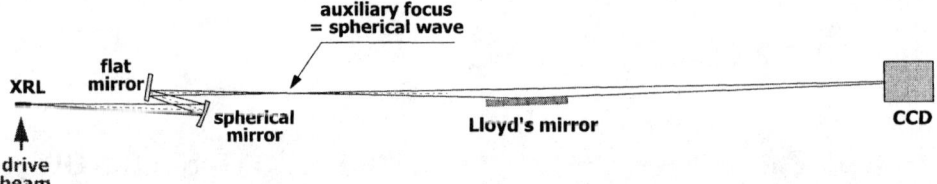

FIGURE 3. Experimental setup (not to scale). A Mo:Si multilayer concave mirror with a radius of curvature of 1060 mm is used to re-focus the XRL beam down to a point, located ≈ 80 mm in front of the edge of the Lloyd's mirror. The CCD camera (16-bit Roper Scientific, 1024×1024 pixels of $25 \times 25\,\mu m$) was located here 510 cm downstream from the Lloyd's mirror center.

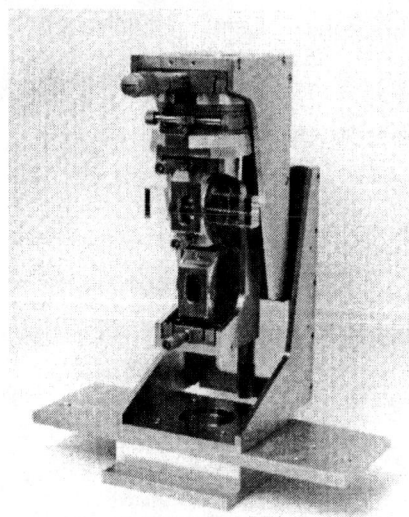

made extremely small (~0.1 mrad) in order to produce fringes with a period extending over a sufficient number of pixels, and amounting to typically 180 μm. Such a small interference angle indeed compromises the spatial resolution for both investigated configurations (cf. Fig.2), but the use of a CCD, on the other hand, allowed to quickly acquire the measured data, which was essential in this early experiment.

The Lloyd's mirror mechanics, shown in Figure 4, consist of a x-y-z-θ-φ mount to position a $100 \times 25 \times 15$-mm glass reflecting the XRL beam under a grazing incidence angle.

FIGURE 4. The Lloyd's mirror assembly, with a block of superpolished BK7 glass constituting the reflective component.

RESULTS AND DISCUSSION

The first holographic arrangement, corresponding to Figure 2(a), was examined exploiting a thin 50-µm tungsten wire as a "2D" object producing an amplitude- and phase-modulated scattered wave. This configuration, represented in Figure 5, is experimentally simple and offers a geometrically well-defined setup which can be numerically analyzed using the mathematical treatment of Fresnel diffraction.

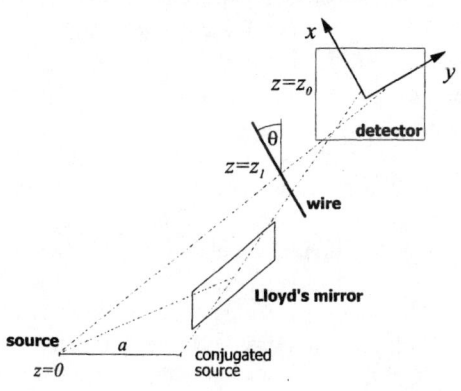

FIGURE 5. Holographic mapping of the diffracted wave produced by a thin tungsten wire, orientated at $\theta \sim 45°$ with respect to the mirror surface and located at $z_1 \sim 38$ cm downstream from the origin of the spherical wavefront.

In the case of an ideally regular diffracting wire of width w, it can be shown that the distribution of intensity $I(x,y)$ on the detector plane will be

$$I(x,y)\, I_0 = \tag{2}$$

$$= 1 + \cos(k_x x + k_y y - \frac{\pi}{4} - k\frac{y^2}{2z_1})\, C(y) + \sin(k_x x + k_y y - \frac{\pi}{4} - k\frac{y^2}{2z_1})\, S(y) + C^2(y) + S^2(y)$$

where $k_x = x_0 k / \sqrt{(x_0 + y_0 + z_0)}$, $k_y = y_0 k / \sqrt{(x_0 + y_0 + z_0)}$, I_0 is the intensity of the unperturbed wave produced by the generating point source, and $C(y)$ and $S(y)$ are Fresnel-type integrals equal to

$$C(y) = \int_{(-y-\frac{w}{2})\beta}^{(-y+\frac{w}{2})\beta} \cos(\frac{\pi}{2} u^2 + k\frac{y\beta}{z_1} u)\, du \quad \text{and} \quad S(y) = \int_{(-y-\frac{w}{2})\beta}^{(-y+\frac{w}{2})\beta} \sin(\frac{\pi}{2} u^2 + k\frac{y\beta}{z_1} u)\, du \tag{3},$$

where $\beta = \overline{2z_1\, \lambda z_1(z_0 - z_1)}$.

The obtained experimental result is displayed in Figure 6. The fringes exhibit a shift along the diffraction pattern cast by the tungsten wire, mapping thereby the phase profile of the diffracted wave. The irregularities of the shift apparent namely in the wire shadow reflect irregularities both of the wire profile and of the conductivity of its surface. The second

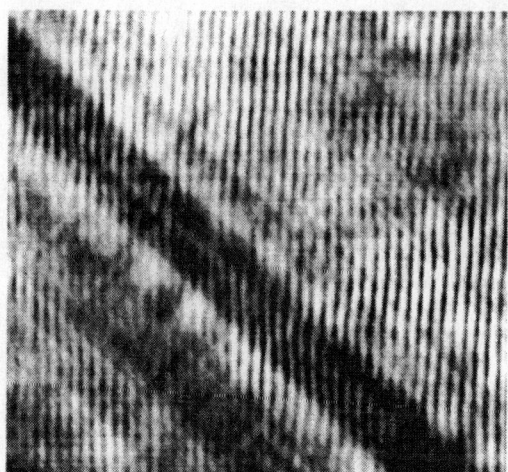

FIGURE 6. Phase profile of the diverging wave diffracted on a 50-µm wire, encoded as a shift of the interferometric fringe system. The experimental setup corresponds to Fig.5.

("ghost") image of the diffracting wire, appearing on the left from the principal image, comes into view as a consequence of the small interference angle mentioned above; it is produced by interaction of the wave reflected from the Lloyd's mirror, casting the interference pattern, with the freely propagating wave. Under these experimental conditions, the spatial resolution on the probed object is ~5 μm. The CCD record furthermore exhibits non-negligible variations of the fringe intensity, which are due to intensity fluctuations of the generating XRL beam.

The numerical analysis of the obtained data, employing the treatment according to Eqs.(2) and (3), is in progress and will be published at a later date.

The surface holography arrangement, corresponding to Figure 2(b), was experimentally investigated by making the corresponding portion of the spherical wave to scatter on microscopic particles deposited on the Lloyd's mirror. These scattering particles were constituted by SnO_2 powder with grains having the characteristic size of ~100 nm, randomly aggregated upon deposition on the glass surface.

The obtained data are shown in Figure 7. The recorded fringe pattern exhibits a complex shift and blur, which are characteristic hologram features and clearly illustrates the X-ray holographic capability of the used arrangement. To our knowledge, this is the first experimental demonstration of soft X-ray holography using an X-ray laser and the side-band (i.e., not inline) arrangement. The surface (elevation) resolution amounts here to ~100 nm, which may be significantly improved in future experiments by employing a larger interference angle in conjunction with a detector having a better spatial resolution. The intensity variations across the record partially reflect the encoded holographic information, but are mainly due to nonuniformities of the XRL beam wavefront.

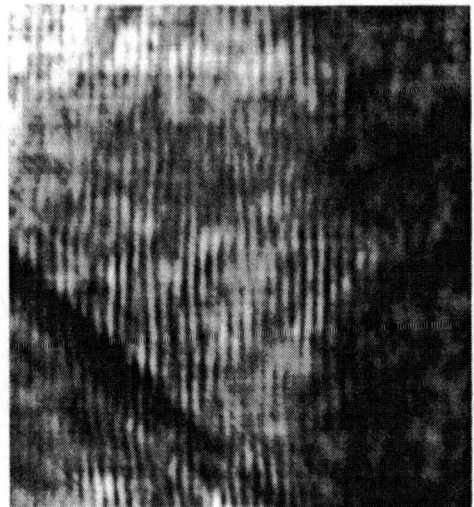

FIGURE 7. Surface hologram of SnO_2 grains deposited on the Lloyd's mirror; a shift equal to one half of the fringe period corresponds to the surface elevation of ~500 nm. The mirror surface lies on the right side of the record.

The numerical reconstruction of the probed object is, in the surface holography configuration, significantly more complex than for a quasi 2-D object investigated in the former case. We are currently developing a numerical model allowing to infer, from the measured data, the appearance of the probed surface, i.e. both its elevation relief and its XUV reflectivity. This model, as well as its sensitivity to imperfections of spatial uniformity of the illuminating radiation, will be experimentally verified using a reference surface with well known characteristics. On the other hand, it is apparent that achieving the full holography potential of the demonstrated arrangement calls for an improvement of the XRL beam wavefront quality.

SUMMARY

We have successfully applied the Lloyd's mirror wavefront-splitting X-ray interferometer to a transient collisional excitation XRL, operating at 13.9 nm. The beam emitted by this laser was found to exhibit a good transverse coherence, producing, under the given experimental configuration, straight fringes with a visibility lying typically between 0.15 and 0.20. The visibility was found to locally vary across the interferograms, significantly decreasing over the high-intensity speckles presented in the beam.

The employed setup made it possible to demonstrate X-ray transmission and reflection Fourier transform (side-band) holography, using ps-class XRL as a source. The application potential especially of the reflection picosecond holography includes, for instance, probing of spatial dynamics of microbiological objects, large organic molecules or fullerene structures, and investigating of phase transitions of surfaces and/or thin layers subjected to e.g. intense radiation or thermal fields. Quantitative measurements associated with these experiments will, however, require an improved beam quality of the transient collisional excitation X-ray lasers.

Based on the prototype of the Lloyd's mirror used in these experiments, we are developing an advanced version making it possible to de-couple the angle of incidence on the mirror from the interference angle (cf. Fig.1). This may be achieved by using two separate mirrors, the first serving to reflect the reference wave, and the second carrying the probed object. This arrangement makes it theoretically possible, in conjunction with an appropriate detector, to achieve a surface resolution of the order of a few nanometers.

REFERENCES

1. Svatos, J., Joyeux, D., Phalippou, D., and Polack, F., *Opt.Lett.* **18**, 1367-1369 (1993)
2. Albert, F., Joyeux, D., Jaeglé, P., Carillon, A., Chauvineau, J.P., Jamelot, G., Klisnick, A., Lagron, J.C., Phalippou, D., Ros, D., Sebban, S., and Zeitoun, P., "Interferograms obtained with an X-ray laser by means of a wavefront division interferometer", in *X-Ray Lasers 1998, Proc. 6th Int. Conf. on X-Ray Lasers*, ed. by Y.Kato, H.Takuma, and H.Daido, IOP Conf. Series No.159, Bristol, 1999, pp. 673-676
3. Rocca, J.J., Moreno, C.H., Marconi, M.C., and Kanizay, K., *Opt.Lett.* **24**, 420-422 (1999)
4. Rus, B., Mocek, T., Präg, A.R., Kozlová, M., Jamelot, G., Carillon, A., Ros, D., Joyeux, D., and Phalippou, D., accepted for publication in *Phys.Rev. A* (2002)
5. Albert, F., Zeitoun, P., Jaeglé, P., Joyeux, D., Boussoukaya, M., Carillon, A., Hubert, S., Jamelot, G., Klisnick, A., Phalippou, D., Ros, D., and Zeitoun-Fakiris, A., *Phys. Rev. B* **60**, 11089-11094 (1999)
6. Mocek, T. *et al.*, these Proceedings
7. Goodman, J.W., *Introduction to Fourier Optics*, Second Edition, McGraw-Hill (New York, 1996)
8. Daido, H. *et al.*, these Proceedings

Exploring the Potential of Table - top X-ray Lasers and Capillary Discharges for Applications

V.N.Shlyaptsev[1], J.Dunn[2], R.F.Smith[2], S.J. Moon[2], K.B.Fournier[2], J.Nilsen[2],
A.L.Osterheld[2], J.Kuba[2], R.London[2], A.J.Wootton[2], R.W.Lee[2], J.J.Rocca[3],
A.Rahman[3], E. Hammarsten[3], J.Filevich[3], E. Jankowska[3], M.C.Marconi[3],
N.Fornaciari[4], D.Buchenauer[4], H.A.Bender[4], S.Karim[4], M.Kanouff[4],
J.Dimkoff[4], G.Kubiak[4], G.Shimkaveg[5], W.T.Silfvast[5]

[1]*UC Davis-Livermore, ILSA / LLNL, Livermore, CA 94551*
[2]*Lawrence Livermore National Lab, Livermore, CA 94551,*
[3]*Colorado State University, Ft.Collins, CO 80523*
[4]*Sandia National Labs, Livermore, CA 94551*
[5]*School of Optics/CREOL, University of Central Florida, Orlando FL 32816*

Abstract. The advantages of using of table top x-ray lasers (XRLs) for different applications have been described. Examples of the first successful use of XRLs, the current efforts in applying them and the potential applications where an XRL can be used in future have been discussed. Modeling results showing the possibility of 3-4 times shorter wavelength capillary discharge x-ray lasers and calculated spectrum of Xe capillary EUV source are presented.

INTRODUCTION

The high efficiency of short pulse duration transient and electric current-driven capillary discharge x-ray lasers have allowed the scaling of these lasers to table-top dimensions. This achievement has opened new possibilities for many different applications. We discuss some past, present and future applications using table-top x-ray lasers.

The applications require more efficient and powerful x-ray lasers with shorter wavelengths and in some cases shorter pulse durations. Specifically, in this paper, we summarize nowadays achievable applications, such as shadowgraphy, soft-x-ray imaging, x-ray and photoelectron spectroscopy and some others. The interferometry and study of XRL interactions with matter will be treated in detail.

In the last part of the paper we present the gain calculations for capillary discharges driven by large, up to 300 kA currents which allow to achieve shorter wavelengths and hence enable new applications. We also discuss EUV sources based on capillary discharge for such an important application as microlithography.

TABLE-TOP X-RAY LASERS AND APPLICATIONS

During the last decade there was substantial progress in scaling down x-ray lasers in dimensions, pumping energy, and cost. XRLs have achieved improved characteristics of peak power, efficiency, coherence, higher repetition rates of operation and dramatically decreased pulse duration. But the major goal of scaling to affordable, table-top dimensions during the last decade was driven by the need to utilize these unique x-ray

CP641, *X-Ray Lasers 2002: 8th International Conference on X-Ray Lasers*, edited by J. J. Rocca et al.

sources for a spectrum of new potential applications and, therefore, substantially increase the range of experimental tools available to a larger user base. Another path is the usage of a large, central, multi-user facilities, utilizing accelerator-based ASE x-ray laser sources which have simultaneously many record parameters, e.g. brightness, tunability and short pulse duration, but at a substantially higher cost and size. In some respects, these resemble the past of plasma x-ray lasers when they were using large, expensive, high power laser facilities. Table 1 compares current plasma-based table top XRL parameters with one of the future DESY VUV- Free Electron Laser (FEL) sources planned to become operational in the next few years. Many achieved parameters of plasma-based x-ray lasers heated by powerful lasers and electrical discharges compare favorably to large scale present and near future particle acceleration x-ray sources. Definitely, plasma XRLs will continue to improve during the next years with the potential to approach the characteristics of FELs. In some cases to keep the cost and size down, it would be reasonable to chose the required type of a table-top laser having just those parameters which are needed by a given application since it is clear that applications in science, industry, metrology, inspection, active and passive diagnostics, heating, surface modification, and plasma ionization are very different. On the way of further increasing the peak brightness which is the most important parameter of XRL, improvement of several parameters will be needed including:

- increasing the output energy,
- decreasing the divergence,
- shortening the pulse duration.

These are possible and will boost brightness by 3-4 orders of magnitude, elevating XRLs to match some FELs. Also, high repetition rates are achievable for laser-produced plasma XRL similar to proposed FEL or capillary XRL. This will boost average power and brightness of laser plasma XRL by orders of magnitude. Additionally, wavelength of

TABLE 1. COMET/Capillary/DESY VUV-FEL x-ray laser source parameters

Source Parameters	COMET x-ray Laser	Capillary [b] x-ray laser	DESY VUV-FEL[c]
Pump Energy – laser or condensor (for capillary) (J)	5 – 10	40-100	-
X-ray Laser Energy (μJ)	25	100-900	300
Photons/Shot	2×10^{12}	2×10^{14}	10^{13}
Shot Rate (Hz)	0.004	1-10	10
Wavelength (nm)	12 – 47	47, 53, 61	6
Δλ/λ	10^{-4}	10^{-4}	6×10^{-3}
Source Dimensions (μm)	25 by 100	3×10^{-2}	150
Divergence (mrad)	2.5 by 10	2-5	0.06
XRL Pulse Duration (ps)	2 – 25	700 – 1500	0.1
Peak Brightness, B [a]	1.6×10^{25}	3×10^{25}	10^{29}
Average Brightness, B [a]	1.3×10^{11}	5×10^{14}	-

[a] Units of ph. mm^{-2} $mrad^{-2}$ s^{-1} (0.1% BW) $^{-1}$; [b] parameters shown are for the 46.9nm Ne-like Ar laser; [c] Expected start operation 2005

table-top XRLs will continue to decrease which will further, and dramatically, increase the brightness. This makes table-top lasers attractive and viable alternatives to other x-ray sources except for applications that really require continuous tunability.

Recently a number of interesting applications was identified for use with future FELs which will operate at 60 Å within 3 years and will reach 1 Å within 10-12 years. The Joint Proposal for Peak Brightness Experiments on the TTF-FEL (LLNL) covers a very wide range of scientific areas from plasma physics to finite temperature condensed matter physics to biological-related imaging. Table 2 cites the main topics in this Proposal together with a brief description.

TABLE 2 Summary of the Peak Brightness Beamline Experiments

Experiment	Brief Description
Warm Dense Matter	Using the x-ray laser to uniformly warm solid density samples. Isochoric heating of a thin foil
Equation of State Measurements	Use an optical laser to heat a sample and the x-ray laser to provide a diagnostic of the bulk conditions
Femtosecond Ablation Studies	Probe the nature of the ablation process on the sub-ps time scale
Near Edge Absorption	Use an optical laser to heat a solid and the x-ray laser to probe the structural changes that occur
Trapped Strongly Coupled Plasmas	Use an EBIT / laser-cooled trap and probe highly charged strongly coupled Coulomb systems
Diagnostic Development	Develop the FEL for Thomson scattering, interferometry, and radiographic imaging
Gas Jet Interaction	Create exotic, long-lived highly perturbed electron distribution functions in dense plasmas
Solid Interactions	Use laser directly to create extreme states of matter at high temperature and density
Plasma Spectroscopy	Use the FEL as a pump to move bound state populations and study radiation redistribution
Coulomb Explosion	Study the effects of the Coulomb Explosion process with emphasis on Biological imaging problems
Diffraction Imaging Studies	Validate imaging techniques. Perform microscopy on living systems beyond the current resolution limited by radiation damage
Optics Damage	Study structural changes and disintegration processes of solids as a function of laser intensity

All of them were inspired by the attractive properties of future FEL sources. Part of these proposed experiments require direct interaction of FEL with matter (e.g. the experiments on creating warm dense matter, plasma physics of photoionized gases, Coulomb explosions of biological samples, diffraction imaging of biological samples, optics damage and high-field effects). Others use this x-ray source just for diagnostics purposes. The proposal on optics damage experiment, for example, combines both heating and probing together. Even a short analysis of these proposals shows that most of them can be successfully realized with smaller and cheaper table-top XRLs. It should be noted that an optics damage experiment is already under way utilizing the transient Comet XRL at LLNL (see below) as well as some others like photoelectron spectroscopy studies [2]. Some experiments will require additional solid state lasers synchronized with XRL pulse of fs, ps or ns duration (as it is the case in above proposal for development of diagnostics of Thomson scattering, interferometry and radiography, equation of state measurements, femtosecond ablation studies, near-edge absorption studies, plasma spectroscopy and strongly coupled plasma studies using a laser-cooled trap). An obvious advantage of table-top XRLs is that this additional laser will be easier to implement since transient XRLs are pumped with the optical lasers and multiple synchronized beams are available.

Utilized currently mostly for scientific applications, table-top x-ray lasers are already expanding the field of traditional scientific diagnostic methods allowing to extract more detailed information, in some cases with an unprecedented accuracy. Among them are:

- **Shadowgraphy:** XRL can be used here because of high peak brightness, and in principle, can be applied as diagnostics of NIF hohlraum and other fusion experiments. In addition, this method may have exceptional sensitivity because of the exponential dependence of transparency on the plasma absorption coefficients. This has been confirmed by modeling and experiments with the argon 46.9nm capillary x-ray laser beam through an elongated plasma formed by an another capillary discharge [3]. Due to plasma inhomogeneity, the extremely sharp boundaries observed in these experiments transform into precise position of specific ion stages and are a demonstration of the mentioned sensitivity.

- **Soft x-ray imaging:** The images of plasmas were recorded in many spectral regions for numerous experiments for a wide range of conditions in capillary and laser plasmas. Of special interest was near- and far-field imaging of the capillary output and transient XRL. The experiments and numerical modeling allowed to reveal detailed data about plasma evolution, gain formation and amplification dynamics. Like with shadowgraphy, due to the same exponential dependence (this time on gain coefficient) it can be arranged as a very sensitive method for plasma diagnostics [4]. Another example of successful application of diagnostics of small variations in distribution function responsible for changes in laser radiation absorption coefficients at shorter pulse durations is demonstrated in [5].

- **X-ray spectroscopy:** This method is the most routine but when combined with XRL it can have exponential sensitivity too. It can be used to find the temperature and density of plasma column, evaluate amplification of x-ray laser etc. It can be applied to find extremely weak plasma processes like Zeeman effect [6] or isotopic hyperfine splitting in spectral lines [7] due to their influence on linewidth and hence gain.

- **Photoelectron spectroscopy:** The comparison of using transient XRL with laser plasma x-ray source shows clear advantage of XRL which provides 1-2 orders of magnitude more photons in selected spectral band and is monochromatic which is important for accuracy. The experiments are currently in progress on COMET [2].

- **Interferometry:** This promises to be the key application of x-ray lasers because it utilizes the unique property of their coherence. The coherence of the first generation of table-top x-ray lasers was sufficient to demonstrate plasma interferometry [8, 9]. Given the very high brightness of all x-ray lasers, the substantially reduced refraction and inverse bremsstrahlung absorption, and the simple straightforward relationship between electron density and index of refraction, interferometry will bring incredible precision to applications in material science, x-ray holography, metrology, dense plasma and fusion applications.

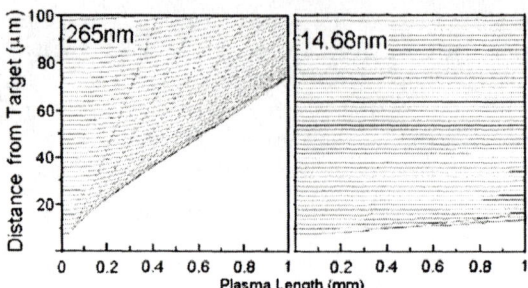

Fig. 1 RADEX ray-tracing of the probe beams of 14.68nm and 265nm wavelength through the calculated density profile. A 1mm long Al plasma is probed at 500ps after the peak of the 600ps plasma forming beam. The plasma was generated with a 6x0.04 mm line focus with an intensity of 4×10^{11} W/cm^2.

The x-ray laser interferometry, first reported on LLNL kJ-scale lasers [10], is expected to have a similar impact on the range of applications as was achieved many years ago with famous x-ray tubes. Figure 1 shows ray-tracing on a typical 1 mm long laser plasma density profile and demonstrates how challenging it is to probe plasmas of important, practical sizes [11]. This is one of the reasons, which may seem surprising, that laser plasma density profiles even at very low laser fluxes accessible for decades were not precisely characterized. Even longer plasmas, e.g. a NIF capsule could be 0.5 - 1 cm, or Z-pinches up to several cm, and in some cases an x-ray laser medium may be tens of cm long, require accurate probing to better understand the plasma conditions. In the case of [12, 13] which attempted to probe laser produced plasma using optical lasers, the refraction restricted the maximum density to around $\sim10^{19}$ cm^{-3}. As a result the density structures resembling very pronounced side lobes, which represent a general property of laser-produced plasmas, were missed by researchers (see ref. [14] for explanation). The inability to probe sufficiently high densities above a certain length with optical or UV wavelengths is clearly demonstrated by Fig. 1. This shows the detrimental ray tilt due to refraction as they propagate through the target plasma. In another calculation shown in Fig. 2 we have included both the effects of the density dependent fringe shifts and refraction-related tilting due to density gradient. The solid line is the initial density profile as was obtained from the RADEX simulations in cylindrical geometry with R_0=40 μm. The other two RADEX curves are the densities deduced from

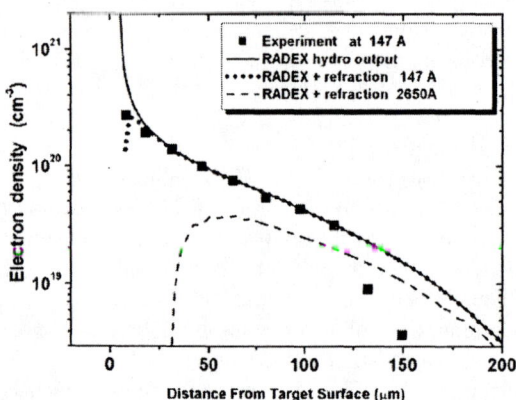

Fig. 2 RADEX calculated density profile (solid line) in comparison with experiment [squares] at 500ps after the peak of the 600ps plasma forming beam 1×10^{12} W/cm^2. Similar values were also obtained with 1.5D LASNEX. Two additional curves show reconstructed density by ray-tracing of the probe beams of 14.68nm and 265nm wavelength through 2mm long Al plasma as they would be deducted by ideal interferometer.

the simulated interference patterns based on this profile if the plasma would be probed by 147 Å or the 4ω higher harmonic (2650 Å) of a Nd-glass laser. The 'ideal' interferometer was implied in this calculation which assumes the ray path in the instrument arms is not affected if they come with different angles which is usually taken into account in real situations [9]. Fig. 2 demonstrates the advantage of XRL over UV interferometry for probing long plasmas. But at such relatively large plasma lengths even 147 Å XRL probes may give an inaccurate picture if there exist very short lateral micron-scale density gradients. As a result of very high ratio of plasma length to gradient length ~ 1000, in this particular case the probing can not come higher than approximately 3×10^{20} cm^{-3}. Larger magnifications, shorter wavelengths or smaller plasma sizes are needed to get into the regions of such small scale length at the critical surface. In all other areas where the gradients are not as severe, this method works extremely well and accurately reproduces the density profiles up to ~10^{22} cm^{-3} for a 0.1 - 0.2 cm long plasma. Higher densities >10^{23} cm^{-3} almost up to XRL critical density can be probed if the gradients are negligible and the plasma length is appropriately short not to absorb the signal and to ensure a reasonable number of fringe shifts. We believe that future experiments will clearly demonstrate that with table-top XRLs we get extremely powerful and sensitive diagnostic tool with great potential.

OPTICS DAMAGE BY HIGH X-RAY FLUENCES

Opposite to numerous kinds of plasma x-ray sources, the predictable parameters of radiation sources which utilize high-energy particles allow to plan the development of such sources far in advance. During the next decade the 4th generation x-ray sources promise to provide the parameters that are far beyond the reach of current 3rd generation synchrotron light sources. It has been expected that a decade from now these sources will produce unprecedented levels of peak and average brightness of monochromatic and spontaneous x-ray radiation. With brightness and fluences of this level there will be no materials which can withstand these fluxes without destruction. As a result, one of the important application in this case is investigation of damage thresholds of x-ray optics.

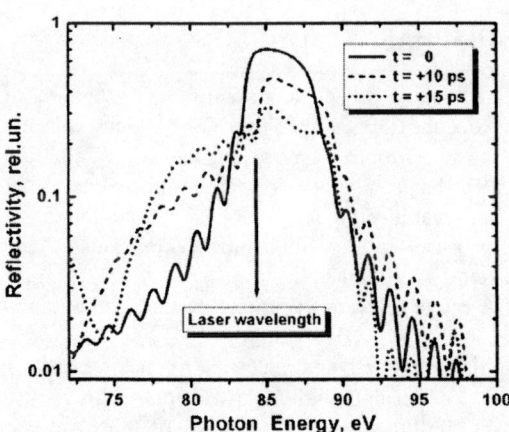

Fig. 3. Modeling of the X-ray reflectivity of a Mo/Si multilayer mirror under the incidence angle of 8 degrees (50 pairs, 30Å Mo/45Å Si on a silicon substrate), where t=0 is the beginning of the incident laser pulse: duration 5 ps at FWHM, intensity 1.7 10^{11} W/cm^2, photon energy 84.4 eV.

One of such x-ray SASE sources will be the Linac Coherent Light Source (LCLS) at Stanford while another at DESY laboratory in Hamburg, Germany is currently constructing a 60 Å source as part of the TESLA linear collider project. The BESSY laboratory at Berlin, Germany, is also designing a 12 Å SASE FEL User Facility.

To emulate optics damage effects we are using laboratory sources of optical and x-ray radiation including femtosecond solid state lasers, K-α plasma sources and x-ray lasers. The experiments with optical wavelengths of similar fluences were performed at LLNL which produced unexplainable results due to potentially non-linear interaction of radiation with matter [15] though optical wavelength interaction physics is very different from x-rays. Given the same nature and reliable values for absorption which will not leave linear regimes, and given the increased accessibility of table-top x-ray lasers the project is under development to utilize the COMET x-ray laser at LLNL. The main interest with this experiment is to emulate the interaction of 10 keV photons of future FEL x-ray sources similar to the LCLS. With the proposed design the 84.4 eV COMET XRL fits well to this goal because the damage layer caused by 10 keV radiation is defined by photoelectron stopping length, 0.5 - 1 μm, and is similar to the photo-absorption length of 84.4 eV XRL radiation for certain materials. To increase sensitivity of the method and at the same time reflectivity which in x-ray region for solid materials at normal incidence is typically very small we proposed to use multilayer mirrors as a test object. The substantial increase of sensitivity of the method will be due to resonance properties of multilayer mirrors.

RADEX and LASNEX simulations were used to predict the heating and expansion of the multilayers under the influence of radiation from the COMET XRL. The reflection coefficient was obtained as a function of photon energy, incidence angle, laser flux and time. Figure 3, shows the dynamic changes in reflectivity of a Mo/Si multilayer mirror modeled with the codes RADEX and XOP. These changes are seen on the hydrodynamics timescale when several top layer pairs will expand into the vacuum. Other physical effects, such as phase transitions, are expected to be observed which may occur with potentially different timescales (from sub-ps to ns-scales) based on their different reflectivity temporal, spectral and angular behavior.

CAPILLARY DISCHARGE STUDIES

The capillary discharge as one of variants of Z-pinches attracted attention of plasma physics researchers for almost two decades. It has been used for hot dense plasma formation and x-ray lasers [16, 17], for transportation of laser beams and XUV radiation generation in x-ray lithography [18, 19], basic Z-pinch research and some other applications. Due to its geometry, and achievable high densities and temperatures, Z-pinches represent a natural medium for x-ray lasers. Substantial experimental and theoretical efforts are now devoted to extend this kind of XRL to shorter wavelengths because of obvious benefit of 100 – 150 Å range relevant to projection EUV lithography and other practical needs.

The progress in parameters of capillary materials and electric drivers for capillary discharges allowed the achievement of 200 kA currents with 10 ns risetime. This in turn enabled to reach higher temperatures $T_e \sim 200\text{-}400$ eV, densities $N_e > 10^{20}$ cm^{-3}, and as large as 300 times density compression ratios in high-Z plasma needed for next generation of capillary discharge XRLs [17, 20]. With such temperatures and densities the collisional XRL scheme on Ni-like ions becomes feasible. The atomic numbers of elements suitable for lasing are in the range 42 - 50 which are lasing at wavelengths down

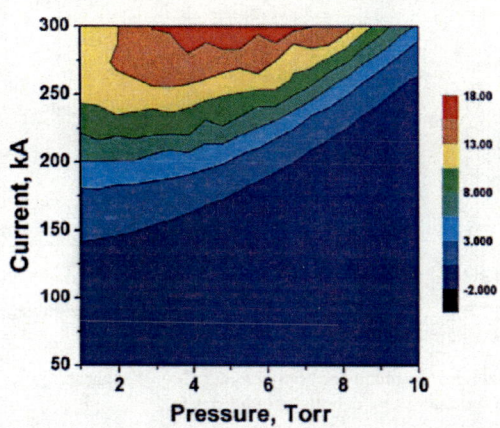

Fig. 4   Small signal gain on 4d-4p transitions in Ni-like AgXX as a function of current and pressure. Current pulse duration is set 20ns, capillary diameter is 3.3 mm.

to ~100 Å. For Ni-like ions e.g. Cd XXI, the gain is calculated to be ~1-2 cm^{-1}, which is in qualitative agreement with the current experimental data [20]. Figure 4 shows the results of numerical calculations of small signal gain for the Ni-like ion AgXX obtained with over 400 hydro/atomic kinetics runs. The numerical model, as usual, involves plasma heating, ionization, radiation transport and material ablation physics. The calculated gain is substantially larger than using Cd. We plan the experiments with the Ni-like Ag ion in the future.

Z-pinches are naturally efficient x-ray and EUV sources so that they can be used for many important applications in science and technology. In particular, the capillary discharge can appear as powerful potential candidate for emerging EUV microlithography. We will shortly describe this potentially very important application of capillary discharges. It is not a coherent or even monochromatic source like an x-ray laser but their physics and numerical models are very strongly related so that the new data and knowledge obtained in either case is equally beneficial. Hence, it is possible to consider x-ray conversion studies as a very useful side product of x-ray laser research. In fact, the RADEX code can treat the hydrodynamics, atomic kinetics and radiation transport for these two cases in almost the same way. EUVL sources must match the specific practical, technological and environmental requirements of complex processes of microchip production. Among the requirement are the achievement of efficient x-ray conversion in the specific spectral range of interest, high average power, pulse-to-pulse stability, large source lifetime etc. The extremely high spatial and temporal stability (radial position jitter of dense plasma column can be comparable to laser plasma, ~20 microns or less), relatively large temperatures achieved with small currents, simplicity and efficiency attracted the attention of researchers to develop capillary plasmas as a radiation source.

Numerous experiments with different capillary materials, currents and gas fills were performed in CREOL and Sandia National Labs [19, 21, 22]. The parameters of this discharge are somewhat intermediate between capillary XRL case and one of 300 μm microcapillary investigated previously [3] since the discharge source has been driven by much smaller (3 - 6 kA) and longer (1 - 2 μs) current pulses. We compared our simulations with the optical interferometry of capillary discharge plasma performed at CREOL and found good agreement in source size, density spatial temporal behavior. We also reproduced spectra obtained in Sandia National Labs capillary discharge source [23].

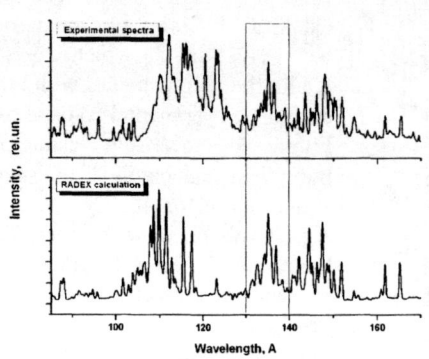

XeVIII	117A: 4p5s-g, 4p5p–g
XeIX	90A: 4d5f-g, 4d6p-g,
	117A: 4d4f-g,
	123A: 4p5s-4d5s
	135A: 4p5s-4d5s, 4p5p-4d5p,
	145A: 4d4f-g,
	163A: 4d5p-g
XeX	90A: 4d5f-g
	110A: 4d4f-g, 4p4d-g,
	145A: 4d5p-g, 4d4f-g
XeXI	80A: 4d5f-g,
	110A: 4d4f-g, 4p4d-g,
	135A: 4d5p-g, 4d4f-g, 4d5f-g

Fig. 5 The experimental (top) and RADEX calculated spectra (bottom) of capillary discharge. Table3 summarizes transitions which contribute to these spectra. g here denoted ground state(s)

To describe spectra correctly, the atomic kinetics model data size for high-Z gases like Xe used in this source reach tremendous dimensions of the order of 1 GB due to the number of ion stages XeVIII - XeXI involved, the atomic levels of the order of 10^4 and spectral lines approaching 10^6. The main source of radiation in such systems is due to numerous spectral lines often many times overlapped over instrumental resolution of ~0.2 Å. The spectra obtained from 5 kA 1.4 μs diamond capillary discharge in Xe at 2 Torr in comparison with modeling spectra are shown in Fig. 5. Table 3 marks the major transitions contributing into different spectral bands. Some missing transitions in RADEX spectra around 120 Å are due to omission in the calculations of the lower Z stages Xe V - VII radiating at much lower temperatures. The calculations done with a wide range of parameters clearly support the observations that Ru-like Xe ions (Xe XI) are contributing the most into the 135 Å region of interest for microlithography. We will continue modeling this source for further optimization for lithography.

ACKNOWLEDGMENTS

This work was performed under the auspices of the US Department of Energy by the University of California Lawrence Livermore National Laboratory under Contract No. W-7405-Eng-48. Part of this work was supported by DARPA grant DE-FG 03-00ER15084 and by the National Science Foundation.

References

1. Th.Tschentscher, TESLA Technical Design Report (2001).
2. J. Dunn *et al.*, in these proceedings (2002).
3. M.C. Marconi, C.H. Moreno, J.J. Rocca, V.N. Shlyaptsev and A.L. Osterheld, *Phys. Rev. E* **62**, 7209 (2000)

4. C.H.Moreno, M.C.Marconi, V.N.Shlyaptsev, B.R.Benware, C.D.Macchietto, J.I..A.Chilla, J.J.Rocca, A.L.Osterheld, *Phys. Rev.A* **58**(2), 1509 (1998).
5. V.N.Shlyaptsev, J.Dunn, K.B.Fournier, S.Moon, A.L.Osterheld, J.J.Rocca, F.Detering[4], W.Rozmus, F.Alouani-Bibi, J. P. Matte, H. Fiedorowicz, A. Bartnik, M. Kanouff, Proc.SPIE Vol.**4505** "X-ray lasers and applications", 14 (2001).
6. F.G. Tomasel, V.N. Shlyaptsev and J.J. Rocca, Phys.Rev.A, **54** ,2474, (1996).
7. J. N.Nilsen, J.Koch, J.H.Scofield, B.J.MacGowan, J.C.Moreno, L.B.DaSilva, *Phys. Rev. Lett.* **70**, 3713 (1993).
8. J. Filevich *et al.*, Opt. Lett. **25**, pp. 356 (2000).
9. R.F. Smith, J. Dunn, J. Nilsen, V.N. Shlyaptsev, S. Moon, J. Filevich, J.J. Rocca, M.C. Marconi, J.R. Hunter, and T.W. Barbee, Jr. *Phys. Rev. Lett.* **89**(6), 065004 (2002).
10. L.B. Da Silva *et al*, *Phys.Rev.Lett.* **74**, pp. 3991 (1995).
11. D.T. Attwood, D.W. Sweeney, J.M. Auerbach, P.H.Y. Lee, *Phys. Rev. Lett.* **40**, 184 (1978).
12. Yu.A. Zakharenkov, N.N. Zorev, O.N. Krokhin, Yu.A. Mikhailov, A.A. Rupasov, G.V. Sklizkov, and A.S. Shikanov, *Sov. Phys. JETP* **70**, 547 (1976).
13. L.A. Bol'shov, I.N. Burdonskii, A.L. Velikovich *et al.*, *Sov. Phys. JETP* **65**(6), 1160 (1987).
14. J. Filevich, J.J. Rocca, E. Jankowska, E.C. Hammarsten, M.C. Marconi, S.J. Moon, and V.N. Shlyaptsev, submitted to *Phys.Rev.Lett.* (2002).
15. J. Kuba *et al.* 'X-ray Optics Research for the Linac Coherent Light Source: Interaction of Ultra-short X-ray Laser Pulses with Optical Materials' in these proceedings (2002).
16. J.J.Rocca, V.Shlyaptsev, F.G.Tomasel, O.D.Cortazar, D.Hartshorn, J.L.A.Chilla, Phys. Rev. Lett., **73**, 2192 (1994).
17. J.J. Gonzalez, M. Frati, J.J. Rocca, V.N. Shlyaptsev, and A.L.Osterheld, *Phys.Rev.E* **65**(2), 026404 (2002)
18. Y. Ehrlich, C. Cohen, and A.Zigler, J.Krall, P. Sprangle, and E. Esarey, *Phys. Rev. Lett.* **77**, 4186 (1996).
19. M.A. Klosner, H. Bender, W.T. Silfvast and J.J. Rocca, *Opt. Lett.* **22**, 34 (1997).
20. S. Sakadzic, A.Rahman, M. Frati , F.G.Tomasel, J.J.Rocca, V.N.Shlyaptsev, A.L.Osterheld. Proc. SPIE Vol.**4505** "X-ray lasers and applications" , 35 (2001).
21. N.R. Fornaciari, H. Bender, D. Buchenauer, M.P. Kanouff, S. Karim, C.D. Moen, K.D. Stewart, W.T. Silfvast, and G.M.Shimkaveg, Proc. SPIE Vol.**4688**, "Microlithography-2002" (in press, 2002).
22. J. Dimkoff, N. Fornaciari, D. Buchenauer, S. Karim, and H. Bender, Proc.SPIE Vol.**4688**, "Microlithography-2002" (in press).
23. V.Shlyaptsev *et al.*, Proc. 5th Int.Conference on Dense Z-pinches, Albuquerque, 2002 (in press).

Interferometric Diagnosis of Two-Dimensional Plasma Expansion

Raymond F. Smith[1], Steven Moon[1], James Dunn[1], Joseph Nilsen[1], Vyacheslev N. Shlyaptsev[1], James R. Hunter[1], Jorge Rocca[3], Jorge Filevich[3], Mario C. Marconi[3, 4]

[1]Lawrence Livermore National Laboratory, Livermore, CA 94551
[2]University of California Davis-Livermore, Livermore, CA 94551
[3]Dept. of Electrical and Computer Engineering, Colorado State University, Fort Collins, CO 80523
[4]Physics Dept., University of Buenos Aires, Argentina

Abstract. Recent advances in interferometry has allowed for the characterization of the electron density expansion within a laser produced plasma to within 10 μm of the target surface and over picosecond timescales. This technique employs the high brightness output of the transient gain Ni-like Pd collisional x-ray laser at 14.7 nm to construct an effective moving picture of the two-dimensional (2-D) expansion within the plasma. In this paper we present experimentally measured density profiles from an Al plasma and make comparisons with 1.5-D and 2-D code simulations. The results are discussed along with an analysis of the underlying mechanisms driving the plasma expansion.

1. INTRODUCTION

Interferometry is a powerful tool for accurately diagnosing the two-dimensional (2-D) evolution of dense laser-produced plasmas. For fast evolving plasmas it is desirable that the duration of the probe pulse is short to obtain an effective snapshot of the density profile while reducing the effects of plasma motion blurring at the ablation front. The picosecond duration and short wavelength of the 14.7 nm Ni-like Pd laser mitigates effects associated with motion blurring and refraction through millimeter scale plasmas. This enables direct measurement of the electron density profile to within 10 μm of the target surface [1]. A series of high quality 2-D density measurements provide unambiguous characterization of the time evolution in a fast evolving plasma suitable for validation of existing 1-D and 2-D hydrodynamic codes. The electron density evolution of a laser-heated Al plasma is measured using a diffraction grating interferometer (DGI) [2] at different times, relative to the peak of the plasma forming pulse. The experimental results are compared with 1.5-D and 2-D hydrodynamic simulations in order to further our understanding of the mechanisms driving plasma expansion.

CP641, *X-Ray Lasers 2002: 8th International Conference on X-Ray Lasers*, edited by J. J. Rocca et al.
© 2002 American Institute of Physics 0-7354-0096-2/02/$19.00

2. EXPERIMENTAL RESULTS

The Ni-like Pd 14.7 nm x-ray laser probe beam and the plasma to be studied were generated using three laser beams at 1054 nm wavelength from the COMET facility at LLNL [1]. Single pass saturated x-ray laser output of a few 10's of μJs was achieved with an optical pumping combination of a 600 ps long pulse (2 J, 2×10^{11} W cm^{-2}) and a 6 ps (5 J, 7×10^{13} W cm^{-2}) main heating pulse. The x-ray laser output was imaged and routed into a diffraction grating interferometer for plasma probing experiments. For details of the instrumentation see ref [1, 2].

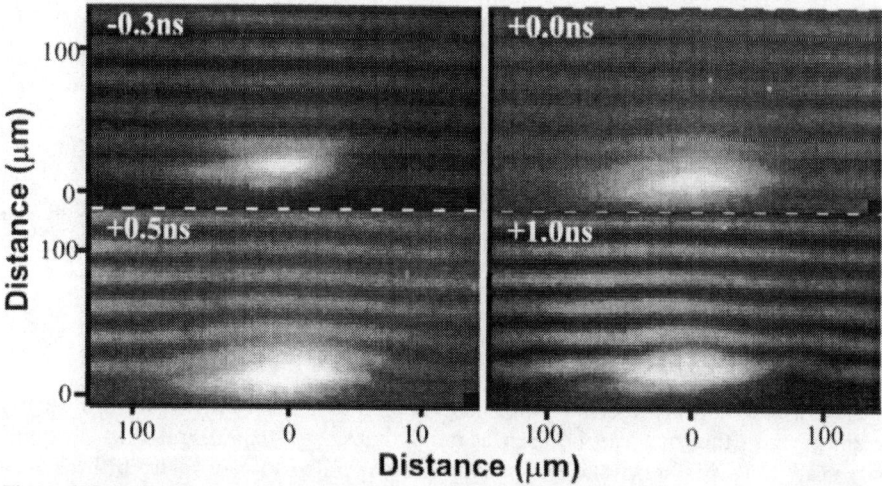

Figure 1. Interferograms representing 2-D electron density profiles for different times relative to the peak of the 600ps plasma forming pulse.

A plasma heated by up to 2 J of energy in a 600 ps, 1054 nm pulse, corresponding to a maximum intensity of 3×10^{12} W cm^{-2}, was produced in one arm of the interferometer. A 6 mm long line focus with a 20 μm focal width, was generated on a polished 1mm long Al slab target using a combination of a cylindrical lens, $f = -200$ cm, and an off-axis paraboloid, $f = 30$ cm. The relative delay between the arrival of the x-ray laser probe pulse, represented by the short pulse beam, to the peak of the plasma forming beam was measured to within 100 ps with a fast diode. The x-ray laser could probe the plasma in the temporal range of -1 ns to $+2$ ns relative to the peak of the 600 ps plasma forming pulse by adjusting a delay arm in the plasma laser beam. The line focus plasma was probed longitudinally by the x-ray laser, thereby minimizing uncertainties in the interpretation of the interferograms arising from plasma gradients along the probe path. The plasma was imaged by a 25 cm focal length spherical multilayer mirror and relayed to a thinned back-illuminated 1024 × 1024 CCD detector with 24 × 24 μm^2 pixels. A 2000 Å Zr/1000 Å Polyimide ($C_{22}H_{10}N_2O_5$) filter was placed in front of the CCD to block visible and UV light. The magnification of the imaging system was determined to be 9.94, by imaging a fine mesh at the target plane, giving a pixel-limited spatial resolution of 2.55 μm. The target angle was determined to be parallel to the x-ray laser beam to better than ± 0.25°. Using the x-ray laser beam with no plasma present, high quality fringes, with visibility $V = (I_{max} - I_{min})/(I_{max} + I_{min})$ of 0.72 ± 0.12, were observed for a 700×500 μm^2 (H × V) region indicating excellent spatial coherence in the laser beam. Figure 1 shows a series of interferograms at different probing times relative to the peak of the plasma forming pulse. The electron density, n_e in cm^{-3}, is related to the measured fringe shifts as $N_{fringe} = 6.68 \times 10^{-20} n_e L$, where L is the length (cm) of the plasma being probed by the 14.7 nm x-ray laser [3].

Significant lateral expansion from the initial 20 μm focal width is observed at all times with the expansion velocity parallel to the target surface measured to be 7.5×10^6 cm/s over the interval of probing. This is in agreement with the expected sound speed for a plasma electron temperature of $T_e \sim 80$ eV. For all the interferograms the sideways expansion is symmetrical about the center of the focusing region. A lot of structure is observed less than 10 μm from the target surface at the limit of the spatial resolution of the imaging system. At 1 ns after the peak of the plasma forming pulse we observed off-axis density enhanced lobes close to the target surface and at approximately 80 μm either side of the central region. In addition, perturbations in the fringes away from the target surface indicated an expansion angle of ± 35° at t > 0.5 ns.

3. SIMULATIONS

Figure 2(a) shows the experimentally observed on-axis electron density profile for different probing times. It can be seen at early times, t = -0.3 ns and 0 ns, that the density scale-length is short and the plasma has expanded to 60 μm from the initial target surface. At later times, t = +0.7 ns and 1.0 ns when the plasma heating pulse is effectively off, the plasma scale-length has relaxed and the plasma has continued to expand to 160 μm. Analysis considering the effects of refraction [4] has shown a 20% uncertainty in the highest density measurements. The lowest measured electron density for this target length is limited to 2×10^{18} cm^{-3} and is dependent on the minimum detection of the fringe shift. The error bars on the density measurement in Figure 2(a) represent the uncertainty in determining the position of the fringes.

Alongside these measurements are the predicted electron density values from the 1.5-D LASNEX [5] plasma physics code. The LASNEX simulations are one-dimensional but include an expansion angle of 15 degrees in the direction perpendicular to the primary expansion (1.5-D) so as to simulate the 2-D effects associated with the narrow width of the line focus on the target. The code predictions give qualitatively good agreement in reproducing the plasma density profile evolution. The overall trend, however, is for the simulations to show somewhat higher density at all times. The predicted n_e values increase rapidly within 5 μm of the target surface. Experimental images indicate very fine fringe structure in this region, which are not well resolved by our present imaging setup. Within our current experimental setup, this has proved to be the limit on the maximum diagnosable electron density. This however does not represent the limit of the technique. The critical density for the 14.7

nm probe wavelength is 5×10^{24} cm^{-3} and gives the ultimate limit to the density with which the x-ray laser can probe. By optimizing the instrumentation and plasma conditions we expect to closer approach this figure in future experimental work.

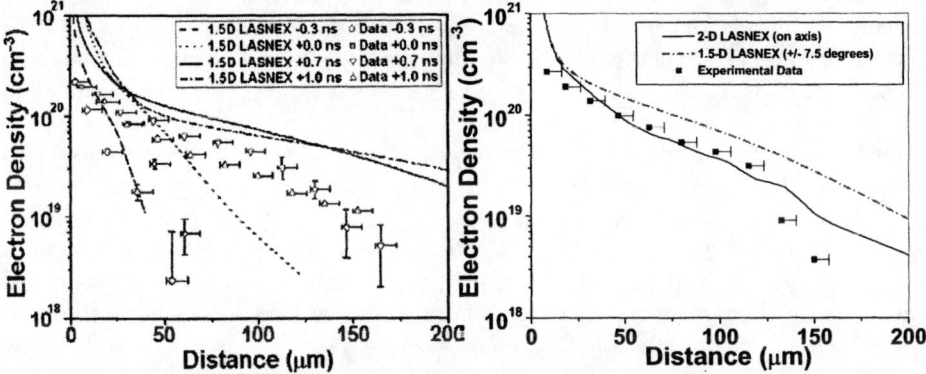

Figure 2 (a) 1-D on axis density slices from experimentally obtained interferograms compared with profiles predicted by 1.5-D LASNEX simulations. (b) Comparison of on axis density from 2-D and 1.5-D ($\pm$ 7.5°) LASNEX simulations with experimentally obtained electron density measurements at t = 0.5 ns.

Figure 2(b) shows the experimentally determined electron density as a function of distance away from the target surface taken 0.5 ns after the peak of the plasma forming pulse. Also shown is the predicted on axis electron density from the LASNEX 1.5-D ($\pm$ 7.5°) and 2-D simulations. It is clear that there is better agreement for the latter. For the 1.5-D case, all the ablated mass is artificially confined to a fixed cone angle of expansion, an approximation which, in the case of significant lateral expansion, will increase the predicted on-axis density. It has been found that by increasing the cone angle within the 1.5-D simulations improved agreement with experiment is obtained.

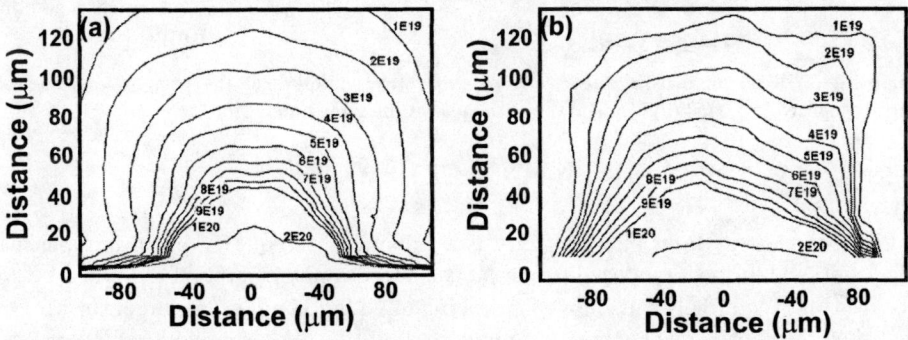

Figure 3 (a) Simulated 2-D electron density profile using LASNEX. (b) Measured 2-D electron density profile 0.5 ns after the peak of the 600ps duration plasma forming beam. The error bars for distance away from the target are -2 μm, $+8$ μm.

Figure 3 shows a contour plot of a 2-D LASNEX [4] simulation and the corresponding experimental data at $\Delta t = 0.5$ ns with good quantitative agreement. Such modeling covers all phases of expansion leading to a plasma size much larger than the focal spot. To understand the mechanisms driving the expansion of the plasma it is necessary to track the evolution of the plasma parameters. Figure 4(a) shows the calculated 2-D electron temperature and density profile at the peak of the 600 ps plasma forming beam. It is predicted that there exists a hot (~150 eV) on-axis region close to the critical surface of the 1ω driver. The temperature gradients relax slowly in the plane of primary expansion. However there are steep gradients in the lateral direction. The plasma close to the target is calculated to be relatively cold (~40 eV) just outside the 20 μm FWHM of the laser driver. An electron density contour map is also shown on the same plot. It is seen that at the peak of the pulse a rippling of the critical density surface is predicted, with side lobes developing at low temperatures at approximately 20 μm off-axis. In figure 4(b) a momentum vector map illustrates that there is a significant lateral push of mass from these side lobe regions in the direction away from the center of the focusing beam.

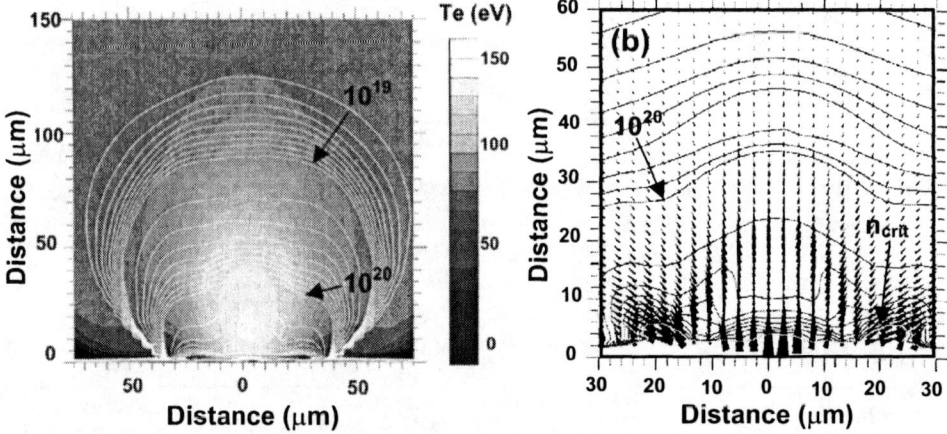

Figure 4 (a) Electron density and electron temperature contours at the peak of the plasma forming pulse [Δt = 0 ns] (b) Momentum vector plot at the same time.

4. CONCLUSIONS

Picosecond x-ray laser interferometry is a valuable technique in diagnosing plasma evolution within laser produced plasmas. The short sampling time of the probe beam reduces blurring effects and allows for probing to within 10μm of the target surface. In addition, the short wavelength probe minimizes effects associated with refraction and free-free absorption making this diagnostic well suited for studying large, fast evolving, dense plasmas. It has been shown that for the experimental conditions reported within this paper lateral expansion is a significant effect. Two-dimensional plasma physics codes are therefore necessary to model such experiments. Preliminary simulations have indicated that cool dense and relatively slow moving regions, which lie outside of the full width at half maximum (FWHM) of the laser focus, drive the lateral expansion. Future work will focus on understanding the dominant energy transport mechanisms, which drive the plasma expansion.

ACKNOWLEDGEMENTS

The support of Al Osterheld and Andy Hazi is greatly appreciated. The authors are pleased to acknowledge the technical contributions from Carl Bruns and Al Ellis. This work was performed under the auspices of the U.S. Dept. of Energy by the University of California Lawrence Livermore National Laboratory, through the Institute for Laser Science and Applications, under Contract No. W-7405-Eng-48 and by US Department of Energy Grant No. DE-FG03-98DP00208. The development of the interferometer gratings was supported with a grant from the State of Colorado Photonics and Optoelectronics program.

REFERENCES

1. R.F. Smith, J.Dunn, J. Nilsen, V.N. Shlyaptsev, S. Moon, J. Filevich, J.J. Rocca, M.C. Marconi, J.R. Hunter, and T.W. Barbee, Jr., Phys. Rev. Lett., **89** (6), 065004-1 (2002).

2. J. Filevich, K. Kanizay, M. C. Marconi, J. L. A. Chilla, and J. J. Rocca, Opt. Lett. **25**, 356-357 (2000).

3. L. B. Da Silva, T. W. Barbee, Jr., R. Cauble, P. Celliers, D. Ciarlo, S. Libby, R. A. London, D. Matthews, S. Mrowka, J. C. Moreno, D. Ress, J. E. Trebes, A. S. Wan, and F. Weber, Phys. Rev. Lett. **74**, 3991-3994 (1995).

4. R.F. Smith, V.N. Shlyaptsev, J. Dunn, J. Nilsen, J.R. Hunter, J.J. Rocca and J. Filevich, M.C. Marconi, "Refraction Effects on X-ray and UV Interferometric Probing of Laser Produced Plasmas" submitted for publication to JOSAB (2002).

5. G.B Zimmerman and W.L. Kruer, Comments Plasma Phys. Controlled Fusion **2**, 51-61 (1975).

Four Wave Sum Difference Non-Linear effects with X-ray Lasers in a Gas Jet Plasma

S. J. Topping*, R. Keenan*, C. L. S. Lewis*, A. M. McEvoy* and S. Hubert*

* School of Mathematics and Physics, Queen's University of Belfast, Belfast, BT7 1NN

Abstract. Progress in a four-wave sum difference frequency generation scheme to convert quasi-coherent soft X-ray radiation to shorter wavelengths, using a plasma medium is discussed. Experimental results and calculations involving the Ne-like Ni XRL (λ_{XRL} = 0.023 µm) mixed in phase-matched mode with a Nd:glass laser at the second harmonic (λ_{OPT} = 0.52 µm) in a Na-like Argon plasma are presented. A 25 mm single slab target of Ni pumped with ~ 200 J of energy from 5 optical beams has routinely produced ~ 5 mJ of XRL energy at 23.1 nm. The XRL was split and recombined with the Nd:glass laser in the Na-like Ar plasma. Considerations for phase matching the XRL and optical beams in the ionized gas jet are also discussed.

INTRODUCTION

Non-linear optics provides us with the opportunity to extend the operating wavelengths of lasers currently available to us. In this experiment we attempt to near frequency double a Ne-like Ni X-ray laser (XRL) at 23.1 nm to 11.5 nm using the third order non-linear process of Four Wave Sum Difference Mixing (FWSDM). Essentially the **k**-vectors of the three input beams, two XRL and one optical are added to give a new output frequency. The choice of our mixing medium is somewhat limited due to the strong absorption of soft X-ray wavelengths in most conventional environments. We are therefore led to consider plasma as a suitable mixing medium.

We focus in particular on a system where we have two XRL photons at the same frequency for a $2\omega_{xrl} - \omega_{opt}$ process. We choose to study Na-like ions for the simplicity of their level structure where the lines are known to high precision. We also choose to study Na-like ions for the relatively high line strengths of the transitions involved and because the energy level structure is appropriate to allow a $2\omega_{xrl} - \omega_{opt}$ process to occur.

Preliminary experiments have been carried out at the Rutherford Appleton Laboratories using the Vulcan laser [1]. The experiments were carried out using the Ne-like Ni XRL at 23.1 nm and the fundamental of Vulcan at ~ 1 µm for the non-phase matched case. Modeling has indicated that using the second harmonic of Vulcan at ~ 0.5 µm and using the phase-matched case should yield a higher output of our non-linear signal compared to previous experiments.

CP641, X-Ray Lasers 2002: 8th International Conference on X-Ray Lasers, edited by J. J. Rocca et al.

MODELLING

In developing an expression for the third order susceptibility (χ^3) a quantum mechanical process known as the density matrix approach is employed. Also as we work near resonance it is feasible to extract the resonant terms of the susceptibility and neglect all others, greatly simplifying the expression [2], equation (1). Figure 1 (a) shows the Na-like scheme and the transitions of interest.

$$\chi^3 = \frac{1}{6\hbar^3} \sum \mu_{ab}\mu_{bc}\mu_{cd}\mu_{da}\rho_a \left[\frac{2}{\omega_{ba}-\omega_{xrl}-i\Gamma_{ba}}\right] x \left[\frac{1}{\omega_{ca}-2\omega_{xrl}-i\Gamma_{ca}}\right] x \left[\frac{1}{\omega_{da}-(2\omega_{xrl}-\omega_{opt})-i\Gamma_{da}}\right] \quad (1)$$

where μ_{abcd} are the dipole matrix elements, ρ_a the thermal population of level a, ω_{abcd} the transition frequencies between levels, ω_{xrl} and ω_{opt} are the XRL and optical frequencies respectively and $i\Gamma_{abcd}$ represents broadening mechanisms in the plasma. We assume an electron temperature of 30 eV so there is a sizeable thermal population in the 3p level. The dipole matrix elements are calculated from tabulated values of the absorption oscillator strengths [3] and by finding the reduced matrix elements and applying the Wigner-Eckart theorem.

In dilute media such as plasma and gases, we have few contributors to the non-linear signal. The dipole matrix elements also scale poorly with shorter wavelengths and the result is a low coupling efficiency at XUV wavelengths. To increase the coupling efficiency we require small frequency difference components between the incident XRL and optical frequencies and the transition frequencies between levels. Working close to the transition frequencies in this way is known as resonance enhancement. However we need to avoid exact resonance or single photon absorption of the pump beams will occur. Equation (2) represents the converted intensity of the non-linear signal (NLS).

$$I_{nls} = \frac{576\pi^4}{c^4} \omega_{nls} I^2_{xrl} I_{opt} \left|\chi^3\right|^2 N^2 L^2 \frac{\sin^2\left(\Delta kL/2\right)}{\left(\Delta kL/2\right)^2} \quad (2)$$

where I_{nls} is the generated intensity at frequency ω_{nls}, I_{xrl} and I_{opt} are the intensities of the XRL and optical beams respectively and Δk represents the wave vector mismatch over an interaction distance L in a medium of number density N.

Figure 1 (b) shows the converted intensity for ArVIII as a function of the incident XRL frequency, for the non-phase matched case. The optical frequencies are set to the fundamental and second harmonic of Vulcan with an intensity of $\sim 10^{12}$ W/cm^2. The XRL has an intensity of $\sim 10^{10}$ W/cm^2 and is assumed to have a thermal line-width $(\Delta\omega/\omega)$ of 10^{-4}. For the non-phase matched case we assume a density-interaction length product NL corresponding to $\Delta kL = \pi$. The conversion efficiency was calculated to be $\sim 10^{-5}$. However the experiment was run in the phase matched case and we would expect higher conversion efficiency $\sim 10^{-3}$.

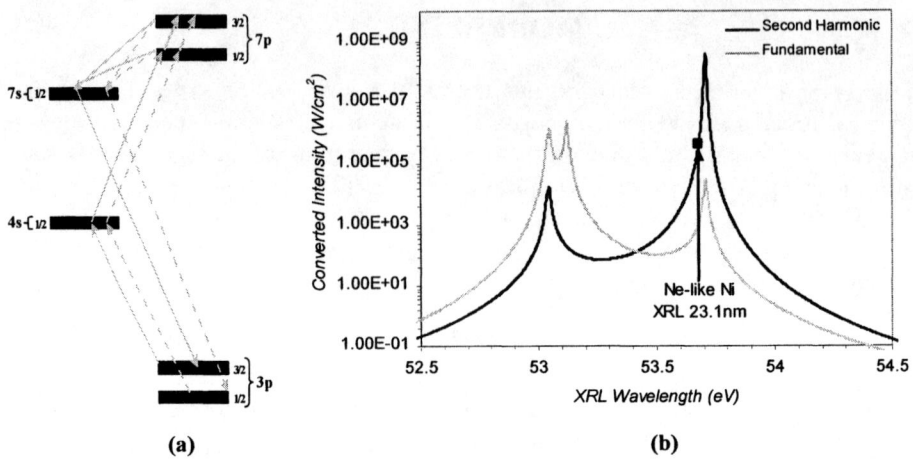

FIGURE 1. a) Na-like scheme and b) converted intensity for ArVIII

In dense plasma the line-widths will be both collisionaly and Doppler broadened due to the temperatures and pressures involved. However at the densities at which we operate at, $\sim 10^{18}$ cm^{-3}, and to simplify matters we assume that Doppler broadening dominates. Figure 2 shows how the broadening mechanism affects the susceptibility of ArVIII as a function of the incident XRL frequency with the optical frequency set to the second harmonic. The first peak represents a two-photon resonance to the 7p level. The second peak represents single-photon resonance to the 4s level. As we move closer to the natural line-width $\sim 10^{-7}$ we can see the peaks become sharper. Alternatively the peaks are smoothed and eventually become featureless as the line-width is broadened.

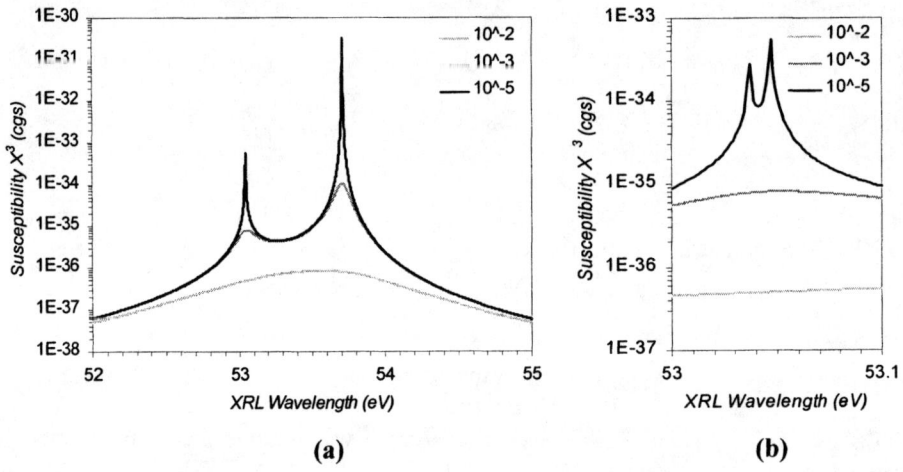

FIGURE 2. Susceptibility of ArVIII a) for various line-widths b) showing two-photon resonance

EXPERIMENT

The Ne-like Ni XRL is created from a plasma column produced from 5 beams at 1.054 µm in a line focus set-up. The beams are first focused to a point (surrogate focus) using a series of f/2.5 lenses. A set of spherical mirrors, are then used to image the surrogate focus to the target plane. The spherical mirrors operate off axis and as a result a large astigmatism is introduced into the beams producing two line foci. One focus is a line, which is set within the plane of incidence and the beams are overlapped along the axis to create a uniform line focus. The beams have a pulse duration of 80 ps with an average energy of ~ 40 J per beam and when distributed across a 100 µm x 25 mm line focus gave an on target intensity of the order of ~ 10^{13} W/cm^2. A 10% pre-pulse was set 2 ns ahead of the main drive beams and the solid Ni slab target retracted 100 µm off axis to maximize the XRL output. Figure 3 shows the experimental set-up used.

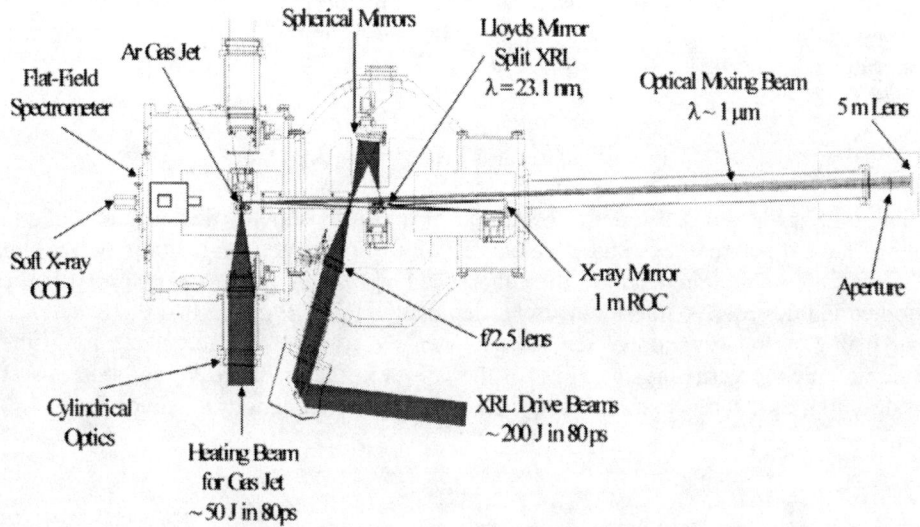

FIGURE 3. Experimental set-up for FWSDM

The illumination uniformity and ionization balance were monitored on a shot to shot basis using a crossed slit camera and Bragg spectrometer. The images showed reasonably good uniformity along the line focus and indicated our Ni plasma was in the desired Ne-like state. The Ne-like Ni XRL was driven into saturation for a 25 mm long target. The output was monitored with an on axis grazing incidence flat-field spectrometer with a 1200 lines/mm aperiodic ruled grating and a soft X-ray CCD. The XRL was found to have a maximum output of ~ 5 mJ, a FWHM divergence of ~ 10 mrad and a refraction angle of ~ 8 mrad which is consistent with previous results [1].

The XRL exit point from the Ni target was imaged using a pair of spherical X-ray mirrors (XRMs) at high magnification (x40). The XRMs had a reflectivity of ~ 20% and a narrow bandpass of ~ 30 Å centered around 23.1 nm. A series of baffles were used to reduce the amount of scattered light and obtain a clean signal with low background. The XRL exit point was found to extend to ~ 200 μm FWHM in the horizontal and ~ 200 μm FWHM in the vertical directions [4].

The XRL was split using a Lloyds mirror, the edge of which is imaged with a 1 m ROC XRM and focused into the gas jet at ~ x2 magnification. The sixth Vulcan beam was used as the optical mixing component, frequency doubled to ~ 0.5 μm. The beam was focused down using a 5 m lens into the gas jet plane and apertured to form a spot size comparable to the XRL spot size in the overlap region.

The Ar gas jet was ionized to the desired Na-like ArVIII state using beam 7 focused with cylindrical optics to form a line focus of ~ 0.4 mm x 3 mm with an intensity of ~ 6×10^{13} W/cm^2. Modeling from Medusa has suggested optimum Na-like densities are created for neutral atom densities of ~ 10^{18} cm^{-3} and irradiation intensities of between $6 - 8 \times 10^{13}$ W/cm^2. Further experimental work including time resolving the gas jet emission is needed to verify this. Characterization of the gas jet was carried out through emission spectroscopy and optical interferometry during a preliminary experimental phase [5].

Phase Matching

The NLS propagates through the plasma at a phase velocity tied to those of the pump waves. As the two phase velocities of the generated and pump waves are different, after a certain distance, the coherence length, they fall out of phase with one another and destructive interference will deplete the NLS. We find that the coherence length is a strong function of the electron density, as it is the free electrons in the system, which governs the velocities of the waves in the medium. When operating at an optimum electron density of ~ 10^{19} cm^{-3} the coherence length is reduced to ~ 100 μm.

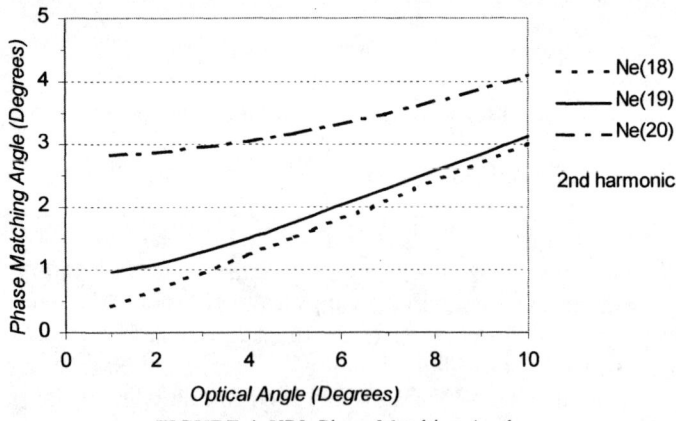

FIGURE 4. XRL Phase Matching Angle

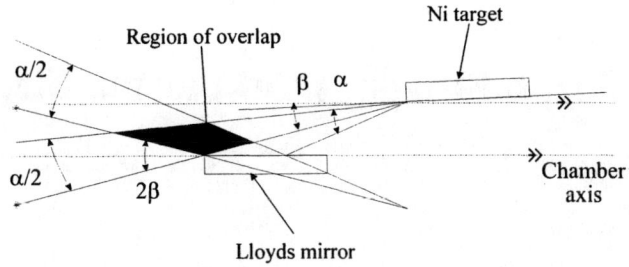

FIGURE 5. Split XRL at Lloyds mirror plane

There are a number of ways in which to phase match, but in this particular case it is convenient to change the input angles of the pump beams, angular phase matching [6]. The phase matching angle between the XRL beams is a function of the electron density N_e, λ_{XRL}, λ_{OPT} and the input angle of the optical beam, see Fig. 4. The angle required between the XRL beamlets is secured by using the natural refraction of the XRL (~8 mrad) and by tilting the Ni slab target, see Fig. 5.

The XRL was split into two separate beamlets by a Lloyds mirror positioned parallel to the axis. This consisted of a thin piece of glass that reflects a portion of the XRL while allowing the remainder to travel through. The flat field spectrometer was used to diagnose the split ratio between the beamlets.

CONCLUSIONS

The FWSDM scheme was tried using the second harmonic of Vulcan, for the phase matched case and with the input beams overlapped spatially and temporally at the gas jet plane. Only a limited parameter scan was made and no non-linear signal was detected. The detailed coherence properties of the XRL beam are still unknown and higher resolution interferometry is needed to diagnose the XRL beam. A TG could also be used to split the XRL beam with the correct co-angle for phase matching. An XRL beam split with an amplitude divider instead of a wave front divider may prove more efficient in the long term and this is under development.

REFERENCES

1. Topping, S.J., Keenan, R., et al "Progress in FWSD non-linear effects with XRLs" in *Applications of X-ray lasers and other bright X-ray sources*, edited by G. Jamelot et al., J. Phys. IV 11, St Malo, France, 2000, pp. 487-490.
2. Muendel, M.H., and Hagelstein, P.L., *Phys. Rev. A* **44**, 7573-7579 (1991)
3. Biemont E., *Astron. Astrophys. Suppl.* **31**, 285 (1978)
4. Keenan, R., Lewis, C.L.S., et al "Development of X-ray lasers for radiographic and other applications" *in Applications of X-ray lasers and other bright X-ray sources*, AIP Conference Proceedings, Melville, New York, 2002, (in these proceedings).
5. O'Rourke, N., "The Development of X-ray Lasers for Non-Linear Optics", PhD Thesis, Queens University Belfast, 1999, pp. 120-126.
6. Reintjes, J.F., Nonlinear Parametric Optical Processes in Liquids and Gases, Orlando Academic Press, 1984.

Experimental Diagnosis of Plasma Electron Density with an X-Ray Laser Probe

Wang Chen, Wang Wei, Gu Yuan, Wang Shiji

Shanghai Institute of Laser Plasma, P.O.Box 800-229, Shanghai, 201800, China

Abstract An application demonstration of diagnosing electron density gradient distributions of CH laser-plasma by x-ray laser was reported. Using an saturated Ni-like Ag x-ray laser as the probe beam and the moiré deflection method, A legible moiré deflectogram including the plasma electron density information was recorded. Analysis of the deflectogram gave a peak electron density of $1.1 \times 10^{21} cm^{-3}$, and this value had been higher than the critical density.

1. INTRODUCTION

The first demonstration of x-ray laser emission was given in 1984[1]. From then on, the x-ray lasers have shown widespread application foregrounds in many fields with its short wavelength, high intensity and high coherence. Ni-like Ag x-ray laser has better benefits for applications than others, because its wavelength at 13.9 nm is most close to the maximum reflection wavelength of Mo:Si multilayer x-ray optics components. Some experiments have achieved the saturated Ni-like Ag x-ray laser [2-4]. Plasma electron density measured by x-ray laser probe is very significative for studies in some fields. There are papers have reported this application by several ways[5-7].

In this paper, we report an experimental study of measuring plasma electron density distribution by the moiré deflection method with an saturated Ni-like Ag x-ray laser as the probe beam. This method was developed by Ress et.al. in reference [5], but there are some differences in our experiment. In the experiment, the aim plasma was created by 80ps-duration short pulse laser and the probe passed it at about 1 ns after the drive pulse terminal.

CP641, *X-Ray Lasers 2002: 8th International Conference on X-Ray Lasers,* edited by J. J. Rocca et al.
© 2002 American Institute of Physics 0-7354-0096-2/02/$19.00

2. EXPERIMENTAL SETUP

The experiments were performed with the ShenguangII laser facility at Chinese National Laboratory on High Power Laser and Physics. The experimental setup was shown in Fig.1. Two beams of drive laser at 1.053μm oppositely irradiated a couple of double 16mm-long Ag target by through a cylindrical lens array (CLA) line-focus system to produce the Ni-like Ag x-ray laser. Each drive laser beam was divided into pre- and main- pulses with the intensity ratio being about 4% and the pre-pulse preceding 3ns. The energy of each beam was about 80J for the duration of 80ps and the irradiating intensity on the target surface was about $6 \times 10^{13} \text{Wcm}^{-2}$. Under this condition, a saturated Ni-like Ag x-ray laser was produced. The divergence angle was about 2.5mrad and the output energy was about 250μJ.

Then the x-ray laser was used as a probe to diagnoses an aim plasma created by another drive laser irradiating a C_8H_8 flat target. The duration(FWHM) and energy of this drive laser beam was 80ps and 20J respectively. With spot-focus diameter of 0.5mm, the irradiating intensity was about $1 \times 10^{14} \text{Wcm}^{-2}$. The probe first traveled 500mm and passed through the aim plasma, when there were about 1 ns after the plasma being created. The aim plasma was imaged by a Mo:Si spherical multilayer mirror focused at the center of the plasma. The image formed by this mirror was relayed by a Mo:Si flat multilayer mirror to the CCD detector plane at a magnification of 9.4. A Zr filter by an attenuated factor of about 150 for 13.9nm was used before the detector to eliminate the effects of background lights. The moiré pattern was formed by a pair of gold bar rulings with period 28.6μm and rotational offset angle 4.5° to give a fringe spacing of 365μm in the image plane (39μm at the object plane). The bars paralleled the target surface, and the rulings were separated by a single Talbot distance 58.7mm to give a sensitivity of 4.6mrad per fringe shift at the object plane. A soft x-ray sensitive CCD camera with sized 24μm×24μm per pixel was tightly close to the second ruling to record the moiré deflectogram.

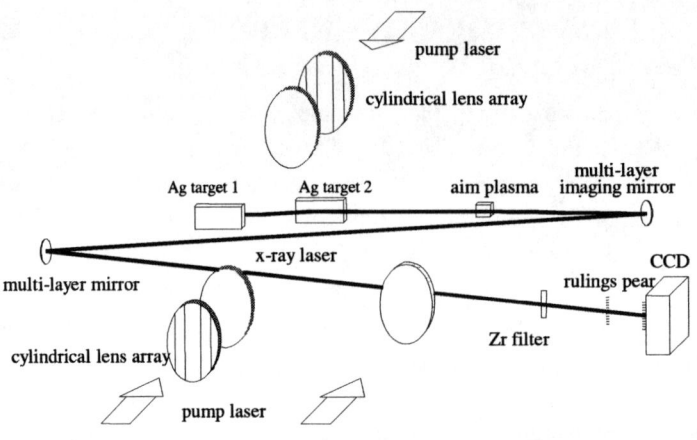

Fig.1 Experimental setup of moiré deflection method

3. RESULTS

We recorded the legible moiré deflectograms in experiments, shown in Fig. 2.
Fig. 2 (a) is the static deflectogram without being perturbed by plasma, and (b) is the dynamic deflectogram being perturbed by the CH plasma. The fringe contrast $(I_{max}-I_{min})/(I_{max}+I_{min})$ is about 0.6, and close to the theoretical value 0.7. It is evidence that the stripe at the top of Fig. 2(b) is deflected by plasma's perturbing, and it includes the information of electron density distribution. The clear edge between stripe and target's shadow at the bottom of Fig. 2(b) exactly shows the target surface original position.

We choose a single stripe at the middle of the aim plasma radiation area to deal with. In Fig. 3, the solid square symbols show the electron density gradation distribution along the target normal direction calculated from experimental data. The circle's show the results from theoretical simulation in which the limiting parameter fe of the hydromechanical heat conduction was assumed as 0.1. The good accordance indicates the 0.1 might be the suitable value in the plasma hydromechanical theory. In the Fig. 3, between 30μm and 60μm distance from target surface, the electron density grad was steep; and where farther than 60μm, the stripe deflection was not evident, showing the density distribution was very flat. Assuming the original electron density at 100μm was 1×10^{20} or 7×10^{20}cm^{-3}(Theory value), we calculated the electron density as a function distance from target surface, shown in Fig. 4. The maximum electron density in experiment was about 1.1×10^{21}cm^{-3}, which was higher than the critical density(1×10^{21}cm^{-3}).

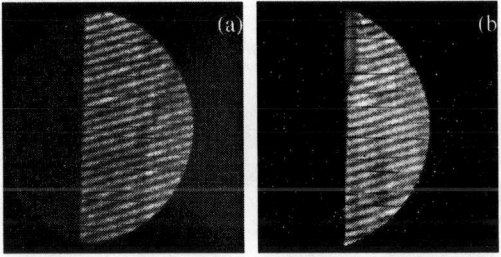

FIGURE 2. Moiré deflectograms recorded by the experiments. (a) without plasma perturbed, (b) with plasma perturbed.

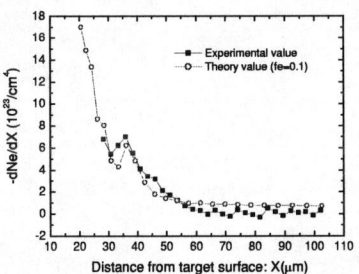

FIGURE 3. Electron density gradation distribution along the target normal direction as a function of distance from target surface.

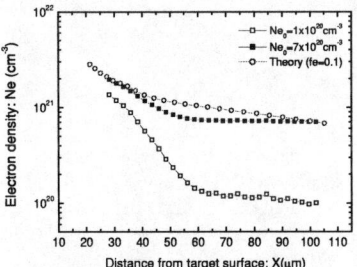

FIGURE 4. Electron density distribution along the target normal direction as a function of distance from target surface.

4. CONCLUSION

We report our experiments about x-ray laser application. Using a gain saturated Ni-like Ag x-ray laser at 13.9nm as probe beams, we measured the electron density distribution of CH plasmas with moiré deflection method. The electron density gradation around the critical density area was measured.

REFERENCES

1. D.L. Matthews et.al., *Phys.Rev.Lett.* **54**, 110 (1985).
2. J. Zhang et.al., *Phys.Rev.Lett.* **78**, 3856 (1997).
3. R. Tommasini et.al., *J.Opt.Soc.Am. B* **16**, 1664 (1999).
4. S. Sebban et.al., *Phys. Rev. A* **61**, 043810 (2000).
5. D. Ress et.al., *Science,* **265**, 514 (1994).
6. L.B. DaSilva et.al., *Phys.Rev.Lett.*, **74**, 3991 (1995).
7. J. Filevich et.al., *Opt.Lett.*, **25**, 356 (2000)

DEVELOPMENT OF X-RAY LASER DRIVERS

Raman Amplification of Laser Pulses in Microcapillary Plasmas

Y. Ping, I. Geltner, A. Morozov, N. J. Fisch and S. Suckewer

Princeton University, Princeton, NJ 08540

Abstract. Raman amplification of ultrashort pulses is demonstrated in microcapillary plasmas. Experiments in very short microcapillaries (0.2 - 0.5 mm) with a broadband seed pulse show that the amplification factor is in agreement with the linear growth rate.

INTRODUCTION

Raman amplification is based on the three-wave interaction between two counterpropagating electromagnetic waves and a plasma wave, whose frequencies and wave vectors satisfy the energy and momentum conservation relations:

$$\omega_1 = \omega_2 + \omega_p, \ \vec{k}_1 = \vec{k}_2 + \vec{k}_p. \tag{1}$$

where ω_1, ω_2, ω_p are the frequencies of the pumping pulse, seed pulse, and plasma wave, respectively, and k_1, k_2, k_p are the corresponding wave vectors. The idea of using the Raman instability as a means to amplify and compress laser pulses has been the object of study for considerable time [1]-[3]. Early work on the compression of laser pulses in gases was reported in [1], and recently, Raman compression from tens of nanoseconds to tens of picoseconds has been achieved in gas mixtures [2]. The maximum power of laser pulses that can be compressed in gases is limited by ionization. The advantage of using plasma as the medium is that there is no theoretical limit on the laser power since plasma can handle ultrahigh power without any "damage" to itself.

In the linear regime of Raman amplification, the intensities of the pump and the seed pulses are moderate and there is no significant pump depletion. The seed pulse grows exponentially with a growth rate, γ_{RBS}, independent of the seed intensity [4]:

$$\gamma_{RBS} = a_1 (\omega_1 \omega_p / 4)^{1/2} \tag{2}$$

where a_1 is the normalized vector potential of the pump; $a_1 = 0.85 \times 10^{-9} \lambda_1 I_1^{1/2}$ (the pump wavelength λ_1 is in μm and the intensity I_1 is in W/cm²). Recently, new effects were identified in the nonlinear behavior of plasma when interacting with ultra-intense laser pulses [5]-[7]. Theoretical studies have shown that the amplified pulse duration decreases inversely proportional to the pulse amplitude and a compression from picoseconds to femtoseconds is achievable in the nonlinear regime [8]. The Raman amplification in combination with the compression effect in plasma provides a potential of overcoming the power limit of current chirped-pulse-amplification (CPA)

CP641, *X-Ray Lasers 2002: 8th International Conference on X-Ray Lasers*, edited by J. J. Rocca et al.

techniques, set by the thermal damage threshold of the optics. Such Raman amplifiers can be useful to produce ultra-intense laser pulses for pumping soft x-ray lasers.

EXPERIMENTAL RESULTS

Our experiments were initially designed for the linear regime. In this paper we present the results of Raman amplification in microcapillary plasmas. The analysis of the data suggests that the observed amplification in the initial experiment was obtained from a plasma with a length much shorter than the microcapillary length [9]. The second experiment was performed in much shorter microcapillaries to check this hypothesis, and the results show that the amplification agrees well with the estimated linear growth rate.

Experiment with a 745 nm, 200 fs Seed Pulse in LiF Microcapillaries

The experimental setup with a 745 nm, 200 fs seed pulse is shown in Fig. 1. The pumping pulse was provided by the second harmonic of a Nd:YAG laser at $\lambda = 532$ nm with 150 mJ in 5 ns (FWHM). The seed pulse and the pumping pulse were both focused onto the center of the microcapillary by an F/25 lens L_1 and an F/8 lens L_2, respectively. The initial plasma was created inside the LiF microcapillary through wall ablation by a low power nanosecond KrF laser (50-60 mJ in 20 ns at $\lambda = 248$ nm). The amplified seed pulse, propagating in the opposite direction to the pumping pulse, passed through mirrors M_2 and M_3 and was focused by a lens L_4 onto the entrance slit of a spectrometer with a photomultiplier detector.

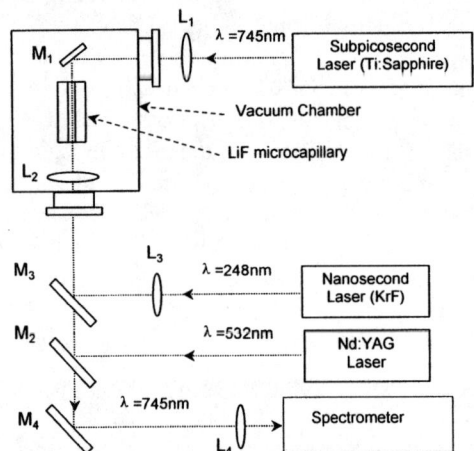

FIGURE 1. Experimental setup for Raman amplification with a 745 nm seed in a LiF microcapillary.

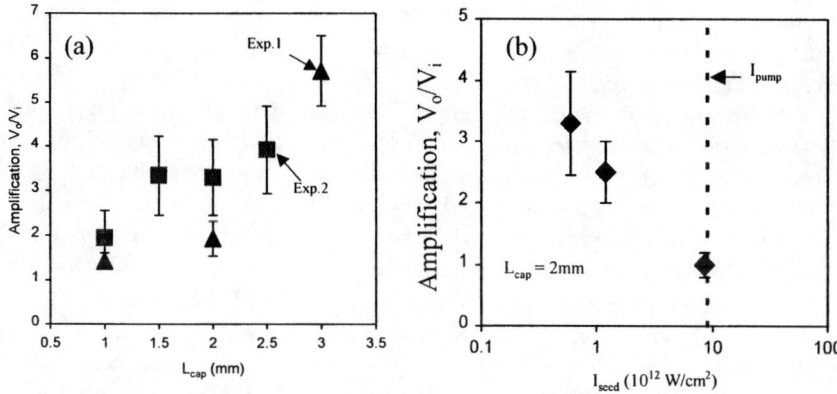

FIGURE 2. Amplification as a function of the microcapillary length (a) and initial intensity of the seed pulse (b).

Figure 2 (a) shows the amplification ratio V_o/V_i vs. the microcapillary length at a delay of 60 ns. Each point in Fig. 2 (a) is an average of 20 shots, with the error bar calculated from the square root of standard variance. The results were obtained in two series of measurements (Exp. 1 and Exp. 2). The relatively large error was mainly due to the irreproducibility of the plasma and the intensity fluctuations (~ 10%) of the seed and the pumping pulses. These results show the amplification increasing gradually with the microcapillary length. A ratio V_o/V_i of ~ 5 - 6 was reached inside the L_{cap} = 3 mm microcapillary. However, the increment of the amplification as a function of the microcapillary length is much less than the exponential growth predicted by the linear theory. The intensity of the seed pulse was also varied by inserting filters between L_1 and the chamber window. The amplification ratio for three initial seed pulse intensities, 0.5, 1.2, and 8.5×10^{12} W/cm^2 (corresponding energies: 2, 4 and 30 µJ) in a 2 mm microcapillary are shown in Fig. 2 (b). It was found that the amplification ratio drops as the intensity of the seed pulse (I_2) approaches the intensity of the pumping pulse (I_1). When I_2 was significantly smaller than I_1, the seed pulse was amplified more than 3 times. As I_2 approached I_1 the ratio dropped to ~ 1.

For the 532 nm pumping pulse focused to 10^{13} W/cm^2, the vector potential is $a_1 \approx$ 0.0014, therefore the growth rate, calculated by Eq. (2), is $\gamma_{RBS} \approx 1.3 \times 10^{12}$ s^{-1}. If there is no energy loss and the amplification occurs through the whole Rayleigh length of the pumping beam ($Z_R \approx 0.6$ mm), for a low intensity seed pulse (no pump depletion - linear regime) one would expect an amplification of $\exp(2\gamma_{RBS}Z_R/c) \approx 180$. However, given an estimated preplasma temperature $T_e \approx 20$ eV and $n_e = 3 \times 10^{20}$ cm^{-3}, the inverse Bremsstrahlung absorption length is ~100 µm, i.e. the penetration distance of the pumping pulse into the plasma is only 100-200 µm. The actual situation is more complicated since the plasma is heated and the plasma density is changed by the pumping pulse. In later density measurements it was found that the plasma does not fill the whole microcapillary at short delays, hence the microcapillary length is not the actual interaction length. If the effective interaction length is $L_{eff} \sim 200$ μm, the amplification would be $\exp(2\gamma_{RBS}L_{eff}/c) \sim 6$, which would be in good agreement with the observed amplification.

The short effective interaction length can also explain the decrease of the amplification with increasing seed pulse intensity. During the measurement of Raman amplification as a function of seed pulse intensity (at constant pumping intensity $I_1 \approx 10^{13}$ W/cm²) the amplification dropped as the intensity of the seed pulse (I_2) approached that of the pumping pulse (I_1). The explanation in view of the short L_{eff} could be as follows. The available pumping energy within the short interaction time (for $L_{eff} = 0.2$ mm the interaction time is $\tau_{eff} = 2L_{eff}/c \approx 1.5$ ps) of the total 150 mJ in 5 ns is only ~ 40 μJ. When $I_2 \approx I_1$, the energy of the seed pulse is ~ 30 μJ, which is comparable to the available pumping energy. Even with total pump depletion one cannot expect the amplification to exceed a factor of 2. While taking into account absorption and no full depletion, there should be no visible amplification.

Experiment with a broadband Seed Pulse in Copper Microcapillaries

In order to check our hypothesis that in the 3 mm long microcapillary, only ~ 200 μm of the plasma had the matched high density, we continued the experiment with much shorter microcapillaries [10]. This allowed us to have a more precise knowledge of the interaction length and to avoid any refractive effects.

The experimental setup was the same as in Fig. 1, except that the seed pulse was provided by an optical parametric oscillator (OPO), which was pumped by the 745 nm subpicosecond pulse and produced a very broadband output pulse in the visible region. Since the broadband seed pulse has a spectrum with some random peaks and the spectral distribution varies from shot to shot, a small portion of the seed pulse was split out and focused into the spectrometer to provide a reference spectrum. Fig. 3 (a) displays the reference spectrum (top) and the seed pulse spectrum after passing through the empty microcapillary (bottom), demonstrating a good correspondence to each other. In order to minimize the spatial mismatch of the seed and the pumping pulses, as encountered in the earlier experiment, the microcapillary was imaged at the entrance slit of the spectrometer. The related geometry is shown in Fig. 3 (c). The pumping pulse was focused onto the center of the microcapillary with ~ 20 μm FWHM ($Z_R \sim 0.6$ mm $> L_{cap}$). The seed pulse was slightly defocused to make sure that part of the seed overlaps with the pump. The entrance slit was closed to a width of 200 μm, corresponding to 30 μm in the microcapillary plane, to cover the interaction area between the seed and the pump pulses. With such an arrangement, the spectrum displayed by the CCD camera had a spectral resolution of ~ 0.6 nm and a spatial resolution of ~ 10 μm in the vertical direction [marked as the y-axis in Fig. 3 (b) and (c)].

The prepulse for creating the preplasma inside the microcapillary and the pumping pulse were the same as in the previous experiment, i.e. about 60 mJ at 248 nm in 20 ns and 150 mJ at 532 nm in 5 ns, respectively. The spectrally resolved amplification was calculated as $I_p (\lambda) / I_0 (\lambda)$, where $I_p (\lambda)$ is the intensity of the amplified pulse spectrum and $I_0 (\lambda)$ is the intensity of the reference spectrum. Without the plasma the ratio I_p /I_0

is approximately a flat curve. When the plasma is created in the microcapillary, the seed pulse spectrum remains almost the same, with small spatial modifications, probably due to nonuniformities of the plasma density. When the pumping pulse is fired into the plasma, the seed pulse spectrum demonstrates a significant difference between the center, I_p (λ, y = 0), and the edge, I_p (λ, y = 100 μm), as shown in Fig. 3 (d) (for a copper microcapillary with L_{cap} = 200 μm, D_{cap} = 150 μm and the delay between the prepulse and the pumping pulse was 40 ns). The ratio I_p /I_0 at the central part has a distinguished peak at λ = 645-650 nm with a maximum of ~ 7, while the ratio at the edge of the plasma channel is much flatter and does not exceed 2 for most wavelengths. The observed maximum amplification is 3-4 for copper microcapillaries with L_{cap} = 200 μm and D_{cap} = 150 μm. The intensity of the seed pulse was below 10^{11} W/cm^2 thus there was no significant pump depletion, i.e., we were in the linear regime for this experiment. The linear growth rate of Raman backscattering is calculated to be $\gamma_{RBS} \approx 1.0 \times 10^{12}$ s^{-1} for $I_{pump} = 1 \times 10^{13}$ W/cm^2 and $n_e = 1.3 \times 10^{20}$ cm^{-3} (the density corresponding to 645 nm). The amplification predicted by the linear theory for L_{cap} = 200 μm is then exp ($2L_{cap}$ γ_{RBS}/c) $\approx$ 3.8, which is in good agreement with the experimental observation.

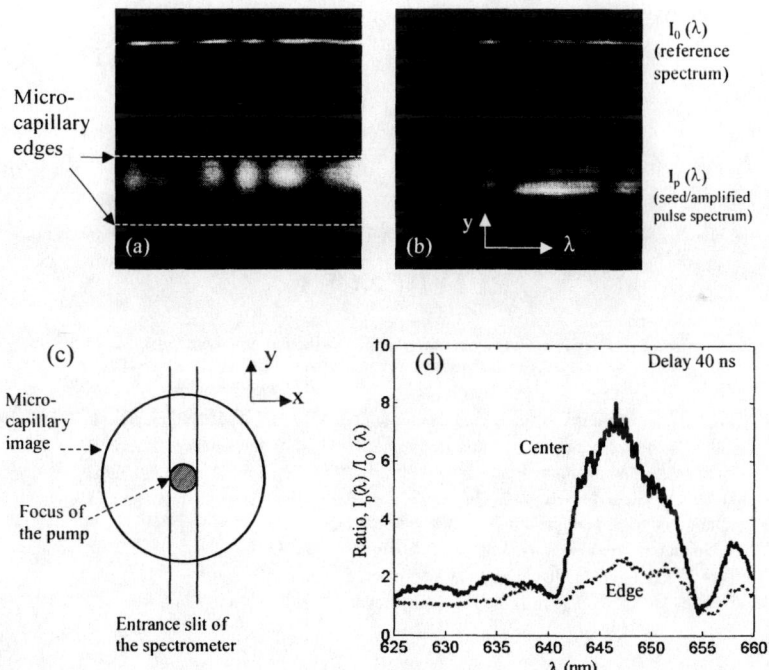

FIGURE 3. (a) Reference spectrum (top, I_0) and the seed pulse spectrum (bottom, I_p). (b) Reference spectrum and the amplified pulse spectrum in a copper microcapillary with L_{cap} = 200 μm and D_{cap} = 150 μm at a delay of 40 ns. (c) Geometry relation at the entrance slit of the spectrometer. (d) Ratio of I_p (λ) /I_0 (λ) at the center area (y = 0) and at the edge (y = 100 μm), both integrated over Δy = 30 μm, for the spectra shown in (b).

CONCLUSION

In conclusion, we have shown Raman amplification of ultrashort laser pulses by a factor of ~ 6 in microcapillary plasmas. The analysis of the experimental data in LiF microcapillaries, together with simulation results [10], indicates that the length of the plasma with the matched density, $n_e = 3 \times 10^{20} cm^{-3}$, was much shorter than the microcapillary. We conclude that due to this short effective length the observed amplification in our initial experiment was significantly less than the linear theory predicted, and the amplification dropped as the energy of the seed pulse increased since the available pumping energy within this narrow time window was very limited. Further investigations in 200 μm-long microcapillaries support the conclusion of short interaction lengths and the results show an agreement between the observed amplification and the linear growth rate.

In order to obtain larger amplification a longer plasma column with proper density is desired. Longer copper microcapillaries (L_{cap} = 500 μm) were found unsuitable since the plasma could not fill the whole microcapillary, due to the illumination geometry of the prepulse. An alternative method for creating longer plasma is a high-pressure gas jet, which allows for 2-D density measurements since it is an open system. The experiment of Raman amplification in a gas jet plasma has confirmed Raman resonance by simultaneous density measurements.

ACKNOWLEDGMENTS

We would like to thank N. Tkach for technical support. This work was supported by DARPA and NSF (PHYS).

REFERENCES

1. Murray, J. R., Goldhar, J., Eimerl, D., and Szoke, A., *IEEE J. Quantum Electronics* **15**, 342 (1979).
2. Nishioka, H., *et al.*, *IEEE J. Quant. Electr.* **29**, 2251 (1993); Takahashi, E., *et al.*, *Fusion Eng. Des.* **44**, 133 (1999).
3. Maier, M., Kraiser, W., and Giordmaine, J. A., *Phys. Rev. Lett.* **17**, 1275 (1966); *Phys. Rev.* **177**, 580 (1969).
4. Kruer, W. L., *The Physics of Laser Plasma Interaction*, Reading, MA, Addison-Wesley, 1988.
5. Shvets, G., Fisch, N. J., Pukhov, A. and Meyer-ter-Vehn, J., *Phys. Rev. Lett.* **81**, 4879 (1998).
6. Malkin, V. M., Shvets, G., and Fisch, N. J., *Phys. Rev. Lett.* **82**, 4448 (1999).
7. Malkin, V. M., and Fisch, N. J., *Phys. Plasmas* **8**, 4698 (2001).
8. Malkin, V. M., Shvets, G., and Fisch, N. J., *Phys. Plasmas* **7**, 2232 (2000).
9. Ping, Y., Geltner, I., Fisch, N. J., Shvets, G., and Suckewer, S., *Phys. Rev. E* **62**, R4532 (2000).
10. Ping, Y., Geltner, I., Morozov, A., Fisch, N. J., and Suckewer, S., submitted to *Phys. Rev. E*.

Present state and future of laser-driven X-ray lasers in France : LASERIX.

G. Jamelot[1], D. Ros[1], A. Klisnick[1], Ph. Zeitoun[1], J. Dubau[1],
J-C. Lagron[1], L. Vanbostal[1], J-P. Chambaret[2] and M. Pittman[2]

[1] *Laboratoire d'Interaction du rayonnement X Avec la Matière (former LSAI), bâtiment 350,
Avenue Jean Perrin, 91405 Orsay Cedex, France. E-mail : gerard.jamelot@lixam.u-psud.fr*
[2] *Laboratoire d'Optique Appliquée, ENSTA, CNRS UMR 7639, F-91761 Palaiseau cedex, France*

Abstract : From the experience of several French teams involved in X-ray laser research, a new concept of X-ray laser facility is presented. This idea, based on a high power Ti:sapphire CPA driver would give a large versatility to the future facility "LASERIX".

INTRODUCTION

Laser-driven X-ray laser (XRL) research has been initiated in France by the former LSAI group (presently LIXAM) in the early seventies [1,2]. We studied extensively the recombination scheme in lithium-like ions [3,4]. In 1993 we have ascertained that the 6-beam laser of the LULI (450 J in 600 ps.) was powerful enough to pump neon-like zinc by electron-ion collisions, producing an intense XRL at 21.2 nm [5]. Collisional XRLs driven by the LULI laser are very intense and bright; they are robust too and can be used routinely to perform application experiments. Besides the quasisteady state (QSS) scheme, we develop studies of transient XRLs [6].

Two other French groups are working on XRLs in France. They use ultrashort femtosecond driver pulses to produce optical field ionized (OFI) plasmas. Collisional XRLs in Pd-like and Ni-like ions are under investigation at the LOA [7], whereas the CEA-DRECAM performs studies of the recombination scheme in H-like nitrogen to make short wavelength lasers [8].

Transient pumping and OFI schemes have changed the state of XRL research. Former quasisteady state (QSS) XRLs driven by subnanosecond pulses need large scale facilities that are mainly intended for inertial confinement fusion (ICF) studies. The shot rates of such facilities are low (3 shots per hour max), and a small time is devoted to non-ICF experiments. On the contrary new schemes imply small pumping energy that can be delivered in small laser facilities that may be entirely devoted to XRL studies and applications. After a brief description of the current XRL studies in France, this paper will present the French project of XRL facility : LASERIX.

CP641, *X-Ray Lasers 2002: 8th International Conference on X-Ray Lasers,* edited by J. J. Rocca et al.
© 2002 American Institute of Physics 0-7354-0096-2/02/$19.00

QSS COLLISIONAL XRLs DRIVEN BY ns-RANGE PULSE.

First actual XRLs needed a large energy driver of several kilojoules in a laser pulse of typically 0.5 ns in duration [9]. Lowering of the pumping energy has been obtained later by the use of a very low energy prepulse that smoothes density gradients and reduces refraction. Then the XRL beam is able to propagate through the whole plasma length before leaving the small amplifying zone of the plasma near the target surface.

Another factor that made the QSS XRL efficient has been a multilayer mirror set at a few millimeters from the plasma edge and used as a half-cavity [10] ; indeed the duration of population inversion is long enough for a double pass of the X-ray beam through the plasma, but too short for making efficient a complete cavity. XRL intensity is increased by almost 2 orders of magnitude by double-pass with respect to single-pass operation. Plasma jets or debris destroy the multilayer after each shot, but only a small part of the mirror surface is burnt owing to a shield with a small diaphragm. The mirror is moved after each shot to present a fresh area for the following shot.

With the 6–beam laser of LULI (420-450 J in 600 ps), a prepulse of about 0.7 J 10 ns before the main pulse, a zinc-target length of 2 cm and a Mo:Si half-cavity, we can make a very bright XRL (~ 1 mJ in 80 ps) emitting at 21.2 nm in a solid angle of 3 x 10 mrad^2 [5,11]. The pointing angle may be slightly adjusted by the half-cavity.
This QSS XRL is very robust, and shot-to-shot variations are small; then it can be currently used for application experiments.

The main defect of this QSS XRL is that it works only with neonlike ions. Attempts to make XRL at shorter wavelength with nickellike ions have failed with such long pulses, because the gain coefficient was low [12].

Recent experiments at the PALS have shown that it is possible to improve optical properties of such XRLs and especially their collimation by using two separate focusing devices for the prepulse and the main pulse [13]. Further developments at the LULI will be made in order to separate geometrically the two beans.

QSS XRLs DRIVEN BY PULSE IN THE RANGE OF 100 ps.

Shorter driver pulses, of about 0.1 ns, give rise to larger gain values. Energy delivered by the LULI laser for such pulse duration is too small to pump zinc efficiently, but a gain value as large as 15 cm^{-1} has been obtained at 25.5 nm with neonlike iron [14]. However the main interest of 0.1-ns pulses is their ability to pump nickellike ions that lase at shorter wavelengths. For example, Ni-like silver emits at 13.9 nm, a wavelength close to the maximum efficiency of the Mo:Si multilayer optics, i.e., very attractive for many applications. We obtained at LULI a gain value of 8 cm^{-1}. Then, saturation can be approached at 13.9 nm in a single pass operation through the 2-cm long plasma filament.

The XRL pulse duration is too short for supporting double-pass, as for 0.5 ns drivers. As we have proved the large efficiency of half-cavity to improve the optical properties of XRLs, we planned to use at LULI two pulses in the same driver shot in order to regenerate the population inversion after a few hundreds of picoseconds. The idea is to use a large scale half-cavity in order that the first XRL beam pass through the plasma when the second pulse re-heats the plasma and regenerates the inversion.

With this aim, we have observed that a second pulse interacting with the plasma 800 ps after the first one gives rise to a slightly lower amplification than the first, i.e., 11 cm^{-1} at 25.5 nm for iron and 6 cm^{-1} at 13.9 nm in the case of silver [15]. One can also conceive an unstable full-cavity for a high-quality XRL that should be sequentially pumped by three pulses or more (figure 1). The delay between 2 successive pulses is to be matched with the cavity length, c being the velocity of light, i.e., $T_i = 2L_i/c$.

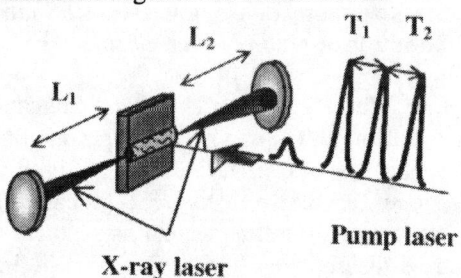

Figure 1 : Concept of a full-cavity X-ray laser sequentially pumped by a series of ~ 100 ps pulses

TRANSIENT COLLISIONAL EXCITATION PUMPING.

A 100 TW chirped pulse amplification laser has been developed at LULI since 1997. It has given to the LIXAM team the opportunity to study the transient-collisional-excitation-pumping scheme (TCE) that needs a nanosecond pulse to produce plasma followed by a picosecond one to pump inversion. Series of collaborative experiments between several European teams at CEA Limeil (P102), LULI and Rutherford Appleton Laboratory (RAL) have lead to a good understanding and expertise of the transient XRL [16].

In TCE plasmas, local gain is large, and saturation arises after propagation of a few millimeters through the plasma. The duration of population inversion is in the range of 1 ps, much smaller than the transit time of radiation ; so it is necessary to use traveling wave [17]. Far-field and near field records of the X-ray laser beam present indications that several amplification channels do exist in the plasma and they can interfere in the far field [18].

XRLs IN OPTICAL FIELD IONISED (OFI) PLASMAS.

OFI plasmas produced by ultrashort CPA laser pulses (20 fs) of high intensity ($> 10^{15}$ Wcm^{-2}) are able to amplify soft X-rays. Circular polarization gives rise to high electron temperature and collisional pumping, whereas electron temperature is expected to remain low if the laser is linearly polarized, which is the condition of recombination pumping. Impressive results were obtained years ago for both cases [19,20]. Two French groups are now working on XRLs in OFI plasmas, the LOA on the collisional scheme with a 2 J, 20 fs driver [21], whereas the CEA-D RECAM use their UHI 10 system to study the recombination scheme [22].

The LOA group has confirmed the Lemoff's experiment in using the same Pd-like xenon line at 41.8 nm [23]. In addition, they realized an OFI X-ray laser in Ni-like krypton at 32.6 nm [21]. The CEA group wanted to make a short wavelength laser at 2.5 nm in H-like nitrogen recombining plasma. They use a focused laser flux of 10^{19} W.cm^{-2} in a 70 fs pulse. Results did not show any indication of amplification. On one hand, the plasma filament appears as not linear and strongly inhomogeneous, as shown in figure 2, probably due to autofocusing. On the second hand, there was a large discrepancy between the experimental (400 eV) and the theoretical (10 eV) values of electron temperature. Plasma was probably collisionally heated by the pedestal of the CPA pulse and overheated by parametric instabilities, which prevent plasma cooling and creation of population inversion [24].

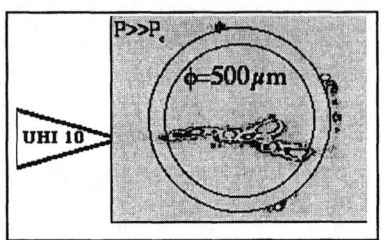

Figure 2 : Pinhole image of a recombining plasma produced by the UHI 10 laser in a gas-jet of 500 μm in diameter. Non-uniformity of recombining plasmas is a major cause that prevents population inversion, in addition to high temperature. (Courtesy of T. Auguste, CEA-DRECAM).

APPLICATIONS OF XRLS.

A major objective of the research devoted to XRLs is to develop their use as a tool for new investigations in several areas such as dense plasma or surface physics, non-linear interaction with matter, biology, chemistry, etc. With respect to other soft X-ray sources, XRLs are remarkable by their high brightness. This property is important for studies of interaction between matter and high-intensity radiation, and for recording images with high spatial and temporal resolution (single shot records).

The LIXAM group has developed several techniques of XUV interferometry and imagery, most of them in collaboration with LCFIO, and has performed experiments needing the properties of XRLs. In particular, we have observed new features in the luminescence spectra of ionic crystals excited by the Ne-like zinc laser at 21.2 nm with respect to spectra classically obtained with synchrotron radiation [25]. We have observed also in situ, for the first time, nanoscale deformations of metallic surfaces by high electric field [26,27]. Other examples of XRL applications are described in ref [28].

LASERIX PROJECT : A LASER FACILITY DEVOTED TO XRL RESEARCH AND APPLICATIONS.

The major difficulty and limitation to these application experiments comes from the fact that XRL drivers are generally high power lasers that are mainly intended for physics of inertial confinement fusion. In addition to the short time conceded to non-fusion studies, the shot rate of these lasers is low, namely 3 shots per hour for the LULI lasers (6-beams and CPA). Then, all the XRL application experiments were made with very few shots and their results must be considered as preliminary in the best case. We have shown that experiments are possible, but complete studies would need long series of identical XRL shots.

Development of XRL studies and applications calls for the creation of compact laser facilities specifically devoted to XRLs. A few such facilities do exist, one for QSS pumping at the University of Bern [29], the others based on CPA lasers for the TCE scheme : COMET at the LLNL [30] and the XRL facility of the APRC (JAERI Kansai) [31].A few others are under construction. We have elaborated the project of LASERIX, a French laser facility intended to

> realization and optimization of actual transient XRLs
> development of new types of XRLs
> development of XRL applications. This facility will be considered as a "light line" and open to the scientific community.

The typical energy of a "compact" CPA laser for transient XRLs is two times 10 J, in 0.5 ns and 1 ps pulses respectively. According to these values, it is possible to make XRLs in the wavelength range comprised between 10 and 40 nm, by using either Ne-like 3p-3s line with 20 < Z < 27 or Ni-like 4d-4p line with 38 < Z < 50 (figure 3). Proposals for new types of XRLs must take these energy values into account.

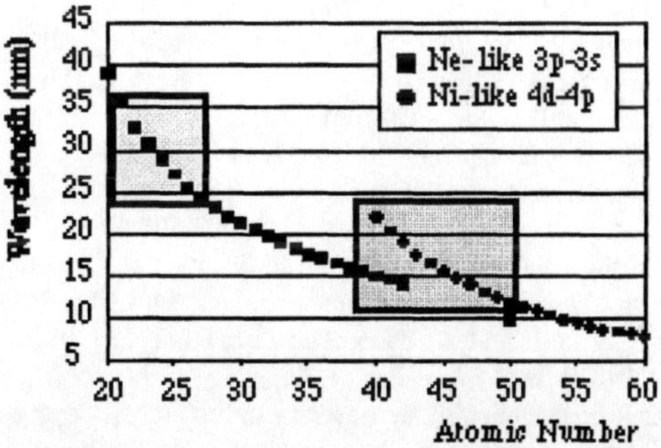

Figure 3 : Wavelength range of transient X-ray lasers that can be produced by the pump CPA laser planned for LASERIX.

The actual XRL facilities based on CPA laser uses the titane:sapphire technology for the front-end. Amplification of the stretched pulse is based on neodymium:glass rods. The wavelength of these lasers is 1.06 µm; the shot rate is limited by the thermal properties of the Nd:glass at 6 to 10 shots per hour. Relative experiences of LIXAM and LOA, supported by recent experiments at LLNL that used a large Ti:Sa crystal [32], have lead to modify the initial project and to conceive a new architecture of the LASERIX CPA driver, entirely based on titane:sapphire. The driving wavelength should be 0.8 µm in this case, but a few preliminary simulations have shown that XRL emission is not strongly affected by this difference. Ti:Sa driver should present many advantages with respect to the "old" Nd:glass technology: larger shot rate, easier optical alignments, larger versatility of the pulse duration, extension of LASERIX toward OFI plasmas and innershell XRLs.

A Ti:Sa driver for XRLs seem to be technologically possible. We are at present studying if such a solution is also economically possible.

ACKNOWLEDGMENTS

Pierre Jaeglé has lead the X-ray laser research in France since the origin and has played a major role in the genesis of the LASERIX project. The award conferred by his colleagues at the 8th ICXRL in Aspen was the mark of reconnaissance of his contribution to our research field. Authors warmly thank him for his action and his support to the project.

Contributions of Bedrich Rus and Antoine Carillon to the birth of the "LULI collisional X-ray laser" are deeply acknowledged. Authors thank many colleagues of several French Laboratories (LIXAM, LOA, CEA-DSM, CEA-DAM, GREMI, CELIA, LPGP, LPTP) who contribute to a number of studies concerning this project. This work is supported by the French Ministry of Research and the "Département de l'Essonne" in the frame of the "Contrat de Plan Etat-Région". The scientific support of the Université Paris-Sud is deeply appreciated.

REFERENCES

1 Jamelot G., *Thèse de 3ème cycle*, Orsay (France), 1971 (in French).
2 Jaeglé P., Jamelot G., Carillon A., Sureau A., and Dhez P., *Phys. Rev. Letters*, **33**,1070(1974)
3 Jamelot G., Klisnick A., Carillon A., Guennou H., Sureau A., Jaeglé P., *J. Phys B* **18**, 4647-63, (1985)
4 Zeitoun P. et al., AIP Conf. Proc.332, p. 55 (1994)
5 Rus B et al., *J. Opt. Soc. Am.* B **11**, 564 (1994)
6 Ros D. et al., these Proceedings
7 Sebban S. et al., *Phys. Rev. Lett.* **86**, 3004 (2001)
8 Hulin S., Auguste T., Monot P., Jacquemot S., Bonnet L., Lefebvre E., *Phys. Rev. E* **61**, 5693 (2000)
9 Matthews D.L. et al., *Phys. Rev. Lett.* **54**, 110 (1985)
10 Carillon A. et al., *Phys. Rev. Lett.* **68**, 2917 (1992)
11 Rus B. et al *Phys. Rev. A*. **55**, 3858 (1997)
12 Ros, D. et al., X-ray Laser 1996 Inst. Phys. Conf. Ser. N° 151, IOP, Bristol, 50 (1996)
13 Rus, B. et al, these Proceedings
14 Sebban S et al., Proceedings of the SPIE, Vol. 3156, 11 (1997)
15 Klisnick A. et al., X-ray Lasers 1998, Inst. Conf. Ser.N° 159, IOP, Bristol, 107 (1999)
16 Klisnick et al, Meas.Sci.Technol.**12**, 1(2001)
17 Klisnick et al, J.O.S.A. B **17**, 1093 (2000)
18 Ros, D. et al., these Proceedings
19 Lemoff B.E. et al., Phys. Rev. Lett. **74**, 1574 (1995).
20 Nagata Y. et al., Phys. Rev. Lett. **71**, 3774 (1993)
21 Sebban S. et al., these Proceedings
22 Hulin S., Thèse de l'Université Paris XI, Orsay, 2000 (in French)
23 Sebban S. et al., Phys. Rev. Lett. **86**, 3004 (2001)
24 Dobosz S. et al., to be published in Phys. Rev. E.
25 Belsky A.N. et al., X-Ray Lasers 2000, J. Phys. IV France **11**, Pr2-495(2001)
26 Zeïtoun-Fakiris A.et al., IEEE transactions on dielectrics and electrical insulation **6**, 418 (1999)
27 Mocek T. et al., these Proceedings
28 Jamelot G. et al., IEEE Journal of Selected Topics in Quantum Electronics, **5**, 1486-94 (1999)
29 Balmer J. et al., these Proceedings
30 Dunn J. et al., X-ray Lasers 1998, Inst. Phys. Conf. Ser. N° 159, IOP Publishing, Bristol, 51 (1999)
31 Kato Y. et al., J. Phys. IV France **11**, Pr2-3, (2001)
32 Bonlie J.D. et al., Appl.Phys. B, **70** (2000) S155-60

A multi-terawatt OPCPA laser system

Yuxin Leng, Zhizhan Xu, Xiaodong Yang, Haihe Lu, Lihuang Lin,
Zhengquan Zhang, Ruxin Li, Wenqi Zhang, Dingjun Yin, Bing Tang

*Laboratory for High Intensity Optics, Shanghai Institute of Optics and Fine Mechanics, Chinese
Academy of Sciences
P.O.Box 800-211, Shanghai 201800, China*

Abstract The novel optical parametric chirped pulse amplification (OPCPA) scheme is
promising for generating ultra-intense and ultra-short laser pulse. Based on the OPCPA
technique, a compact 3.67TW tabletop laser system was developed with an output of the
570mJ/155fs per-pulse, which is the highest power output in the existing laser systems based
on OPCPA, to the best of our knowledge.

INTRODUCTION

The development of ultra-intense and ultra-short laser is always an important field
in optics. Based on chirped pulse amplification (CPA) technique, the pulse of several
hundred femtoseconds with a peak power above 1PW has been obtained from
Nd:glass laser system [1]. And the Ti:sapphire CPA laser system can supply
peak-power above 100TW with <20fs output pulse width [2]. But, limited by the size
of Ti:sapphire crystal or the gain bandwidth of Nd:glass, it is difficult to achieve
broadband high gain to supply the ultra-intense pulse with high energy and short pulse
width in CPA laser system.

In recent years, the novel concept [3] of optical parametric chirped pulse
amplification (OPCPA), is regarded as an efficient means that can promote the output
power of laser. The OPCPA concept can supply broadband high gain, which can
support pulse duration about 10fs. OPCPA possesses several advantages over CPA [4].
First, in OPCPA process, the level of the pre-pulse intensity can be significantly
reduced. Second, the OPCPA-based laser system can obtain good beam quality. Third,
the OPCPA-based laser system may be genuinely tabletop in size. According to these
advantages, it is possible that the laser system based on OPCPA can generate
sub-10-fs pulse with high energy, or acts as an alternative of regenerative or
multi-pass Ti:sapphire amplifier in ultra-high power laser system.

CP641, *X-Ray Lasers 2002: 8th International Conference on X-Ray Lasers*, edited by J. J. Rocca et al.
© 2002 American Institute of Physics 0-7354-0096-2/02/$19.00

The studies on OPCPA at low energy level have been reported in several papers [3-5]. In this paper, we present an OPCPA laser system with two OPA pre-amplifier stages and a final OPA power amplifier. The energy of the signal is amplified from ~20pJ to 900mJ with the energy gain of above 4×10^{10}. After a grating compressor, a 3.67-TW pulse with the energy of 570mJ and pulse duration of 155fs is obtained, which is the highest output from an OPCPA laser, to the best of our knowledge.

EXPERIMENT

The experimental setup is shown in Fig.1. The system consists of a femtosecond Ti:sapphire oscillator, a pulse stretcher, a Nd:YAG-Nd:glass amplifiers chain as pumping laser system supplying pumping pulse, an OPA chain for amplifying chirped signal pulse, and a compressor. The pumping pulse and the chirped signal pulse are come from the same fs Ti:sapphire oscillator. The Ti:sapphire oscillator generates ~140mW and 76MHz mode-locking pulses train at ~1064nm with the ~150fs pulse width and ~8nm bandwidth (shown in Fig.2). Then the pulses are expanded to about 150ps by an öffner stretcher. By a 50%-50% splitter, one part of the chirped pulses after being stretched is reflected as the signal pulse of the OPA stages. And another fraction is induced into the stretcher again to be stretched to ~300ps, which is used as the seed of the pumping laser for OPCPA.

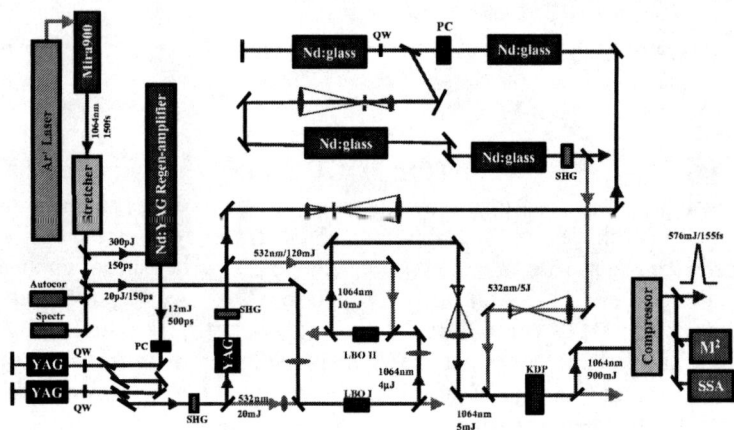

Figure 1 The experiment scheme of OPCPA laser system

The pumping laser is a Nd:YAG-Nd:glass amplifiers chain consisting of a regenerative amplifier, three Nd:YAG pre-amplifiers and four Nd:silicate-glass amplifiers. The amplified pulse ejected from the regenerative amplifier is injected to three Nd:YAG pre-amplifiers. Using three frequency doublers in the three position in the pumping laser, the 532nm pulses with ~20mJ, ~100mJ and ~5J are generated as the pumping pulses for three OPA stages.

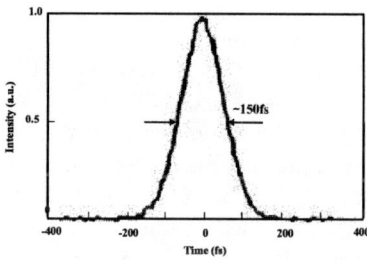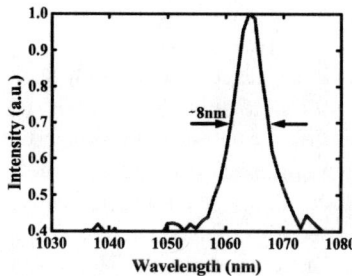

Figure 2 The autocorrelation trace and the spectrum of the pulse output from the oscillator. The pulse

duration is ~150fs (FWHM). The bandwidth is ~8nm (FWHM) centering at 1064nm.

The OPA chain consists of two OPA preamplifiers with 15-mm-long and 18-mm-long LBO crystals, and a final large aperture OPA power amplifier with a 30-mm-long KDP crystal. All the OPA stages operate in degeneracy for type I OPA process. The second OPA stage operates in near gain saturation and the final OPA stage in gain saturation.

In the OPCPA process, since the signal pulse is chirped pulse and can be amplified only in the time window of the pumping pulse, the timing jitter between the signal and pumping pulse can affect gain and spectrum of the signal pulse significantly. In this laser system, the OPA stages and the pumping laser system share the same oscillator as the seed source. Then the timing jitter between the signal and pumping laser can be reduced to near 10ps.

In order to obtain the optimum OPA, the width of the pumping pulse should wider than the chirped signal pulse suitably to avoid the time jitter effect and amplify the whole signal pulse. First, a Nd:glass regenerative amplifier is installed. The amplified pulse from the Nd:glass regenerative amplifier can adjusted continually from 0.3ns to ~3ns according to the ~300ps injected chirped pulse [6]. So the optimum OPA can be obtained when the pumping pulse width is adjusted to ~500ps. Second, a Nd:YAG regenerative amplifier is used to replace the Nd:galss regenerative amplifier for the longer using lifetime of Nd:YAG. By inserting a 5-mm glass etalon in the regenerative amplifier cavity, the bandwidth of the pulse amplified in cavity is narrowed. So that the width of the pulse amplified in the regenerative amplifier is stretched to generate the ~500ps pumping pulse width in OPA stages.

Considering the destroy threshold of the crystal, the pumping intensities in all the OPA stages should be ~4GW/cm^2 to obtain high signal gain and high efficiency of energy extraction.

The input signal pulse energy of the first OPA stage is about 20pJ and its bandwidth is ~8nm. The sizes of the signal and pumping beam in the first LBO crystal are about 0.9mm and 1.1mm. With pumping intensity of ~4.2GW/cm^2, the measured gain is 2×10^5. The amplified signal pulse is then injected to the second OPA stage with beam size of 2.2mm. The pumping beam size in the second LBO crystal is about 2.5mm. With a pumping intensity of ~4GW/cm^2 in the second LBO crystal, the signal pulse energy rises to ~10mJ with signal gain of ~2×10^3, with an energy conversion of

~10%. Then the signal is injected into the final OPA stage and the energy decreased to about 5mJ. The beam sizes of the signal and the pumping beam in KDP crystal are 1cm and 1.2cm, respectively. With a pumping intensity of ~4.1GW/cm^2, the signal pulse is amplified to about 900mJ, with an energy conversion of 18%.

The amplified pulses are collimated and then sent to a grating compressor, which compression efficiency is 63%, resulting in the compressed pulse energy of 570mJ.

The spectrum of the unamplified and amplified signal pulse is measured. The measured spectrum of seed pulse and amplified signal pulse of the first and the second OPA stages have bandwidth of ~8nm centering at 1064nm. After the final OPA stage, although the bandwidth is still ~8nm, the center wavelength shifts from 1064nm to 1066nm. Fig.3 gives the measured amplified signal pulse spectrum of the final OPA stage. The pulse duration of the recompressed pulse is measured with a single short autocorrelator. The autocorrelation trace for the amplified signal pulse is shown in Fig.4. The amplified pulse has a pulse duration of 155fs (FWHM) resulting in a time-bandwidth product of 0.34. The major factor of the shoulder on the autocorrelation trace is possibly due to an uncompensated spectral phase.

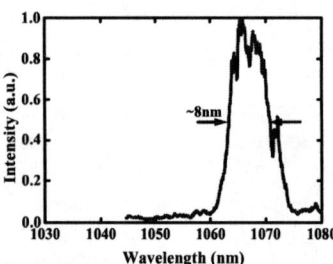

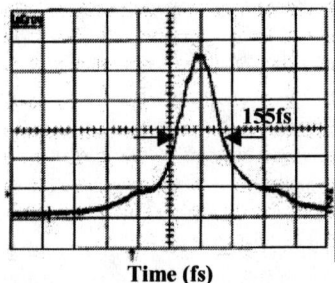

Figure 3 The spectrum of the amplified signal pulse output from the final OPA stage.

Figure 4 The autocorrelation trace of the amplified signal pulse after compressor.

CONCLUSION

We demonstrated a compact terawatt laser system based on the scheme of OPCPA. Chirped pulse has been amplified from 20pJ to 900mJ in two LBO pre-amplifiers and a final KDP power amplifier with the pumping energy of 5J at 532nm. An energy gain of above 4×10^{10} has been obtained. After compression, we obtained a final output as 570-mJ/155-fs pulse with a peak power of 3.67TW.

REFERENCES

1.   M.D.Perry, D.Pennington, C.Rouyer, et. al., *Opt. Lett.* 24, 160 (1993).

2.   K.Yamakawa, M.Aoyama, S.Matsuoka, T.Kase, Y.Akahane, and II.Takuma, *Opt. Lett.* 23, 1468 (1998)

3.   Dubietis, G.Jonušauskas, and A.Piskarskas, *Opt. Comm.* 88, 437 (1992).

4.   I.N.Ross, P.Matousek, M.Towrie, A.J.Langley, and J.L.Collier, *Opt. Commun.* 144, 125 (1997).

5.   I.N.Ross, J.L.Collier, P.Matousek, et. al., Appl. Optics. 39, 2422 (2000).

6.   Yuxin Leng, Haihe Lu, Lihuang Lin, Wenqi Zhang, Xu Zhizhan. *CLEO/Pacific Rim 2001,* P1-10.

XUV OPTICS, INSTRUMENTATION, AND CHARACTERIZATION OF X-RAY BEAMS

Thermoresistive multilayer mirrors with antidiffusion barriers for work at the wavelengths 40-50 nm

Dmitriy L. Voronov[a], Evgeniy N. Zubarev[a], Valeriy V. Kondratenko[a], Alexey V. Penkov[a], Yuriy P. Pershin[a], Alexander G. Ponomarenko[a], Igor A. Artioukov[b], AlexanderV. Vinogradov[b], Yuriy A. Uspenskii[b], and John F. Seely[c]

[a] National Technical University "Kharkiv Polytechnical Institute" , Frunze str. 21, Kharkiv 61002, Ukraine
[b] Lebedev Physical Institute, Leninski prosp. 53, Moscow 119991, Russia
[c] Naval Research Laboratory, Washington DC 20375, USA

Abstract. To improve the thermal stability of Si/Sc multilayer mirrors, thin layers of W were deposited at interlayer boundaries. Using X-ray scattering and transmission electron microscopy, we studied the interaction of Si and Sc layers at elevated temperatures. It was shown that the W layers of 0.5-0.8 nm thickness form dense WSi_2 barriers, which prevent a direct contact between Si and Sc and greatly slow down the formation of scandium silicides. Presented measurements show that Si/W/Sc/W multilayers fabricated by dc-magnetron sputtering possess long thermal stability up to 250° C and the normal incidence reflectivity of 24 %.

1. INTRODUCTION

Multilayer EUV and soft X-ray optics offers wide opportunities for numerous researches, the creation of novel equipment and technology. Its key components are reflective multilayer coatings constructed from alternate nanolayers of two materials with different optical constants. As roughness significantly reduces multilayer performance, layers of amorphous materials are widely used to smooth interfaces. The existence of interfaces and amorphous layers makes multilayer mirrors sensitive to heating, intense radiation and streams of particles. The action of these factors enhances diffusion and chemical reactions at interfaces, as well as crystallization in amorphous layers, and so causes the degradation of mirrors. This hampers the use of multilayer optics in applications with intense fluxes: space studies of the Sun, synchrotron beam manipulation, soft X-ray lasers, and, especially, free-electron lasers. To overcome this impediment, special methods of multilayers protection should be developed.

The present work applies antidiffusion W barriers to enhance the thermal stability of Si/Sc multilayer mirrors. This research was stimulated by the development of optics for mission studies of the Sun at the wavelength λ=40 nm [1] and the EUV laser with λ=46.9 nm [2]. In Section 2 we describe the fabrication of Si/Sc and Si/W/Sc/W

CP641, X-Ray Lasers 2002: 8th International Conference on X-Ray Lasers, edited by J. J. Rocca et al.
© 2002 American Institute of Physics 0-7354-0096-2/02/$19.00

coatings. Processes going in the multilayers at elevated temperatures are considered in Sections 3 and 4. The reflectivity of Si/W/Sc/W coatings is studied in Section 5. Section 6 summarizes results and discusses their application to other coatings.

2. FABRICATION AND TESTING OF MULTILAYERS

A number of Sc/W/Si/W multilayers with a period H=20-35 nm was prepared by dc-magnetron sputtering at $3 \cdot 10^{-3}$ Torr of Ar. All multilayers were designed as having 20 periods, the equal thicknesses of Sc and Si and capped by the Si layer of 5 nm. The thickness of W barriers was varied from 0 nm to 0.8nm.

The period and structure of fabricated multilayers were controlled by CuK_α (λ=0.154 nm) X-ray scattering at small and large angles, as well as by cross-sectional transmission electron microscopy (TEM). To test the thermal stability, the annealing of varied duration at 210°C and 600°C in vacuum of 10^{-5} Torr was used. The reflectivity of coatings was measured at the incidence angle 5° in the interval 25-50 nm at Brookhaven synchrotron.

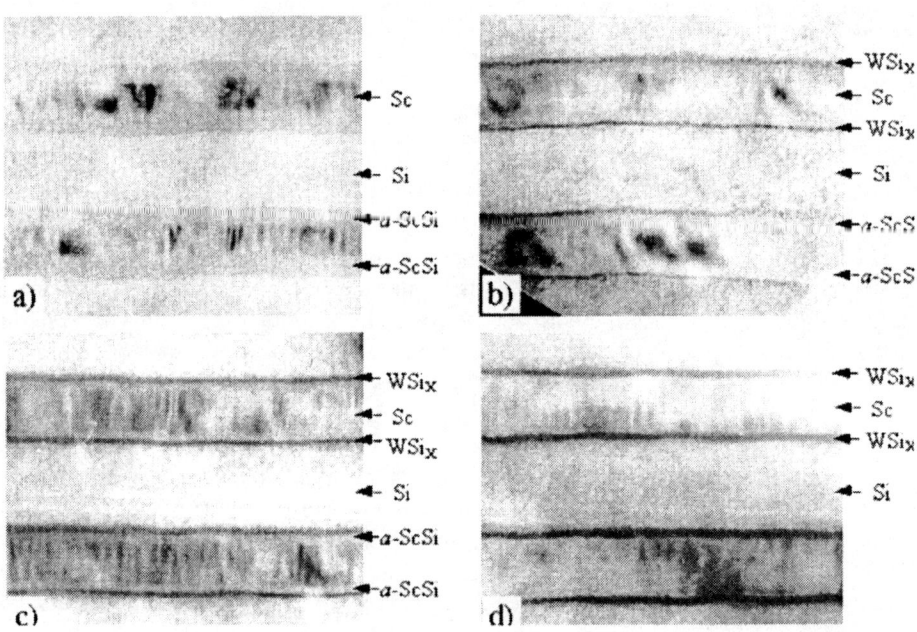

FIGURE 1. TEM pictures of Sc/Si (*a*) and Sc/W/Si/W (*b*) ML coatings with nominal thickness of tungsten layers of 0.2, 0.4, and 0.8 nm (*b, c,* and *d,* correspondingly). Growth of barrier layer thickness is accompanied by decrease of amorphous silicide thickness from 3 nm down to 0 nm.

3. PROCESSES IN SC/SI MULTILAYERS AT ELEVATED TEMPERATURES

Earlier studies [2, 3, 7] have shown that the amorphous scandium-silicon alloy interlayers of 3 nm are formed at interfaces between polycrystalline Sc and amorphous Si layers during Sc/Si multilayer deposition. The composition of amorphous alloy is close to composition of one of crystalline scandium silicide, namely ScSi [9]. When the temperature increases beyond ~130 °C solid state amorphisation starts in Sc/Si coating resulting in growth of amorphous ScSi interlayer thickness and contraction of Sc and Si layers. After total consumption of Sc layers the process of silicide formation abruptly slows down but doesn't stop. Under further heating Sc-based layers consume additional amount of silicon, and their composition shifts to the more rich of silicon silicide, Sc_3Si_5. Within the temperature range of 430-450 °C the amorphous alloy crystallizes through the mechanism of polymorphous crystallization into metastable crystal Sc_3Si_5 silicide which later undergoes eutectoid decomposition into equilibrium mixture of ScSi silicide and silicon.

Silicide formation processes are accompanied by decrease of specific volume and corresponding shrinkage of ML period. Character of shrinkage allowed to ascertain that the solid state amorphization proceeds along diffusion kinetics, i.e. rate of amorphous ScSi silicide growth is limited by the rate of diffusion mass transfer of Si atoms through an amorphous silicide layer toward ScSi/Sc interface where chemical reaction, $Sc+Si \rightarrow ScSi$, takes place.

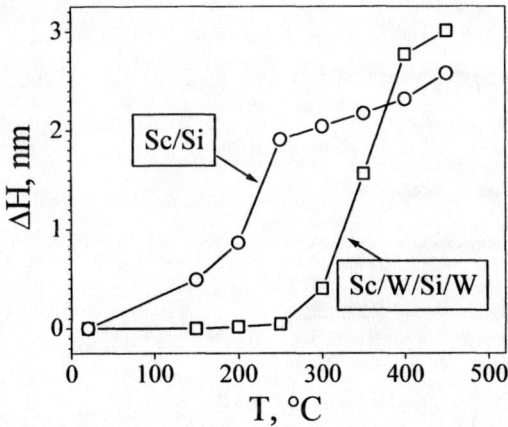

FIGURE 2. Curves of Sc/Si and Sc/W/Si/W ($hw^{nom.} = 0.8$ nm) ML shrinkage during annealing. Diffusion barriers retard silicide formation process in Sc/W/Si/W ML coatings shifting it to higher temperature range by 100-150 °C.

4. STRUCTURE AND THERMAL RESISTANCE OF SC/W/SI/W MULTILAYERS

As it was mentioned above, silicide formation processes change both phase composition and period of ML bringing to a shift of resonance wavelength and some reduction of reflectivity. To block the degradation process we used thin tungsten layers as diffusion barriers placed at the Sc/Si interfaces. High melting temperature for tungsten make possible a formation of its ultra-thing continuous layers. Tungsten does not practically react with Sc. W/Si multilayers fabricated by dc-magnetron sputtering had the dense symmetrical regions of the WSi_2 composition with the thickness of 0.9-1.4 nm [10]. These facts determined our choice for W as diffusion barrier between Sc and Si.

Fig. 1 shows TEM pictures of Sc/W/Si/W ML coating with diffusion barriers of different thickness. It is clear visible that on tungsten thickness growing from 0 (Fig.1a) up to 0.8 nm (Fig.1d) the amorphous silicide thickness gradually decreases. Intermixed zones are not observable at $h_W^{nom.} = 0.6$-0.8 nm (Fig. 1b). At the same time observed thickness of barrier interlayers exceeds the nominal noticeably. It is looks

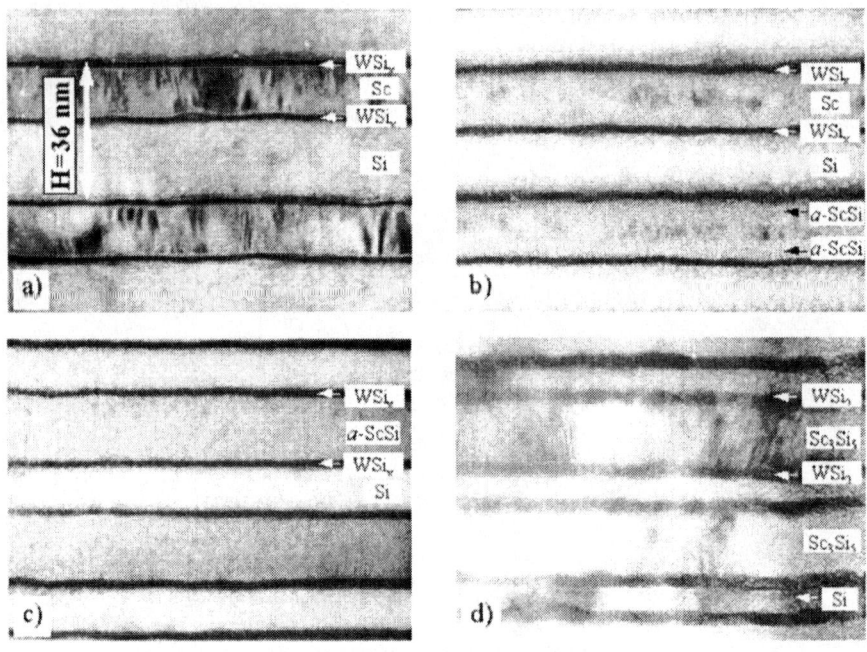

FIGURE 3. Evolution of Sc/W/Si/W ML structure ($h_W^{nom} = 0.6$ nm) under thermal heating: *a*) as-deposited state; *b*), *c*) annealing for 10 and 35 hours correspondingly at temperature 300 °C – reaction of solid state amorphisation takes place; *d*) annealing at 600 °C – appearance of metastable Sc_3Si_5 silicide and crystallization of tungsten disilicide WSi_2.

like to be connected with a possibility for tungsten to react with silicon forming WSi$_x$ silicide layers of larger volume.

So, the diffusion W-based barriers inhibit silicide formation processes during ML fabrication as well as subsequent heating, as shown in Fig. 2. Here it is visible that heating above 150°C results in substantial shrinkage of Sc/Si ML period while noticeable changes in Sc/W/Si/W ML coating is revealed at the temperature of ~ 300°C only. Therefore diffusion barriers shift silicide formation process to a higher temperature range improving thermal stability of Sc/W/Si/W mirror at least by 100-150 degrees.

Formation of scandium silicides in Sc/W/Si/W MLs occurs under the similar to Sc/Si MLs scheme. At the beginning, reaction of solid-state amorphisation takes place (Figs. 3b and 3c), forming amorphous Sc-Si alloy with a composition close to ScSi silicide and then metastable crystal Sc$_3$Si$_5$ silicide arises.

During annealing noticeable changes in W-based layers are also observed. Low-angle X-ray diffraction reveals substantial increase of barrier layer thickness (Fig. 4) at the initial stages of annealing at 300°C. Growth of barrier layers obviously originates from further progress in silicide formation within these layers finishing with appearance of amorphous tungsten disilicide: WSi$_x$ + Si → WSi$_2$. After 10-15 hours the formation of WSi$_2$ is completed, growth of barrier layers stops, and their thickness stabilizes. Within the temperature range of 500-600°C amorphous WSi$_2$ crystallizes (Fig. 3d).

Amorphisation kinetics in Sc/W/Si/W MLs appreciably differs from that in Sc/Si MLs. Fig. 5 shows curves of period shrinkage in ML with nominal tungsten thickness of 0.6 nm during isothermal annealing at 300 °C. Both formation of a-ScSi silicide and tungsten disilicide WSi$_{2-x}$ are responsible for this shrinkage. The latter process gives relatively small contribution (roughly 10 %), and in the first approximation we

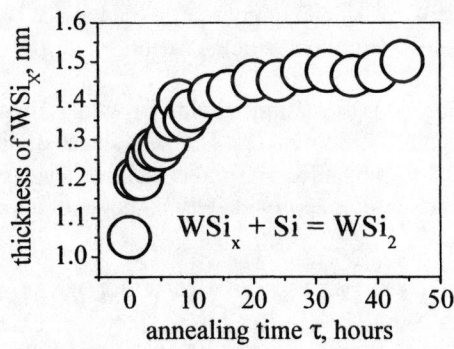

FIGURE 4. Changes of barrier layer thickness during annealing of Sc/W/Si/W ML ($h_W{}^{nom}$ = 0.6 nm) at temperature of 300°C in accordance with small-angle diffraction (λ=0.154 nm). As a result of reaction WSi$_x$ + Si → WSi$_2$ the thickness W-based layers increases. After 10-15 hours of annealing the reaction of tungsten disilicide formation is completed, and the thickness of barrier layers is stabilized.

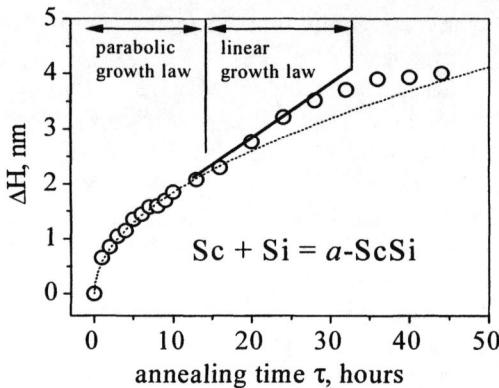

FIGURE 5. Kinetics of amorphisation reaction in Sc/W/Si/W ML coating ($h_W^{nom} = 0.6$ nm). At the stage of barrier thickness growth (see Fig. 4) the growth of amorphous silicide a-ScSi follows to the parabolic law (dotted line). After stabilization of diffusion barrier thickness the parabolic law is replaced by linear one (solid line).

can suppose that shrinkage $\Delta H \sim h_{ScSi}$, with h_{ScSi} being the thickness of amorphous scandium silicide. Fig. 5 shows the initial stage growth of a-ScSi obeys the parabolic law, and the experimental points will deviate from parabolic after 10-15 hours of annealing (dotted line) when linear part of a-ScSi growth can be observed.

Obviously such change of amorphization kinetics is caused by tungsten layers transformations described above (Fig. 4) since diffusion of silicon atoms through diffusion barrier is the process that limits the amorphous scandium silicide growth rate. Since thickness of barrier layers increases during first 10-15 hours (Fig. 5) then silicon diffusion through these layers becomes slower, and that decreases a-ScSi growth rate (the part of parabolic growth in Fig. 5). After stabilization of barrier thickness a diffusion flux of silicon atoms through a barrier layer ceases depending on time, and growth rate of amorphous silicide becomes steady (linear part of growth in Fig. 5).

Linear low of amorphous scandium silicide growth is maintained up to complete consumption of scandium layers (Fig. 3c). The next stages of silicide formation proceed with lower rate and are accompanied by smaller volume changes. That is why a bend of the contraction curve arises after approximately 35 hours of annealing (Fig. 5).

REFLECTIVITY OF SC/W/SI/W MIRRORS

Our earlier studies [1, 2] have shown that the reflectivity of Sc/Si coatings fastly decreases with the growth of interaction between Sc and Si layers. The deposition of W barriers affects on the reflectivity of Sc/Si multilayers by two ways: removing the Sc-Si mixed regions and providing more EUV absorption, as W has high absorption at the wavelengths of interest. These factors significantly balance each other. For example, the calculated reflectivity of the Sc/W/Si/W multilayer mirror with H=20.5 nm and $h_W^{nom.}$=0.55 nm is approximately equal to that of the as-deposited Sc/Si mirror with the same period. In this calculation the dielectric function of Si was taken from the handbook [11] and that of Sc from our recent measurement [12]. Our measurements of R excellently agree with the calculations (Fig. 6) and evidence that the Sc/Si multilayers with W barriers allow to obtain both the good thermal stability and rather high performance of reflectors.

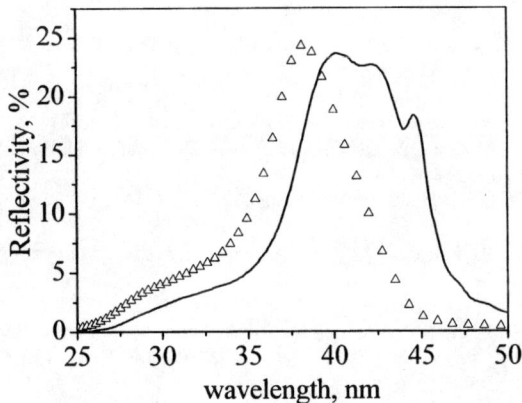

FIGURE 6. Calculated (line) and measured (symbols) reflectivity of the Sc/W/Si/W coating with H=20.5 nm and $h_W^{nom.} = 0.6$ nm.

6. SUMMARY AND DISCUSSION

The present studies showed that W barriers reliably block the formation of scandium silicides at the Si/Sc and Sc/Si interfaces. This block works equally well at the stage of multilayer deposition and under heating of a multilayer mirror. The deposition of W layers of 0.6 nm leads to the formation of dense WSi_2 barriers, which prevent a direct contact between Si and Sc layers and so greatly slow down the kinetics of their mixing. Because of a small thickness of W layers, the increase of thermal stability is not attended by a significant reduce of multilayer performance. It is probable that a similar compromise between the heat or flux resistance and the performance of a multilayer mirror may be obtained for many other coatings stabilized by interlayer barriers.

ACNOWLEDGEMENTS

This work was supported by the CRDF grant RP2-2267 and the "Integration" Grant IO 859. One of us (YAU) is grateful to the RFBR grant 01-02-17432a.

REFERENCES

1. Seely J.F., Uspenskii Yu.A., Pershin Yu.P, Kondratenko V.V., and Vinogradov A.V., Applied Optics **41**, 1846–1851 (2002).
2. Uspenskii Yu.A., Levashov V.E., Vinogradov A.V., Fedorenko A.I., Kondratenko V.V., Pershin Yu.P., Zubarev E.N., and Fedotov V.Yu., *Optics Letters* **23**, 771–773 (1998).
3. Uspenskii Yu.A, Levashov V.E., Vinogradov A.V., Fedorenko A.I., Kondratenko V.V., Pershin Yu.P, Zubarev E.N., Mrowka S, and Schaefers F., *Nucl. Instrum. Methods* **A 448**, 147–151 (2000).
4. Moreno C.H., Maarconi M.C., Kanizay K., Rocca J.J., Uspenskii Yu.A., Vinogradov A.V., and Pershin Yu.P., *Phys. Rev.* **E60**, 911–917 (1999).
5. Artioukov I.A., Benware B.R., Rocca J.J., Forsythe M., Uspenskii Yu.A., and Vinogradov A.V., *IEEE J. Sel. Topics in Quantum Electronics* **5**, 1495–1501 (1999).
6. Benware B.R, Seminario M., Lecher A.L., Rocca J.J., Uspenskii Yu.A., Vinogradov A.V, Kondratenko V.V., Pershin Yu.P., and B.Bach, *JOSA* **18**, 1041–1045 (2001).
7. Voronov D.L., Zubarev E.N., KondratenkoV.V., Penkov A.V., Pershin Yu.P., and Fedorenko A.I., *Functional Materials* **6**, 856–859 (1999).
8. Bugaev E.A., Fedorenko A.I., Kondratenko V.V., and Zubarev E.N., *J.X-Ray Sci. Tech.* **5**, 295–298 (1995).
9. Fedorenko A.I., Pershin Yu.P., Poltseva O.V., Ponomarenko A.G., Sevryukova V.S., Voronov D.L., and Zubarev E.N., *J.X-Ray Sci. Tech.* **9**, 1–6 (2000).
10. Shin W.S., and Stobbs W.M., *Ultramicroscopy* **32**, 219-226 (1990).
11. *Handbook of Optical Constants of Solids*, edited by E.D. Palik ,Academic Press, Orlando, 1998.
12. Uspenskii Yu.A, Seely J.F., Popov N.L., Vinogradov A.V., Pershin Yu.P., and Kondratenko V.V., to be published.

A New Method for X-ray Spatial Coherence Measure

Hongyi Gao, Jianwen Chen, Honglan Xie, Peiping Zhu and Zhizhan Xu

Shanghai Institute of Optics and Fine Mechanics , The Chinese Academy
of Sciences, P.O. Box 800-211, Shanghai 201800, China

Abstract. In this paper, a new method for X-ray spatial coherence measure is suggested by means of a quasi-equal optical path interference device with a Fresnel zone plate. It is also feasible to take Gabor in-line holograms of higher resolution using a lower resolution recording medium on this device.

1. INTRODUCTION

The spatial and temporal coherence is one of the most important parameter that describes the source quality for either X-ray holography and micro-tomography, respectively. However, the result from usual method used to be a total effect mixed with both of the spatial and the time coherence. It's difficult to measure them separately.

In general, the measure for X-ray spatial coherence is carried out by Yang's double slits [1] or multi-slits interference setup [2], which is usually performed by repetitive times.

For an X-ray lasers and syncrotron radiation with a finite source and a finite spectral range, it is necessary to specify the correlation between the vibrations from two arbitrary points in the wave field. It is difficult to obtain the spatial coherence by a single exposure.

In this paper, a new method for X-ray spatial coherence measure is suggested using a quasi-equal optical path interference device. The factor from the temporal coherence can be neglected, then the visibility of the interference fringes recorded on the imaging plane reveals an independent spatial coherence of the source.

2. QEOPI(QUASI-EQUAL OPTICAL PATH INTERFEROMETER)

The quasi-equal optical path interferometer (QEOPI) is simply composed of an X-ray zone-plate, a pinhole diaphragm and a square aperture. The geometric scheme of the quasi-equal optical path interferometer is shown in Fig.1.

CP641, *X-Ray Lasers 2002: 8th International Conference on X-Ray Lasers,* edited by J. J. Rocca et al.

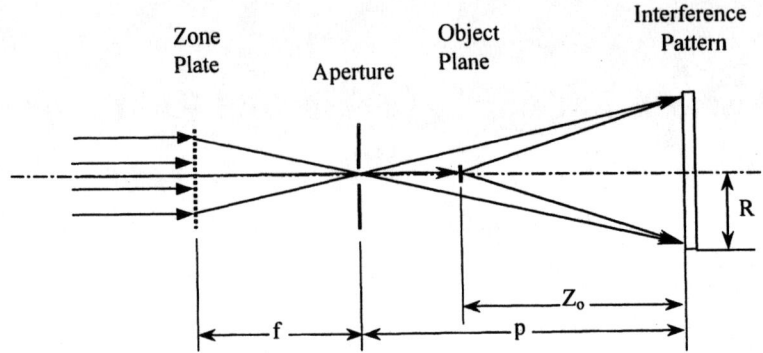

Figure 1. Schematic diagram for spatial coherence measuring setup

The focal length of Fresnel zone-plate is f, the distance from the focus to the pattern plane is p, and Z_o is the distance from the object plane to the interference pattern plane, R the maximum radius of the interference pattern.

The physical principle of QEOPI is, at a certain place behind the square aperture there exists an area overlapped by two parts of the diffraction wave, one part is the first order diffraction wave coming from the zone plate and the other is zero order wave diffracted by the square aperture, of which the optical path difference approaches to zero.

The verification to this issue is derived from the following calculation. From figure 1 we get the optical path of the first order diffraction wave at the edge of the pattern:

$$L_1 = (p^2 + R^2)^{1/2}(f + p)/p$$
$$= (f + p)(1 + R^2/p^2)^{1/2}$$

Extend above formula with binomial expression, then

$$L_1 \approx (f + p)(1 + R^2/2p^2 - R^4/8\,p^4)$$
$$= f + p + (f + p)R^2/2p^2 - (f + p)R^4/8p^4 \qquad (1)$$

The optical path of zero order wave diffracted by the square aperture at the edge of the pattern L_2 is:

$$L_2 = f + (p - Z_o) + (Z_o^2 + R^2)^{1/2}$$
$$\approx f + p + R^2/2Z_o - R^4/8Z_o^3 \qquad (2)$$

From expression (1) and (2), the optical path difference between L_1 and L_2 is:

$$\Delta L = |L_1 - L_2|$$
$$= |(f + p)R^2/2p^2 - (f + p)R^4/8p^4 - R^2/2Z_o + R^4/8Z_o^3|$$
$$\approx |[\,Z_o(p + f) - p^2\,]\,R^2/2Z_o\,p^2 - [(p + f)/\,p^4 - Z_o^{-3}]\,R^4/8| \qquad (3)$$

584

If expression (3) satisfies the following condition:

$$Z_o(p+f) = p^2 \qquad (4)$$

Then it will be

$$\Delta L = |[Z_o^{-3}-(p+f)/p^4]R^4/8|$$
$$=(p^2- Z_o^2) R^4/8Z_o^3 p^2 \qquad (5)$$

Suppose the resolvable voxels radius of the object is v, hence the maximum interference pattern radius should be

$$R = 0.61Z_o\lambda/ v \qquad (6)$$

Substitute expression (6) into (5), then we get

$$\Delta L = (p^2- Z_o^2) (0.61Z_o\lambda/ r)^4/8Z_o^3 p^2$$
$$= Z_o (p^2- Z_o^2) (0.61\lambda/ r)^4/8 p^2 \qquad (7)$$

The zone plate radius r_z and its focus spot radius r_f follows such relation:

$$r_z = 0.61f \lambda/ r_f \qquad (8)$$

The voxels radius v and the focus spot radius r_f satisfy the relationship as following (see Fig. 1):

$$v = (Z_o/p)r_f$$

Substitute it into (8), we obtain

$$r_z = 0.61f \lambda Z_o/vp \qquad (9)$$

Apply these theoretical results to the actual. Suppose $\lambda=3.3$nm(water-window wavelength), v=60nm, f=1.25mm, Z_o= 4mm,

From expression (4) and (9) we obtain p=5mm and r_z = 33.55µm.

From expression (7) we get:
$$\Delta L = 0.23\text{nm}.$$

The calculation shows that the temporal coherent length required by the X-ray quasi-equal optical path interference is so small near to the size of an atom that it can be completely neglected in spatial coherence measure.

As a result, the visibility of the interference pattern observed on the recording plane represents the spatial coherence of the X-ray source.

3. A NOVEL X-RAY IN-LINE HOLOGRAPHY

When the square aperture is substituted with a semi-opaque specimen, the setup becomes a novel system of an X-ray in-line holography. Compare with normal Gabor in-line holography it has the merits as following: First, the requirement for the temporal coherence is greatly decreased. Second, the imaging resolution exceeds the limit of the recording medium. Since the reference wave comes from a point source, the imaging resolution is decided by the angle between the reference wave and the object wave rather than the recording medium resolution just like traditional Gabor in-line holography. Third, the disturb come from the twin images is mostly eliminated, because the recording medium is far away from the specimen.

4. CONCLUSIONS

The quasi-equal optical path interference is capable to measure the spatial coherence of the source by a single experiment, which is very important for large-scale X-ray laser systems. It is also feasible to record a novel Gabor in-line hologram with higher resolution.

ACKNOWLEDGMENTS

This project was partly supported by Chinese High-Tech Program, Chinese National Major Basic Research Development Program (No. G1999075200).

REFERENCES

1. Max Born and Emil Wolf, *Principles of Optics*, Second (revised) Edition, Pergamon Press, 1964.
2. Jianwen Chen, Zhizhan Xu et al., An estimate of time coherent length for the soft X-ray beam in Hefei National Synchrotron Radiation Laboratory, *Acta Optica Sinica* (in Chinese), Vol.17, No.8, 1997.

Simple In-Line EUV Pulse Characterization

Emily A. Gibson[*], Thomas Weinacht[†], Sterling Backus[*], Margaret M. Murnane[*] and Henry C. Kapteyn[*]

*JILA, University of Colorado, Campus Box 440, Boulder, CO 80309
†Dept. of Physics, SUNY Stony Brook, NY 11794-3800

Abstract. We have developed a simple design for an in-line apparatus to measure the spectral and temporal characteristics of an EUV pulse. The apparatus can be used in front of an experiment without disrupting the beam.

In our experiment, EUV pulses are generated by focusing 20 fs, 800nm pulses with 0.7mJ of energy at 1kHz into a hollow core fiber filled with Argon. The high harmonic generation process yields odd orders of the fundamental 800nm light [1]. The fiber allows phase-matching of harmonic orders 21 to 27 (29.6 to 38 nm), increasing the brightness at these wavelengths.

The spectrum is measured using a compact magnetic-bottle time-of-flight (TOF) spectrometer. Figure 1 shows a schematic of the experimental set-up. Argon or Neon is continuously leaked into the vacuum chamber through a nozzle at the interaction region of the spectrometer. The magnetic-bottle TOF is designed to collect photoelectrons emitted over initial trajectories within a solid angle of 2π [2]. The permanent magnets produce a strong field (3000G) at the interaction region directed along the TOF tube. A solenoid is constructed around the flight tube providing a uniform weaker field (30G) in the same direction. Electrons generated in the interaction region follow the field lines to the detector. As they pass from the high field region to the low field, their velocities are parallelized so that even electrons initially traveling in a direction away from the flight tube end up with most of their velocity redirected along axis. The count rate is typically around 2000 to 3000 counts per minute corresponding to 2 to 3 electrons detected per pulse. A typical spectrum can be acquired in 60s. The EUV spectrum is simply the spectrum of the emitted photoelectrons after correcting for the ionization potential and absorption of the detection gas used. The resolution of our spectrometer is constant over a large energy region (10 to 100eV). Figure 2 shows spectra taken with different detection gases.

CP641, *X-Ray Lasers 2002: 8th International Conference on X-Ray Lasers*, edited by J. J. Rocca et al.
© 2002 American Institute of Physics 0-7354-0096-2/02/$19.00

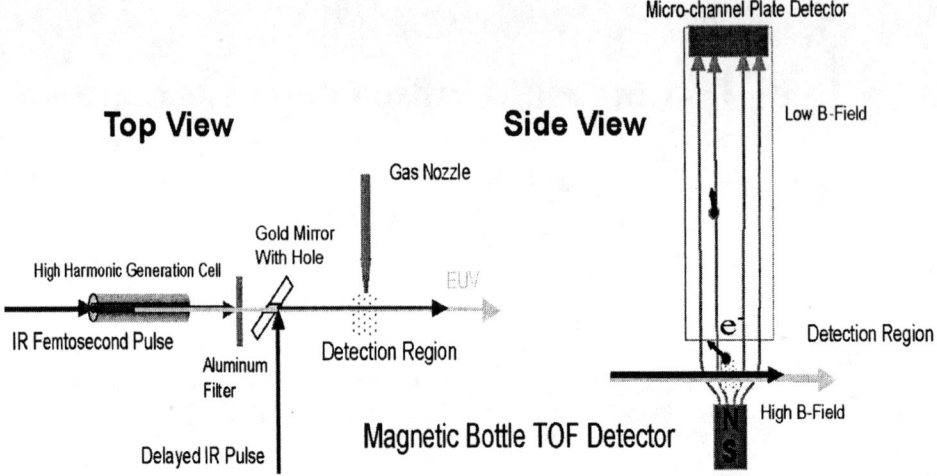

Top View　　　　　　　　**Side View**

Gas Nozzle

Micro-channel Plate Detector

Low B-Field

Gold Mirror
With Hole

High Harmonic Generation Cell

EUV

IR Femtosecond Pulse

Detection Region

Aluminum
Filter

Detection Region

Delayed IR Pulse

Magnetic Bottle TOF Detector

High B-Field

FIGURE 1. Experimental set-up.

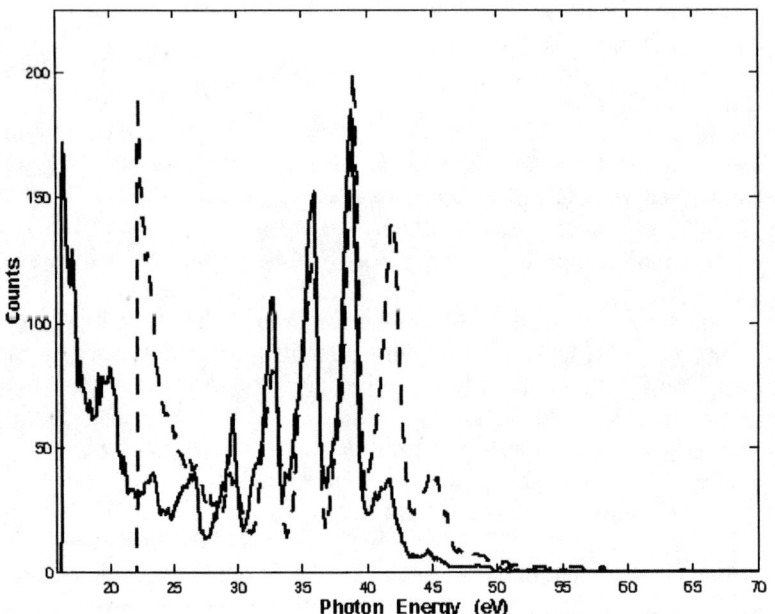

FIGURE 2. EUV spectrum with Argon (solid) and Neon(dashed) in detection region, not corrected for absorption.

The time duration of the EUV pulse is measured by cross-correlation with a weak 800nm probe pulse. Approximately 10% of the 800nm light used to drive high harmonic generation is split off, delayed, and recombined co-linearly with the EUV beam at the interaction region. There is no time smearing effect due to the co-linear geometry of the two beams. When both the EUV and probe are overlapped spatially and temporally, the resulting photoelectron spectrum contains sidebands to the main harmonic peaks as a result of laser-assisted ionization (see Figure 3) [3,4]. The sidebands have energies equal to the harmonic peaks plus or minus one photon of the fundamental. The cross-correlation signal is obtained by integrating the intensity of the sidebands for each delay (see Figure 4).

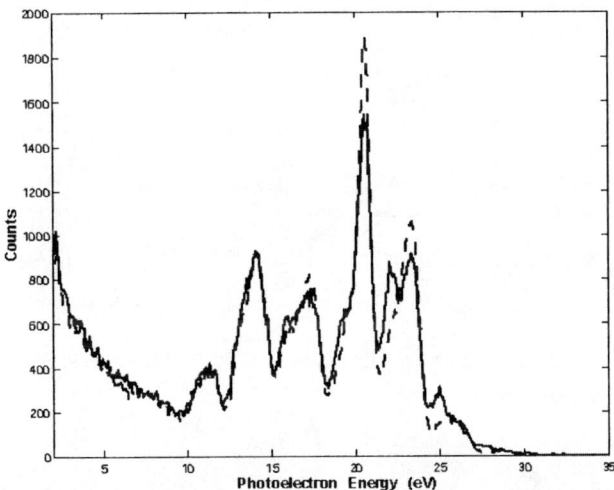

FIGURE 3. Photoelectron spectra for zero probe delay (solid) and 35 femtosecond delay (dashed).

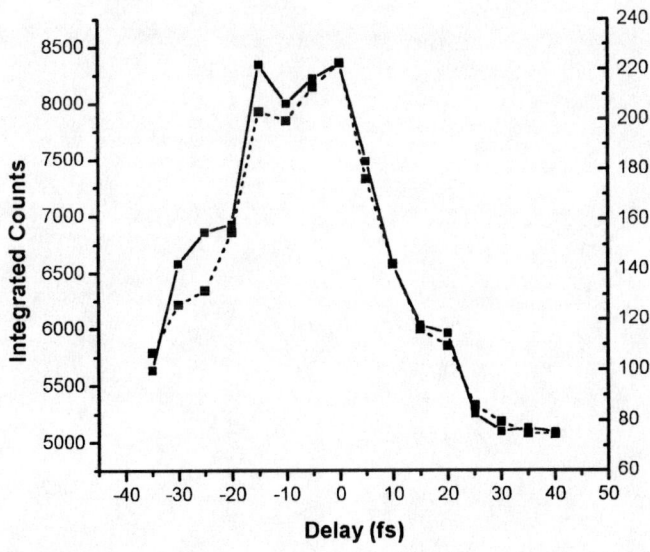

FIGURE 4. Sideband intensity vs. 800nm probe delay for sidebands at 22eV(solid) and 25eV(dashed).

To first order, the time duration of our EUV pulse is given by the de-convolution of the cross-correlation time width, obtained from the sideband signal, and the time duration of the probe pulse. Preliminary measurements of the cross-correlation give an EUV pulse duration that is instrument limited. Several possible future experiments using this set-up include the direct measurement of Auger lifetimes in atoms and molecules, measurement of attosecond pulses from quasi-phase matched modulated fiber [5], and observation of molecular dynamics.

REFERENCES

1. Rundquist, A. et al., *Science* **280**, 1412 (1998).
2. Kruit, P. and Read, F. H., *J. Phys. E. Sci. Instrum.* **16**, 313-324 (1983).
3. Glover, T. E. et al., *Phys. Rev. Lett.* **76**, 2468 (1996).
4. Schins, J. M. et al., *J. Opt. Soc. Am. B* **13**, 197 (1996).
5. Christov, I.P. et al., *Optics Express* **7**, 362-367 (2000).

Measurements of spatial coherence of X-ray lasers by diffraction on haircross wires

M. Kozlová, B. Rus, T. Mocek, and A.R. Präg

Gas Lasers Department / PALS Centre, Institute of Physics, Academy of Sciences of the Czech Republic, Na Slovance 2, 18221 Prague 8, Czech Republic

Abstract. Diffraction of a partially coherent beam on narrow elongated obstacles, such as thin haircross wires in an XRL experiment, is numerically investigated. The generated diffraction pattern is employed to evaluate the spatial coherence across the incident beam. The investigated geometry corresponds to a configuration frequently encountered in XRL experiments where the X-ray beam, upon intersecting a cross-wire, produces diffraction fringes within the footprint record. The diffraction pattern is modelled here with the help of angular spectrum of plane or spherical waves having an arbitrary degree (including zero) of mutual coherence. We compare the calculated diffraction pattern with the experimental data obtained using a neon-like Zn laser at 21.2 nm, operated in double-pass configuration.

INTRODUCTION

One of the key parameters of a XRL beam is its spatial coherence across a given plane situated at a distance of interest from the amplifying plasma. A number of coherence measurements have been performed using Young's slits structures [1,2], in which the fringe visibility is examined by several pairs of slits making it possible to sample the XRL beam in a few points on a single shot. These techniques, however, employ a device blocking the XRL beam and precluding its use for other measurements (e.g., footprint), and does not present a satisfactory approach in experiments with a repetition rate of typically 20 minutes and housed in vacuum environment.

In this paper we present an alternative method for measuring the XRL spatial coherence, based on diffraction on a thin wire. This technique is an "inverse problem" in which parameters of a partially coherent source are adjusted to produce a fringe pattern matching the experimentally measured fringes. The fringe pattern produced by a partially coherent wavefront are modelled assuming the output laser plane to be composed of an array of mutually incoherent spherical emitters, with the individual source strengths corresponding to the overall intensity profile across the source.

The pattern produced by the individual emitters is calculated using Fresnel diffraction treatment [3]. Firstly, diffraction pattern for a point source is calculated for the given experimental parameters (wavelength, distances, wire diameter) for a single emitter sitting on the axis (see Figure 1). In the second step, the diffraction pattern thus obtained is shifted by a distance $\Delta y r_0/\rho_0$, where Δy is the assumed coherence area on the source, and is multiplied by the intensity factor I_G corresponding to the transverse

CP641, *X-Ray Lasers 2002: 8th International Conference on X-Ray Lasers*, edited by J. J. Rocca et al.

profile of the source intensity (see Fig. 1). The intensity of the resulting diffraction pattern is then a sum of the diffraction patterns generated by the individual points.

The summing up of the individual diffraction patterns to I_{total} may be represented by

$$I_{total} = \sum_{m=-n}^{m=n} I_m\left(y - m\Delta y \, \frac{r_0}{\rho_0}\right) I_G(m\Delta y) \qquad (1),$$

where I_m is the diffraction pattern corresponding to a point emitter sitting at the transverse distance $m\Delta y$ on the source aperture. In this work, the intensity distribution I_G of the source is assumed gaussian or super-gaussian.

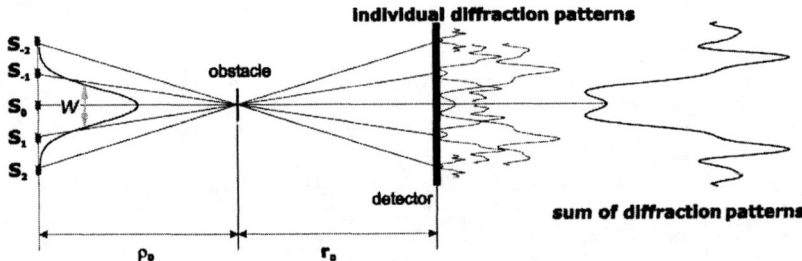

FIGURE 1. Calculation of the diffraction pattern generated by a partially coherent source exhibiting a gaussian intensity distribution; ρ_0 is the distance between the source and the diffracting object (wire), and r_0 is the distance wire-detector. The individual point sources S_m, constituting the XRL output aperture and being mutually incoherent, are separated by a distance Δy (see Eq.1). The resulting diffraction pattern is the sum of intensities of the individual patterns.

NUMERICAL MODELLING

The diffraction patterns were calculated with the help of a developed Fortran program efficiently treating the Fresnel integrals, serving to evaluate the diffraction pattern on a slit or a wire, for an incident spherical (or alternatively, plane) wave. For the given experimental parameters (see below) r_0=1500 mm, λ=21.2 nm, and the wire diameter of 120 μm, Fig. 2(a) shows the calculated diffraction pattern for a plane wave, whereas Fig.2(b) corresponds to a spherical wave with a radius ρ_0=1400 mm.

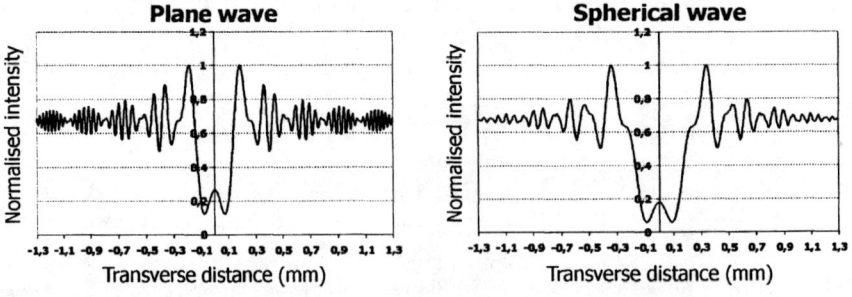

FIGURE 2. Interference pattern for a fully coherent (a) plane wave, (b) spherical wave.

By employing the method described above we calculated the diffraction patterns for different Δy and different gaussian sizes W (measured at the full-width at half-maximum, cf. Fig.1) of the emitting source. The results are shown in Figures 3, 4, and 5. Figure 6 corresponds to a super-gaussian source (of degree 4) with $W=50$ μm.

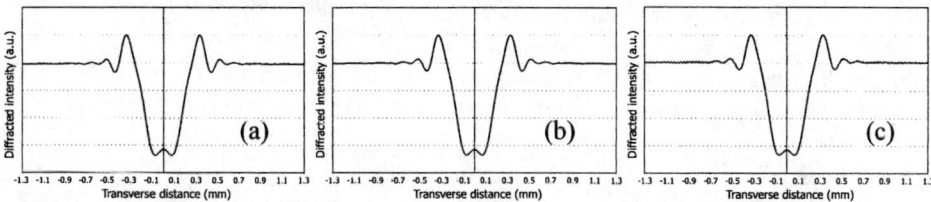

FIGURE 3. The sum diffraction pattern for $W=100$ μm gaussian; (a) $\Delta y=1$μm, (b) $\Delta y=10$μm, (c) $\Delta y=27$μm.

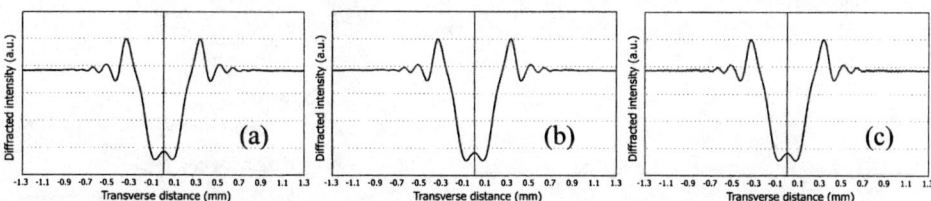

FIGURE 4. The sum diffraction pattern for $W=80$ μm gaussian; (a) $\Delta y=1$μm, (b) $\Delta y=10$μm, (c) $\Delta y=27$μm.

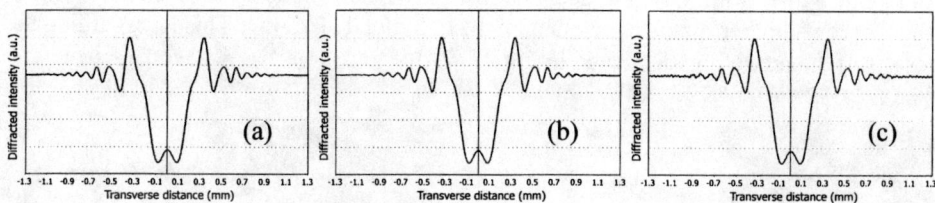

FIGURE 5. The sum diffraction pattern for $W=50$ μm gaussian; (a) $\Delta y=1$μm, (b) $\Delta y=10$μm, (c) $\Delta y=27$μm.

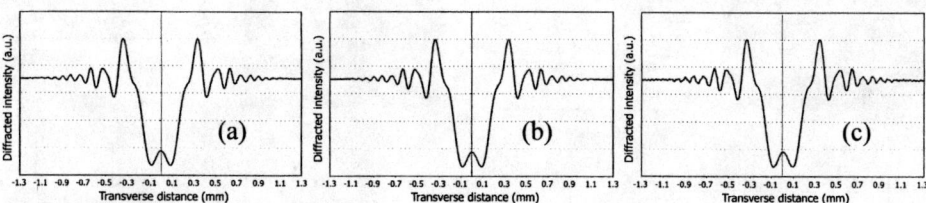

FIGURE 6. The sum pattern for $W=50$ μm super-gaussian; (a) $\Delta y=1$μm, (b) $\Delta y=10$μm, (c) $\Delta y=27$μm.

EXPERIMENTAL RESULTS

Figure 7 shows the experimental setup of a Ne-like zinc XRL [4], implemented at the PALS facility. The laser, employing a 30-mm long plasma amplifier and a half-cavity, delivers a saturated emission at 21.2 nm. A haircross wire system, positioned along the XRL beam path at an appropriate distance, generated a diffraction pattern detected on the XUV spectrometer; this was configured to a low spectral resolution, i.e. with the entrance slit open and with the detector plane rotated with respect to the Wadsworth (=Rowland configuration adjusted for a plane incoming wave) focusing circle.

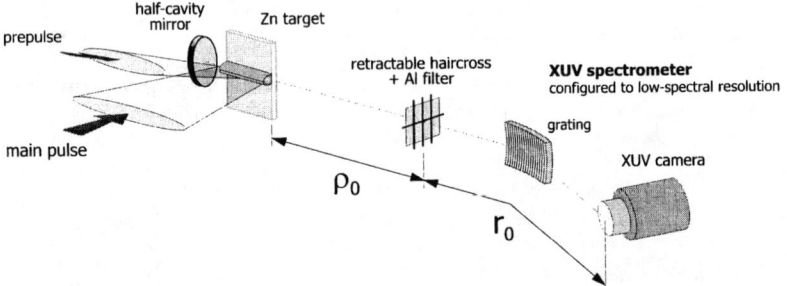

FIGURE 7. The experimental setup: $\rho_0 = 1400$ mm, $r_0 = 1500$mm, wire diameter $w = 120$ μm.

Examples of the recorded footprints, obtained as a "by-product" of the half-cavity optimization, are shown in Figure 8. By comparing the fringes produced in the horizontal and vertical directions, it may be seen that the XRL beam yields more contrasted fringes, and thereby a higher spatial coherence, in the horizontal direction. This is, regarding the results of the modelling displayed in Figs. 3-6, due to a larger vertical source size.

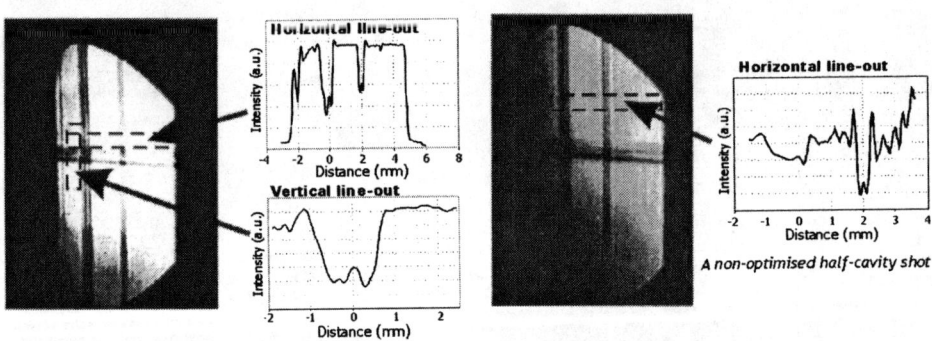

FIGURE 8. Diffraction patterns generated on 120-μm haircross wires by the half-cavity XRL beam.

CONCLUSIONS

In this paper we outlined a method for investigation of the spatial coherence of X-ray lasers. It is based on the assumption that the exit aperture of the XRL may be considered as a superposition of mutually incoherent spherical emitters. The approach makes it possible to deduce both the intrinsic coherence and the size of the exit aperture of the investigated XRL. Both the numerical results (see Figs. 5 and 6) and the measurements (see the non-optimized shot in Fig.8) indicate that the XRL spatial coherence is mainly due to propagation of the radiation in free space, which allows in most practical situations to assess the coherence of the XRL beam by the van Cittert-Zernike theorem.

REFERENCES

1. Trebes, J.E., Nugent, K.A., Mrowka, S., London, R.A., Barbee, T.W., Carter, M.R., Koch, J.A., MacGowan, B.J., Matthews, D.L., DaSilva, L.B., Stone, G.F., and Feit, M.D., *Phys. Rev. Lett.* **68**, 588-591, 1992.
2. Burge, R.E., Browne, M.T., and Slark, G.E., *"Coherence experiments with the Ne-like Ge X-ray laser"*, in Rutherford Appleton Laboratory Annual Report 1994, RAL-94-042, 1994, p.19.
3. Hecht, E., *Optics, Second Edition*, Addison-Wesley Publishing Company, Inc., Massachusetts, 1987.
4. Rus, B., Mocek, T., Präg, A.R., Kozlová, M., Jamelot, G., Carillon, A., Ros, D., Joyeux, D., Phalippou, D., in print in *Phys. Rev. A.*; also Rus, B., *et al.*, these Proceedings

X-ray Optics Research for Linac Coherent Light Source: Interaction of Ultra-short X-ray Pulses with Matter

Jaroslav Kuba[♣], *Alan Wootton, Richard M. Bionta, Ronnie Shepherd,*
Ernst E. Fill[1], *James Dunn, Raymond F. Smith, Todd Ditmire*[2],
Gilliss Dyer[2], *Richard A. London, Vyacheslav N. Shlyaptsev,*
Sasa Bajt, Michael D. Feit, Rick Levesque,
Mark McKernan and Ronald H. Conant

Lawrence Livermore National Laboratory, 7000 East Ave, Livermore, CA 94550, USA
[1]*Max-Planck Institut für Quantenoptik, D-85748 Garching, Germany*
[2]*Department of Physics, The University of Texas at Austin, Austin, TX 78712, USA*

Abstract. Free electron lasers operating in the 0.1 to 1.5 nm wavelength range have been proposed for the Stanford Linear Accelerator Center (USA) and DESY (Germany). The unprecedented brightness and associated fluence predicted for pulses <300 fs pose new challenges for optical components. A criterion for optical component design is required, implying an understanding of x-ray – matter interactions at these extreme conditions. In our experimental effort, the extreme conditions are simulated by currently available sources ranging from optical lasers, through x-ray lasers (at 14.7 nm) down to K-alpha sources (~0.15 nm). In this paper we present an overview of our research program, including (a) Results from the experimental campaign at a short pulse (100 fs – 5 ps) power laser at 800 nm, (b) K–α experiments, and (c) Computer modeling and experimental project using a tabletop high brightness ps x-ray laser at the Lawrence Livermore National Laboratory.

1. INTRODUCTION

The development of modern high power laser facilities in XUV regime brings new opportunities for the research of basic physical processes involved in the x-ray – matter interaction. In turn, it also poses new challenges on optical components. For example, at the Linac Coherent Light Source (USA), the projected brightness and associated non-focused fluence (up to 30 J cm^{-2}) in <300 fs will be available in 2008 and, at the TESLA source (DESY, Germany), the focused intensities of 10^{19} W/cm^2 in 100 fs pulses will be achievable in 2011.

[♣] E-mail: kuba1@llnl.gov

CP641, *X-Ray Lasers 2002: 8th International Conference on X-Ray Lasers,* edited by J. J. Rocca et al.
© 2002 American Institute of Physics 0-7354-0096-2/02/$19.00

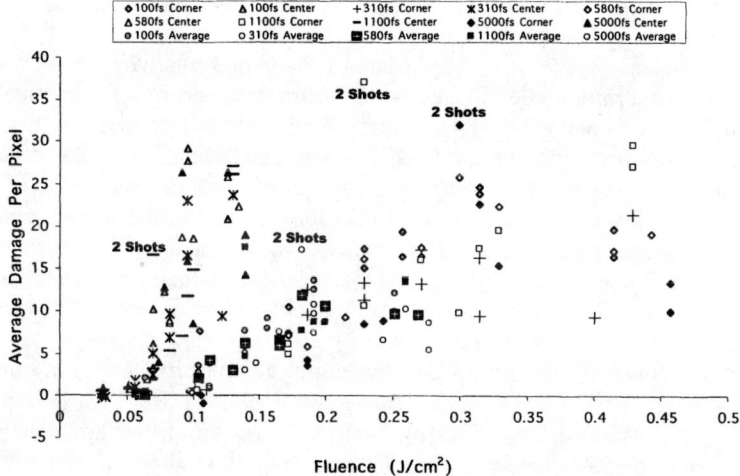

FIGURE 1. Irreversible damage at fluences as low as ~0.1 J cm^{-2} with no dependence on pulse length was observed. High power short pulse duration laser experiment at 800 nm was carried out in 2001.

Interaction experiments have been performed at many laser facilities, e.g. [1–4] in the (near-) visible wavelength range or in the XUV range [5] using a quasi-steady-state long pulse x-ray laser (XRL) at the NOVA and Janus facilities. At the LLNL the research involves the interaction of x-ray pulses with various materials from sources available nowadays with the aim to simulate the expected parameters of the 4th generation x-ray sources. In this paper we present our results, numerical modeling and experimental projects on the three major paths being pursued at the LLNL.

2. EXPERIMENTS AND SIMULATIONS
A. Damage by 800-nm Laser Pulses

Surrogate experiments using infrared pulses provided by the USP facility at the LLNL were carried out in 2001. A single Si crystal (111) was irradiated by 800-nm converging beam laser pulses under normal incidence, with the footprint defined by a 1 mm x 1 mm aperture. The sample was placed 5.5 cm beyond the aperture. The footprint on the sample was measured to be 0.321 x 0.359 mm. Both incident and reflected energies were measured, allowing the absorbed energy to be deduced. The measured reflectivity was 0.12 ± 0.06. Pulse lengths were varied between 100 fs and 5000 fs, with fluences up to ~0.3 J cm^{-2}.

After irradiation the samples were examined by an optical microscope. White light was reflected at normal incidence off the Si wafer, and the resulting pictures recorded at each spot heated by the laser. Presumably, the darker areas in the pictures correspond either to increased white light absorption, or to areas where the white scattering has been out of the microscope objective, in either case due to surface damage.

For each pixel in the image a background value representative of no damage was subtracted. Furthermore, the images were numerically corrected for the Fresnel diffraction pattern on the square aperture. Where obvious extraneous darkening appears (e.g. striations running diagonally across the wafer) the data was excluded from the analysis. Three values were taken for each picture: (a) an average darkening across each area inspected (i.e. across the 0.0156 mm^2), (b) in the central region, and (c) in the four areas of maximum irradiation due to the diffraction.

The data (Fig. 1) show clearly observable irreversible damage at fluences as low as ~ 0.1 J cm^{-2}, with no dependence on intensity (the pulse duration was varied by a factor of more that 100 while keeping the energy constant; the damage onset fluence remained unchanged). Recently, a German group has repeated our experiment and found the same results [6]. This onset fluence is 50 times lower than that predicted from the linear absorption coefficient (800 cm^{-1}) assuming the threshold damage process is melting of the surface. We infer that an additional absorption mechanism must be operating for these rather intense pulses.

Experiments at 625 nm [3] see a melting threshold of 0.17 J cm^{-2}, similar to ours, which they attribute to 2-photon absorption. However, at 800 nm, the 2-photon energy is below the direct band gap for Si and therefore the absorption coefficient is expected to be fairly small [7]. Furthermore, we observe that the damage fluence is insensitive to pulse length, arguing against 2- or multiphoton processes. It is possible that the enhancement is due to free-free absorption. In this case, linear absorption seeds an avalanche breakdown, which generates sufficient free electrons to increase the strength of the free-free absorption coefficient. This is similar to the multi-photon seeded avalanche breakdown inferred in ultra-short pulse interaction with fused silica [8].

R. London, S. Rubenchik, and M. Feit of the LLNL are currently developing a theoretical model.

B. K-α source Interaction with Matter

The second type of x-ray – matter interaction experiments involves the use of a K-α source generated by a short (~ 100 fs) laser pulse. The target materials range from Cu (8.0 keV) to Al (1.5 keV). The projected experiment will use such a K-α source to irradiate (in back-emission) a sample material (Fig. 2). The advantages of this scheme consist in a short wavelength comparable with the projected X-FEL facilities, and short pulse duration achievable. The scheme, however, requires basic research to produce reliable and well-described K-α sources.

Preliminary experiments at the ATLAS 10 (Max-Planck-Institut, Germany) facility were carried out in order to develop and characterize various K-α sources. The ATLAS 10 laser facility delivered typically 200-600 mJ (on-target energy) in 130-fs pulses at 10-Hz repetition rate or in a single pulse mode. The beam was focused by an off-axis parabola to a minimum of 9.4-µm-in-dimeter spot on a solid target [50-75 µm by 4 cm x 4 cm] supported by an aluminum substrate. Copper, iron, and titanium targets were irradiated with P-polarized laser light at a 45 deg angle of incidence. The intensity on target was controlled by varying the target position to change the focal spot size. The best focus position was optimized by the Hartman plate method. The resulting intensities on target and spot size were measured after the original experiment using a Coherent LaserCam CCD camera with the help of SPIRICON Laser Beam Diagnostics software. The measurement showed that the maximum irradiation intensities were around 2×10^{18} W/cm^2.

The resulting spectra were diagnosed by a 512x512, 16-bit, 12.3x12.3-mm Photometrics CCD under an angle of ~45 degrees. The CCD irradiation was reduced by transmission filters (5 μm of Ti and 10 μm of Ni). The K-alpha source vertical dimension was measured by means of a knife-edge input horizontally into the emission in front of the CCD (~7 cm away from the target).

In the first part of the experiment, the Cu K-alpha measurements reproduced well the results obtained during previous experiments by Eder et al. [9]. The optimization with respect to the focal spot yielded the maximum K-alpha emission of 3.7×10^9 photons/sterrad (supposing an isotropic emission in this angle). Our maximization of Cu K-alpha emission corresponds to the source diameter of 75 μm (even though of a slightly oval shape due to the parabola focusing), which is consistent with 70-μm-diameter reported in [9].

Comparison in Fig. 3 shows the optimization of the laser intensity to maximize the K-α output. The emission from a Ti target appears more intense than that of Fe or Cu. The results suggest that the emission rises for lower Z number of the target (cf. also [10]). A possible explanation would be that the energy conversion (laser to K-alpha) remains similar within the observed Z interval and hence only a smaller number of high energetic photons can be released in comparison with those at lower photon energy for low-Z materials.

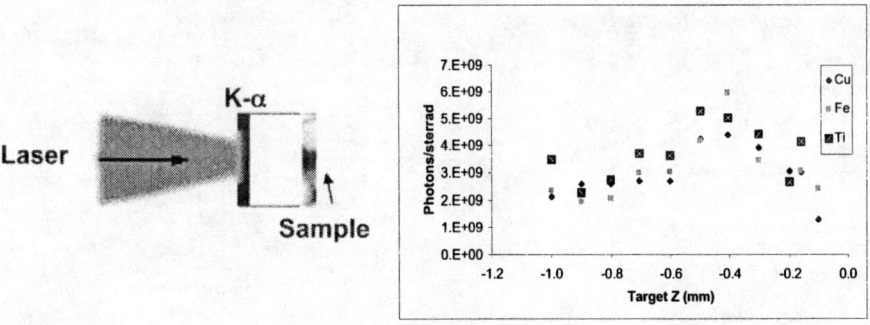

FIGURE 2. In a proposed experiment x-ray radiation from a short pulse laser driven K–α source will interact with an investigated sample.

FIGURE 3. K–α source optimization with respect to the laser intensity (varied by means of defocusing the irradiation laser).

C. X-ray Laser Interaction with Matter

The proposed experimental campaign will concentrate on the interaction of the focused XRL beam provided by the COMET tabletop facility [12] with multi-layer mirrors. These will include Si- and C-based optics, such as Mo-Si or Ru-C, operated under near-normal incidence. The damage will be studied by time-resolved measurements of the reflected beam using an ultra-fast streak camera with resolution up to 300 fs. The reflected pulse will be compared on a shot-to-shot basis with the original pulse, a small portion of which will be sent directly to the streak camera without the focusing and interaction with the target (Fig. 4). This scheme will allow us to evaluate the intensity profile on the target, and to calibrate the spot size (and hence intensity) as a function of the distance between the focusing optics and the target. The experimental work will provide data on reflectivity changes and then by interpretive modeling on material changes both during and after the x-ray pulse.

The hydrodynamic changes that affect mirror properties on intermediate to later times can be currently modeled by the LASNEX and RADEX codes [11]. Preliminary results demonstrating the high sensitivity to small material property changes are shown in Fig. 5, where dynamic photon absorption and subsequent expansion in a Mo-Si multilayer mirror are modeled with the code RADEX. The resulting density profiles were then input into a multilayer optics code to calculate the dynamic reflectivity, which is what will be experimentally measured. By varying the multilayer period and incident beam angle, substantial changes in reflectivity are predicted. The changes shown take place on hydrodynamic timescales and produce distinct 'signatures' in the multilayer reflectivity as a function of photon energy (or incidence angle) arising from material expansion into vacuum. Other physical effects

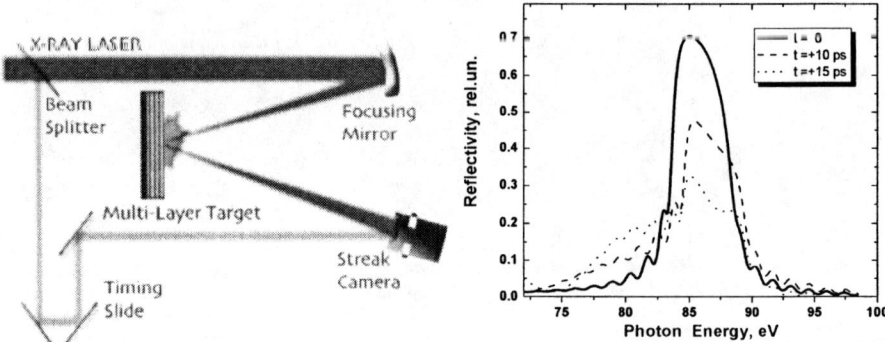

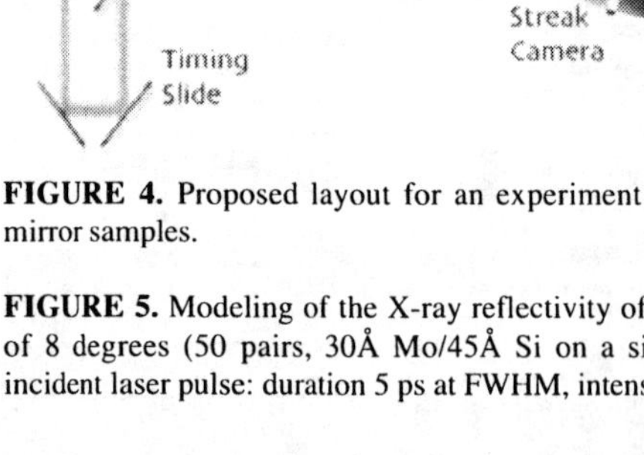

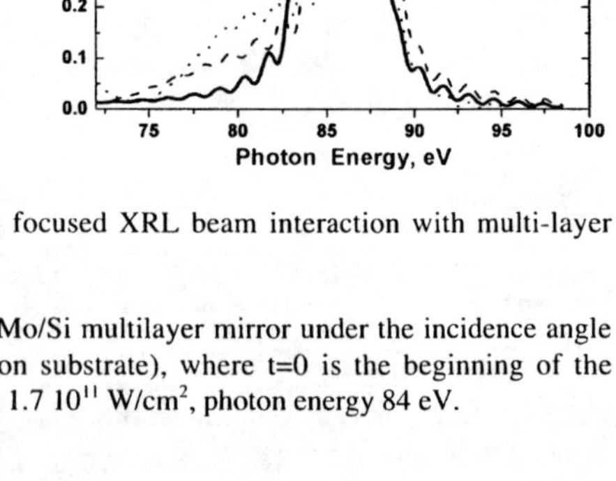

FIGURE 4. Proposed layout for an experiment on focused XRL beam interaction with multi-layer mirror samples.

FIGURE 5. Modeling of the X-ray reflectivity of a Mo/Si multilayer mirror under the incidence angle of 8 degrees (50 pairs, 30Å Mo/45Å Si on a silicon substrate), where t=0 is the beginning of the incident laser pulse: duration 5 ps at FWHM, intensity 1.7 10^{11} W/cm^2, photon energy 84 eV.

with different timescales lead to different reflectivity temporal behavior, including changes seen only after the x-ray pulse is over. For example, one might find improvements in reflectivity on subsequent shots. The preliminary theoretical analysis predicts in some conditions the occurrence of phase transitions like melting on (sub-) ps timescales.

3. CONCLUSIONS

In this paper we presented the experimental program and associated simulations on x-ray – matter interaction that is being carried out at the Lawrence Livermore National Laboratory. In a surrogate experiment with a short pulse (100 fs – 5 ps) power laser at 800 nm we demonstrated irreversible changes in a silicon target at fluences as low as 0.1 J cm^{-2}, which is 50 times lower than that predicted from the linear absorption coefficient assuming the threshold damage process is melting of the surface, which suggest that other processes are involved.

In frame of the K-α program, we optimized K-α sources from different targets. In our further work the K-α sources will be used to investigate the interaction of a sample with this short x-ray pulse. The experimental program will continue both at the ATLAS 10 facility in Germany and at the JanUSP facility at the LLNL.

Finally, the XRL project was presented where the focused short pulse transient XRL will interact with a multi-layer mirror target. The simulations predict substantial changes in reflectivity of this resonance system.

REFERENCES

[1] J. Larsson et al., Appl. Phys. **A 66**, 587 (1998)
[2] A. M. Lindenberg et al., Phys. Rev. Lett. **84**, 111 (2000)
[3] K. Sokolowski-Tinten et al., Phys. Rev. **B 61**, 2643 (2000)
[4] D. A. Reis, Phys. Rev. Lett. **86**, 3072 (2001)
[5] B. J. MacGowan et al., Journal of X-Ray Science and Technology 3 (1993) 231
[6] J. Bonse et al., Appl. Phys. **A 74**, (2002) 19
[7] Reitze, et al., J. Opt. Soc. Am. **B 7**, 84 (1990)
[8] B.C. Stuart et al, Phys. Rev. **B 53**, 1749 (1996)
[9] D. C. Eder, G. Pretzler, E. Fill, K. Eidmann, A. Saemann, Appl. Phys. **B 70**, 211 (2000)
[10] Ch. Reich, Phys. Rev. Lett. **84**, 4846 (2000)
[11] Yu.V. Afanasiev, V.N. Shlyaptsev et al., J. of Sov. Laser Research **10**, 1 (1989)
[12] J. Dunn et al., Phys. Rev. Lett. 84, 4834 (2000)

This work was performed under the auspices of the U.S. Department of Energy by the University of California, Lawrence Livermore National Laboratory under contract No. W-7405-Eng-48.

Measurements of soft X-ray lasers wavefront

S. Le Pape [(1)], Ph. Zeitoun [(1)], M. Idir [(1)], P.Dhez [(1)], D. Ros [(1)], A. Carillon [(1)] J. J. Rocca [(2)], M. Francois [(3)], S.Sebban [(4)]

[(1)] LIXAM, Bât. 350, Université Paris-Sud, 91405 Orsay, France
[(2)] Department of Electrical and Computer Engineering, Colorado State University, Fort Collins CO 80523-1373, USA
[(3)] Institut d'Electronique et de Microélectronique du Nord, Université des Sciences et Technologie de Lille, avenue Poincaré BP 69, 59652 Villeneuve d'Ascq, France
[(4)] Laboratoire d'Optique Appliqué, chemin de la Huniére F-91761 Palaiseau France

Abstract; A Shack-Hartmann wavefront sensor has been developed at the Laboratoire de l'Interaction du Rayonnement X avec la Matière, for a characterization of soft x-ray laser beam. A mapping of the wave-vectors pointing of laser-pumped soft x-ray laser (Optical Field induced and 100 ps laser pumped lasers) has been achieved. Capillary discharge soft X-ray laser has been also investigated. The electromagnetic field has been reconstructed and the influence of the argon pressure in the capillary on the wave front regularity has been investigated.

The emergence of these table-top XUV-ray sources [1,2,3] represents a major development that allows many small laboratories to experiment with a source of their own, which in turn is leading to new applications. Nonetheless, application experiments require a new forward in the knowledge and the optimization of the x-ray laser beam (XRL) characteristics. The XUV optic group of the LIXAM has developed a Shack-Hartmann wavefront sensor dedicated to the study of XUV sources. In the reminding paper, results on laser pumped x-ray laser are presented: optical field induced x-ray laser pumped by femtosecond pulses in collaboration with Laboratoire d'Optique Apliquée,iron single pass x-ray laser pumped by 100 ps pulses in collaboration with the x-ray laser team from LIXAM. The last section is dedicated to the study of the capillary discharge beam in collaboration with Colorado State University.

1. STUDY OF THE WAVEFRONT OF LASER PUMPED X-RAY LASERS

A first experiment was conducted on the optical field induced (OFI) x-ray laser emitting at 41.8 nm. In this scheme a 500 mJ, 30 fs infrared pulse is focused in an Xenon gaz cell. Electrons are ionized by tunneling effect. The population inversion is caused by the collisions of the free electrons with the Xe^{8+} ions [4]. The Shack-Hartmann wavefront sensor was placed at 2 m from the end of the cell. The residual defects of the lenses array used during this experiment was not previously calibrated

CP641, X-Ray Lasers 2002: 8th International Conference on X-Ray Lasers, edited by J. J. Rocca et al.
© 2002 American Institute of Physics 0-7354-0096-2/02/$19.00

with a spherical wavefront. Its defaults were analyzed with a "zygo" interferometer in the visible range providing an accuracy of about λ_{XUV}. As the spatial coherence of the OFI XRL has not yet been studied, the wavefront shape can not be reconstructed. So a map of the wave vector's tilt is realized. Results presented on figure 1 shows a map of the difference of the OFI XRL wavefront to a perfect spherical wavefront.

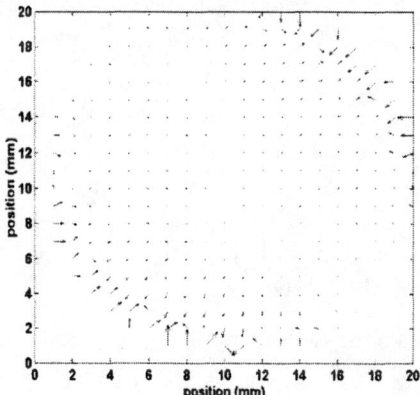

FIGURE 1. Map of the difference of the OFI XRL wavefront to a spherical wave.

The wave vector's norm is globally equal to zero and their orientation is homogeneous. It means that the XRL wavefront is almost spherical. Defaults that appear on the beam side are probably induced by the sensor substrate default. During this experiment the Shack-Hartman resolution was not sufficient to detect the amplitude of the wavefront aberrations, still we can conclude that they are smaller than λ_{XUV}.

A second experiment was realized on a single pass x-ray laser at 25.5 nm pumped by a 120 J, 100 ps IR pulse [5]. A plasma column is then created on a 100 μm × 2 cm focal line. The Shack-Hartmann wavefront sensor was placed at 2 m from the end of the plasma column. Some previous experiments have shown that single pass XRL are not spatially coherent over the full beam dimension [6]. In this configuration we have investigated the influence of the prepulse level on the wave vectors distribution. The prepulse level is the ratio between the energy in the first pulse and the energy in the second pulse. This prepulse has a strong influence on the electron density gradient and therefor on the angular aperture of the wave vectors. The time delay between the two pulses is 6 ns, which corresponds to the higest XRL intensity previously observed [7]. As in the previous OFI x-ray laser experiment, we present a map of the single pass x-ray laser wave difference to a spherical wavefront for two different prepulse levels 1% (fig 2.a) and 24% (fig 2.b).

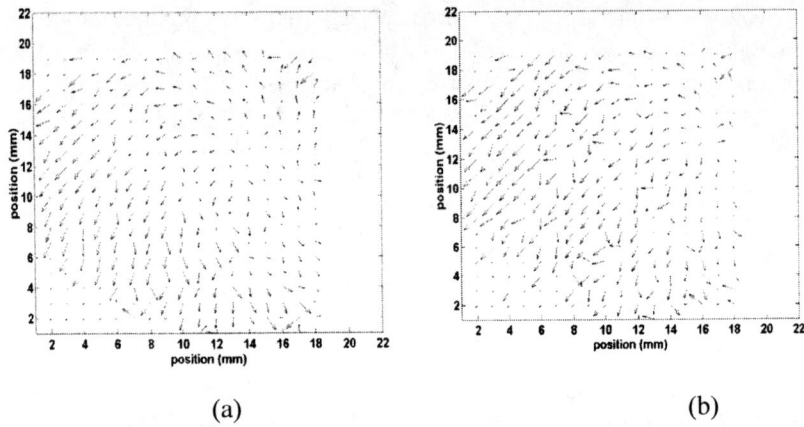

(a)   (b)

FIGURE 2. Map of the difference of the single pass XRL wave-vectors to a spherical wave for a 1% prepulse level. (a) for a 24 % prepulse level (b).

First of all we can notice the high amplitude of the wave-vectors norm as well as the high degree of disorder of the pointing angle. It shows that the wave is far from being spherical. In addition the orientation and the norm of the wave vectors do not exhibit any particular symmetry. The prepulse increase from 1% to 24 % seems to reduce the vectors homogeneity. The amplitude of the depart from a spherical wave is in the order of 10^{-4} rad, which is equivalent to 5 λ_{xuv}.

2. STUDY OF THE CAPILLARY DISCHARGE X-RAY LASER

The measurements reported herein allowed the full characterization of the electromagnetic field distribution of a table-top capillary discharge pumped x-ray laser operating at a wavelength of 46.9 nm (λ_{XUV}) in the 3s $^1P_1$- 3p $^1S_0$ line of Ne-like Ar [8]. The laser has been described in previous publications [3,9]. The gain medium is an elongated plasma column generated in a capillary channel (35 centimeter long) by a fast discharge current pulse. Since the X-ray laser under study is fully coherent [10], there is a unique phase relation between every points on the detector, and consequently the wavefront shape can be reconstructed from the focal spot positions. A program was developed to estimate each focal spot centroïd with an accuracy of about 10^{-2} pixel. The wavefront was subsequently reconstructed with a second program based on the algorithm written by Southwell [11]. The phase in each point was calculated by using a Gauss-Seidel method, which has been chosen for its convergence speed and its accuracy. The accuracy of the wavefront reconstruction is 10^{-3} λ_{XUV}. We have investigated the variation of the wavefront characteristics as a function of the argon pressure in the capillary discharge laser channel. The results of measurements corresponding to discharge pressures of 200 mtorr, 420 mTorr (near the optimum for maximum laser power), and 660 mTorr are shown in figure 3.a-b-c.

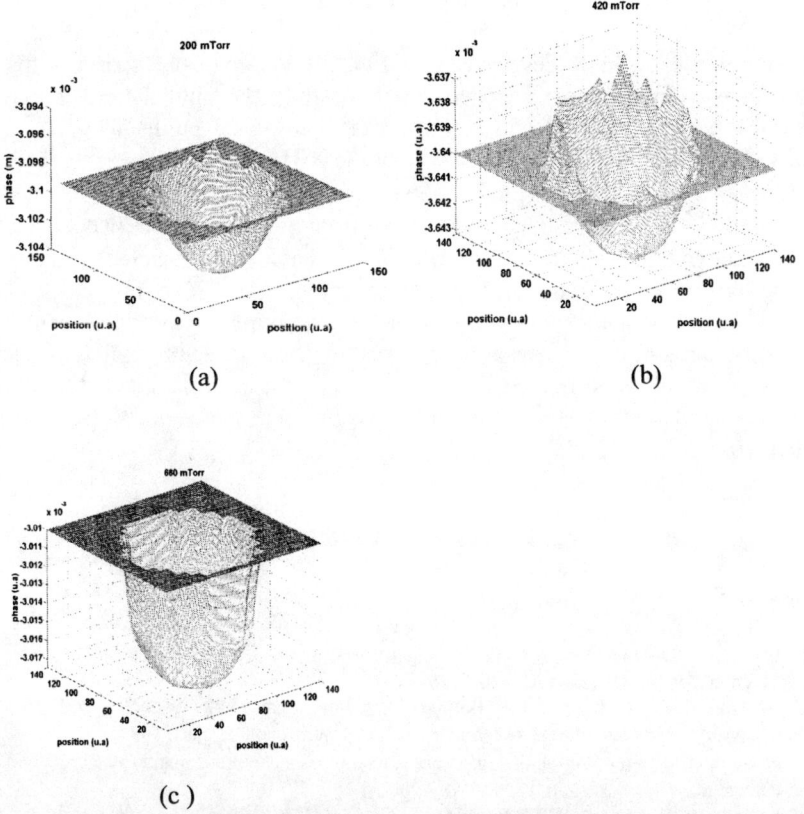

FIGURE 3. Wavefront shape for different discharge pressures (a) 200 mtorr, (b) 420 mtorr and (c) 660 mtorr. The image dimension correspond to 20 millimeters. The radius of curvature of the wavefront is 6.5 meters.

For all pressures the wavefront is nearly a divergent spherical wave with a curvature radius of almost 6.5 meters. The wavefront shape and radius are observed to be dominantly determined by beam refraction in the discharge-pumped amplifier induced by the radial electron density gradient in the capillary plasma column [12]. It is also refraction what dominantly determines the annular beam shape. Two general trends are observed for every pressure: i) the wavefront is smoother in the center of the beam, ii) the amplitude of wavefront defects decreases as the pressure is increased (from $7\lambda_{XUV}$ rms at 200 mtorr, to $3\lambda_{XRL}$ rms at 660 mtorr). The defects are evaluated by subtracting the XRL wavefront from a best-fit spherical wavefront. This decrease in wavefront aberration at increased pressure results from the smoother electron density gradients and reduced refraction that are a consequence of reduced plasma column compression at the higher discharge pressures. The XRL shot to shot wavefront stability was also investigated making several measurements for each pressure. The shot-to-shot wavefront variation was determined to be $0.22\lambda_{XRL}$.

CONCLUSION

In summary, we have developed the first XUV wavefront sensor. This Shack-Hartmann wavefront sensor has been used to study the optical field induced x-ray laser (OFI XRL) wavefront and a single pass x-ray laser pumped by 100 ps pulse. Results obtained on the OFI XRL show that the XRL wavefront presents difference to a spherical wavefront having an amplitude less than λ_{XUV}. Results obtained on the single pass XRL show that the XRL wavefront presents difference to a spherical wavefront having an amplitude up to $5\lambda_{XUV}$. We have also measured for the first time the absolute electromagnetic field of a soft x-ray laser. A soft x-ray Shack-Hartmann sensor was used to gather in a single shot the data required for the reconstruction of the electromagnetic field corresponding to a capillary discharge soft x-ray laser. The results suggest it should be possible to focus the output of capillary discharge lasers onto near diffraction limited spots, achieving unprecedented x-ray power densities, about 4×10^{13} Wcm^{-2}.

REFERENCES

[1] P.V Nikles et al., *Phys. Rev. Lett*, **78** (1997) 2748
[2] A. L'Huillier and Ph. Balcou, *Phys. Rev. Lett*. **59**, 56 (1987)
[3] B.R. Benware, C.D. Macchietto, C. Moreno and J.J. Rocca, *Phys. Rev. Lett*. **81**, 5804, (1998)
[4] S. Sebban et al., *Phys. Rev. Lett*. **86**, 3004 (2001)
[5] R. Tommasini, F. Löwenthal, and J. E. Balmer, *Phys. Rev. A*, **59**, 1577 (1999)
[6] F. Albert et al., 6 th international Conference on x-ray laser, Kyoto, Japan (1998)
[7] Ph. Zeitoun et al., 6 th international Conference on x-ray laser, Kyoto, Japan (1998)
[8] S. Le Pape et al., *Phys. Rev. Lett*. **88**, 183901 (2002)
[9] C. Macchietto, B. Benware, and J.J. Rocca, *Opt. Lett*., **24**, 115, (1999)
[10] Y. Liu, M. Seminario et al.,*Phys. Rev. A*. **63**, 2001, 033802-1
[11] W.H. Southwell, *J. Opt. Am.* **70**, 8, 998 (1980)
[12] C.H. Moreno et al, *Phys. Rev. A*, **58**, 1509 (1998)

Spatial Coherence of Currently Available EUV/Soft X-Ray Sources

Yanwei Liu[1], David T. Attwood[1], Jorge J. Rocca[2],
Henry C. Kapteyn[3], and Margaret M. Murnane[3]

[1]*Applied Science and Technology Graduate Group, University of California, Berkeley, CA 94720*
Center for X-Ray Optics, Lawrence Berkeley National Lab, Berkeley, CA 94720
[2]*Department of Electrical and Computer Engineering, Colorado State University, Fort Collins, CO 80523*
[3]*Department of Physics and JILA, University of Colorado and NIST, Boulder, CO 80309*

Abstract. Spatial coherence properties of 3 different kinds of coherent sources in extreme ultraviolet/soft x-ray wavelength region, namely undulator, x-ray laser and high harmonic generation source, are introduced. Advantages of each source type and their potential applications are also discussed.

INTRODUCTION

Extreme ultraviolet (EUV) and soft x-ray (SXR) wavelengths, covering the electromagnetic spectrum region of ~1-50nm, is attracting more and more research interests and attentions (1). At these wavelengths, there are ample atomic and molecular resonances that can be used to identify elements and understand structural changes. Various soft x-ray spectroscopy techniques have revealed a lot of new information to chemists and material scientists. In addition, EUV/SXR have much shorter wavelengths (compared with infrared/visible/ultraviolet light), making it possible to extend the resolution limits of current optical systems. Two examples are high-resolution x-ray microscopy and EUV lithography for next generation IC manufacturing.

For many applications, spatial coherence of the radiation plays a critical role. High spatial coherence is necessary for diffraction-limited focusing (to achieve smallest possible spot) and recording high contrast fringes in interfere experiments. Currently there're mainly three kinds of coherent sources in EUV/SXR wavelengths: undulator, x-ray lasers and high-order harmonic generation sources. Undulators provide tunable coherent radiation covering the whole EUV/SXR region and have been widely used by scientists working at synchrotron beamlines. On the other hand, EUV lasers (2) and high harmonic generation sources (3) are two examples of the efforts to develop alternative 'table-top' sources. In this paper, we'll examine the spatial coherence properties of these sources, together with discussions of their applications.

CP641, X-Ray Lasers 2002: 8th International Conference on X-Ray Lasers, edited by J. J. Rocca et al.
© 2002 American Institute of Physics 0-7354-0096-2/02/$19.00

UNDULATOR

Undulator radiation is generated by relativistic electrons traversing a periodic magnet structure. (See Fig. 1a) The relativistic motion of electrons helps to generate very short wavelength radiation and compress the radiation into a very small forwarding cone in the order of $1/\gamma$, where γ is the Lorentz factor. The interference effect between the individual periods of the undulator serves to enhance the radiation whose wavelength satisfies undulator equation, generating very bright radiation at those wavelengths. Generally the useful portion of undulator radiation is contained in the central radiation cone, whose angle is determined by $\theta_{cen} = 1/\gamma\sqrt{N}$, where N is the undulator periods. Power in the central radiation cone is generally in the order of watts, with a relative spectral bandwidth of 1/N, all contained in a small cone angle in the order of µrad.

Coherent power from undulator can be obtained by applying adaquate filtering techniques. For spatial coherence, a spatial filter (limiting source size or collection angle, or both) can be used to limit the phase-space of radiation to the coherence limit, i.e. $d\cdot\theta = \lambda/2\pi$. This is the phase-space of a perfect TEM$_{00}$ mode laser, also representing the minimal phase-space any radiation would occupy. From van Cittert-Zernike theorem (4), even with an incoherent source, a spatial filter satisfing this criteria would also ensure essentially full spatial coherence. In 3rd generation synchrotrons, electron beam emittance (the product of beam size and divergence angle) are tightly controlled to minimize the phase-space of the radiation. As a result, a good portion of the power in the central radiation cone will go through the spatial filtering. A monochrmator can be further used to improve the temporal coherence, if a relative spectral bandwidth better than 1/N is required. (More details about coherent undulator radiation can be found at Chapter 5 and 8 in Ref. 1.)

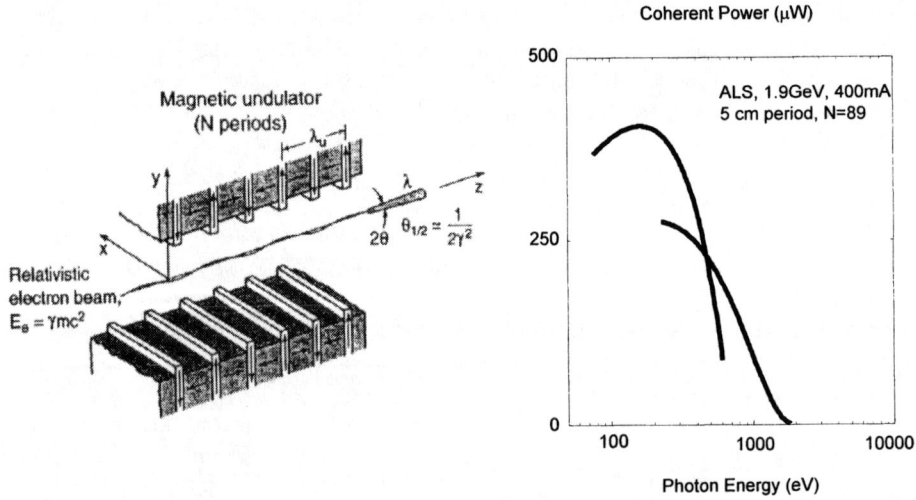

FIGURE 1. a. Undulator b. Coherent power for ALS U5 undulator. Both 1st and 3rd harmonic are shown.

As an example, Fig. 1b shows the coherent power from a 5-cm period undulator at Advanced Light Source (ALS) in Lawrence Berkeley National Lab. The coherent power is spatially coherent, with relative spectral bandwidth of 1/1000. An efficiency of 10% is also included for power losses on the beamline optics. It can be seen from the figure that ALS U5 undulator provides good coverage of EUV/SXR region, from below 100eV to over 1keV. Higher energy machines, such as APS in Argonne National Lab, ESRF in Europe and Spring-8 in Japan with electron beam energy of 6-8GeV, can extend the coverage of undualator radiation to hard x-ray region.

SOFT X-RAY LASERS

Due to relatively high photon energies involved, x-ray lasers generally have to operate on the transition between energy levels of ions, rather than atoms as in optical lasers. As a result, x-ray lasing (population inversion) usually only happens in plasmas produced by high intensity lasers or fast discharges. Short gain duration (picosecond to nanosecond scale) and plasma debris damage to mirrors generally limit x-ray laser to amplification of spontaneous emission (ASE) through the plasma, even at wavelengths where multilayer mirrors have relatively high reflectivity. The lack of cavity, thus the lack of efficient mode selection mechanism, greatly limits the spatial coherence of x-ray lasers. Early experiments of x-ray lasers indeed showed limited spatial coherence (5).

From the van Cittert-Zernike theorem (4), a high degree of spatial coherence from an ASE-based laser can be achieved with a gain medium has a Fresnel number ($N_F = a^2/\lambda L$, where a and L are plasma column width and length) less than unity. Given the very short wavelength, this will require very long and narrow plasma, hard to achieve with current plasma excitation techniques. Theoretical works (6,7,8) have suggested refractive anti-guiding a plasma with sharp density gradients can serve to improve the spatial coherence by reducing the number of guided modes. This is not demonstrated until recently in a series of two-pinhole interference experiment with the 46.9nm discharge capillary laser at Dr. Rocca's group in Colorado State University (9). In the experiment, fast capillary discharge produced plasma columns with both very high uniformity and length-to-diameter ratio exceeding 1000:1. Very rapid coherence buildup is observed, as an evidence of refractive anti-guiding. Essentially full spatial coherence is obtained with the longest capillary length (36cm).

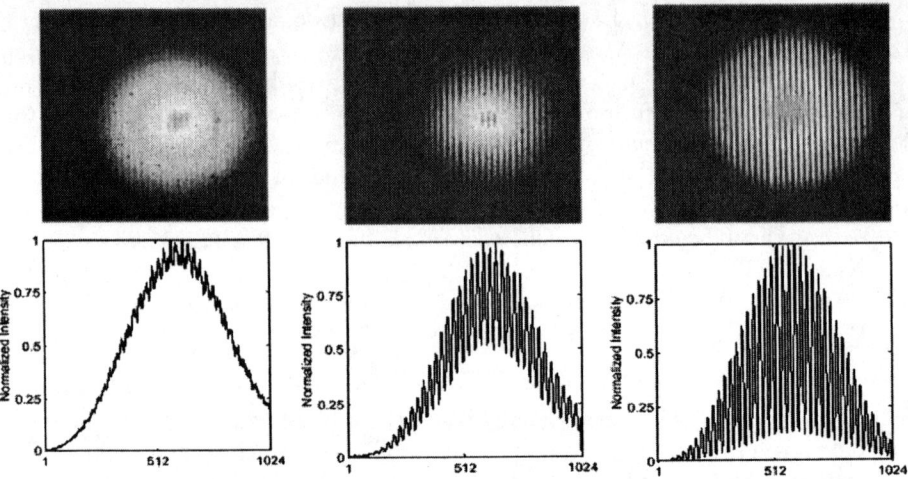

FIGURE 2. Rapid coherence buildup in the 46.9nm Neon-like Ar laser at Colorado State University. Interferograms are taken with a same pinhole pair at same position. Capillary lengths are 18, 27 and 36 cm. See detailed experiment description and results in ref. 9.

HIGH HARMONIC GENERATION

In HHG, EUV radiation is produced by focusing a high-intensity femtosecond laser into a gas. With the progress of ultrafast laser technique (10), compact table-top lasers can now readily deliver TW-order of peak power. The availability of extremely high intensity greatly accelerated the HHG research and helps to push the HHG EUV photon energies to > 500eV (corresponding to harmonic orders >300). By selecting different harmonic peaks or changing the wavelength of pump laser, HHG can also provide 'tunable' radiation in good part of EUV spectrum.

HHG has its unique advantage in generating coherent EUV radiation. While both undulator radiation and x-ray lasers begin with incoherent radiation from uncorrelated electrons and ions, HHG coherently up-shifting the infrared/visible laser to EUV region. Since the whole process is driven by the coherent laser beam, high degree of coherence is also expected from the generated EUV beam. However, past experiments have shown that the HHG radiation has only partial spatial coherence (11,12). Electron formation by ionization and associating refractive index change has been identified as one possible reason for degraded coherence (11). More importantly, HHG process is an extremely high order nonlinear process where perturbation analysis can no longer be used. As a result, the nonlinear polarization is not simply proportional to the q-th power of fundamental field (q is the harmonic order). A consequence of this is that HHG not necessarily retains the beam quality and coherence of the driving laser. Theoretical analysis has shown that the limiting factor on coherence of HHG is the intensity-dependent phase of dipole momentum responsible for generating harmonics (12). For typical HHG experiments, the intensity of the pump laser varies in both space (focused beam) and time (ultrashort pulse), thus degrading both the spatial and temporal coherence. Moreover, even exposed to the same intensity, two separate electron trajectories (corresponding to different ionization and recollision times within the optical cycle) could also generate the same photon energy (13, 14). These all could degrade the coherence of HHG.

In a recent experiment carried at University of Colorado at Boulder, we demonstrated that HHG in a hollow core fiber can generate spatially coherent beam with very good beam mode quality (15). The hollow core fiber geometry had been used in past work to demonstrate phase-matching technique over a long interaction region (16). The pump laser propagates in the hollow fiber in guided EH_{11} mode, eliminating the Guoy phase change associated with focused Gaussian beam as in free-space HHG setup. This quasi-plane wave interaction helps to achieve phasematching in a long propagation distance and select a single trajectory. The observed high contrast fringes even after more than a million shots demonstrated the ability of HHG to generate stable, coherent radiation with this extended phasematching.

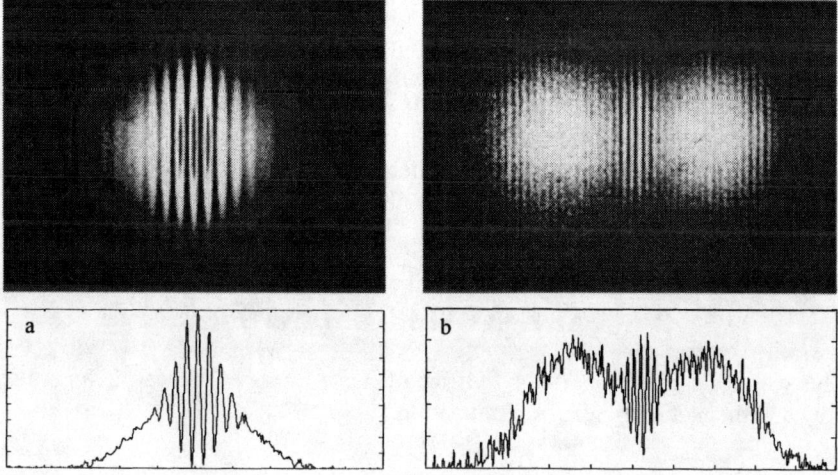

Figure 3. Observation of full spatial coherence in EUV beam generated by HHG in a hollow core fiber filled with argon gas. The interferograms are taken using two pinholes separated by a) 384 μm and b) 779 μm, while the beam diameter ($1/e^2$) is only 1.0mm. See detailed experiment description and results in ref. 9.

Generating fully coherent radiation by HHG is a demonstration of its coherent process nature. This unique property makes HHG a controllable process. Past work showed sophisticated method can be used to control the HHG process with unprecedented precision (17). HHG is an especially promising source in ultrafast x-ray applications. Thanks to its broad spectrum and coherent nature, HHG holds the promise of generating sub-fs pulses.

COMPARISON OF THE SOURCES

Of the three kinds of source discussed here, undulator radiation has the best spectrum coverage and flexibility to satisfy different experiment requirements. To the long wavelength end, low energy machines can go as far as infrared. High energy machines, on the other hand, can provide coherent radiation in hard x-ray region. For much of the EUV/Soft X-ray region, undulator is still the only reliable source providing continuously tunable radiation with high photon flux. With adequate filtering techniques, desired coherence can be readily obtained.

Both EUV/soft x-ray laser and HHG source are very useful complement sources for their smaller size. For many situations, it's inconvenient to conduction experiments in synchrotron facilities. The progresses in these sources have already seen their applications in interferometry, holography and spectroscopy. As the researches in both fields continue, improvements such as higher photon flux and better spectrum coverage will make it possible to perform more experiments, once only be possible with synchrotron radiation, in a laboratory scale. Moreover, both of them have some unique properties that are not available in undulators. For example, both of them provide much higher peak powers, promising for nonlinear optics and strong light-matter interaction in EUV/SXR region. For lasers, high energy per pulse and good temporal coherence (monochromatic light) make them the suitable sources for experiments such as single-shot interferometry and holography. For HHG, the femtosecond and even shorter pulse width will open new fields in ultrafast phenomenon research.

ACKNOWLEDGMENTS

This work is partly supported by Office of Basic Energy Science of Department of Energy and the National Science Foundation.

REFERENCES

1. Attwood, D. T., Soft X-rays and Extreme Ultraviolet Radiation, Cambridge University Press, New York, 1999
2. Rocca, J. J., Rev. Sci. Instrum. 70, 3799 (1999) and the references therein.
3. McPherson, A. et al., J. Opt. Soc. Am. B 4, 595-601 (1987).
4. Born, M. and Wolf, E., Principles of Optics, 7th ed. Cambridge University Press, Cambridge, 1999
5. Trebes, J. E. et al., Phys. Rev. Lett. 68, 588 (1992)
6. London, R. A., Phys. Fluids 31, 184 (1988)
7. London, R. A., Strauss, M. and Rosen, M. D., Phys. Rev. Lett. 65, 563(1990)
8. Gasparyan, P. D., Starikov, F. A., and Starostin, A. N., Phys. Usp. 41, 761(1998)
9. Liu, Y. et al, Phys. Rev. A 63, 033802 (2001)
10. Backus, S. et al., Review of Scientific Instruments 69, 1207 (1998).
11. Ditmire, T. et al., Phys. Rev. Lett. 77, 4756 (1996).
12. Salieres, P., L'Huillier, A., Lewenstein, M., Phys. Rev. Lett. 74, 3776 (1995).
13. Lewenstein, M. et al, Phys. Rev. A 49, 2117 (1994).
14. Lewenstein, M., Salieres, P., L'Huillier, A., Phys. Rev. A 52, 4747(1995).
15. Bartels, R. et al, Science 297, 376 (2002).
16. Rundquist, A. et al., Science 280, 1412-1415 (1998).
17. Bartels, R. et al., Nature 406, 164 (2000).

Beam Properties of a Saturated, Half Cavity Zinc Soft X-ray Laser

T. Mocek[1], B. Rus[1], A.R. Präg[1], M. Kozlová[1], G. Jamelot[2], A. Carillon[2], D. Ros[2], and D. Joyeux[3]

[1]*Department of Gas Lasers/PALS Research Center, Institute of Physics, Na Slovance 2, Prague 8, Czech Republic*

[2]*Laboratoire de Spectroscopie Atomique et Ionique (LSAI), Bâtiment 350, Université Paris-Sud, 91405 Orsay, France*

[3]*Institut d'Optique Théorique et Appliquée, Laboratoire Charles Fabry, Bâtiment 503, Université Paris-Sud, 91403 Orsay, France*

Abstract. We report on beam quality of the multi-millijoule, double-pass Ne-like zinc soft X-ray laser (XRL) operating in deep saturation. The ASE (amplified spontaneous emission) single-pass XRL output emitted by a 30-mm long plasma column delivers a smooth, highly symmetric ellipsoidal beam with horizontal divergence of 3 ($\pm$0.5) mrad and vertical divergence of 5 ($\pm$0.5) mrad. Deep saturation was achieved by implementing a half-cavity consisting of a Mo/Si flat multilayer mirror. The XRL output was boosted up by a factor of ~11 providing narrowly collimated beam with horizontal and vertical divergence of 3.8 ($\pm$0.5) mrad and 5.8 ($\pm$0.5) mrad, respectively. The refraction angle is close to 5 mrad in both regimes of operation, being nearly independent on the re-injection angle in the case of half-cavity.

INTRODUCTION

To date, the most robust and versatile approach to achieve lasing action in the soft X-ray spectral region has proven to be the collisional excitation scheme. In this contribution we report on detailed characterization of the 21.2 nm Ne-like Zn X-ray laser beam produced in a double-pass regime and compare with the single-pass output. For the first time we have experimentally explored the use of a *flat* Mo:Si mirror which would be ideal for an automated regime of the XRL operation, due to extremely easy alignment and fast re-positioning. In contrast to the widely used pumping by 1.06-μm lasers we use rather longer wavelength of 1.315 μm of the iodine laser which is expected to be more beneficial for the XRL performance [1]. To produce a plasma with reduced lateral gradients and to suppress the detrimental effects of lateral beam refraction, we introduced a novel approach in which the main pumping pulse is focused into a much wider preplasma column. The spatial characteristics of the emitted X-ray laser beam were measured for different prepulse conditions and discussed.

CP641, X-Ray Lasers 2002: 8th International Conference on X-Ray Lasers, edited by J. J. Rocca et al.
© 2002 American Institute of Physics 0-7354-0096-2/02/$19.00

EXPERIMENTAL SETUP

The experiment was performed with the iodine laser system at PALS (Prague Asterix Laser System) [2,3]. The laser, operating at wavelength of 1.315 µm, provides two beams of 450-ps (FWHM) pulse duration. The temporal delay between the two pulses was adjusted at 10 ns. The main beam is focused on the target by a composite system consisting of a matrix of 10 concave cylindrical lenses and a single aspherical lens. It provides a highly uniform 130 µm x 30 mm line focus. The auxiliary beam is sent onto a 30-mm long, optically polished zinc slab target at an angle of 25 deg. from above, being focused by a tandem of cylindrical and spherical lens to form a ~700 µm x 40 mm line focus. The typical net irradiances on the target surface were ~2.8 x 10^{13} W/cm^2 and ~1.6 x 10^{10} W/cm^2 (for optimum prepulse energy of 1.6 J) for the main pulse and the prepulse, respectively. The saturated operation of the zinc XRL has been achieved by implementing the half-cavity. It consists of a flat, 25-mm-diameter Mo:Si multilayer mirror with reflectivity of about 30 %, positioned at 8.5 mm from the plasma column. The reflecting surface was protected from plasma debris by a stainless-steel shield with a 1-mm-diameter hole which allows to pass the X-ray beam. The generated XRL beam was analyzed either by a footprint monitor consisting of a fine-grain phosphor screen coated by a 40-nm-thick Al layer, or by an X-ray spectrometer operating in Wadsworth geometry [4].

RESULTS

Figure 1 shows a typical far-field pattern of the single-pass ASE X-ray beam produced by a 30-mm long plasma column recorded with the footprint monitor. The net energies delivered by the main-pulse and prepulse onto the target were typically 500 J and 1.6 J, and the pulse duration was 485 ps in both cases. A smooth, highly symmetric X-ray beam of ellipsoidal profile is generated with a maximum of intensity at the center. From the averaged horizontal and vertical profiles we inferred the respective divergences of 3 (±0.5) mrad and 5 (±0.5) mrad. The XRL beam emerges from the plasma under the pointing angle of 5.5 (±0.5) mrad with respect to the target surface.

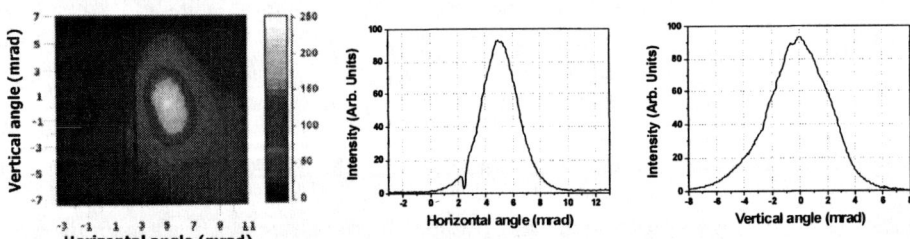

FIGURE 1. Far-field pattern (footprint) of the single-pass Ne-like Zn XRL beam, with its horizontal and vertical lineout, respectively. The target surface is on the left side of the footprint (at zero angle).

Figure 2(a) shows the measured divergence of the single-pass XRL as a function of prepulse energy. Apparently, the vertical divergence extends with prepulse while the horizontal one which is controlled by plasma expansion in the direction perpendicular to the target surface, remains similar. As the width of the prepulse focus is significantly larger than the focus of the main driving beam, varying the prepulse level modifies the vertical electronic density gradients (parallel to the target surface) present in the amplifying medium along with its dimension. To clearly understand these results, however, a ray-tracing code would be necessary.

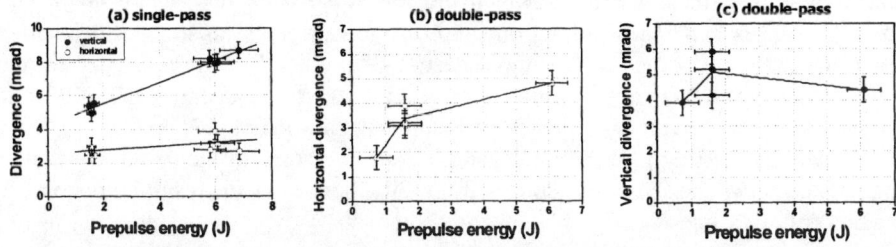

FIGURE 2. (a) Horizontal and vertical divergences of the single-pass XRL are shown as a function of the prepulse energy compared with the double-pass (b), (c).

Figure 3 displays a set of footprint records of the double-pass XRL beam for prepulse energies ranging from 0.7 up to 6.1 J. The azimuth angle of the half-cavity mirror is 3 mrad, which corresponds to the re-injection angle of 1 mrad. It is seen that both the spatial quality and the intensity of the XRL beam are optimized for a prepulse energy of 1.6 J, corresponding to a net irradiance of ~1.6 x 10^{10} W/cm^2. The double-pass XRL beam emerges from the plasma under an angle of ~5 mrad, having its horizontal and vertical divergence of 3.8 (±0.5) mrad and 5.8 (±0.5) mrad, respectively, very close to those of the single-pass ASE.

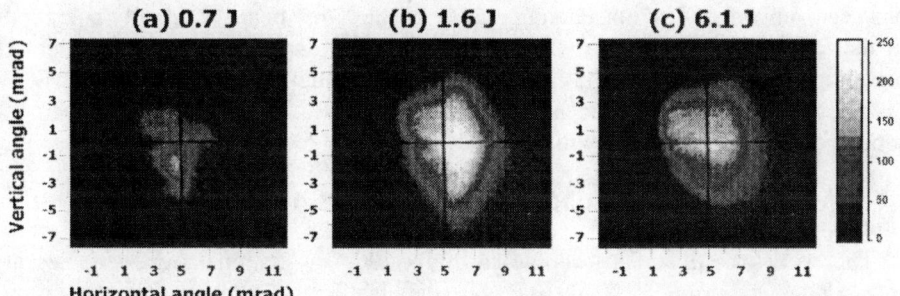

FIGURE 3. Recorded footprints of the double-pass XRL beam for prepulse energies of (a) 0.7 J, (b) 1.6 J, and (c) 6.1 J, respectively.

The dependency of horizontal and vertical divergence on the prepulse energy for a double-pass XRL is shown in Fig. 2(b),(c). In this case, the prepulse enlarges exclusively the horizontal divergence but not the vertical one. The experimental data demonstrate that the second propagation through amplifying medium strongly modifies the spatial profile of the XRL beam in both directions. This interesting feature is rather different from the single-pass case (Fig. 2(a)) and can be attributed to a combined effect of larger gain region and smaller refraction of the re-injected XRL beam in the case of stronger prepulse.

The optimization of half-cavity has also been performed with respect to the re-injection angle under which the single-pass ASE emission is reflected back to the plasma for return amplification. During this measurement we kept the prepulse energy at 1.6 J. The XRL beam divergence was optimized for re-injection angle of 1 mrad. On the other hand, we found that the refraction angle is nearly independent of the re-injection angle, retaining a value of about 5 mrad which is just same as in the case of single-pass. The trajectory along which the X-ray emission reflected from the half-cavity mirror is sent back for return amplification presents an important parameter determining the properties of the output beam. When the ASE beam is reflected into a low density plasma far away from the target, the emerging beam is just moderately amplified, having a refraction-unaffected duplicate profile of the ASE seed. On the other hand, as the reflected ASE emission is injected closer to the target, it samples a higher gain but at the same time suffers from refraction. Thus the resulting output beam profile is unlike the ASE beam. However, our data show that the pointing angle is essentially independent of the re-injection conditions which implies that the intensity of XRL reflects the gain magnitude at the sampled distance from the target. Using the experimental data we estimate that the transverse size of the gain region is ~50 μm.

CONCLUSION

In conclusion, the output beam of a deeply saturated Ne-like Zn X-ray laser has been characterized. We found that the ASE as well as the double-pass XRL are both optimized for the same level of prepulse, exhibiting maximum output signal together with low divergence. The wide prepulse focus helped to minimize the lateral refraction and as a result a highly symmetric ASE X-ray beam of ellipsoidal profile has been generated. The outstanding quality of the X-ray beam is actually comparable to that of common optical lasers. This justifies us to speak about a true laser. The unique parameters of this half-cavity XRL such as high output energy (~4 mJ), low divergence (3.8 x 5.8 mrad^2), and very symmetric beam shape together with robustness and reproducibility make this XRL an ideal tool for applications.

ACKNOWLEDGMENTS

This work was financially supported by the EU Transnational Access to Research Infrastructures grant HPRI-00108, by the National Research Centers project LN00A100, and by the Czech Academy of Sciences grant A1010014.

REFERENCES

1. Holden, P.B., and Pert, G.J., J. Phys. B: At. Mol. Opt. Phys. **29**, 2151-2157 (1996).
2. Rus, B., Rohlena, K., Skála, J., Králiková, B., Jungwirth, K., Ullschmied, J., Witte, K.J., and Baumhacker, H., Laser Part. Beams **17**, 179-194 (1999).
3. Jungwirth, K., Cejnarová, A., Juha, L., Králiková, B., Krása, J., Krouský, E., Krupičková, P., Láska, L., Mašek, K., Mocek, T., Pfeifer, M., Präg, A., Renner, O., Rohlena, K., Rus, B., Skála, J., Straka, P., and Ullschmied, J., Phys. Plasmas **8**, 2495-2501 (2001).
4. Carillon, A., Dhez, P., Gauthé, B., Jaeglé, P., Jamelot, G., Klisnick, A., and Lagron, J.C., "Experimental device for XUV experiments especially adapted to the directional emission from a plasma column," in *X-Ray Instrumentation in Medicine and Biology, Plasma Physics, Astrophysics, and Synchrotron Radiation*, edited by R. Benattar, Proceedings of SPIE Vol. 1140, Bellingham, Washington, 1989, pp. 271-278.

Longitudinal Coherence Measurements of the Transient Collisional X-ray Laser

Raymond F. Smith[1], Sebastian Hubert[3], Marta Fajardo[2], Philippe Zeitoun[2],
James Dunn[1], James R. Hunter[1], Christian Remond[3], Laurent Vanbostal[2],
Silvie Jaquemot[3], Joseph Nilsen[1], Ciaran L.S. Lewis[4], Remy Marmoret[3]

[1]*Lawrence Livermore National Laboratory, Livermore, CA 94551*
[2]*Laboratoire d'Interaction du rayonnement X avec la Matiére (LIXAM), Université
Paris-Sud, Bât 350, 91405 Orsay, France.*
[3]*Commisariat á l'Ènergie Atomique, BP 2, 91680 Bruyeres-le-Chatel, France.*
[4]*School of Mathematics and Physics, The Queen's University of Belfast, Belfast BT7
1NN, UK.*

Abstract. The first longitudinal coherence measurement of the transient inversion collisional x-ray laser is presented. The scheme under study is the picosecond output of the Ni-like Pd x-ray laser at 14.68 nm generated by the COMET laser facility at LLNL. Interference fringes were generated using a Michelson interferometer setup in which a thin multilayer membrane was used as a beam splitter. Longitudinal coherence measurements were made for this transition by changing the length of one interferometer arm and measuring the resultant variation in fringe visibility. The nature of this dependence also allows for an estimation of the linewidth of the lasing transition to be made. Analysis indicates a linewidth of ~0.3 pm which is a factor of four less than previous measurements on quasi-steady state x-ray laser schemes.

1. INTRODUCTION

The rapid development of x-ray lasers in recent years combined with the availability of optics in the XUV has led to several applications, such as interferometry [1] and holography [2], which depend on the coherence properties of the source pulse. Accurate measurements of the longitudinal coherence are important for determining feasibility and design parameters of instrumentation for such experiments. The longitudinal coherence is quantitatively related to the spectral bandwidth, $\Delta\lambda$, of the source and is a measure of the temporal separation along a beam in which the different spectral components maintain a phase relationship. In amplitude division interferometry the phase front of one arm is spatially overlapped and co-propagated with the other arm at the output of the interferometer. If the arm lengths are equalized to an accuracy better than the coherence length, L_c, of the source, interference fringes will be generated. The additional measurement of the spectral linewidth provides valuable insights into the gain dynamics within the lasing medium.

Saturated lasing at XUV wavelengths is typically generated through single pass

CP641, *X-Ray Lasers 2002: 8ᵗʰ International Conference on X-Ray Lasers*, edited by J. J. Rocca et al.
© 2002 American Institute of Physics 0-7354-0096-2/02/$19.00

amplification along an extended laser-produced plasma column. Previous experimental measurements of L_c have been made on the NOVA generated Ne-like Y x-ray laser at 15.5 nm [3]. This lasing scheme was generated with a 350 ps pulse, which generated a quasi-steady state population inversion on the main lasing transition. The duration of the x-ray laser pulse was comparable to the pulse width of the optical driver with gain conditions ultimately extinguished through plasma expansion and radiative cooling. The longitudinal coherence of this lasing scheme was measured to be ~100 μm, using a Mach-Zehnder type interferometer [3] of a long ~ns prepulse with a short ~ps main heating pulse. A large population inversion is created for a period before collisional redistribution of the excited state populations can take place. The attractiveness of this lasing scheme lies in the picosecond output, which makes it particularly suitable for characterizing fast evolving events. With this paper we present the first measurements of the longitudinal coherence of the transient collisional x-ray laser.

2. EXPERIMENTAL SETUP

The Ni-like Pd 14.7 nm x-ray laser probe beam was generated using two laser beams at 1054 nm wavelength from the COMET facility at LLNL [5]. Single pass saturated x-ray laser output of a few 10's of μJs was achieved with an optical pumping combination of a 600 ps long pulse (2 J, 2×10^{11} W cm^{-2}) and a 13 ps (5 J, 3×10^{13} W cm^{-2}) main heating pulse [6]. Traveling wave irradiation using a 7-step reflection echelon along the 1.6 cm line focus geometry was employed on a 1.25 cm polished slab Pd target to produce amplification in the axial direction. Approximately ten shots can be taken on the same target position before the output of the x-ray laser degrades significantly. Over the lifetime of a target position the output mode structure of the x-ray laser was found to be repeatable.

The x-ray laser output was imaged (Figure 1) and partially collimated by a normal incidence Mo/Si multilayer spherical mirror (S1), with $f = 11.75$ cm, and routed via a 45° multilayer mirror, to the input path of the Michelson interferometer. The measured reflectivity (assuming unpolarized light) of these two Mo/Si multilayer optics was 0.62 (S1) and 0.30 at the laser wavelength with a bandpass of 0.62 nm and 1.44 nm, respectively. The x-ray laser pointing stability along the input path of the interferometer was better than 100 μm from shot-to-shot. Upon entering the interferometer the x-ray laser beam interacts with a thin foil beamsplitter (BS), which is partially transmissive and partially reflective at the 14.7 nm wavelength. The beamsplitter consists of a 89nm thick Si_3N_4 membrane (0.5 × 0.5 cm^2) coated either side with 4.5 bi-layers of Mo/Si. The reflectivity at the x-ray laser wavelength is 14.4 % with a bandwidth of 2.7 nm.

One arm of the interferometer is directed to 0° multilayer optic, M1, via a reflection from BS. The beam is then returned along the same optical path and is transmitted through the beamsplitter. The other interferometer arm, which is initially transmitted through BS, reflects off the 0° mirror M2 and reflects off BS to the output collection optics. Both arms are co-propagating after the beamsplitter en route to M3. If sufficient spatial and temporal overlap of the two arms is achieved interference fringes will be generated. The two arms of the interferometer (BS→M1, BS→M2) were set to have the same nominal length with a mechanical alignment rod, to an estimated accuracy of ~100 μm. Rough spatial overlap of the two arms was done using a charged coupled device camera (CCD1) placed before the imaging system (Fig. 1). Optimization of the spatial overlap is possible under vacuum with remote adjustment of the M2 tilt stage.

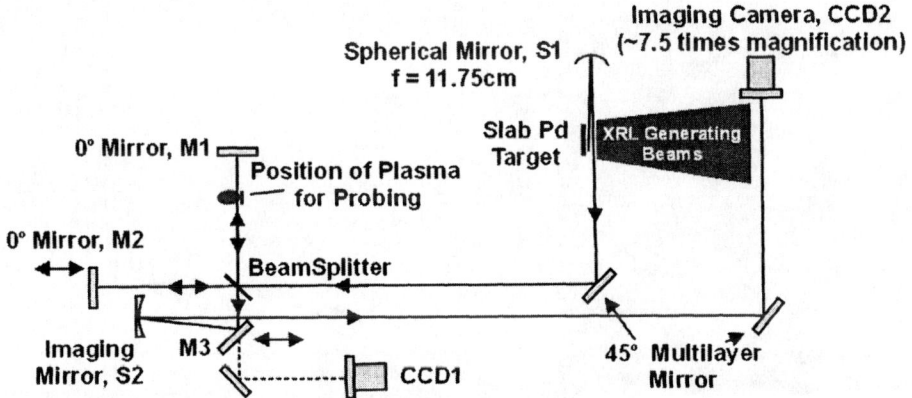

Figure 1. Setup for Michelson Interferometer Experiment.

The output signal from the interferometer reflects off M3 onto a spherical multilayer imaging mirror, S2, which has a focal length of 50 cm. The signal is routed from the imaging optic, via a 45° multilayer optic, to a thinned back-illuminated CCD2 detector with a 1024×1024 24 μm^2 pixel array. A 2000 Å Zr/1000 Å Polyimide ($C_{22}H_{10}N_2O_5$) filter was placed in front of the CCD to block visible and UV light. The imaging mirror was set to image the output plane of an Al target, within the M1 arm, from which a secondary plasma can be generated for pump-probe experiments. The total magnification of the imaging system was estimated to be ~7.5 giving a pixel limited resolution of 3.2 μm at the target plane. The reflectivity of the multilayered mirrors in the interferometer (also M3 and S2) is measured to be ~35% at the x-ray laser wavelength (s-polarization). The throughput of the interferometer is approximately 0.014.

Alignment of all the optics within the interferometer was done with the aid of a telescope, placed in the CCD2 camera position, and set on axis with the aid of alignment crosshair fiducials. A 5 mW HeNe diode laser was injected into the entrance aperture of the telescope. Absolute positioning of the optics tilt stages was achieved by retro-reflection of the HeNe alignment beam, off a given optic, back down to the telescope entrance aperture (autocollimation).

A sample interferogram is shown in Fig. 2 (a). The image captured by the CCD2 camera represents a region of ~1700 x 1700 μm^2 at the target plane. The curvature associated with the fringes is due to stresses across the beamsplitter causing local variations in the angles of the phase front within the x-ray beam. Figure 2(b) shows an integrated lineout (over box in 3(a)) which illustrated good levels of coherence across several hundred microns. Fringe visibilities, $V = (I_{max} - I_{min})/(I_{max} + I_{min})$, up to 70% have been measured. The large scale intensity variations across the image are due to mode structure within the x-ray laser. It has been observed that after equalization of the arms is achieved, there exists good coherent relationship between the modes. Structure along individual fringes may be due to non-uniformities within the beamsplitter or multilayer optics.

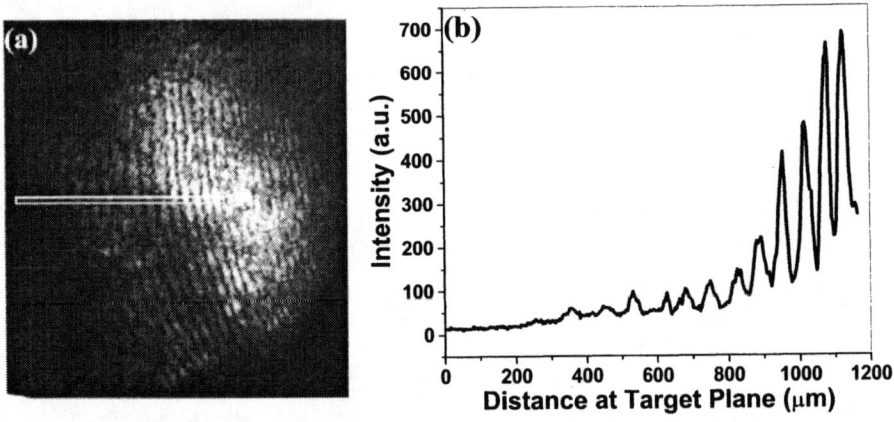

Figure 2 (a) Interferogram taken at $\Delta L = 0$ for a 13 ps CPA heating pulse (b) Integrated lineout over box in (a) shows good visibility over several hundred microns.

3. LONGITUDINAL COHERENCE MEASUREMENTS

To determine the longitudinal coherence of the source the differential path length between the two arms of the interferometer was systematically varied and the fringe visibility in the interferograms monitored. This was achieved by incrementing the M2→BS length with encoded translation of M2 while keeping the distance from the beamsplitter to M1 fixed (Fig. 3). At least two shots were taken at each position to ensure repeatability. Each interferogram can be characterized in a number of ways: overall fringe continuity, number of adjacent fringes, number of counts on CCD, uniformity of intensity along a fringe and visibility. Sufficient quality must be achieved in all of these categories to ensure the interferogram is suitable for plasma probing experiments. It was found that there existed a high degree of speckle in the x-ray intensity pattern which was considered to be unrelated to the larger scale intensity variations associated with the inherent transverse mode structure of the x-ray laser. This speckle, which is possibly a result of inhomogeneities within the beamsplitter, was superimposed onto the interferograms giving rise to random high spatial frequencies in the intensity of the fringes. To obtain visibility measurements intensity averaging was carried out over a 30-pixel region, which is equivalent to 225 μm at the target plane. This had the effect of negating the speckle contribution whilst maintaining the accuracy of the visibility measurement. The results are shown in Fig.

3(b). Each data point signifies the average of nine measurements across a single interferogram. Ideally fringe visibilities up to 100% would be observed. In practice this was not achieved due to a small degree of spatial misalignment and slight variations between the optics. It was observed that for shots with increasing distance away from the nominal overlap position the regions over which fringes were visible became smaller. This may be due to adjacent transverse modes in the x-ray laser profile with, for example, the same coherence length but being temporally offset. Fringe visibilities below 0.2 were difficult to measure because at larger values of ΔL the fringe continuity was on a scale less than that of the integration window. Also shown in a Gaussian fit to the data points with the 1/e half-width yielding a coherence length measurement of 400 μm ± 35 μm. This is equivalent to a temporal coherence, τ_c, of ~1.33 ps.

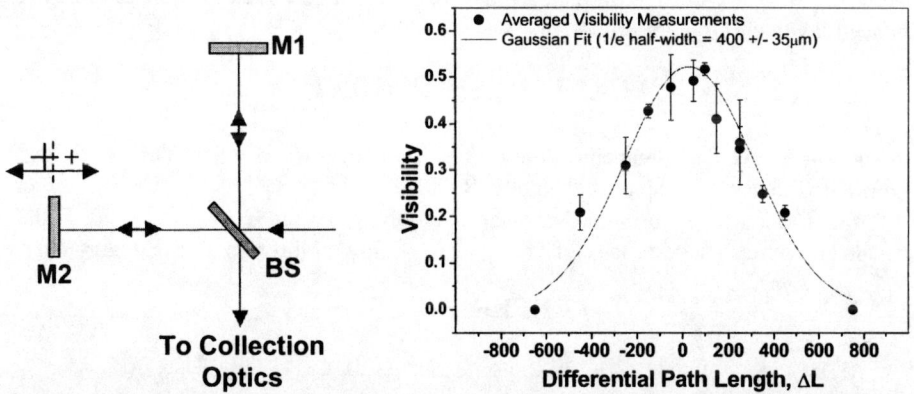

Figure 3. Variation of fringe visibility with path difference between two arms of the interferometer. Each data point represents the average visibility for a given shot taken from nine separate measurements across the interferogram. Standard deviations are shown via the error bars. Also shown is the best fit Gaussian curve with a 1/e half width of ~400 μm ± 35 μm.

Following [4] the measured fringe visibility is equal to the magnitude of the complex degree of coherence within the illumination beam. The latter is a measure of the degree of correlation between the two combining phase fronts at the output of the interferometer. By incrementing the length of the M2 arm by a distance L the phase front associated with that arm is delayed by a time $\tau = 2L/c$. By using a Fourier transform relationship (Wiener-Khintchine theorem) the complex degree of coherence is shown to be related to the power spectral density of the x-ray laser pulse, $I(\upsilon)$. As a result of this relationship, the linewidth $\Delta\upsilon$ and the coherence time τ_c are inversely related. We assume the x-ray laser line has a Gaussian power spectrum such that,

$$I(\upsilon) \propto \exp\left[-2\sqrt{\ln(2)}\left(\frac{\upsilon-\upsilon_0}{\Delta\upsilon}\right)^2\right], \qquad (1)$$

where $\Delta\upsilon$ is the linewidth of the lasing transition, $\upsilon_0 = c/\lambda$. The measured fringe visibility dependence on τ, $V(\tau)$, is the envelope of the autocorrelation function of the power spectrum varying as,

$$V(\tau) = \exp\left[-\left(\frac{\pi\Delta\upsilon\tau}{2\sqrt{\ln 2}}\right)^2\right].\qquad(2)$$

The coherence time, τ_c, is defined as the value of τ at which the visibility has decreased to 1/e the maximum value. From figure 4(b) we have a $\tau_c \sim 1.33 \pm 0.12$ ps, which yields an equivalent Gaussian FWHM linewidth of 0.29 ± 0.03 pm. For an expected ion temperature of 80eV [5] within the gain medium the Doppler broadened lineshape would be equivalent to ~1 pm. Our lower measurement of ~0.3 pm is an expected result of gain narrowing effects. The measured linewidth is a factor of four less than the 1.3 pm previously measured for the Ni-like Y quasi-state state x-ray laser scheme [3]. In that work the predicted ion temperature was 600eV, which gave rise to enhanced levels of Doppler broadening.

ACKNOWLEDGEMENTS

The support of Al Osterheld and Andy Hazi is greatly appreciated. This work was performed under the auspices of the U.S. Dept. of Energy by the University of California Lawrence Livermore National Laboratory, under Contract No. W-7405-Eng-48. The optics used in the interferometer were funded through the CEA, France.

REFERENCES

1. L. B. Da Silva, T. W. Barbee, Jr., R. Cauble, P. Celliers, D. Ciarlo, S. Libby, R. A. London, D. Matthews, S. Mrowka, J. C. Moreno, D. Ress, J. E. Trebes, A. S. Wan, and F. Weber, Phys. Rev. Lett. **74**, 3991-3994 (1995); R.F. Smith, J.Dunn, J. Nilsen, V.N. Shlyaptsev, S. Moon, J. Filevich, J.J. Rocca, M.C. Marconi, J.R. Hunter, and T.W. Barbee, Jr., Phys. Rev. Lett., **89** (6), 065004-1 (2002).

2. J.E. Trebes, S.B. Brown, E.M. Campbell, D.L. Matthews, D.G. Nilson, G.F. Stone, D.A. Whelan, Science, **238** (4826), p.517 (1987); B. Rus, H. Daido, H. Tang, M. Nishiuchi, M. Kishimoto, M. Tanaka, T. Kawachi, N. Hasegawa, K. Nagahima, T. Arisawa, Y. Kato, , *these proceedings.*

3. P. Celliers, F. Weber, L.B. DaSilva, T.W. Barbee, Jr., R. Cauble, A.S. Wan, and J.C. Moreno, Opt. Letts. **20** (18), 1907 (1995).

4. M. Born and E. Wolf, Principles of Optics, Chap. 7, 7th ed. (Cambridge University Press, 1999).

5. J. Dunn, R.F. Smith, J. Nilsen, J.R. Hunter, T.W. Barbee, Jr., V.N. Shlyaptsev, H. Fiedorowicz, A.Bartnik, Soft X-ray Lasers and Applications IV meeting, SPIE Proc. Vol. 4505, p.62 (2001).

6. J. Dunn et al., Journal de Physique IV (Proceedings), vol.11, (no.2), (X-Ray Lasers 2000. 7th International Conference on X-Ray Lasers, Saint-Malo, France, 19-23 June 2000.), (2001).

Spatial Coherence of Recombination X-ray Laser Pumped by Pulse-Train Laser

Naohiro Yamaguchi, Yuuji Okamoto, Hideki Yamaguchi, and Tamio Hara

Toyota Technological Institute
2-12-1 Hisakata, Tempaku, Nagoya 468-8511, Japan

Abstract. We have started measurements of spatial coherence of x-ray radiation from recombination x-ray laser source that is pumped by pulse-train laser. The dispersing coherence diagnostic system has been constructed. In the preliminary experiments using double-slits of 24 μm and 50 μm separations, fringe patterns representing the interference from the pair of Young's slits have been observed for the Al XI 3d-4f line. The measured fringe visibility was 0.3 for 50 μm separated two points at 370 mm distant from the source. On the other hand, there have not been observed any clear fringe visibility for the other non-lasing lines.

INTRODUCTION

High brightness coherent radiation at soft-x-ray wavelengths can open many applications such as microscopy, lithography, interferometry and holography. The degree of spatial coherence of radiation plays a critical role in many of the above applications. Experimental studies on spatial coherence of x-ray laser radiation have been reported for the laser pumped [1, 2] or the capillary discharge pumped x-ray lasers [3, 4] in the collisional excitation scheme. Soft x-ray laser beams were limited to single-pass amplification of spontaneous emission (ASE) in these experiments.

We performed x-ray laser cavity experiments for Al XI 3d-4f transition line (15.47 nm) [5], that were conducted in the recombining plasma scheme, because our x-ray laser medium pumped by the pulse-train YAG laser could maintain x-ray laser gain with a duration of about 1 ns [6]. It would be interesting to observe development of the degree of coherence of x-ray laser beam in the x-ray laser cavity.

In this report, we present the results of Young's interference experiments for the x-ray laser radiation in a single-pass and double-target configuration. Results of comparison between experiment and calculation are also shown.

EXPERIMENTAL SETUP

A 100 ps laser pulse from a mode-locked YAG oscillator, 1.064 μm, was transformed into a linearly polarized 16-pulse-train through an optical stacker and a delay line

CP641, *X-Ray Lasers 2002: 8th International Conference on X-Ray Lasers*, edited by J. J. Rocca et al.
© 2002 American Institute of Physics 0-7354-0096-2/02/$19.00

component. The interval of each pulse was 200 ps. The shaped pulse-train YAG laser has a 1.5-2 J energy. An Nd:glass amplifier having a 25 mm diameter rod was added to the YAG laser system. The output laser energy was increased up to 6 J. The laser beam was divided into two beams through a half mirror and each beam impinged onto an Al slab target via a segmented lens system. Each lens assembly forms an 11-mm-long focused-line on a target. Therefore, plasmas produced in this experiment had a total length of 22 mm in the double-target configuration.

The diagnostic system was based on the dispersing coherence diagnostic method [1]. It consisted of a Young's double-slit, a transmission grating, and an x-ray CCD camera. Each of the two slits is itself crossed by a narrow diffraction grating with lines and spaces perpendicular to the slit length. This provides wavelength dispersion by diffraction through the transmission grating in a direction perpendicular to the interference direction of the double-slit. Thus, this diagnostic system can separate the x-ray laser spectral lines from the other emission lines of the laser-produced-plasma, and gives us the fringe pattern representing the interference from the pair of Young's slits for individual spectral lines.

Several pairs of Young's slits were fabricated on a stainless steel plate of 30 μm thickness, whose center-to-center separations were 24 μm, 50 μm, 74 μm, 100 μm and 125 μm. The transmission grating (X-OPT, F002HA0805:FS-210) had free standing parallel gold bars with a 5000 lines/mm grating. A single slit of 0.1 mm width and 2 mm height was settled parallel to the grating bar in front of the transmission grating. The source-to-slit or source-to-grating distance was 370 mm, and the slit-to-camera or grating-to-camera distance was about 500 mm. Therefore the reciprocal dispersion on the detection plane was 0.4 nm/mm for the first order light. The detector used was an x-ray CCD camera (SX-TE/CCD 512TKB, Princeton Instrum.) using a back-illuminated CCD of 512×512 pixels with 24×24 μm pixel size.

METHOD OF CALCULATION

A theoretical calculation model has been developed which is based on two-beam interference with partially coherent and quasimonochromatic light to analyze measured interference patterns. We take a generalized geometric framework for a two-slit interference experiment in our calculation as shown in Fig. 1. X-ray source is assumed to have a Gaussian intensity distribution. Each slit opening has a rectangular shape with a width of $2a$ and a length of $2b$. Radiation intensity at a point on the screen, the detector plane, is calculated by substituting diffracted intensities from each slit opening into the general interference law for stationary optical fields [7]. The diffracted intensity is given by the Fraunhofer formula for diffraction by a rectangular aperture, because the Fraunhofer condition is satisfied in our experimental geometry. Then, the intensity at a point $Q(x, y) = (R_1 \tan\theta_X, R_1 \tan\theta_Y)$ on the screen is expressed as

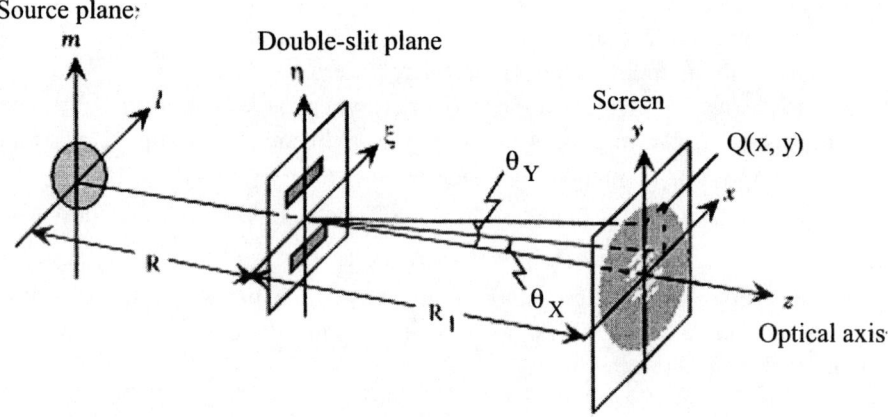

FIGURE 1. Schematic diagram for calculation of the two-slit interference.

$$I(Q) = 2\left(\frac{\sin X}{X}\right)^2\left(\frac{\sin Y}{Y}\right)^2\left\{1 + e^{-\frac{v^2}{2}} \cdot \cos(\phi + \psi + \delta)\right\} \tag{1}$$

where $X = \frac{2\pi}{\lambda}a\sin\theta_x$, $Y = \frac{2\pi}{\lambda}b\sin\theta_y$, $v = \frac{2\pi\sigma}{\lambda R}\sqrt{\Delta\xi^2 + \Delta\eta^2}$, σ is the rms

spatial measure of a Gaussian distribution of the source. Phase differences, ϕ, ψ and δ, are denoted as

$$\phi = \frac{2\pi}{\lambda R}(\Delta\xi\Delta\ell + \Delta\eta\Delta m)$$

$$\psi = \frac{\pi}{\lambda R}(\rho_2^2 - \rho_1^2) \tag{2}$$

$$\delta = \frac{\pi}{\lambda R_1}(\rho_2^2 - \rho_1^2 - 2\Delta\xi \cdot x - 2\Delta\eta \cdot y)$$

where $\rho_1 = \sqrt{\xi_1^2 + \eta_1^2}$, $\rho_2 = \sqrt{\xi_2^2 + \eta_2^2}$, (ξ_1, η_1) or (ξ_2, η_2) is the position of slit 1 or slit 2, respectively, therefore $\Delta\xi = \xi_1 - \xi_2$, $\Delta\eta = \eta_1 - \eta_2$. $\Delta\ell$ and Δm are the source displacement from the optical axis.

The spatial coherence is characterized by the cross correlation of fields through the double slit, which can be described in terms of the normalized complex degree of coherence μ_{12}. According to the van Cittert-Zernicke theorem, the complex degree of coherence corresponds to the Fourier transform of the source intensity distribution [8], that is described as

$$\mu_{12} = e^{-\frac{v^2}{2}} \cdot e^{i\psi} \cdot e^{i\phi} \tag{3}.$$

In a double-slit interference experiment, the modules of μ_{12} is proportional to the fringe visibility V, that is defined as $V = (I_{max} - I_{min})/(I_{max} + I_{min})$, where I_{max} and I_{min} are the maximum and minimum intensities of the fringe pattern.

Calculations have been carried out in the case when the source displacement exists only with respect to the m-axis, $\Delta m \neq 0; \Delta \ell = 0$, the center of slit opening is on the η-axis, $\Delta \xi = 0$, and the double-slit is placed symmetrically to the optical axis z, $\eta_1 = -\eta_2$. Calculated interference patterns are compared with an observed interference pattern, where we take the source size, [FWHM in Gaussian profile]$=2\sqrt{2\ln 2}\sigma$, and source displacement along the interference direction, Δm, as the fitting parameters. It was found that the source displacement along the interference direction was sensitive to the asymmetry of the fringe pattern.

RESULTS

Emission spectra were observed mainly in the wavelength range from 12 nm to 16.5 nm including the Al XI 3d-4f transition line. The x-ray source configuration was the double-target configuration and the single-pass case. In this preliminary experiment the double-slit used had a 24 μm or 50 μm separation. An example of spectrophotograph taken by the CCD camera is shown in Fig.2, where the slit separation was 50 μm. Intensity profiles in the interference direction for the two spectral lines are also shown. Fringe patterns were observed for the Al XI 3d-4f line, with the fringe separations which is consistent with the calculated value. On the other hand, fringe patterns were not clear for the non-lasing lines around 15.1 nm including the Al XI 3p-4d transition line.

The measured interference pattern was analyzed with comparing the calculation described in the previous section. One of the results of calculation for the Al XI 3d-4f

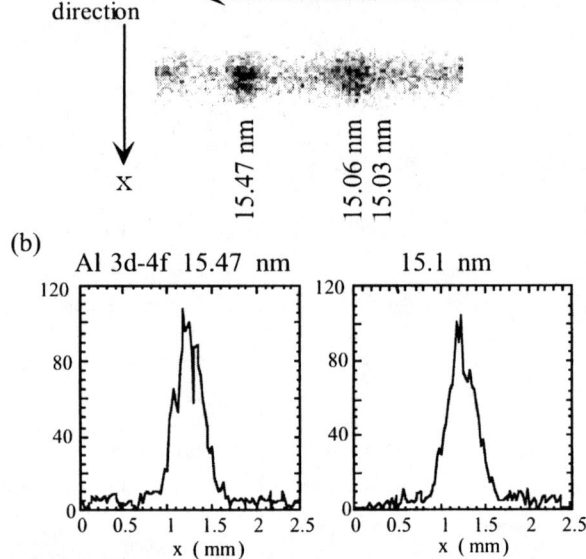

FIGURE 2. (a) Part of an output image of the dispersing coherence diagnostic system. The slit separation of double-slit is 50μm.
(b) Intensity profiles of interference fringe for the two different spectral lines shown in the above image.

626

line, when the slit separation was 50 μm, is shown in Fig. 3 along with the measured result shown in Fig. 2. From this analysis, we can determine the fringe visibility as 0.32, the source size as 65 μm, and the source displacement as 17 μm. While the fringe visibility was 0.09 - 0.13 for the 15.1 nm line, where the source size was about 80 μm and the source displacement was 17 μm.

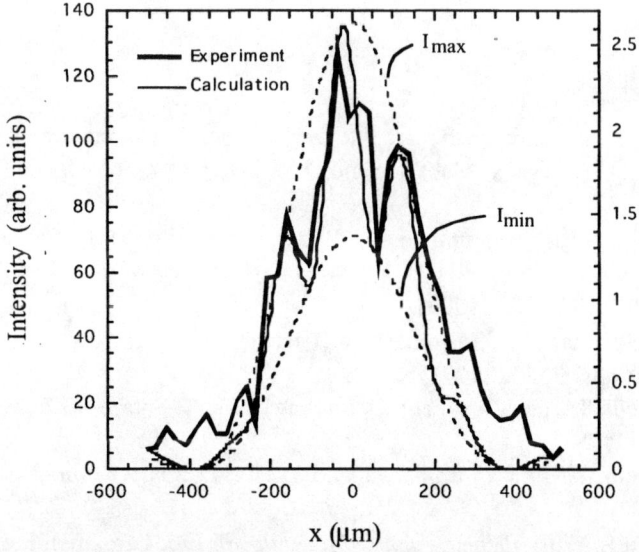

FIGURE 3. Calculated intensity curve from Gaussian source model (thin curve), where the FWHM of the source is 65 μm and its displacement from the optical axis is 17 μm. I_{max} is an envelope of maxima in the fringe pattern and I_{min} is that of minima. The measured intensity curve for the Al XI 3d-4f line is also shown (thick curve).

CONCLUSIONS

The dispersing coherence diagnostic system has been constructed to measure the spatial coherence of our x-ray laser source. In the preliminary experiment, the fringe pattern representing the interference from the pair of Young's slits has been observed for the Al XI 3d-4f line, while there has not been observed clear fringe visibility for the other non-lasing lines. Our calculation for double-slit experiment can reproduce measured interference patterns including asymmetry of the intensity distribution, assuming Gaussian intensity distribution and displacement of the source position.

Measurements will be continued using other slit pairs and under the various configurations of x-ray laser amplification. The data accumulated would clarify the spatial coherence of our x-ray laser source and its build-up in an x-ray laser cavity.

ACKNOWLEDGMENTS

This work was supported in part by a Grant-in-Aid for Scientific Research from the Ministry of Education, Science, Sports and Culture (No. 1265044).

REFERENCES

1. R. E. Burge, G. E. Slark, M. T. Browne, X.-C. Yuan, P. Charalambous, X.-H. Cheng, C. L. S. Lewis, A. MacPhee and D. Neely, *J. Opt. Soc. Am. B* **14**, 2742-2751 (1997).
2. P. Lu, E. Fill, Y. Li, J. Maruhn, and G. Pretzler, *Phys. Rev. A* **58**, 628-635 (1998).
3. M. C. Marconi, J. L. A. Chilla, C. H. Moreno, B. R. Benware, and J. J. Rocca, *Phys. Rev. Lett.* **79**, 2799-2802 (1997).
4. Y. Liu, M. Seminario, F. G. Tomasel, C. Chang, J. J. Rocca, and D. T. Attwood, *Phys. Rev. A* **63**, 033802 (2001).
5. N. Yamaguchi, T. Hara, T. Ohchi, C. Fujikawa, and T. Sata, *Jpn. J. Appl. Phys.* **38**, 5114-5116 (1999).
6. N. Yamaguchi, T. Hara, C. Fujikawa, and Y. Hisada, *Jpn. J. Appl. Phys.* **36**, L1297-L1300 (1997).
7. M. Born and E. Wolf, *Principles of Optics (6th edition)*, Pergamon Press, Oxford, 1980, p. 501.
8. D. T. Attwood, *Soft X-rays and Extreme Ultraviolet Radiation*, Cambridge University Press, New York, 1999, pp. 321-330.

List of Participants

8th International Conference on X-Ray Lasers

Aspen, Colorado, USA

May 27th – 31st, 2002

Kanti Aggarwal
Queen's University
Department of Pure and
Applied Physics
University Road
Belfast,
N. Ireland   BT7 INN
K.Aggarwal@qub.ac.uk

Igor Artiokov
Lebedev Physical Institute
53, Leninsky Prospekt
Moscow,
Russia 119991
iart@sci.lebedev.ru

David Attwood
University of California-
Berkeley
Lawrence Berkeley National
Laboratory
One Cyclotron Road
Berkeley, CA 94720
attwood@eecs.berkeley.edu

Yoav Avitzour
Princeton University
MAE Department, E-Quad.
Princeton, NJ 08544
avitzour@princeton.edu

Juerg Balmer
University of Berne
Institute of Applied Physics
Sidlerstrasse 5
Berne,
Switzerland   CH-3012
balmer@iap.unibe.ch

Nir Bar-Gill
Technion
38 Shivtei-Israel Street
Kiryat-Haim, 26275
Israel
bargill@tx.technion.ac.il

Djamel Benredjem
LIXAM
Bat 350 Universite Paris-Sud
Orsay,   Essonne
France   91405
djamel.benredjem@lsai.u-psud.fr

Sandra Biedron
Argonne National Laboratory
Advanced Photon Source
9700 South Cass Avenue
Argonne, IL 60439
biedron@aps.anl.gov

Maite Braud
University of Berne
Institute of Applied Physics
Sidlerstrasse 5
Berne,
Switzerland   3012
maite.braud@iap.unibe.ch

Jianwen Chen
Shanghai Inst. of Optics &
Fine Mechanics
The Chinese Academy
P. O. Box 800-211
Shanghai,
China
j-w-chen@yahoo.com

Boris Chichkov
Laser Zentrum Hannover e.v.
Hollerithallee 8
Hannover,
Germany   D-30419
ch@lzh.de

Hiroyuki Daido
Advanced Photon Research
Center
K. R. E. - Japan Atomic
Energy Research Inst.
8-1, Umemidai, Kizu
Kyoto,
Japan 619-0215
daido@apr.iaeri.go.jp

Sam Dalhed
Lawrence Livermore National
Laboratory
7000 East Avenue
Mail Stop L-015
Livermore, CA 94550
dalhed1@llnl.gov

Arif Demir
University of Kocaeli
Department of Physics
Kocaeli,
Turkey 41200
arifd@kou-edu.tr

James Dunn
University of California
Lawrence Livermore National
Laboratory
P. O. Box 808 - I-251
Livermore, CA 94551
dunn6@llnl.gov

Jorge Filevich
Colorado State University
Dpt ERC
Room B308
Fort Collins, CO 80521
rage@engr.colostate.edu

Ernst Fill
Max-Panck-Institut
Hans-Koptermann - Str. 1
Garching,   D-85748
Germany
ernst.fill@mpg.mpg.de

John Galayda
Stanford Linear Accelerator
Center
2575 Sand Hill Road
MS 69
Menlo Park, CA 94025
galayda@slac.stanford.edu

Hongyi Gao
Shanghai Inst. of Optics &
Fine Mechanics
The Chinese Academy
P. O. Box 800-211
Shanghai,
China
gaohy@mail.shcnc.ac.cn

David Gaudiosi
University of Colorado
JILA, Campus Box 440
Boulder, CO 80309-0440
gaudiosi@colorado.edu

Iddo Geltner
Princeton University
MAE Department, E-Quad.
Princeton, NJ 08544
Igeltner@princeton.edu

Emily A. Gibson
University of Colorado
JILA
Campus Box 440
Boulder, CO 80309
emily.gibson@colorado.edu

Olivier Guilbaud
LIXAM
Bat 350 Universite Paris-Sud
Orsay, Essonne
France 91405
olivier.guilbaud@lsai.u-
psud.fr

Mike Hentschel
Inst. PHOTONIK/TU-WIEN
Gusshausstr. 27/387
Vienna, 1040
Austria
michael.hentschel@tuwien.ac.
at

Kuroda Hiroto
University of Tokyo
Institute for Solid State
Physics - 5-1-5
Kashiwa-no-ha, Kashiwa City
Kashiwa,
Japan 277-8581
kuroda@issp.u-tokyo.ac.jp

Eiki Hotta
Tokyo Institute of Technology
Dept. of Energy Sciences
4259, Nagatsuta - Midori-ku
Yokohama, Kanagawa
Japan 226-8502
ehotta@es.titech.ac.jp

Elena Ivanova
Institute of Spectroscopy of
RAS
Troitsk Moscow Region
Troitsk, 142190
Russia
eivanova@isan.troitsk.ru

Sylvie Jacquemot
CEA/DIF
BP 12
Bruyeres-Le-Chatel,
France 91680
sylvie.jacquemot@cea.fr

Pierre Jaegle
LIXAM
Bat 350 Universite Paris-Sud
Orsay,
France 91405
pierre.jaegle@lixam.u-psud.fr

Gerard Jamelot
LIAM Universite Paris Sud
Batiment 350, av. Jean Perrin
Orsay,
France 91405
gerard.jamelot@lixam.u-
psud.fr

Alexandr Jancarek
Czech Technical University
Brehova 7
Prague, 11519
Czech Republic
jancarek@troja.fjfi.cvut.cz

Elizabeth Jankowska
Wroclaw University of
Technology
Wroclaw,
Poland 50-370

Karol Janulewicz
Institut Berlin
Max Born Institute
Max-Born-Str.2A
Berlin,
Germany D-12489
kaj@mbi-berllin.de

Henry Kapteyn
University of Colorado
JILA
440 UCB
Boulder, CO 80309-0440
kapteyn@jila.colorado.edu

Shosuke Karashima
Science University of Tokyo
Dpt of Electrical Engineering
1-3 Kagurazaka
Shinjuku-Ku, Tokyo
Japan 162-8601
kara@ee.kagu.sut.ac.jp

Yoshiaki Kato
JAERL - Kansai Laboratory
8-1 Umemidai, Kizu
Soutaku, Kyoto
Japan 619-0215
ykato@apr.jaerl.go.jp

Midorikawa Katsumi
RIKEN
Hirosawa 2-1
Wako, Saitama
Japan 351-0198
kmidori@postman.riken.go.jp

Tetsuya Kawachi
Japan Atomic Energy
Research Institute - 8-1
Umemidai, Kizu
Kyoto,
Japan 619-0215
kawachi@apr.jaeri.go.jp

Roisin Keenan
Queen's University
Dpt of Pure - Applied Physics
University Road
Belfast, BT7 1NN
N. Ireland
r.keenan@qub.ac.uk

Robert King
University of York
Heslington
York, Yorkshire
United Kingdom
reki02@york.ac.uk

Daniel Klir
Czech Technical University
Dept. of Physical Electronics
Brehova 7
Prague, 11519
Czech Republic
klir@troja.fjfi.cvut.cz

Annie Klisnick
CNRS/LIXAM
Av. Jean Perrin, bâtiment 350
Orsay,
France 91405
annie.klisnick@lixam.u-psud.fr

Takahisa Koike
Int. Univ. Health & Welfare
Dpt of Radiological Sciences
2600-1 Kitakanamaru -
Otawara-shi, Tochigi
Otawara, 324-8501
Japan
takahisa@iuhw.ac.jp

Karel Kolacek
Institute of Plasma Physics
Academy of Sciences
Za Slovankou 3 - PO Box 17
18221 PRAGUE 8
Czech Republic
kolacek@ipp.cas.cz

Misha Kozlova
Institute of Physics
PALS Research Center
Na Slovance 2
Prague, 18221
Czech Republic
kozlova@fzw.cz

Jaroslav Kuba
Lawrence Livermore National
Laboratory
7000 East Avenue
L-251
Livermore, CA 94550
kuba1@llnl.gov

Kitae Lee
Korea Atomic Energy
Research Inst.
P. O. Box 105
Yuseong
Daejeon,
Korea 305-600
klee@kaeri.re.kr

Sébastien Le Pape
CNRS/LIXAM
Av. Jean Perrin, bâtiment 350
Orsay,
France 91405
Sebastien.Le-Pape@lsai.u-psud.fr

Ciaran Lewis
Queen's University
University Road
Belfast, BT9 5ED
N. Ireland
c.lewis@qub.ac.uk

Yanwei Liu
Univ. of California-Berkeley
Lawrence Berkeley Nat. Lab.
MS 2-400, LBL - 1 Cyclotron
Road
Berkeley, CA 94720
ywliu@lbl.gov

Pierre Loiseleur
Laboratoire LPTP
Ecole Polytechnique
Palaiseau,
France 91128
ploisel@lptp.polytechnique.fr

Brad Luther
Colorado State University
ECE Department
Fort Collins, CO 80523

Carmen S. Menoni
Colorado State University
ECE Department
Fort Collins, CO 80523
carmen@engr.colostate.edu

Hamed Merdji
CEA/SPAM –
Centre d'Etudes de Saclay
522 piece 114D
Gif-sur-Yvette,
France 91191
merdji@drecam.cea.fr

Stephen Milton
Argonne National Laboratory
Advanced Photon Source
9700 South Cass Avenue
Argonne, IL 60439

Tomas Mocek
Institute of Physics
PALS Research Centre
NA Slovance 2
Prague, 182 21
Czech Republic
mocek@fzu.cz

Stephen Moon
Lawrence Livermore National
Laboratory
7000 East Avenue
L-041
Livermore, CA 94551
moon1@llnl.gov

Margaret Murnane
University of Colorado
JILA
440 UCB
Boulder, CO 80309-0440
nurnane@jjila.colorado.edu

Chang Hee Nam
Dpt Physics and Coherent X-
Ray
Research Center
KAIST
Daejeon 305-701
Korea
chnam@mail.kaist.ac.kr

Art Nelson
Lawrence Livermore National
Laboratory
7000 East Avenue
Livermore, CA 94550
nelson63@llnl.gov

Peter Viktor Nickles
Institut Berlin
Max Born Institute
Max-Born-Str. 2A
Berlin,
Germany D-12489
kaj@mbi-berlin.de

Gohta Niimi
Nat.l Ins. Advanced Indust.
Sci
Adv. Semiconductor Res.
Center
AIST Tsukuba Central-2 -
Umezono 1-1-1
Tsukuba,
Japan 305-8568

Joseph Nilsen
Lawrence Livermore National
Laboratory
7000 East Avenue
Livermore, CA 94551-0808
jnilsen@llnl.gov

Tadashi Nishikawa
NTT Basic Research
Laboratories
3-1, Morinosato Wakamiya
Atsugi-shi, Kanagawa
Japan 243-0198
nisikawa@will.bri.nttco.jp

Yuuji Okamoto
Toyota Technological Institute
2-12-1 Hisakata, Tempaku
Nagoya, Aichi
Japan 468-8511
y_okamoto@toyota.ti.ac.jp

Albert Osterheld
Lawrence Livermore National
Laboratory
P.O. Box 808
Livermore, CA 94550
osterheld@llnl.gov

Geoff Pert
University of York
Heslington
York, Yorkshire
United Kingdom Y010 5DD
gjp1@york.ac.uk

Jerome Pons
GREMI/ Universite d Orleans
14, rue d Issoudun
BP 6744
Orleans,
France 45067
jerome.pons@univ-orleans.fr

Jean-Michel Pouvesle
GREMI/ Universite d Orleans
14, rue d Issoudun
BP 6744
Orleans,
France 45067
*jean-michel.pouvesle@univ-
orleans.fr*

Ansgar R. Praeg
Inst. Physics - PALS R. Cent.
Czech Academy of Sciences
Na Slovance 2
Prague 8, 18221
Czech Republic
praeg@fzu.cz

Abdur Rahman
Colorado State University
ECE Department
Fort Collins, CO 80523

Antonio Ritucci
Phys. Dept. of Univ. L'Aquila
via Vetoio
67010 Coppito
L'Aquila,
Italy
antonio.ritucci@aquila.infn.it

Jorge J. Rocca
Colorado State University
ECE Department
Fort Collins, CO 80523
rocca@engr.colostate.edu

David Ros
LIXAM
Bat 350 Universite Paris-Sud
Orsay, Essonne
France 91405
david.ros@lsai.u-psud.fr

Bedrich Rus
Institute of Physics
PALS Research Centre
Na Slovance 2
Prague 8, 18221
Czech Republic
rus@fzu.cz

Stéphane SEBBAN
Laboratoire d'Optique
Appliquée - ENSTA
Chemin de la Huniere
Palaiseau Cedex
France 91761
sebban@enstay.ensta.fr

Vyacheslav Shlyaptsev
University of California-Davis
Lawrence Livermore National
Laboratory - L-411
7000 East Avenue
Livermore, CA 94550
slava@llnl.gov

Raymond Smith
Lawrence Livermore National
Laboratory
L-251, 7000 East Avenue
Livermore, CA 94550
smith248@llnl.gov

Szymon Suckewer
Princeton University
MAE Department, E-Quad.
Room D-246
Princeton, NJ 08544
suckewer@princeton.edu

Greg Tallents
University of York
Department of Physics
York, Yorkshire
United Kingdom Y010 5DD
gjt5@york.ac.uk

Kai Tiedtke
HASYLAB at DESY
Notkestrasse 85
Hamburg,
Germany D-22603
kai.tiedtke@desy.de

Raanan Tobey
University of Colorado
Campus Box 440
JILA Bldg.
Boulder, CO 80309
tobey@ucsub.colorado.edu

Simon J. Topping
Queen's University
Department of Pure and
Applied Physics
University Road
Belfast, BT7 1NN
N. Ireland
s.j.topping@qub.ac.uk

Ozaki Tsuneyuki
NTT Basic Research
Laboratories
3-1, Morinosato Wakamiya
Atsugi, Kanagawa
Japan
ozaki@wlll.brl.ntt.co.jp

Lee Upcraft
Laboratoire d'Optique
Appliquee
Ensta-Loa
Chemin de la Huniere
Palaiseau, 91761
France
upcraft@enstay.ensta.fr

Alexander Vinogradov
Lebedev Physical Institute
53, Leninsky prospekt
Moscow,
Russia 119991
vinograd@sci.lebedev.ru

Pavel Vrba
Institute of Plasma Physics
Czech Academy of Science
Za Slovankou 3
Prague 8, 18221
Czech Republic
vrba@jpp.cas.cz

Miroslava Vrbova
Czech Technical University
Brehova 7
Prague 1, 11519
Czech Republic
vrbova@troja.fjfl.cvut.cz

Chen Wang
Shanghai Inst. Laser Plasma
P.O. Box 800-229
Shanghai, 201800
P.R. CHINA
wangch@mail.shcnc.ac.cn

Tim Whiteaker
Colorado State University
124 N. Sherwood
Fort Collins, CO 80521
whiteake@lamar.colostate.edu

Naohiro Yamaguchi
Toyota Technological Institute
2-12-1 Hisakata, Tempaku
Nagoya,
Japan   468-8511
yamagch@toyota.ti.ac.jp

Goupling Zhang
Institute of Applied Physics &
Comp. Math
P.O. Box 8009
Beijing,   100088
P.R. CHINA
zheng_wudi@mail.iapcm.ac.cn

Zhengquan Zhang
Shanghai Inst. of Optics &
Fine Mechanics
The Chinese Academy
P. O. Box 800-211
Shanghai,
China   201800

AUTHOR INDEX

A

Abou-Ali, Y., 3, 15, 291
Aggarwal, K. M., 227, 233
Agostini, P., 406
Anderson, E., 461
Andrejczuk, A., 504
Arisawa, T., 473, 522
Artioukov, I. A., 575
Attwood, D. T., 461, 607
Avitzour, Y., 193, 212

B

Baba, M., 397
Backus, S., 401, 587
Bajt, S., 596
Bakshaev, Y. L., 504
Balcou, P., 204
Ballester, F., 166, 518
Balmer, J. E., 147, 154
Barbee Jr., T. W., 271
Bar-Gill, N., 301
Bartels, R. A., 401
Bartnik, A., 21, 31, 504
Bender, H. A., 528
Benredjem, D., 46, 52, 239
Biedron, S. G., 357
Bionta, R. M., 596
Bobrova, N. A., 133, 139
Boháček, V., 91
Boody, F. P., 504
Botton, M., 301
Bouhouch, K., 69
Boussoukaya, M., 166, 518
Braley, D. A., 125
Brandl, F., 305
Braud, M., 147, 154
Breger, P., 406
Brunner, S., 193
Buchenauer, D., 528

C

Cachoncinlle, C., 107, 133

Carillon...

Carillon, A., 69, 166, 182, 204, 518, 602, 613
Carré, B., 406
Cejnarová, A., 504
Celliers, P. M., 342
Cha, Y. H., 373
Chambaret, J.- P., 563
Chanteloup, J.-C., 69
Chen, H., 318
Chen, J., 469, 495, 583
Chen, W., 550
Cheng, C., 318
Chernenko, A. S., 504
Chichkov, B. N., 441
Christov, I. P., 401
Chvostová, D., 504
Clarke, R., 15, 291
Conant, R. H., 596
Constant, E., 406

D

Daido, H., 31, 40, 69, 166, 473, 522
Demir, A., 312
Denbeaux, G., 461
Dhez, P., 602
Dimkoff, J., 528
Ding, P., 265
Ditmire, T., 596
Drescher, M., 389
Drska, L., 46, 52
Dubau, J., 239, 563
Dunn, J., 9, 21, 271, 481, 528, 538, 596, 617
Dyer, G., 596

E

Eder, D. C., 342
Edwards, M. H., 3, 15, 69, 166, 291
Egbert, A., 441

F

Faenov, A. Y., 9, 119
Fajardo, M., 617
Feit, M. D., 596

637

Fiedorowicz, H., 21, 31, 504
Filevich, J., 125, 481, 489, 498, 528, 538
Fill, E. E., 305, 596
Fisch, N. J., 557
Fleurier, C., 107, 133
Flora, F., 119
Fojtik, A., 139
Fornaciari, N., 528
Fournier, K. B., 528
Francois, M., 602
Freeman, R. R., 434
Frolov, O., 91
Fu, P., 434

G

Gaeta, G., 119
Galayda, J. N., 365
Ganeev, R. A., 65, 397
Gao, H., 469, 495, 583
Gauthier, J.-C., 52
Geltner, I., 199, 212, 557
Gibson, E. A., 587
Goldberg, K., 461
Gonthiez, T., 133
Grillon, G., 204
Grisham, M., 125
Guilbaud, O., 3, 15, 69, 116, 291
Guo, D., 434

H

Habs, D., 305
Hammarsten, E. C., 113, 125, 489, 498, 528
Han, J. M., 373
Hara, T., 623
Haroutunian, R., 204
Hasegawa, N., 31, 40, 473, 522
Hayashi, Y., 103
Hentschel, M., 389
Hervé du Penhoat, M. A., 166
Hong, K. H., 338, 418, 451
Horioka, K., 103
Hotta, E., 103
Hubert, S., 15, 510, 544, 617
Hudeček, M., 182
Hulin, D., 204

Hunter, J. R., 21, 481, 538, 617

I

Idir, M., 602
Ishizawa, A., 65, 397
Ivanov, M. I., 504
Ivanova, E. P., 247, 253

J

Jacques, E., 518
Jaeglé, P., 166
Jamelot, G., 69, 166, 182, 204, 518, 563, 613
Jančárek, A., 91, 139
Jankowska, E., 125, 489, 498, 528
Janulewicz, K. A., 26, 58
Jaquemot, S., 617
Jarocki, R., 21
Jia, T., 318
Joyeux, D., 69, 166, 182, 518, 613
Juha, L., 504
Jurek, M., 504

K

Kado, M., 31, 40, 473
Kaiser, J., 119
Kanai, T., 65, 397
Kanouff, M., 528
Kapteyn, H. C., 401, 587, 607
Karashima, S., 322
Karim, S., 528
Kato, Y., 31, 40, 473, 522
Kawachi, T., 31, 40, 473, 522
Keenan, F. P., 227, 233
Keenan, R., 3, 15, 160, 166, 291, 510, 544
Kienberger, R., 389
Kilpio, A. V., 40
Kim, B. H., 373
Kim, C. M., 338, 451
Kim, D., 97, 373
Kim, H. T., 338, 418, 451
Kim, J. H., 97, 418
King, R. E., 52, 259, 291, 510

638

Kishimoto, M., 31, 40, 473, 522
Klinger, D., 504
Klir, D., 447
Klisnick, A., 3, 15, 52, 69, 166, 204, 239, 291, 563
Koike, M., 31, 40
Koike, T., 322
Koláček, K., 91
Kondratenko, V. V., 575
Korolev, V. D., 504
Kovacev, M., 406
Kozlová, M., 174, 182, 518, 591, 613
Králiková, B., 504
Krása, J., 504
Krausz, F., 389
Kravárik, J., 447, 504
Krzywinski, J., 504
Kuba, J., 46, 52, 69, 166, 528, 596
Kuba, K., 239
Kubeš, P., 447, 504
Kühl, T., 69, 166
Kukhlevsky, S. V., 119
Kuroda, H., 65, 397

L

Lagron, J.-C., 166, 182, 563
Lan, K., 107
Lee, D. G., 338, 418, 451
Lee, K., 97, 373
Lee, R. W., 528
Lei, C., 401
Leng, Y., 569
Le Pape, S., 602
Levesque, R., 596
Lewis, C. L. S., 3, 15, 160, 166, 291, 510, 544, 617
Li, R., 318, 328, 428, 569
Li, X., 434
Li, Y., 77
Limongi, T., 119
Limpouch, J., 139
Lin, L., 569
Liu, J., 328, 328
Liu, Xi., 265
Liu, Xu., 265
Liu, Y., 607
London, R. A., 528, 596
Lu, H., 569

Lu, P., 31, 40, 473
Lu, X., 77
Lucianetti, A., 26, 58
Luther, B. M., 125, 332

M

Mader, B., 441
Mairesse, Y., 406
Marconi, M. C., 125, 332, 481, 489, 498, 528, 538
Marmoret, R., 617
Masuda, H., 455
McEvoy, A. M., 544
McKernan, M., 596
Merdji, H., 406
Mével, E., 406
Mezi, L., 119
Midorikawa, K., 412
Mocek, B., 287
Mocek, T., 166, 174, 182, 204, 287, 338, 451, 518, 591, 613
Möller, C., 46, 52, 239
Montchicourt, P., 406
Moon, S. J., 125, 342, 481, 489, 528, 538
Moribayashi, K., 31
Morozov, A., 199, 212, 557
Msezane, A. Z., 233
Murnane, M. M., 401, 587, 607

N

Nabekawa, Y., 412
Nadvornikova, L., 139
Nagashima, K., 31, 40, 473, 522
Nakajima, M., 103
Nakano, H., 424, 455
Nakao, M., 455
Nam, C. H., 338, 418, 451
Namikawa, K., 31, 473
Naulleau, P., 461
Neely, D., 15, 291
Nelson, A. J., 481
Neumeyer, P., 69, 166
Nickles, P. V., 26, 58
Niimi, G., 103